P9-CCC-593

Viral Insecticides for Biological Control

Academic Press Rapid Manuscript Reproduction

Viral Insecticides for Biological Control

Edited by

Karl Maramorosch

Waksman Institute of Microbiology
Rutgers—The State University of New Jersey
Piscataway, New Jersey

K. E. Sherman

George Washington University
School of Medicine
Washington, D.C.

1985

ACADEMIC PRESS, INC.

(Harcourt Brace Jovanovich, Publishers)

Orlando San Diego New York London
Toronto Montreal Sydney Tokyo

ACADEMIC PRESS, INC.
Orlando, Florida 32887

United Kingdom Edition published by
ACADEMIC PRESS INC. (LONDON) LTD.
24–28 Oval Road, London NW1 7DX

LIBRARY OF CONGRESS CATALOG CARD NUMBER: 84-45187

ISBN 0-12-470295-3

PRINTED IN THE UNITED STATES OF AMERICA

85 86 87 88 9 8 7 6 5 4 3 2 1

Contents

Contributors *ix*
Preface *xi*

I
Taxonomy and Identification

Taxonomy and Nomenclature of Insect Pathogenic Viruses 3
T. W. Tinsley and D. C. Kelly

Classification, Identification, and Detection of Insect Viruses by Serologic Techniques 27
Loy E. Volkman

Viral Proteins for the Identification of Insect Viruses 55
Sally B. Padhi

II
Pathology

Pathology Associated with Baculovirus Infection 81
H. M. Mazzone

Pathology Associated with Cytoplasmic Polyhedrosis Viruses 121
Tosihiko Hukuhara

Pathobiology of Invertebrate Icosahedral Cytoplasmic Deoxyriboviruses (Iridoviridae) 163
Donald W. Hall

Pathology Associated with Densoviruses 197
Shigemi Kawase

Pathology Associated with Small RNA Viruses of Insects 233
Norman F. Moore

III
Ecology and Environmental Biology

Natural Dispersal of Baculoviruses in the Environment 249
D. L. Hostetter and M. R. Bell

Stability of Insect Viruses in the Environment 285
Robert P. Jaques

Development of Viral Resistance in Insect Populations 361
D. T. Briese and J. D. Podgwaite

The Safety of Insect Viruses as Biological Control Agents 399
Gabriele Döller

The Role of Viruses in the Ecosystem 441
W. J. Kaupp and S. S. Sohi

IV
Physical, Biological, and Chemical Characteristics

Structure and Physical Characteristics of Baculoviruses 469
D. C. Kelly

The Nature of Polyhedrin 489
Just M. Vlak and George F. Rohrmann

V
Replication

The Replication of Iridoviruses in Host Cells 545
Peter E. Lee

The Replication of Baculoviruses 569
James L. Vaughn and Edward M. Dougherty

The Replication Schemes of Insect Viruses in Host Cells 635
Norman F. Moore

Quantitation of Insect Viruses 675
W. J. Kaupp and S. S. Sohi

Receptors in the Infection Process 695
H. M. Mazzone

Multiple Virus Interactions 735
Kenneth E. Sherman

VI
Production and Field Application

Considerations in the Large-Scale and Commercial Production of Viral Insecticides 757
Kenneth E. Sherman

Strategies for Field Use of Baculoviruses 775
J. D. Podgwaite

Index *799*

Contributors

Numbers in parentheses indicate the pages on which the authors' contributions begin.

M. L. Bell (249), *USDA, ARS, Western Cotton Research Laboratory, Phoenix, Arizona*

D. T. Briese (361), *CSIRO Division of Entomology, Canberra, 2601, Australia*

Gabriele Döller (399), *Department for Medical Virology and Epidemiology of Virus Diseases, Hygiene-Institute, University of Tübingen, D-7400 Tübingen, Federal Republic of Germany*

Edward M. Dougherty (569), *Insect Pathology Laboratory, Plant Protection Institute, BARC, ARS, USDA, Beltsville, Maryland 20705*

Donald W. Hall (163), *Department of Entomology and Nematology, University of Florida, Gainesville, Florida 32611*

D. L. Hostetter (249), *USDA, ARS, Biological Control of Insects Research Laboratory, University of Missouri, Columbia, Missouri 65201*

Tosihiko Hukuhara (121), *Faculty of Agriculture, Tokyo University of Agriculture and Technology, Tokyo 183, Japan*

Robert P. Jaques (285), *Research Station, Agriculture Canada, Harrow, Ontario NOR 1GO, Canada*

W. J. Kaupp (441, 675), *Canadian Forestry Service, Forest Pest Management Institute, Sault Ste. Marie, Ontario P6A 5M7, Canada*

Shigemi Kawase (197), *Faculty of Agriculture, Nagoya University, Nagoya 464, Japan*

D. C. Kelly (3, 469), *Natural Environment Research Council, Institute of Virology, Oxford OX1 3SR, England*

Peter E. Lee (545), *Department of Biology, Carleton University, K1S 5B6 Ottawa, Canada*

H. M. Mazzone (81, 695), *U.S. Department of Agriculture, Forest Service, Center for Biological Control of Northeastern Forest Insects and Diseases, Hamden, Connecticut 06514*

Norman F. Moore *(233, 635), Natural Environment Research Council, Institute of Virology, Oxford OX1 3SR, England*

Sally B. Padhi *(55), Waksman Institute of Microbiology, Rutgers—The State University of New Jersey, Piscataway, New Jersey 08854*

J. D. Podgwaite (361, 775), *U.S. Department of Agriculture, Forest Service, Center for Biological Control of Northeastern Forest Insects and Diseases, Hamden, Connecticut 06514*

George F. Rohrmann (489), *Department of Agricultural Chemistry, Oregon State University, Corvallis, Oregon 97331*

K. E. Sherman (735, 757), *George Washington University School of Medicine, Washington, D.C. 20052*

S. S. Sohi (441, 675), *Canadian Forestry Service, Forest Pest Management Institute, Sault Ste. Marie, Ontario P6A 5M7, Canada*

T. W. Tinsley (3), *Natural Environment Research Council, Institute of Virology, Oxford OX1 3SR, England*

James L. Vaughn (569), *Insect Pathology Laboratory, Plant Protection Institute, BARC, ARS, USDA, Beltsville, Maryland 20705*

Just M. Vlak (489), *Department of Virology, Agricultural University, 6709 PD Wageningen, The Netherlands*

Loy E. Volkman (27), *Department of Entomology and Parasitology, University of California, Berkeley, California 94720*

Preface

Pesticide chemicals have created complex problems in the environment and in the food chain. Food, wildlife, and even human breast milk have become contaminated with organic pesticides such as DDT. Recently, important advances have been made in combating insect pests and vectors of disease agents by nonchemical, biological means. Several biological approaches have been developed to deal with pests that compete for food supplies and natural resources as well as with arthropods that act as carriers of pathogens.

During the past several years, we have become aware that viral insecticides have the potential to help cope with pest control problems without endangering man and the environment. Since the production of food and fiber and the prevention of vector-borne disease are among the most urgent problems that mankind must face in the near future, it seemed appropriate to devote a treatise to viral insecticides and to discuss the many factors and considerations involved in the use of such biological control agents. While an impressive wealth of information has been assembled in journals and books on microbial control, no single volume has been devoted entirely to viral insecticides until now.

There are many different types of viruses associated with insects. Several cause epizootics under natural conditions. The introduction of viral insecticides has the potential to significantly reduce the current reliance on chemical pesticide technology. The use of integrated pest control in which insect viruses are used in combination with small amounts of chemical pesticides thus promises to provide a desirable alternative method of pest management.

The host specificity of most insect–pathogenic viruses, especially baculoviruses, permits the control of certain undesirable insects without damaging beneficial insects or the ecosystem. While no detectable undesirable effects of baculoviruses have been observed, safety of each baculovirus has to be established before wide-scale use, including tests on mammals and other vertebrates, inhalation tests, and skin and eye sensitivity tests. In addition, it may be necessary to apply cytogenic analysis, DNA hybridization, and radioimmunoassay, as well as tests for the possible persistence of the insect virus genome in nontarget systems, before new large-scale applications are undertaken. Baculoviruses have been identified as having the best potential for microbial pest control. On the other hand, in certain conditions and parts of the world, naturally occurring insect disease outbreaks caused by viruses other than baculoviruses have already been utilized on a smaller scale to control pests efficiently and at a much lower cost than with the use of chemicals. Should the use of field-collected insect viruses continue or should it be halted because such viruses may pose a potential biohazard? The information assembled for the first time in this volume by foremost experts will enable others to develop the means to identify hazardous problems and prevent them. It supports the view that a number of already licensed viral pesticides, as well as several others that are now being developed, seem to be safe biological control agents with promising potential effectiveness. In contrast to most chemical pesticides, the use of viral insecticides does not seem to result in environmental contamination or disruption of the ecosystem.

We hope that this comprehensive volume, containing basic as well as applied aspects of this vast and multidisciplinary area, will benefit a wide audience of readers, including researchers, students, and those working directly in crop protection. The book will be of value to agronomists, horticulturists, entomologists, virologists, as well as health officials because it collates information that until now has been scattered throughout the literature. We also hope that this work will stimulate new environmental approaches to pest and vector control and will help elucidate new concepts, applications, and techniques.

We wish to thank all contributors for the effort and care with which they have prepared their chapters, and the staff of Academic Press for their part in the production of the volume.

KARL MARAMOROSCH
KENNETH E. SHERMAN

I.

Taxonomy and Identification

TAXONOMY AND NOMENCLATURE OF INSECT PATHOGENIC VIRUSES

T. W. TINSLEY
D. C. KELLY

Natural Environment Research Council
Institute of Virology, Oxford

I. INTRODUCTION

This chapter deals with those viruses that are able to cause disease in the insect host. It does not include viruses that cause disease in vertebrates or in plants even though virus replication occurs in the arthropod vectors.

The nomenclature of insect viruses is closely linked with the history of their discovery and to their subsequent characterization. Much of the earlier work on the isolation and characterization of insect viruses was by entomologists who were well versed in the taxonomy of the host but not necessarily of the pathogens. This led them to assume that the viruses were very host specific. It has become routine to add the Latin generic and/or the specific name of the host to form a descriptive name for the virus, e.g., the nuclear polyhedrosis virus of the silkworm became *Bombyx mori* nuclear polyhedrosis virus. Such composite names are still found in the literature and this may indicate either that they still have some value or that old systems take time to change. The problems arise when additional but quite different viruses of the same group are found in the same host. Thus the use of the host Latin binomial name for the virus leads to considerable confusion in the literature. The position becomes even more complicated when a vernacular name of an insect is added to virus groups.

ISBN 0-12-470295-3

For example, "mosquito iridescent virus" is used to describe viruses isolated from a wide range of different mosquito species. To add to this point, subsequent work has shown that, in fact, a complex of unrelated viruses is involved in these blood-sucking flies. The confusion with iridescent viruses grew until it became necessary to propose a system of interim type numbers for the various isolates of the iridescent viruses (Tinsley and Kelly, 1970). However, it must be emphasized that the problems of virus nomenclature and their subsequent taxonomy were considerable. The measures taken to solve at least some of these problems are emphasized in this chapter.

The importance of diseases in populations of beneficial insects, such as, the silkworm and the honeybee (*Apis mellifera*) were recognized early by workers in the field. Full accounts of diseases now known to be caused by viruses have been provided by Nysten (1808), Cornalia (1956), Maestri (1856), Langstroth (1857), Bolle (1894), von Prowazek (1907) and White (1917). Reviews of much of the early work have been presented by McKinley (1929) and Steinhaus (1949). There were several early attempts to provide systems of nomenclature which, it was hoped, would bring some form of order to the rapidly growing confusion in the literature. Holmes (1948) in his large and general scheme of virus nomenclature introduced two generic names, "Borrelina" for the polyhedrosis viruses associated with inclusion bodies and "Morator" for all the other known insect viruses that were not associated with inclusion bodies. Unfortunately, this system did little to clarify the situation and, therefore, was not widely adopted.

Bergold et al. (1960) put forward a system of insect nomenclature derived from the names of various prominent workers in the field. The proposed names included "Bergoldiavirus," "Smithiavirus," "Vagoiavirus," and, although recongizing human merit, did not provide any information about the nature and properties of the viruses. Further, it ignored the fact that such systems were not in accord with the current thinking in other branches of virology. Thus there was an indication that insect virology was developing in isolation. This was a distinct disadvantage and it became clear that the problems would multiply rather than decrease if a major effort was not forthcoming to prevent further confusion and divergence of opinion. To this end, a significant event occurred at the International Congress for Microbiology held in Moscow, USSR, in 1966--the launching of the International Committee on the Nomenclature of Viruses (ICNV). This body set up subcommittees to consider the entire field of virology, first independently and later together. It is very likely that the establishment of this Committee will come to be regarded as perhaps the most important single event in the general study of virus taxonomy, having effectively brought insect virology into the mainstream of events. The development of this international body and its influence on policy and on virus research will be discussed later. It is also significant that the work of the various subcommittees and study groups has had a profound stimulating effect on the development of the study of comparative virology; this has been an unexpected bonus as it was certainly not part of the original plan.

One problem that soon becomes apparent in any study or consideration of insect pathogenic viruses is the actual state of the virus infection. The expression of frank infection and concomitant disease are perhaps atypical of virus infections in field populations of insects. Inapparent or latent infections with viruses are so common in natural populations that it has been suggested (Tinsley, 1975) that perhaps these must be regarded as the normal expression. However, natural epizootics of insect viruses do occur and can be so extensive as to cause the collapse of the host population. However, detailed investigations usually reveal that adverse climatic or biotic factors have provided the stimuli to activate occult infections. Such latent infections make the problem of virus isolation very difficult and also have a direct relevance to the difficulties in trying to fulfill Koch's Postulates, particularly when insects are used as the test material.

The development of more sensitive and reliable methods of virus identification has helped considerably to distinguish between virus isolates. These methods have also done much to put the questions of specificity of insect viruses on a quantifiable basis. Such studies have indicated that isolates of certain viruses previously thought to be distinct were, in fact, related strains of one and the same virus. This has naturally reduced the confusion in the literature to a certain extent but there is still much to be done before a clear picture can emerge.

II. GROUPS OF PATHOGENIC VIRUSES FOUND IN INSECTS

There are at least ten groups of viruses known from insects. The present classification of these groups is shown in Table I. This list must be regarded as incomplete since many more viruses need to be characterized and described. To date, only about 1% of the known insect species have been examined for virus infection. This figure, given present workers and resources, is not likely to increase appreciably in the next decade. A further complication is that some of the methods in current use for the isolation of insect pathogenic viruses may, in fact, be counterproductive. For example, many laboratories in which insect pathogenic viruses are studied, use various

TABLE I. *Groups of Viruses Associated with Insects*

Family	Genus	Nucleic acid	Particle shape	Association with inclusion body	Biochemical and Biophysical similarities to plants found in	
					Vertebrates	Plants
Baculoviridae	Baculovirus	DNA	Rod	+	None	None
Poxviridae	Entomopoxvirus	DNA	Ovoid	+	Poxviruses	None
Reoviridae	Cytoplasmic polyhedrosis virus	RNA	Isometric	+	Reovirus blue-tongue	Rice dwarf, wound tumor
Iridoviridae	Iridovirus	DNA	Isometric	-	African swine fever Frog viruses 1-3	None
Parvoviridae	Densovirus	DNA	Isometric	-	Parvoviruses	None
Picornaviridae	Unassigned	RNA	Isometric	-	Picornaviruses	Many small RNA viruses
Caliciviridae	Unassigned (Amyelois virus)	RNA	Isometric	-	Caliciviruses	None
Nodaviridae	Nodavirus	RNA	Isometric	-	[a]	None
Rhabdoviridae	Sigmavirus	RNA	Bullet-shaped bacilliform	-	Rhabdoviruses	Rhabdoviruses
Nudaurelia β virus group	Unassigned	RNA	Isometric	-	None	None

[a] Nodamura virus also replicates in various vertebrate cells and in suckling mice.

organic solvents in the initial extraction process and such materials would be destructive to any virus that contains lipid in the outer membranes or layers. Furthermore, viruses that are also labile at alkaline pH may not survive the initial extraction. Viruses prone to aggregation without special treatment may also be eliminated in the initial low-speed centrifugation cycles. Therefore, the extraction and purification protocols in current use may have exerted a certain degree of selection pressure on the type or group of insect viruses that would be present in the final preparations. Unfortunately, there are not yet enough permissive cell systems available for the direct recovery of viruses from diseased material.

A. *BACULOVIRUSES (Baculoviridae)*

Baculoviruses have been recorded from the following Orders of insects: Neuroptera, Trichoptera, Lepidoptera, Diptera, Hymenoptera and Coleoptera. Again, this must be regarded as a minimum list as little work has been done on the more primitive orders particularly if the individuals hosts are small, such as, Collembola, Psoptera, Mallophaga, or Thysanoptera. Another obstacle is provided by the difficulty in collecting diseased aquatic larvae of species in the Ephemeroptera or Odonata or in catching the very actively flying adults. Baculoviruses (or Baculovirus-like particles) have been isolated from various Crustacea, mites, and a fungus (Matthews, 1982).

Baculovirus, derived from the Latin *baculum* (a rod or stick) are rod-shaped of variable size but within the range of 40-70 nm × 250-400 nm. They are composed of a lipoprotein envelope surrounding a protein/DNA core (nucleocapsid). The virions are usually occluded within a proteinaceous lattice crystal known as the inclusion body.

The properties of the viruses included in the four groups have remarkably similar basic properties. These characteristics provided for logical grouping. However, relationships within the groups are much more complex and are therefore more difficult to assess. However, the application of modern techniques has done much to clarify the situation.

The virions of baculoviruses possess 11-25 polypeptides (Payne, 1974; Harrap et al., 1977; Brown et al., 1977; Summers and Smith, 1978; Harrap and Payne, 1979) of which 4-11 are associated with the nucleocapsids; the remainder are most probably associated with the capsid envelopes. The genome is composed of a single molecule of double-stranded, circular, supercoiled DNA with a molecular weight of 50 to 100 × 106 (Harrap and Payne, 1979). Most baculoviruses appear to replicate in the cell nucleus although some viruses in subgroups B and C can be seen to replicate in the cytoplasm (Kelly, 1976; Summers, 1977).

Signs of gross pathology in infected insects is of little use in the identification of baculoviruses or most insect viruses. Insect larvae react to infection with any virus in roughly the same manner: the integument changes color, usually becoming pale white or grey in color, general flaccidity occurs, and finally the integument ruptures. Moribund larvae of the Lepidoptera and Hymenoptera usually hang from the host plant in an inverted position. There are marked tissue tropisms in that viruses in subgroups A and B affecting Lepidoptera replicate in most tissues whereas viruses in these groups affecting the Hymenoptera apparently replicate only the epithelial cells of the gut.

It is possible to stain preparations of baculoviruses so that they can be seen with the light microscope. These viruses can often be identified by differential staining with Giemsa. However, such methods are obviously only useful in determining the major virus group to which the stained inclusion bodies belong. It is evident that although the inclusion body proteins of baculoviruses are distinguishable, it is the properties of the nucleic acid and associated proteins in the virion that are likely to provide the more sensitive systems of identification.

B. *ENTOMOPOX VIRUSES (Poxviridae)*

Entomopox viruses have been recorded in the Lepidoptera, Orthoptera, Coleoptera, and Diptera (Kurstak and Garzon, 1977) and have many morphological and biochemical similarities to the poxviruses found in vertebrates. However, the somewhat atypical entomopox virions are occluded within a proteinaceous inclusion body which has a different molecular weight from that associated with baculoviruses, although the protein matrix has a similar structure.

The dsDNA has a molecular weight of approximately 110-200 $\times 10^6$ and there are at least four associated enzymes. The virus replicates exclusively in the cytoplasm of infected cells, mainly in the fat body (Bergoin and Dales, 1971; Kurstak and Garzon, 1977).

Histological examination with the light microscope of diseased tissue reveals little that would be useful in diagnosis. Examination with the electron microscope is essential in the detection of the virions and to determine core structure.

C. *CYTOPLASMIC POLYHEDROSIS VIRUSES (Reoviridae)*

This group of viruses (CPVs) is very common in the Lepidoptera and Diptera, although CPV-like particles also have been recorded by Federici and Hazard (1975) in a crustacean (*Simocephalus espinosus*). These viruses exhibit some biochemical similarities to viruses within the family Reoviridae, which affect vertebrates and plants. The viruses are icosahedral and have a diameter of 50 to 65 nm. The particles have spikes at their vertices (Hosaka and Aizawa, 1964; Payne and Harrap, 1977) which are thought to be hollow, some 20 nm in length, and appear to originate within the core of the particle (Miura et al., 1969).

The virions can be occluded either singly or in large numbers within the inclusion body. The protein lattice of this body comprises a single major polypeptide (mol. wt. 25,000-31,000). The virus particles contain from three to six polypeptides of which two (sometimes three) have molecular weights in excess of 100,000 (Payne and Rivers, 1976). The genome has 10 segments of double-stranded RNA and, by analogy with other viruses in the Reoviridae, it can be assumed that each segment represents a single gene. A provisional classification of the insect viruses within the group has been made on the basis of the different sizes of the RNA segments. This system proposed 12 distinct virus types from the 40 isolates of CPV examined (Payne and Rivers, 1976; Payne et al., 1977).

Most CPVs are thought to replicate in the cytoplasm of the epithelial cells of the gut tissue, although Kawase et al. (1973) observed inclusion bodies in the nuclei of the silkworm (*Bombyx mori*). Inclusion bodies of CPVs can be distinguished from those of baculoviruses by differential staining with Giemsa (NPV) and naphthalene black (CPV).

Serological methods confirm the typing suggested by the analysis of RNA segments, with ELISA and radioimmunoassay being particularly sensitive. However, as Payne and Churchill (1977) have pointed out, it is essential that intact virus particles are used as antigens in comparative tests, because there is a danger of spurious reactions as a result of antibodies being produced against the dsRNA as well as the virus-specific proteins. A further safeguard would be to absorb out the dsRNA-specific antibodies with a nonviral dsRNA antigen.

D. *IRIDOVIRUSES (Iridoviridae)*

These large DNA viruses are associated with the cytoplasm and can be grouped into two sizes, the smaller being about 130 nm and the larger about 190 nm in diameter. The nucleic acid is in an inner core which is surrounded by an internal capsid and lipid membrane which, in turn, is covered by an outer capsid layer (Kelly and Vance, 1973). The DNA is linear and has molecular weights of 114-150 × 10^6 (small IV) and 240-288 × 10^6 (large IV) (Bellett and Inman, 1967; Kelly and Avery, 1974; Wagner and Paschke, 1977). There are reports of several enzymes in these virions including: DNA-RNA polymerase (Kelly and Tinsley, 1973), nucleotide phosphohydrolase (Kelly and Robertson, 1973; Monnier and Devauchelle, 1978), and a protein kinase (Monnier and Devauchelle, 1978; Kelly et al., 1980b).

The iridoviruses replicate in the cytoplasm of cells in a wide range of tissues in the natural order Lepidoptera, Diptera, Coleoptera, Hemiptera, and Hymenoptera of the Class Insecta.

Infection by many iridoviruses can be easily recognized by the iridescence caused by the high concentration of virus particles in the cytoplasm packing into quasicrystalline arrays. However, this is not a constant feature and reflects very high concentrations of virus and the physical morphology of the particles rather than the intrinsic pathology. Biochemical methods including SDS gel electrophoresis, restriction enzyme analysis, and homology experiments have not proved as useful as was hoped in providing adequate data for identification. Kelly et al. (1979) have summarized what is known of relationships of iridoviruses based on biological techniques. Two of the isolates in the classification of Tinsley and Kelly (1970) (Types 6 and 24) appeared to be quite distinct from each other and from all the other isolates. Types 1, 2, 9, 10, 16, 18, 21, 22, 23, 25, and 28 all shared common antigens and at least three groups of serotypes could be distinguished by sensitive analytical methods.

E. DENSOVIRUSES (Parvoviridae)

These small DNA viruses have many features in common with parvoviruses associated with vertebrates. Only three of the various insect parvoviruses described have been studied in any detail, namely, the viruses from the greater wax moth, *Galleria mellonella* (DNV1) (Meynadier et al., 1964), the Buckeye moth, *Junonia coenia* (DNV2) (Rivers and Longworth, 1972), and the silver spotted flambeau, *Agraulis vanillae* (Kelly et al., 1980a). The isometric particles contain linear single-stranded DNA (Barwise and Walker, 1970) and complementary strands are packaged in separate virions. The ssDNA has a molecular weight of 1.6-2.2 × 10^6 (Barwise and Walker, 1970; Kurstak, 1972; Kelly et al., 1977) and is closely associated with polyamines in the virion (Kelly and Elliott, 1977).

Reactions in gel diffusion tests show that the three viruses share common antigen(s) and must be regarded as related strains or serotypes. This relationship has been further demonstrated by REN analysis and DNA homology (Kelly et al., 1977; Kelly et al., 1980c). On the other hand, at least two of these related strains have very different host ranges in that DNV1 is known to infect only its original host *Galleria mellonella*, whereas DNV2 will not only infect the host of origin, *Junonia coenia*, but several other members of the *Nymphalidae* as well as certain species in the family *Noctuidae*; DNV2 will not infect *G. mellonella*.

F. SMALL RNA VIRUSES

Large numbers of small RNA viruses have been isolated from a wide range of insect species. Most of them are regarded as being potential members of the family Picornaviridae, although only a few have been given possible assignments to genera. Longworth (1978) has pointed out that many of the insect RNA viruses may be so different from members of the established groups that new groups may have to be delineated and new terminology set up with which to describe them.

The viruses, so far described, seem to have marked associations with the cytoplasm and also with gut epithelial cells. Certain viruses also invade and replicate within the brain and nervous system which leads to paralysis. The general pathological picture is one of acute enteric disease often followed by varying degrees of paralysis. It is interesting to note the overall similarity to picornavirus, particularly enterovirus, infection in vertebrates. The viruses in this diffuse group can be readily detected using gel diffusion methods. Some of the viruses also grow readily in cell culture derived from the fruit fly *Drosophila melanogaster*.

Longworth (1978) has proposed an interim classification of these viruses similar to that proposed for vertebrate picornaviruses. Matthews (1982) presented the view of the ICTV, which was more cautious in its assessment and although three insect viruses were recognized as being members of the Picornaviridae (but not yet assigned to genera) another 30 viruses were designated as "unclassified small RNA viruses of invertebrates." Two other groups recognized by Longworth [*Nudaurelia* β viruses and Nodamura virus (Nodaviridae)] were accorded Family status by Matthews but were clearly outside the Picornaviridae.

G. SIGMAVIRUSES (Rhabdoviridae)

The sigma virus of the fruit fly *D. melanogaster* is a hereditary infectious agent often present in populations of these flies. They can be detected by the sensitivity of the infected flies to carbon dioxide. Berkaloff et al. (1965) and Teninges (1968) have produced photographic evidence of rhabdovirus-like particles in tissues of infected flies. The particles measured 140-180 × 7 nm and cross-striation was visible in negatively stained preparations. Since the sigma virus has not been purified, little is known of its properties. Present evidence points to an agent which multiplies in its host, is transmissible, and will confer sensitivity to carbon dioxide; rhabdovirus-like particles can be seen in infected tissues. Unfortunately, until the viruslike particles are purified and thereby concentrated, it is not possible to state unequivocally that they are responsible for the manifestation of CO_2 sensitivity. It is for this reason that the actual viral nature of this disease is frequently questioned.

H. OTHER GROUPS AND POSSIBLE MEMBERS

The class Insecta comprises about three-quarters of the known animal species and it is also quite possible that insects harbor the largest number of viruses of all known groups. Furthermore, it is very likely that insects and other invertebrates have viruses that do not fit into this present classification and so would require new genera and perhaps the creation of new families.

Recent examples of these and unclassified viruses from or associated with insects are given below.

1. The X virus of *Drosophila melanogaster* (Teninges et al., 1979) which is a double-stranded RNA virus in two segments of molecular weight 2.5×10^6 and 2.6×10^6. The particle is ca. 60 nm in diameter and is similar to viruses found in chickens (Nick et al., 1976), mollusks (Underwood et al., 1977), and fish (Dobos et al., 1979). Recently, another virus in this tentative group was found in Lake Thirlmere (UK) (Kelly et al., 1982). This virus has probable invertebrate connections in view of the fact that the virus replicates in insect cell lines. There is also circumstantial evidence of large populations of such animals in the lake.

2. A small RNA virus isolated from diseased larvae of the navel orange worm (*Amyelois transitella*) which has certain similarities to viruses in the Caliciviridae (Kellen and Hoffman, 1981; Hillman et al., 1982).

3. A group of very small viruses (13-17 nm) and with low S values (42-45 S). It has been suggested that two of these viruses isolated from *Antheraea eucalypti* (Bellett et al., 1973) and the honeybee *Apis mellifera* (Bailey, 1976), many have the function of satellites for larger viruses found in these host species. However, more work is required since this sug-

gestion is based on analogies only. A very small virus (13 nm) has been reported from the grasshopper *Melanoplus bivattatus* (Jutila et al., 1970) which, atypically, is highly pathogenic for its host and appears to function in its own right and not as a satellite. Reinganum (1975) has described another small virus (14 nm) from preparations of the cricket paralysis virus (CrPV) from infected larvae of *A. eucalypti*. These viruslike particles were not related serologically to CrPV or to a 34 nm virus also found in *A. eucalypti*. A similar sized virus is present in preparations of CrPV grown in larvae of the wax moth, *Galleria mellonella*. Such particles are not found in extracts of noninfected larvae. It is not known if this small virus is a contaminant or if it has a more specific role as satellite to the larger CrPV (Tinsley and Robertson, 1983).

The more members that can be obtained of established groups of viruses associated with insects will facilitate comparisons. It follows that basic biochemical data is also required on newly isolated viruses because simple records of their occurrence in new hosts add little to our general knowledge. Further, the discovery of entirely new groups of viruses, even though restricted to insects or perhaps invertebrates generally, would make the classification of viruses more comprehensive. Such information would make a valuable contribution to general virus taxonomy. Therefore, the need to make detailed searches for new viruses and new virus groups is clearly indicated and it is hoped that such projects will be encouraged.

II. CRITERIA AND METHODS TO DETERMINE RELATIONSHIPS BETWEEN VIRUSES

There is no single reliable method to provide either unequivocal identification of a virus or evidence of relationship of the test virus to others that have similar properties. Therefore, we have to rely on the sum total of results obtained from many biochemical and biological methods if a balanced assessment is to be made.

Serological methods are by far the most commonly used methods and it is usually wise to use the simplest and least complicated methods first.

It is obvious that preparations of pure viral antigens are essential and the methods by which these can be obtained are to be found in standard textbooks of virology and immunology. The specificity of the antiserum that is produced depends largely on the purpose of the investigations. If accurate diagnosis is required then freedom from any contaminating viruses and/or inhibitors of antigen/antibody reactions is essential. When detailed knowledge of component antigenic groups is called for then the highest degree of antigen purity must be provided with the elimination of any host protein contaminants.

Rabbits are the usual animals in which specific antibodies to insect viruses are produced, although guinea pigs are also used. Less commonly, the ascitic fluid from immunized mice is used and this is a potent and convenient source of specific antibodies. The immunization procedures vary but are usually based on a series of intramuscular injections of viral antigens. The initial injection is made with addition of a complete adjuvant, while subsequent injections use incomplete adjuvant. Such intramuscular routes may be interspersed with one or more intravenous injections of antigen alone without adjuvant. The importance of using groups of animals for the production of any one antisera must be stressed as individual animals vary so widely in their level of hyperimmune responses to antigens. Subsequent selection from the serial bleedings taken from groups of animals then provides the most potent antisera.

The development of a technique to produce monoclonal antibodies from the fusion of specific antibody-producing cells with malignant cells (Köhler and Millstein, 1975) represents a significant advance. There is every likelihood that this technique will increase the accuracy with which the antigens and/or antigenic groups of insect viruses can be studied. There is the additional advantage that once the various hybridomas have been produced, the need to use living animals for the production of specific antibodies will be reduced.

Most of the serological methods used to study insect viruses are based on the precipitation of antigen by specific antibody. The degree of combination can be measured by various forms of visual assessment, e.g., the development of opalescent bands by diffusion of antigen and antibody in agar gels or by the volume of precipitate in aqueous media. The process of the binding of specific antibody to antigen can be speeded up by immunoelectrophoretic techniques. The antigens are separated first by electrophoresis in one direction and then allowed to diffuse in a second direction toward a source of antiserum. This system provides a measurement of electrophoretic mobility as well as antigenic specificity. A further refinement is to use cross-over (counter) electrophoresis which is based on the fact that most antigens at pH 7.0-8.0 carry a high negative charge and in a suitable media, as buffered agar gels or cellulose acetate, move toward the anode. On the other hand, antibodies in such systems have a lower negative charge and move toward the cathode. Antigen and antibody thus can be made to move toward each other and precipitation lines are formed that are visible after treatment with suitable strains. The principal advantage of the cross-over electrophoresis system is not simply speed of reaction but also that much smaller quantities of antigen can be detected.

Enzyme-labeled antibodies have proved very useful in quantitative determination of antigens and antibodies. Specific modifications for the serodiagnosis of such viruses as measles, rubella, mumps, and Newcastle disease have been described by Voller and Bidwell (1977). Kelly et al. (1978) have reported on the use of the enzyme-linked immunosorbent assay (ELISA) for insect viruses. The detection of small quantities of antigen by this method is well documented (Clark et al., 1976; Kelly et al., 1978; Devergne et al., 1978) and unequivocal results with as little as 1-10 ng/ml virus have been reported.

The ELISA test is capable of discriminating between closely related virus strains and its specificity is much greater than that of precipitin tests. Indeed, it is claimed to equal that of neutralization *in vitro* (Mills et al., 1978). A pool of antisera, specific for various serotypes, may be required when a broad range of virus strains is involved.

The radioimmunoassay test involves the use of a radioisotope linked either to the antigen or to the antibody. The radioactivity associated with the antigen/antibody complex can be measured after any of the unbound labeled material has been removed. The solid phase system involves the bonding of one of the reagents to the surface of a particular compound and all reactions take place at this specific site; in the liquid phase system, all the reagents are in suspension. The methods to be used and the applications of radioimmunoassays in virology have been presented by Daugherty and Ziegler (1977). The development of additional solid-phase techniques with the objectives of additional sensitivity have been described by Middleton et al. (1977).

The relative ability of direct versus indirect solid-phase radioimmunoassays to detect polyhedral protein of an insect baculovirus has been studied by Crawford et al. (1977). The direct assay using rabbit IgG labeled with ^{125}I was able to detect 200 μg of polyhedral protein whereas the indirect assay using labeled sheep anti-rabbit IgG was able to measure as little as 50 μg of protein.

The complement fixation test has been the method of choice in the routine diagnosis of mammals in viruses in many laboratories but the use of other more sensitive methods, such as ELISA, have rather overshadowed its use. The basis of the test is provided by the visual assessment of the degree of binding or "fixing" of a proteinaceous compound known as complement. This substance is present in sera unless destroyed by heating to 56°C for 10 min. The actual test involves two stages: one is the actual test virus antigen and antibody reaction and the other is the indicator system of sheep red blood cells (RBC) and homologous antibody. The first step is to inactivate complement in the test antiserum and also in the anti-sheep RBC serum. Then, a quantity of guinea pig serum complement, calculated to be able to cause total hemolysis of the RBC in the indicator system is added to the test antiserum. The viral antigen and homologous antibody plus complement are incubated overnight. Differing amounts of complement will be fixed depending on the proportions of antigen and antibody present in the test system. At this stage, the sheep RBC plus lytic serum are added and the degree of hemolysis which occurs is a direct measure of the amount of free complement present. This, in turn, is inversely proportional to the extent to which the test antigen combined with its homologous antisera. Perhaps the main disadvantage of this system is the exhibition of anticomplementarity whereby certain sera and/or antigen preparations can bind complement in the absence of a true antigen/antibody reaction.

Serological methods provide valuable criteria for the initial diagnosis of viruses and also can provide varying amounts of information on the relationship between virus isolates. Ideally, serological methods should be used in conjunction with other chemical techniques so that an overall picture of relationship can be accumulated. Analyses of the structural proteins of a virus provide an indication of the genetic information contained in the viral genome. Variations in the number and proportion of virus structural proteins can provide valuable markers with which to differentiate between virus strains. The biochemical techniques used to investigate the proteins of insect viruses include gel electrophoresis and mapping of the peptides. Polyacrylamide gel electrophoresis of proteins treated with anionic detergents has been used to examine isolates of all the major groups of insect viruses. Valuable comparative analyses can be made of viruses and virus strains when tested under the same conditions. A virus contains virus-specific proteins in the same concentration in each and every virus particle and so will present a unique and reproducible pattern on each examination. However, different viruses within the same group, such as, baculoviruses (Harrap et al., 1977; Summers and Smith, 1978) and iridescent viruses (Kelly and Tinsley, 1973; Elliott et al., 1977; Carey et al., 1978; Kelly et al., 1979), exhibit different profiles for different isolates. Differences of mobility and even concentration between viral polypeptides are significant parameters. However, it must be remembered that even though two proteins exhibit similar molecular weight this does not imply identity. In such cases, additional biochemical evidence would be required to establish structural and chemical identity.

Perhaps the most significant application of peptide mapping is to examine viral proteins that appear to be identical using other methods. The method is based on the enzymatic cleavage of the proteins followed by chromatographic and/or electrophoretic separation of the resulting peptides. The comigration of individual peptides strongly suggests that they are, in fact, identical. This technique has been applied to the analysis of baculoviruses inclusion body proteins by Summers and Smith (1976). Maruniak and Summers (1978) found that, although the proteins from different viruses showed a general relationship, there were distinguishing peptides.

Naturally, differences between viral proteins are reflections of basic differences between the composition of viral nucleic acids. Virus groups can be easily identified by the nature of the nucleic acids or by their strandedness (single or double). Finer differences have to be established by comparisons of molecular weights, base compositions, and buoyant densities of the nucleic acids. However, the most sensitive and therefore perhaps most useful techniques are those employed to compare various base sequences of the nucleic acids. For example, restriction endonucleases are used to recognize specific base sequences which are then cleaved at these sites. The fragments can then be separated on a size basis by electrophoresis on agarose gels. This separation then produces a characteristic electrophoretic profile for each virus. Common patterns would suggest genetic relatedness whereas completely different profiles would reflect either very distinct relatedness or nonidentity.

The use of several restriction enzymes improves this identification technique not just by providing more data but by considerably reducing the chance of obtaining identical profiles. The method has been used with great success in the analysis of baculovirus genomes (Miller and Dawes, 1978, 1979; Lee and Miller, 1978; Smith and Summers, 1978). The fragmentation of baculovirus DNA's by restriction enzymes will provide a very sensitive method with which to evaluate the accuracy and sensitivity of simpler serological techniques.

The natural segmentation of the dsRNAs from the cytoplasmic nuclear polyhedrosis viruses (CPV) can be used to classify, at least provisionally large numbers of virus isolates in this group (Payne and Rivers, 1976). Unfortunately, a range of enzymes with which to cleave to segments is not yet available and so identity in electrophoretic mobility does not necessarily imply identical base sequence.

Homology studies provide an additional parameter to distinguish between virus isolates by determining the degree of hybridization between the viral nucleic acids. An accurate assessment of homology also requires a knowledge of the reassociation kinetics of double stranded nucleic acids. Such analyses have been applied only to a few insect viruses in the Baculovirus, Iridovirus, and Densovirus groups (Kelly and

Avery, 1974; Kelly, 1977; Kelly et al., 1977; Wagner and Paschke, 1977; Rohrmann et al., 1978; Jurkovicova et al., 1979). There is no doubt that homology studies can make a contribution to the analyses of closely related viruses, but at the moment the techniques are very time consuming and are therefore not suited to routine diagnostic procedures.

The criteria of relationships are vital parameters in any system of taxonomy. It is clear from this brief review of the identification procedures available that a range of techniques must be used in order to provide data for a dependable assessment. The selection of the techniques must be determined by the type or group of virus involved, e.g., double diffusion in agar gels is obviously not indicated for large nondiffusible virions of the Baculoviruses or Iridoviruses. Further, hemagglutination and/or hemagglutination inhibition are obviously not applicable to the many insect viruses that fail to attach to erythrocytes. Finally, neutralization of infectivity tests in whole animals is tedous and often produces equivocal results but this may be the only alternative when tissue culture cell lines are not available. It is hoped that this sensitive technique will have greater application to insect viruses when more cell lines are available.

III. DEVELOPMENT OF TAXONOMY AND NOMENCLATURE OF INSECT VIRUSES

The advances made in the taxonomy of viruses since the establishment of International Committees have already been mentioned. However, it should not be forgotten that the original function was not taxonomy but simple nomenclature. This was realistic and expedient because the confusion over names was growing and it was necessary first to attempt a system of nomenclature. The question of relationships between viruses was a much thornier problem to tackle particularly because it raised questions of dubious phylogeny and of latinized binomials for viruses in the minds of those more acquainted with the classic systems of the taxonomy for animals and plants.

The policy laid down in Moscow in 1966 was that (1) groups (genera) of viruses would be defined and listed; (2) type members (species) representative of the groups would be provided; (3) that names for the groups (genera) would be proposed; and (4) that the use of the cryptogram or taxonomic device would be explored (Wildy, 1971). Many problems were faced by the original committees--not the least of which were the widely differing attitudes toward virus taxonomy which often seemed quite irreconcilable. Knowledge of vertebrate viruses was very much greater than the rest of virology and expertise varied widely. It was difficult to use this specialized knowledge and to still maintain a universal approach. Perhaps the most difficult problem to resolve was to produce names of viruses that working virologists actually wanted to use.

Three reports on the International Committee's work have been published since 1971 (Fenner, 1976; Matthews, 1979; 1982). It is remarkable that rapid progress has been made in a decade. A glance at the most recent report reveals the wide adoption of latinized names for families and genera. Most significant is the fact that descriptions of type species are beginning to appear under defined genera. Evidently, the fears and misgivings expressed so forcefully at the early meetings have been dispelled or at least allayed. This is a good augury for the future.

Viruses of insects and other invertebrates are now being considered, in conjunction with those of vertebrates, by joint working parties. This is an important step in the comparative study of viruses since it emphasizes that viruses of animals should be considered in totality. This is substantiated by the fact that, to date, only one family, the Baculoviridae, can be considered as being unique to invertebrates. However, as was pointed out early in this chapter, there are many unknown viruses in invertebrates waiting to be described and it may be wiser to reserve a detailed taxonomic consideration until these are described.

REFERENCES

Bailey, L. (1976). Viruses attacking the honeybee. *Advan. Virus. Res. 20*, 271-304.

Barwise, A. H. and Walker, I. O. (1970). Studies on the DNA of a virus from *Galleria mellonella*. *FEBS Lett. 6*, 13-16.

Bellett, A. J. D. and Inman, R. B. (1967). Some properties of deoxyribonucleic acid preparations from *Chilo, Sericesthis* and *Tipula*. *J. Mol. Biol. 25*, 425-432.

Bellett, A. J. D., Fenner, F., and Gibbs, A. J. (1973). The Viruses. *In* "Viruses and Invertebrates" (A. J. Gibbs, ed.), pp. 41-88. North Holland, Amsterdam.

Berkaloff, A., Breghano, J. C., and Ohanessian, A. (1965). Mise en évidence de virions dans des drosophiles inféctees par le virus héréditaire σ. *C.R. Acad. Sci. D260*, 5956-5959.

Bergoin, M. and Dales, S. (1971). Comparative observations on poxviruses of invertebrates and vertebrates. *In* "Comparative Virology" (K. Maramorosch and E. Kurstak, eds.), pp. 169-205. Academic Press, New York.

Bergold, G. H., Aizawa, K., Smith, K., Steinhaus, E. A., and Vago, C. (1960). The present status of insect virus nomenclature and classification. *Intern. Bull. Bacteriol. Nomencl. Taxon. 10*, 259-262.

Bolle, J. (1894). Il Giallume del baco da seta: notizia preliminare. *Atti. Mem I.R. Soc. Agr. Gorizia 33*, 193.

Brown, D. A., Bud, H. M. and Kelly, D. C. (1977). Biophysical properties of the structural components of a granulosis virus isolated from the cabbage white butterfly (*Pieris brassicae*). *Virology 81*, 317-327.

Carey, G. P., Lescott, T., Robertson, J. S., Spencer, L. K., and Kelly, D. C. (1978). Three African isolates of small iridescent viruses: type 21 from *Heliothis armigera* (Lepidoptera: Scaraebidae), and type 28 from *Letriocerus columbiae* (Hemiptera Heteroptera: Belostomatidae). *Virology 85*, 307-309.

Clark, M. F., Adams, A. N., and Barbara, D. J. (1976). The detection of plant viruses by enzyme-linked immunosorbent assay (ELISA). *Acta Horticult. 67*43-47.

Cornalia, E. (1856). Monografia del bombice del gelso (*Bombyx mori* Linneo). *Mem. I.R. Inst. Lombardo Sci., Lett. Arti 6*, 1-387.

Crawford, A. M., Kalmakoff, J., and Longworth, J. F. (1977). A solid-phase radioimmunoassay for polyhedron protein from baculoviruses. *Intervirology 8*, 117.

Daugherty, H. and Ziegler, D. W. (1977). Radioimmune assay in viral diagnosis. *In* "Comparative Diagnosis of Viral Diseases" (E. Kurstak and C. Kurstak, eds.), Vol. II, pp. 459-487. Academic Press, New York.

Devergne, J. C., Cardin, L., and Quiot, J. B. (1978). Détection et identification sérologique des infections naturelles par le virus de la mosaique du Concombre. *Ann. Phytopathol. 10*, 253.

Dobos, P., Hill, B. J., Hallett, R., Kells, D. T. C., Becht, H., and Teninges, D. (1979). Biophysical and biochemical characterization of 5 animal viruses with bisegmented double stranded RNA genomes. *J. Virol. 32*, 593-605.

Elliott, R. M., Lescott, T., and Kelly, D. C. (1977). Serological relationships of an iridescent virus (Type 25) recently isolated from *Tipula* sp. with two other iridescent viruses (Types 2 and 22). *Virology 81*, 309-316.

Federici, B. A. and Hazard, E. I. (1975). Iridovirus and cytoplasmic polyhedrosis virus in the freshwater daphnid *Simocephalus expinosus*. *Nature (London) 254*, 327-328.

Harrap, K. A. and Payne, C. C. (1979). The structural properties and identification of insect viruses. *Advan. Virus. Res. 25*, 273-355.

Harrap, K. A., Payne, C. C., and Robertson, J. S. (1977). The properties of three baculoviruses from closely related hosts. *Virology 79*, 14-31.

Hillman, B., Morris, T. J., Kellen, W. R., Hoffman, D., and Schlegel, D. E. (1982). An invertebrate/calici-like virus: Evidence for partial virion disintegration in host excreta. *J. Gen. Virol. 60*, 115-123.

Holmes, F. O. (1948). Borrelinaceae. "Bergey's Manual of Determinative Bacteriology," 6th Ed., pp. 1225-1228. Williams & Wilkins, Baltimore, Maryland.

Hosaka, Y. and Aizawa, K. (1964). The fine structure of the cytoplasmic-polyhedrosis virus of the silkworm, *Bombyx mori* (Linnaeus). *J. Insect Pathol. 6*, 53-77.

Jurkovicova, M., van Touw, J. H., Sussenbach, J. S., and Terschegget, J. (1979). Characterization of the nuclear polyhedrosis virus DNA of *Adoxophyes orana* and of *Barathra brassicae*. *Virology 93*, 8-19.

Jutila, J. S., Henry, J. E., Anachev, R. L., and Brown, W. R. (1970). Some properties of a crystalline array virus (CAV) isolated from the grasshopper *Melanoplus bivattatus* (Say) (Orthoptera: Acvididae). *J. Invertebr. Pathol. 15*, 225-231.

Kawase, S., Kawamoto, F., and Yamaguchi, K. (1973). Studies on the polyhedrosis virus forming polyhedra in the midgut. Cell nucleus of the silkworm, *Bombyx mori*. I. Purification procedure and form of the virion. *J. Invertebr. Pathol. 22*, 266-272.

Kellen, W. R. and Hoffman, D. F. (1981). A pathogenic non-occluded virus in hemocytes of the navel orange worm, *Amyelois transitella* (Pyralidae: Lepidoptera). *J. Invertebr. Pathol. 38*, 52-66.

Kelly, D. C. (1976). "Oryctes" virus replication: electron microscope observations on infected moth and mosquito cells. *Virology 69*, 596-606.

Kelly, D. C. (1977). The DNA contained by nuclear polyhedrosis viruses isolated from four *Spodoptera sp*. (Lepidoptera, Noctuidae): genome size and homology assessed by DNA reassociation kinetics. *Virology 76*, 468-471.

Kelly, D. C. and Avery, R. J. (1974). The DNA content of four small iridescent viruses: genome size, redundancy, and homology determined by renaturation kinetics. *Virology 57*, 425-435.

Kelly, D. C. and Elliott, R. M. (1977). Polyamines contained by two densonucleosis viruses. *J. Virol. 21*, 408-410.

Kelly, D. C. and Robertson, J. S. (1973). Icosahedral cytoplasmic deoxyriboviruses. *J. Gen. Virol. (Suppl.) 20*, 17-41.

Kelly, D. C. and Tinsley, T. W. (1973). Ribonucleic acid polymerase activity associated with particles of iridescent virus types 2 and 6. *J. Invertebr. Pathol. 22*, 199-202.

Kelly, D. C. and Vance, D. E. (1973). The lipid content of two iridescent viruses. *J. Gen. Virol. 21*, 417-423.

Kelly, D. C., Barwise, A. H., and Walker, I. O. (1977). DNA contained by two densonucleosis viruses. *J. Virol. 21*, 396-407.

Kelly, D. C., Edwards, M. L., and Robertson, J. S. (1978). The use of enzyme-linked immunosorbent assay to detect, and discriminate between small iridescent viruses. *Ann. Appl. Biol. 90*, 369-374.

Kelly, D. C., Ayres, M. D., Lescott, T., Robertson, J. S., and Happ, G. M. (1979). A small iridescent virus (type 29) isolated from *Tenebrio molitor*: a comparison of its proteins and antigens with six other iridescent viruses. *J. Gen. Virol. 42*, 95-105.

Kelly, D. C., Ayres, M. D., Spencer, L. K., and Rivers, C. F. (1980a). Densonucleosis virus 3: a recent insect parvovirus isolated from *Agraulis vanillae* (Lepidoptera: Nymphalidae). *Microbiologica 3*, 455-460.

Kelly, D. C., Elliott, R. M., and Blair, G. E. (1980b). Phosphorylation of iridescent virus polypeptides *in vitro*. *J. Gen. Virol. 48*, 205-211.

Kelly, D. C., Moore, N. F., Spilling, C. R., Barwise, A. H., and Walker, A. H. (1980c). Densonucleosis virus structural proteins. *J. Virol. 36*, 224-235.

Kelly, D. C., Ayres, M. D., Howard, S. C., Lescott, T., Arnold, M. K., Seeley, N. D., and Primrose, S. B. (1982). Isolation of a bisegmented double-stranded RNA virus from Thirlmere Reservoir. *J. Gen. Virol. 62*, 313-322.

Köhler, G. and Millstein, C. (1975). Continuous cultures of fused cells secreting antibody of predefined specificity. *Nature (London) 256*, 495.

Kurstak, E. (1972). Small DNA densonucleosis virus (DNV). *Advan. Virus Res. 17*, 207-241.

Kurstak, E. and Garzon, S. (1977). Entomopoxviruses (Poxviruses of invertebrates). *In* "The Atlas of Insect and Plant Viruses" (K. Maramorosch, ed.), pp. 29-66. Academic Press, New York.

Langstroth, L. L. (1857). A practical treatise on the hive and honey bee. 2nd Ed. C. M. Saxton & Co., New York.

Lee, H. H. and Miller, L. K. (1978). Isolation of genotypic variants of *Autographa californica* nuclear polyhedrosis virus. *J. Virol. 27*, 754-767.

Longworth, J. F. (1978). Small isometric viruses of Invertebrates. *Advan. Virus Res. 23*, 103-157.

McKinley, E. B. (1929). Filterable virus and *Rickettsia* diseases. *Philipp. J. Sci. 39*, 1-413.

Maestri, A. (1856). Frammenti anotomiei, fisiologiei e patologici sul baco da seta (*Bombyx mori* Linn). *Fratelli Fusi, Pavia*, 1-172.

Maruniak, J. E. and Summers, M. D. (1978). Comparative peptide mapping of baculovirus polyhedrins. *J. Invertebr. Pathol.* *32*, 196-201.
Matthews, R. E. F. (1982). Classification and nomenclature of viruses. *Intervirology 17*, 1-200.
Meynadier, G., Vago, C., Plantevin, G., and Atger, P. (1964). Viruse d'une type inhabituel chez le lépidoptère *Galleria mellonêlla* L. *Rev. Zool. Agr. Appl.* *63*, 207-208.
Middleton, P. J., Holdaway, M. D., Petrie, M., Szymanski, M. T., and Tam, J. S. (1977). Solid-phase radioimmunoassay for the detection of rotavirus. *Infect. Immun.* *16*, 439.
Miller, L. K. and Dawes, K. P. (1978). Restriction endonuclease analysis to distinguish two closely related nuclear polyhedrosis viruses: *Autographa californica* and *Trichoplusia ni* MNPV. *Appl. Environ. Microbiol.* *35*, 1206-1210.
Miller, L. K. and Dawes, K. P. (1979). Physical map of the DNA genome of *Autographa californica* nuclear polyhedrosis virus. *J. Virol.* *29*, 1044-1045.
Mills, K. W., Gerlach, E. H., Bell, J. W., Farkas, M. E., and Taylor, R. J. (1978). Serotyping herpes simplex virus isolates by enzyme-linked immunosorbent assays. *J. Clin. Microbiol.* *7*, 73-76.
Miura, K., Fujii-Kawata, I., Iwata, H., and Kawase, S. (1969). Electron-microscopic observations of a cytoplasmic polyhedrosis virus from the silkworm. *J. Invertebr. Pathol.* *14*, 262-265.
Monnier, C., and Devauchelle, G. (1978). Protein kinase activity in *Chilo* iridescent virus. *Abstr. 4th Intern. Congr. Virol., The Hague*, p. 542.
Nick, H., Cursiefen, D., and Becht, H. (1976). Structural and growth characteristics of infectious bursal disease virus. *J. Virol.* *18*, 227-234.
Nysten, P. H. (1808). Recherches sur les maladies des Vers à Soie. *Impr. Impêriale*, Paris.
Payne, C. C. (1974). The isolation and characterization of a virus from *Oryctes rhinoceros*. *J. Gen. Virol.* *25*, 105-116.
Payne, C. C. and Churchill, M. P. (1977). The specificity of antibodies to double-stranded (ds) RNA in antisera prepared to three distinct cytoplasmic polyhedrosis viruses. *Virology 79*, 251-258.
Payne, C. C. and Harrap, K. A. (1977). Cytoplasmic viruses. *In* "The Atlas of Insect and Plant Viruses" (K. Maramorosch, ed.), pp. 105-129. Academic Press, New York.
Payne, C. C. and Rivers, C. F. (1976). A provisional classification of cytoplasmic polyhedrosis viruses based on the sizes of the RNA genome segments. *J. Gen. Virol.* *33*, 71-85.
Payne, C. C., Piasecka-Serafin, M., and Pilley, B. (1977). The properties of two recent isolates of cytoplasmic polyhedrosis viruses. *Intervirology 8*, 155-163.

Reinganum, C. (1975). The isolation of cricket paralysis virus from Emperor gum moth *Antherea eucalypti* (Scott) and its infactivity towards a range of insect species. *Intervirology* *5*, 77-102.

Rivers, C. F. and Longworth, J. F. (1972). A non-occluded virus of *Junonia coenia* (Nymphalidae: Lepidoptera). *J. Invertebr. Pathol.* *20*, 369-370.

Rohrmann, G. F., McParland, R. H., Martignoni, M. E., and Beaudrean, G. S. (1978). Genetic relatedness of two nucleopolyhedrosis viruses pathogenic for *Orgyia pseudotsugata*. *Virology* *84*, 213-217.

Smith, G. E. and Summers, M. D. (1978). Analysis of baculovirus genomes with restriction endonucleases. *Virology* *89*, 517-527.

Steinhaus, E. A. (1949). "Principles of Insect Pathology." McGraw Hill, New York.

Summers, M. D. (1977). Baculoviruses (Baculoviridae). *In* "The Atlas of Insect and Plant Viruses (K. Maramorosch, ed.), pp. 3-27. Academic Press, New York.

Summers, M. D. and Smith, G. E. (1976). Comparative studies of baculovirus granulins and polyhedrins. *Intervirology* *6*, 168-180.

Summers, M. D. and Smith, G. E. (1978). Baculovirus structural polypeptides. *Virology* *84*, 390-402.

Teninges, D. (1968). Mise en évidence de virions sigma dans les cellules de la lignée germinale mâle de drosophile stabilisées. *Arch. Gesamte Virusforsch.* *23*, 378-387.

Teninges, D., Obanessian, A., Richard-Molard, C., and Contamine, D. (1979). Isolation and biological properties of *Drosophila* X virus. *J. Gen. Virol.* *42*, 241-254.

Tinsley, T. W. (1975). Factors affecting virus infection of insect gut tissue. "Invertebrate Immunity--Mechanisms of Invertebrate Vector-parasite relations," pp. 55-63. Academic Press, New York.

Tinsley, T. W. and Kelly, D. C. (1970). An interim nomenclature system for the iridescent group of insect viruses. *J. Invertebr. Pathol.* *16*, 470-472.

Underwood, B. O., Smale, C. J., Brown, F., and Hill, B. J. (1977). Relationship of a virus from *Tellina tenuis* to infectious pancreatic necrosis. *J. Gen. Virol.* *36*, 93-110.

Voller, A. and Bidwell, D. E. (1977). Enzyme immunoassays and their potential in diagnostic virology. *In* "Comparative Diagnosis of Viral Diseases" (E. Kurstak and C. Kurstak, eds.), Vol. II, pp. 449-457. Academic Press, New York.

von Prowazek, S. (1907). Chlamydozoa. 2. Gelbsucht der Seidenraupen. *Arch. Protistenk.* *10*, 358-364.

Wagner, G. W. and Paschke, J. D. (1977). A comparison of the DNA of the "R" and "T" strains of mosquito iridescent virus. *Virology* *81*, 298-308.

White, G. F. (1917). Sacbrood. *U.S. Dept. Agr. Bull. No. 431.*
Wildy, P. (1971). Classification and nomenclature of viruses. *Monogr. Virol. 5.*
Yung, L. L. L., Loh, W., and Ter Meulen, V. (1977). Solid-phase indirect radio-immunoassay: standardization and applications in viral serology. *Med. Microbiol. Immunol. 163*, 111.

CLASSIFICATION, IDENTIFICATION, AND DETECTION OF INSECT VIRUSES BY SEROLOGIC TECHNIQUES

LOY E. VOLKMAN

Department of Entomology and Parasitology
University of California
Berkeley, California

I. INTRODUCTION

A casual glance at the table of contents of almost any immunology textbook will reveal that the study of immunology is traditionally divided into two broad categories: humoral immunity and cell-mediated immunity (Thaler et al., 1977). It is the humoral immunity portion that is of concern in serology and serodiagnostics. A basic understanding of how the production of antibodies capable of binding foreign (non-self) antigens can best be managed and manipulated to produce the type of antiserum that is optimal for the purpose at hand, whether it be classification, identification, or detection, is fundamental to the success of serologic testing. The purpose of this chapter is to review factors affecting antibodies and serologic tests that are relevant to serologic studies of insect viruses. The results of published insect virus serologic studies have been reviewed elsewhere (Mazzone and Tignor, 1976; Harrap and Payne, 1979), although some results of a few of the more recent studies are mentioned here.*

**Key references containing extensive background information on serologic methodology or specific techniques are indicated in the text and in the reference list by an asterisk (*).*

ISBN 0-12-470295-3

A. CLASSIFICATION, IDENTIFICATION, AND DETECTION

Although classification, identification, and detection are common terms familiar to everyone associated with insect viruses, it is important to be reminded that for purposes of classification, the similarities of viral characters are emphasized, while identification requires that viruses be distinguishable by at least one stable character difference. Thus, in concept, similarities are emphasized in classification, and differences in identification (Rypka et al., 1978). Implicit in the concept of detection is that identification is not an issue; generally, the target of detection is a known entity.

Primary characters used to classify viruses into families are the kind and strandedness of the nucleic acid comprising the genome, and the presence or absence of a lipoprotein envelope (Matthews, 1982). Family membership is confirmed and further classification is determined according to characters which may pertain only to certain families, such as replication strategy, numbers of segments comprising the genome, size of genome, shape of viral particles, host from which the virus was initially isolated, etc. By using these characters, insect viruses have been classified into families, subgroups, and genera. A guideline for further classification to the species level is established by rule 11 of *The Rules of Nomenclature of Viruses*: "A virus species is a concept that will normally be represented by a cluster of strains from a variety of sources, or a population of strains from a particular source, which have in common a set or pattern of correlating stable properties that separates the cluster from other clusters of strains" (Matthews, 1982).

The justifying premise for the use of serologic relatedness as a tool for virus classification is that two viruses that react with the same populations of antibodies to a significant extent have structural components that are similar in tertiary (spatial), if not in primary, structure. Further, it is implied that similar structural components reflect genetic similarity. By the same logic, unique serologic reactions, indicative of unique physical structures, can be used for virus identification.

It is evident that the identification of a virus is highly dependent upon the completeness of the data base by which viruses can be distinguished from one another. Absolute identification requires a complete nucleotide sequence of the viral genome, or the amino acid sequences of virus-coded proteins (Vlak, 1982). Because of the technical difficulties involved, however, virus identification is rarely absolute.

Identification generally is established only to the level needed for the study at hand, or to the level determined by the ease and sophistication of the technique(s) being used.

B. *CATEGORIES OF INSECT VIRUSES*

Viruses that infect insects can be grouped broadly according to their ability to infect and replicate in vertebrates as well. Those that can infect vertebrates are referred to as arthropod-born viruses or arboviruses. Arbovirus replication in hematophagous insect vectors is generally accompanied by little or no pathogenic effect, while their replication in vertebrate hosts may cause varying degrees of disease. Nodamura, a small RNA virus of the family Nodaviridae, is an exception to the rule in that it can be highly pathogenic in both vertebrate (mice) and invertebrate (honeybees and greater wax moths) hosts (Knudson and Buckley, 1977). The arboviruses, with the exception of a very few, are RNA viruses, and the majority have been classified in the families of Togaviridae, Bunyaviridae, Rhabdoviridae (all ssRNA, enveloped), or Reoviridae (dsRNA, nonenveloped). In contrast, viruses that are restricted to an insect host range are commonly, but not necessarily, pathogenic to their insect host. These viruses are generally more diverse than arboviruses, and may be members of the families Picornaviridae, Caliciviridae (both ssRNA, nonenveloped), Reoviridae (dsRNA, nonenveloped), Parvoviridae (ssDNA, nonenveloped), Iridoviridae (dsDNA, nonenveloped), Poxviridae, or Baculoviridae (both dsDNA, enveloped) (Matthews, 1982).

1. *Serology and Arboviruses*

Because of their medical importance, arboviruses have been more extensively studied than insect-restricted viruses. Serologic methodology played a major role in the grouping of arboviruses before the current system of classification (based on physical properties of the virion) was instituted (Casals, 1971). The validity of the premise that serologic cross-reactivity is a reflection of structural similarity was demonstrated as the viruses of serologically defined Group A were all subsequently classified in the genus Alphavirus, while the serogroup B viruses, structurally different from and serologically unrelated to serogroup A viruses, were placed in the genus Flavivirus (Matthews, 1982; Berge, 1975). In view of this correlation, antigenic relatedness continues to be an important character in arbovirus classification.

2. *Serology and Insect-Restricted Viruses*

The success of serology as a tool in arbovirus research is partially due to several advantages not available to scientists studying insect-restricted viruses. One of these is that antibodies are produced as a consequence of natural infection of vertebrates, while infection of invertebrates does not lead to antibody formation. Thus arbovirologists can probe for either specific antibodies or viral antigens in diagnostic tests, while serologic studies with insect-restricted viruses are limited to testing for viral antigen. In addition, most arboviruses replicate in brains of newborn mice. Unpurified virus generated in this way may be used to immunize adult mice which respond by producing antibodies only to viral antigens, and not to self (mouse) antigens (Casals, 1967*). This procedure has two obvious advantages: antisera to arboviruses may be generated by a standardized protocol, and antigen purification is unnecessary. Insect-restricted viruses, on the other hand, must be highly purified before use as an antigen if inadvertent immunization with host-derived material is to be avoided.

II. ANTIBODIES AS REAGENTS FOR INSECT VIRUS CLASSIFICATION AND IDENTIFICATION

A. *PREPARATION OF ANTISERA*

The lack of a standard protocol for preparing antisera for insect-restricted viruses is one significant reason that a reliable serologic data base has not been generated for these viruses. It has been well documented that repeated inoculations of a viral antigen into an animal will elicit a different antibody response than that obtained with a single, or few injections of the same antigen (Casals, 1967*). The former antiserum will have a higher titer, show greater affinity for the stimulating antigen and be more cross-reactive with related antigens than the latter antiserum, which will be more specific.

Variability in antisera also arises because many different viral antigenic determinants (epitopes) may be recognized by the immunized animal, and up to five distinct antibodies can be produced to each one. As a consequence, antibody response may differ from animal to animal (Oxford, 1982). The specificity of antisera is also dependent upon the time elapsed between the final inoculation and bleeding. Animal species differences also contribute to variability in antisera.

Cross-reacting antisera can be used to advantage in systems where contaminating nonviral antigens are not a problem, such as in arbovirus serology. The development of the concept of antigenic groups of arboviruses came from taking advantage of natural cross-reactions (Casals, 1967*). The cross-reactivity of sera can be enhanced by hyperimmunizing animals with several antigenically related viruses in succession, and then pooling the sera of several animals immunized in this fashion. For classification purposes, where antigenic similarities are stressed, these group-polyvalent sera can be used in tests to maximize chances of detecting cross-reacting antigens of unknown viruses.

Whereas the enhancement of cross-reactivity can be advantageous for classification purposes, it can lead to trouble in situations where antigen purity is a concern, or when differences between related viruses are of interest. In this situation specific antisera that highlight antigenic differences are appropriate. If antigen purity is a problem for immunization, then the concentration of immunizing antigen should be reduced to a minimum in an effort to dilute the extraneous materials to such an extent that they are no longer antigenically active (Casals, 1967*).

1. *Antisera to Individual Viral Proteins*

Antiserum specificity can sometimes be enhanced by immunization with individual viral proteins. One procedure is to prepare virus to a fairly high degree of purity, and then subject it to preparative sodium dodecyl sulfate-polyacrylamide gel electrophoresis (SDS-PAGE). Viral proteins, recognized by predetermined molecular weights, can be cut out of the gel and either extracted from the acrylamide, or used complete with acrylamide for inoculating an animal (Barta and Issel, 1978; Summers and Smith, 1975; Emini et al., 1982). It should be noted, however, that at least some of the epitopes of the natural protein can be altered or destroyed by the SDS-PAGE procedure. Nevertheless, antisera prepared in this manner can be used for a number of purposes depending on the viral protein used as the antigen. Antiserum to a protein carrying group specific determinants can be useful for classification purposes, and antiserum with type specific or neutralizing activity can be useful for identification purposes.

2. *Monoclonal Antibodies*

Factors such as antigen impurity, variability of antisera, and reaction of antisera with contaminating, nonviral antigens have led to problems in interpretation, reproducibility, and standardization of serologic assays. Since these problems can be overcome by using monoclonal antibodies, the advent of lymphoyte hybridoma technology and monoclonal antibody production has been welcomed as a valuable addition to the immunologic arsenal for viral study and identification (Oxford, 1982). Monoclonal antibodies are obtained from cloned hybrids produced by the fusion of antigen-stimulated lymphocytes and myeloma cells (Chan and Mitchison, 1982*). Monoclonal antibodies are chemically homogeneous and react with constant avidity to single antigenic determinants. For this reason, standardization of reagents is not a problem. Since hybridomas inherit immortality from their myeloma parent, it is possible to bank and ship cell lines producing the same characterized monoclonal antibodies all over the world. Further, immunization can be made with a complex mixture of impure antigens, and still the final product is a homogeneous antibody specific for one component of the mixture. Antibodies reacting with contaminating host antigens can be recognized and screened out.

Although monoclonal antibodies have the potential of being highly specific in that they may react only with a unique epitope, they are not necessarily so. Any given antibody may react with a particular epitope as well as structurally similar ones (Yelton and Scharff, 1981*). This capability of recognizing individual cross-reacting epitopes may prove to be a valuable tool in identifying evolutionarily conserved macromolecular structures with apparently important functions.

III. SEROLOGIC TESTS

It is evident that by indirectly manipulating the humoral response, i.e., by immunizing repeatedly or just once with antigenically related viruses, single whole viruses, or isolated viral structural proteins, antibodies can be elicited that differ with regard to their titer, affinity, and specificity. These qualities can be modified further by pooling the sera from several immunized animals. Epitope-specific monoclonal antibodies also can be pooled to make reagents that react with several epitopes, and thus standardized, defined, cross-reacting reagents can be generated (de Macario and Macario, 1983).

While the reagents play a crucial role in establishing antigenic relatedness among viruses, the choice of assay used often affects the information obtained. A survey of the standard immunologic tests applied to the study of viruses reveals that there are several general methodological approaches with many variations.

A. TUBE PRECIPITATION, GEL IMMUNODIFFUSION, AND IMMUNOELECTROPHORESIS

When antibodies and macromolecular antigens react, they frequently form insoluble complexes (lattices) that visibly precipitate from solution. This phenomenon is the basis of many classic serologic tests. If the reaction occurs in liquid media in test tubes, it is called tube precipitation. If it is done by introducing antigen and antibody into different regions on an agar or agarose gel, allowing them to diffuse toward each other and form a band of precipitate at the junction of their diffusion fronts, it is called gel immunodiffusion. A double diffusion variation of this technique, developed mainly by Ouchterlony (1962) in Sweden, can give information as to whether two (or more) substances are antigenically related by the various patterns of reaction bands obtained (Crowle, 1961*). The bands frequently overlap, however, and complex mixtures of antigens cannot be resolved. If the mixture of antigens is first subjected to electrophoresis, the resolution of the reaction bands can be greatly improved. This is done in immunoelectrophoresis. The protein in an individual band can be identified by comparison with patterns developed by using monospecific antisera to known protein markers (Nisonoff, 1982).

Tests performed based on precipitin reactions played a major role in the development of serology. They can be used for either antigen or antibody detection. They are relatively insensitive, however, requiring approximately 50-1000 μg antibody per ml (Nisonoff, 1982). In addition, some antigens are too large to diffuse through agar gels.

In insect virology, gel immunodiffusion tests have been used principally with the small, spherical DNA and RNA viruses because they move easily in the gel and antigen concentration usually is not a problem (Harrap and Payne, 1979). Since precipitation is dependent upon antigen cross-linking to form insoluble lattices, monoclonal antibodies can not be used effectively in this procedure if they bind to only one determinant on a monomeric antigen (Yelton and Scharff, 1981*). This can be overcome by mixing monoclonal antibodies that are reactive to different sites on an antigen. Kawanishi and Huang (1983) used this approach to determine whether monoclonal antibodies elicited to *Heliothis zea* M nuclear polyhedrosis virus polyhedron protein (polyhedrin) reacted with single or multiple epitopes of that molecule, and to determine which pairs of monoclonal antibodies reacted with different epitopes, sufficiently separated to avoid antibody steric hindrance.

B. HEMAGGLUTINATION INHIBITION

Many intact viruses or viral-coded proteins have the capacity to agglutinate erythrocytes of certain species (hemagglutination). In some cases, the adsorption of viruses to host cells has been shown to involve the same receptors that mediate hemagglutinating activity (Howe and Lee, 1970). This hemagglutination (HA) reaction can be inhibited if specific antibodies attach to the effector proteins (the hemagglutinins) and prevent their attaching to receptors on the red blood cells. The hemagglutination inhibition (HI) test has been used extensively for virus classification and identification (Clarke and Casals, 1958*). With some viruses, the HI test is capable of identification to a subtype, group of strains, or even a strain level, while with other viruses it is useful for classification to the group level only (Casals, 1967*). The specificity difference, in this case, is not necessarily a function of the antiserum used, but rather reflects the nature of the hemagglutinating antigens and the degree to which they are conserved evolutionarily. The more highly conserved the antigen(s), the less specific the test.

Some insect-restricted viruses have demonstrated hemagglutinating activity (Cunningham et al., 1966; Anderson et al., 1981). Of these, the most extensive testing has been done with baculoviruses, but there is considerable disagreement as to which of the viral components is involved in mediating the hemagglutinating activity. In some instances the polyhedrins have been implicated (Reichelderfer, 1974; Norton and DiCapua, 1975), while in others, enveloped nucleocapsids were found to be responsible (Anderson et al., 1981).

HI tests can be performed with monoclonal antibodies if single epitopes of the viral proteins are involved in the hemagglutination reaction. HI tests are more sensitive than precipitin tests because of the particulate nature of the erythrocytes (Nisonoff, 1982). It should be kept in mind, however, that the HI test only measures the interaction of viral hemagglutinins and antibodies specific for them. Other viral proteins could be reacting extensively with antibodies, and these reactions would go undetected in this test. (The same qualification applies to neutralization. See Section III,D.)

C. COMPLEMENT FIXATION

Complement is a complex system of serum proteins that has two properties pertinent to the complement fixation test (CF): it has the ability to cause lysis of red blood cells in the presence of antibodies reactive with red blood cells, and it combines with antigen-antibody complexes if it is present during an antigen-antibody reaction. The CF test is based on the observation that complement bound up in antigen-antibody complexes is not free to react with red blood cell-antibody complexes causing cytolysis and release of hemoglobin into the medium. The "sequestering" of complement in the primary antigen-antibody reaction is called complement fixation, and can be used to determine whether a specific or cross-reacting antigen is present in some unknown sample. This test has been used extensively, along with HI and neutralization tests in the grouping of arboviruses and other viruses. The CF test has been used by Kelly and collaborators (1979) to measure the antigenic relatedness of various isolates of small iridescent viruses.

It is now recognized that complement fixing activity is mediated by the Fc (nonantigen binding) portion of certain subclasses of IgG antibodies, and by IgM (Thaler et al., 1977). One molecule of pentameric IgM can fix complement whereas at least two adjacent molecules of IgG appear to be required. The requirement for proximity between IgG molecules for this test to work makes the spatial relationship of antibody-specific epitopes a factor, especially if monoclonal IgG antibody is used.

D. NEUTRALIZATION

Virus neutralization historically has been regarded as the most sensitive and specific serologic test available. Now radioimmunoassay (RIA) and enzyme-linked immunosorbent assay (ELISA) are contenders for this claim, but the outcome depends upon the system. The neutralization test is based on the fact that many viruses are inactivated (neutralized) by antibodies that bind to critical sites, usually on the viral surface (Mandel, 1979*). In the neutralization test, the remaining infectious activity is determined after virus exposure to antiserum. If the remaining activity is assessed *in vivo*, the antiserum concentration usually is held constant and the virus concentration is varied; if it is assessed *in vitro*, then the opposite is usually the case (Casals, 1967*; Martignoni et al., 1980). The sensitivity (in terms of amount of antigen required to do the test) is closely associated with the infectious to physical particle ratio of the virus in the host system used. For example, it has been calculated that the physical to infectious particle ratio of the budded phenotype of *A. californica* nuclear polyhedrosis virus (AcMNPV) infecting TN-368 cells *in vitro* is 1.28×10^2:1, while the ratio for the occluded phenotype (LOVAL) is 2.4×10^5:1 (Volkman et al., 1976). To perform a neutralization test in this system, starting with 100 plaque forming units (PFU) in the absence of any inactivation, 1.4 ng of budded virus and 2.8 μg of LOVAL is required (Volkman et al., 1976). The latter concentration is easily within the range of RIA and ELISA (see Section III,E).

The specificity of the neutralization test is based on the "critical site" binding feature. In a sense, cross-reactivity is a measure of shared "critical sites." Differences in the rate of binding of antibody to cross-reactive critical sites can be measured in kinetic neutralization experiments, which is useful in discriminating among virus strains. To date, Kelly and collaborators (1979) are the only ones who have used kinetic neutralization experiments to detect strain differences in insect-restricted viruses.

It is thought that antibody usually neutralizes the virus by interfering with the initial virus-host cell interaction either directly or indirectly (Mandel, 1979*). The neutralization assay can, therefore, be useful in detecting differences in specific interactions of viruses and host cell surfaces. This may be the case with the budded and occluded phenotypes of AcMNPV, which are neutralized by different populations of antibodies *in vitro* (Volkman et al., 1976). It has been known for some time that the budded and occluded phenotypes of subgroup A and B baculoviruses (nuclear polyhedrosis viruses and granulosis viruses) are morphologically different from each other (Summers and Volkman, 1976; Adams et al., 1977). Differences in the mechanism of cell entry have been described, as well, although not as a complete *in vivo* and *in vitro* study for any one virus. A survey of the literature reveals that the occluded form seems to enter cells by fusion (Summers, 1971; Kawanishi et al., 1972; Tanada et al., 1975; Granados and Lawler, 1981) while the budded form enters by viropexis, both *in vitro* (Raghow and Grace, 1974; Hirumi et al., 1975) and *in vivo* (Kislev et al., 1969; Kamamoto et al., 1977). While this observation needs to be tested further, it is consistent with the neutralization data.

Neutralizing monoclonal antibodies have been generated for several viruses, and have been used to better understand the mechanism of neutralization (Yewdell and Gerhard, 1981*; Yelton and Scharff, 1981*; Oxford, 1982; Massey and Schochetman, 1981). Monoclonal antibodies have been used in neutralization tests to show clear-cut antigenic differences between street and vaccine strains of rabies which were previously thought to be antigenically indistinguishable (Wiktor and Koprowski, 1978; Flamand et al., 1980). In this instance, minor antigenic changes were shown to be of major importance in virus-host interactions.

Recently, neutralizing monoclonal antibodies were elicited to the budded phenotype of AcMNPV (Hohman and Faulkner, 1983). Subsequently, one neutralizing antibody was shown to bind specifically to envelop surface antigens of the budded virus. That same antibody did not neutralize or bind the occluded phenotype of AcMNPV (Volkman et al., 1984).

E. RIA AND ELISA

The sensitivity of the neutralization test is due to the fact that any remaining infectious virus is greatly amplified in the course of the assay. The intensity of the signal that the reaction has (or has not) taken place is a critical factor in assay sensitivity. This principle was well demonstrated in the development of the RIA, and subsequently, the ELISA. The RIA makes use of a radiolabeled (usually ^{125}I) antigen or antibody, and the ELISA, an enzyme-labeled (usually alkaline phosphatase) antigen or antibody to signal the occurrence of an antigen-antibody reaction (Weir, 1978*; Voller et al., 1979*). Both are exceedingly sensitive assays and frequently antigen can be detected at a concentration of a few nanograms per milliliter. Furthermore, the antigen does not have to be infectious for the assay to work as it does for neutralization.

In principle, RIA and ELISA are precisely the same (Fig. 1). There are multiple variations of both; the method of choice being dependent upon the nature of the experiments (Halonen and Meurman, 1982*; Voller et al., 1979*; 1982*). Antibodies used in RIA and ELISA procedures should be of high titer and high affinity to remain attached throughout the multiple washes. Crawford et al. (1978*) and Crook and Payne (1980*) tested some of the RIA and ELISA variations, respectively, in insect virus systems. They agreed that the most sensitive assay for antigen detection in the absence of plentiful extraneous matter was the indirect test, and in the presence of plentiful extraneous matter, the sandwich test (Fig. 1). ELISA has been used more extensively than RIA in insect virology, probably because of expense and safety considerations involved in using ^{125}I, its relatively short half life (60 days), and the cost of a gamma counter.

1. The Indirect ELISA

Recent uses of the indirect ELISA in insect virology include host range studies, assessment of antigen purification techniques, and viral relatedness determinations. Some examples of these are given below.

Indirect ELISA has been used as a sensitive test for viral antigen production in nonpermissive cells. Rubenstein et al. (1982) used it to detect *Estigmene acrea* granulosis virus antigen increase when fat body cells in culture were exposed to the virus. (No productive infection of a granulosis virus in cell culture has ever been reported.)

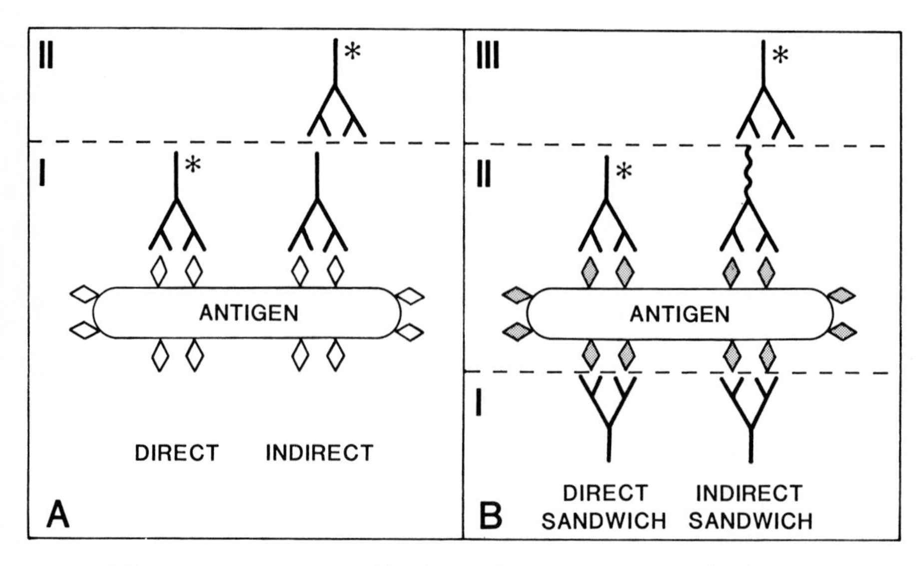

FIGURE 1. Four solid phase immunoassay variations are shown (from left to right). In the direct and indirect variations (A), the antigen is attached to the solid phase first, and then is incubated with either labeled (*) (direct) or unlabeled antibody (indirect). In the indirect variation, a second incubation is performed with a labeled second antibody specific for the first antibody. The direct and indirect sandwich methods (B), are conducted in nearly the same way except antibody specific for the antigen of interest is first attached to the solid substrate. The antigen-containing substance is then incubated with the attached antibody followed by an incubation with antigen-specific antibody that is either labeled (direct sandwich) or unlabeled (indirect sandwich). In the latter variation, the second antibody should be from a different species, but of the same specificity as the first. A labeled third antibody, specific for the species and type of the second antibody, is used in the final incubation of the indirect sandwich assay.

An increase in AcMNPV antigens in AcMNPV-exposed codling moth cells (cell line Cp169) was detected by Langridge et al. (1981a) using indirect ELISA, while Summers et al. (1978), using an indirect immunoperoxidase staining method, were unable to do so. The difference in results was presumably due to the difference in sensitivity of the techniques.

Langridge et al. (1981b) also used the indirect ELISA when they repeated a published study to determine whether alkali-liberated virus of the occluded phenotype of a nuclear polyhedrosis virus could be separated completely from polyhedrin by sequential sucrose gradient and sepharose column purification techniques (Bell and Orlob, 1977). By immunodiffusion, their results agreed with those of Bell and Orlob; virus purified in such a manner appeared to be free of polyhedrin. The ELISA results, however, showed that even these highly purified virions had elicited a considerable antibody titer to polyhedrin, indicating that they, in fact, were not free of the protein.

Cole and Morris (1980) used the indirect ELISA to determine that a new iridovirus isolated from terrestrial isopods was serologically related (slightly) to *Tipula* iridescent virus and *Phyllophage* iridescent virus.

Crook (1981) found he was able to discriminate between the granulosis virus of *Pieris brassicae* and *Pieris rapae*, which he determined to be 97.7% homologous at the DNA level, using indirect ELISA. In contrast, using Western blots (Section III,F), Harvey and Volkman (1983) were unable to discriminate between variants of *Cydia pomonella* granulosis virus related to a similar degree at the DNA level.

Brown et al. (1982) recently reported using a variation of the indirect ELISA wherein enzyme-linked protein A was substituted for the second antibody in a study demonstrating antigenic relatedness of four baculoviruses from *Spodoptera* species. This variation was no more or less sensitive than the standard technique.

Roberts and Nasar (1982) used monoclonal antibodies to probe for differences between *in vivo* and *in vitro* generated AcMNPV polyhedrin using direct and indirect ELISA. No differences were detected.

2. *The Sandwich ELISA*

While generally not as sensitive as the indirect ELISA, the sandwich ELISA can be used to detect antigen in a preparation which is highly contaminated with host tissue components. It is, therefore, the method of choice for virus detection in crude virus extracts. There are two major variations of the sandwich ELISA; the direct sandwich and the indirect sandwich (Fig. 1B). The basic principle underlying both of them is that the antigen of interest is selectively concentrated and bound to the plate by its specific interaction with preadsorbed antibody. After this step, the contaminating extraneous host antigens are washed away, and the remainder of the test is the same as the basic direct or indirect ELISA, with one exception. The species of the second antibody in the indirect sandwich test should be different from the primary antibody unless only the Fab (antigen-binding) portion of the primary antibody is used (Fig. 1B). This restriction is necessary to avoid a spurious reaction between the tertiary (labeled) antibody and the primary antibody.

The direct sandwich ELISA was very effectively used by Morris et al. (1981) to quantitatively monitor the degree of contamination of AcMNPV preparations with a small RNA virus, TRV.

Kelly et al. (1978a) found that the direct sandwich ELISA was an effective tool for discriminating among purified preparations of five small iridescent viruses. In another study, Kelly and collaborators (1978b) used the same assay for detecting *Heliothis armiger* NPV in *H. armiger* larval extracts. They reported a sensitivity of 1 ng virus per ml of extract.

Langridge et al. (1981a) used the direct sandwich ELISA to monitor possible AcMNPV antigen increase in *L. dispar* larvae fed AcMNPV occlusion bodies (in view of their positive results with AcMNPV exposed Cp169 cells), but found no evidence for viral activity.

An indirect sandwich ELISA assay has been established and used for monitoring the presence of *Oryctes rhinoceros* baculovirus in field populations (Longworth and Carey, 1980; Young and Longworth, 1981).

Payment et al. (1982) recently developed an indirect sandwich ELISA for *Euxoa scandens* cytoplasmic polyhedrosis virus that is sensitive enough to detect 10 ng virus per ml of larval extract.

Volkman and Falcon (1982) examined the possibility of using a monoclonal antibody as one component of an indirect sandwich ELISA for the detection of *Trichoplusia ni* S NPV in infected *T. ni* larvae. They determined that the assay worked well if the monoclonal antibody was used as the primary antibody, but not if it was used as the secondary antibody.

F. WESTERN BLOTS

Although RIAs and ELISAs are highly sensitive assays and can demonstrate serologic cross-reactivity and differences among viruses, no information is obtained about which of the virus proteins is involved in the reactions. A recently developed variation of immunoelectrophoresis that combines SDS-PAGE and RIA or ELISA can be used to determine which viral structural proteins are involved in serologic cross-reactions (Towbin et al., 1979*). This technique has been called Western blotting and has the potential of emerging as a powerful tool for understanding the basis of serologic groupings of complex insect viruses (with more than a few structural proteins). Smith and Summers (1981) demonstrated the power of Western blotting by comparing the antigenic relatedness of 17 different species of baculoviruses from lepidopteran hosts. They found that subgroup A baculoviruses had conserved interspecies antigenic determinants both between and among MNPVs and SNPVs. They were able to determine the molecular weights of the structural proteins retaining these conserved epitopes. They also found that some subgroup B baculoviruses (GVs) were antigenically related to some subgroup A baculoviruses. They determined unequivocally that all the polyhedrins and granulins used in the study were serologically related. More recently, Knell et al. (1983) expanded these studies to find additional common antigenic determinants among different baculovirus subgroups.

Smith and Summers (1981) also used the Western blotting technique to compare the antigenic relatedness of the occluded and budded phenotypes of AcMNPV that they found to be considerably different. Volkman (1983) explored this further by doing reciprocal Western blots of the two phenotypes.

Roberts and Naser (1982) and Hohman and Faulkner (1983) used Western blotting as a method of determining which of the AcMNPV structural proteins were reactive with monoclonal antibodies elicited to that virus. Furthermore, Naser and Miltenburger (1982) explored the possibility of using Western blots in conjunction with a specific monoclonal antibody for the identification of AcMNPV.

The Western blot technique is still too new for a complete understanding of all its limitations, but loss of some epitopes due to the SDS-PAGE procedure is an expected drawback (Erickson et al., 1982*).

G. *IMMUNOFLUORESCENT AND IMMUNOPEROXIDASE STAINING*

The first immunoassay developed by tagging an antibody with a substance to signal its reaction with an antigen was the fluorescent antibody (FA) technique. Coons and collaborators (1942) determined that fluorescein isothiocyanate could be used to label antibodies without destroying their specificity. Antibodies to viruses labeled in this manner will react with viral epitopes in infected cells, and the complexes may be viewed with a microscope equipped with special FA filters and an ultraviolet light source. The FA technique has long been used for viral identification and for determining the location of viral antigens in infected cells during the course of replication (Casals, 1967*; Schmidt and Lennette, 1973). Then, by monitoring infections of cell cultures with monoclonal antibodies, the time of production and location of specific proteins can be determined.

Since the development of FA, immunocytochemical techniques have expanded to overcome some of the inherent limitations in using fluorescein as the signal, such as the requirement for a

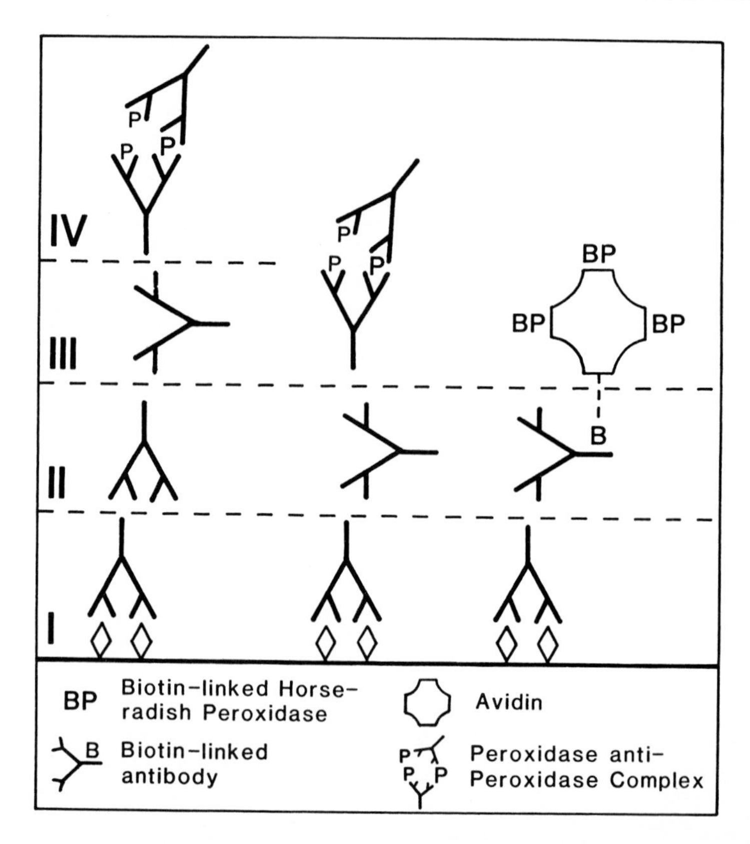

FIGURE 2. Three immunohistochemical methods are shown (from left to right): the four-layer PAP method, the three-layer PAP method, and the avidin-biotin method (Naritoku and Taylor, 1982). In the PAP methods, the second antibody of the four-layer variation, and the first antibody of the three-layer variation, must be of the same species as the antibodies complexing the horseradish peroxidase. The bridging antiserum, specific for that species, is used at a high concentration to ensure the "bridge" will be formed. In the avidin-biotin method, the biotin-linked primary antibody is detected by the avidin-biotin-peroxidase complex added as the third layer.

special microscope, signal fading, and incompatibility with histological staining operations (Volkman, 1982). One very successful method of overcoming these limitations, in addition to achieving much greater sensitivity and specificity, is the use of horseradish peroxidase-labeled antibodies. In addition to the usual direct and indirect strategies of staining used with FA (Fig. 1A), three- and four-layered peroxidase anti-peroxidase (PAP) techniques have been developed which have enhanced sensitivity significantly (Fig. 2) (Naritoku and Taylor, 1982*). One important difference between the direct/ indirect and the peroxidase anti-peroxidase staining methods is that in the latter, the enzyme is not covalently linked to the Fc portion of the antibody, but rather is held in position as an antigen in an antigen-antibody complex, and thus retains full activity. The high sensitivity of the PAP method is due to the virtual absence of background, which results in a high signal-noise ratio. The antigen-antibody reaction usually is marked by the use of hydrogen peroxide as enzyme substrate, and diaminobenzidine as electron donor, resulting in the deposition of a brown, highly insoluble polymeric reaction product.

Another promising immunocytochemical variation makes use of the extraordinary affinity between avidin and biotin. In this method, the secondary antibody is labeled with biotin, which is detected by a biotin-peroxidase-avidin complex (Fig. 2). This method is relatively recent but there have been reports that it is slightly more sensitive (fourfold) than the three-layered PAP technique (Hsu et al., 1981*; Lloyd and Fruhman, 1982*).

Detection and discrimination between two different antigens can be achieved by employing a double immunoenzyme labeling technique using alkaline phosphatase and horseradish peroxidase (Mason and Sammons, 1978*; Malik and Daymon, 1982*). Although dependent upon the substrate used, the alkaline phosphatase reaction is usually located by the appearance of a blue or deep red reaction product.

Both immunofluorescent and immunoperoxidase staining procedures have been used to detect insect viruses. Krywienczyk (1963) used immunofluorescence to detect an NPV in *B. mori*, and Kurstak and Kurstak (1974) reported the use of immunoperoxidase to detect infections of *Tipula* iridescent virus and densonucleosis virus in *Galleria mellonella*. More recently, the indirect immunoperoxidase technique was used to monitor time course studies of AcMNPV in cell culture (Summers et al., 1978). Peroxidase anti-peroxidase staining has been used to develop a quantitative assay for AcMNPV, for studying the kinetics of viral replication in single cells and to determine the location (surface or intracellular) of viral antigens (Volkman and Goldsmith, 1981*; 1982; Volkman, 1983). An extensive host range study of AcMNPV in vertebrate cell lines was conducted using the PAP technique to monitor intracellular increase in viral antigen (Volkman and Goldsmith, 1983). The assay used in this study was estimated to be sensitive to about 940 fg of virus per area equivalent to the nucleus of an average TN-368 cell, 35 μm in diameter.

The direct and indirect peroxidase and PAP staining techniques can be used with electron microscopy as well as with light microscopy (Andres et al., 1973*). In electron

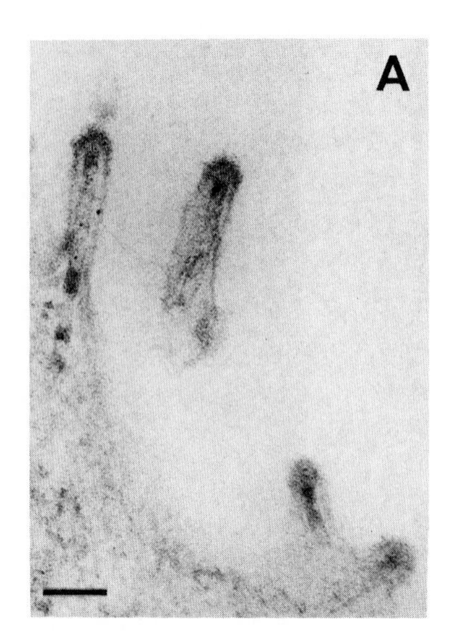

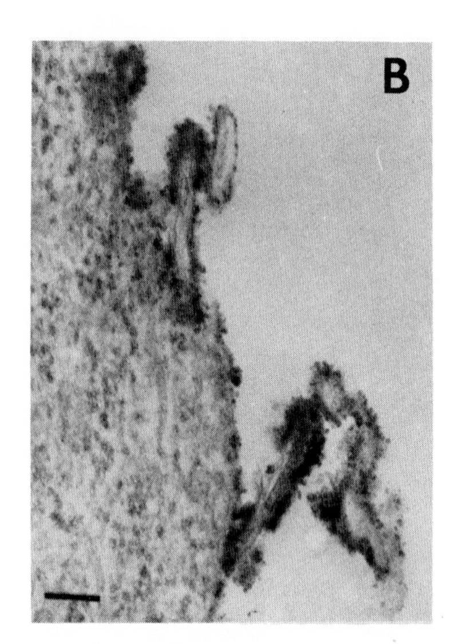

FIGURE 3. Electron micrographs of AcMNPV budding from the surface of infected Spodoptera frugiperda IPLB-SF-21 cells. Viral surface antigens are stained by the indirect peroxidase method using antiserum to the budded form of AcMNPV (B). Unstained budding virus is shown for comparison (A). The bars represent 100 nm.

microscopy, the antigen is localized when the brown, highly insoluble, osmiophilic reaction product reduces OsO_4 to insoluble osmium. Figure 3 shows an example of budding AcMNPV stained using homologous antiserum and the indirect immunoperoxidase technique. Unstained budding virus also is shown for comparison.

H. IMMUNOPRECIPITATION AND IMMUNOAFFINITY CHROMATOGRAPHY

Immunoprecipitation is a widely used method for identifying the specificities of monoclonal antibodies (Yewdell and Gerhard, 1981*). It can be used also for the identification of viral antigens when the specificity of the antisera or antibodies is known. The procedure involves three steps.

1. Formation of antigen-antibody complexes. Antibodies are incubated with dissociated, usually radiolabeled, viral components or *in vitro* translation products. In the case of dissociated virus, care must be taken that conditions do not interfere with antigen-antibody interaction, or denature critical epitopes.

2. Precipitation of antigen-antibody complexes. Complex precipitation can be achieved by using anti-immunoglobulin antiserum to form an insoluble immunoprecipitate, or by using protein A, either in its "natural" state bound to formalin-fixed, heat-killed *Staphylococcus aureus*, or coupled to Sepharose beads (Jones, 1980*). Protein A binds to immunoglobulins via their Fc region, but does not bind to all immunoglobulins equally well (Goding, 1978). Species of origin, antibody isotype, and pH all effect binding efficiency (Yewdell and Gerhard, 1981*) and must be taken into consideration.

3. Analysis of precipitated viral components. Usually precipitates are analyzed by SDS-PAGE, followed by autoradiography if radiolabeled viral preparations are used.

Immunoprecipitation has been used recently in insect virology to identify *in vitro* translated protein products of AcMNPV mRNA (Vlak et al., 1981; Adang and Miller, 1982). In addition, it was used to identify the specificity of AvMNPV-generated monoclonal antibodies (Roberts and Naser, 1982; Volkman et al., 1984).

Immunoaffinity chromatography is a technique that will become more and more popular as monoclonal antibodies gain wider usage. Roberts and Naser (1982) used it to show specific binding of some of their monoclonal antibodies with AcMNPV polyhedrin. In this technique, Sepharose or protein A-Sepharose is covalently linked with purified monoclonal antibody, or monospecific polyclonal antibodies. A column made of this material can then be used to bind, purify, and concentrate the specifically reacting antigen (Schneider et al., 1982*), thereby providing the first necessary step in the characterization of individual viral components. In addition, protein antigens purified in this way can be used for the identification and cloning of their structural genes. This task can be accomplished by first determining a partial amino acid sequence of the gene product (the antigen), synthesizing DNA probes that cover all the possible codes for that sequence, and using these probes to screen cDNA libraries or to generate cDNA probes specifically by binding appropriate mRNA's and acting as primers. The specific cDNA probes then can be used to identify and clone the structural genes (Hunkapiller and Hood, 1983).

IV. SUMMARY AND CONCLUSIONS

The choice of serologic reagents and techniques may have profound effects on the credibility of the results obtained in a serologic study. Clearly significant advances in techniques and reagents in the past few years have made it possible to use antibodies as tools for detection, classification, and identification with much greater confidence than was ever possible before because of increased specificity and sensitivity. We can now answer questions raised by Mazzone and Tignor in their review of serologic relationships of insect viruses (1976): "... is it presently possible to construct a serologic differentiation or characterization of insect viruses of different types? If so, on which virus-specific antigens may these differences be based and which serologic method commonly used is the most sensitive and which is the most specific for differentiating group specific and virus-strain-specific antigens?" The answer is clearly "yes" to the first question. We have seen the power of Western blots and ELISA as an answer to the second question. Several studies cited in this chapter demonstrate the sensitivity of ELISA for virus detection in crude biological preparations. This capability makes ELISA ideal for use in microbial control for the assessment of contamination of the agents used, and for environmental monitoring. In addition, the use of Western blots and ELISA in conjunction with well-characterized antisera and monoclonal antibodies could provide a serologic data base needed for the unambiguous identification and classification of insect viruses. Harrap and Payne (1979) discussed the need for establishing such a data base several years ago. They suggested that, as a measure to aid in the identification of new isolates, standard reference strains of viruses (cloned, if possible) should be made available to all workers for the calibration and assessment of various techniques. It is now possible to provide standardized monoclonal antibody reagents, individually and in pools, for confirming the uniqueness or identity of new isolates (de Macario and Macario, 1983).

ACKNOWLEDGMENTS

The author thanks P. Faulkner, J. Harvey, J. Morris, and Y. Arcus for commenting on the manuscript. Thanks also go to Pam Carroll for typing.

REFERENCES

Adams, J. R., Goodwin, R. H., and Wilcox, T. A. (1977). Electron microscopic investigations on invasion and replication of insect baculoviruses *in vivo* and *in vitro*. *Biol. Cell.* *28*, 261-268.

Adang, M. J. and Miller, C. K. (1982). Molecular cloning of DNA complementary to an RNA of the baculovirus *Autographa californica* nuclear polyhedrosis virus: location and gene products of RNA transcripts found late in infection. *J. Virol. 44*, 782-793.

Anderson, D. K., Bulla, L. A., Jr., and Consigli, R. A. (1981). Agglutination of vertebrate erythrocytes by the granulosis virus of *Plodia interpunctella*. *Virology 113*, 242-253.

*Andres, G. A., Hso, K. C., and Seegal, B. C. (1973). Immunologic techniques for the identification of antigens or antibodies by electron microscopy. *In* "Handbook of Experimental Immunology, Vol. 2: Cellular Immunology" (D. M. Weir, ed.), Chapter 34, pp. 34.1-34.45. Blackwell, London.

Barta, V., and Issel, C. J. (1978). Anemia virus proteins P26 and P16. *Amer. J. Vet. Res. 39*, 1856-1859.

Bell, C. D. and Orlob, G. B. (1977). Serological studies on virions and polyhedron protein of a nuclear polyhedrosis virus of the cabbage looper, *Trichoplusia ni*. *Virology 78*, 162-172.

Berge, T. O. (1975). "International Catalogue of Arboviruses Including Certain other Viruses of Vertebrates." DHEW Publication No. (CDC) 75-8301.

Brown, D., Allen, C., and Bignell, G. (1982). The use of the protein A conjugate in an indirect enzyme-linked immunosorbent assay (ELISA) of four closely related baculoviruses from *Spodoptera* species. *J. Gen. Virol. 62*, 375-378.

*Casals, J. (1967). Immunological techniques for animal viruses. *In* "Methods in Virology" (K. Maramorosch and H. Koprowski, ed.), Vol. 3, pp. 13-199. Academic Press, New York.

Casals, J. (1971). Arboviruses: incorporation in a general system of virus classification. *In* "Comparative Virology" (K. Maramorosch and E. Kurstak, eds.), pp. 307-333. Academic Press, New York.

*Chan, W. L. and Mitchison, N. A. (1982). The use of somatic cell hybrids for the production of monospecific viral antibodies. *In* "New Developments in Practical Virology: Laboratory and Research Methods in Biology and Medicine" (C. R. Howard, ed.), Vol. 5, pp. 125-141. Alan R. Liss, New York.

*Clarke, D. H. and Casals, J. (1958). Techniques for hemagglutination and hemagglutination-inhibition with arthropod-borne viruses. *Amer. J. Trop. Med. Hyg. 7*, 561-573.

Cole, A. and Morris, T. J. (1980). A new iridovirus of two species of terrestrial isopods, *Armadillidium vulgare* and *Porcellio scaber*. *Intervirology 14*, 21-30.

Coons, A. H., Creech, H. J., Jones, R. N., and Berliner, E. (1942). The demonstration of pneumococcal antigen in tissues by the use of fluorescent antibody. *J. Immunol. 45*, 159-170.

*Crawford, A. M., Faulkner, P., and Kalmakoff, J. (1978). Comparison of solid-phase radioimmunoassays for baculoviruses. *Appl. Environ. Microbiol. 36*, 18-24.

Crook, N. E. (1981). A comparison of the granulosis viruses from *Pieris brassicae* and *Pieris rapae*. *Virology 115*, 173-181.

*Crook, N. E., and Payne, C. C. (1980). Comparison of three methods of ELISA for baculovirus. *J. Gen. Virol. 46*, 29-37.

*Crowle, A. J. (1961). "Immunodiffusion." Academic Press, New York.

Cunningham, J. C., Tinsley, T. W., and Walker, J. M. (1966). Haemagglutination with plant and insect viruses. *J. Gen. Microbiol. 42*, 397-401.

de Macario, E. G. and Macario, A. J. L. (1983). Monoclonal antibodies for bacterial identification and taxonomy. *ASM News 49(1)*, 1-8.

Emini, E. A., Elzinga, M., and Wimmer, E. (1982). Carboxy-terminal analysis of poliovirus proteins: termination of poliovirus RNA translation and location of unique poliovirus polyprotein cleavage sites. *J. Virol. 42*, 194-199.

*Erickson, P. F., Miniez, L. N., and Gasher, R. (1982). Quantitative electrophoretic transfer of polypeptides from SDS polyacrylamide gels to nitrocellulose sheets: A method for their re-use in immunoautoradiographic detection of antigens. *J. Immunol. Methods 51*, 241-249.

Flamand, A., Wiktor, T. J., and Koprowski, H. (1980). Use of hybridoma monoclonal antibodies in the detection of antigenic differences between rabies and rabies-related virus proteins. II. The glycoprotein. *J. Gen. Virol. 48*, 105-109.

Goding, J. W. (1978). Use of staphylococcol protein A as an immunological reagent. *J. Immunol. Methods 20*, 241-253.

Granados, R. R. and Lawler, K. A. (1981). *In vivo* pathway of *Autographa californía* baculovirus invasion and infection. *Virology 108*, 297-308.

*Halonen, P. and Meurman, O. (1982). Radioimmunoassay in diagnostic virology. *In* "New Developments in Practical Virology: Laboratory and Research Methods in Biology and Medicine" (C. R. Howard, ed.), Vol. 5, pp. 83-124. Alan R. Liss, New York.

Harrap, K. A. and Payne, C. C. (1979). The structural properties and identification of insect viruses. *Advan. Virus Res. 25*, 273-355.

Harvey, J. P. and Volkman, L. E. (1983). Biochemical and biological variation of *Cydia pomonella* (Codling moth) granulosis virus. *Virology 124*, 21-34.

Hirumi, H., Hirumi, K., and McIntosh, A. H. (1975). Morphogenesis of a nuclear polyhedrosis virus of the alfalfa looper in a continuous cabbage looper cell line. *Ann. N.Y. Acad. Sci. 226*, 302-326.

Hohmann, W. and Faulkner, P. (1983). Monoclonal antibodies to baculovirus structural proteins: determination of specificities by Western blot analysis. *Virology 125*, 432-444.

Howe, C. and Lee, L. T. (1970). Virus-erythrocyte interaction. *Advan. Virus Res. 17*, 1-50.

*Hsu, S. M., Raine, L., and Fanger, H. (1981). A comparative study of the peroxidase-antiperoxidase method and an avidin-biotin complex hormones with radioimmunoassay antibodies. *Amer. J. Clin. Pathol. 75*, 734-738.

Hunkapiller, M. W. and Hood, L. E. (1983). Protein sequence analysis: Automated microsequencing. *Science 219*, 650-659.

*Jones, P. P. (1980). Analysis of radiolabeled lymphocyte proteins by one- and two-dimensional polyacrylamide gel electrophoresis. *In* "Selected Methods in Cellular Immunology" (B. B. Mishell and S. M. Shiigi, eds.), pp. 398-440. Freeman, San Francisco.

Kawamoto, F., Suto, C., Kumade, N., and Kobayashi, M. (1977). Cytoplasmic budding of a nuclear polyhedrosis virus and comparative ultrastructural studies of envelopes. *Microbiol. Immunol. 21*, 255-265.

Kawanishi, C. Y. and Huang, Y. S. (1983). Manuscript in preparation.

Kawanishi, C. Y., Summers, M. D., Stoltz, D. B., and Arnott, H. J. (1972). Entry of an insect virus *in vivo* by fusion of viral envelope and microvillus membrane. *J. Invertebr. Pathol. 20*, 104-108.

Kelly, D. C., Edwards, M. L., and Robertson, I. S. (1978a). The use of enzyme-linked immunosorbent assay to detect and discriminate between small iridescent viruses. *Ann. Appl. Biol. 90*, 369-374.

Kelly, D. C., Edwards, M. L., Evans, H. F., and Robertson, J. S. (1978b). The use of the enzyme-linked immunosorbent assay to detect a nuclear polyhedrosis virus in *Heliothis armigera* larvae. *J. Gen. Virol. 40*, 465-469.

Kelly, D. C., Ayres, M. D., Lescott, T., Robertson, J. S., and Happ, G. M. (1979). A small iridescent virus (type 29) isolated from *Tenebrio molitor*: a comparison of its proteins and antigens with six other iridescent viruses. *J. Gen. Virol. 42*, 95-105.

Kislev, N., Harpaz, I., and Zelcer, A. (1969). Electron microscopic studies on hemocytes of the egyptian cotton worm, *Spodoptera littoralis* (Boisduval) infected with a nuclear polyhedrosis virus as compared to noninfected hemocytes. II. Virus-infected hemocytes. *J. Invertebr. Pathol. 14*, 245-257.

Knell, J. D., Summers, M. D., and Smith, G. E. (1983). Serologic analysis of 17 baculoviruses from subgroups A and B using protein blot immunoassay. *Virology 125*, 381-392.

Knudson, D. L. and Buckley, S. M. (1977). Invertebrate cell culture methods for the study of invertebrate--associated animal viruses. *In* "Methods in Virology" (K. Maramorosch and H. Koprowski, eds.), Vol. 6, pp. 323-391. Academic Press, New York.

Krywienczyk, J. (1963). Demonstration of nuclear polyhedrosis in *Bombyx mori* (Linngeus) by fluorescent antibody technique. *J. Invertebr. Pathol.* *5*, 309-317.

Kurstak, E. and Kurstak, C. (1974). Application of the immunoperoxidose technique in virology and cancer virology: light and electron microscopy. "Viral Immunodiagnosis." Academic Press, New York.

Langridge, W. M. R., Granados, R. R., and Greenberg, J. F. (1981a). Detection of baculovirus protein in cell culture and insect larvae by enzyme-linked immunosorbent assay (ELISA). *J. Gen. Virol.* *54*, 443-448.

Langridge, W. H. R., Granados, R. R., and Greenberg, J. F. (1981b). Detection of *Autographa californica* and *Heliothis zea* baculovirus proteins by enzyme-linked immunoassay (ELISA). *J. Invertebr. Pathol.* *38*, 242-250.

*Lloyd, R. V. and Fruhman, J. (1982). Comparison of peroxidase-antiperoxidase and avidin-biotin complex methods with radioimmunoassay antibodies. *Amer. J. Clin. Pathol.* *78*, 795-796.

Longworth, J. F. and Carey, G. P. (1980). The use of an indirect enzyme-linked immunosorbent assay to detect baculovirus in larvae and adults of *Oryctyes rhinoceros* from Tonga. *J. Gen. Virol.* *47*, 431-438.

*Malik, N. J. and Daymon, M. E. (1982). Improved double immunoenzyme labelling using alkaline phosphatase and horseradish peroxidase. *J. Clin. Pathol.* *35*, 1092-1094.

*Mandel, B. (1979). Interaction of viruses with neutralizing antibodies. *In* "Comprehensive Virology" (H. Frankel-Conrat and R. R. Wagner, eds.), Vol. 15, pp. 37-121. Academic Press, New York.

Martignoni, M., Iwai, P. J., and Rohrmann, G. F. (1980). Serum neutralization of nucleopolyhedrosis viruses (*baculovirus* subgroup A) pathogenic for *Orgyia pseudotsugata*. *J. Invertebr. Pathol.* *36*, 12-20.

*Mason, D. Y. and Sammons, R. (1978). Alkaline phosphatase and peroxidase for double immunoenzymatic labelling of cellular constituents. *J. Clin. Pathol.* *31*, 454-460.

Massey, R. J. and Schochetman, G. (1981). Viral epitopes and monoclonal antibodies: isolation of blocking antibodies that inhibit neutralization. *Science* *213*, 447-449.

Matthews, R. E. F. (ed.) (1979). Classification and nomenclature of viruses. *Intervirology* *12*(3-5), 132-281.

Matthews, R. E. F. (ed.) (1982). Classification and nomenclature of viruses. *Intervirology* *17*(1-3), 1-200.

Mazzone, H. M. and Tignor, G. M. (1976). Insect viruses: serological relationships. *Advan. Virus Res. 23*, 237-270.

*Naritoku, W. Y. and Taylor, C. R. (1982). A comparative study of the use of monoclonal antibodies using three different immunohistochemical methods. *J. Histochem. Cytochem. 30*, 253-260.

Morris, T. J., Vail, P. V., and Collier, S. S. (1981). An RNA virus in *Autographa californica* nuclear polyhedrosis virus preparations: detection and identification. *J. Invertebr. Pathol. 38*, 201-208.

Naser, W. L. and Miltenburger, H. G. (1982). A rapid method for the selective identification of *Autographa californica* nuclear polyhedrosis virus using a monoclonal antibody. *FEMS Microbiol. Lett. 15*, 261-264.

Nisonoff, A. (1982). "Introduction to Molecular Immunology." Sinauer, Sunderland, Massachusetts.

Norton, P. W. and Dicapua, R. A. (1975). Serological relationship of nuclear polyhedrosis virus. I. Hemaglutination by polyhedral inclusion body protein from the nuclear polyhedrosis virus of *Porthetria (=Lymantria) dispar*. *J. Invertebr. Pathol. 25*, 185-188.

Ouchterlony, Ö. (1962). Diffusiton-in gel methods for immunological analysis II. *Progr. Allergy 6*, 30-154.

Oxford, J. (1982). The use of monoclonal antibodies in virology. *J. Hyg. (Cambridge) 88*, 361-368.

Payment, P., Arora, J. S., and Belloncik, S. (1982). An enzyme-linked immunosorbent assay for the detection of cytoplasmic polyhedrosis virus. *J. Invertebr. Pathol. 40*, 55-60.

Raghow, R. and Grace, T. D. C. (1974). Studies on a nuclear polyhedrosis virus in *Bombyx mori* cells *in vitro*. *J. Ultrastr. Res. 47*, 384-399.

Reichelderfer, C. F. (1974). A hemagglutinating antigen from a nuclear polyhedrosis virus of *Spodoptera frugiperda*. *J. Invertebr. Pathol. 23*, 46-50.

Roberts, P. L. and Naser, W. (1982). Characterization of monoclonal antibodies to the *Autographa californica* nuclear polyhedrosis virus. *Virology 122*, 424-430.

Rubenstein, R., Lawler, R., and Grandos, R. (1982). Use of primary fat body cultures for the study of baculovirus replication. *J. Invertebr. Pathol. 40*, 226-273.

Rypka, E., Volkman, L. E., and Kinter, E. (1978). Construction and use of an optimized identification scheme. *Lab. Med. 9*, 32-41.

Schmidt, N. J. and Lennette, . (1973). Advances in the serodiagnosis of viral infections. *Progr. Med. Virol. 15*, 244-308.

*Schneider, C., Newman, R. A., Sutherland, D. R., Asser, U., and Greaves, M. F. (1982). A one-step purification of membrane proteins using a high efficiency immunomatrix. *J. Biol. Chem. 257*, 10766-10769.

Smith, G. E. and Summers, M. D. (1978). Analysis of baculovirus genomes with restriction endonucleases. *Virology 89*, 517-527.

Smith, G. and Summers, M. D. (1981). Application of a novel radioimmunoassay to identify baculovirus structural proteins that share interspecies antigenic determinants. *J. Virol. 39*, 125-137.

Summers, M. D. (1971). Electron microscopic observations on granulosis virus entry, uncoating and replication processes during infection of the midgut cells of *Trichoplusia ni*. *J. Ultrastr. Res. 35*, 606-625.

Summers, M. D. and Smith, G. E. (1975). *Trichoplusia ni* granulosis virus granulin: a phenol-soluble phosphorylated protein. *J. Virol. 16*, 1108-1116.

Summers, M. D. and Volkman, L. E. (1976). Comparison of biophysical and morphological properties of occluded and extracellular nonoccluded baculovirus from *in vivo* and *in vitro* systems. *J. Virol. 17*, 962-972.

Summers, M. D., Volkman, L. E., and Hsieh, C. H. (1978). Immunoperoxidase detection of baculovirus antigens in insect cells. *J. Gen. Virol. 40*, 545-557.

Tanada, Y., Hess, R. T., and Omi, E. M. (1975). Invasion of a nuclear polyhedrosis virus in midgut of the armyworm, *Pseudoletia unipuncta*, and the enhancement of the synergistic enzyme. *J. Invertebr. Pathol. 26*, 99-104.

Thaler, M. S., Klausner, R. D., and Cohen, H. J. (1977). "Medical Immunology." Lippincott, Philadelphia.

*Towbin, H., Staehelin, R., and Gordon, J. (1979). Electrophoretic transfer of proteins from polyacrylamide gels to nitrocellulose sheets: procedure and some applications. *Proc. Natl. Acad. Sci. U.S. 76*, 4350-4354.

Vlak, J. M. (1982). Restriction endonucleases as tools in baculovirus identification. *Proc. 3rd Intern. Colloq. Invertebr. Pathol. Microbial Control, Brighton*, pp. 218-225.

Vlak, J. M., Smith, G. E., and Summers, M. D. (1981). Hybridization selection and *in vitro* translation of *Autographa californica* nuclear polyhedrosis virus. *J. Virol. 40*, 762-771.

*Voller, A., Bidwell, D. E., and Bartlett, A. (1979). "The Enzyme-Linked Immunosorbent Assay (ELISA)." Dynatech laboratories, Inc.

*Voller, A., Bidwell, D. E., and Bartlett, A. (1982). ELISA techniques in virology. *In* "New Developments in Practical Virology: Laboratory and Research Methods in Biology and Medicine" (C. R. Howard, ed.), Vol. 5, pp. 59-82. Alan R. Liss, New York.

Volkman, L. E. (1982). The application of immunofluorescence and immunoperoxidase light microscopy techniques to problems in invertebrate pathology. *Proc. 3rd Intern. Colloq. Invertebr. Pathol. Microbial Control, Brighton*, pp. 248-251.
Volkman, L. E. (1983). Occluded and budded *Autographa californica* nuclear polyhedrosis virus: immunological relatedness of structural proteins. *J. Virol. 46*, 221-229.
Volkman, L. E. and Falcon, L. A. (1982). Use of monoclonal antibody in an enzyme-linked immunosorbent assay to detect the presence of *Trichoplusia ni* (lepidoptera: noctuidae) S nuclear polyhedrosis virus polyhedrin in *T. ni* larvae. *J. Econ. Entomol. 75*, 868-871.
*Volkman, L. E. and Goldsmith, P. A. (1981). Baculovirus bioassay not dependent upon polyhedra production. *J. Gen. Virol. 56*, 203-206.
Volkman, L. E. and Goldsmith, P. A. (1982). Generalized immunoassay for *Autographa californica* nuclear polyhedrosis virus infectivity *in vitro*. *Appl. Environ. Microbiol. 44*, 227-233.
Volkman, L. E. and Goldsmith, P. A. (1983). *In vitro* survey of *Autographa californica* nuclear polyhedrosis virus interaction with nontarget vertebrate host cells. *Appl. Environ. Microbiol. 45*, 1085-1093.
Volkman, L. E., Goldsmith, P., Hess, R., and Faulkner, P. (1984). Neutralization of budded *Autographa californica* NPV by a monoclonal antibody: Identification of the target antigen. *Virology 133*.
Volkman, L. E., Summers, M. D., and Hsieh, C. H. (1976). Occluded and nonoccluded nuclear polyhedrosis virus grown in *Trichoplusia ni*: neutralization, comparative infectivity, and *in vitro* growth studies. *J. Virol. 19*, 820-832.
*Weir, D. M. (ed.) (1978). "Handbook of Experimental Immunology," 3rd ed. Blackwell, Oxford.
Wiktor, T. J. and Koprowski, H. (1978). Monoclonal antibodies against rabies virus produced by somatic cell hybridization: detection of antigenic variants. *Proc. Natl. Acad. Sci. U.S. 75*, 3938-3942.
*Yelton, D. E. and Scharff, M. D. (1981). Monoclonal antibodies: a powerful new tool in biology and medicine. *Annu. Rev. Biochem. 50*, 657-680.
*Yewdell, J. W. and Gerhard, W. (1981). Antigenic characterization of viruses by monoclonal antibodies. *Annu. Rev. Microbiol. 35*, 185-206.
Young, E. C. and Longworth, J. F. (1981). The epizoology of the baculovirus of the coconut palm rhinoceros beetle (*Oryctes rhinoceros*) in Tonga. *J. Invertebr. Pathol. 38*, 362-369.

VIRAL PROTEINS FOR THE IDENTIFICATION OF INSECT VIRUSES

SALLY B. PADHI

Waksman Institute of Microbiology
Rutgers-The State University of New Jersey
New Brunswick, New Jersey

I. INTRODUCTION

In recent years a great deal of progress has been made in the characterization of the structural proteins of insect viruses, mostly through the use of sodium dodecyl sulfate polyacrylamide gel electrophoresis (SDS-PAGE). The insect viruses receiving most of the attention have been those that are presently in use or under consideration for use as biological control agents. One could not overexaggerate the importance of the correct identification of an insect virus before it is dispersed into the environment for the control of an insect pest(s). In order to do this there must be methods of ascertaining the identity of the insect virus that is being used as a biological control agent. One must be certain that during the mass production of the virus, no mutations have occurred and/or no viral contaminants are present in the preparation.

II. VIRUS IDENTIFICATION BY PROTEIN CHARACTERIZATION

A study of the virus structural proteins are a means by which viruses can be characterized. One should bear in mind that accurate identification of a virus will possibly require

ISBN 0-12-470295-3

serological and other biochemical studies in addition to methods such as PAGE, since PAGE is only comparing electrophoretic mobilities of proteins. Other methods which have been used for studies of insect virus proteins are: peptide mapping and amino acid analysis, and the study of enzymes associated with viruses. Although there are many nonstructural proteins coded for during insect-virus infections, characterization of these proteins is unlikely to provide a direct means of insect virus identification. Thus, most attention should be given to a study of the structural proteins of insect viruses (Harrap and Payne, 1979).

A. ELECTROPHORESIS

Basically, electrophoresis is a method which is used to separate or analyze proteins, nucleic acids, or other charged molecules by their differing electrophoretic mobilities. Separation is brought about by propelling charged molecules through a porous gel in an electrical field generated in a buffer which permeates the gel. By making variations in the gel and buffer it is possible to separate proteins not only according to charge, but also according to molecular weight and isoelectric point (pI).

The two types of commonly used gel materials are agarose and polyacrylamide. The agarose gel separates proteins according to charge density (i.e., the higher the ratio of charge to mass the faster the protein will move) and contains pores large enough for even the largest protein molecules to pass through. Agarose is also used to separate large DNA or RNA molecules (according to size). Polyacrylamide gels have smaller pores so that movement of the larger proteins is slowed down and there is a sieving of proteins. This increases the resolution of proteins so that a protein sample that might produce 5 bands on an agarose gel could produce 15 bands on a polyacrylamide gel.

Polyacrylamide gels are formed by the polymerization of acrylamide monomers in the presence of a comonomer such as *N*,*N*'-methylene-bis-acrylamide, which acts as a cross-linker. The pore sizes are regulated by varying the relative concentrations of these two monomers in the gel mixture.

1. *Types of Electrophoretic Separations*

a. Separation of proteins by size: SDS-PAGE and gradient electrophoresis. It was first observed by Shapiro et al. (1967) that proteins reduced with mercaptoethanol and heated with the detergent, sodium dodecyl sulfate (SDS) disaggregates into protomers, and their electrophoretic mobilities are inversely related to the logarithm of their molecular weight if measured in cross-linked polyacrylamide. This finding was further supported by Weber and Osborn (1969). The SDS binds to the protein molecules, denaturing the proteins and masking their native charge with its own negative charge. When subjected to electrophoresis, the proteins which are now equal in charge density are separated solely according to size by the molecular sieving effects of the gel.

In gradient electrophoresis, the pore size of the gel decreases as the concentration of acrylamide increases so that when the electrical field is applied to the gel, molecules begin moving at a speed determined by charge. As the pore size decreases in the gradient gel, the larger protein molecules begin to slow down and eventually are immobilized at the point where the pores are too small for the large protein molecules to pass through (Fawcett and Morris, 1966; Margolis and Kenrick, 1968).

b. Separation by isoelectric point. The separation of proteins by isoelectric point takes advantage of the fact that each protein has a different pH at which it is electrically neutral (its isoelectric point). With this technique, proteins are separated on a gel in which a pH gradient has been generated, extending from a low pH at the anode to a high pH at the cathode (Wrigley, 1968).

c. Two-dimensional electrophoresis. With two-dimensional electrophoresis, two electrophoretic techniques are applied in succession and are capable of resolving over 1000 proteins (O'Farrell, 1975). The technique can be used for the identification of genetically altered or modified polypeptides that are separable by minor charge differences. The proteins are separated by isoelectric focusing in the first dimension and by SDS-PAGE in the second dimension.

2. *Electrophoresis of Insect Virus Proteins*

Electrophoretic studies of insect virus proteins have been mainly on acrylamide gels in the presence of SDS. A review of these studies will comprise the major portion of this chapter. Studies published prior to 1979 will be found in a review by

Harrap and Payne (1979). For information on the details of SDS-PAGE procedures refer to Stegemann (1979).

By utilizing SDS-PAGE, one can compare the polypeptide profiles of insect viruses when they are examined under identical conditions. As pointed out by Harrap and Payne (1979), electrophoresis of proteins can show whether viruses have similar or distinct protein compositions but the detection of proteins with similar electrophoretic mobilities does not necessarily mean they are identical, since the actual amino acid composition of the proteins can be different. Also, one protein band on a gel does not necessarily mean one protein. Even when separation is based on molecular weight, the virus could contain more than one protein having similar molecular weights. For this reason, two-dimensional electrophoresis is a very useful tool as it can resolve a greater number of proteins. To date there is one report utilizing O'Farrell's (1975) method of two-dimensional electrophoresis in the study of insect virus proteins (Singh et al., 1983). Yamamoto and Tanada (1979) also analyzed baculovirus proteins with two-dimensional electrophoresis but they used agarose gel electrophoresis in the first dimension and either SDS-PAGE or immunoelectrophoresis in the second dimension.

B. PEPTIDE MAPPING AND AMINO ACID ANALYSIS

Peptide mapping has been of limited use in insect virus characterization (Harrap and Payne, 1979) but has been used in the comparative analysis of *Baculovirus* granulins and polyhedrins (Summers and Smith, 1976; Cibulsky et al., 1977a; Maruniak and Summers, 1978; Griffith, 1982). Peptide mapping was used in studies which determined the first complete amino acid sequence o an insect virus protein--the polyhedrin of *Bombyx mori* NPV (Serebryani et al., 1977). Total amino acid analysis is of little use for virus identification as it is the least sensitive technique available for comparing proteins (Harrap and Payne, 1979).

C. VIRUS-ASSOCIATED ENZYMES

Comparative studies of virus-associated enzymes are considered unlikely to be useful for virus identification (Harrap and Payne, 1979). Baculoviruses contain an alkaline protease activity which is associated with and degrades the matrix protein of the polyhedra of NPVs, the capsules of granulosis viruses, or the inclusion bodies of entomopoxviruses and cytoplasmic polyhedrosis viruses (Eppstein and Thoma, 1975, 1977; Kozlov et al., 1975a,b; McIntosh and Padhi, 1976; Summers and

Smith, 1976; Tweeten et al., 1978; Bilimoria and Arif, 1979; Rubinstein and Polson, 1983). It was first documented by Wood (1980b) (using the *Autographa californica* NPV) that this alkaline protease activity can also have an effect on the structural proteins of the virions. More recently, Stiles et al. (1983) reported that the alkaline protease has a pronounced effect on the envelope proteins of another baculovirus (*Lymantria dispar* NPV). As a result of these studies, one can see that it is important to consider the activities of this enzyme(s) when studying the proteins of these insect viruses.

III. RESULTS OF PROTEIN STUDIES FOR INSECT VIRUSES IDENTIFICATION

A. *BACULOVIRUSES*

1. *Nuclear Polyhedrosis Viruses*

Nuclear polyhedrosis viruses (NPVs) are rod-shaped double-stranded DNA viruses. They may either be occluded in a proteinaceous inclusion body (this protein is called polyhedrin) or they may occur as nonoccluded viruses (NOV). A number of enveloped nucleocapsids are included in a single inclusion body. Some NPVs consist of only one nucleocapsid per envelope (SNPV) and some NPVs have one or several nucleocapsids per envelope (MNPV).

a. Polyhedrin. Polyhedrin from protease-inactivated polyhedra of ten different NPVs has been found to be composed of a single major polypeptide having a molecular weight in the range of 28,000 to 33,000 (McCarthy and Liu, 1976; Summers and Smith, 1978; Brown et al., 1979; Zethner et al., 1979; Dobos and Cochran, 1980; Maruniak and Summers, 1981; Maskos and Miltenburger, 1981; Smith and Summers, 1981). Polyhedrin from *Autographa californica* MNPV and *Porthetria (Lymantria) dispar* MNPV were analyzed by two-dimensional electrophoresis (Singh et al., 1983). They report that the major polyhedrin polypeptide produced in one-dimensional electrophoresis produced six polypeptides in the second dimension all having the same molecular weight with the pI ranging from 5.3 to 5.9 for *A. californica* MNPV and 5.7 to 6.2 for *P. dispar* MNPV.

Griffith (1982) suggests that studying the dissolution kinetics of polyhedral inclusion bodies (PIB) of NPV or capsules of granulosis viruses may at times be a simple and rapid method for the identification of baculoviruses. He found the rate of dissolution of each baculovirus in alkaline buffer to be reproducible and also to be different even in four closely related NPVs (*Spodoptera exigua*, *S. frugiperda*, *S. littoralis*, and *S. exempta*).

Griffith (1982) also did peptide mapping of polyhedrin and again found a great deal of sequence homology with the polyhedrin from three NPVs of three related insects. He suggests that the use of automated methods of peptide analysis, together with a computerized file of information on tryptic peptide fingerprints might be useful for baculovirus identification.

The alkaline protease activity which is known to be present in the proteinaceous inclusion bodies of NPV when harvested from diseased insects is absent in NPV propagated *in vitro* (Zummer and Faulkner, 1979). Even though this enzyme is only associated with baculoviruses and not synthesized by them, it is still important to consider the effect it has on the virion proteins (Wood, 1980b). When the proteins from occluded virus particles are analyzed, the alkaline protease is first inactivated by heat or chemical treatment to eliminate the protease activity; or the alkali treatment to release the virions from the inclusion bodies is conducted at 0-4°C.

b. Viral proteins. When one-dimensional SDS-PAGE is used, studies have shown that the virions of baculoviruses contain 10-25 polypeptides having molecular weights ranging from 10,000 to 160,000 (Young and Lovell, 1973; Padhi, 1974; Padhi et al., 1974; McCarthy and Liu, 1976; Cibulsky et al., 1977b; Harrap et al., 1977; Summers and Smith, 1978; Maruniak et al., 1979; Zethner et al., 1979; Singh et al., 1980; Maskos and Miltenburger, 1981; Brown et al., 1981).

Wood (1980b) used SDS-PAGE to determine the polypeptide profile obtained from virions liberated in the presence of active and heat-inactivated endogenous protease; and also in the absence of an endogenous protease utilizing inclusion bodies propagated *in vitro*. Polypeptide analysis of *A. californica* virions from active protease inclusion bodies showed both qualitative and quantitative differences when compared with the virions from the heat-inactivated protease samples. The nucleocapsids themselves were also analyzed by releasing the virions in the presence of protease and treating them with Nonidet P-40 (NP-40) to remove the viral envelope. These nucleocapsids were shown to have three major proteins with molecular weights of 16,000, 19,000, and 39,000 and four minor proteins of 18,000, 32,000, 33,000, and 35,000.

With increased knowledge of invertebrate tissue culture (and the absence of alkaline protease in the inclusion bodies), more studies of NPV proteins are being done with *in vitro*-propagated virions. *In vitro*-produced *Galleria mellonella*

NPV was determined to have 24 polypeptides by SDS-PAGE (Fraser and Hink, 1982). Carstens et al. (1979), Dobos and Cochran (1980), Wood (1980a), and Maruniak and Summers (1981) all analyzed *A. californica* NPV-induced protein synthesis *in vitro*. With the use of autoradiography, which improved resolution of proteins in PAGE, 22-36 polypeptides were detected in *A. californica* NPV infected cells. Similar results were obtained with virions isolated from the extracellular fluid of infected cells (Vlak, 1979).

Properties of the major nucleocapsid protein of *Heliothis zea* SNPV were determined by Kelly et al. (1983). It was found to be a low molecular weight (estimated to be 12,600), basic protein comprising 64% of the nucleocapsid proteins. Amino acid analysis of this protein showed it to be rich in serine, glycine, and arginine and have only small amounts of aspartic and glutamic acid, cysteine, histidine, and methionine. They report that there are 15 polypeptides in the virus particles and eight of these are in the nucleocapsid. The presence of this extremely basic, low molecular weight protein in baculoviruses was initially reported by Tweeten et al. (1980b) in granulosis viruses. Kelly et al. (1983) compared some features of this protein with the major nucleocapsid protein of the following three other baculoviruses: *T. ni* MNPV, *Spodoptera litura* granulosis virus, and *Oryctes rhinoceros* nonoccluded baculovirus. They found that they were all low molecular weight, arginine-rich, lysine-poor, DNA-binding proteins.

Some studies have been made of the structural glycoproteins in *A. californica* NPV (Maruniak, 1979; Dobos and Cochran, 1980) and a NOV of *Heliothis zea* (Burand et al., 1983). Stiles and Wood (1983) identified eleven structural glycoproteins in *A. californica* NPV and eight of these were identical in both virus preparations (i.e., virus was propagated in two different cell lines: TN-368 and lPLB-21).

Stiles et al. (1983) characterized the proteins of *Lymantria dispar* NPV in the absence of alkaline protease and 29 viral structural proteins were identified by SDS-PAGE. They report that 14 are surface envelope proteins (determined by lactoperoxidase iodination) and 11 proteins are associated with the nucleocapsids. They also reported distinct alterations of viral proteins when virions were purified in the presence of active alkaline protease. The patterns they obtained were then considered to be similar to previous reports on this virus (Padhi et al., 1974; Zethner et al., 1979; Maskos and Miltenburger, 1981).

Singh et al. (1983) analyzed the proteins of *Porthetria (Lymantria) dispar* NPV and *Autographa californica* NPV by two-dimensional electrophoresis. The enveloped virions of *A. californica* NPV had 81 acidic polypeptides ranging in molecular weight from 13,500 to 86,000 daltons. The major capsid pro-

TABLE I. SDS-PAGE of Nuclear Polyhedrosis Virus Structural Proteins[a]

Host species	*Treatment for alkaline protease in PIB*[b]
Anticarsa gemmatalis	0-4°C, 1.0 hr
Autographa californica	0-4°C, 1.0 hr[d]
	0-4°C, 1.0 hr[d]
	NOV (*in vitro*)
	In vitro
	NOV (*in vitro*)
	In vitro
	In vitro
	0.4°C, 1.0 hr or 30°C, 0.5 hr[d]
	0-4°C, 1.0 hr
Galleria mellonella	NOV (*in vitro*)
Heliothis armigera	0-4°C, 1.0 hr
Heliothis zea (SNPV)	0-4°C, 1.0 hr
Lymantria dispar (*Lymantria=Porthetria*)	80°C, 2 hr[f]
	0-4°C, 1.0 hr
Rachiplusia ou	0-4°C, 1.0 hr[d]
Trichoplusia ni (SNPV)	0-4°C, 1.0 hr

[a]Results of studies in which the alkaline protease enzyme is absent or inactive.

[b]Method used to eliminate the action of alkaline protease on the virions or polyhedrin during alkali dissolution. If "in vitro" is written in this column, the PIB are from tissue culture and therefore no alkaline protease is present.

PIB protein M.W.[c]	*+ Virus proteins*		*Reference*
	Number	*M.W. range*[c]	
29.0	*15*	*17-97*	*Summers and Smith (1978)*
30	*24*	*16-150*	*Summers and Smith (1978)*
30	*23*	*16-125*	*Maruniak et al. (1979)*
	35	*≤11.6-126*	*Vlak (1979)*
30	*22*	*15-90*	*Dobos and Cochran (1980)*
	22	*16-115*	*Dobos and Cochran (1980)*
33	*26*	*15-115*	*Wood (1980b)*
30	*26*	*16-125*	*Maruniak and Summers (1981)*
28.9	*22*	*12.4-92.5*	*Maskos and Miltenburger (1981)*
31.2	*81*[e]	*13.5-86*	*Singh et al. (1983)*
	24	*14-142*	*Fraser and Hink (1982)*
28.0	*25*	*16-90*	*Summers and Smith (1978)*
27.0	*19*	*15-89*	*Summers and Smith (1978)*
30	*29*	*13-109*	*Stiles et al. (1983)*
30.6	*95*[e]	*13.5-85.5*	*Singh et al. (1983)*
30	*16*	*16-160*	*Summers and Smith (1978)*
31	*15*	*16-90*	*Summers and Smith (1978)*

[c]*All protein molecular weights (M.W.) are given* × 10^{-3}.

[d]*Also analyzed in vitro PIBs.*

[e]*The number of acidic polypeptides obtained from two-dimensional PAGE.*

[f]*Treatment used before alkali treatment of PIBs; also assayed for alkaline protease activity.*

tein was determined to have a molecular weight of 41,500. Nucleocapsids (envelope removed) had 64 polypeptides, and at least 11 of the polypeptides, including most of the high molecular weight proteins, that were not resolved in nucleocapsids were considered to be envelope proteins. The *P. dispar* NPV had 95 acidic polypeptides with molecular weights ranging from 13,500 to 85,500. The predominant polypeptide had a molecular weight of 46,500 daltons. The number of polypeptides resolved are over twice the number previously determined with one-dimensional electrophoresis for any baculovirus. It is thought that with further manipulation of this technique, there will be even more virion polypeptides resolved. A compilation of the SDS-PAGE analyses of NPV proteins in the absence of alkaline protease activity is presented in Table I.

2. *Granulosis Viruses*

Granulosis viruses (GV) are enveloped rod-shaped DNA viruses. They are embedded singly in a proteinaceous inclusion body called a capsule and each contain one virion. The inclusion body protein is called granulin.

a. Granulin. Granulosis virus capsules have also been established as containing an alkaline protease which degrades the granulin (Summers and Smith, 1975; Tweeten et al., 1980b). Summers and Smith (1978) determined that the granulin of *Trichoplusia ni* GV and of *Spodoptera frugiperda* GV has molecular weights of 25,000 and 26,000, respectively.

Boucias and Nordin (1980) reported that the polyhedrin of *Hyphantria cunea* and *Diacrisia virginica* GVs have a molecular weight of approximately 28,000 when analyzed on SDS-PAGE.

Langridge and Balter (1981) determined the granulin of *Estigmene acrea* GV to be 30,000 daltons. Hotchkin (1981) determined the molecular weight of granulin from a *Pseudaletia unipuncta* GV produced in two different host species (*P. unipuncta* and *Spodoptera exigua*). The granulin produced in either host was found to have the same molecular weight (around 28,500).

Crook (1981) determined the molecular weight of *Pieris brassicae* and *P. rapae* GV granulin to be 30,000. Crook and Brown (1982) then analyzed a GV from *Lacanobia oleracea* with SDS-PAGE. The molecular weight of the granulin was determined to be 29,000.

Griffith (1982) mapped tryptic peptides of the granulin of *P. rapae*, *Phthorimaea operculella*, and *Heliothis punctiger* and a comparison of the maps from the three GVs showed very little sequence homology.

b. *Viral proteins*. The structural proteins of *P. brassicae* GV (Brown et al., 1977), *Trichoplusia ni* GV, and *Spodoptera frugiperda* GV (Summers and Smith, 1978) were the first GVs to be characterized by PAGE. They were reported to have 12-18 polypeptides with molecular weights from 12,000 to 90,000. Tweeten et al. (1980a) analyzed the proteins of *Plodia interpunctella* GV and found them to consist of 15 polypeptides ranging from 12,600 to 97,300 daltons. Five of these proteins were identified as envelope proteins. In further studies of the GV, Tweeten et al. (1980b) isolated the nucleocapsids and found them to consist of eight structural polypeptides. The two major proteins are 12,500 and 31,000. It was determined that the 12,500 molecular weight protein is a very basic protein which is associated with the viral DNA in the form of a DNA protein complex. Similar basic polypeptides were extracted from other NPVs and GVs which indicates that these basic proteins are common to baculoviruses.

Boucias and Nordin (1980) analyzed the virion proteins of *Hyphantria cunea* and *Diacrisia virginica* GV using SDS-PAGE. The protein patterns of the GVs from these two closely related insects were found to be very similar; *H. cunea* GV had 13 polypeptides and *D. virginica* had 12 polypeptides. Langridge and Balter (1981) determined the *Estigmene acrea* GV to have 24 polypeptides ranging in molecular weight from 11,000 to 98,000 daltons.

The GVs of *P. brassicae* and *P. rapae* were found to be very similar in their protein pattern (Crook, 1981). Each GV had nine major and ten minor polypeptides and only slight differences were determined in molecular weight for three of the polypeptides in the viral envelope. Nucleocapsids, obtained by NP-40 treatment, had four major polypeptides that remained unaltered. More recently, Harvey and Volkman (1983) analyzed the proteins of three isolates of *Cydia pomonella* GV by SDS-PAGE. They were found to have 23-25 polypeptides. Glycoproteins were found present in the virions (and capsule protein) of *Pseudaletia unipuncta* (Hotchkin, 1981). They had previously been reported to be in the capsule proteins of *T. ni* GV (Summers and Smith, 1975).

A compilation of the results of SDS-PAGE can be found in TABLE II. As no tissue culture system has yet been devised for the propagation of GV, one must always deal with the possible presence of alkaline protease in GV capsules. Various methods for the release of virions from the capsules have been used. In some studies the alkaline protease was inactive (Summers and Smith 1978; Tweeten et al., 1978; Boucias and Nordin, 1980; Langridge and Balter, 1981; Hotchkin, 1981) while in others the enzyme was still active (Brown et al., 1977; Tweeten et al., 1980a; Crook, 1981; Crook and Brown, 1982; Harvey and Volkman, 1983). Of those studies in which the

TABLE II. *Properties of Granulosis Virus Structural Proteins*

Host species	Inclusion body protein granulin M.W.[a]	Virus proteins		Reference
		Number	M.W. range[a]	
Cydia pomonella	--	23-25	14-140	Harvey and Volkman (1983)
Diacrisia virginica	28	12	12-80	Boucias and Nordin (1980)
Estigmene acrea	30	24	11-98	Langridge and Balter (1981)
Hyphantria cunea	28	13	12-94	Boucias and Nordin (1980)
Lacanobia oleracea	29	15+	--	Crook and Brown (1982)
Pieris brassicae	27.5	12	12-91	Brown *et al.* (1977)
Pieris brassicae	30	19	12.5-96	Crook (1981)
Pieris rapae	30	19	12.5-98	Crook (1981)
Plodia interpunctella	28	--		Tweeten *et al.* (1978)
Plodia interpunctella	--	15	12.6-97.3	Tweeten *et al.* (1980a)
Pseudaletia unipuncta	28.5	10	13-125	Hotchkin (1981)
Spodoptera frugiperda	26	18	16-160	Summers and Smith (1978)
Trichoplusia ni	25	19	16-90	Summers and Smith (1978)

[a]All protein molecular weights (M.W.) are given $\times 10^{-3}$.

protease was active the time period for alkali treatment for the release of virions was variable. Therefore, it is important to remember that it is difficult to make comparisons between the viruses as all of these studies were not done under the same conditions or with identical electrophoretic methods.

A. ENTOMOPOXVIRUSES

Entomopoxviruses (EPV) are large brick-shaped DNA viruses and they have morphological similarities to vertebrate poxviruses. They are also found in crystalline proteinaceous inclusion bodies. The inclusion body protein is called spheroidin.

1. Spheroidin

Bilimoria and Arif (1979) characterized the spheroidin of *Choristoneura biennis* EPV and found the molecular weight to be 102,000. They also determined that the inclusion bodies contain alkaline protease activity. *Amsacta moorei* EPV spheroidin was found to have a molecular weight of 110,000 (Langridge and Roberts, 1982; Langridge, 1983).

2. Viral Proteins

There have been comparatively few reports on the viral proteins of EPVs. Bilimoria and Arif (1980) characterized the proteins of *C. biennis* EPV with SDS-PAGE and determined that the alkali-released virions contained at least 40 polypeptides ranging from 12,000 to 250,000 in molecular weight. There are 12 major proteins which constitute approximately 95% of the total virus protein, with the four most abundant ones having molecular weights of 12,000, 46,000, 59,000, and 89,000. They further studied the core structure of this EPV by treating intact virions with NP-40 which removed the shell and lateral body structures. Their studies determined that one of the four major viral polypeptides is contained in the core structure (59,000 daltons).

Langridge (1983) studied *A. moorei* EPV protein synthesis in *Estigmene acrea* cells. He reported 37 virus structural proteins ranging in molecular weight from 13,000 to 208,000 for both occluded and nonoccluded forms of the virus. More than 35 structural proteins were also found in *Euxoa auxilaris* and *Melanoplus sanguinipes* EPVs (Langridge and Roberts, 1982).

C. CYTOPLASMIC POLYHEDROSIS VIRUSES

Cytoplasmic polyhedrosis viruses (CPV) are spherical double-stranded RNA viruses which are occluded in proteinaceous inclusion bodies. The protein is referred to as polyhedral protein.

1. Polyhedral Protein

Langridge (1983) determined the molecular weight of the polyhedral protein from *Estigmene acrea* CPV to be 28,500. Rubinstein (1983) determined the polyhedral protein of *Heliothis armigera* CPV to be 28,700. These are in agreement with previous studies which determined that the major polypeptide of CPV inclusion bodies has a molecular weight in the range of 25,000 to 30,000 (Lewandowski and Traynor, 1972; Payne and Rivers, 1976). Also, as has been determined with the other groups of occluded insect viruses, Rubinstein and Polson (1983) reported that there is alkaline protease activity associated with inclusion bodies of *H. armigera* CPV.

2. Viral Proteins

Using SDS-PAGE, Langridge (1983) reported there are nine virus proteins ranging in molecular weight from 14,000 to 128,000 in the *Estigmene acrea* CPV. Rubinstein (1983) characterized the proteins of *Heliothis armigera* CPV. SDS-PAGE revealed five proteins in the virus particles ranging in molecular weight from 32,000 to 156,000. In another study, a comparison was made of three closely related CPVs from *Dendrolimus spectabilis*, *Bombyx mori*, and *Lymantria dispar*. SDS-PAGE analysis showed five proteins for each CPV with no significant differences between them (Payne et al., 1978).

D. IRIDOVIRUSES (IRIDESCENT VIRUSES)

Iridoviruses are large DNA viruses which are divided into two groups on the basis of size: the smaller ones are approximately 130 nm in diameter and the larger ones are around 180 nm in diameter.

Moore and Kelly (1980) demonstrated the presence of one major and more than 20 minor polypeptides in iridescent viruses Type 2, 6, and 9. The major polypeptide and the majority of the minor ones from each virus had distinct molecular weights. These viruses were all grown in *Galleria mellonella* larvae. The major structural polypeptide of each virus was purified from polyacrylamide gels and N-terminal analysis of the polypeptide gave a single N-terminal, suggesting the protein was homogeneous. Amino acid analysis of the major polypeptide from each virus indicated that the total amino acid content was very similar; the only difference found in the major polypeptide from each of these iridescent viruses was the molecular weight.

Iridovirus Type 29, isolated from *Tenebrio molitor*, contains 20-25 proteins ranging in molecular weight from 15,000 to 72,000 as analyzed by SDS-PAGE and also by two-dimensional electrophoresis (Black et al., 1981).

Previous reports (Carey et al., 1978; Kelly et al., 1979) characterizing the proteins of iridovirus Types 2, 6, 21, 22, 23, 25, 28, and 29 stated that each of these viruses had 20 or more polypeptides.

E. DENSOVIRUSES (INSECT PARVOVIRUSES)

Densoviruses (DNV) are small spherical DNA viruses. Insect viruses recognized as belonging to this group are DNV 1 of *Galleria mellonella*, DNV 2 of *Junonia coenia*, and DNV 3 of *Agraulis vanillae*. A study characterizing the proteins of a virus isolates from *Bombyx mori* infected with the flacherie virus has tentatively placed this virus as a member of the densovirus group (Nakagaki and Kawase, 1980). It was determined that the virus contained four proteins: the major one had a molecular weight around 50,000, and the three minor proteins had molecular weights of 57,000, 70,000, and 77,000. The *Galleria* DNV has four structural proteins--one major protein with a molecular weight of 49,000 and three minor proteins (Tijssen et al., 1976).

F. PICORNAVIRUSES

Two small RNA viruses, cricket paralysis virus and *Drosophila* C virus, have been tentatively classified as picornaviruses (Scotti et al., 1981). The cricket paralysis virus is composed of three major polypeptides around 30,000 daltons and one minor polypeptide (Moore et al., 1980). The *Drosophila* C virus contains three major polypeptides of 31,000, 30,000, and 28,000 daltons and two minor components of 37,000 and 8,500 daltons (Jousset et al., 1977). Another possible member of this group of viruses is an infectious flacherie virus of *Bombyx mori*. According to Hashimoto and Kawase (1983) there are four major structural proteins in this virus (molecular weights 35,000, 33,000, 31,200, and 11,600). High resolution of these four proteins was obtained with the use of isoelectric focusing and two-dimensional electrophoresis. This is the first report on insect picornaviruses that presents discriminative properties of each of the viral proteins and these were compared to mammalian picornaviruses, in addition to the other insect picornaviruses.

G. NUDAURELIA β VIRUS

Another group of small RNA viruses is the nudaurelia β viruses. They are characterized as having only one polypeptide species with a molecular weight of 61,000-63,000 (Struthers and Hendry, 1974; Reinganum et al., 1978).

SDS-PAGE of a virus isolated from *Daschira pudibunda* was found also to have only one polypeptide with a molecular weight of 66,000 (Greenwood and Moore, 1981). Based on this and other characteristics of this new virus isolate, it is being considered as a new member of the nudaurelia β virus group.

H. UNCLASSIFIED VIRUS

A filamentous virus of the honey bee is presently not classified in any of the currently described groups of insect viruses. SDS-PAGE of this double-stranded DNA enveloped virus showed that there are 12 structural proteins with molecular weights ranging from 13,000 to 70,000 (Bailey et al., 1981). This resembles the protein profile of the *Oryctes rhinoceros* virus which also has 12 proteins with molecular weights from 10,000 to 76,000 (Payne et al., 1977). This virus is considered to be a nonoccluded baculovirus. The similarity in protein profiles of these two viruses indicates that this virus of the honey bee might also be a nonoccluded baculovirus.

IV. CONCLUSIONS

From the results of these studies one can see that SDS-PAGE has been the most commonly used tool for the identification of insect virus proteins. Using this information alone to identify insect viruses would, at present, be difficult. It is first necessary to establish standard methods to be used by all investigators so that more accurate comparisons of viruses can be made. It is also important to have a well-characterized virus from each group of insect viruses that can be used as a "standard" for comparison.

The group of viruses receiving the most attention have been the baculoviruses. Two of the NPVs (*A. californica* NPV and *L. dispar* NPV) have been studied to a great extent probably since these are important candidates for use in biological control of economically important insect pests. Thus, with the use of two-dimensional electrophoresis and the knowledge already obtained of NPVs, it appears possible to identify small changes in an NPV due to mutations, or to identify the presence of contaminating viruses.

It is now quite well documented that the inclusion body material of baculoviruses (polyhedrin or granulin) contains one protein and all that have been studied have been shown to have similar molecular weights. In contrast to this, the virions themselves have a large number of proteins and the resulting profile of each virus will be useful in virus identification.

The inclusion body protein of the other occluded viruses (CPVs and EPVs) also appear to be composed of a single protein. The molecular weight of CPV polyhedral protein is similar to that found in granulin or polyhedrin of the baculoviruses while the spheroidin of EPVs is a much larger protein with a molecular weight greater than 100,000. The polypeptide profile that have been obtained of EPVs show that this group of insect viruses is the most complex in its protein structure.

It is possible to carry out further detailed studies on viral protein synthesis and identification using tissue culture for the propagation of NPVs. Hopefully, methods for the successful propagation of GVs, the other group of baculoviruses, will also soon be determined. The use of tissue culture also eliminates the interference of the action of alkaline protease on the virus proteins as this enzyme has not been found in insect viruses produced *in vitro*.

All the other groups of insect viruses have received attention in the area of protein analysis. The results have been useful in classifying these viruses since the protein profiles are different for each known group of insect viruses.

In order to use protein characterization studies for the identification of insect viruses, the most important prerequisites are: to be sure the virus preparation is free from contaminating proteins and that no enzymes are present which will alter the protein profile. Once standards are agreed upon for protein analysis and they are used by all investigators, the identification of insect viruses by protein analysis will indeed be even more useful than it is at present.

ACKNOWLEDGMENTS

I would like to thank B. Stiles and H. A. Wood for providing me with their manuscript prior to its publication.

REFERENCES

Bailey, L., Carpenter, J. M., and Woods, R. D. (1981). Properties of a filamentous virus of the honey bee (*Apis melifera*). *Virology 114*, 1-7.

Bilimoria, S. L., and Arif, B. M. (1979). Subunit protein and alkaline protease of entomopoxvirus spheroids. *Virology 96*, 596-603.

Bilimoria, S. L., and Arif, B. M. (1980). Structural polypeptides of *Choristoneura biennis* entomopoxvirus. *Virology 104*, 253-257.

Black, P. N., Blair, C. D., Butcher, A., Capinera, J. L., and Happ, G. M. (1981). Biochemistry and ultrastructure of iridescent virus type 29. *J. Invertebr. Pathol. 38*, 12-21.

Boucias, D. G., and Nordin, G. L. (1980). Comparative analysis of the alkali-liberated components of the *Hyphantria cunea* and the *Diacrisia virginica* granulosis viruses. *J. Invertebr. Pathol. 36*, 264-272.

Brown, D. H., Bud, H. M., and Kelly, D. C. (1977). Biophysical properties of the structural components of a granulosis virus isolated from the cabbage white butterfly (*Pieris brassicae*). *Virology 81*, 317-327.

Brown, D. A., Evans, H. F., Allen, C. J., and Kelly, D. C. (1981). Biological and biochemical investigations of the European isolates of *Mamestra brassicae* nuclear polyhedrosis virus. *Arch. Virol. 69*, 209-217.

Brown, S. E., Kaczmarek, F. S., Dubois, N. R., Zerillo, R. T., Holleman, J., Breillatt, J. P., and Mazzone, H. M. (1979). Comparative properties of the inclusion body proteins of the nucleopolyhedrosis viruses of *Neodiprion sertifer* and *Lymantria dispar*. *Arch. Virol. 59*, 319-329.

Burand, J. P., Stiles, B., and Wood, H. A. (1983). Structural and intracellular proteins of the non-occluded baculovirus Hz-1. *J. Virol. 46*, 137-142.

Carey, G. P., Lescott, T., Robertson, S. J., Spencer, L. K., and Kelly, D. C. (1978). Three African isolates of small iridescent viruses: Type 21 from *Heliothis armigera* (Lepidoptera: Noctuidae), Type 23 from *Heteronychus arator* (Coleoptera: Scarabaeidae), and Type 28 from *Lethocerus columbiae* (Hemiptera Heteroptera: Belostomidae). *Virology 85*, 307-309.

Carstens, E. B., Tjia, S. T., and Doerfler, W. (1979). Infection of *Spodoptera frugiperda* cells with *Autographa californica* nuclear polyhedrosis virus. I. Synthesis of intracellular proteins after virus infection. *Virology 99*, 386-398.

Cibulsky, R. J., Harper, J. D., and Gudauskas, R. T. (1977a). Biochemical comparison of polyhedral proteins from five nuclear polyhedrosis viruses infecting plusiine larvae (Lepidoptera: Noctuidae). *J. Invertebr. Pathol. 29*, 182-191.

Cibulsky, R. J., Harper, J. D., and Gudauskas, R. T. (1977b). Biochemical comparison of virion proteins from five nuclear polyhedrosis viruses infecting plusiine larvae (Lepidoptera: Noctuidae). *J. Invertebr. Pathol. 30*, 303-313.

Crook, N. E. (1981). A comparison of the granulosis viruses from *Pieris brassicae* and *Pieris rapae*. *Virology 115*, 173-181.

Crook, N. E., and Brown, J. D. (1982). Isolation and characterization of a granulosis virus from the tomato moth, *Lacanobia oleracea*, and its potential as a control agent. *J. Invertebr. Pathol. 40*, 221-227.

Dobos, P., and Cochran, M. A. (1980). Protein synthesis in cells infected by *Autographa californica* nuclear polyhedrosis virus (Ac-NPV): the effect of cytosine arabinoside. *Virology 103*, 446-464.

Eppstein, D. A., and Thoma, J. A. (1975). Alkaline protease associated with the matrix protein of a virus infecting the cabbage looper. *Biochem. Biophys. Res. Commun. 62*, 478-484.

Eppstein, D. A., and Thoma, J. A. (1977). Characterization and serology of the matrix protein from a nuclear polyhedrosis virus of *Trichoplusia ni* before and after degradation by an endogenous proteinase. *Biochem. J. 167*, 321-332.

Fawcett, J. C., and Morris, C. J. O. R. (1966). Molecular-sieve chromatography of proteins on granulated polyacrylamide gels. *Separation Sci. 1*, 9-26.

Fraser, M. J., and Hink, W. F. (1982). The isolation and characterization of the MP and FP plaque variants of *Galleria mellonella* nuclear polyhedrosis virus. *Virology 117*, 366-378.

Greenwood, L. K., and Moore, N. F. (1981). A single protein Nudaurelia-β-like virus of the pale tussock moth, *Dasychira pudibunda*. *J. Invertebr. Pathol. 38*, 305-306.

Griffith, I. P. (1982). A new approach to the problem of identifying baculoviruses. *In* "Microbial and Viral Pesticides" (E. Kurstak, ed.), pp. 507-527. Dekker, New York.

Harrap, K. A., and Payne, C. C. (1979). The structural properties and identification of insect viruses. *Advan. Virus Res.* *25*, 273-355.

Harrap, K. A., Payne, C. C., and Robertson, J. S. (1977). The properties of three baculoviruses from closely related hosts. *Virology 79*, 14-31.

Harvey, J. P., and Volkman, L. E. (1983). Biochemical and biological variation of *Cydia pomonella* (codling moth) granulosis virus. *Virology 124*, 21-34.

Hashimoto, Y., and Kawase, S. (1983). Characteristics of structural proteins of infectious flacherie virus from the silkworm, *Bombyx mori*. *J. Invertebr. Pathol.* *41*, 68-76.

Hotchkin, P. G. (1981). Comparison of virion proteins and granulin from a granulosis virus produced in two host species. *J. Invertebr. Pathol.* *38*, 303-304.

Jousset, F.-X., Bergoin, M., and Revet, B. (1977). Characterization of the *Drosophila* C virus. *J. Gen. Virol.* *34*, 269-285.

Kelly, D. C., Ayres, M. D., Lescott, T., Robertson, J. S., and Happ, G. M. (1979). A small iridescent virus (type 29) isolated from *Tenebrio molitor*: a comparison of its proteins and antigens with six other iridescent viruses. *J. Gen. Virol.* *42*, 95-105.

Kelly, D. C., Brown, D. A., Ayres, M. D., Allen, C. J., and Walker, I. O. (1983). Properties of the major nucleocapsid protein of *Heliothis zea* singly enveloped nuclear polyhedrosis virus. *J. Gen. Virol.* *64*, 399-408.

Kozlov, E. A., Sidorova, N. M., and Serebryani, S. B. (1975a). Proteolytic cleavage of polyhedral protein during dissolution of inclusion bodies of the nuclear polyhedrosis viruses of *Bombyx mori* and *Galleria mellonella* under alkaline conditions. *J. Invertebr. Pathol.* *25*, 97-101.

Kozlov, E. A., Levitina, T. L., Sidorova, N. M., Radavski, Y. L., and Serebryani, S. B. (1975b). Comparative chemical studies of the polyhedral proteins of the nuclear polyhedrosis viruses of *Bombyx mori* and *Galleria mellonella*. *J. Invertebr. Pathol.* *25*, 103-107.

Langridge, W. H. R. (1983). Virus DNA replication and protein synthesis in *Amsacta moorei* entomopoxvirus-infected *Estigmene acrea* cells. *J. Invertebr. Pathol.* *41*, 341-349.

Langridge, W. H. R. (1983). Characterization of a cytoplasmic polyhedrosis virus from *Estigmene acrea* (Lepidoptera). *J. Invertebr. Pathol.* *42*, 259-263.

Langridge, W. H. R., and Balter, K. (1981). Protease activity associated with the capsule protein of *Estigmene acrea* granulosis virus. *Virology 114*, 595-600.

Langridge, W. H. R., and Roberts, D. W. (1982). Structural proteins of *Amsacta moorei*, *Euxoa auxiliaris*, and *Melanoplus sanguinipes* entomopoxviruses. *J. Invertebr. Pathol. 39*, 346-353.

Lewandowski, L. J., and Traynor, B. L. (1972). Comparison of the structure and polypeptide composition of three double-stranded ribonucleic acid-containing viruses (Diplornaviruses): cytoplasmic polyhedrosis virus, wound tumor virus, and reovirus. *J. Virol. 10*, 1053-1070.

McCarthy, W. J., and Liu, S.-Y. (1976). Electrophoretic and serological characterization of *Porthetria dispar* polyhedron protein. *J. Invertebr. Pathol. 28*, 57-65.

McIntosh, A. H., and Padhi, S. B. (1976). *In vitro* and *in vivo* comparative studies of several nuclear polyhedrosis viruses (NPVs) by neutralization, immunofluorescence and polyacrylamide gel electrophoresis. *In* "Invertebrate Tissue Culture: Applications in Medicine, Biology, and Agriculture" (E. Kurstak and K. Maramorosch, eds.), pp. 331-338. Academic Press, New York.

Margolis, J., and Kenrick, K. G. (1968). Polyacrylamide gel electrophoresis in a continuous molecular sieve gradient. *Anal. Biochem. 25*, 347-362.

Maruniak, J. E. (1979). Biochemical characterization of baculovirus structural and infected TN-368 cell polypeptides, glycoproteins, and phosphoproteins. Ph.D. Thesis, Univ. Texas, Austin, Texas.

Maruniak, J. E., and Summers, M. D. (1978). Comparative peptide mapping of baculovirus polyhedrins. *J. Invertebr. Pathol. 32*, 196-201.

Maruniak, J. E., and Summers, M. D. (1981). *Autographa californica* nuclear polyhedrosis virus phosphoproteins and synthesis of intracellular proteins after virus infection. *Virology 109*, 25-34.

Maruniak, J. E., Summers, M. D., Falcon, L. A., and Smith, G. E. (1979). *Autographa californica* nuclear polyhedrosis virus structural proteins compared from *in vivo* and *in vitro* sources. *Intervirology 11*, 82-88.

Maskos, C. B., and Miltenburger, H. G. (1981). SDS-PAGE comparative studies on the polyhedral and viral polypeptides of the nuclear polyhedrosis viruses of *Mamestra brassicae*, *Autographa californica*, and *Lymantria dispar*. *J. Invertebr. Pathol. 37*, 174-180.

Moore, N. F., and Kelly, D. C. (1980). Comparative study of the polypeptides of three iridescent viruses by N-terminal analysis, amino acid analysis and surface labeling. *J. Invertebr. Pathol. 36*, 415-422.

Moore, N. F., Kearns, A., and Pullen, J. S. K. (1980). Characterization of cricket paralysis virus-induced polypeptides in *Drosophila* cells. *J. Virol. 33*, 1-9.

Nakagaki, M., and Kawase, S. (1980). Structural proteins of densonucleosis virus isolated from the silkworm, *Bombyx mori*, infected with the flacherie virus. *J. Invertebr. Pathol. 36*, 166-171.

O'Farrell, P. H. (1975). High resolution two-dimensional electrophoresis of proteins. *J. Biol. Chem. 250*, 4007-4021.

Padhi, S. B. (1974). Studies on the nuclear polyhedrosis virus of *Porthetria dispar*. Ph.D. Thesis, Rutgers--The State University, New Brunswick, New Jersey.

Padhi, S. B., Eikenberry, E. F., and Chase, T., Jr. (1974). Electrophoresis of the proteins of the nuclear polyhedrosis virus of *Porthetria dispar*. *Intervirology 4*, 333-345.

Payne, C. C., and Rivers, C. F. (1976). A provisional classification of cytoplasmic polyhedrosis viruses based on the sizes of the RNA genome segments. *J. Gen. Virol. 33*, 71-85.

Payne, C. C., Compson, D., and DeLooze, S. M. (1977). Properties of the nucleocapsids of a virus isolated from *Oryctes rhinoceros*. *Virology 77*, 269-280.

Payne, C. C., Mertens, P. P. C., and Katagiri, K. (1978). A comparative study of three closely related cytoplasmic polyhedrosis viruses. *J. Invertebr. Pathol. 32*, 310-318.

Reinganum, C., Robertson, J. S., and Tinsley, T. W. (1978). A new group of RNA viruses from insects. *J. Gen. Virol. 40*, 195-202.

Rubinstein, R. (1983). Characterization of the proteins and serological relationships of cytoplasmic polyhedrosis virus of *Heliothis armigera*. *J. Invertebr. Pathol. 42*, 292-294.

Rubinstein, R., and Polson, A. (1983). Midgut and viral associated proteases of *Heliothis armigera*. *Intervirology 19*, 16-25.

Scotti, P. D., Longworth, J. F., Plus, N., Croizier, G., and Reinganum, C. (1981). The biology and ecology of strains of insect small RNA virus complex. *Advan. Virus Res. 26*, 117-143.

Serebryani, S. B., Levitina, T. L., Kautsman, M. L., Radavski, Y. L., Gusak, N. M., Ovander, M. N., Sucharenko, N. V., and Kozlov, E. A. (1977). The primary structure of the polyhedral protein of nuclear polyhedrosis virus (NPV) of *Bombyx mori*. *J. Invertebr. Pathol. 30*, 442-443.

Shapiro, A. L., Vinuela, E., and Maizel, J. V. (1967). Molecular weight estimation of polypeptide chains by electrophoresis in SDS-polyacrylamide gels. *Biochem. Biophys. Res. Commun. 28*, 815-820.

Singh, S. P., Gudauskas, R. T., and Harper, J. D. (1980). Gel electrophoresis of sonicated virions of two-nuclear polyhedrosis viruses. *J. Ala. Acad. Sci. 51*, 249-255.

Singh, S. P., Gudauskas, R. T., and Harper, J. D. (1983). High resolution two dimensional gel electrophoresis of structural proteins of baculoviruses of *Autographa californica* and *Porthetria (Lymantria) dispar*. *Virology 125*, 370-380.

Smith, G. E., and Summers, M. D. (1981). Application of a novel radioimmunoassay to identify baculovirus structural proteins that share interspecies autigenic determinants. *J. Virol. 39*, 125-137.

Stegemann, H. (1979). SDS-gel-electrophoresis in polyacrylamide, merits and limits. *In* "Electrokinetic separation methods" (P. G. Righetti, C. J. van Oss, and J. W. Vanderhoff, eds.), pp. 313-336. Elsevier/North-Holland Biomedical Press.

Stiles, B., and Wood, H. A. (1983). A study of the glycoproteins of *Autographa californica* nuclear polyhedrosis virus (AcNPV). *Virology 131*, 230-241.

Stiles, B., Burand, J. P., Meda, M., and Wood, H. A. (1983). Characterization of gypsy moth (*Lymantria dispar*) nuclear polyhedrosis virus. *Appl. Environ. Microbiol. 46*, 297-303.

Struthers, J. K., and Hendry, D. A. (1974). Studies of the protein and nucleic acid components of *Nudaurelia capensis* β virus. *J. Gen. Virol. 22*, 355-362.

Summers, M. D., and Smith, G. E. (1975). *Trichoplusia ni* granulosis virus granulosis virus granulin: a phenol-soluble, phosphorylated protein. *J. Virol. 16*, 1108-1116.

Summers, M. D., and Smith, G. E. (1976). Comparative studies of baculovirus granulins and polyhedrins. *Intervirology 6*, 168-180.

Summers, M. D., and Smith, G. E. (1978). Baculovirus structural polypeptides. *Virology 84*, 390-402.

Tijssen, P., van den Hurk, J., and Kurstak, E. (1976). Biochemical, biophysical, and biological properties of densonucleosis virus. I. Structural proteins. *J. Virol. 17*, 686-691.

Tweeten, K. A., Bulla, L. A., Jr., and Consigli, R. A. (1978). Characterization of an alkaline protease associated with a granulosis virus of *Plodia interpunctella*. *J. Virol. 26*, 702-711.

Tweeten, K. A., Bulla, L. A., and Consigli, R. A. (1980a). Structural polypeptides of the granulosis virus of *Plodia interpunctella*. *J. Virol. 33*, 877-886.

Tweeten, K. A., Bulla, L. A., and Consigli, R. A. (1980b). Characterization of an extremely basic protein derived from granulosis virus nucleocapsids. *J. Virol. 33*, 866-876.

Vlak, J. M. (1979). The proteins of nonoccluded *Autographa californica* nuclear polyhedrosis virus produced in an established cell line of *Spodoptera frugiperda*. *J. Invertebr. Pathol. 34*, 110-118.

Weber, K., and Osborn, M. (1969). The reliability of molecular weight determinations by dodecyl sulfate-polyacrylamide gel electrophoresis. *J. Biol. Chem. 244*, 4406-4412.

Wood, H. A. (1980a). *Autographa californica* nuclear polyhedrosis virus-induced proteins in tissue culture. *Virology 102*, 21-27.

Wood, H. A. (1980b). Protease degradation of *Autographa californica* nuclear polyhedrosis virus proteins. *Virology 103*, 392-399.

Wrigley, C. W. (1968). Gel electrofocusing--a technique for analyzing multiple protein samples by isoelectric focusing. *Sci. Tools 15*, 17-23.

Yamamoto, T., and Tanada, Y. (1979). Comparative analysis of the enveloped virions of two granulosis viruses of the armyworm, *Pseudaletia unipuncta*. *Virology 94*, 71-81.

Young, S. Y., and Lovell, J. S. (1973). Virion proteins of the *Trichoplusia ni* nuclear polyhedrosis virus. *J. Invertebr. Pathol. 22*, 471-472.

Zethner, O., Brown, D. A., and Harrap, K. A. (1979). Comparative studies on the nuclear polyhedrosis viruses of *Lymantria monacha* and *L. dispar*. *J. Invertebr. Pathol. 34*, 178-183.

Zummer, M., and Faulkner, P. (1979). Absence of protease in baculovirus polyhedral bodies propagated *in vitro*. *J. Invertebr. Pathol. 33*, 383-384.

II.

Pathology

PATHOLOGY ASSOCIATED WITH BACULOVIRUS INFECTION

H. M. MAZZONE

U.S. Department of Agriculture--Forest Service

I. INTRODUCTION

The pathology of baculoviruses has both an old and a modern heritage. Silkworm larvae were of economic importance even in the 16th century when Vida noted in his poem that infection thinned the ranks of their population (in Bergold, 1953). The particular disease observed by Vida was referred to as "jaundice." Such larvae, we now know, had died from nucleopolyhedrosis, a disease caused by certain baculoviruses.

In the 19th century, as the light microscope was becoming indispensable in pathology, the large, many-sided crystals associated with nucleopolyhedrosis, the polyhedra, were being studied. The idea that viruses were the cause of the disease expressed in silkworm larvae and other insects was realized through progressive experimentation on these crystals. In the early decades of the 20th century, it was demonstrated that when polyhedra were treated with weak alkali, minute objects were exposed and showed Brownian movement in dark field optical microscopy (Komárek and Breindl 1924). This observation led to the idea that the causative agent of the disease was contained within the polyhedra. With the advent of the transmission electron microscope, Bergold (1947) repeated the experiment and confirmed the presence of virus in polyhedra.

ISBN 0-12-470295-3

The baculoviruses, in addition to having the unique characteristic of viral-containing inclusions for many of their representatives, have other important qualities. A large number of the baculoviruses can be grown in considerable quantity with relatively little expenditure in terms of effort and cost. Moreover, the baculoviruses do not appear to have counterparts among the viruses that infect vertebrates.

The baculoviruses, for all their interest, did not receive the attention accorded viruses having a more personal relationship with humans. It was not until the importance shed on the continued use of chemical insecticides, catalyzed by Rachel Carson (1962) in her classic book, "Silent Spring," that alternatives to chemical pesticides were demanded. Quite naturally, attention was now turned to insect viruses as one approach. The idea became imminent that insect viruses, particularly the baculoviruses, should be used to control insects. Historically, the idea of using viruses to regulate insect populations is a relatively old one. The use of such agents against insects was suggested in the first decade of the present century (Smith, 1967, p. 6).

Modern pathologists use a variety of disciplines in the hope of answering questions including the relatively recent scientific procedures such as serologic techniques, *in vitro* cell culture, high-voltage electron microscopy, and genetic engineering. Such methodologies are essential both in the characterization of baculoviruses and to understand their mode of action in current and potential hosts, particularly if they are to properly serve their role as viral insecticides.

This chapter will review the pathological features of baculoviruses. Studies will also be presented which are both relevant and indispensable to a proper understanding of the complex and fascinating pathology of the baculoviruses.

II. THE BACULOVIRUSES

A. *CLASSIFICATION*

The Invertebrate Virus Subcommittee of the International Committee on Nomenclature of Viruses classifies the baculoviruses as rod-shaped (*baculo=rod*), enveloped viruses containing double-stranded DNA. The free viruses are sensitive to heat and lipid solvents, and are often occluded in protein crystals (Wildy, 1971; Fenner, 1976).

The baculoviruses include several types listed below.

1. Numerous virus particles occluded in a polyhedral-shaped proteinic crystal. The occlusion of viruses in this

group occurs in the nuclei of certain types of infected cells. This type of baculovirus is referred to as a nucleopolyhedrosis virus (NPV).

2. One virus particle, or rarely two, occluded in a granular- or capsular-shaped protein crystal. In many cases, the occlusion of the virus particles occurs in the nucleus, but the cytoplasm is also involved. This type of baculovirus is referred to as a granulosis virus (GV).

3. Virus particles which occlude in other types of proteinic crystals.

4. Virus particles which are not or are only slightly occluded.

For the most part, baculoviruses infect insects. The typical representatives of NPVs are those infecting the silkworm, *Bombyx mori* (L.) and the gypsy moth, *Lymantria dispar* (L.) (Bergold, 1947). Examples of the GVs are those infecting the pine shoot roller, *Choristoneura murinana* (Hübner) (Bergold, 1948) and the Indian meal moth, *Plodia interpunctella* (Hübner) (Arnott and Smith, 1968).

Rod-shaped viruses may also be occluded in protein črystals not having polyhedral or capsular shapes. Representatives of baculoviruses infecting some mosquitoes include the non-typical inclusions found in *Aedes sollicitans* (Clark et al., 1969) and in *A. triseriatus* (Federici and Lowe, 1972). The nonoccluded baculoviruses are represented by those infecting the palm rhinoceros beetle, *Oryctes rhinoceros* (L.) (Huger, 1966), the parasitoid wasp, *Apanteles melanoscelcus* (Krell and Stoltz, 1979), and the ichneumonid parasitoid, *Mesoleius tenthredinis* (Stoltz, 1981).

Baculoviruses appear to have been found outside the class Insecta. The rod-shaped virus particle infecting the citrus red mite, *Panonychus citri* (Reed and Hall, 1972), is very likely to be classified as a baculovirus. This virus is discussed further in Section I,B.

Another possible member of this group is the occluded virus infecting a crustacean, the pink shrimp, *Penaeus duorarum* (Couch, 1974a,b). This latter example is a less desirable addition to the baculoviruses. While many hosts of the baculoviruses are invertebrate pests, the pink shrimp, as a crustacean, represents a link in man's food chain. The significance of this report is further discussed in Section V,A.

Bạculovirus infection outside the class Insecta will have to be considered in order to fully classify the baculoviruses. Revision of the baculovirus designation to include viruses found in several classes of invertebrates, in addition to Insecta, is thus very likely to occur. Other examples are given in this chapter.

B. STRUCTURAL RELATIONSHIPS

1. *Occluded Baculoviruses*

The baculoviruses may be occluded within a polyhedron or a granule (capsule). In the literature these bodies are variously referred to as inclusions, inclusion bodies, or occlusions. In this chapter, the terms polyhedron, polyhedron inclusion body (PIB), granule, and capsule will be used.

The occluded baculoviruses have three distinct components: the inclusion body, the polyhedron or capsule, the virus or virions contained therein, and the deoxyribonucleic acid contained within each virus particle. These components will be discussed below.

a. Inclusions. Polyhedral inclusions can be observed with both the light and electron microscope. They generally range in size from 1 to 15 μm, although, on occasion, there have been reports of larger dimensions. Although their name indicates the many sideness of their structure, polyhedra are variable in size and shape.

As noted above, polyhedra of the baculovirus class are formed in the nucleus of the infected cell, occluding the completely formed virions which, in turn, contain DNA. The term cytoplasmic polyhedrosis virus (CPV) appears in the literature in association with insect viruses. The CPVs are not baculoviruses. Their inclusions are formed in the cytoplasm of infected cells and they occlude spherical virions which contain RNA (Smith and Wyckoff, 1950; Xeros, 1952; Smith and Xeros, 1953; Wildy, 1971).

The nuclear polyhedra possess an envelope believed to be composed of carbohydrate (Minion et al., 1979) on which a paracrystalline protein matrix is arranged in a cubic lattice (Bergold, 1963 a,b). Contained within the nuclear polyhedron are a number of enveloped rod-shaped virions, occurring as single virions (single enveloped virion or SEV) or as groups of single virions contained within an envelope. The groups of single virions are called bundles (multiply enveloped virions or MEV). In some cases the nuclear polyhedron may contain a mixture of single enveloped virions and enveloped bundles.

The inclusions referred to as granules are capsular or ovocylindrical in shape and are smaller than most polyhedra. The granules have been reported to average 300-500 nm in length by 120-350 nm in width (Bergold, 1963a; Huger, 1963). As for polyhedral inclusions, a protein matrix with a paracrystalline lattice comprises the granule and an electron-dense layer, distinct from the protein matrix, exists at the surface (Arnott and Smith, 1968; Kawanishi et al., 1972). This layer has been reported in the baculoviruses to be a highly condensed protein (Harrap, 1972) or a lipoprotein (Beaton and Filshie,

1976; Hess and Falcon, 1978). A granule contains one rod-shaped virus, although on rare occasions, it contains two.

The inclusion bodies, polyhedra or granules, protect the virions from disrupting forces in the environment. Inclusion bodies have relatively good storage qualities, particularly in the dried state and at refrigeration temperatures (Steinhaus, 1954, 1960). The inclusion bodies are sensitive to alkaline pH, their matrix structure dissociating at a pH of about 10.5 (Bergold, 1947). When inclusion bodies are ingested as contaminants of food, the matrix structure disassociates in the alkaline environment of the insect gut. Most likely, this reaction is aided by proteolytic enzymes with the result that free virions are liberated. Chemical dissolution or physical thin sectioning demonstrates the presence of rod-shaped virions within the inclusion bodies, as does the use of high-voltage electron microscopy (Mazzone et al., 1980a,b). For laboratory and field studies, inclusion bodies can be prepared in highly purified, kilogram quantities by zonal rotor centrifugation (Anderson, 1966; Martignoni et al., 1968; Mazzone et al., 1970; Breillatt et al., 1972).

b. Virions. The single virus particles found in polyhedra and in granules are enveloped. The virus envelope is believed to be composed of protein (Bergold and Wellington 1954). Within the viral envelope is a nucleocapsid composed of the protein capsid and DNA core. In some GVs protein may also be associated with the DNA core, forming a nucleoprotein complex within the capsid (Tweeten et al., 1980).

The bundles, enveloped groups of single viruses, found within nuclear polyhedra vary in length and width depending upon the number of single viruses contained within each bundle. Each single virus within a polyhedron or a granule generally has the same length and width, with only slight variation. Occasionally, anomalies do occur as in the case of a long single rod sometimes seen in the granules of some GVs (Smith and Brown, 1965).

c. Deoxyribonucleic Acid. The baculoviruses are known to contain DNA (Breindl and Jirovec, 1936; Wyatt, 1952a,b), and infectious DNA has been isolated from these microorganisms (see Section IV,A,2). Thus far, no baculoviruses have been found where RNA is considered as the infectious entity. Some reports indicate that RNA is associated with the protein matrix of baculovirus polyhedra (Faulkner, 1962; Aizawa and Tida, 1963) and RNA from the host insect has been implicated in the formation of a baculovirus (Gershenson et al., 1963). However, overwhelming evidence points to DNA as the nucleic acid which directs replication of baculoviruses.

It appears that the baculoviruses have a wide range of molecular weights. Although the maximum value is given as over 100 million daltons (Summers, 1976; Tweeten et al., 1981), the minimum value cannot be established with certainty. Onodera et al. (1965) reported a molecular weight of 2 million for infectious DNA of the NPV of the silkworm. One may speculate on this and other reports of relatively low molecular weights for infectious DNA whether the high values found for some baculoviruses necessarily coincides with the portion of DNA needed for infectivity. In this regard, apparently many of the high molecular weight DNAs isolated from the baculoviruses have not been tested for infectivity.

2. *Nonoccluded Viruses*

Huger (1966) observed enveloped rod-shaped viruses in the nuclei of cells of the Indian (also called palm) rhinoceros beetle larvae. The virus particle was found to consist of a nucleocapsid surrounded by an envelope; in this form, the average length × width dimensions were 235 nm × 110 nm. The DNA was double stranded, supercoiled with a molecular weight of 60 to 92 million daltons (Monsarrat et al., 1973a,b; Revet and Monsarrat, 1974; Payne, 1974; Payne et al., 1977). The virus is normally nonoccluded, although inclusion bodies containing virions have been seen in the cytoplasm of cells of fat body, ovarian sheaths, inner spermathecal walls, and midgut epithelium (Monsarrat et al., 1973a). The inclusion bodies were considered to be unlike polyhedra (Bedford, 1981). The characteristics of this virus satisfy the definition of baculoviruses, and it is likely to be included in this class (Bedford, 1981).

Another nonoccluded virus and one outside the class Insecta is that infecting the citrus red mite, *Panonychus citri* (Smith et al., 1959). Recently, Reed and Hall (1972) have characterized the virus as rod shaped, 58 × 194 nm, and enclosed in envelopes, 111 × 266 nm. These structures are formed within the nuclei of midgut epithelial cells (Reed, 1981). Viruses isolated by density gradient centrifugation are elongate to oval in shape 81 × 206 nm, usually with a rounded projection or knob, at one end, and remnants of the envelope, at the other (Reed, 1981). The morphology is reported to be very similar to the rod-shaped viruses which infect the Indian (palm) rhinoceros beetle described above.

Other viruses which are nonoccluded and bear a close structural resemblance to baculoviruses have been reported in the ovaries of many parasitoid braconid wasps (Vinson and Scott, 1975; Poinar et al., 1976; Stoltz et al., 1976; Stoltz and Vinson, 1977). There are some interesting features noted for these viruses. Krell and Stoltz (1979) observed that in the case of the parasitoid wasp, *Apanteles melanoscelus*, the nucleocapsids of the virus particles were not of uniform length. The diameter of the particle is constant, approximately 40 nm, while the length of the nucleocapsid varies from 28 nm to 100 nm. The double-stranded, circular DNA had a variable molecular weight ranging from 2 to 25 million daltons. The variable DNA size may represent a multipartite genome (Faulkner, 1982). On the basis of acceptable morphological criteria, Krell and Stoltz (1979) have provisionally assigned such particles to the baculovirus class.

An apparent baculovirus was also found in the ovarial calyx tissue of an ichneumonid parasitoid, *Mesoleius tenthredinis*. The virions were morphologically similar to typical baculoviruses, and viral nucleocapsids measuring approximately 65 × 235 nm were found. The viral DNA is believed to be circular and to have a molecular weight of about 100 million daltons (Stoltz, 1981).

III. PATHOLOGY

A. *HOST RANGE*

Baculoviruses appear to be restricted to the Invertebrates. The class Insecta has been notably represented as a host area and occluded baculoviruses have been isolated from Lepidoptera, Hymenoptera, Diptera, Neuroptera, and Tricoptera. An occluded baculovirus has been reported in a crustacean (Couch, 1974a,b).

The nonoccluded baculoviruses observed in Insecta have been found in Coleoptera, Homoptera, Phasmatodea, Lepidoptera, Hymenoptera, and Diptera. Nonoccluded baculoviruses have been reported in Arachnida and Crustacea (Crawford and Granados, 1982).

Generally, a host specificity by a baculovirus has been expressed within a genus (Ignoffo, 1968). However, the NPV isolated from the alfalfa looper, *Autographa californica*, has been shown to have a wide host range within the Lepidoptera (Vail and Jay, 1973); Vail et al., 1973a,b). In comparison, the GVs appear to be restricted to the Lepidoptera (David, 1975).

Of particular interest have been reports of baculovirus infection in mosquitoes. Clark et al. (1969) reported an NPV infecting the gastric ceca and midgut of *Aedes sollicitans*, obtained from collections in the field. Inclusions ranged from 0.1 to slightly over 1.0 μm. Rod-shaped virions were about 250 mμ long and 75 mμ in diameter. The nuclei of cells first infected were found in a portion of the midgut extending from the Malphigian tubule anteriorly about one-third the length of the midgut. Attempts at laboratory transmission yielded infections in only about 5% of exposed larvae, but all larvae died before pupation. Several other mosquito species were reported to be infected with this NPV (Clark and Fukuda, 1971), and Federici and Lowe (1972) studied its pathology in *Aedes triseriatus*. In larvae, the virus infected the cardia, gastric ceca, and the entire stomach of larval midgut epithelium. Progression of the disease was described as being similar to that of other baculoviruses of the NPV type. Rod-shaped nucleocapsids, approximately 50 nm in diameter, by 200 nm in length, were formed within the nucleus. The nucleocapsids were enveloped by a membranous material and occluded in small irregular and polyhedral protein inclusion bodies. The disease differed from other NPV infections in that the small inclusions gradually coalesced in this host forming fusiform bodies. The size of the inclusions in the nuclei varied with the number as a result of the coalescing process. Examination of the stained preparations indicated inclusions ranging from 3 to 8 μm in diameter to 6 to 20 μm in length.

The investigators noted that this baculovirus did not fit into the NPV or GV groups because of the fusiform-shaped inclusions. However, the viral nomenclature of baculoviruses allows for other shapes of inclusion bodies: "virus particles occluded in other proteinic crystals" (Wildy, 1971).

B. *In Vivo INFECTION*

The larval insect can be used as a model to investigate the events occurring during an infection by baculovirus. What is described below is a plausible explanation of various processes involved in infection, although other explanations may also be possible.

1. *Ingestion of Virus*

The usual mode of entry of a baculovirus is via ingestion of food contaminated with the virus. While oral ingestion is the most common route of entry of virus, other routes are also possible. Virus may enter the host by any of the following alternate pathways: contamination of the egg surface (the transovum route), passage within the egg (transovarially), and as a result of injury (a likely occurrence would be injury of an insect by a parasite. These alternative modes of entry are discussed in Section V,B.

2. *Events in the Insect Gut*

In the lumen of the gut, the virus, if occluded in the form of a polyhedron or capsule is affected by an alkaline pH environment. The alkalinity in the gut breaks down the matrix protein of the inclusion body thereby releasing free virions. It is very likely that enzymes such as proteases are involved in the breakdown of the inclusion bodies; this topic is discussed in Section IV,C. In lepidopteran hosts the larval gut does not appear to deteriorate in the presence of virus, but rather serves as the vehicle for allowing the virus to enter the hemocoel, where susceptible cells are present; destruction will occur in such cells. In hymenopteran hosts, such as sawflies, baculoviruses apparently multiply only in midgut tissue.

Primary infection occurs in the gut. The envelope of free virions from NPVs or GVs associate with columnar cell microvilli. The virus envelope interacts and fuses with the plasma membrane of the microvillus in such a manner as to permit the naked nucleocapsid, which has lost its envelope, to enter the cytoplasm of columnar cells (Harrap and Robertson, 1968; Harrap, 1969, 1970; Summers, 1969).

In a GV infection, Summers (1969) observed that while virions were in an apparently nonspecific association with the nuclear envelope of gut columnar cells, some viruses appeared directly associated end-on with the nuclear pore. Moreover, structures similar to empty capsid membranes of the virus were also associated in a like manner with the nuclear pore. Summers (1969) suggests that in such cases the virus genome is released into the nucleus without the virion entering the nuclear region, resembling a mechanism similar to phage infection of bacteria.

As the primary infection progresses changes in the nucleoplasm appear and virus progeny are observed. The nuclear envelope loses its structural integrity, establishing a continuity with the cytoplasm. The nuclear envelope breaks down and enveloped virus particles become engulfed in membrane vesicles in the cytoplasm of the columnar cells.

For GV primary infections envelope acquisition of the viruses may occur in a number of ways: from intracytoplasmic membranes which had proliferated after virus infection (Tanada and Leutenneger, 1970); the nuclear membrane may contribute to the formation of the envelope (Robertson et al., 1974; Hunter et al., 1975; Tweeten et al., 1980); envelope acquisition may consist of budding of free nucleocapsid through the plasma membrane (Summers, 1971; Robertson et al., 1974).

An important observation, at least for Lepidoptera, is that for viruses originating from inclusions, there is no occluding of the virus particles in the gut cells. The viroplasm, if present, is not as obvious as it is in cells of hemocoelic tissues. The events described above are believed to be the means by which the virus is transported through the columnar cells to the basement membrane. The passage of virus through the basement membrane into the hemocoelic cavity, where secondary infection occurs, is not clear.

Tanada and Leutenegger (1970) have postulated the passage of viruses dissociated from GV inclusion bodies in the gut, directly into the hemocoel of the insect host. Uptake of free virus particles was reported as being accomplished by pinocytosis and intercellular passage of particles through the zonula adherentes and zonula occludentis of the gut into the hemocoel.

As an example of an infection caused by a nonoccluded baculovirus, we may consider the virus disease of the Indian rhinoceros beetle. Huger (1966) noted in ultrathin sections that the nuclei of fat body cells were the principal sites of virus reproduction; however, virogenic foci could also be frequently observed in the cytoplasm.

The virus rods have a highly differentiated structure, and are not occluded in polyhedra or capsules of microcrystalline proteinaceous material. Yet, clusters of rods were observed by Huger to be densely packed in a two- or three-dimensional pseudocrystalline pattern.

The disease was easily transmissable to third instar larvae by contaminated food as well as by intrahemocoelic inoculation. The period of lethal infection was from about 6 to 30 days.

3. *Events in the Hemocoel*

In the hemocoel the virus is involved in more dramatic events. The NPVs infect a number of cell types including hemocytes, tracheal cells, and fat body cells. The GVs appear to be especially attracted to fat body cells. Many cells are infected for each cell type and the yield of virus per cell is greater than that observed for the midgut epithelial cells. Moreover, a complete cycle of replication takes place, including the occlusion of virus particles.

The mode of attachment of an NPV enveloped virus and its penetration into a susceptible cell appears to be by specific attachment of the end of the viral envelope to the host cell membrane, followed by viropexis (Hirumi et al., 1976; Adams et al., 1977; Kawamoto et al., 1977). For GVs the attachment and penetration of enveloped nucleocapsids into cells of the fat body and possibly other hemocoelic tissues is not known (Tweeten et al., 1980).

After viral penetration into the cell, the nucleus enlarges. A dense network of chromatinlike material, viroplasm, is formed. This "virogenic stroma" consists of aggregated strands reported to be different from host chromatin (Xeros, 1956). As events proceed, the viroplasm network apparently contracts, or the nucleus enlarges further, giving the infected nuclei a characteristic appearance, described as a "ring zone" (Xeros, 1956).

The stroma, apparently, is the site of viral DNA synthesis and nucleocapsid assembly because rod-shaped virus particles can be observed from the substance of the stromal strands (Tweeten et al., 1980). As morphogenesis continues the nucleocapsids acquire envelopes and matrix protein formation proceeds as the stroma is depleted. Occlusion of virus particles occurs only with enveloped nucleocapsids (Tinsley, 1976).

In the advanced stages of NPV infection, the cells are completely packed with virus particles present in the nucleus. Eventually, the nucleus or cell ruptures, liberating occluded viruses. Some virus particles not occluded may be responsible for the infection of adjacent cells. In some GV infections nucleocapsid assembly and occlusion may continue in the cytoplasm after the nuclear membrane is ruptured (David, 1975; Tweeten et al., 1980).

4. *Gross Appearance of the Host*

When inclusions are ingested and infection results, the disease is referred to as a nuclear polyhedrosis, in the case of an NPV infection, or as a granulosis, in the case of a GV infection. Infected larvae, in either case, become sluggish and a reduction in growth results. As the disease progresses, the integument of the larva may assume a color change, and/or a glossy sheen as in the case of NPV-infected larvae of the gypsy moth. In time the integument becomes very prone to damage. If it is ruptured, a whitish or greyish-white fluid, is liberated. The fluid contains masses of viral inclusion bodies. In many cases, as death occurs, the infected larva may be seen hanging in an inverted V-shaped position from a leaf or branch.

5. *Changes in Hemolymph*

In many insects, as in the Lepidoptera, hemolymph is a principle vehicle for observing the progress of baculovirus pathology. The hemolymph yields information on the course of virus disease: hemocytes, hemolymph proteins, carbohydrates, and lipids, phagocytosis, and rank infection of hemolymph in bioassay may reveal the progress of infection.

Raheja and Brooks (1971) studied the infectivity of hemolymph of larvae of the forest tent caterpillar, *Malacosoma disstria* (Hübner). After larvae fed on a diet containing polyhedra of the insect's NPV, hemolymph was withdrawn at various intervals, centrifuged to remove the hemocytes, and bioassayed. It was observed that hemolymph was infectious as early as 12 hr after ingestion of polyhedra by larvae. The infectivity increased up to 96 hr after which there was a sharp decline, followed by a plateau 108-144 hr after feeding.

In NPV infections, disc electrophoretic techniques demonstrated a general hypoproteinemia in hemolymph in the variegated cutworm, *Peridroma saucia* (Hübner) (Martignoni and Milstead, 1964; Van der Geest and Craig, 1967), the fall webworm, *Hyphantria cunea* (Drury) (Watanabe, 1967), the cabbage moth *Mamestra brassicae* (L.) (Van der Geest and Wassink, 1969), the cabbage looper *Trichoplusia ni* (Hübner), (Young and Scott, 1970; Young and Lovell, 1971), and the lawn armyworm *Spodoptera maurita* (Boisduval) (Takei and Tamishiro, 1975).

Kislev et al. (1969) studied the types of hemocytes which take part in an NPV infection of larvae of the Egyptian cottonwood, *Spodoptera littoralis* (Boisduval). Of the four major types of hemocytes differentiated in the blood of the insect, virus formation was found to take place mainly in the plasmatocytoids and only to a much lesser extent in the granular hemocytes and oenocytoids. Adipohemocytes were not seen to sustain virus development. Plastmatocytoids were observed phagocytosing free virus particles as well as several whole polyhedra. Virus particles, originating presumably from the nucleus of the cell, were found also in cytoplasmic extensions of infected plasmatocytoids.

Wittig (1968) studied the process of phagocytosis using hemocytes of larvae of the armyworm, *Pseudaletia unipuncta* (Haworth), fed a granulosis virus or a nuclear polyhedrosis virus. In the case of the GV infection she concluded that virus capsules observed in the hemocytes were phagocytized and that the capsules were treated by the blood cells as inert particles. In armyworms treated with NPV, a substantial percentage of hemocytes, up to 20% of the total circulating cells, phagocytized polyhedra after they had appeared in the hemolymph. Other blood cells, up to 20% of the total circulating cells, developed the disease in their nuclei.

Mazzone and McCarthy (1981) reported a number of changes in larval hemolymph of the gypsy moth as a result of NPV infection. Such changes were based on a comparison of the same parameters observed in hemolymph of healthy larvae (Brown and Mazzone, 1977; Brown et al., 1977). In addition to hypoproteinemia, lipids and carbohydrates decreased during the course of infection. The blue-green color of the hemolymph changed with viral infection to a whitish or greyish white. The pH of the hemolymph (6.57), did not change appreciably with infection. One of the parameters which persisted during infection was the ability of hemolymph to melanize. Polyphenol oxidase, the enzyme responsible for the melanization of hemolymph, is very active in the gypsy moth (Mazzone and Brown, 1975; Mazzone, 1976). Polyacrylamide gel electrophoresis, while showing a decrease in protein bands and concentration over uninfected hemolymph, consistently demonstrated the presence of the enzyme.

C. *In Vitro* OBSERVATIONS

To a certain extent pathology of baculoviruses *in vitro* mimics that observed *in vivo*. There are differences, however, and these should be noted.

To initiate baculovirus infections in cell cultures, the optimal medium appears to be that of infectious hemolymph from diseased larvae. If cells and inclusion bodies are centrifuged, the hemolymph can serve as a source of infectious, free virus, but also, apparently, other infectious forms as well (see Section IV,A,3).

Raghow and Grace (1974) studied the multiplication of the NPV in a cell line of the silkworm. The course of infection was followed by electron microscopy of ultrathin sections. Attachment of enveloped virus particles was observed along with their invagination into the cell by viropexis, up to 8 hr postinfection. Disruption of the virus envelope in the cytoplasm and alignment of the transverse and of the naked virus rods with the nuclear membrane pore was observed between 4 and 8 hr after infection. Some of the naked virus rods associated with the nuclear pores were partially empty. Nuclear changes were first observed 16 hr after infection. By 24 hr there was an obvious virogenic stroma with naked virus particles. Virus envelopes apparently surrounded individual and groups of virus particles by 36 hours. By 48 hr small polyhedra could be detected. Polyhedra formation was complete at 72 hr.

Knudson and Harrap (1976) studied the multiplication of an NPV of the fall armyworm, *Spodoptera frugiperda* (J. L. Smith), in host cell culture. Both enveloped naked virus particles could be observed in contact with the plasma membrane by 1 hr postinfection. Naked virus particles could be found in the cytoplasm, and enveloped, partially enveloped, and naked virus particles could be found in the nucleus as early as 3 hr after infection. After 9 hr, nuclear changes were apparent, with the development of a virogenic stroma and some naked virus rods. By 12 hr, the ring zone stage of development was obvious and virus envelope formation could be observed. Polyhedron formation occurred between 18 and 48 hr postinfection.

The viruses which replicate in cell cultures are also infectious in insects. Vail et al. (1973b) infected cabbage looper cells *in vitro* with the NPV of the alfalfa looper. The polyhedra which resulted were as infectious for larvae of several species as those produced in the insects themselves.

However, in some case, anomalies are found in *in vitro* infections. Sohi and Cunningham (1972a,b) studied the NPV of the western oak looper, *Lambdina fiscellaria somniaria* (Hulst) in two cell lines of hemocytes of the forest tent caterpillar. The infected cells of one cell line showed the typical events observed in *in vivo* virus infection of hemocoelic tissues: an extensive virogenic stroma, nuclocapsid formation, viral acquisition, and occlusion into polyhedra. In the other cell line, few cells showed polyhedra formation, and the polyhedra present contained no virus particles. Although virogenic stroma could be detected in the cells, few virus particles were evident. In these two hemocyte cell lines from the same source virus efficiency was normal in one cell line and virtually absent in the other.

Ignoffo et al. (1971) observed no virus particles or inclusion bodies in an ovarian cell line of *Heliothis zea* (Boddie) challenged with an NPV. However, the challenged cells fed to larvae, after seven passages, produced typical NPV infection.

McKinnon et al. (1974) found that the capacity for *in vivo* replication of cabbage looper NPV that had been repeatedly passaged in cabbage looper cell cultures decreased with time. In early passages, virus morphogenesis was very similar to that in insect host tissues. Virogenic stroma could be observed followed by polyhedra formation. A few polyhedra contained a few or no virus particles. In 24 to 48 hr postinfection nucleocapsids could be seen acquiring envelopes within infected nuclei. At the periphery of the nuclei, some nucleocapsids could be seen budding through the nuclei envelope into the cisterna of the endoplasmic reticulum. Budding through the plasma membrane was also seen between 48 and 72 hr when the nuclear envelope had ruptured and the nucleocapsids were released directly into the cytoplasm.

After 15 serial passages, a reduction in polyhedra occurred and continued. After 20 passages, abnormalities of morphogenesis were noticeable and became pronounced from 40 passages onward. Anomalies such as nuclei with virogenic stroma but not nucleocapsids were noted, and virus particles and nuclei with virus present, but no or few polyhedra. The cell line remained fully susceptible to infection by NPV-infected insect hemolymph resulting in normal NPV morphogenesis. Repeated passage of the NPV had resulted in the accumulation of defective interfering particles. There is the possibility of aberrant virus formation if the virus is repeatedly passaged.

The study of infection of cell cultures with GVs has been minimal. Vago and Bergoin (1963) inoculated gypsy moth primary ovarian cultures with GV-infected fat bodies from the large white butterfly (also called the cabbage white butterfly) *Pieris brassicae* (L.). Virus production was irregular and incomplete, with nuclear hypertrophy and formation of occluded virus detected in only some of the cells.

In other studies, cell lines from other insects were not able to replicate GVs. In cabbage looper cells aberrant infection occurred, in which numerous membrane structures and long cylindrical forms resembled capsids, but no normal nucleocapsids were synthesized (Tweeten et al., 1980).

Kelly (1976) noted that the nonoccluded baculovirus of the Indian rhinoceros beetle replicated efficiently in continuous moth (*Spodoptera frugiperda*) and mosquito (*Aedes albopictus*) cell cultures. Progeny virus was detected by bioassay and electron microscopy 15 hr after infection in both cell cultures. Replication in moth cells was morphologically similar to that observed *in vivo*--occurring in the cell nucleus. However, replication of the virus in mosquito cells occurred exclusively in the cytoplasm. Kelly observed a virogenic stromalike arrangement in cells infected in the nuclei (moth cells), but no such arrangement in the cytoplasm (mosquito cells). However, Crawford (1981) suggested that infection in these two cell lines may have resulted from a contaminating virus.

IV. RELATED STUDIES OF PATHOLOGICAL INTEREST

A. *MORPHOLOGICAL FORMS AND INFECTIVITY*

1. *Free Virions*

A number of observations concerning free virions remain unsettled. Most investigators agree that free virions of baculoviruses are infectious. However, whether an envelope is required for infectivity is a matter which lacks universal agreement. Another subject of consideration is the relative infectivity of the virion from infectious hemolymph versus that freed from inclusion bodies. Comparisons have also been made on the routes in which the free virus is administered to the host: orally versus hemocoelic injection. Parallel observations are being made in *in vitro* experiments in an attempt to resolve these questions.

Bergold (1958) reported that a virus with an envelope is the infectious unit of an NPV. However, a number of investigators assert that virions from NPVs are not surrounded with membranes when found in hemolymph (Bird and Whalen, 1954; Bird, 1959; Stairs and Ellis, 1971). Kawarabata (1974) conducted studies on infectious hemolymph of the silkworm. After density gradient centrifugation, only nucleocapsids, accounting for 80% of the infectivity, were present in a highly infectious fraction. Comparing the nucleocapsid infection to that of enveloped virions liberated from polyhedra by alkaline dissolution, the former were nearly a hundred times more infectious than the latter.

Apparent differences between the virions in infected hemolymph and those occluded within inclusion bodies have been noted (Vaughn, 1965; Summers and Volkman, 1976). Kawarabata and Aratake (1978) differentiated between a peroral infectious unit isolated from polyhedra and a hemocoelic infectious unit. The peroral infection was largely attributed to a virion with an envelope, while that of the hemocoelic infection was the result of a virion without an envelope.

Virions liberated from NPVs by alkaline dissociation are known to be infectious when injected into the hemocoel of insects (Bergold, 1958; Shapiro and Ignoffo, 1969; Mazzone et al., 1973). In this regard, Khosaka and Himeno (1972) reported that enveloped and unenveloped virions freed from polyhedra were almost equally infectious by hemocoelic injection into silkworm pupae. They concluded that an envelope was not required for viral infection.

Summers and Volkman (1976) suggested that hemolymph-derived viruses are loosely enveloped. They draw support for their observations from the report of Summers (1971) that the GV which buds from midgut cells into the hemolymph of infected larvae of the cabbage looper is loosely enveloped. Tweeten et al. (1980) also report that the infectious unit in GV infection is the enveloped virion.

In vitro infectivity studies involving nucleocapsids of NPVs showed that they were highly infectious to cell cultures and were morphologically different from viruses with envelopes (Henderson et al., 1974; Knudson and Tinsley, 1974; Raghow and Grace, 1974; Dougherty et al., 1975; Knudson and Harrap, 1976; Summers and Volkman, 1976). Dougherty et al. (1975) observed that unenveloped virus from an NPV of the alfalfa looper derived from cell culture or from infectious hemolymph was not infectious when injected orally into larvae of the fall armyworm, but was infectious when injected into the hemocoel of these larvae.

2. *DNA*

The infectivity of purified baculovirus DNA has been investigated only to a limited extent. In some cases the DNA reported to show infection in susceptible insects has been of relatively low molecular weight. In this regard the infectious DNA of the NPV of the silkworm, *Bombyx mori* would appear to have a variable molecular weight value.

Bergold (1963b) obtained DNA from the silkworm NPV. He reported the DNA to have a sedimentation coefficient of 14.5 S and a molecular weight ranging to 10 million daltons. The DNA was about 1/10,000 times as infectious relative to the DNA content of the enveloped particle. Yamafugi et al. (1966) obtained an infectious DNA from the NPV of the silkworm. Their values for sedimentation coefficient and molecular weight were 35 S and 24 million daltons. Kok et al. (1972) isolated an infectious DNA from silkworm NPV with a similar sedimentation coefficient, 37 S, but a somewhat higher molecular weight value, 31.5 million daltons. As noted above (Section II,B), Onodera et al. (1965) reported an infectious DNA with a molecular weight of 2 million daltons, isolated from the NPV of the silkworm.

Some DNAs from other NPVs have been reported to show infectivity. Ignoffo et al. (1971) challenged a cell line from adult ovaries of the cotton bollworm with DNA extracted from the NPV of the bollworm. These cells produced infection when fed to larvae of the insect.

An infectious DNA was isolated from the NPV of the gypsy moth. Polyhedra were dissociated at alkaline pH to release the virions. The virus fraction obtained by centrifugation was subjected to sodium dodecyl sulfate and mild heating. The DNA extracted had a 260/280 ratio between 1.75 and 2.0, a sedimentation coefficient of 43 S, and a molecular weight value determined by analytical centrifucation of about 50 million daltons. The DNA molecules were mostly circular. Adjusting a portion of the DNA preparation to an optical density of 0.3, and injecting an aliquot into the hemocoel of third instar larvae, 39% of the larvae died in 18 days. At the same optical density value, the virus fraction from which the DNA was extracted killed 100% of the larvae in 18 days. When the DNA was injected under the same conditions, but at full strength (optical density of 2.8), 72% of the larvae died in the same time period. The death of the larvae was the result of nucleopolyhedrosis, as evidenced by the formation of polyhedra in the blood cells and in the hemolymph of inoculated larvae (Mazzone et al., 1973).

A report by Gershenson (1956) remains to be corroborated. DNA-free protein and protein-free DNA were isolated from the NPV of the *Antheraea pernyi*. Neither preparation was infectious alone, although mixtures of the DNA and protein caused 40% infection in pupae of the insect.

In an *in vitro* study, transfection of a continuous cell line of the cabbage looper was attempted with DNA's from the GVs of the fall armyworm and the cabbage looper. Plaque formation was not observed (Burand et al., 1980).

3. *Smaller Entities*

The studies discussed here emphasize the possibility of subviral and smaller genomic entities being involved in the infection process of the baculoviruses.

Aizawa (1967) observed that the hemolymph of silkworms with nuclear polyhedrosis contained DNase-resistant particles with a diameter of about 20 nm. These particles induced the development of nuclear polyhedrosis upon injection into healthy recipients. It was reasoned that the DNA of the NPV would have to be quite small to fit into such particles, if the infectivity attributed to them is in terms of a nucleoprotein entity.

Barefield and Stairs (1970) studied the infectivity of suspensions collected by making lateral incisions in larvae of the codling moth, *Carpocapsa pomonella*, heavily infected with a GV. The initial suspension was centrifuged at 3020 *g* for 30 minutes to remove the GV granules (capsules). The supernatant was then centrifuged at 110,000 *g* for 2 hr. Tests on the second supernatant showed that infectious material was present. After centrifugation of the second supernatant at 190,000 *g* for 3 hr the resulting supernatant was still infectious. The authors suggest that the infectious entity remaining may be a small particle of very low density or a fine strand of viral DNA.

An infectious factor, differing from mature virus rods was detected in tissue homogenates from larvae of the greater wax moth *Galleria mellonella* (L.) infected with an NPV. The infectivity was believed to possibly be connected with a fraction containing spherical particles of uniform appearance and diameter measuring 10-12 nm. This fraction was obtained by centrifugation and precipitation techniques considered adequate to remove cells, polyhedra, and free virions. The 10-12 nm particles isolated from infected tissues were indistinguishable from those particles found in healthy insect homogenates and possessing no infectivity (Zherebtsova et al., 1972).

Kok et al. (1968) in their study on the DNA of the NPV of the silkworm suggested the presence of several genomes in the virus. They postulate that viral DNA is composed of subunits, which are capable of replicating independently of the viral DNA.

B. MIXED INFECTIONS

There have been observations in the field on insects infected with two viruses (Steinhaus, 1957; Vago, 1959). Some of these infections involved either a baculovirus and a non-baculovirus, or two baculoviruses: an NPV and a CPV, an NPV and a nonoccluded virus (a non-baculovirus), and an NPV and a GV (Smith, 1967).

Bird (1959) observed that in the case of an NPV- and a GV-infecting larvae of the spruce budworm, *Choristoneura fumiferana* (Clemens), prior infection by one virus interfered with infection by the second virus. Generally, a cell would accept one virus but not both.

Infection of silkworm larvae, with the NPV of the silkworm and the NPV of *Samia cynthia pryeri* produced a different result. Each type of NPV was formed in the cell nucleus, but both kinds of polyhedra were not observed in the same nucleus. Large triangular polyhedra, different from either of the two types of polyhedra employed in the study, were formed in a few nuclei of the cells. It was not known whether the triangular polyhedra represented a hybridization of polyhedra of the original two types (Ishikawa and Asayama, 1961).

Tanada (1959) observed a synergistic effect in studies involving two baculoviruses. He infected larvae of the armyworm with a GV and an NPV both of which had been isolated, in turn, from the armyworm. The fat body was susceptible to both viruses. However, there was a more rapid rate of development for the NPV, 4-13 days, as compared to 10-34 days for the GV. As a result, the larvae died of nuclear polyhedrosis and not granulosis. Tanada noted that the two viruses acting together produced a more severe disease than each acting alone. Larvae were more susceptible when fed both viruses at the same time. An initial feeding of the GV followed by a feeding of the NPV increased the susceptibility of the armyworm larvae more than the reverse procedure. If the GV was heat-inactivated, it was still capable of enhancing the virulence of unheated NPV. However, heat-inactivated NPV had no apparent effect on unheated GV. The synergism was believed to reside with the GV.

In one case, as many as five virus infections were reported arising as a result of initial infection with one baculovirus. Hess et al. (1978) fed polyhedra of the NPV of the alfalfa looper (*Autographa californica*) to homologous larvae hosts and to larvae of the cabbage looper. The polyhedra were from the multiply virus type (AcMNPV). Two types of virus particles were simultaneously observed in the nuclei of cells of the midguts of the two types of larvae: the AcMNPV and icosahedral unenveloped particles, approximately 40 nm in diameter. Within the same cell, two other types of viruses were observed in the cytoplasm: cytoplasmic polyhedrosis virus, 60 nm in diameter, and dense icosahedral unenveloped particles, approximately 28 nm in diameter. Occasionally, a fifth type of virus particle was observed: icosahedral cytoplasmic deoxyriboviruses, approximately 80 nm in diameter. These observations were made in the two types of larvae infected. Control insects did not appear to have any recognizable forms of virus.

An interesting observation on mixed infections was made by Croizier and co-workers (1980). Larvae of the greater wax moth were infected with a mixture of the homologous NPV and the NPV of the alfalfa looper in a 1:1 ratio. Subsequently, the multiplied viruses were serially passaged four times in the larvae by hemolymph transfer. DNA analysis of the resulting virus population with restriction endonuclease, EcoRl, indicated that new types of virus were formed. These new viruses were believed to be selected with the elimination of the parental viruses. The formation of the new virus types was attributed to genetic recombination.

In vitro observations have also noted mixed baculovirus infections in insect cell cultures. Garzon and Kurstak (1972) demonstrated dual infection of a cell line of the greater wax moth with its NPV and with the Tipula iridescent virus.

Kimura and McIntosh (1976) employed two DNA viruses, the NPV of the alfalfa looper and the cytoplasmic Chilo iridescent virus (CIV) (not a baculovirus) to infect a cell line from the cabbage looper. When cultures of the cabbage looper cell line were inoculated with the alfalfa NPV at 0, 2, 4, 6, and 18 hr and challenged with CIV, polyhedra of the former were produced 48 hr postinfection. When the process was reversed, inoculation first with CIV, followed 24 hr later by superinfection with alfalfa looper NPV, no polyhedra of the latter were observed when cultures were held for 96 hr postinfection. If cells were infected with CIV for 6 hr and then with the alfalfa looper NPV, polyhedra of the alfalfa looper NPV were not inhibited. Dual infection in this situation resulted in production of alfalfa looper NPV in the nucleus and production of the CIV in the cytoplasm.

C. ALKALINE PROTEASE

Proteases undoubtedly operate in the alkaline environment of the insect gut to aid in the dissociation of ingested virus inclusion bodies releasing free virions. An alkaline protease has been described as being associated with the matrix protein of inclusion bodies (Eppstein and Thoma, 1975). Yamafugi et al. (1958) isolated an alkaline protease from inclusion bodies of the NPV from virus-infected larvae of the silkworm. They also isolated an enzyme with similar properties from healthy host tissue. Based on these observations, it may be possible to suggest that the enzyme is being carried over in the purification of inclusion bodies.

The enzyme appears to be absent in GV inclusion bodies isolated from infected larvae of the cabbage white butterfly (S. Brown et al., 1977), and also from NPVs of infected larvae of the European pine sawfly, *Neodiprion sertifer Geoffroy* (Brown et al., 1979). It was reported to be present in NPVs derived from *in vivo* sources of the gypsy moth (McCarthy and Liu, 1976; Brown et al., 1979) but was lacking in *in vitro* cultures of the gypsy moth and the alfalfa looper (DiCapua and Norton, 1976).

It remains unclear how such protease activity could be a function of inclusion bodies. Inclusion bodies derived from *in vivo* and *in vitro* production, and lacking the enzyme, are highly infectious when fed to host insects (Vail et al., 1973 ; Vaughn and Goodwin, 1977; Brown et al., 1977).

D. LATENCY

The pathology of baculovirus infection also includes the phenomenon whereby an insect is infected with a virus but shows no apparent signs of infection. This latent state is such that the infectious entity is not precisely known, and in this regard could exist as naked nucleic acid or as a structural subviral unit. Because of this uncertainty, this phenomenon is referred to as a latent insect infection, rather than as a latent virus infection (Smith, 1967). Latent infections are then defined as inapparent infections which are chronic, and in which a certain virus-host equilibrium is established. The adjective "latent" is used to qualify infection, the term "latent virus" being avoided (Smith 1967). In this connection Lwoff (1958) reminds us that an infection, whether apparent or not, should not be recognized as viral until infectious particles have been detected and identified as a virus. Another term that may add confusion to the phenomenon of latency is "occult virus." Smith (1967) defines occult virus as occurring in such cases where virus particles cannot be detected and in which the actual state of the virus cannot be definitely ascertained.

Latency in the virus infection of insects is quite frequent. Some examples include a latent GV infection in the cabbage worm, *Pieris rapae* (L.) and in the cabbage white butterfly. NPV latent virus infections have also been observed in the silkworm, the spruce budworm, the clothes moth, *Tineola bissellbiella* Hummel, and in larvae of the puss moth, *Cerusa vinula* (L.).

Latency may occur in insects when the initial virus dose is very small as appears to be the case in latent virus infections in plants (Bennett, 1958). If latency continues in the next generation, then it almost surely involves a transovarial (or transovum) transmission.

Latent infections involving viruses in insects are stimulated into virulence by external or stress factors (Steinhaus, 1958a,b), which are believed to aid the penetration of the infectious entity into the cell or to accelerate the multiplication of the virus in the infected cell (Aruga et al., 1963). Examples of stress include the crowding of insects and excessive temperature, with low more effective than high temperatures (Smith, 1967). In the change from latency to virulence, a true virus infection is produced with attendant symptoms as is frequently observed in epizootics in the field. Latent infections which turn virulent are thus believed to be a natural way, if at times not efficacious, in controlling insect populations.

E. WASP BACULOVIRUSES AND INSECT IMMUNITY

The nonoccluded baculoviruses indigenous to parasitoid wasps (Stoltz and Vinson, 1979; Krell and Stoltz, 1979) appear to be capable of modulating other parameters of larval host pathology when transferred to target insects. Specifically, such baculoviruses are capable of interfering with the ability of host insects to elicit an immune response to parasitoid eggs or larvae (Stoltz, 1982).

As a component of the calyx fluid of the parasitoid wasps, they are injected, together with an egg, into target caterpillars. Successful development of parasitoid eggs depends on the presence of the baculovirus which acts to suppress the immune response of the target host insect, and thus prevent encapsulation and inactivation of the egg (Faulkner, 1982). Edson et al. (1981) refer to this phenomenon as "obligatory mutualism" between a virus and a eukaryotic organism.

When virus particles were separated from other calyx fluid components, and the remaining components injected with parasitoid eggs into target tobacco budworm *Heliothis virescens* (Fab.) caterpillars, all of the eggs were encapsulated by the target insects, which then became protected from the parasitoid. Injection of virus separated from other calyx fluid components effectively suppressed the encapsulation. The presence of the living virus was apparently necessary, since when it was inactivated by uv radiation, encapsulation occurred (Stoltz and Vinson, 1979; Faulkner, 1982).

F. THE POTENTIAL OF GENETIC ENGINEERING

Miller et al. (1983) cite the ways in which genetic engineering may offer a number of benefits when using baculoviruses as insecticides: (1) increasing the virulence of the baculoviruses registered for use by the Environmental Protection Agency (EPA); (2) increasing the tolerance of baculoviruses for physical and chemical conditions so that their persistence in the field could be lengthened; (3) broadening the host range of viral insecticides.

In following the above recommendations there may have to be compromises. A more persistent, broader ranged, more virulent viral insecticide would have to be monitored and tested to ensure its safety in the environment. However, the record of presently registered baculoviruses, as insect control agents, is not very spectacular. Of the baculoviruses which have been approved by the EPA or are in the process of being approved, only the NPV of the European pine sawfly appears to be efficacious in the field (refer to the chapter by J. D. Podgwaite in this volume). Although there are registered viral insecticides against the gypsy moth and the Douglas fir tussock moth, these important forest pest insects persist and have made significant advances in the United States and elsewhere.

While the biodegradability of baculoviruses is an asset over persistent chemical insecticides, not having to apply viral insecticides every year would realize a significant savings for the user, e.g., homeowners. In terms of broadening the host range of baculoviruses, the NPV of the alfalfa looper, *Autographa californica*, referred to as AcNPV, infects some 28 different lepidopteran species (Miller et al., 1983). The AcNPV is being extensively studied to determine the reasons for its broad host range.

The DNA genome of AcNPV has been extensively mapped by restriction endonucleases (Miller and Dawes, 1979; Smith and Summers, 1979; Cochran et al., 1982; Vlak and Smith, 1982). This technique and that of plaque purification of a wild population of the AcNPV revealed the presence of many closely related variants differing in the number and size of the DNA fragments produced by restriction endonuclease digestion (Lee and Miller, 1978). The use of a virulent, plaque-purified virus for pesticide use has been recommended to minimize the presence of defective interfering viruses in baculovirus candidates as viral insecticides (Andrews et al., 1980). Restriction endonuclease analysis has also been recommended as a means of quality control and detecting genetic variations in commercial baculovirus preparations for use in the environment.

V. ENVIRONMENTAL CONSIDERATIONS

A. *EPIDEMIOLOGY*

In vivo studies have thus far shown that the baculoviruses are restricted to invertebrates (Wildy, 1971; Ignoffo, 1973; Podgwaite and Mazzone, 1981). Generally, they appear to be confined to the target insect species or to closely related species. Moreover, no morphologically similar counterpart to the baculoviruses has been detected in vertebrates.

As a prerequisite to registration of the NPVs of the gypsy moth and the European pine sawfly, Mazzone et al. (1976) demonstrated that these baculoviruses are not serologically related to the arboviruses, a highly infectious class of viruses having an intimate association with some insects. They also noted the absence of circulating antibodies to the NPVs in the blood of laboratory workers heavily involved in the purification of these baculoviruses. Continuing the study on these NPVs, Tignor et al. (1976) determined their effect in immunodepressed animals, adult mice, and guinea pigs. The results were negative. These studies, among many others, point to the suitability of the baculoviruses as biological control agents for pest insects. Toward this end the Food and Agricultural Organization has recommended only the baculoviruses, among the various classes of insect viruses, for field use [Food and Agricultural Organization, "The Use of Viruses for the Control of Insect Pests and Disease Vectors," *FAO Agr. Study 91*, 1-39 (1973)].

If reports on baculovirus infection of mosquitoes are borne out (Clark et al., 1969; Clark and Fukuda, 1971; Federici and Lowe, 1972), this would be a boon to the use of these agents as regulators of pest insects. On the other hand, the report, of an NPV baculovirus present in pink shrimp (Couch, 1974a,b) should warn against any assumption that baculoviruses *must* be confined only to invertebrate pests.

B. *TRANSMISSION OF VIRUS*

Insects which become infected by eating food contaminated with baculoviruses make a major contribution to the dissemination of virus in the environment. As the disease progresses, the larval epidermis ruptures resulting in the liberation of virus material. The polyhedra granules and nonoccluded viruses are then transmitted by climatic factors, such as, wind and rain. As noted baculoviruses may also be transmitted in other ways: transovarially (in the egg), the transovum route (egg surface), and through parasitism.

Transovarial transmission of baculoviruses has been discussed by Longworth (1973). He concluded that the demonstration of transovarial transmission is difficult to reconcile, because the possibility of contamination of the egg surface by meconium or feces cannot be ruled out. One of the few valid cases of transovarial transmission cited by Longworth is that of the NPV of *Prodena litura* (Fab.) (Harpaz and Ben Shaked, 1963). Although the study referred to transovum transmission, what was really observed was not a mechanical contamination of egg surface or contents, i.e., transovum, but a generation to generation transmission of the NPV involving a genetic mechanism, a transovarian transmission.

As an example of transovum transmission, Hamm and Young (1974) demonstrated the surface contamination of *H. zea* with polyhedra of the NPV of this host. The PIBs had passed through the digestive tract of adult *H. zea* that had been fed these inclusions. Surface sterilization of the eggs eliminated the transmission. The percentage progeny infected increased with increased dose of PIBs to the female, in the range 10^6-10^8 PIBs/female, but transmission generally declined with increasing post-treatment. No adverse effect on adult moths were observed in terms of mating, oviposition, or egg hatching.

Parasites have been shown to transmit virus in insects from larva to larva. Steinhaus (1954), in addition to presenting a number of examples of transovum transmission of microorganisms in several families of insects, also pointed out that parasites, such as Apanteles, could also transmit virus from diseased to healthy insects. In this connection, David (1965) demonstrated the transmission of granulosis virus from infected to healthy larvae of the large white butterfly by the hymenopterous parasite *Apanteles glomeratus*.

C. BACULOVIRUSES AS INSECTICIDES

The NPVs and GVs, when used in the role of insecticides, are dispersed in the environment as inclusion bodies. The inclusion bodies containing viruses represent the most stable form of the baculoviruses. These structures protect the occluded virions from environmental forces, which degrade the more susceptible virions.

The inclusion bodies themselves are eventually degraded by environmental forces, such as uv irradiation. To retard such degradation, inclusion bodies are formulated with a uv protectant, and other agents which enhance storage, wetting, suspension, flow, and dispersal (Miller et al., 1983).

Because they must be ingested to produce infection of the host insect, baculoviruses need to be applied in the pest area at the time the target insect begins feeding. The stage normally infected is the larval rather than the egg, pupal, or adult stages.

Some of the baculoviruses cause natural epizootic diseases within pest insect populations and some have been commercially developed as pesticides. In the United States, the EPA has registered or is considering for registration, several baculoviruses including the NPVs that infect the cotton boll worm, the Douglas Fir tussock moth, the gypsy moth, the alfalfa looper, and the European pine sawfly. A GV that infects the codling moth, *Cydia pomonella*, is also a candidate for EPA registration.

Baculoviruses are being monitored for their possible effects on the health of humans and other living forms, as well as for their effects on the environment. As commercially available microbial insecticides, baculoviruses have met the safety and environmental criteria established by the United States Environmental Protection Agency.

Some of the EPA tests required include: oral feeding, acute dermal toxicity, inhalation, eye sensitivity, dermal sensitivity, subacute oral toxicity, carcinogenicity test, teratological tests, intraperitoneal injection, intracerebral injection, intradermal injection, subdermal injection, intramuscular injection, and intravenous injection. The test animals include: rodents, fish, birds, wild ungulates, estuarian animals (shrimp and oysters), and monkeys. The test results were negative (Ignoffo, 1973; Heimpel, 1976).

In addition, baculoviruses have been tested in tissue cultures of a number of cell lines from humans and other animals. The results of these tests were negative, although there were some exceptions. Himeno et al. (1967) extracted DNA from the silkworm NPV and introduced it into a culture of human amniotic Fogh-Lund (FL) cells. They subsequently observed the formation of polyhedra in the nuclei. The polyhedra were serologically similar to polyhedra purified from the insect. The virions from the polyhedra produced *in vitro* were infectious when injected into silkworm larvae.

McIntosh and Shamy (1975) treated a viper cell line culture (VSW) with virus rods of the alfalfa looper NPV, obtained from insect tissue culture. The virus was labeled with tritiated thymidine. At 72 hr postinfection, autoradiography demonstrated the presence of grains over the cytoplasm and nuclei of inoculated cells. Immunofluorescence confirmed the presence of viral antigens in the inoculated culture, although viral replication was not demonstrated. The investigators concluded that the virus entered the VSW cells and induced the synthesis of virus-specific proteins.

McIntosh and Maramorosch (1973) tested the retention of baculovirus virus activity in mammalian cell cultures. There was some evidence for persistence of viable *H. zea* NPV in cell cultures from human lung, leukocyte, and amnion tissues.

These *in vitro* experiments aimed at observing effects at the cellular and molecular levels point out the need for continual surveillance of the baculoviruses in their function as insecticides. The observation that baculoviruses kill insects should not be our only concern (Axelrod, 1975).

VI. CONCLUSIONS

The baculoviruses offer great potential as biological insecticides and such potential puts them in the forefront of microorganisms of interest. The lure of researchers from various disciplines to take up their study ensures the accumulation of new data concerning their properties, host range, and mode of action.

In order to effectively control the spread of pest insects, the baculoviruses may have to be altered genetically in order to make them more virulent and more persistent in the field. These actions may give the baculoviruses a broader host range. Obviously, such tradeoffs will have to be carefully considered, necessitating a far greater surveillance of these control agents than is currently warranted.

ACKNOWLEDGMENTS

For their valuable discussion and suggestions, I am grateful to Professor Marion A. Brooks, Department of Entomology, University of Minnesota at St. Paul, and to Dr. Arthur H. McIntosh, U.S. Department of Agriculture, Columbia, Missouri.

REFERENCES

Adams, J. R., Goodwin, R. H., and Wilcox, T. A. (1977). Electron microscopic investigations on invasion and replication of insect baculoviruses *in vivo* and *in vitro*. *Biol. Cell. 28*, 261-268.

Aizawa, K. (1967). Mode of multiplication of the nuclear-polyhedrosis virus of silkworm. *J. Sericult. Sci. Japan 36*, 327-331.

Aizawa, K. and Tida, S. (1963). Nucleic acids extracted from the virus polyhedra of the silkworm, *Bombyx mori* (Linnaeus). *J. Insect Pathol. 5*, 344-348.

Anderson, N. G. (1966). The development of zonal centrifuges and ancillary systems for tissue fractionation and analysis. *Nat. Cancer Inst. Monogr. 21*. U.S. Govt. Printing Office, Washington, D.C.

Andrews, R. E., Jr., Spence, K. D., and Miller, L. K. (1980). Virulence of cloned variants of *Autographa californica* nuclear polyhedrosis virus. *J. Appl. Environ. Microbiol. 39*, 932-933.

Arnott, H. J. and Smith, K. M. (1968). An ultrastructural study of the development of a granulosis virus in the cells of the moth *Plodia interpunctella* (Hbn.). *J. Ultrastr. Res. 21*, 251-268.

Aruga, H., Yoshitake, N., and Owada, M. (1963). Factors affecting the infection per os in the cytoplasmic polyhedrosis of the silkworm, *Bombyx mori*, L. *Nippon Sanshigaku Zasshi 32*, 41-50.

Axelrod, L. R. (1975). The Environmental Protection Agency's mandate to evaluate the hazard associated with pesticide uses. *In* "Baculoviruses for Insect Pest Control: Safety Considerations" (M. Summers, R. Engler, L. A. Falcon, and P. Vail, eds.), pp. 12-13. EPA-USDA Working Symp., Bethesda, Maryland. Amer. Soc. Microbiol. Publ.

Barefield, K. P. and Stairs, G. R. (1970). Infectious components of granulosis virus of the codling moth, *Carpocapsa pomonella*. *J. Invertebr. Pathol. 15*, 401-404.

Beaton, C. D. and Filshie, B. K. (1976). Comparative ultrastructural studies of insect granulosis and nuclear polyhedrosis viruses. *J. Gen. Virol. 31*, 151-161.

Bedford, C. O. (1981). Control of the rhinoceros beetle by baculovirus. *In* "Microbial Control of Pests and Plant Diseases" (H. D. Burges, ed.), pp. 409-426. Academic Press, New York.

Bennett, C. W. (1958). Masked plant viruses. *Proc. 7th Intern. Congr. Microbiol., Stockholm*, pp. 218-223. Almqvist and Wiksell, Uppsala.

Bergold, G. (1947). Die isolierung des polyeder-virus and die natur der polyeder. *Z. Naturforsch. 2b*, 122-143.

Bergold, G. H. (1948). Über die kapselvirus-krankheit. *Z. Naturforsch. 3b*, 338-342.

Bergold, G. H. (1953). Insect viruses. *Adv. Virus Res. 1*, 91-139.

Bergold, G. H. (1958). Viruses of insects. *In* "Handbuch der Virusforschung" (C. Hallauer and K. F. Meyer, eds.), Vol. 4, pp. 60-142. Springer, Vienna.

Bergold, G. H. (1963a). The molecular structure of some insect virus inclusion bodies. *J. Ultrastr. Res. 8*, 360-378.

Bergold, G. H. (1963b). The nature of nuclear polyhedrosis viruses. *In* "Insect Pathology" (E. A. Steinhaus, ed.), pp. 413-456. Academic Press, New York.

Bergold, G. H. and Wellington, E. F. (1954). Isolation and chemical composition of the membranes of an insect virus and their relation to the virus and polhedral bodies. *J. Bacteriol. 67*, 210-216.

Bird, F. T. (1959). Polyhedrosis and granulosis viruses causing single and double infections in the spruce budworm, *Choristoneura fumiferana* (Clemens). *J. Insect. Pathol. 1*, 406-430.

Bird, F. T. and Whalen, M. M. (1954). Stages in the development of two insect viruses. *Can. J. Microbiol. 1*, 170-174.
Breillatt, J., Brantley, J. N., Mazzone, H. M., Martignoni, M. E., Franklin, J. E., and Anderson, N. G. (1972). Mass purification of nucleopolyhedrosis virus inclusion bodies in the K-series centrifuge. *Appl. Microbiol. 23*, 923-930.
Breindl, V. and Jirovec, O. (1936). Polyeder und polyedervirus im lichte der ferelgenschen nuclearraktion. *Vest. Kesk. Spolecnosti Zool., Praze 3*, 9-11.
Brown, D. A., Bud, H. M., and Kelly, D. C. (1977). Biophysical properties of the structural components of a granulosis virus isolated from the cabbage white butterfly (*Pieris brassicae*). *Virology 81*, 317-327.
Brown, S. E. and Mazzone, H. M. (1977). Electrophoretic studies on proteins in the egg and hemolymph of the gypsy moth with reference to isoenzymes. *J. N.Y. Entomol. Soc. 85*, 26-35.
Brown, S. E., Patton, R. L., Zerillo, R. T., Douglas, S. M., Breillatt, J. P., and Mazzone, H. M. (1977). Comparative properties of hemolymph of the gypsy moth and the European pine sawfly. *J. N.Y. Entomol. Soc. 85*, 36-42.
Brown, S. E., Kaczmarek, F. S., Dubois, N. R., Zerillo, R. T., Holleman, J., Breillatt, J. P., and Mazzone, H. M. (1979). Comparative properties of the inclusion body proteins of the nucleopolyhedrosis viruses of *Neodiprion sertifer* and *Lymantria dispar*. *Arch. Virol. 59*, 319-329.
Burand, J. P., Summers, M. D., and Smith, C. E. (1980). Transfection with baculovirus DNA. *Virology 101*, 286-290.
Carson, Rachel (1962). "Silent Spring." Houghton-Mifflin, New York.
Clark, T. B. and Fukuda, T. (1971). Field and laboratory observations of two viral diseases in *Aedes sollicitans* (Walker) in southwestern Louisiana. *Mosquito News 31*, 193-199.
Clark, T. B., Chapman, H. C., and Fukuda, T. (1969). Nuclear-polyhedrosis and cytoplasmic polyhedrosis virus infections in Louisiana mosquitoes. *J. Invertebr. Pathol. 14*, 284-286.
Cochran, M. A., Carstens, E. F., Eaton, B. T., and Faulkner, P. (1982). Molecular cloning and physical mapping of restriction endonuclease fragments of *Autographa californica* nuclear polyhedrosis virus DNA. *J. Virol. 41*, 940-946.
Couch, J. A. (1974a). Free and occluded virus, similar to baculovirus, in hepatopancreas of pink shrimp. *Nature (London) 247*, 229-231.
Couch, J. A. (1974b). An enzootic nuclear polyhedrosis virus of pink shrimp: ultrastructure, prevalence, and enhancement. *J. Invertebr. Pathol. 24*, 311-331.
Crawford, A. M. (1981). Attempts to obtain Oryctes baculovirus replication in three insect cell cultures. *Virology 112*, 625-633.

Crawford, A. M. and Granados, R. R. (1982). Nonoccluded baculoviruses. *Proc. Invertebr. Pathol. Microbial Control, 3rd Intern. Colloq. Invertebr. Pathol.*, pp. 154-159. U. of Sussex, Brighton, U.K.

Croizier, G., Godse, D., and Vlak, J. (1980). Sélection de types viraux dans les infections doubles à baculo virus chez les larves de lepidoptere. *C.R. Acad. Sci. (Paris) D290*, 579-582.

David, W. A. L. (1965). The granulosis virus of *Pieris brassicae* L. in relation to natural limitation and biological control. *Ann. Appl. Biol. 56*, 331-334.

David, W. A. L. (1975). The status of viruses pathogenic for insects and mites. *Annu. Rev. Entomol. 20*, 97-117.

DiCapua, R. A. and Norton, P. W. (1976). Immunological characterization of the baculoviruses: present status. *In* "Invertebrate Tissue Culture: Applications in Medicine, Biology, and Agriculture (E. Kurstak and K. Maramorosch, eds.), pp. 317-330. Academic Press, New York.

Dougherty, E. M., Vaughn, J. L., and Reichelderfer, C. F. (1975). Characterization of the nonoccluded form of a nuclear polyhedrosis virus. *Intervirology 5*, 109-121.

Edson, K. M., Vinson, S. B., Stoltz, D. B., and Summers, M. D. (1981). Virus in a parasitoid wasp: Suppression of the cellular immune response in the parasitoid's host. *Science 211*, 582-583.

Eppstein, D. A. and Thoma, J. A. (1975). Alkaline protease associated with the matrix protein of a virus infecting the cabbage looper. *Biochem. Biophys. Res. Commun. 62*, 478-484.

Faulkner, P. (1962). Isolation and analysis of ribonucleic acid from inclusion bodies of the nuclear polyhedrosis of the silkworm. *Virology 16*, 479-484.

Faulkner, P. (1982). A novel class of wasp viruses and insect immunity. *Nature (London) 299*, 489-490.

Federici, B. A. and Lowe, R. E. (1972). Studies on the pathology of a baculovirus in *Aedes triseriatus*. *J. Invertebr. Pathol. 20*, 14-21.

Fenner, F. (1976). Classification and nomenclature of viruses. *Intervirology 7*, 1-115.

Garzon, S. and Kurstak, E. (1972). Infection double inhabituelle de celleles d'un arthropode par le virus de la polyédrie nucléaire (VPN) et le virus irisant de tipula (TIV). *C.R. Acad. Sci. 275*, 507-508.

Gershenson, S. M. (1956). Reconstitution of an active polyhedral virus from nucleic acid protein outside the organism. *Dopo. Akad. Nauk. Ukr. RSR No. 5*, 492-493.

Gershenson, S. M., Kok, I. P., Vitas, K. I., Dobrovolskaya, G. M., and Skuratovskaya, I.N. (1963). Formation of a DNA-containing virus by host RNA. *Proc. 5th Intern. Congr. Biochem. Moscow, 1961, Vol. 9*, p. 150. Pergamon Press, Oxford.

Hamm, J. J. and Young, J. P. (1974). Mode of transmission of nuclear polyhedrosis virus to progeny of adult *Heliothis zea*. *J. Invertebr. Pathol. 24*, 70-81.

Harpaz, I. and Ben Shaked, Y. (1963). Generation-to-generation transmission of the nuclear-polyhedrosis virus of *Prodena litura* (Fabricius). *J. Invertebr. Pathol. 6*, 127-130.

Harrap, K. A. (1969). Viruses of invertebrates. *Proc. 1st Intern. Congr. Virol., Helsinki, 1968, Vol. 1*, p. 281.

Harrap, K. A. (1970). Cell infection by a nuclear polyhedrosis virus. *Virology 42*, 311-318.

Harrap, K. A. (1972). The structure of nuclear polyhedrosis viruses. I. The inclusion body. *Virology 50*, 114-123.

Harrap, K. A. and Robertson, J. S. (1968). A possible infection pathway in the development of a nuclear polyhedrosis virus. *J. Gen. Virol. 3*, 221-225.

Heimpel, A. L. (1976). Practical applications of insect viruses. *In* "Virology in Agriculture" (J. A. Romberger, ed.). Universe Books, New York.

Henderson, J. F., Faulkner, P., and MacKinnon, E. A. (1974). Some biological properties of virus present in tissue cultures infected with the nuclear polyhedrosis virus of *Trichoplusia ni*. *J. Gen. Virol. 22*, 143-146.

Hess, R. T. and Falcon, L. A. (1978). Electron microscopic observations of the membrane surrounding polyhedral inclusion bodies of insects. *Arch. Virol. 56*, 169-176.

Hess, R. T., Summers, M. D., and Falcon, L. A. (1978). A mixed virus infection in midgut cells of *Autographa californica* and *Trichoplusia ni* larvae. *J. Ultrastr. Res. 65*, 253-265.

Himeno, M., Sakai, E., Onodera, K., Nakai, H., Fukuda, T., and Kawada, Y. (1967). Formation of nuclear polyhedra bodies and nuclear polyhedrosis virus of the silkworm in mammalian cells infected with viral DNA. *Virology 33*, 507-512.

Hirumi, H., Hirumi, K., and McIntosch, A. H. (1976). Morphogenesis of a nuclear polyhedrosis virus of the alfalfa looper of a continuous looper cell line. *Ann. N.Y. Acad. Sci. 266*, 302-326.

Huger, A. (1963). Granuloses of insects. *In* "Insect Pathology" (E. A. Steinhaus, ed.), Vol. 1, pp. 531-575. Academic Press, New York.

Huger, A. M. (1966). A virus disease of the Indian rhinoceros beetle, *Oryctes rhinoceros* (Linnaeus), caused by a new type of insect virus, *Rhabdionvirus oryctes* gen. n. sp. n. *J. Invertebr. Pathol. 8*, 38-51.

Hunter, D. K., Hoffman, D. F., and Collier, S. J. (1975). Observations on a granulosis virus of the potato tuberworm, *Phthorimaea operculella*. *J. Invertebr. Pathol. 26*, 397-400.

Ignoffo, C. M. (1968). Specificity of insect viruses. *Bull. Entomol. Soc. Amer. 14*, 265-276.
Ignoffo, C. M. (1973). Effects of entomopathogens on vertebrates. *Ann. N.Y. Acad. Sci. 217*, 141-172.
Ignoffo, C. M., Shapiro, M., and Hink, W. F. (1971). Replication and serial passage of infectious *Heliothis zea* cells. *J. Invertebr. Pathol. 18*, 131-134.
Ishikawa, Y. and Asayama, T. (1961). Studies on the relation between the polyhedroses of the wild insects and the silkworm, *Bombyx mori* L. II. On the double infection of the nuclear polyhedroses in silkworm larvae. *Nippon Sanshigaku Zasshi 30*, 201-205 (Engl. summary).
Kawamoto, F., Suto, C., Kumada, N., and Kobayashi, M. (1977). Cytoplasmic budding of a nuclear polyhedrosis virus and comparative ultrastructural studies of envelopes. *Microbiol. Immunol. 21*, 255-265.
Kawanishi, C. Y., Egawa, K., and Summers, M. D. (1972). Solubilization of *Trichoplusia ni* granulosis virus protein crystal. II. Ultrastructure. *J. Invertebr. Pathol. 20*, 95-100.
Kawarabata, T. (1974). Highly infectious free virions in the hemolymph of the silkworm (*Bombyx mori*) infected with a nuclear polyhedrosis virus. *J. Invertebr. Pathol. 24*, 196-200.
Kawarabata, T. and Aratake, Y. (1978). Functional differences between occluded and nonoccluded viruses of a nuclear polyhedrosis of the silkworm, *Bombyx mori*. *J. Invertebr. Pathol. 31*, 329-336.
Kelly, D. C. (1976). "Oryctes" virus replication: electron microscopic observations on infected moth and mosquito cells. *Virology 69*, 596-606.
Khosaka, T. and Himeno, M. (1972). Infectivity of the components of a nuclear polyhedrosis virus of the silkworm, *Bombyx mori*. *J. Invertebr. Pathol. 19*, 62-65.
Kimura, M. and McIntosh, A. H. (1976). Dual infection of the *Trichoplusia ni* cell line with the Chilo iridescent virus (CIV) and *Autographa californica* nuclear polyhedrosis virus. *In* "Invertebrate Tissue Culture: Applications in Medicine, Biology, and Agriculture" (E. Kurstak and K. Maramorosch, eds.), pp. 391-394. Academic Press, New York.
Kislev, N., Harpaz, I., and Zelcer, A. (1969). Electron microscopic studies on hemocytes of the Egyptian cottonworm, *Spodoptera littoralis* (Boisduval) infected with a nuclear polyhedrosis virus, as compared to noninfected hemocytes. II. Virus-infected hemocytes. *J. Invertebr. Pathol. 14* 245-257.

Knudson, D. L. and Harrap, K. A. (1976). Replication of a nuclear polyhedrosis virus in a continuous cell culture of *Spodoptera frugiperda*. Microscopy study of the sequence of events of the virus infection. *J. Virol. 17*, 254-268.

Knudson, D. L. and Tinsley, T. W. (1974). Replication of a nuclear polyhedrosis virus in a continuous cell culture of *Spodoptera frugiperda*. Purification, assay of infectivity, and growth characteristics of the virus. *J. Virol. 14*, 934-944.

Kok, I. P., Chistyakova-Ryndich, A., and Gudz'-Gorban' (1968). Infectivity and structure of DNA of virus of nuclear-polyhedrosis of the silkworm. *Trans. 13th Intern. Entomol. Congr., Moscow,* . *Nanka, Leningrad,* p. 127.

Kok, I. P., Chistyakova-Ryndich, A., Gudz'-Gorban', A., and Solomko, A. (1972). Macromolecular structure of the DNA of the *Bombyx* nuclearpolyhedrosis virus. *Mol. Biol. 6*, 323-331.

Momárek, J. and Breindl, V. (1924). Die Wippelkrankheit-der nonne und der erreger derselben. *Z. Angew. Entomol. 10*, 99-162.

Krell, P. J. and Stoltz, D. B. (1979). Unusual baculovirus of the parasitoid wasp *Apanteles melanoscelus*: Isolation and preliminary characterization. *J. Virol. 29*, 1118-1130.

Lee, A. H. and Miller, L. K. (1978). Isolation of genotypic variants of *Autographa californica* nuclear polyhedrosis virus. *J. Virol. 27*, 754-767.

Longworth, J. F. (1973). Viruses and Lepidoptera. *In* "Viruses and Invertebrates" (A. J. Gibbs, ed.), Chapter 21. North-Holland, Amsterdam.

Lwoff, A. (1958). Recent progress in microbiology. Symp. IV. Latent and masked virus infections. *Proc. 7th Intern. Congr. Microbiol., Stockholm,* p. 211.

McCarthy, W. J. and Liu, S.-Y. (1976). Electrophoretic and serological characterization of *Porthetria dispar* polyhedron protein. *J. Invertebr. Pathol. 28*, 57-65.

McIntosh, A. H. and Maramorosch, K. (1973). Retention of insect virus infection in mammalian cell cultures. *J. N.Y. Entomol. Soc. 81*, 175-182.

McIntosh, A. H. and Shamy, R. (1975). Effects of the nuclear polyhedrosis virus (NPV) of *Autographa californica* on a vertebrate viper cell line. *Ann. N.Y. Acad. Sci. 266*, 327-331.

MacKinnon, E. A., Henderson, J. F., Stoltz, D. B., and Faukner, P. (1974). Morphogenesis of nuclear polyhedrosis virus under conditions of prolonged passage *in vitro*. *J. Ultrastr. Res. 49*, 419-435.

Martignoni, M. E. and Milstead, J. E. (1964). Hypoproteinemia in a noctuid larva during the course of nucleopolyhedrosis. *J. Insect. Pathol. 6*, 517-531.

Martignoni, M. E., Breillatt, J. P., and Anderson, N. G. (1968). Mass purification of polyhedral inclusion bodies by isopycnic banding in zonal rotors. *J. Invertebr. Pathol. 11*, 507-510.

Mazzone, H. M. (1976). Influence of polyphenol oxidase on hemocyte cultures of the gypsy moth. *In* Invertebrate Tissue Culture: Applications in Medicine, Biology, and Agriculture" (E. Kurstak and K. Maramorosch, eds.). Academic Press, New York.

Mazzone, H. M. and McCarthy, W. T. (1981). Gypsy moth nucleopolyhedrosis virus. Biochemistry and Biophysics. *In* "The Gypsy Moth. Research Toward Integrated Pest Management" (C. C. Doane and M. L. McManus, eds.). U.S. Dept. Agr. Tech. Bull. 1584.

Mazzone, H. M., Breillatt, J. B., and Anderson, N. G. (1970). Zonal rotor purification and properties of a nuclear polyhedrosis virus of the European pine sawfly (*Neodiprion sertifer* Geoffroy). *Proc. 4th Intern. Colloq. Insect. Pathol., College Park, Maryland*, pp. 371-379.

Mazzone, H. M., Breillatt, J., and Bahr, G. (1973). Studies on the rod forms and isolated deoxyribonucleic acid from the nucleopolyhedrosis virus of the gypsy moth (*Porthetria dispar* L.). *Abstr. 5th Intern. Colloq. Insect Pathol. Microbial Control*, p. 42.

Mazzone, H. M., Tignor, G. H., Shope, R. E., Pan, I. C., and Hess, W. R. (1976). A serological comparison of the nuclear polyhedrosis viruses of the gypsy moth and the European pine sawfly with arthropod-borne and other viruses. *Environ. Entomol. 5*, 281-282.

Mazzone, H. M., Wray, G., Engler, W. F., and Bahr, G. F. (1980a). High voltage electron microscopy of cells in culture and viruses. *In* "Invertebrate Systems *In Vitro*" (E. Kurstak, K. Maramorosch, and A. Dübendorfer, eds.), pp. 511-515. Elsevier/North-Holland, Amsterdam.

Mazzone, H. M., Engler, W. F., Wray, G., Szirmae, A., Conroy, J., Zerillo, R., and Bahr, G. F. (1980b). High voltage electron microscopy of viral inclusion bodies. *Proc. 38th Annu. Meet. Electron Microscopy Soc. America, Reno, Nevada* (G. W. Bailey, ed.). Claitor's Publishing Div., Baton Rouge, Louisiana.

Miller, L. K. and Dawes, K. P. (1979). Physical map of the DNA genome of *Autographa californica* nuclear polyhedrosis virus. *J. Virol. 29*, 1044-1055.

Miller, L. K., Lingg, A. J., and Bulla, L. A., Jr. (1983). Bacterial, viral, and fungal insecticides. *Science 219*, 715-721.

Minion, F. C., Coons, L. B., and Broome, J. R. (1979). Characterization of the polyhedral envelope of the nuclear polyhedrosis virus of *Heliothis virescens*. *J. Invertebr. Pathol.* *34*, 303-307.

Monsarrat, P., Meynadier, G., Croizier, G., and Vago, C. (1973a). Recherches cytopathologiques sur une maladie virale du coléoptère *Oryctes rhinoceros*. *C.R. Acad. Sci.* *D276*, 2077-2080.

Monsarrat, P., Veyrunes, J. C., Meynadier, G., Croizier, G., and Vago, C. (1973b). Purification et etude structurale du virus du coleoptere *Oryctes rhinoceros*. *C.R. Acad. Sci.* *D277*, 1413-1415.

Onodera, K., Komano, T., Himeno, M., and Sakai, F. (1965). The nucleic acid of nuclear polyhedrosis virus of the silkworm. *J. Mol. Biol.* *13*, 532-539.

Payne, C. C. (1974). The isolation and characterisation of a virus from *Oryctes rhinoceros*. *J. Gen. Virol.* *25*, 105-116.

Payne, C. C., Compson, D., and de Looze, S. M. (1977). Properties of the nucleocapsids of a virus isolated from Oryctes rhinoceros. *Virology* *77*, 269-280.

Podgwaite, J. D. and Mazzone, H. M. (1981). Development of insect viruses as pesticides: The case of the gypsy moth (*Lymantria dispar*, L.) in North America. *Protection Ecol.* *3*, 219-227.

Poinar, G. O., Jr., Hess, R., and Caltagirone, L. E. (1976). Virus-like particles in the calyx of *Phanerotoma flavitestacea* (Hymenoptera: Braconidae) and the transfer into host tissue. *Acta Zool. (Stockholm)* *57*, 161-165.

Raghow, R. and Grace, T. D. C. (1974). Studies on a nuclear polyhedrosis virus in *Bombyx mori* cells *in vitro*. I. Multiplication kinetics and intrastructural studies. *J. Ultrastr. Res.* *47*, 384-399.

Raheja, A. K. and Brooks, M. A. (1971). Infective hemolymph from forest tent caterpillars diseased by nuclear polyhedrosis virus. *J. Invertebr. Pathol.* *17*, 286-287.

Reed, D. K. (1981). Control of mites by non-occluded viruses. *In* "Microbial Control of Pests and Plant Diseases 1970-1980 (H. D. Burges, ed.), pp. 427-432. Academic Press, New York.

Reed, D. K. and Hall, I. M. (1972). Electron microscopy of a rod-shaped noninclusion virus infecting the citrus red mite. *J. Invertebr. Pathol.* *20*, 272-278.

Revet, B., and Monsarrat, P. (1974). L'acide nucleique du virus du coleoptere *Oryctes rhinoceros* L.; un ADN superhelicoidal de haut poids moleculaire, *C.R. Acad. Sci.* *D278*, 331-334.

Robertson, J. S., Harrap, K. A., and Longworth, J. F. (1974). Baculovirus morphogenesis: the acquisition of the virus envelope. *J. Invertebr. Pathol.* *23*, 248-251.

Shapiro, M. and Ignoffo, C. M. (1969). Nuclear polyhedrosis of Heliothis: Stability and relative infectivity of virions. *J. Invertebr. Pathol. 14*, 130-134.

Smith, G. E. and Summers, M. D. (1979). Restriction maps of five *Autographa californica* MNPV variants. *J. Virol. 30*, 828-838.

Smith, K. M. (1967). "Insect Virology," pp. 172-175. Academic Press, New York.

Smith, K. M. and Brown, R. M., Jr. (1965). A study of the long virus rods associated with insect granuloses. *Virology 27*, 512-519.

Smith, K. M. and Wyckoff, R. W. G. (1950). Structure within polyhedra associated with virus diseases. *Nature (London) 166*, 861-862.

Smith, K. M. and Xeros, N. (1953). Studies on the cross transmission of polyhedral viruses: Experiments with a new virus from *Pyrameis cardui*, the painted lady butterfly. *Parasitology 43*, 178-185.

Smith, K. M., Hill, G. J., Munger, F., and Gilmore, J. E. (1959). A suspected virus disease of the citrus red mite *Pananychus citri* (McG). *Nature (London) 184*, 70.

Sohi, S. S. and Cunningham, J. C. (1972a). Replication of a nuclear polyhedrosis virus in serially transferred hemocyte cultures. *J. Invertebr. Pathol. 19*, 51-61.

Sohi, S. S. and Cunningham, J. C. (1972b). Replication of a nuclear polyhedrosis virus in serially transferred hemocyte cultures. *Monogr. Virol. 6*, 35-42.

Stairs, G. R. and Ellis, B. J. (1971). Electron microscopy and microfilter studies on infectious nuclearpolyhedrosis virus in *Galleria mellonella* larvae. *J. Invertebr. Pathol. 17*, 350-353.

Steinhaus, E. A. (1954). Duration of infectivity of the virus of silkworm jaundice. *Science 120*, 186-187.

Steinhaus, E. A. (1957). New records of insect-virus diseases. *Hilgardia 26*, 417-430.

Steinhaus, E. A. (1958a). Stress as a factor in insect disease. *Proc. 10th Intern. Congr. Entomol., Montreal, 1956*, Vol. 4, pp. 725-730. Intern. Congr. Entomol., Ottawa, Canada.

Steinhaus, E. A. (1958b). Crowding as a possible stress factor in insect disease. *Ecology 39*, 503-514.

Steinhaus, E. A. (1960). The duration of viability and infectivity of certain insect pathogens. *J. Insect Pathol. 2*, 225-229.

Stoltz, D. B. (1981). A putative baculovirus in the ichneumonid parasitoid, *Mesoleius tenthrednis*. *Can. J. Microbiol. 27*, 116-122.

Stoltz, D. B. (1982). Viruses of parasitoid hymenoptera. *Proc. 3rd Intern. Colloq. Invertebr. Pathol. Microbial Control*, pp. 160-161. U. of Sussex, Brighton, U.K.

Stoltz, D. B. and Vinson, S. B. (1977). Baculovirus-like particles in the reproductive tracts of female parasitoid wasps. II. The genus *Apantales*. *Can. J. Microbiol. 23*, 28-37.

Stoltz, D. B. and Vinson, S. B. (1979). Viruses and parasitism in insects. *Advan. Virus Res. 24*, 125-171.

Stoltz, D. B., Vinson, S. B., and MacKinnon, E. A. (1976). Baculovirus-like particles in the reproductive tracts of female parasitoid wasps. *Can. J. Microbiol. 22*, 1013-1023.

Summers, M. D. (1969). Apparent *in vivo* pathway of granulosis virus invasion and infection. *J. Virol. 4*, 188-190.

Summers, M. D. (1971). Electron microscopic observations on granulosis virus entry, uncoating, and replication processes during infection of the mid-gut cells of *Trichoplusia ni*. *J. Ultrastr. Res. 35*, 606-625.

Summers, M. D. (1976). Deoxyribonucleic acids of baculoviruses. *In* "Virology in Agriculture" (J. A. Romberger, ed.), pp. 233-246. Universe Books, New York.

Summers, M. D. and Volkman, L. E. (1976). Comparison of biophysical and morphological properties of occluded and extracellular nonoccluded baculoviruses from *in vivo* and *in vitro* host systems. *J. Virol. 17*, 962-972.

Takei, G. H. and Tamishiro, M. (1975). Changes observed in hemolymph proteins of the lawn armyworm, *Spodoptera maurita* acronyctoides, during growth, development, and exposure to a nuclear polyhedrosis virus. *J. Invertebr. Pathol. 26*, 147-158.

Tanada, Y. (1959). Synergism between two viruses of the armyworm *Pseudaletia unipuncta* Haworth (Lepidoptera Noctuidae). *J. Insect Pathol. 1*, 215-231.

Tanada, Y. and Leutenegger, R. (1970). Multiplication of a granulosis virus in larval midgut cells of *Trichoplusia ni* and possible pathways of invasion into the hemocoel. *J. Ultrastr. Res. 30*, 589-600.

Tignor, G. H., Mazzone, H. M., and Shope, R. E. (1976). Serologic studies with the baculovirus of *P. dispar* and *N. sertifer*. *Proc. 1st Intern. Colloq. Invertebr. Pathol.*, pp. 13-14. Queens Univ., Kingston, Ontario, Canada.

Tinsley, T. W. (1976). Properties and replication of insect baculoviruses. *In* "Virology in Agriculture" (J. A. Romberger, ed.), pp. 117-133. Universe Books, New York.

Tweeten, K. A., Bulla, L. A., Jr., and Consigli, R. A. (1980). Characterization of an extremely basic protein derived from granulosis virus nucleocapsids. *J. Virol. 33*, 866-876.

Tweeten, K. A., Bulla, L. A., Jr., and Consigli, R. A. (1981). Applied and molecular aspects of insect granulosis viruses. *Microbiological Rev. 45*, 379-408.

Vago, C. (1959). On the pathogenesis of simultaneous virus infections in insects. *J. Insect Pathol. 1*, 75-79.

Vago, C. and Bergoin, M. (1963). Developpement des virus a corps d'inclusion du lepidoptere *Lymantria dispar* en cultures cellulaires. *Entomophage 8*, 253-261.

Vail, P. V. and Jay, D. L. (1973). Pathology of a nuclear polyhedrosis virus of the alfalfa looper in alternate hosts. *J. Invertebr. Pathol. 21*, 198-204.

Vail, P. V., Jay, D. L., and Hunter, D. K. (1973a). Infectivity of a nuclear polyhedrosis virus from the alfalfa looper, *Autographa californica*, after passage through alternate hosts. *J. Invertebr. Pathol. 21*, 16-20.

Vail, P. V., Jay, D. L., and Hink, W. F. (1973b). Replication and infectivity of the nu-lear polyhedrosis virus of the alfalfa looper, *Autographa californica*, produced in cells grown *in vitro*. *J. Invertebr. Pathol. 22*, 231-237.

Van der Geest, L. P. S. and Craig, R. (1967). Biochemical changes in the larvae of the variegated cutworm, *Peridroma saucia*, after infection with a nuclear polyhedrosis virus. *J. Invertebr. Pathol. 9*, 43-54.

Van der Geest, L. P. S. and Wassink, H. J. M. (1969). Hemolymph proteins of the cabbage armyworm, *Mamestra brassicae*, after infection with a nucleopolyhedrosis virus. *J. Invertebr. Pathol. 14*, 419-420.

Vaughn, J. L. (1965). Chromatographic evidence of differences between the virus in the hemolymph and in the polyhedra from diseased *Bombyx mori*. *J. Invertebr. Pathol. 7*, 524-525.

Vaughn, J. L. and Goodwin, R. H. (1977). Large-scale culture of insect cells for virus production. *In* "Virology in Agriculture" (J. A. Romberger, ed.), pp. 109-116. Universe Books, New York.

Vinson, S. B. and Scott, J. R. (1975). Particles containing DNA associated with the oocyte of an insect parasitoid. *J. Invertebr. Pathol. 25*, 375-378.

Vlak, J. M. and Smith, G. E. (1982). Orientation of the genome of *Autographa californica* nuclear polyhedrosis virus: a proposal. *J. Virol. 41*, 1118-1121.

Watanabe, H. (1967). Electrophoretic separation of the hemolymph proteins in the fall webworm, *Hyphantria cunea*, infected with a nuclear polyhedrosis virus. *J. Invertebr. Pathol. 9*, 570-571.

Wildy, P. (1971). Classification and nomenclature of viruses. First Report of the international committee on nomenclature. *Monogr. Virol. 5*, 1-81.

Wittig, G. (1968). Phagocytosis by blood cells in healthy and diseased caterpillars. III. Some observations concerning virus inclusion bodies. *J. Invertebr. Pathol. 10*, 211-229.

Wyatt, G. R. (1952a). Specificity in the composition of nucleic acids. *Exptl. Cell Res. Suppl. 2*, 201-217.

Wyatt, G. R. (1952b). The nucleic acids of some insect viruses. *J. Gen. Physiol. 36*, 201-205.

Xeros, N. (1952). Cytoplasmic polyhedral virus disease. *Nature (London) 170*, 1073.

Xeros, N. (1956). The virogenic stroma in nuclear and cytoplasmic polyhedroses. *Nature (London) 128*, 412-413.
Yamafugi, K., Yoshihara, F., and Hirayama, K. (1958). Protease and desoxyribonuclease in viral polyhedral crystal. *Enzymologia 19*, 53-58.
Yamafugi, K., Hashinaga, F., and Fuji, T. (1966). Isolation and identification of polyhedral pre-viral deoxyribonucleic acid from healthy silkworm cells. *Enzymologia 3*, 92-164.
Young, S. Y. and Scott, H. A. (1970). Immunoelectrophoresis of hemolymph of the cabbage looper, *Trichoplusia ni*, during the course of a nuclear polyhedrosis infection. *J. Invertebr. Pathol. 16*, 57-62.
Young, S. Y. and Lovell, J. S. (1971). Hemolymph proteins of *Trichoplusia ni* during the course of a nuclear polyhedrosis infection. *J. Invertebr. Pathol. 17*, 410-418.
Zherebtsova, E. N., Strokovskaya, L. I., and Gudz'-Gorban' (1972). Subviral infectivity in nuclear polyhedrosis of the greater wax moth (*Galleria mellonella* L.) *Acta Virol. 16*, 427-431.
Zummer, M., and Faulkner, P. (1979). Absence of protease in baculovirus polyhedral bodies propagated *in vitro*. *J. Invertebr. Pathol. 33*, 383-384.

NOTE ADDED IN PROOF

Recently, Stiles et al. (1983) reported that the NPV from the mosquito *A. sollicitans* infected four additional species of mosquito: *A. epactius*, *Wyeomia smithii*, *Toxorhynchites brevipalpus*, and *Eretmapodites quinquevitiatus*. Infection was restricted to midgut and gastric caeca cells.

Potter and Miller (1980) reported transfection of two invertebrate cell lines with DNA of *A. californica* NPV. The cells of *Spodoptera frugiperda* were pretreated with DEAE-dextran for 30 min at room temperature. After removal of the DEAE-dextran, viral DNA was inoculated onto the cells for 30 min at room temperature. The cells were washed and overlaid with TC-100 medium containing 0.5% Seaken agarose and incubated at 27°C. The DEAE-dextran transfection method resulted in plaque production with an efficiency of 550-1000 pfu/μg DNA in *S. frugiperda* cells. For *T. ni* cells, the efficiency of plaque production was 220 pfu/μg DNA.

An alkaline protease associated with the GV of the Indian meal moth was characterized by Tweeten et al. (1978). The protease was located within the protein matrix of the occluded virus. The enzyme had a molecular weight of 14,000, a pH optimum of 10.5, and a temperature optimum of 40°C. The protease was capable of hydrolyzing the major constituent of the capsular matrix, a 28,000 dalton protein, to a mixture of polypeptides ranging in molecular weight from 10,000 to 27,000.

Potter, K. N. and Miller, L. K. (1980). Transfection of two invertebrate cell lines with DNA of *Autographa californica* nuclear polyhedrosis virus. *J. Invertebr. Pathol. 36*, 431-432.

Stiles, B., Dunn, P. E., and Paschke, J. D. (1983). Histopathology of a nuclear polyhedrosis infection in *Aedes epamius* with observation in four additional mosquito species. *J. Invertebr. Pathol. 41*, 191-202.

Tweeten, K. A., Bulla, L. A., and Consigli, R. A. (1978). Characterization of an alkaline protease associated with a granulosis virus of *Plodia interpunctella*. *J. Virol. 26*, 702-711.

PATHOLOGY ASSOCIATED WITH CYTOPLASMIC POLYHEDROSIS VIRUSES

TOSIHIKO HUKUHARA

Faculty of Agriculture
Tokyo University of Agriculture and Technology
Fuchu, Tokyo 183, Japan

I. INTRODUCTION

Up to the 1930s, most of our knowledge of virus infections in insects had come from studies of the nuclear polyhedroses of Lepidoptera. At that time it was generally believed that the midgut epithelial cells were refractory to virus infection, because of the usual absence of polyhedral inclusion bodies (Paillot, 1928). Ishimori (1934) discovered a new type of silkworm polyhedrosis, which was characterized by the formation of polyhedral inclusion bodies only in the cytoplasm of midgut epithelial cells. The new type was later found prevalent not only in the silkworm but also in other lepidopterous insects (Kitajima, 1936; Lotmar, 1941) and the name "cytoplasmic polyhedrosis" was coined to represent this group of polyhedroses (Xeros, 1952). The fact that the causative viruses of the new type were entirely different from the nuclear polyhedrosis viruses was gradually realized as their morphology and the nucleic acid type were elucidated (Smith and Wykloff, 1950; Bird and Whalen, 1954; Krieg, 1956).

Cytoplasmic polyhedroses have been recorded in 173 Lepidoptera, 11 Hymenoptera, 33 Diptera, 2 Coleoptera, and 2 Neuroptera (Martignoni and Iwai, 1981). Outside of the Insecta, a freshwater daphnid (Crustacea) is affected by a cytoplasmic polyhedrosis (Federici and Hazard, 1975). It is not possible

VIRAL INSECTICIDES
FOR BIOLOGICAL CONTROL

ISBN 0-12-470295-3

to say whether the many cytoplasmic polyhedroses recorded are all due to different viruses, but this seems unlikely in view of the cross-transmissibility of many of these viruses. Twelve distinct virus "types" have been proposed using a provisional classification system, where viruses with similar RNA gel profiles are included within the same "type" (Payne and Rivers, 1976; Matthews, 1979).

The cytoplasmic polyhedrosis virus (CPV) of the silkworm, *Bombyx mori*, is the prototype and has been most extensively studied on account of its economic importance in sericulture. The studies on this virus as well as other CPVs were reviewed in a comprehensive monograph (Aruga and Tanada, 1971). This chapter will concentrate on advances made since that time and will focus primarily on aspects of the structural, and partly functional, changes in cells and tissues of insect which cause or are caused by cytoplasmic polyhedroses.

A recent advance in cytopathology is the isolation of several virus strains of *Bombyx* CPV which produce polyhedral inclusion bodies in cell nuclei. The inclusion bodies are not "polyhedra" in the well-established sense, because they do not occlude virus particles. "Polyhedron" is defined in the field of invertebrate pathology as "crystallike inclusion body (enclosing a number of polyhedrosis-virus particles) produced in the cells of tissues affected by certain insect viruses" (Steinhaus and Martignoni, 1970). The term "inclusion body," as used in this chapter, may be defined as abnormal structures which occur in virus-infected cells and are readily visible with the light microscope. The disease caused by the strains is often described as "midgut-nuclear-polyhedrosis." The term will not be used in this chapter in order to avoid confusion. "Cytoplasmic polyhedrosis" will be used, instead, to mean "disease caused by a family of insect viruses that contain double-stranded RNA."

II. PATHOLOGY IN LARVAL STAGE

A. *VIRUS STRAINS PRODUCING DIFFERENT CYTOPATHOLOGY*

CPV strains of insects other than the silkworm have been little studied. The entire midgut of an infected larva of *Danaus plexippus* was found to contain only cubic polyhedra, as contrasted with the normal round polyhedra (Arnott et al., 1968). The "mutant" virus from the midgut, when administered to healthy larvae, produced only cubic polyhedra. A virus from *Triphaena pronuba* produced cubic and irregularly shaped polyhedra in addition to the normal "hexagonal" polyhedra in

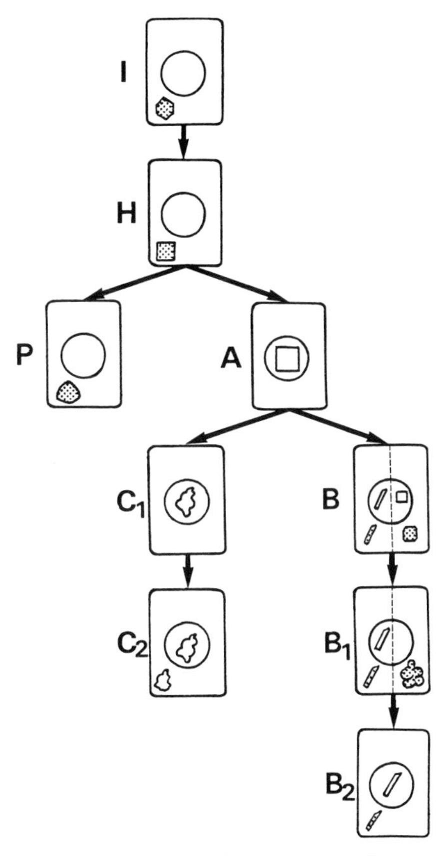

FIGURE 1. Diagram showing the order of establishment of strains of Bombyx CPV and their inclusion bodies. Arrows indicate the origin of new strains. Rare forms of inclusion bodies are not shown. Dots within the inclusion bodies represent virus particles. Where the cells are divided by a central broken line, the left and right halves represent early and late stages of infection, respectively.

infected larvae of *Scotogramma trifolii* (Lipa, 1977), which was attributed to the occurrence of at least three strains in spite of the absence of virus passage data. Two types of polyhedra were found in infected chironomid larvae: cubic polyhedra and polyhedra showing hexagonal outline (Federici et al., 1973). It is yet to be determined whether these are produced by different viruses or strains of the same virus.

During the routine diagnosis of infected silkworm larvae, variation in the shape and/or intracellular location of the inclusion bodies was occasionally noted. Successive passages of these virus isolates through silkworm larvae led to the establishment of nine virus strains in Japan (Fig. 1). The first three strains have been named after the polyhedron shape. Some authors adopt the two-dimensional description such as "HC" (hexagonal cytoplasmic) and "TC" (tetragonal cytoplasmic), whereas some adopt the three-dimensional "I" (icosahedral), "H" (hexahedral), and "P" (pyramidal), etc., and the nomenclature of the same strain differs according to the author. The matter is also confused by the fact that the shapes of polyhedra have not been determined with certainty. The signs, I, H, and P, will be used in this chapter without any connotation with regard to the tridimensional shapes. The other strains, which were later isolated, have been designated by letters and subscript numbers according to the derivation and the order of establishment.

Inclusion bodies of all the strains except for C_1 and C_2, have a lattice structure (Hukuhara and Midorikawa, 1983). The lattice space is in the range of 50 to 65 Å. X-ray diffraction diagrams of inclusion bodies of two strains (H, A) show a number of reflections, which indicates that the lattice structure reflects the crystalline molecular organization (Fujiwara et al., 1984).

The cytoplasm of midgut epithelial cells infected with the first three strains (I,H,P) show similar ultrastructural changes except for the shape of polyhedra. The following account is taken from a work with H strain (Kobayashi, 1971). The virogenic stromata are found in the cytoplasm just below the brush border at an early stage of infection. Dense spherical bodies occur on the surface of the virogenic stroma and release small granular material. The material condenses into small particles, which are then enclosed within a thin layer of membrane to form the virus core. Subsequently, the capsidal protein and the projections are attached to the core. The lysosomes increase in number and enclose virus particles. As the infection progresses, virogenic strmata grow out toward the base of the columnar cell, and numerous free virus particles and small polyhedra are distributed irregularly over their surfaces. The rough-surfaced endoplasmic reticulum is so degraded that only fragments of flattened reticulum occur among the numerous free ribosomes. Mitochondria generally appear normal at this stage but a few corpulent mitochondria with opulent matrices and labyrinthine cristate are present. The process of polyhedron formation is believed to proceed as follows. Many threads of polyhedron proteins diffuse into a mass of virus particles on the surface of the virogenic stroma. Loosely assembled proteins are, then, rearranged and crystallize tightly to form a polyhedron. The polyhedron grows by a process of accretion. In rare cases, the surface of a polyhedron is partly covered by a crystallogenic matrix.

Newly isolated strains induce the formation of not only virogenic stromata and virions in the cytoplasm but also inclusion bodies and associated histopathological changes in the nucleus. Polyhedron formation in the cytoplasm is absent or rare in three strains (A, C_1, C_2). Although crystalline nuclear inclusions were observed in a small proportion of epithelial cells infected with typical strains, the associated nuclear changes were overlooked because of the scarcity of such cells (Xeros, 1966; Hukuhara and Hashimoto, 1966a; Kobayashi, 1971).

In the following sections, the basic observations on these strains will be described and an attempt will be made to correlate them on the basis of a working hypothesis concerning the process of inclusion body formation. Unpublished results of our electron microscopic study will be included.

1. I Strain

The virus strain that predominantly occurs in Japanese sericulture is characterized by the formation of cytoplasmic polyhedra showing hexagonal outline. The outline and the shadows cast by the polyhedron suggest that the tridimensional shape of the polyhedron is that of an icosahedron (Hukuhara and Hashimoto, 1966a), an assumption based on the premise that the facets of a polyhedron are flat. If this is not the case, the shape can be octahedral, rhombic dodecahedral, or a fourteen-faced polyhedral shape (Hukuhara, 1971). Scanning electron microscopy has shown that the typical shape is a rhombic dodecahedron but a dodecahedron and cube-shaped body also occurs (Rao, 1973).

The formation of nuclear inclusions was studied by Miyagawa (1975a), from whose paper this information is derived. Infected columnar cells rarely contain nuclear inclusions at the normal rearing temperature (25°C), but the proportion of such cells is high at lower rearing temperatures (15° or 20°C). Nuclear inclusions show tetragonal outline at early stages of infection and hexagonal outline in advanced stages of infection. However, Hukuhara and Hashimoto (1966a) describe the shape of nuclear inclusions to be the same as that of cytoplasmic polyhedra. The discrepancy may be ascribed to the differences in the isolation of the virus (Yamaguchi, 1972a). On the other hand, the shape of cytoplasmic polyhedra is affected by excessively high rearing temperature (34°C): the hexagonal outline changes to tetragonal outline with rounded corners (Okino and Ishikawa, 1971).

According to Miyagawa (1978) the ultrastructural changes in infected cells takes place in the following order: aggregation of "nucleoli" which have been dispersed within the nucleus, occurrence of virogenic stroma in the cytoplasm, formation of nuclear inclusions in association with the "nucleoli," and formation of cytoplasmic polyhedra. The nuclear inclusions exhibit a crystalline lattice structure and do not occlude virus particles. The structure described by him as nucleoli may be equivalent to dense reticulum observed in the nucleus of columnar cells infected with A strain (Hukuhara and Yamaguchi, 1973).

2. *H Strain*

There is little doubt that polyhedra of this strain are cubic. Observations using light microscopy reveal every edge of the cube. This conclusion is further confirmed by double shadowing and scanning electron microscopy (Hukuhara and Hashimoto, 1966a; Hukuhara, 1972; Rao, 1973). In ultrathin sections, the polyhedra show triangular, square, rectangular, trapezoidal, rhombic, parallelgrammic, pentagonal, and hexagonal outlines, depending on the plane at which they are sectioned (Hukuhara, 1971).

Aberrant polyhedra, composed of two or three cubes, are rarely found in infected silkworm cells (Hukuhara et al., 1972). However, they occur in considerable numbers when the virus strain is cross-transmitted to larvae of *Hyphantria cunea*. In terms of crystallography some of the aberrant polyhedra exhibit parallel growth and some are twin crystals. Exposure of infected silkworm larvae to excessively high temperatures (30° or 34°C) has no effect on the polyhedron shape (Okino and Ishikawa, 1971; Yamaguchi, 1973).

3. *P Strain*

The polyhedron shape of this strain is believed to be similar to that of a noncentrosymmetric square pyramid on the basis of the outlines and shadows cast by the polyhedra (Hukuhara, 1971). Scanning electron micrographs show that the pyramidal form of the polyhedra resembles a tetragonal tristetrahedron (Rao, 1973). This type of solid is also called a deltoid dodecahedron.

According to our electron microscopic study, infected columnar cells show ultrastructural changes typical of CPV (Fig. 2). In rare cases, the nuclei of infected cells contain crystalline inclusions showing the same outline as the cytoplasmic polyhedra (Hukuhara and Midorikawa, 1983).

An isolate from diseased larvae in Japanese sericulture produces inclusion bodies showing similar outlines mainly in the nucleus of infected cells (Miyagawa and Sakai, 1980).

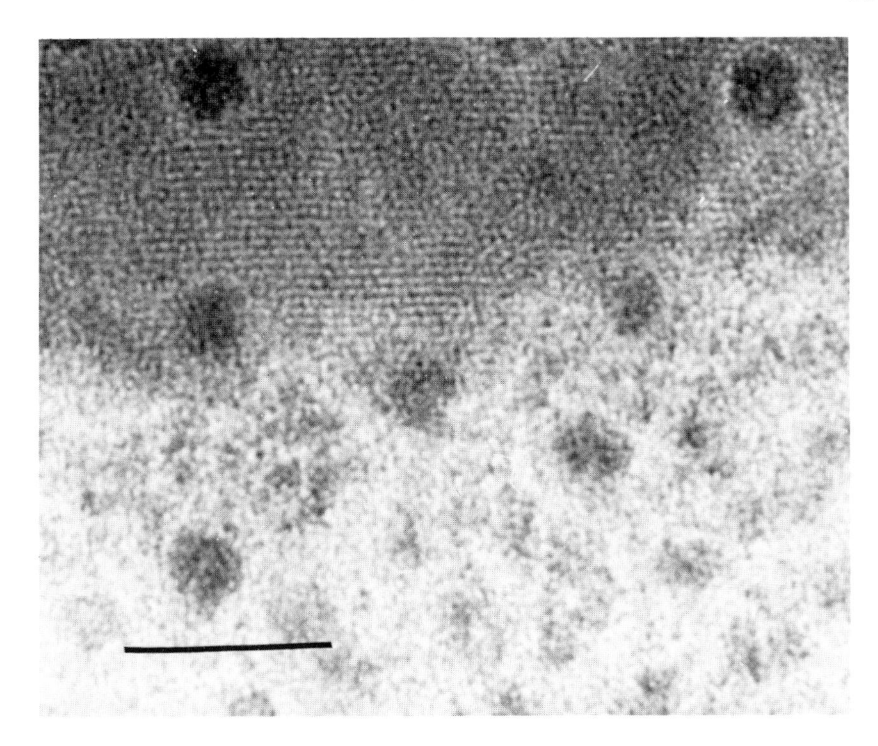

FIGURE 2. The margin of a polyhedron of P strain illustrating its crystalline nature and the distribution of the occluded virus particles. The polyhedron is associated with a dense fibrous area (crystallogenic matrix) containing virus particles; bar, 0.1 μm.

If the characteristic is maintained during serial passage, this isolate may be regarded as a new strain.

4. *A Strain*

Tanaka et al. (1967) discovered that one of the silkworm larvae which had been inoculated with H strain contained large cubic inclusions in the nucleus of columnar cells. The isolate "bred true" in the successive passages through silkworm larvae. Yamaguchi (1968a), who independently obtained a similar isolate from H strain, reported that a few of the infected columnar cells, especially those in the posterior portion of the midgut, contained both intranuclear and intracytoplasmic cubic inclusions. The usual number of nuclear inclusions per infected cell is 1-5, and that of cytoplasmic inclusion is 0-1. Comparative work of the two isolates has shown that they belong to the same strain, which causes the formation of crystalline inclusions almost entirely in the nucleus and only rarely in the cytoplasm (Hukuhara and Yamaguchi, 1973). Diseased larvae which are infected with both I and A strains have been observed in Japanese sericulture (Miyagawa and Sakai, 1971; Miyagawa, 1972; Tomita et al., 1978).

Pathological changes in infected midgut cells have been studied by Iwashita et al. (1968a) and Hukuhara and Yamaguchi (1973), on whose paper the following account is based. The first recognizable manifestation of infection is the enormous hypertrophy of the nucleoli and the aggregation of dense reticulum at multiple foci within the nuclei. The two structures occur primarily in the central portion of the nucleus. The dense reticulum lacks the characteristic granules of nucleonemata and is more electron opaque. At a later stage the nuclei are seen to contain clusters of spherical bodies 60-200 nm in diameter and a few cubic inclusions. They have a crystalline lattice structure. No virus particles are seen in the nuclear inclusions or in the nuclei containing them. The ultrastructural changes in the cytoplasm resemble those of the typical strains except that only a small proportion of the infected cells contain polyhedra. The rough endoplasmic reticula show the most apparent changes in their vesiculate cisternae. Our study shows that the masses of dense reticulum are composed of small granules linked together by fine filaments (Fig. 3), which look like polyhedron protein granules of H strain (Kobayashi, 1971). The dense reticulum is reduced in amount as the inclusions increase in size and in number.

Nuclear inclusions are deformed at high temperatures but resume the former shape when the infected larva is returned to the normal rearing temperature (Yamaguchi, 1968b; Yamaguchi et al., 1969). Electron microscopic observations by Hukuhara and Yamaguchi (1973) show that nuclear inclusions break up into particles of 70 to 250 nm in diameter at 30° and 35°C. Cytoplasmic polyhedra are also decomposed into particles at 100 to 125 nm in diameter, each consisting of virus particles and thick coat of polyhedron protein. The inclusion body proteins may assume the thermodynamically stable states, cubic inclusion or small particle, depending on the temperatures of the protoplasm.

5. *B Strain*

According to Yamaguchi and Ayuzawa (1970), a large number of rod-shaped inclusions are formed in the nuclei of midgut epithelial cells at the early stages of the disease. Late in the development of the disease cubic inclusions are formed both in the nucleus and cytoplasm and the rod-shaped inclusions de-

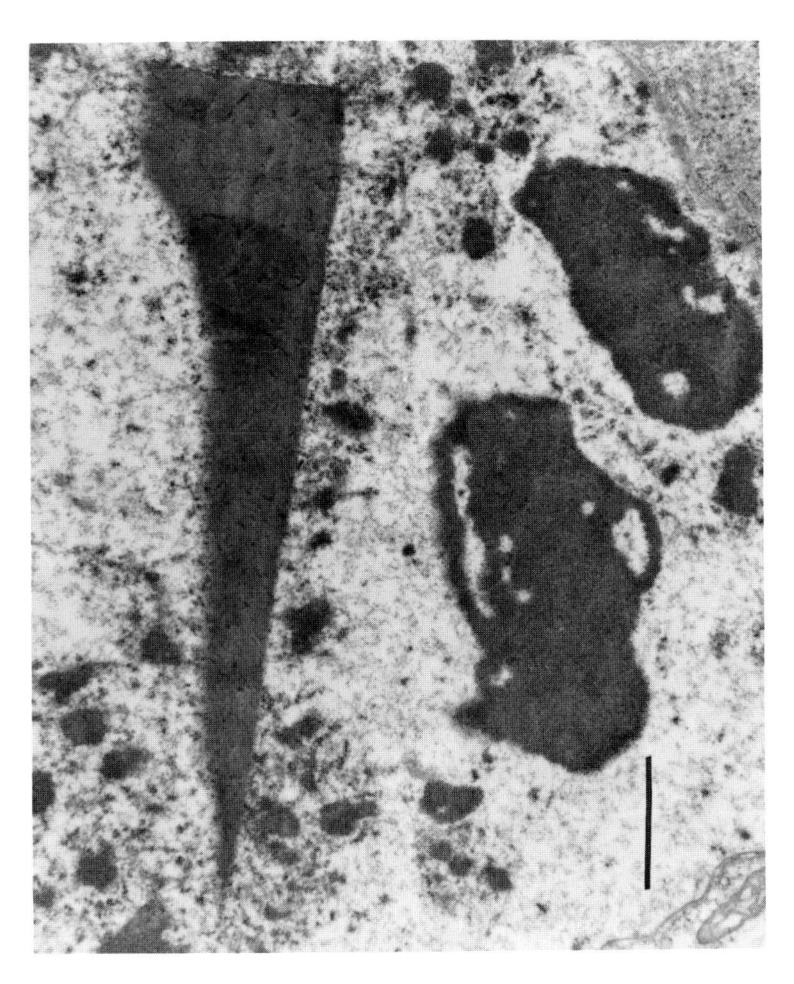

FIGURE 3. Silkworm columnar cell infected with A strain. A nuclear inclusion is seen in an area where many masses of dense reticulum conglomerate. Two nucleoli are present near the area; bar, 1 μm.

crease in number. Some nuclei contain a disklike structure that is readily dissolved by alkali. Rearing of infected larvae at 30°C has no effect on the shape of inclusion bodies (Yamaguchi, 1973).

Our study shows that rod-shaped inclusions are formed not only in the nucleus but also in the cytoplasm. Nuclear changes in ultrastructure are very similar to those induced by A strain. Nuclear inclusions first appear in small number as short, thin rods within the areas where many masses of dense reticulum conglomerate. This area may correspond to the disklike structures

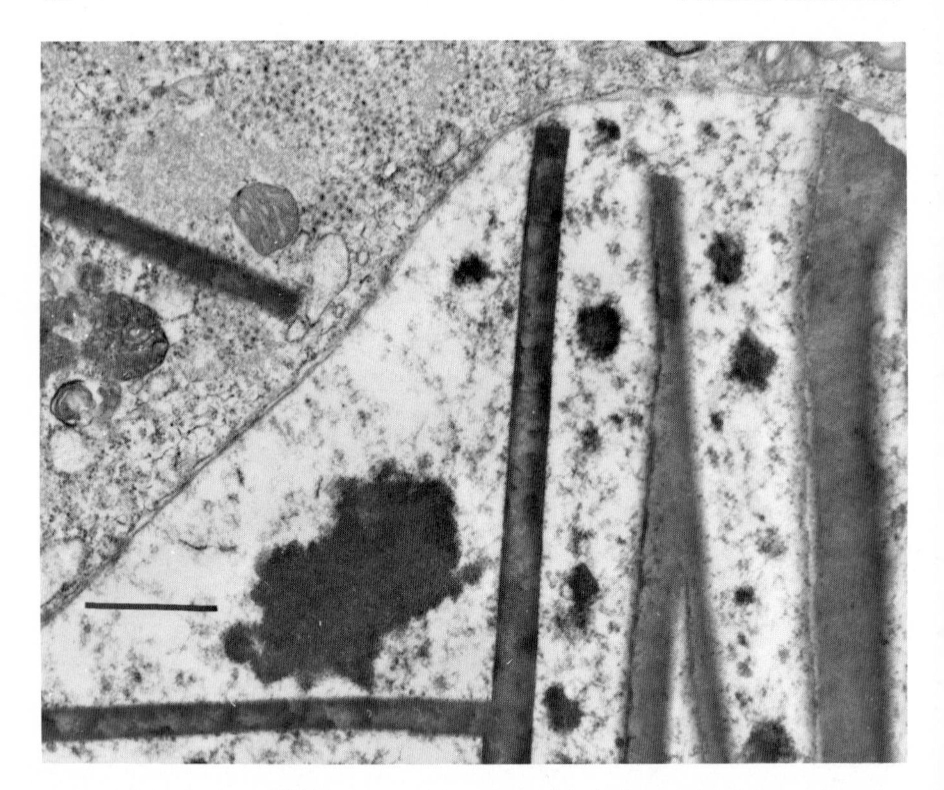

FIGURE 4. Silkworm columnar cell infected with B strain. The nucleus contains masses of dense reticulum and nuclear inclusions. Virogenic stroma, free virus particles, and a rod-shaped polyhedron are seen in the cytoplasm; bar, 1 μm.

as observed by light microscopy. The rod-shaped inclusions increase in size (up to 26 μm × 1 μm) and in number (up to 40 rods/plane of section), and anastomose to produce dendrolic forms (Fig. 4). In an advanced stage of infection the rod-shaped inclusions appear to be transformed to cubic inclusions with the change in the growth pattern. Cytoplasmic polyhedra also change from rod-shaped to cubic forms. Both the nuclear and cytoplasmic inclusions have a crystalline lattice structure. Only the latter occlude virus particles.

6. B_1 Strain

According to Yamaguchi (1973), individual cytoplasmic polyhedra of this strain consist of a number of polyhedral bodies less than 3 μm in diameter, and resemble a bunch of grapes in appearance. Rod-shaped inclusions are formed in the nucleus of the same cells in the early stages of infection but

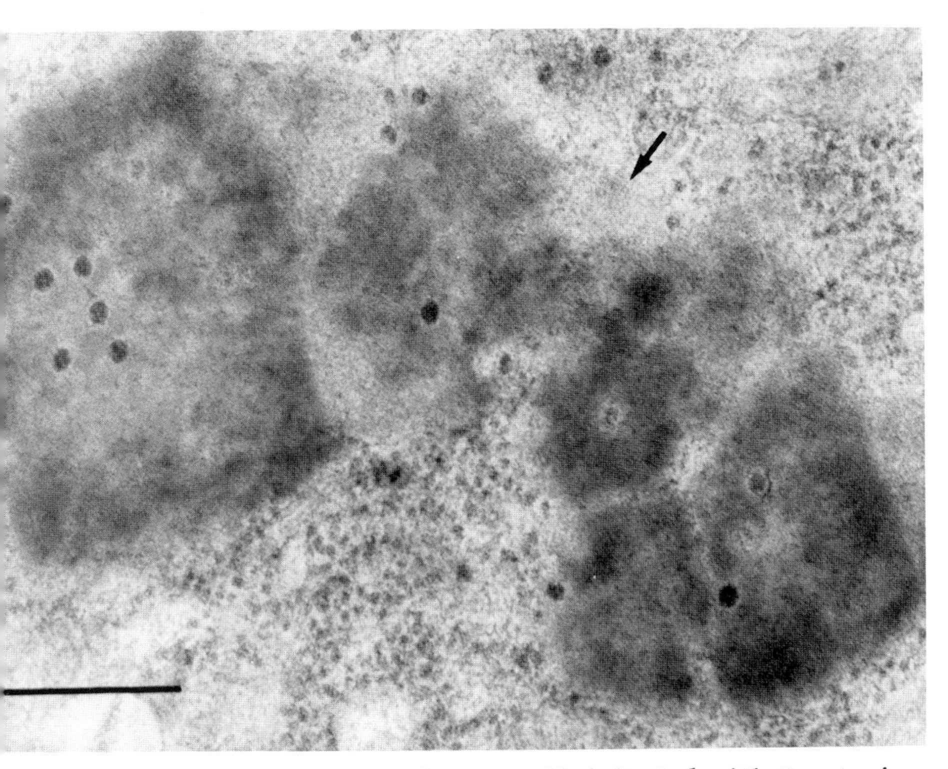

FIGURE 5. Silkworm columnar cell infected with B_1 strain. Remnant of crystallogenic matrix (arrow) fills the spaces between the component crystals. The lattice patterns on both sides of the spaces do not coincide between each other. The crystals occlude virus particles; bar, 0.3 µm.

decrease in number in the late stage. Rearing of infected larvae at different temperatures (20°, 25°, 30° or 35°C) has no effect on the shape of polyhedra.

Our study shows that ultrastructural changes in the nucleus resemble those induced by B strain. The changes in the cytoplasm are unlike the typical picture in that crystallogenic matrices appear as spherical dense areas of 1 to 2 µm in diameter at the early stages of infection. Acicular growth of crystalline inclusions occurs at multiple foci within the matrices.

Later the growth pattern changes resulting in the formation of many polyhedral bodies, which meet during the crystal growth to produce conglomerate forms (Fig. 5). The crystallogenic matrix is absent or greatly reduced at this stage. The internal structure of these polyhedra is similar to that of aberrant polyhedra of H strain (Hukuhara et al., 1972).

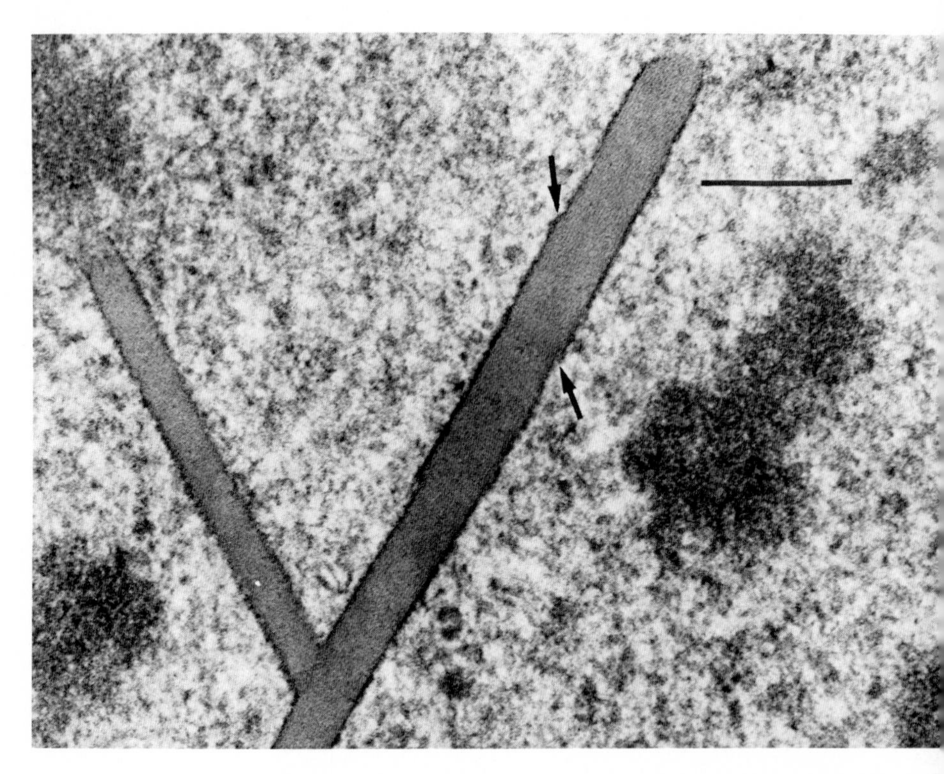

FIGURE 6. Nucleus of a silkworm columnar cell infected with B_2 strain. The side of the acicular nuclear inclusion is more electron opaque than the proximal ends, which are growing more rapidly. Note the presence of steps on the crystal faces (arrows); bar, 0.3 μm.

7. B_2 Strain

This strain is characterized by the presence of many rod-shaped inclusions not only in the nucleus, but also in the cytoplasm, of columnar cells (Yamaguchi and Hukuhara, 1973a). Our study shows that the histopathological changes in the nucleus are the same as described for B and B_1 strains (Fig. 6) and that crystallogenic matrices occur in the cytoplasm prior to polyhedron formation. Acicular crystal growth within the matrices results in the formation of many rod-shaped inclusions which are united to produce a dendritic form (Fig. 7). In most cases, the component rods are independent crystals and the lattice planes are discontinuous at the point of juncture. The growth pattern does not change during the development of the disease.

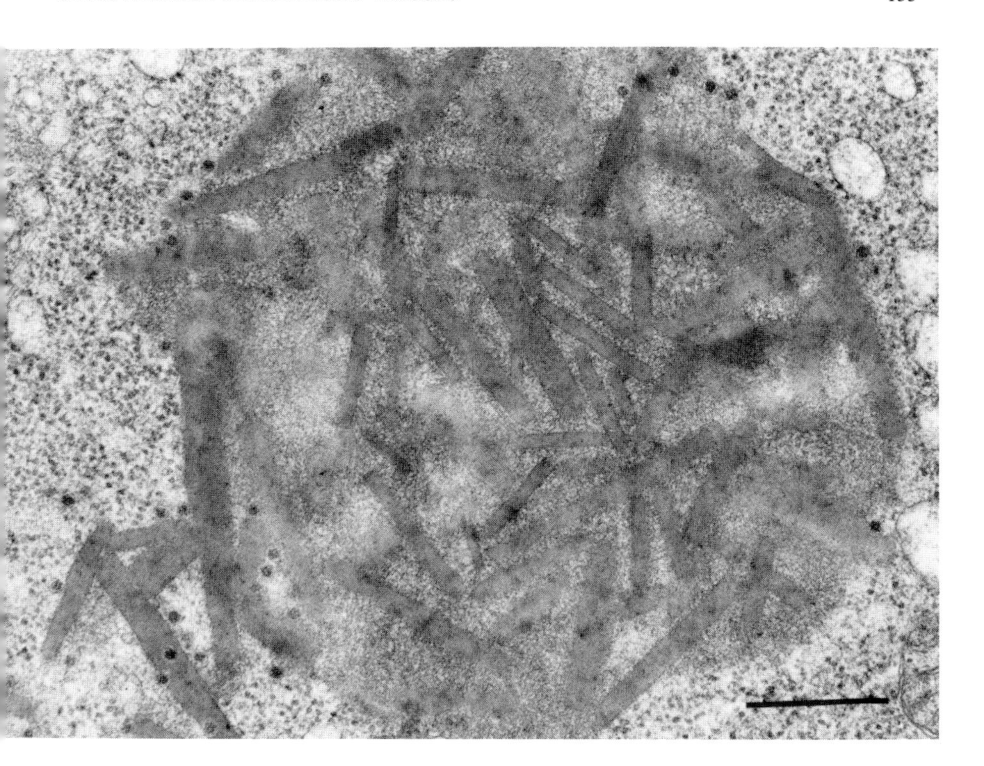

FIGURE 7. Silkworm columnar cell infected with B_2 strain. Rod-shaped inclusions occur in a round area of crystallogenic matrix. Virus particles are present in the cytoplasm and within the rod-shaped inclusions. Those within the crystallogenic matrix are difficult to observe because of the superimposition of fibrils; bar, 0.5 μm.

8. C_1 Strain

A large number of minute inclusion bodies are formed in the hypertrophied nuclei of midgut epithelial cells infected with this strain (Yamaguchi and Ayuzawa, 1970). Electron microscopic studies by Yamaguchi and Hukuhara (1973b) and by the author show that the inclusion bodies are amorphous and consist of tightly packed granules of 30 to 40 Å in diameter, which are never observed in a crystalline structure even with a high level of resolution. As the disease develops, the inclusions grow in size and are united with one another to form large masses of up to 14 μm in diameter. Amorphous inclusions rarely occur in the cytoplasm at the normal rearing temperature (25°C) but more frequently at a lower rearing temperature (20°C) (Yamaguchi, 1970; Yamaguchi and Hukuhara, 1973b).

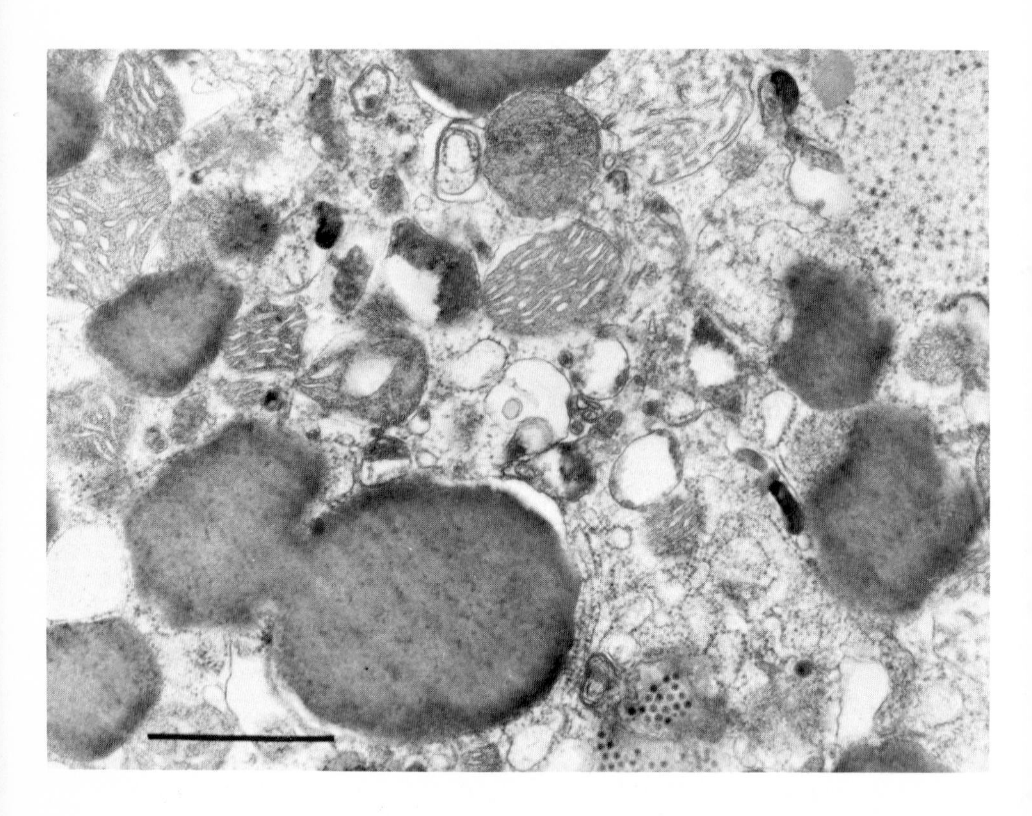

FIGURE 8. Silkworm columnar cell infected with C_2 strain. The cytoplasm contains amorphous inclusion bodies, masses of vesiculated endoplasmic reticulum, corpulent mitochondria, a lysosome enclosing virus particles, virogenic stroma, and virus particles; bar, 1 μm.

9. C_2 Strain

Minute inclusions are formed not only in the nucleus, but also in the cytoplasm, of infected midgut cells (Yamaguchi, 1970). The shape and size of the inclusion bodies depend on the rearing temperature of infected silkworm larvae (Yamaguchi, 1972b). The inclusions are amorphous at 25° and 30°, but spherical at 20°C. The lower the temperature, the larger the inclusions. The spherical inclusion bodies transform to amorphous ones, when the infected larvae are transferred from 20° to 25°C. Our study shows that the inclusions formed at the normal rearing temperature are not crystalline. Although the cytoplasm abounds in virus particles, the cytoplasmic inclusion bodies are never seen to occlude virus particles (Fig. 8).

10. *Strain Relationship*

The virion of A strain contains double-stranded RNA and is very similar to that of H strain in many features such as shape, size, sedimentation coefficient, and amino acid composition (Kawase et al., 1973; Kawase and Yamaguchi, 1974; Tomita and Ebihara, 1980). Autoradiographs with [^{3}H]uridine show that the pattern of nucleic acid synthesis in cells infected with A strain is similar to that described in infection with H strain (Watanabe et al., 1967; Iwashita et al., 1968a). Purified virions of I, H, and A strains are serologically identical or closely related (Hukuhara and Hashimoto, 1966b; Miyajima, 1976; Tomita and Ebihara, 1980). The virus titer of homogenates of midguts infected with H, A, B, B_1, B_2, C_1 or C_2 strain decreased in neutralization tests using antisera against alkali-dissolved inclusion bodies of A, B, C_1 or H strains (Yamaguchi and Ayuzawa, 1970; Yamaguchi, 1970, 1973; Yamaguchi and Hukuhara, 1973a). Although the results appear to indicate the serological relationship, the possibility of cross reaction between proteins of inclusion bodies and virions can not be excluded until tests using purified virions are conducted. Mixed infection of silkworm larvae with two strains results in the formation of two types of inclusions usually in separate midgut cells and rarely in a single cell (Aruga et al., 1961; Tanaka and Aruga, 1967; Yamaguchi, 1974; 1975a,b). A cell infected with one strain may become insusceptible to the other strain shortly after infection. On the other hand, a densonucleosis virus or a nuclear polyhedrosis virus does not interfere with a CPV and often coexist in a single columnar cell (Sato and Inoue, 1978; Inoue, 1981). The interaction between CPV strains may be analogous to "cross-protection" between closely related plant viruses (Matthews, 1981). However, there is also the possibility that the strains multiply in a single cell but the inclusion body is formed at the direction of one strain. In a mixed infection of the tiger moth, *Arctia caja*, with two serological distinct CPVs (types 2 and 3) a polyhedron protein characteristic of only one of these viruses (type 3) is produced (Payne, 1976).

Nuclear and cytoplasmic inclusions of the nine virus strains stain well with basic and acidic dyes after pretreatment with normal hydrochloric acid (Iwashita, 1971; Yamaguchi and Hukuhara, 1973a; Abe, 1973). Inclusion body proteins of I, H, and A strains are very similar in amino acid composition (Kawase and Yamaguchi, 1974; Hukuhara and Midorikawa, 1983). The major polypeptides of inclusion bodies of the nine strains have similar molecular weight (approximately 30,000) and share at least one common antigenic determinant (Hukuhara and Midorikawa, 1983; Hukuhara and Ito, unpublished). These facts suggest that inclusion bodies of different shapes and locations are derived from the same or similar inclusion body protein. The major polypeptide of polyhedron protein of *Bombyx* CPV (probably H strain) is coded at the direction of segment 10 of the genome RNA (McCrae and Mertens, 1983).

The characteristics of the inclusion body are inherited and are governed by the virus genome. We may ask the following basic question concerning the process which leads from subunits to the formation of a final product in its characteristic shape and intracellular location. How many bits of genetically independent information are involved in the process and how they are expressed? The preceding subsections show that the information contained in the inclusion body protein is not likely to be fully sufficient to build a final product by means of a system of self-assembly. Supplementary genetic information is apparently contributed through other pathways. The molecules which control assembly of subunits need not be incorporated into the complete structure, but may include specific gene products which serve a catalytic role in construction.

The general sequence of inclusion formation may be visualized as follows: production of inclusion body protein in the cytoplasm (Kawase and Miyajima, 1969), transport of protein to the site of morphogenesis, nucleation process, crystal growth, incorporation of virus particles, and formation of the mature inclusion body. The state of macromolecular association is determined by the total protein concentration and the equilibrium constants for formation of the various possible aggregates which are a function of environmental conditions (Caspar, 1966). Nuclear inclusions may be formed as the result of transport of the inclusion body protein to the nucleus. Such modification may occur through change in the permeability barrier of the nucleus. The transport, although exceptional with the typical strains, may occur normally with the other strains.

The different shapes of inclusion bodies of the typical strains (I, H, P) may be ascribed to the genetically determined change in the protoplasm, which results in different ways of terminating the same crystalline structure (Hukuhara, 1971). In term of crystallography, the shapes of mature inclusion bodies of these strains may be the equilibrium forms, which are defined by the condition of minimum surface-free energy (Kern, 1969). Before the equilibrium has been reached, the inclusion bodies may exhibit the growth forms. Their shapes may change during the crystal growth because the rapidly growing crystal faces are replaced by slowly growing ones. It has long been known that the speed of crystal growth can be affected by certain substances, which may be present in very small amounts and which do not necessarily have to enter the crystal (Buckley, 1951). The effect is not uniform to all crystal faces but restricted to specific faces, which results in a different habit of a growing crystal. The rod-shaped inclusions of B-series strains may be growth forms of this kind. The growth pattern may be changed if the substances involved disappear.

In protein association reactions where the activation energy required to initiate assembly is relatively high, crystal growth requires a nucleation process to overcome this energy barrier (Caspar, 1966). Noncrystalline inclusion bodies of C_1 and C_2 strains may be the result of defective nucleation. If a crystal is held in a highly supersaturated solution, it frequently ceases to grow in its normal form: dendrites or a polycrystalline mass grow out of the body of the crystal, and these may break off and give rise to further crystals (Strickland-Constable, 1972). This phenomenon, called "secondary nucleation," appears analogous to the process observed in cytoplasmic inclusion formation of B_1 and B_2 strains. Crystallogenic matrix may be equivalent to a highly supersaturated solution of polyhedron protein.

Virus particles which are fortuitously present in the area of crystal growth may be incorporated into the inclusion body. Physical attraction between virus particles and inclusion body protein have to be assumed to explain why the virus particles do not move away as the crystal grows (Arnott et al., 1968). Other cellular components are restricted from occlusion but virus particles of a nuclear polyhedrosis virus are incorporated into nuclear inclusions of CPV in mixed-infected columnar cells (Inoue, 1981).

It is possible that some of the CPV strains have arisen as the result of genetic reassortment of RNA segments, as demonstrated in mammalian reovirus strains (Fields, 1981). For example, B strain differs from the original strain (A) in several traits such as the growth parameter and the site of morphogenesis. If these traits are controlled by independent genes, the mutational event would have been very rare as to make it almost impossible to find a larva infected with a variant virus in infection experiments involving several hundred larvae.

B. SYMPTOMATOLOGY AND HISTOPATHOLOGY

Infected larvae become sluggish, cease to feed, are sometimes diarrheic, and often vomit. Heavily infected larvae emit white fecal pellets and frequently exhibit a color change in the central region of the body where the midgut shows through the larval skin. Upon dissecting the larvae, and exposing the alimentary tract, the midgut is found to assume an opaque yellow to white appearance as compared with the normally clear gut wall. The color change is due to the presence of large numbers of polyhedral inclusion bodies and, therefore, is not so conspicuous in larvae infected with a virus strain producing relatively few inclusion bodies. The wall of the diseased midgut is more fragile than that of the normal gut, and when punctured a whitish fluid containing large numbers of inclusion bodies emerges.

The site of virus multiplication is restricted to the midgut. The tissues in the hemocoel are not infected in spite of the fact that infectious virus can be detected in the hemolymph of diseased larvae (Yamamasu and Kawakita, 1962; Sikorowski et al., 1971; Miyajima, 1975). The reported presence of polyhedra in the fore- and hindgut of the saltmarch caterpillar, *Estigmene acrea* (Smith, 1963), is not confirmed by other workers (Stairs et al., 1968) and may be attributable to an infection by a nuclear polyhedrosis virus (Ishimori, 1936). Reports of the presence of cytoplasmic polyhedra in the fat body of coleopterous insects (Sidor, 1970, 1971) and in the epidermis and imaginal buds of the mosquito, *Culex tarsalis* (Kellen et al., 1966), need further investigation because of the possibility that these tissues are attacked by viruses other than CPVs (Federici, 1974). Of the four types of cells that constitute the lepidopterous midgut epithelium, columnar cells are most commonly infected. Infection of goblet cells is rare (Xeros, 1966; Iwashita, 1971; Iwashita and Seki, 1973). Regenerative and basal-granulated cells have never been observed to be infected.

In blackflies, the anterior and posterior thirds of the midgut are whitened and, more rarely, the entire midgut is grossly infected (Bailey et al., 1975; Weiser, 1978). In most mosquitoes, infection are confined to the anterior (cardia and gastric ceca) and posterior portions of the midgut (Federici, 1977; Andreadis, 1981). The mid and hind portions of the midgut are found to contain polyhedra in a hymenopterous insect, *Anoplonyx destructor* (Longworth and Spilling, 1970), and in most lepidopterous insects. This type of infection occurs in the majority of heavily infected silkworm larvae in the advanced instars (Mitani et al., 1958). At early stages of the disease, however, most larvae show the whitening of only the posterior portion. Thus, infection is believed to proceed from the posterior to the anterior region (Aruga, 1957). Aruga et al. (1963) showed that the progress of whitening was hampered at the region where the midgut epithelial cells had been injured by temporary ligation of the silkworm larva with fine thread. They conclude that the inability of these cells to support virus replication terminates the cell-to-cell virus spread from the initial infection site.

The results of interference experiments by Aruga et al. (1961) support this interpretation. After peroral administration of a mixture of two virus strains (I, H), the majority of infected silkworm midguts contain polyhedra of only one strain. The authors hypothesize the specific receptor sites as follows. There may be only a limited number of "receptor cells" which can be infected by the ingested virus. Due to the interference at the cellular level, one of the two ingested viruses multiplies in these cells and produces the progeny virus which, in turn, infects other midgut cells. In young silkworm larvae the process of midgut infection appears to differ: infection proceeds from the anterior to the posterior region (Miyagawa, 1975b). Moreover, the midgut epithelium of young larvae does not have specific receptor sites, as shown by the lack of virus interference at the organismal level (Aruga et al., 1961; Yamaguchi, 1974, 1975a,b). Inclusion bodies of two virus strains coexist in a mixed-infected midgut, although an individual cell contains only one type. The difference in infection process according to the larval age may be ascribed to the functional differentiation of the midgut epithelial cells (Iwashita and Seki, 1973).

The surface structure of infected midgut epithelium is distinctly different from that of healthy epithelium. The epithelial cells of the fall webworm, *Hyphantria cunea*, are enlarged at the lumenal surface, where the microvilli are partially or completely absent (Boucias and Nordin, 1978). The underlying cell bodies are packed with polyhedra. At the final stage of infection the cells rupture and release polyhedra into the gut lumen. In the silkworm, the apical cytoplasm of infected columnar cells bulges into the gut lumen and the plasma membrane is ruptured because of internal pressure (Iwashita, 1971; Kobayashi, 1971). The infected columnar cells are often discharged into the lumen and the cells are replenished with newly differentiated cells (Iwashita and Aruga, 1957; Yamaguchi, 1962, 1968a; Yamaguchi and Ayuzawa, 1970; Yamaguchi and Hukuhara, 1973a). Autoradiographs of the midgut at this stage of the disease show the acceleration of DNA synthesis in the regenerative cells, which are in the process of active cell division to replace the damaged gut epithelium (Watanabe et al., 1967). The replacement occurs most frequently at the larval molt (Yamaguchi, 1976a). Large numbers of newly formed cells appear just before larval molt and, as the result of their growth and expansion, the infected columnar cells are discharged during the molt (Inoue, 1977; Inoue and Miyagawa, 1978). The newly differentiated cells are susceptible to infection. The polyhedra produced in these cells are smaller than those in the old columnar cells although they coexist in the same epithelium. In some cases, uninfected goblet cells are discharged and the entire midgut epithelium is renewed (Yamaguchi, 1977). Such marked midgut regeneration also occurs in larvae containing mixed infections with a CPV and a flacherie

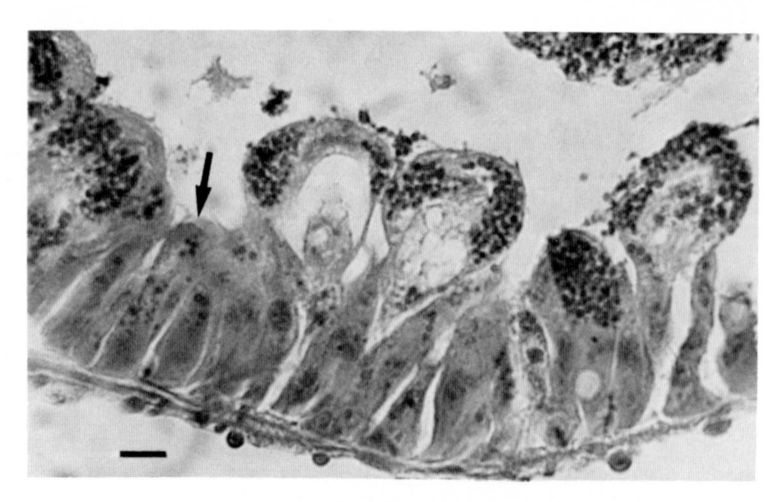

FIGURE 9. Midgut epithelium of a fall webworm larva infected with Hyphantria CPV. Infected columnar cells containing numerous polyhedra are extruded by the development of newly formed cells. Note recent infection in the latter cells (arrow); bar, 10 μm. (From Noguchi and Yamaguchi, 1982.)

virus, which replicates in the goblet cell (Inoue and Miyagawa, 1978).

Infection of fall webworm larvae with *Hyphantria* CPV results in the similar process and many polyhedra are produced in the renewed epithelium (Fig. 9) (Akutsu, 1971; Noguchi and Yamaguchi, 1982). Most of the infected larvae are so severely affected that they are unable to grow into the next molt. Fall webworm larvae infected with *Bombyx* CPV, however, recover naturally from the virus disease and eventually grow into adults (Yamaguchi, 1976a,b, 1979). In this case, newly formed cells which replace the infected columnar cells are refractory to infection. No infectious virus can be detected in the renewed midgut epithelium. The recovered larvae are resistant to a challenge dose of *Bombyx* CPV sufficient to infect unexposed larvae (Noguchi and Yamaguchi, 1978, 1979). Noguchi and Yamaguchi (1982) compared the process of the midgut regeneration in several lepidopterous species infected with three CPVs. Active regeneration is observed in all virus-host combinations but the fate of infected larvae differs according to the response of renewed midgut epithelium to infection. Larvae of *Orgyia thyellina* and *Mamestra brassicae* recover from infection with *Bombyx* CPV. These larvae as well as those of *Spodoptera litura* and the silkworm do not recover from infection with either *Hyphantria* CPV or *Dendrolimus* CPV. Larvae of *Orgyia thyellina* and the silkworm die as the result of frank infection with *Dendrolimus* CPV. In other combinations, however, most larvae survive because polyhedron formation in the newly formed cells is confined to the distal portion of the infected cell or to the posterior portion of the midgut.

The active regeneration of larval midgut epithelium has been recorded in *Amphimallon majalis* invaded by *Bacillus popilliae* (Splittstoesser et al., 1973, 1978), and in the silkworm infected with a flacherie virus (Inoue, 1974; Kurisu and Matsumoto, 1974). The process is closely analogous to vertebrate reactions under inflammatory conditions. The accelerated extrusion of cells, increased mitosis of replacement cells, and subsequent appearance of numerous immature cells are associated with *Salmonella* and *Shigella* infection of gut mucosa (Takeuchi, 1971).

Fecal pellets from infected fall webworm larvae contain relatively few polyhedra at the beginning of midgut epithelial degradation, but consist almost entirely of polyhedra at the advanced stage of the disease (Boucias and Nordin, 1978). Feces from infected silkworm larvae are infectious (Saito and Yamaguchi, 1960). The infectivity is not lost after the polyhedra are removed from the feces by low-speed centrifugation, which indicates that they contain not only polyhedra but also free virions (Ishikawa et al., 1961; Furuta, 1963; Watanabe, 1968). During the natural infection of lepidopterous species, polyhedra are solubilized by the action of alkaline gut juice to release the occluded virions into the midgut lumen. The solubilization proceeds from the polyhedron surface to the interior (Watanabe, 1974). The lowering of the gut pH at an advanced stage of the disease may be responsible for the appearance of large numbers of polyhedra in feces (Vago et al., 1959; Watanabe, 1971a). Polyhedra in fecal pellets can be detected by an enzyme-linked immunosorbent assay following purification with an aqueous two-phase separation system (Hukuhara et al., 1981a; Hukuhara and Midorikawa, 1981).

C. EFFECTS OF INFECTION ON THE HOST

CPV infections have little or no adverse effect on most dipterous insects, and infected larvae usually pupate and emerge as apparently healthy adults (Federici, 1974, 1977; Bailey, 1977; Weiser, 1978). Larvae of other insect orders suffer from lethal infections but, because of the chronic nature of the diseases, some larvae survive especially when infected late in the larval stage. In the following sections, deleterious effects of CPV infections in these species, particularly in species of Lepidoptera, will be discussed.

Infected larvae lag behind uninfected insects in their development. For example, larvae of *Pectinophora gossypiella* treated with a CPV are significantly smaller than untreated larvae (Bullock et al., 1970) and infected larvae of *Heliothis virescens* weigh one-half as much as healthy larvae at 7 and 12 days of age (Simmons and Sikorowski, 1973). Fall webworm larvae which have recovered from infection are of diminutive size and weight (Yamaguchi, 1976a). The time required for pupation is prolonged in infected larvae of *Trichoplusia ni*, *Lymantria dispar*, and *Heliothis virescens* (Vail et al., 1969; Magnoler, 1974; Sikorowski and Thompson, 1979). When the virus dose is increased, larval weights at a given age are progressively decreased in *Pectinophora gossypiella* and the duration of larval stage of survivors is increased in this insect and *Euxoa scandens* (Bell and Kanavel, 1976; Bellemare and Belloncik, 1981). The increased developmental periods of infected larvae of the silkworm and *Lymantria dispar* are occasionally associated with an increased number of larval molts (Kurata, 1971; Magnoler, 1974).

Slow growth, increased molts, diminutive size, high or complete mortality of the immature stages, and prolonged survival are familiar pathologies of most nutritional defects in insects (House, 1963). It is possible that the failure of infected larvae to achieve normal growth and development is a manifestation of dystrophy. CPV infection must impair the midgut functions, such as, the secretion of digestive enzymes and the absorption of nutrients. It also causes the weakening of the rhythmic contractile movements of the midgut in silkworm larva (Hukuhara et al., 1981b, unpublished). The midgut dysfunction must bring about reduced feeding of the infected larvae by some unknown feedback mechanism.

Although few biochemical and physiological studies have been conducted on tissues of infected larvae other than the midgut, almost every parameter measured shows the same change as exhibited by starved larvae. The parameters include reduced hemolymph content of free amino acids (Kawase, 1965), protein (Kawase and Hayashi, 1965), and lipid (Bell, 1977), lower hemolymph refractive index (Miyajima, 1982), lower gut pH, increased excretion of uric acid (Watanabe, 1971b), and underdeveloped fat body (Ignoffo and Adams, 1966). An exception to this generalization is the increased oxygen uptake per body weight as shown by Wiygul and Sikorowski (1978) in infected larvae of *Heliothis virescens*. They speculate that the difference in oxygen uptake is due to extensive metabolism associated with the production of viral bodies in the midgut.

For a better understanding of the effects of infection more comprehensive data must be assembled with the use of improved techniques, possibly involving axenic rearing. The period of lethal infection of infected silkworm larvae is prolonged in an axenic condition and a higher proportion of individuals survive to adulthood (Kurata, 1971). Under ordinary rearing conditions, some of the intestinal bacteria may proliferate at lowered gut pH of CPV-infected larvae and act synergistically to induce the deleterious effects on the host insects (Kodama and Nakasuji, 1969).

III. PATHOLOGY IN POSTLARVAL STAGES

A. *HISTOPATHOLOGY*

The midgut of a normal lepidopterous larva is completely renewed at the larval-pupal ecdysis. The regenerative cells proliferate and form an epithelial layer round the outside of the larval cells which thus come to lie in the lumen of the new alimentary canal. Sometimes partial or complete renewal occurs again when the adult tissues are forming (Chapman, 1982).

Ishikawa and Asayama (1959) studied the histopathology of infected silkworm midguts at the time of metamorphosis. Numerous polyhedra (I strain) and the disintegrated larval epithelium are discharged into the lumen at the larval-pupal ecdycis. The midgut of diseased pupae and adults is larger than the normal midgut and contains an amorphous greyish-brown clump. Infection occurs in the newly formed epithelium. Polyhedra are present in the cytoplasm of columnar cells in late-pupal and adult stages. According to Iwashita et al. (1968b), the polyhedra are more concentrated in the distal areas of the cytoplasm in pupal epithelium infected with H strain. Their size is generally smaller than those in the old larval epithelium. The nuclei of the infected cells are atrophied, the chromatin granules aggregated, and the nucleoli indistinct. They occasionally contain relatively large crystalline inclusions of the same shape as that of the cytoplasmic polyhedra. Some pupal epithelia are not infected notwithstanding the presence of numerous polyhedra in the lumen.

Neilson (1965) reports the histopathology of infected adults of *Alsophila pometaria*, *Nymphalis antiopa*, *Operophtera brumata*, and *Paleacrita vernata* as follows. Polyhedra are produced only in the midgut cells. A typical infected midgut becomes milky-white and fragile, and in advanced stages of infection may break down completely into a milky fluid. Polyhedra are sometimes found in the hemolymph, but these are most likely released into the hemocoel from the ruptured gut cells. Irregularly shaped bodies of 1 to 3 mm^3 in size occur in the abdomens. Usually these bodies are found inside the midgut, but occasionally some are found unattached and free-floating in the hemocoel. They are hard, usually dark in color, and composed of a hard central core of tightly packed polyhedra covered by several disorganized layers of cells all surrounded by an apparently noncellular layer about 10 to 20 μm thick. Similar bodies are found in infected adults of *Lymantria dispar* (Magnoler, 1974). Neilson suggests that the bodies are formed from a larval midgut not undergoing histolysis. The author suspects that they may be masses of disintegrated larval epithelium enveloped within peritrophic membrane. Because of the development of a much smaller gut, they may occasionally be pushed out through the fragile gut wall and come to lie in the hemocoel. Otherwise, they must be transported to the rectal sac. Polyhedra are detected in the meconium from infected adults of *Pseudaletia unipuncta* and *Euxoa scandens* (Tanada and Chang, 1960; Bellemare and Belloncik, 1981).

B. *EFFECTS OF INFECTION ON THE HOST*

Pupae from infected larvae are generally smaller than healthy controls but are normal in external morphology. Malformations in *Euxoa scandens* pupae may be an exceptional case (Bellemare and Belloncik, 1981). The weight of infected pupae of *Trichoplusia ni*, *Lymantria dispar*, and *Pectinophora gossypiella* is not related to the virus dose administered to the larvae but to the time of pupation (Vail et al., 1969; Magnoler, 1974; Bell and Kanavel, 1976). Thus, the pupal weight decreases as the duration of larval stage increases. The duration of pupal stage is shortened by infection in *L. dispar* (Magnoler, 1974), but is not affected in *Heliothis virescens* and *P. gossypiella* (Simmons and Sikorowski, 1973; Bell and Kanavel, 1976). Not all the individuals that survive to pupate reach the adult stage. Adult emergence is reduced in infected pupae of *T. ni*, *P. gossypiella*, *H. virescens*, and *L. diapar* (Vail et al., 1969; Bullock et al., 1970; Simmons and Sikorowski, 1973; Magnoler, 1974). In diapausing *H. virescens*, however, infected pupae yield as many moths as healthy pupae (Sikorowski and Thompson, 1979).

Trichoplusia ni moths from infected pupae are usually smaller and are often deformed (Vail et al., 1969; Vail and Gouch, 1970). The proportion of deformed *T. ni* moths increases as pupal weight decreases. No such relationship is observed in *L. diapar* by Magnoler (1974), who reports that a higher percentage of deformed moths occur among diseased females than among males. The malformations include inability to emerge from the pupa, unexpanded wings, and reduction in size. The wing coloration of diseased males is significantly lighter than that of healthy controls. The percentage of light-colored male moths increases as pupal weight decreases. The longevity of diseased adults of *P. gossypiella* and *H. virescens* is less than healthy moths (Bullock et al., 1970; Simmons and Sikorowski, 1973), while there is no difference in longevity between healthy and infected groups of *T. ni* and diapausing *H. virescens* (Vail and Gough, 1970; Sikorowski and Thompson, 1979).

The wings of diseased moths of *E. scandens* are often so seriously deformed as to render the insects unable to fly and copulate (Bellemare and Belloncik, 1981). Vail and Gough (1970) noted a trend to decreased matings of both sexes among *T. ni* moths emerging from small pupae. Production of eggs by normal females mating with virus-treated males are not affected, but virus-treated females from small pupae mating with normal males lay fewer eggs than do females of normal weight from either virus-treated or control groups. Diseased female moths of *P. gossypiella* and *H. virescens* also exhibit decreased egg production, which may be attributed to the lower level of lipid content in infected individuals as compared with the healthy controls (Bullock et al., 1970; Simmons and Sikorowski, 1973; Bell and Kanavel, 1977; Sikorowski and Thompson, 1979). The fecundity of infected silkworm female moths is not affected (Watanabe and Namura 1971).

Some of the deleterious effects observed in the postlarval stages may be a reflection of malnutrition during the larval stage. Accordingly, they may be observed irrespective of whether the midgut is infected or has recovered from infection. They have little diagnostic value, because reduced pupal size induced by means other than CPV infection gives rise to similar effects (Shorey, 1963; Cheng, 1972). The effects of infection in the postlarval stages may vary depending on the insect mode of nutrition and reproduction, as suggested by Neilson (1965). For example, adults of *Alsophila pometaria* live only for a few weeks and do not feed. The dysfunction of the midgut due to infection, therefore, probably has little direct effect upon the insect's ability to fulfill the reproductive roles. Moths of *Nymphalis antiopa*, on the other hand, do feed and must sustain themselves for 7 to 10 months until the spring of the following year, before they begin their reproductive functions. Severely infected moths, with functionless midguts, never live more than 1 month.

IV. INFECTION *IN VITRO*

A. EMBRYO CULTURE

Under natural conditions the embryonic stage is not affected by frank CPV infection. Susceptibility of the embryonic stage to CPV was demonstrated by Kitazawa and Takami (1959). When an extract from infected midgut is smeared over a pinhole made near the micropyle of a silkworm egg, the embryo swallows the virus-contaminated serosa at the head pigmentation stage and its midgut is found to contain many polyhedra 2-3 days postinoculation. Most of the infected cells are columnar cells, although goblet cells are sometimes infected. They are distributed all over the entire midgut. Silkworm embryos cultured *in vitro* are also infected by CPV added to the culture media (Takami et al., 1967). The site of polyhedron formation is reséricted to the midgut epithelium. Only those embryos that have swallowed the serosa are infected. Addition of the virus inoculum to young embryo cultures does not result in precocious infection; polyhedron formation occurs invariably at the final embryonic stage. The midgut primordium of young embryos is not infected even though it is occasionally extruded into virus-contaminated culture medium through an incomplete dorsal closure. Takami and his associates conclude that a certain degree of morphological or physiological differentiation is a prerequisite for the successful infection of midgut cells. No spontaneous development of CPV infection is observed in cultures of embryos from diseased female moths. Both virions from alkali-dissolved polyhedra and extracts from infected midgut are infectious and induce the same histological changes in cultured embryos (Iwashita and Seki, 1973).

B. CELL CULTURE

Cultured insect cells are used to investigate virus infectivity, the process of virus multiplication, the assay of virus titers, and fundamental virus-cell interactions. Cultured midgut epithelium would be a most valuable substrate for CPV study. However, among various insect tissues, this is one of the most difficult to maintain *in vitro*. Kobayashi (1971, 1972) added *Bombyx* CPV released from polyhedra to cultures of silkworm midgut epithelial cells to elucidate the process of virus entry. He suggests that the virions adsorb to the cell surface by their projections and inject the virus core into the cytoplasm. The morphological evidence for such mechanisms of infection is not convincing, particularly in the light of the biochemical evidence that isolated viral RNA is not infectious (Lewandowski et al., 1969) and the viral transcriptase does not function if the virion is degraded (Lewandowski and Traynor, 1972). Moreover, the cultured midgut cells were not maintained long enough to determine if the inoculated cells became infected.

Grace (1962) reported development of CPV in early passage cultures of *Antheraea eucalypti* ovarian cells. Polyhedra in the infected cells are cubic and occlude spherical virions. When the virus is administered to *A. eucalypti* larvae, spherical polyhedra are produced in the midgut. The infection was discovered during the experiments designed to determine the susceptibility of the cultured cells to a nuclear polyhedrosis virus of the silkworm. He attributed the infection to provocation of latent infection by a foreign virus. However, this is not always reproducible (Hukuhara, 1971). *Lymantria dispar* ovarian tissues cultured *in vitro* are infected with *Lymantria* CPV and exhibit the same cytopathology as the infected midgut cells *in vivo* (Vago and Bergoin, 1963). Primary cultures of silkworm ovarian and tracheal tissues are infected with purified virions from diseased midguts of *Malacosoma disstria* (Sohi et al., 1971). Most infected cells show changes typical of CPV infection. Nuclei of some ovarian cells, however, contain crystalline inclusions of the same shape and size as the cytoplasmic polyhedra and much smaller spherical bodies. No virions are seen in these inclusions. *Malacosoma disstria* CPV replicates in an insect cell line but polyhedron formation does not take place (Kawarabata and Hayashi, 1972). The synthesized virus can be detected by the incorporation of [^{3}H]uridine into components of the same size as intact virions. The cell line used was described as Grace's *Aedes aegypti* cell line but may actually have been an *A. eucalypti* ovarian cell line (Krywienczyk and Sohi, 1973). The earlier studies show that tissue specificity of CPVs is not as restrictive *in vitro* as *in vivo* and have led to the use of several established ovarian cell lines, which may have come from cells of the intermediate layer of the ovariole or hemocytes that have adhered to the surface of the ovary.

Recent studies of CPV replication *in vitro* include those on *Trichoplusia ni* cell line infected with *T. ni* CPV (Granados et al., 1974), *Lymantria dispar* cell line infected with *Euxoa scandens* CPV (Quiot and Belloncik, 1977), and *Spodoptera frugiperda* cell line infected with *Chrysodeixis eriosoma* CPV (Longworth, 1980). The first two lines were infected with both extracts of infected midguts and virions released from polyhedra and the third line was infected with virions purified from cadavers. The percentage of infected cells varied depending on the cell-virus systems: 1-40% in *T. ni* cells, more than 90% in *L. dispar* cells, and 8% in *S. frugiperda* cells. CPV development in the three cell lines was similar to that described in a typical CPV infection *in vivo*. Infected cells are often enlarged, assume a granulated appearance, and produce a variable number of cytoplasmic polyhedra. The development of virus particles was restricted to localized areas of virogenic stroma in the cytoplasm. There appears to be little lysis of infected cells. Thus there is no major release of extracellular virus and minimal spread of infection in the cell monolayer. Virus can not be serially passaged by using cell-free medium from infected cultures. However, homogenates of infected cultures are infectious. A majority of the polyhedra produced *in vitro* were cubic, while the polyhedra produced in infected larvae were predominantly spherical.

A high infection rate in the cell-virus system involving *L. dispar* cells and *Euxoa scandens* CPV has allowed more sophisticated study. The process of virus entry appears to resemble phagocytosis because virions are observed in phagocytic vacuoles prior to the first manifestation of virus replication (Quiot and Belloncik, 1977). Viral protein can be detected in a homogenate of cultured cells by an enzyme-linked immunosorbent assay at very early stages of infection (Payment et al., 1982). In γ-irradiated cultured cells, the number of polyhedra per cell increases but the process of polyhedron maturation becomes abnormal (Belloncik and Arella, 1981). For the virus titration, a tissue microculture assay is available (Belloncik and Chagnon, 1980). The procedure is similar to the 50% tissue culture infectious dose assay, but a single infected cell, detected by the presence of polyhedra, is scored rather than the degeneration of cell monolayer. According to Belloncik and Bellemare (1980), the virus in the cell culture maintains the capacity to produce cubic polyhedra during the passages *in vitro* but, when inoculated to *Euxoa scandens* larvae, induces the production of spherical polyhedra. They attribute the change in polyhedron shape to the different crystallization process of polyhedron protein in two distinct cellular systems rather than to the selection of different virus strains from a heterologous virus population. This conclusion is substantiated by the infection of the same cell line with various strains of *Bombyx* CPV. The characteristic intracellular location of inclusion body is retained but its shape is changed *in vitro* (Hukuhara and Midorikawa, 1983, unpublished).

Additional cell-virus systems reported in recent literature are *Trichoplusia ni* CPV in established cell lines derived from *Estigmene acrea* hemocytes (Granados and Naughton, 1976) and *Spodoptera frugiperda* and *Lymantria dispar* ovarian tissues (Granados, 1976) and *Chrysodeixis erisoma* CPV in *Trichoplusia ni* ovarian cell line (Longworth, 1980). However, no detailed accounts of replication processes have been reported on these cell-virus systems.

V. CONCLUSIONS

The nature of the mechanism that controls the formation of crystalline inclusion bodies appears to be a fertile field for investigation of biomolecular organization. It would be most interesting to know at what point the viral genetics are interposed on the situation to bring about the alternate crystal forms of protein molecules.

Histopathological studies of *Bombyx* CPV has led to the isolation of several occlusion-body-deficient strains. Had it not been for their derivation and close relationship with the typical strains in many criteria, they would probably have remained as unclassified nonoccluded viruses as did the viruses from *Musca domestica* and *Drosophila melanogaster* (Moussa, 1980; Haars et al., 1980). The possibility of a polyhedrosis virus mutating to a nonoccluded virus reinforces the desirability of classifying viruses on the basis of the properties of the virions rather than the syndromes. The occlusion body-deficient variants of *Bombyx* CPV are different from those of a nuclear polyhedrosis virus (Wood, 1980) in that they retain the capacity to produce the inclusion body protein and that they are as virulent as the typical strains *in vivo*. Inasmuch as the variant CPV strains occur in farmer's silkworm rearing, the occlusion of the virus particles in polyhedra may not be the prerequisite for the survival of the CPV in nature.

We should make serious attempts to understand the mechanisms of tissue and host specificity. Development of sophisticated techniques such as tissue culture needs increased study and emphasis. Establishment of midgut cell lines would certainly contribute to our understanding of specificity, pathogenesis, and viral genetics. With progress in these areas, we shall be able to explain and predict the harmful results of CPV infection, which may vary widely depending on the virulence of the virus, the regenerative capacity of the midgut epithelium, synergism from mixed infections, host physiological state or mode of life.

REFERENCES

Abe, Y. (1973). Affinity for biological stains of nuclear and cytoplasmic polyhedra of the silkworm, *Bombyx mori* L. *Sansi-Kenkyu* (80), pp. 104-111.

Akutsu, K. (1971). Cytoplasmic polyhedrosis of the fall webworm, *Hyphantria cunea* Drury (Lepidoptera: Arctiidae). *Appl. Entomol. Zool. 6*, 198-205.

Andreadis, T. G. (1981). A new cytoplasmic polyhedrosis virus from the salt-marsh mosquito, *Aedes cantator* (Diptera: Culicidae). *J. Invertebr. Pathol. 37*, 160-167.

Arnott, H. J., Smith, K. M., and Fullilove, S. L. (1968). Ultrastructure of a cytoplasmic polyhedrosis virus affecting the monarch butterfly, *Danaus plexippus*. I. Development of virus and normal polyhedra in the larva. *J. Ultrastr. Res. 24*, 479-507.

Aruga, H. (1957). Mechanism of resistance to virus diseases in the silkworm, *Bombyx mori*. II. On the relation between the nuclear polyhedrosis and the cytoplasmic polyhedrosis. *J. Sericult. Sci. Japan 26*, 279-283.

Aruga, H., and Tanada, Y. (1971). "The Cytoplasmic-Polyhedrosis Virus of the Silkworm." Univ. Tokyo Press, Tokyo.

Aruga, H., Hukuhara, T., Yoshitake, N., and Israngkul Na Ayudhya, A. (1961). Interference and latent infection in the cytoplasmic polyhedrosis of the silkworm, *Bombyx mori*. (Linnaeus). *J. Insect Pathol. 3*, 81-92.

Aruga, H., Yoshitake, N., Hukuhara, T., and Owada, M. (1963). Invasion route of the cytoplasmic polyhedrosis virus in the silkworm, *Bombyx mori*. L. *J. Sericult. Sci. Japan 32*, 58-62.

Bailey, C. H. (1977). Field and laboratory observations on a cytoplasmic polyhedrosis virus of blackflies (Diptera: Simuliidae). *J. Invertebr. Pathol. 29*, 69-73.

Bailey, C. H., Shapiro, M., and Granados, R. R. (1975). A cytoplasmic polyhedrosis virus from the larval blackflies *Cnephia mutata* and *Prosimulium mixtum* (Diptera: Simuliidae). *J. Invertebr. Pathol. 25*, 273-274.

Bell, M. R. (1977). Pink bollworm: effect of infection by a cytoplasmic polyhedrosis virus on diapausing larvae. *Ann. Entomol. Soc. Amer. 70*, 675-677.

Bell, M. R., and Kanavel, R. F. (1976). Effect of dose of cytoplasmic polyhedrosis virus on infection, mortality, development rate, and larval and pupal weights of the pink bollworm. *J. Invertebr. Pathol. 28*, 121-126.

Bell, M. B., and Kanavel, R. F. (1977). The effect of a cytoplasmic polyhedrosis virus on lipid and protein content of pupae of the pink bollworm (Lepidoptera: Gelechiidae). *J. Kansas Entomol. Soc. 50*, 359-362.

Bellemare, N., and Belloncik, S. (1981). Études au laboratoire des effets d'une polyédrose cytoplasmique sur le ver gris blanc *Euxoa scandens* (Lepidoptère: Noctuidae Agrotinae). *Ann. Soc. Entomol. Québec 26*, 28-40.

Belloncik, S., and Arella, M. (1981). Production of cytoplasmic polyhedrosis virus (CPV) polyhedra in a gamma irradiated *Lymantria dispar* cell line. *Arch. Virol. 68*, 303-308.

Belloncik, S., and Bellemare, N. (1980). Polyèdres du CPV d'*Euxoa scandens* (Lep.: Noctuidae) produits *in vivo* et sur cellules cultivées *in vitro*. Études comparatives. *Entomophaga 25*, 199-207.

Belloncik, S., and Chagnon, A. (1980). Titration of a cytoplasmic polyhedrosis virus by a tissue microculture assay: some application. *Intervirology 13*, 28-32.

Bird, F. T., and Whalen, M. M. (1954). A nuclear and a cytoplasmic polyhedral virus disease of the spruce budworm, *Choristoneura fumiferana* (Clem.). *Can. J. Zool. 32*, 82-86.

Boucias, D. G. and Nordin, G. L. (1978). A scanning electron microscope study of *Hyphantria cunea* CPV-infected midgut tissue. *J. Invertebr. Pathol. 32*, 229-233.

Buckley, H. E. (1951). "Crystal Growth." Wiley, New York.

Bullock, H. R., Martinez, E., and Stuermer, C. W., Jr. (1970). Cytoplasmic-polyhedrosis virus and the development and fecundity of the pink bollworm. *J. Invertebr. Pathol. 15*, 109-112.

Caspar, D. L. D. (1966). Design principles in organized biological structures. *In* "Principles of Biomolecular Organization" (G. E. W. Wilstenholme and M. O'Conner, eds.), pp. 7-39. Churchill, London.

Chapman, R. F. (1982). "The Insects. Structure and Function." 3rd ed. Hodder and Stoughton, London.

Cheng, H. H. (1972). Oviposition and longevity of the dark-sided cutworm *Euxoa messoria* (Lepidoptera: Noctuidae) in the laboratory. *Can. Entomol. 104*, 919-925.

Federici, B. A. (1974). Viral pathogens of mosquitoes and their potential use in mosquito control. *In* "Le Contrôle des Moustiques/Mosquito Control" (A. Aubin, J.-P. Bourassa, S. Belloncik, M. Pellissier, and E. Lacoursiere, eds.), pp. 93-135. Les Presses de l'Université du Quebec, Quebec.

Federici, B. A. (1977). Virus pathogens of Culicidae (mosquitoes). *In* "Pathogens of Medically Important Arthropods" (D. W. Roberts and M. A. Strand, eds.), Bull. WHO 55 (Supp. 1), pp. 25-46.

Federici, B. and Hazard, E. I. (1975). Iridovirus and cytoplasmic polyhedrosis virus in the freshwater daphnid *Simocephalus expinosus*. *Nature (London) 254*, 327-328.

Federici, B. A., Hazard, E. I., and Anthony, D. W. (1973). A new cytoplasmic polyhedrosis virus from chironomids collected in Florida. *J. Invertebr. Pathol. 22*, 136-138.

Fields, B. N. (1981). Genetics of reovirus. *Current Topics Microbiol. Immunol. 91*, 1-24.

Fujiwara, T., Yukibuchi, E., Tanaka, Y., Yamamoto, Y., Tomita, K., and Hukuhara, T. (1984). X-ray diffraction studies of polyhedral inclusion bodies of nuclear- and cytoplasmic-polyhedrosis viruses. *Appl. Entomol. Zool. 19*, in press.

Furuta, Y. (1963). Pathogenicity of feces from a larva of the silkworm, *Bombyx mori* L., infected with a cytoplasmic-polyhedrosis virus. *Sanshi-Kenkyu (48)*, pp. 30-34.
Grace, T. D. C. (1962). The development of a cytoplasmic polyhedrosis in insect cells grown *in vitro*. *Virology 18*, 33-42.
Granados, R. R. (1976). Infections and replication of insect pathogenic viruses in tissue culture. *Advan. Virus Res. 20*, 189-236.
Granados, R. R., and Naughton, M. (1976). Replication of *Amsacta moorei* entomopoxvirus and *Autographa californica* nuclear polyhedrosis virus in hemocyte cell lines from *Estigmene acrea*. *In* "Invertebrate Tissue Culture" (E. Kurstak and K. Maramorosch, eds.), pp. 379-389. Academic Press, New York.
Granados, R. R., McCarthy, W. J., and Naughton, M. (1974). Replication of a cytoplasmic polyhedrosis virus in an established cell line of *Trichoplusia ni* cells. *Virology 59*, 584-586.
Haars, R., Zentgraf, H., Gateff, E., and Bautz, F. A. (1980). Evidence of endogenous reovirus-like particles in a tissue culture cell line from *Drosophila melanogaster*. *Virology 101*, 124-130.
House, H. L. (1963). Nutritional diseases. *In* "Insect Pathology. An Advanced Treatise" (E. A. Steinhaus, ed.), Vol. 1. Academic Press, New York.
Hukuhara, T. (1971). Variations in cytoplasmic-polyhedrosis virus. *In* "The Cytoplasmic-Polyhedrosis Virus of the Silkworm" (H. Aruga and Y. Tanada, eds.), pp. 61-78. Univ. Tokyo Press, Tokyo.
Hukuhara, T. (1972). Demonstration of polyhedra and capsules in soil with scanning electron microscope. *J. Invertebr. Pathol. 20*, 375-376.
Hukuhara, T., and Hashimoto, Y. (1966a). Studies of two strains of cytoplasmic-polyhedrosis virus. *J. Invertebr. Pathol. 8*, 184-192.
Hukuhara, T. and Hashimoto, Y. (1966b). Serological studies of the cytoplasmic- and nuclear-polyhedrosis viruses of the silkworm, *Bombyx mori*. *J. Invertebr. Pathol. 8*, 234-239.
Hukuhara, T. and Midorikawa, M. (1981). Assessment of the occurrence of the cytoplasmic polyhedrosis in silkworm populations reared on an artificial diet. *J. Sericult. Sci. Japan 50*, 345-346.
Hukuhara, T. and Midorikawa, M. (1983). Pathogenesis of cytoplasmic polyhedrosis in the silkworm. *In* "Double Stranded RNA Viruses" (D. H. L. Bishop and R. W. Copans, eds.), pp. 405-414. Elsevier, Amsterdam.

Hukuhara, T., and Yamaguchi, K. (1973). Ultrastructural investigation on a strain of a cytoplasmic polyhedrosis with nuclear inclusions. *J. Invertebr. Pathol. 22*, 6-13.

Hukuhara, T., Namura, H., and Kobayashi, M. (1972). Aberrant polyhedra of the cytoplasmic-polyhedrosis virus of the silkworm, *Bombyx mori*. *J. Sericult. Sci. Japan 41*, 187-196.

Hukuhara, T., Midorikawa, M., and Ito, H. (1981a). Purification of a cytoplasmic-polyhedrosis virus from silkworm feces. *J. Sericult. Sci. Japan 50*, 301-305.

Hukuhara, T., Satake, S., and Sato, Y. (1981b). Rhythmic contractile movements of the larval midgut of the silkworm, *Bombyx mori*. *J. Insect Physiol. 27*, 469-473.

Ignoffo, C. M., and Adams, J. R. (1966). A cytoplasmic-polyhedrosis virus, *Smithiavirus pectinophorae* sp. n. of the pink bollworm, *Pectinophora gossypiella* (Saunders). *J. Invertebr. Pathol. 8*, 59-66.

Inoue, H. (1974). Multiplication of an infectious-flacherie virus in the resistant and susceptible strains of the silkworm, *Bombyx mori*. *J. Sericult. Sci. Japan 43*, 318-324.

Inoue, H. (1977). Thermal therapy of virus diseases of the silkworm, *Bombyx mori*. *J. Sericult. Sci. Japan 48*, 306-312.

Inoue, H. (1981). Double infection of midgut epithelial cells with nuclear and cytoplasmic polyhedrosis viruses. *J. Sericult. Sci. Japan 50*, 311-319.

Inoue, H., and Miyagawa, M. (1978). Regeneration of midgut epithelial cell in the silkworm infected with viruses. *J. Invertebr. Pathol. 32*, 373-380.

Ishikawa, Y., and Asayama, T. (1959). On the translocation of the cytoplasmic polyhedral bodies in the mid-gut during metamorphosis in the silkworm, *Bombyx mori*. *J. Sericult. Sci. Japan 28*, 308-312.

Ishikawa, Y., Asayama, T., and Okino, H. (1961). Dissemination of the cytoplasmic polyhedrosis of the silkworm. *Sanshi-Kaiho (821)*, pp. 14-19.

Ishimori, N. (1934). Contribution a l'étude de la grasserie du ver a soie (*Bombyx mori*). *C. R. Soc. Biol. 116*, 1169-1170.

Ishimori, N. (1936). Gut epithelial cells affected by insect polyhedroses. *Shokubutsu oyobi Dobutsu 4*, 26-28.

Iwashita, Y. (1971). Histopathology of cytoplasmic polyhedrosis. *In* "The Cytoplasmic-Polyhedrosis Virus of the Silkworm" (H. Aruga and Y. Tanada, eds.), pp. 79-101. Univ. Tokyo Press, Tokyo.

Iwashita, Y., and Aruga, H. (1957). Mechanism of resistance to virus diseases in the silkworm, *Bombyx mori*. III.

Histological studies on the polyhedrosis in the silkworm. *J. Sericult. Sci. Japan 26*, 323-328.

Iwashita, Y., and Seki, H. (1973). The development of the cytoplasmic-polyhedrosis virus in the midgut of the embryo and the young silkworm, *Bombyx mori* L. *Bull. Coll. Agr. Utsunomiya Univ. 8*, 27-42.

Iwashita, Y., Inoue, K., and Yamaguchi, K. (1968a). Light and electron microscopic investigation of midgut-nuclear-polyhedrosis virus of the silkworm, *Bombyx mori* L. *Bull. Coll. Agr. Utsunomiya Univ. 7*, 81-95.

Iwashita, Y., Kanke, E., and Takahashi, A. (1968b). Histopathological changes in the midgut epithelium of the silkworm infected with the cytoplasmic polyhedrosis virus and the infectious flacherie virus. *Bull. Coll. Agr. Utsunomiya Univ. 7*, 1-10.

Kawarabata, T., and Hayashi, Y. (1971). Development of a cytoplasmic polyhedrosis virus in an insect cell line. *J. Invertebr. Pathol. 19*, 414-415.

Kawase, S. (1965). Free amino acids in the hemolymph and midgut epithelium of the silkworm, *Bombyx mori* (Linnaeus), infected with cytoplasmic-polyhedrosis virus. *J. Invertebr. Pathol. 7*, 113-116.

Kawase, S., and Hayashi, Y. (1965). Nucleic-acid and protein changes in blood and midgut of the silkworm, *Bombyx mori* (Linnaeus), during the course of cytoplasmic polyhedrosis. *J. Invertebr. Pathol. 7*, 49-54.

Kawase, S., and Miyajima, S. (1969). Immunofluorescence studies on the multiplication of cytoplasmic-polyhedrosis virus of the silkworm, *Bombyx mori*. *J. Invertebr. Pathol. 13*, 330-336.

Kawase, S., and Yamaguchi, K. (1974). A polyhedrosis virus forming polyhedra in midgut-cell nucleus of silkworm, *Bombyx mori*. II. Chemical nature of the virion. *J. Invertebr. Pathol. 24*, 106-111.

Kawase, S., Kawamoto, F., and Yamaguchi, K. (1973). Studies on the polyhedrosis virus forming polyhedra in the midgut-cell nucleus of the silkworm, *Bombyx mori*. I. Purification procedure and form of the virion. *J. Invertebr. Pathol. 22*, 266-272.

Kellen, W. R., Clark, T. B., Lindegren, J. E., and Sanders, R. D. (1966). A cytoplasmic-polyhedrosis virus of *Culex tarsalis* (Diptera: Culicidae). *J. Invertebr. Pathol. 8*, 390-394.

Kern, R. (1969). Crystal growth and adsorption. *In* "Growth of Crystals" (N. N. Sheftal', ed.), Vol. 8, pp. 3-23. Plenum, New York.

Kitajima, E. (1936). Insect virus diseases. *Kagoshima Kotonoringakko Koyukai Kaiho (27)*, pp. 5-11.

Kitazawa, T., and Takami, T. (1959). Inoculation of silkworm embryos with the intestinal cytoplasmic polyhedral virus. *J. Sericult. Sci. Japan 28*, 59-64.
Kobayashi, M. (1971). Replication cycle of cytoplasmic-polyhedrosis virus as observed with the electron microscope. *In* "The Cytoplasmic-Polyhedrosis Virus of the Silkworm" (H. Aruga and Y. Tanada, eds.), pp. 103-128. Univ. Tokyo Press, Tokyo.
Kobayashi, M. (1972). Penetration of polyhedrosis viruses into the cultured midgut cells of the silkworm, *Bombyx mori*. *J. Sericult. Sci. Japan 41*, 1-6.
Kodama, R., and Nakasuji, Y. (1969). Pathogenicity of bacteria for silkworm larvae reared aseptically on an artificial diet. *Annu. Report Inst. Fermentation, Osaka 4*, 1-11.
Krieg, A. (1956). Über die Nucleinsäuren der Polyeder-Viren. *Naturwissenschaften 43*, 537.
Krywienczyk, J., and Sohi, S. S. (1973). Serologic characterization of a *Malacosoma disstria* Hübner (Lepidoptera: Lasiocampidae) cell line. *In Vitro 8*, 459-465.
Kurata, K. (1971). On symptoms of silkworm, *Bombyx mori*, infected with a cytoplasmic-polyhedrosis virus under aseptic condition. *J. Sericult. Sci. Japan 40*, 32-36.
Kurisu, K., and Matsumoto, T. (1974). Histopathological observations on the midgut epithelium in the germ-free silkworm larvae infected with flacherie virus. *J. Sericult. Sci. Japan 43*, 283-289.
Lewandowski, L. J., and Traynor, B. L. (1972). Comparison of the structure and polypeptide composition of three double-stranded ribonucleic acid containing viruses (diplornaviruses): cytoplasmic polyhedrosis virus, wound tumor virus, and reovirus. *J. Virol. 10*, 1053-1070.
Lewandowski, L. J., Kalmakoff, J., and Tanada, Y. (1969). Characterization of a ribonucleic acid polymerase activity associated with purified cytoplasmic-polyhedrosis virus of the silkworm, *Bombyx mori*. *J. Virol. 4*, 857-865.
Lipa, J. J. (1977). Electron microscope observations on the development of cytoplasmic polyhedrosis virus in *Scotogramma trifolii* Rott. (Lepidoptera, Noctuidae). *Bull. Acad. Polon. Sci. 25*, 155-158.
Longworth, J. F. (1980). The replication of a cytoplasmic polyhedrosis virus from *Chrysodeixis eriosoma* (Lepidoptera: Noctuiidae) in *Spodoptera frugiperda* cells. *J. Invertebr. Pathol. 37*, 54-61.
Longworth, J. F., and Spilling, C. R. (1970). A cytoplasmic-polyhedrosis of the larch sawfly, *Anoplonyx destructor* (Tenthredinidae: Hymenoptera). *J. Invertebr. Pathol. 15*, 276-280.

Lotmar, R. (1941). Die Polyederkrankheit der Kleidermotte *(Tineoia bisselliella). Mitt. Schweitz. Entomol. Ges. 18*, 372-373.

Magnoler, A. (1974). Effects of a cytoplasmic polyhedrosis on larval and postlarval stages of the gypsy moth, *Porthetria dispar. J. Invertebr. Pathol. 23*, 263-274.

Martignoni, M. E., and Iwai, P. J. (1981). A catalog of viral diseases of insects, mites and ticks. *In* "Microbial Control of Pests and Plant Diseases 1970-1980" (H. D. Burges, ed.), pp. 897-911. Academic Press, New York.

Matthews, R. E. F. (1979). Classification and nomenclature of viruses. *Intervirology 12*, 131-296.

Matthews, R. E. F. (1981). "Plant Virology," 2nd ed. Academic Press, New York.

McCrae, M. A., and Mertens, P. P. C. (1983). *In vitro* translation studies on and RNA coding assignments for cytoplasmic polyhedrosis viruses. *In* "Double Stranded RNA Viruses" (D. H. L. Bishop and R. W. Compans, ed.), pp. 35-41. Elsevier, Amsterdam.

Mitani, K., Kitai, M., and Soda, F. (1958). Study of the midgut polyhedrosis of the silkworm, *Bombyx mori. Bull. Shimane Sericult. Expt. Sta. 36*, 28-75.

Miyagawa, M. (1972). Isolation of large hexahedral polyhedra from larvae affected with cytoplasmic polyhedrosis in sericultural farms. *Bull. Niigata Sericult. Exp. Sta. 12*, 114-118.

Miyagawa, M. (1975a). Effect of rearing temperature on the formation of nuclear inclusion in the midgut of the silkworm, *Bombyx mori*, infected with a cytoplasmic-polyhedrosis virus. *J. Sericult. Sci. Japan 44*, 26-32.

Miyagawa, M. (1975b). Infection site in the midgut epithelium of newly-hatched silkworm larvae affected by cytoplasmic polyhedrosis. *Bull. Niigata Sericult. Exp. Sta. 14*, 139-143.

Miyagawa, M. (1978). Electron microscopic observation on the midgut epithelium at the time of polyhedron and nuclear-inclusion formation in silkworm larva infected with cytoplasmic polyhedrosis. *Bull. Niigata Exp. Sta. 17*, 55-64.

Miyagawa, M., and Sakai, H. (1971). On a cytoplasmic polyhedrosis forming polyhedron-like material in the nuclei of midgut cells. *Bull. Niigata Sericult. Exp. Sta. 11*, 122-129.

Miyagawa, M., and Sakai, H. (1980). On a previously unreported strain of the cytoplasmic-polyhedrosis virus of the silkworm, *Bombyx mori. J. Sericult. Sci. Japan 49*, 169-170.

Miyajima, S. (1975). Changes in virus-infectivity titer in the hemolymph and midgut by peroral infection with cytoplasmic polyhedrosis virus of the silkworm, *Bombyx mori* L. *Res. Bull. Aichi Agr. Res. Ctr. D6*, 19-26.

Miyajima, S. (1976). Serological properties of cytoplasmic-polyhedrosis virus of the silkworm, *Bombyx mori* L. *J. Sericult. Sci. Japan 45*, 245-250.

Miyajima, S. (1982). Refractive index in hemolymph and gut juice of the silkworm infected with some viruses. *J. Sericult. Sci. Japan 51*, 176-181.

Moussa, A. Y. (1980). The housefly virus contains double-stranded RNA. *Virology 106*, 173-176.

Neilson, M. M. (1965). Effects of a cytoplasmic polyhedrosis on adult Lepidoptera. *J. Invertebr. Pathol. 7*, 306-314.

Noguchi, Y., and Yamaguchi, K. (1978). The fate of immune response observed in a virus infection. *Saitama Sericult. Expt. Sta. 50*, 37-38.

Noguchi, Y., and Yamaguchi, K. (1979). An immune response of the fall webworm, *Hyphantria cunea*, to the infection by a cytoplasmic-polyhedrosis virus of the silkworm, *Bombyx mori*. *J. Sericult. Sci. Japan 48*, 15-18.

Noguchi, Y., and Yamaguchi, K. (1982). Development of disease in several species of lepidopterous insects subjected to cross-infection with cytoplasmic-polyhedrosis viruses. *Jap. J. Appl. Entomol. Zool. 26*, 281-287.

Okino, H., and Ishikawa, Y. (1971). Change in the shape of silkworm cytoplasmic polyhedra under high temperature. I. Effect of high temperature on change in the shape of polyhedra. *Res. Bull. Aichi Agr. Res. Ctr. D2*, 65-69.

Paillot, A. (1928). "Les Maladies du Ver à Soie." Éditions du service photographique de L'Université, Lyon.

Payment, P., Arora, D. J. S., and Belloncik, S. (1982). An enzyme-linked immunosorbent assay for the detection of cytoplasmic polyhedrosis virus. *J. Invertebr. Pathol. 40*, 55-60.

Payne, C. C. (1976). Biochemical and serological studies of a cytoplasmic polyhedrosis virus from *Arctia caja*: a naturally-occurring mixture of two virus types. *J. Gen. Virol. 30*, 357-369.

Payne, C. C., and Rivers, C. F. (1976). A provisional classification of cytoplasmic polyhedrosis viruses based on the sizes of the RNA genome segments. *J. Gen. Virol. 33*, 71-85.

Quiot, J. M., and Belloncik, S. (1977). Caracterisation d'une polyédrose cytoplasmique chez le lépidoptère *Euxoa scandens*, Riley (Noctuidae, Agrotinae). Etudes *in vivo* et *in vitro*. *Arch. Virol. 55*, 145-154.

Rao, C. B. J. (1973). Surface topography and shapes of polyhedral inclusion bodies of the cytoplasmic polyhedrosis virus. *J. Ultrastr. Res. 42*, 582-593.

Saito, C., and Yamaguchi, T. (1960). Studies on flacherie of the silkworm, *Bombyx mori* L. I. On the cytoplasmic polyhedrosis. *Bull. Gunma Sericult. Exp. Sta. 36*, 1-79.

Sato, F., and Inoue, H. (1978). Double infection of viruses in the midgut of the silkworm, *Bombyx mori*. *Bull. Sericult. Exp. Sta. 27*, 427-444.

Shorey, H. H. (1963). The biology of *Trichoplusia ni* (Lepidoptera: Noctuidae). II. Factors affecting adult fecundity and longevity. *Ann. Entomol. Soc. Amer. 56*, 476-480.

Sidor, C. (1970). Poliedarno virozno oboljenje topoling krasnika (*Melanophila picta* Pall, Coleoptera, Buprestidae). *Topola 77/78*, 19-21.

Sidor, C. (1971). Virus diseases of some economically important harmful insects. *Topola 83/85*, 64-69.

Sikorowski, P. P., and Thompson, A. C. (1979). Effects of cytoplasmic polyhedrosis virus on diapausing *Heliothis virescens*. *J. Invertebr. Pathol. 33*, 66-70.

Sikorowski, P. P., Andrews, G. L., and Broome, J. R. (1971). Presence of cytoplasmic-polyhedrosis virus in the hemolymph of *Heliothis virescens* larvae and adults. *J. Invertebr. Pathol. 18*, 167-168.

Simmons, C. L., and Sikorowski, P. P. (1973). A laboratory study of the effects of cytoplasmic polyhedrosis virus on *Heliothis virescens* (Lepidoptera: Noctuidae). *J. Invertebr. Pathol. 22*, 369-371.

Smith, K. M. (1963). The cytoplasmic virus diseases. *In* "Insect Pathology. An Advanced Treatise" (E. A. Steinhaus, ed.), Vol. 1, pp. 457-497. Academic Press, New York.

Smith, K. M., and Wyckoff, R. W. G. (1950). Structure within polyhedra associated with insect virus diseases. *Nature (London) 166*, 861-862.

Sohi, S. S., Bird, F. T., and Hayashi, Y. (1971). Development of *Malacosoma disstria* cytoplasmic polyhedrosis virus in *Bombyx mori* ovarian and tracheal tissue cultures. *Proc. 4th Intern. Colloq. Insect Pathol. 1970*, pp. 340-351.

Splittstoesser, C. M., Tashiro, H., Lin, S. L., Steinkraus, K. H., and Fiori, B. J. (1973). Histopathology of the European chafer, *Amphimallon majalis*, infected with *Bacillus popilliae*. *J. Invertebr. Pathol. 22*, 161-167.

Splittstoesser, C. M., Kawanishi, C. Y., and Tashiro, H. (1978). Infection of the European chafer, *Amphimallon majalis*, infected with *Bacillus popilliae*: Light and electron microscope observations. *J. Invertebr. Pathol. 31*, 84-90.

Stairs, G. R., Parrish, W. B., and Allietta, M. (1968). A histopathological study involving a cytoplasmic-polyhedrosis virus of the salt-marsh caterpillar, *Estigmene area*. *J. Invertebr. Pathol. 12*, 359-365.

Steinhaus, E. A., and Martignoni, M. E. (1970). "An Abridged Glossary of Terms Used in Invertebrate Pathology." 2nd ed. Pacif. Northwest Forest Range Exp. Sta. U.S.D.A., Forest Service.

Strickland-Constable, R. F. (1972). The breeding of crystal nuclei--A review of the subject. In "Crystallization from Solution: Nucleation Phenomena in Growing Crystal Systems" (J. Estrin, ed.), pp. 1-7. Amer. Inst. Chem. Eng., New York.

Takami, T., Sugiyama, H., Kitazawa, T., and Kanda, T. (1967). Infection of cultured silkworm embryos with the cytoplasmic-polyhedrosis virus. *Japan. J. Appl. Entomol. Zool. 11*, 182-186.

Takeuchi, A. (1971). Penetration of the intestinal epithelium by various microorganisms. *Current Topics Pathol. 54*, 1-27.

Tanada, Y. and Chang, G. Y. (1960). A cytoplasmic polyhedrosis of the armyworm, *Pseudaletia unipuncta* (Haworth) (Lepidoptera: Noctuidae). *J. Insect Pathol. 2*, 201-208.

Tanada, Y., and Tanabe, A. M. (1964). Response of the adult of the armyworm, *Pseudaletia unipuncta* (Haworth), to cytoplasmic-polyhedrosis-virus infection in the larval stage. *J. Insect Pathol. 6*, 486-490.

Tanaka, S., and Aruga, H. (1967). Interference between the midgut nuclear-polyhedrosis virus and the cytoplasmic-polyhedrosis virus in the silkworm, *Bombyx mori* L. *J. Sericult. Sci. Japan 36*, 169-176.

Tanaka, S., Shimizu, T., and Aruga, H. (1967). Midgut nuclear polyhedrosis forming polyhedra in the nucleus of midgut cell of the silkworm, *Bombyx mori* L. *J. Sericult. Sci. Japan 36*, 1-9.

Tomita, K., and Ebihara, T. (1980). Purification and nature of a midgut-nuclear-polyhedrosis virus of the silkworm, *Bombyx mori* L. *Bull. Ibaraki Sericult. Exp. Sta. 34*, 10-14.

Tomita, K., Ebihara, T., and Iwashita, Y. (1978). Histopathological observations of midgut-nuclear-polyhedrosis virus obtained from sericultural farm. *Bull. Ibaraki Sericult. Exp. Sta. 32*, 28-32.

Vago, C., and Bergoin, M. (1963). Dévelopment des virus a corps d'inclusion du lépidoptère *Lymantria dispar* en cultures cellulaires. *Entomophaga 4*, 253-261.

Vago, C., Croissant, O., and Lepine, P. (1959). Processus de l'infection a virus a partir des corps d'inclusion de "Polyédrie Cytoplasmique" ingèrés par le lépidoptère sensible. *Mikroskopie 14*, 36-40.

Vail, P. V., and Gough, D. (1970). Effects of cytoplasmic-polyhedrosis virus on adult cabbage loopers and their progeny. *J. Invertebr. Pathol. 15*, 398-400.

Vail, P. V., Hall, I. M., and Gough, D. (1969). Influence of a cytoplasmic polyhedrosis on various developmental stages of the cabbage looper. *J. Invertebr. Pathol. 14*, 237-244.

Watanabe, H. (1968). Pathogenic changes in the feces from the silkworm, *Bombyx mori* L., after peroral inoculation with a cytoplasmic polyhedrosis virus. *J. Sericult. Sci. Japan 37*, 385-389.

Watanabe, H. (1971a). Pathophysiology of cytoplasmic polyhedrosis in the silkworm. *In* "The Cytoplasmic-Polyhedrosis Virus of the Silkworm" (H. Aruga and Y. Tanada, eds.), pp. 151-167. University of Tokyo Press, Tokyo.

Watanabe, H. (1971b). Pathophysiology of nitrogen catabolism in the midgut of silkworm, *Bombyx mori* L. (Lepidoptera: Bombycidae), infected with a cytoplasmic-polyhedrosis virus. *Appl. Entomol. Zool. 6*, 163-168.

Watanabe, H. (1974). Electron-microscope investigation on dissolution of polyhedra in the gut juice of the silkworm, *Bombyx mori* L. *J. Sericult. Sci. Japan 43*, 29-34.

Watanabe, H., and Namura, H. (1971). Fecundity of adult of the silkworm, *Bombyx mori*, infected with a cytoplasmic-polyhedrosis virus in the pupal stage. *J. Sericult. Sci. Japan 40*, 145-146.

Watanabe, H., Aruga, H., and Tanaka, S. (1967). Autoradiographic studies on the nucleic-acid synthesis in the midgut of the silkworm, *Bombyx mori* L., infected with midgut-nuclear-polyhedrosis virus. *J. Sericult. Sci. Japan 37*, 34-42.

Weiser, J. (1978). A new host, *Simulium argyreatum* Meig., for the cytoplasmic polyhedrosis virus of the blackflies in Czechoslovakia. *Folia Parasitol. 25*, 361-365.

Wiygul, G., and Sikorowski, P. P. (1978). Oxygen uptake in tabacco budworm larvae (*Heliothis virescens*) infected with cytoplasmic polyhedrosis virus. *J. Invertebr. Pathol. 32*, 191-195.

Wood, H. A. (1980). Isolation and replication of an occlusion body-deficient mutant of the *Autographa californica* nuclear polyhedrosis virus. *Virology 105*, 338-344.

Xeros, N. (1952). Cytoplasmic polyhedral virus diseases. *Nature (London) 170*, 1073.

Xeros, N. (1966). Light microscopy of the virogenic stromata of cytopolyhedroses. *J. Invertebr. Pathol. 8*, 79-87.

Yamaguchi, S. (1962). Studies on the functional localization of digestive system in the silkworm larva, *Bombyx mori* L. VI. Susceptibility and its local difference of the midgut epithelial cell for virus multiplication. *J. Sericult. Sci. Japan 31*, 90-96.

Yamaguchi, K. (1968a). Studies on the midgut-nuclear polyhedrosis in the silkworm, *Bombyx mori* L. I. The formation site and some nature of the polyhedra. *J. Sericult. Sci. Japan 37*, 34-42.

Yamaguchi, K. (1968b). Studies on the midgut-nuclear polyhedrosis in the silkworm, *Bombyx mori* L. II. Temperature effects on the formation of polyhedra. *J. Sericult. Sci. Japan 37*, 462-470.

Yamaguchi, K. (1970). Studies on the midgut-nuclear polyhedrosis of the silkworm, *Bombyx mori* L. (V) A previously undescribed strain, C_2. *J. Sericult. Sci. Japan 39*, 363-370.

Yamaguchi, K. (1972a). On variant strains, B_1 and B_2, of the midgut-nuclear-polyhedrosis virus of the silkworm. *Konchu-Byori-Danwakai Kaiho 21*, 11.

Yamaguchi, K. (1972b). Studies on the midgut-nuclear polyhedrosis of the silkworm, *Bombyx mori* L. (VII) Electron microscopic observation on the inclusion formation of C_2 strain. *Bull. Saitama Sericult. Exp. Sta. 44*, 45-51.

Yamaguchi, K. (1973). Studies on the midgut-nuclear polyhedrosis of the silkworm, *Bombyx mori* L. (VIII) A previously undescribed virus strain, B_1. *J. Sericult. Sci. Japan 42*, 74-78.

Yamaguchi, K. (1974). Studies on the interference between viruses in the silkworm, *Bombyx mori* L. I. Interference between typical and new strains of the cytoplasmic-polyhedrosis virus. *Bull. Saitama Sericult. Exp. Sta. 46*, 78-83.

Yamaguchi, K. (1975a). Studies on the interference between viruses in the silkworm, *Bombyx mori* L. II. Interference between B, C_1 and C_2 strains of the cytoplasmic-polyhedrosis virus. *Bull. Saitama Sericult. Exp. Sta. 47*, 59-61.

Yamaguchi, K. (1975b). Studies on the interference between viruses in the silkworm, *Bombyx mori* L. III. Interference between A and C_1 strains of the cytoplasmic-polyhedrosis virus. *J. Sericult. Sci. Japan 44*, 468-471.

Yamaguchi, K. (1976a). Natural cure of the fall webworm, *Hyphantria cunea*, infected with the cytoplasmic polyhedrosis virus of the silkworm, *Bombyx mori*. *J. Sericult. Sci. Japan 45*, 60-65.

Yamaguchi, K. (1976b). Resistance of the fall webworm, *Hyphantria cunea*, to the cytoplasmic polyhedrosis virus of the silkworm, *Bombyx mori*. *J. Sericult. Sci. Japan 45*, 377-378.

Yamaguchi, K. (1977). Regeneration of the midgut epithelial cells in the silkworm, *Bombyx mori*, infected with the cytoplasmic polyhedrosis virus. *J. Sericult. Sci. Japan 46*, 179-180.

Yamaguchi, K. (1979). Natural recovery of the fall webworm, *Hyphantria cunea*, to infection by a cytoplasmic-polyhedrosis virus of the silkworm, *Bombyx mori*. *J. Invertebr. Pathol. 33*, 126-128.

Yamaguchi, K., and Ayuzawa, C. (1970). Studies on the midgut-nuclear polyhedrosis of the midgut-nuclear polyhedrosis of the silkworm, *Bombyx mori* L. (IV) Two previously undescribed virus strains, B and C_1. *J. Sericult. Sci. Japan 39*, 342-350.

Yamaguchi, K., and Hukuhara, T. (1973a). Studies on the midgut-nuclear polyhedrosis of the silkworm, *Bombyx mori* L. (IX) New virus strain, B_2. *J. Sericult. Sci. Japan 42*, 239-243.

Yamaguchi, K., and Hukuhara, T. (1973b). Studies on the midgut-nuclear polyhedrosis of the silkworm, *Bombyx mori* L. (X) Electron microscopic observation on the inclusion formation of C_1 strain. *Bull. Saitama Sericult. Exp. Sta. 45*, 61-62.

Yamaguchi, K., Iwashita, Y., and Inoue, K. (1969). On the midgut nuclear polyhedrosis in the silkworm, *Bombyx mori* L. III. Effects of high temperature treatment on the shape of polyhedron of the infected larvae. *J. Sericult. Sci. Japan 38*, 157-162.

Yamamasu, Y., and Kawakita, T. (1962). Studies on the grasserie of the silk producing insects. V. On the polyhedrosis of silkworm, *Bombyx mori*. *Bull. Fac. Sci. Fibers, Kyoto Unit. Ind. Arts Text. Febers 3*, 444-467.

PATHOBIOLOGY OF INVERTEBRATE ICOSAHEDRAL CYTOPLASMIC DEOXYRIBOVIRUSES (IRIDOVIRIDAE)

DONALD W. HALL

Department of Entomology and Nematology
University of Florida
Gainesville, Florida

I. INTRODUCTION

In 1954 Xeros reported a new virus disease of the crane fly *Tipula paludosa* in which the fat body of infected individuals appeared iridescent purple when viewed through the intact integument. The virus was shown to contain DNA and was located in the cytoplasm of the infected tissues. When partially purified and pelleted, the virus exhibited the same iridescence as the infected tissue, and it was later given the name *Tipula* iridescent virus (TIV).

Since the discovery of TIV, other viruses with similar characteristics have been reported. Most of these have been reported from insects, but a few have been reported from other invertebrates. In addition, a number of similar viruses (African swine fever virus, lymphocystis virus of fish, and an array of viruses of amphibians) have been reported from vertebrates. Both the invertebrate and vertebrate viruses of this group have been placed into the family Iridoviridae by the International Committee on Taxonomy of Viruses (ICTV), and the subfamily name Invertebrate Icosahedral Cytoplasmic Deoxyribovirus (IICDV) has been proposed for the viruses of invertebrates (Matthews, 1982). This chapter will review the

ISBN 0-12-470295-3

TABLE I. Iridescent Virus Isolates Assigned Type Numbers[a]

Type	Host	References
1	*Tipula paludosa* (Diptera: Tipulidae)	Xeros (1954)
2	*Sericesthis pruinosa* (Coleoptera: Scarabaeidae)	Steinhaus and Leutenegger (1963)
3	*Aedes taeniorhynchus* (Diptera: Culicidae)	Matta and Lowe (1970)
4	*Aedes cantans* (Diptera: Culicidae)	Weiser (1965)
5	*Aedes annulipes* (Diptera: Culicidae)	Weiser (1965)
6	*Chilo suppressalis* (Lepidoptera: Pyralidae)	Fukaya and Nasu (1966)
7	*Simulium ornatum* (Diptera: Simuliidae)	Weiser (1968)
8	*Culicoides* sp. (Diptera: Ceratopogonidae)	Chapman et al. (1968)
9	*Wiseana cervinata* (Lepidoptera: Hepialidae)	Fowler and Robertson (1972)
10	*Witlesia sabulosella* (Lepidoptera: Pyralidae)	Fowler and Robertson (1972)
11	*Aedes stimulans* (Diptera: Culicidae)	Anderson (1970)
12	*Aedes cantans* (Diptera: Culicidae)	Tinsley et al. (1971)
13	*Corethrella brakeleyi* (Diptera: Chaoboridae)	Chapman et al. (1971)
14	*Aedes detritus* (Diptera: Culicidae)	Hasan et al. (1971)
15	*Aedes detritus* (Diptera: Culicidae)	Vago et al. (1969)
16	*Costelytra zealandica* (Coleoptera: Scarabaeidae)	Kalmakoff et al. (1972)

17	*Pterostichus mandidus* (Coleoptera: Carabidae)	J. S. Robertson (unpublished)
18	*Opogonia* (Coleoptera: Scarabaeidae)	Kelly and Avery (1974)
19	*Odontria striata* (Coleoptera: Scarabaeidae)	J. Kalmakoff (unpublished)
20	*Simocephalus expinosus* (Crustacea: Cladocera)	Federici and Hazard (1975)
21	*Heliothis armigera* (Lepidoptera: Noctuidae)	Carey et al. (1978)
22	*Simulium* sp. (Diptera: Sumuliidae)	Batson et al. (1976)
23	*Heteronychus arator* (Coleoptera: Scarabaeidae)	Longworth et al. (1979)
24	*Apis cerana* (Hymenoptera: Apidae)	Bailey et al. (1976)
25	*Tipula* sp. (Diptera: Tipulidae)	Elliott et al. (1977)
26	*Ephemeropteran* (Insecta: Ephemeroptera)	B. Federici (unpublished)
27	*Nereis diversicolor* (Annelida: Polychaeta)	Devauchelle and Durchon (1973a)
28	*Lethocerus columbiae* (Hemiptera: Belostomatidae)	Carey et al. (1978)
29	*Tenebrio molitor* (Coleoptera: Tenebrionidae)	Black et al. (1981)
30	?	
31	*Armadillidium vulgare* (Crustacea: Isopoda)	Federici (1980)
32	*Porcellio dilatatus* (Crustacea: Isopoda)	Federici (1980)

[a] *Modified and updated from Tinsley and Harrap (1978). Used with permission of the authors and Plenum Publishing Company.*

TABLE II. Iridescent Virus Isolates Not Assigned Type Numbers

Invertebrate host	Reference
Aedes fulvus pallens (Diptera: Culicidae)	Chapman et al. (1966)
Aedes dorsalis (Diptera: Culicidae)	Chapman et al. (1966)
Aedes vexans (T type) (Diptera: Culicidae)	Chapman et al. (1966)
Aedes vexans (R type) (Diptera: Culicidae)	Hall and Anthony (1976)
Aedes caspius caspius (R type) (Diptera: Culicidae)	Torybaev (1970)
Aedes cantans (R type) (Diptera: Culicidae)	Popelkova (1982)
Aedes sticticus (Diptera: Culicidae)	Chapman et al. (1969)
Psorophora ferox (Diptera: Culicidae)	Chapman et al. (1969)
Psorophora varipes (Diptera: Culicidae)	Chapman et al. (1969)
Psorophora horrida (Diptera: Culicidae)	Chapman et al. (1969)
Culiseta annulata (Diptera: Culicidae)	Buchatsky (1977)
Culex territans (Diptera: Culicidae)	Buchatsky (1977)
Chironomus plumosus (Diptera: Chironomidae)	Stoltz et al. (1968)

Simulium rubicundulum (Diptera: Simuliidae)	Takaoka (1980)
Simulium earlei (Diptera: Simuliidae)	Takaoka (1980)
Simulium callidum (Diptera: Simuliidae)	Takaoka (1980)
Prosimulium sp. (Diptera: Simuliidae)	Avery and Bauer (1983)
Phyllophaga anxia (Coleoptera: Scarabaeidae)	Cole and Morris (1980)
Formica lugubris (Hymenoptera: Formicidae)	Steiger et al. (1969)
Tenebrio molitor (Coleoptera: Tenebrionidae)	Thomas and Gouranton (1978)
Porcellio scaber (Crustacea: Isopoda)	Cole and Morris (1980)
Armadillidium vulgare (Crustacea: Isopoda)	Cole and Morris (1980)
Octopus vulgaris Gastropoda)	Runnger et al. (1971)

characteristics and pathobiology of the IICDVs, and for simplicity they will be referred to as iridescent viruses even though not all exhibit iridescence.

II. CLASSIFICATION AND LIST OF VIRUSES

Initially, as new iridescent viruses were reported they were given names composed of either the genus or common name of the host plus the name iridescent virus (IV) [e.g., *Sericesthis* iridescent virus (SIV) and mosquito iridescent virus (MIV)]. This system became confusing as new viruses were reported from hosts of the same species or genus as previously described viruses. To avoid this problem, Tinsley and Kelly (1970) proposed an interim classification system in which each new virus would be assigned a type number. If any two virus types were later shown to be the same virus, the earlier type designation would have priority and the additional host record would be noted for it. This system has been adopted and many of the isolates of iridescent viruses have now been assigned numbers (Table I). The individual viruses will henceforth be referred to as "IV" plus the type number designation for the particular virus. A number of viruses have not yet been assigned type designations (Table II) and will be referred to according to their respective hosts.

The ICTV has recently revised the generic classification of the iridescent viruses (Matthews, 1982). The original genus name *Iridovirus* has been reserved for the small iridescent viruses (ca. 130 nm) which includes IV's 1, 2, 6, 9, 10, 16-29, probably *Chironomous* iridescent virus (which does not iridesce), and possibly *Octopus vulgaris* virus. The type species for the genus *Iridovirus* is *Tipula* iridescent virus (IV-1). The genus name *Chloriridovirus* has been assigned to the large iridescent viruses (ca. 180 nm) and includes IV's 3-5, 7, 8, and 11-15. The type species of the genus *Chloriridovirus* is the regular (R) strain of mosquito iridescent virus from *Aedes taeniorhynchus* (IV-3). The name *Chloriridovirus* is somewhat of a misnomer since it is the T strain of IV-3 which produces green iridescence rather than the R strain. There seems to be some validity to this generic grouping since a number (but not all) of the IVs assigned to the genus *Iridovirus* have been shown to be serologically related to each other and some have been shown not to be serologically related to IV-3 (Kelly et al., 1979; Cunningham and Tinsley, 1968; Tinsley and Harrap, 1978); IV-3 and IV-12 are serologically related (Tinsley et al., 1971). However, size alone does not appear to be a completely satisfactory criterion for generic placement, as has been pointed out by Kelly and Robertson (1973). The IV-8 and IV-13 which are currently included in the genus *Chloriridovirus* are both approximately 130 nm in size; IV-7 is reported to be 140-160 nm whereas IV-27, which has been assigned to the genus *Iridovirus*, is reported to be 160-180 nm (Devauchelle, 1977). Further serological comparisons are desirable to clarify the placement of some of the IV isolates.

III. ULTRASTRUCTURE

The IV-1, -2, and -27 virions have been shown to be icosahedral by electron microscopic analysis of the shadows cast when freeze-dried virions were double-shadowed (in two directions differing by 60° in azimuth) (Williams and Smith, 1958; Steinhaus and Lentenegger, 1963; Devauchelle, 1977). The icosahedral nature of the other IVs is inferred from the profiles of virions in ultrathin sections.

The iridescent virus virion is composed of an electron-dense central core consisting of deoxyribonucleoprotein, which is surrounded by a unit membrane (Stoltz, 1973). The unit membrane is in turn surrounded by an icosahedral capsid composed of subunits (Stoltz, 1973). Wrigley (1969, 1970) found that when virions of IV-1 and IV-2 were treated with the nasal decongestant Afrin and negatively stained, morphological subunits of the capsids could be observed. He also found that by storing the viruses at 4°C in distilled water for long periods of time the capsids broke up into fragments including triangular (trisymmetrons), pentagonal (pentasymmetrons), and linear fragments (disymmetrons). Disymmetrons were observed only with IV-2. An analysis of these fragments suggested that the surface of IV-1 is composed of 1472 subunits and that IV-2 is composed of 1562 subunits, although the evidence for the 1562 figure is somewhat weaker, and 1472- or 1292-subunit models could not be excluded for IV-2 or could 1562- or 1292-subunit models be excluded for IV-1. For IV-1, Wrigley (1970) concluded that the icosahedral capsid is composed of 20 trisymmetrons (each composed of 55 hexagonally packed subunits) and 12 pentasymmetrons (each composed of 31 subunits) which comprise the vertices. In some micrographs, trisymmetrons were found with their edges touching each other. These fragments were not lined-up corner to corner but were usually out of register with each other by three subunits, which suggested that the trisymmetrons were skewed about the pentasymmetrons. The conclusions of Wrigley concerning IV-1 have been confirmed by Manyakov (1977) who disrupted the virions by storage at 4°C or by treatment with chloroform and then purified the fragments in sucrose density gradients. Manyakov actually observed a pentasymmetron with five neighboring trisymmetrons, each out of register with its neighbors.

Stoltz (1973) studied trisymmetrons of IV-3 (T strain) and the IV from *Chironomus* (Stoltz et al., 1968). These viruses are larger than IV-1 and IV-2. The subunits were more difficult to resolve in these viruses, but trisymmetrons of *Chironomous* IV were estimated to have 12 subunits per edge, and those of the T strain of IV-3 were estimated to have either 13 or 14 subunits per edge with the latter figure being most likely. Further studies of the capsid subunit structures of the turquoise and regular strains of IV-3 should be interesting since the regular strain of IV-3 is 35 nm larger than the turquoise strain, yet the proteins of the two viruses have been shown to be nearly identical serologically (Hall and Lowe, 1972; Wagner et al., 1974a) and by polyacrylamide gel electrophoresis, amino acid, and tryptic peptide analysis (Wagner et al., 1974a).

Chironomus IV has long fibrils attached to the capsid subunits giving the virions a "whiskered" appearance (Stoltz et al., 1968). These fibrils may be responsible for the absence of iridescence for this virus. Smaller structures which may be homologous have been observed in IV's 1, 3, 6, 29, and isopod iridescent virus (Stoltz, 1973; Willison and Cocking, 1972; Cole and Morris, 1980; Black et al., 1981; Guerillon et al., 1982) and will likely be found in other IV's upon closer examination.

IV. MECHANISM OF IRIDESCENCE

The iridescence of infected tissues and pellets of purified virus is due to constructive interference (Bragg reflections) of visible wavelengths of light by paracrystalline arrays of virions (Fig. 1), and the wavelength reflected is a function of the distance between successive reflecting planes (interparticle distance) in the crystals. Klug et al. (1959) conducted a crystallographic study of IV-1 crystals with monochromatic light and concluded that the virions are packed in face-centered cubic arrays with an interparticle spacing of 250 nm, a distance nearly twice the diameter of the dehydrated virions.

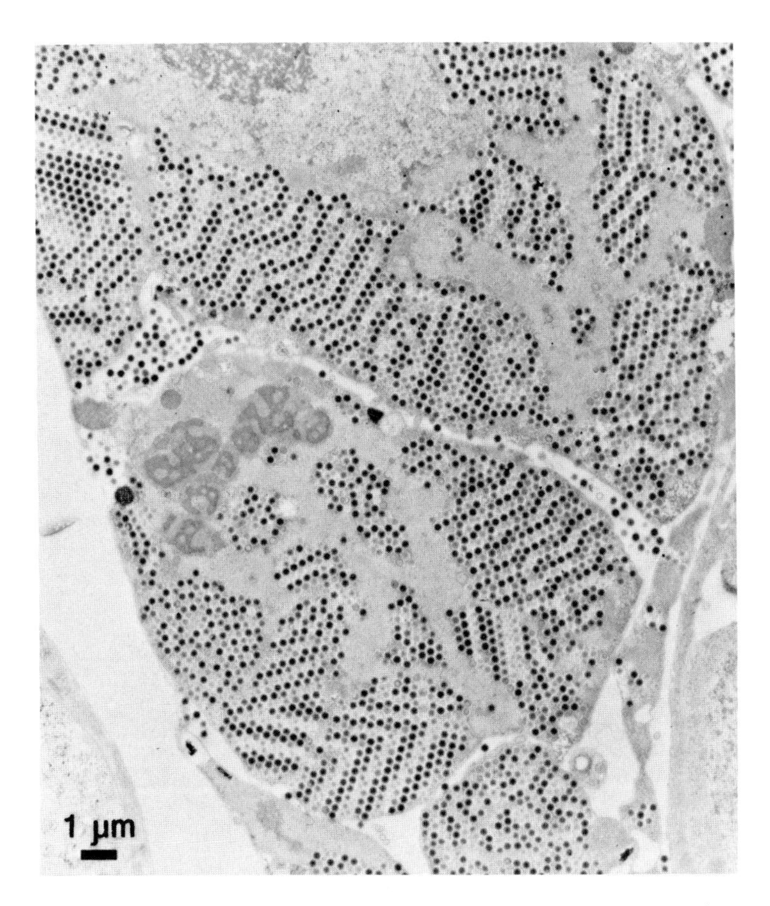

FIGURE 1. Paracrystalline arrays of iridescent virus type 3 in fat body cells of Aedes taeniorhynchus.

V. PHYSIOCHEMICAL CHARACTERISTICS

Prior to consideration of any group of pathogens for biological control, it is important to establish a taxonomic foundation. For the viruses this involves physical and chemical characterization as well as the use of certain approaches such as serological and nucleic acid homology techniques which yield results with more direct phylogenetic implications. There are a number of excellent reviews on the physicochemical characteristics of the invertebrate iridescent viruses (Bellett, 1968; Goorha and Granoff, 1979; Kelly and Robertson,

TABLE III. *Physicochemical Properties of Selected Invertebrate Iridescent Viruses*

	Property	Iridescent virus type			
		1	2	3[q]	6
Whole particle	Sedimentation constant	2200[e]	2200[e]	4460[k] 4440[o]	3300[m]
	Weight ($\times 10^9$) (daltons)	1.28[e]	1.28[e]	2.49[k] 2.75[o]	--
	Diffusion coefficient (cm^2/sec)	0.16×10^{-7}[e]	0.16×10^{-7}[e]	1.49×10^{-8}[k] 1.23×10^{-7}[o]	--
	Optical Density (260 nm/280 nm)	--	1.28[d]	1.2[o]	1.22[m]
	Density	1.33[e]	1.33[e]	1.35[k] 1.32[o]	--
DNA	Percentage	17[e] 19[f]	17[e]	16[k] 17[o]	11[m]
	Molecular wt. ($\times 10^6$)	126[c]	134[b,g]	243[n]	135[c]
	G + C (%)	31.7[c,1]	31[c]	53.9[n]	28.7[c]
Lipid	Percentage	5.2[1]	9[i]	3.9[o]	9[i]

Proteins	Number of polypeptides (polyacrylamide gel) electrophoresis	28[j]	20[h]	9[f] 4[f]	16[a] 26[a] 19[h]
	Range of molecular weights ($\times 10^3$)	17.5 - 300[j]	11 - 213[h]	15 - 98[p] 15 - 97[p]	18 - 115[a] 10 - 230[a] 10 - 213[h]

[a]Barry and Devauchelle (1979)
[b]Bellett (1968)
[c]Bellett and Inman (1967)
[d]Day and Mercer (1964)
[e]Glitz et al. (1968)
[f]Kalmakoff and Tremaine (1968)
[g]Kelly and Avery (1974)
[h]Kelly and Tinsley (1972)
[i]Kelly and Vance (1973)
[j]Krell and Lee (1974)
[k]Matta (1970)
[l]Thomas (1961)
[m]Tojo and Kodama (1968)
[n]Wagner and Paschke (1977)
[o]Wagner et al. (1973)
[p]Wagner et al. (1974a)
[q]Values for IV-3 are for the R strain.

1973; Tinsley and Harrap, 1978) and only a brief summary and update will be given here.

There is considerable variation in some of the published values of certain physicochemical parameters for the same virus. This is most often due to the use of different techniques for measuring these parameters. The most striking example of this variation is the range of published values for the particle sizes of IVs. Much of the variation can be attributed to the degree of hydration of the virions or whether the virions are oriented in the two-, three-, or fivefold axes of symmetry (Wrigley, 1969; Stoltz, 1971) when measurements are made with the electron microscope of freeze-dried, negatively stained, or sectioned virions. Wagner et al. (1973) found that diameters (fivefold axis) of negatively stained virions of the R and T strains of IV-3 were 37 nm and 13 nm larger, respectively, than virions in tissue sections. DeBlois et al. (1978) reported values for IV-1 of 170 nm (negatively stained virions), 180 nm (resistive pulse technique), and 212 nm (laser light-scattering spectroscopy) compared to the earlier published values of 130 nm (Williams and Smith, 1958). If size is to be used as a major criterion for generic classification, a standard method for size determination is desirable. Standardization of techniques is also desirable for measurement of other physicochemical properties.

Detailed physicochemical information is lacking for most of the IVs. A summary of values for some of the better studied viruses is given in Table III. The reader is referred to the original research papers for the techniques involved. In addition, limited information is available for IV-18 (Kelly and Avery, 1974), IV-22 (Hibbin and Kelly, 1981), IV-29 (Black et al., 1981), and IV-31 (Cole and Morris, 1980).

With the exception of IV-3 (R strain), the DNA of the invertebrate iridescent viruses is in the form of single linear duplexes. Redundant sequences have been demonstrated in the DNA of IV-2 and IV-6 (Kelly and Avery, 1974) and IV-3 (R and T strains) (Wagner and Paschke, 1977). Wagner and Paschke (1977) suggested that the R strain of IV-3 contains two identical duplex molecules while the T strain contains a single duplex molecule and that the two duplexes of the R strain are 15% smaller than the single duplex of the T strain. They concluded that the DNA of the two strains contain identical sequences and that the portion of the T strain genome accounting for the higher molecular weight is composed of repeated sequences common to both strains. These results suggested that the T strain is a mutant of the R strain which arose as a result of a recombination event between the two R strain genomes resulting in the larger single genome of the T strain. The T strain appeared spontaneously in a single larva during laboratory studies with the R strain (Woodard and Chapman, 1968) and has never been collected from the field.

At least some of the variation in numbers of polypeptides reported for the different viruses in Table III is due to differing methods of solubilization and maintenance of solubility during electrophoresis. Direct comparisons of the polypeptides of different viruses in the same laboratory under the same conditions like that of Kelly and Tinsley (1972) for IV-2 and IV-6 would be desirable. In addition to structural polypeptides, a number of virion-associated enzymes have been found in iridescent viruses including DNA-dependent RNA polymerases from IV's 2 and 6 (Kelly and Tinsley, 1973), a nucleotide phosphohydrolase from IV-6 (Monnier and Devauchelle, 1976) and protein kinases from IV-6 (Monnier and Devauchelle, 1980), IV-22, and IV-23 (Kelly et al., 1980). The advantage of a virion-associated RNA polymerase to a cytoplasmic DNA virus is obvious. The functions of the other enzymes are not well understood. Monnier and Devauchelle (1980) have suggested that the protein kinases may regulate the activity of a viral enzyme during infection or replication or they may shut down host cell macromolecular synthesis by phosphorylation of host proteins. The function of the nucleotide phosphohydrolase activity is unknown.

VI. RELATIONSHIPS BETWEEN IRIDESCENT VIRUSES

Due to the difficulty in obtaining large quantities of most of the large iridescent viruses, few serological studies have been done with this group. The R and T strains of IV-3 are indistinguishable by gel immunodiffusion (Hall and Lowe, 1972; Wagner et al., 1973). IV-3 and IV-12 are serologically related, but IV-3 shares no common antigens with IV's 1 or 2 (Cunningham and Tinsley, 1968); IV-12 is not related to IV's 1, 2, 6, or 9 (Tinsley et al., 1971).

The small iridescent viruses have been more thoroughly studied, but much remains to be done. Most of the small iridescent viruses (1, 2, 9, 10, 16, 18, 21, 22, 23, 24, 25) are serologically related to some extent (Cunningham and Tinsley, 1968; Glitz et al., 1968; Fowler and Robertson, 1972; Kalmakoff et al., 1972; Kelly and Robertson, 1973; Bailey et al., 1976; Batson et al., 1976; Elliott et al., 1977; Carey et al., 1978; Kelly et al., 1979). IV-29 is related to some, but not all, members of the above serogroup (Kelly et al., 1979), and IV-6 does not appear to be related to any of the small viruses with which it has been compared (Carey et al., 1978; Kelly et al., 1979), with the possible exception of IV-24 (Bailey et al., 1976). Iridescent virus 31, a small virus from a southern California terrestrial isopod is not related to either IV-1 or IV-6 (M. Ohba, personal communication). However, another small iridescent virus isolate from nothern California isopods appears to be distantly related to IV-1 (Cole and Morris, 1980).

Comprehensive testing to compare all isolates of small iridescent viruses is needed. The enzyme-linked immunosorbent assay appears to be the method of choice for future comparisons (Kelly et al., 1978) as it is both sensitive and quantitative. For identification, it is not unreasonable to expect that type-specific antibodies may be produced in the future by gel resection (Krøll, 1981) or monoclonal antibody techniques (Galfre and Milstein, 1981; Zola and Brooks, 1982). Although monoclonal antibodies have the disadvantage of being more expensive and technically more difficult to produce initially, unlimited quantities of completely uniform antibodies can be produced once the hybridoma lines are established.

Four of the small iridescent viruses (types 2, 6, 9, and 18) have also been compared by DNA-DNA hybridization (Kelly and Avery, 1974). Iridescent viruses 9 and 18 were completely homologous. The IV-9 genome was 45% homologous to the genome of IV-2, but in the reciprocal experiment only 25% of the genome of IV-2 was homologous to the genomes of IV's 9 and 18. IV-6 contained no sequences common to IV's 2, 9, and 18. These results support those from serological studies (Kelly et al., 1979).

VII. HOST RANGE

A number of the small iridescent viruses have wide host ranges when injected, but much narrower ones when fed to prospective hosts. Ohba (1975) listed the following experimental hosts for IV-6 (by per os inoculation and/or intrahemocoelic injection): Lepidoptera, 65 species (26 families); Coleoptera, 8 species (7 families); Hymenoptera, 2 species (2 families); Diptera, 2 species (2 families); and Orthoptera, 2 species (1 family). He also noted that many species of the orders Lepidoptera, Orthoptera, Blattaria, Isoptera, Neuroptera, Hemiptera, and Dermaptera are not susceptible. IV-6 has also been transmitted to leafhoppers (Mitsuhashi, 1967; Jensen et al., 1972) and to 13 species of mosquitoes (Fukuda, 1971). In contrast, IV-3 (a large iridescent virus) appears to have a narrow host range. Woodard and Chapman (1968) were able to infect only a single additional mosquito species out of 9 with this virus. The statements by Smith (1976) that larvae of *Corethrella brakeleyi* are voracious predators of first instar mosquito larvae and easily pick up the iridescent virus from infected mosquitoes and that the virus was readily transmissible to both *C. brakeleyi* and *Corethrella appendiculata*, are misinterpretations of statements of Chapman et al. (1971). These workers were reporting on a new virus of *C. brakeleyi* (IV-13) and stated that iridescent viral infections were never observed in larvae of mosquitoes from the same habitat. Furthermore, in the laboratory, they were never able to infect mosquito larvae with this virus.

There are a number of instances of iridescent virus cross-infections in the field. Field-collected isopods of two species have been reported to harbor the same iridescent virus (Cole and Morris, 1980). This virus has also been found to replicate in a mermithid nematode parasite of the isopods (Poinar et al., 1980; Poinar, 1981). The relationship of this isopod virus to two other isopod isolates (IV's 31 and 32) is presently unknown. Moore et al. (1974a) reported IV-16 infections in field-collected larvae of three species of beetles representing two families.

VIII. PATHOLOGY

A. *SYMPTOMS AND SIGNS*

The characteristic iridescence of infected tissues usually becomes evident during the last larval stadium. Iridescent virus infections in insects with thin translucent integuments can usually be diagnosed by examination of the intact animal. The iridescence is often enhanced by viewing the infected individuals against a black background. Most iridescent viruses produce iridescences of green, blue or purple, with the color produced by any given virus varying somewhat according to the intensity of the infection.

Most of the iridescent viruses do not noticeably affect the activity of patently infected hosts immediately, but shortly before death the hosts become moribund. Death usually occurs near the end of the last larval stadium or occasionally during the pupal stage. IV-29 does not produce iridescence in its host, *Tenebrio molitor*, until the pupal or adult stage (Black et al., 1981) and IV-24 produces iridescence in *Apis cerana* during the adult stage. The adult *A. cerana* exhibit clustering and crawling symptoms and eventually paralysis (Bailey and Ball, 1978; Singh, 1979) in the terminal stages of IV-24 infection.

B. *REPLICATION AND CYTOPATHOLOGY*

The events of the replicative cycle of IV's are linked to the pathology produced by these viruses. Most research on IV replication has utilized cell cultures, but the events which take place in cell cultures probably do not differ markedly from those in tissues, with the possible exception of the mechanisms of viral release. The major features of replication are similar for the IV's which have been studied, although the time sequences vary somewhat.

Penetration of virus occurs by viropexis (Kelly and Tinsley, 1974a; Mathieson and Lee, 1981; Webb et al., 1976), although the possibility that some virus also penetrates directly has not been ruled out. Webb et al. (1976) reported that IV-3 is released from the phagocytic vesicles within 60 min postinfection. Iridescent virus type 1 in *Estigmene acrea* cells (Mathieson and Lee, 1981) and types 2 and 6 in *Aedes aegypti* and *Antherea eucalypti* cells (Kelly and Tinsley, 1974a) could be observed in phagocytic vacuoles 6-8 hr postinfection. The IV-1 particles were reported to undergo uncoating in the phagocytic vesicles, and although IV-2 and IV-6 virions were observed both free in the cytoplasm and in vesicles up to 8 hr after infection, they were believed to be uncoated while free in the cytoplasm.

Rapid inhibition (within 1 hr) of host cell macromolecular synthesis has been reported in cells infected with IV-1 and IV-6 (Krell and Lee, 1974; Cerutti and Devauchelle, 1980) allowing diversion of cell resources to production of virus. This inhibition is not dependent on the integrity of the viral genome and in the case of IV-6 occurs in nonpermissive invertebrate and vertebrate cell lines as well as permissive invertebrate lines. The rate of inhibition by IV-6 is dependent on the multiplicity of infection. As mentioned earlier in this chapter (Section V), Monnier and Devauchelle (1980) have suggested that inhibition of host cell macromolecular synthesis may result from phosphorylation of host proteins by iridovirus-associated protein kinases.

Large inocula of IV-6 also produce rapid cytotoxic effects in *A. eucalypti* cell cultures (M. Ohba, personal communication) and *Galleria mellonella* hemocytes *in vivo* (Ohba and Aizawa, 1979). IV-6 also produces lethal toxic symptoms in mice (Ohba and Aizawa, 1982), while IV's 6 and 31 produce lethal toxicity in the frog, *Rana limnocharis* (Ohba and Aizawa, 1981), when injected intraperitoneally. This toxic activity is destroyed by heat (60°C for 20 min) and reaction with homologous antiserum but not by ultraviolet irradiation, which suggests that the toxic factor is a virus-associated enzyme or structural protein. Toxic effects from feeding insects large quantities of virus have not been reported. Therefore, it appears that either the toxic factor is inactivated by gut enzymes or that the gut epithelial cells are resistant. The relationship, if any, of these toxic effects to the presence of protein kinases or to inhibition of host cell macromolecular synthesis is unknown.

Uncoating of virions is followed by formation of viroplasmic centers in the cytoplasm of infected cells. Early replication events are often accompanied by extensive changes in cell size and shape. Infection by IV's 2 and 6 also result in loss of adhesion in *Aedes aegypti* cells (Kelly and Tinsley, 1974a).

IV-6 induces syncytia formation in a wide range of invertebrate cells and also some vertebrate cells (Cerutti and Devauchelle, 1979). Cell fusion occurs in invertebrate cell lines at high multiplicity of infection within 2 to 4 hr postinfection. Syncytia-stimulating capacity of the virus is destroyed by antiserum and heat treatment (55°C, 3 hr) of virus but not by ultraviolet irradiation. IV-1 has also been shown to induce formation of syncytia, but cell fusion in this case rarely occurs before 7 hr postinfection (Mathieson and Lee, 1981). Based on known and hypothesized effects of protein kinases associated with oncogenic viruses (Marx, 1981; Kolata, 1983), it is not unreasonable to suspect that changes in cell size and shape, syncytia formation, and loss of cell adhesion may all result from the action of iridescent virus-associated protein kinases. It would be very interesting to determine whether iridescent virus-associated kinases are capable of phosphorylating tyrosine residues.

Viral DNA is produced in the viroplasmic centers (Yule and Lee, 1973). In studies with IV's 2 and 6 in *A. eucalypti* and *Ae. aegypti*, virus-specific RNA and DNA synthesis were detected 24 and 96 hr postinfection, respectively, and viral antigen could not be detected until 96 hr (Kelly and Tinsley, 1974a,b). Yule and Lee (1973) were able to detect IV-1 antigen and DNA after only 24 hr in *Galleria* hemocytes. The results of Kelly and Tinsley (1974a) suggest that progeny viral DNA condenses to form a core which is then surrounded by capsid protein to form the complete virion. This supports the work of Bird (1961, 1962). Alternative hypotheses for assembly are that once the shell is assembled, the core material is introduced, possibly through an opening left in the shell (Xeros, 1964; Smith, 1958a,b; Yule and Lee, 1973) or that the granular core material is enclosed by the developing shell and condenses after completion of the shell (Federici, 1980). Release of progeny virions is reported to occur by budding into fingerlike protusions of the plasma membrane (Bellett and Mercer, 1964), by extrusion of vacuoles containing clusters of virus particles, or by cell lysis (Kelly and Tinsley, 1974a).

C. *GROSS PATHOLOGY*

Iridescent virus infections *in vivo* often lead to large quantities of intracellular virus and iridescence of the infected tissues (Hukahara and Hashimoto, 1966; Mitsuhashi, 1966). Since the iridescent viruses replicate in the cytoplasm, tissues with large concentrations of virus may also be detected by staining with hematoxylin or the nucleic acid differentiating fluorochromes acridine orange or coriphosphine O (Hall and Anthony, 1976). Detection of light infections may require electron microscopy.

Arthropod iridescent viruses appear to produce systemic infections and to infect almost all tissues (Bird, 1961; Mitsuhashi, 1966; Hall and Anthony, 1971; Federici, 1980). Hall and Anthony (1971) list the following tissues as sites of IV-3 development: fat body, tracheal epithelium, imaginal disc, epidermis, hemocytes, esophagus, nerve, muscle, and ovaries.

The greatest concentration of virus is usually found in the epidermis and fat body, often completely destroying the cytoplasmic structure of the latter tissue. Fat body is somewhat of a misnomer for this tissue which is often compared to the vertebrate liver in function. The fat body is most extensively developed in larval insects where it serves, among other functions, as a storage organ for lipids, protein, and glycogen, and is a particularly crucial organ during molting and metamorphosis (Wigglesworth, 1972). In the adult insect the fat body is the site of synthesis of vitellogenin (Engelmann, 1979), the precursor of egg yolk protein. The extensive destruction of the fat body is probably responsible for the inability of most infected insect larvae to pupate and to successfully develop to the adult stage. Extensive infection of the epidermal cells may also impair their ability to produce molting fluid enzymes and new cuticle. Nerve and muscle do not usually suffer extensive pathology. Therefore, infected individuals may move about in a normal manner until shortly before death when they likely exhaust their reserves for energy and tissue maintenance.

Hosts which become infected late in their larval development may reach the adult stage and transmit the viruses to their progeny. If adult fat body suffers the same degree of destruction as larval fat body, it is likely that the ability to synthesize vitellogenins would be impaired with a concomitant reduction in fecundity of the infected female or reduced fitness of her progeny. Reduced fecundity of infected females would obviously reduce the vertical transmission component of the natural epizootiological cycle.

Juvenile hormonelike effects have been reported in *Bombyx mori* infected with IV-6 (Ono and Fukaya, 1969) and in *Heliothis zea* infected with its own virus (Stadelbacher et al., 1978). In both cases, pupae were produced with patches of larval cuticle, and in *B. mori*, adults were eventually produced which had patches of larval cuticle. Ono et al. (1972) attributed these effects to the action of iridescent virus infection at the cellular level rather than to virus-induced perturbations of juvenile hormone level or cellular sensitivity to juvenile hormone.

Another anomaly of iridescent virus infection of *B. mori* is the production of epidermal "tumors" by IV-1 (Hukuhara, 1964). These tumors result from proliferation of epidermal cells and subsequent formation of a multilayered epidermis. Cells in the tumors are smaller and their cytoplasm is more basophilic than the normal epidermal cells. Iridescent virus-induced tumors have not been reported from other species of insects, although some iridescent viruses of vertebrates and of the octopus are associated with tumors.

IX. EPIZOOTIOLOGY

A. *NATURAL INCIDENCE AND INFECTIVITY*

The natural incidence of iridescent virus infections is typically low. Infection rates in mosquitoes are usually less than 1% (Hall and Lowe, 1971; Hall and Anthony, 1976). However, relatively high infection rates of some of the viruses have been recorded. Chapman et al. (1971) reported infection rates with IV-13 ranging from 0 in August to 70% in November with an average monthly level of 36%. A 50% average infection rate was recorded for IV-8 in *Culicoides arboricola* in two treeholes (Chapman et al., 1969). Infection rates with IV-1 are usually low (Carter, 1976), but Ricou (1975) has reported an epizootic in which a 90% infection rate was observed. The factors responsible for initiation of iridescent virus epizootics are not known. It is possible that in some cases infected larvae survive for long periods of time, as reported by Fukaya and Nasu (1966) for IV-6 infected *Chilo suppressalis*. If most healthy larvae pupated and emerged as adults, this would result in higher apparent infection rates.

There is considerable variation in the dose of virus required to produce infections *in vivo* by injection, even though Brown et al. (1977) have demonstrated with IV-22 that a plaque is initiated by a single infectious unit in tissue culture.

Infectious doses of IV-2 range from little more than a single particle for highly susceptible hosts (Day and Gilbert, 1967) to greater than 2.5×10^7 for less susceptible hosts (Glitz et al., 1968).

Wagner et al. (1974b) isolated a picornavirus from both "healthy" and IV-3 infected *Ae. taeniorhynchus* larvae and found that inocula containing both viruses resulted in higher IV-3 infection rates than inocula containing only IV-3. The natural incidence of the picornavirus is unknown as is the occurrence of similar "helper" viruses for the other IV's.

Careful quantitative per os infection studies with IV's have not been done, but it is evident that infection by feeding is more difficult and requires much larger doses than by injection (Mitsuhashi, 1966; Woodard and Chapman, 1968; Carter, 1973b). This "gut barrier" is probably responsible for the low rates of infection commonly seen in nature.

B. PATHWAY OF INFECTION

The route of infection has been investigated only for IV-3 in its host the black salt marsh mosquito, *Ae. taeniorhynchus*. The alimentary canal of insects is divided into a foregut and a hindgut, each of which is lined with a cuticular intima, and a midgut which is lined with microvilli. The foregut of mosquito larvae invaginates (Fig. 2) and runs forward to a point where it meets the anterior midgut. In the invagination, the midgut cells secrete a chitinous fluid which is pressed into a cylindrical peritrophic membrane. This membrane hardens and forms a continuous barrier between the solid food and the midgut epithelial cells.

The midgut has been considered to be the natural route of infection for the iridescent viruses, but direct evidence for this has been forthcoming only recently. Neither Hall and Anthony (1971) nor Stoltz and Summers (1971) were able to detect IV-3 virions in midgut cells of *Ae. taeniorhynchus* larvae after allowing the larvae to feed in concentrated suspensions of virus. Stoltz and Summers noted that the virions were apparently too large to penetrate the pores of the peritrophic membrane, and that they were degraded in the anterior region of the midgut. In both papers, infection of the midgut by naked nucleic acid was suggested as a possibility. More recently, Hembree and Anthony (1980) have observed intact IV-3 virions in a single midgut epithelial cell (Fig. 3) of a larva which had been exposed to the virus for 5 hours. The infected epithelial cell was in the extreme anterior area opposite the invaginated foregut. It is unknown whether the virions gained access to this area via the unsolidified peritrophic membrane or through a break in the hardened region. If this is the normal path of infection, it would help explain the inability to obtain high infection rates with this virus in the laboratory.

Fukuda and Clark (1975) were able to infect a few adult *Ae. taeniorhynchus* with IV-3 by aerosol application. It is unlikely that these infections resulted from contamination of the mosquitoes' food since no infections occurred in individuals which were fed on a sucrose-virus mixture. Breaks in the integument or in the intima of the tracheal system were considered possible entry sites for the virus. The integument and tracheal system are not believed to be natural sites of entry of virus into larval mosquitoes because the larvae are not usually exposed to a concentrated suspension of virus in nature.

C. NATURAL TRANSMISSION CYCLES

The natural transmission cycles have been studied for IV-1 in the terrestrial crane fly, *Tipula paludosa*, and for IV-3 in the salt marsh mosquito, *Aedes taeniorhynchus*, an aquatic host. The epizootiology of IV-3 has been studied most thoroughly and it will be considered first.

A brief account of the biology of *Ae. taeniorhynchus* is helpful in understanding the transmission cycle of IV-3. The eggs of *Ae. taeniorhynchus* are laid in depressions in salt marshes and, after embryonation, hatching follows the next flooding by either rain or tides. The majority of each egg clutch hatch with the first flooding, but a few may hatch in installments during subsequent floodings. Development through the four larval instars and pupal stage requires 5-7 days under ideal conditions, but may last much longer. Fourth instar *Ae. taeniorhynchus* larvae are gregarious, form clusters at the surface of the water in certain areas of the flood pools, and are very active feeders.

The following account of the major natural transmission cycle of IV-3 has been given by Linley and Nielsen (1968a,b). Infected adult females pass the virus to their progeny by transovarial transmission. The progeny larvae develop patent infections during the fourth instar and most become moribund prior to pupation. The moribund, infected larvae are cannibalized by healthy fourth instars which are infected too late to develop patent infections and survive to later pass the virus transovarially to their progeny. The aggregation behavior and extreme feeding activity of fourth instar larvae may facilitate cannibalization of the infected cadavers. There is apparently no paternal vertical transmission of IV-3 (Hembree, 1979). A possible secondary cycle of transmission might occur if larvae of different ages were present in flood pools at the

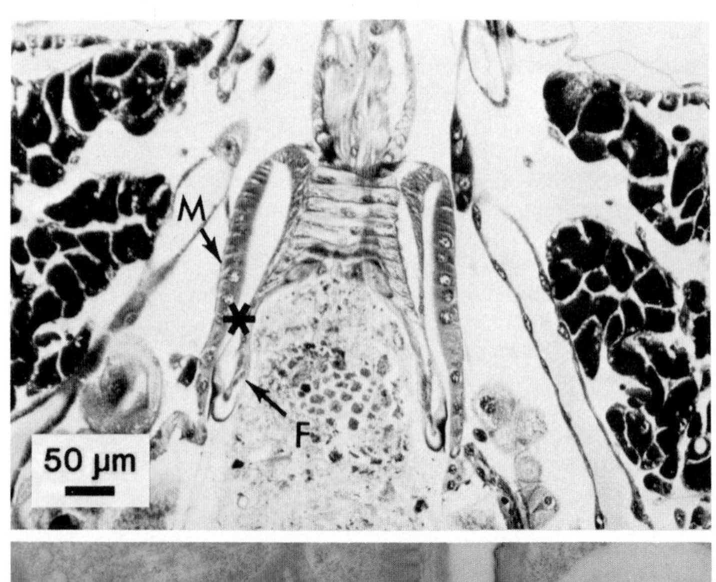

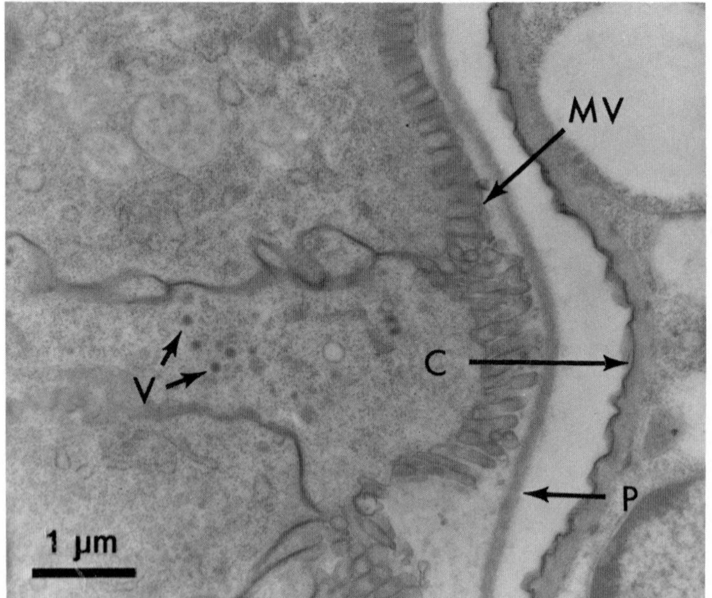

FIGURE 2. Esophageal invagination of larval Aedes taeniorhynchus. (6 µm section, Heidenhain's hematoxylin). F, foregut; M, midgut.

FIGURE 3. Iridescent virus type 3 in a first instar Aedes taeniorhynchus midgut cell adjacent to the esophageal invagination 5 hr after exposure to virus suspension (area marked with an asterisk in Fig. 2). Used with the permission

same time due to inundation of additional eggs after an initial flooding. The younger larvae (first or second instars) might feed on decomposing infected cadavers. These larvae would then develop patent infections as fourth instars, die, and serve as sources of inoculum for healthy fourth instars. This cycle is believed to be relatively insignificant in the epizootiology of IV-33 because of the inability of first instars to effectively cannibalize cadavers, the high concentrations of virus required to produce infections, and the fairly rapid inactivation of free virus in water (Linley and Nielsen, 1968a,b). The epizootiology of IV-3 is represented diagrammatically in Fig. 4. No alternate host is presently known for IV-3 although iridescent viruses occur in algae and lower fungi; Kelly (1981) has suggested the possibility that these may be infective for mosquitoes.

The transmission cycle of IV-1 in its natural hosts, terrestrial *Tipula* spp., appears to be virtually identical to that of IV-3. Both first and fourth instar larvae of *Tipula oleracea* may become infected by feeding on infected cadavers and the older larvae may also kill and cannibalize infected larvae (Carter, 1973a,b). Both IV-1-infected *T. oleracea* and IV-9-infected *Wiseana* larvae are reported to migrate upward in the soil thereby becoming more accessible to cannibalism by the actively feeding healthy larvae (Carter, 1973b; Fowler and Robertson, 1972). Early attempts to demonstrate vertical transmission of IV-1 were unsuccessful (Carter, 1973b; Rivers, 1966), but Carter (personal communication) has recently demonstrated vertical transmission by field-collected females of *Tipula paludosa*. Studies to determine the quantitative contribution of vertical transmission (see Fine, 1975) to the epizootiology of IV-1 and IV-3 are desirable.

Another iridovirid, African swine fever virus, is vertically transmitted in its tick vector, *Ornithodoros moubata* (Plowright et al., 1970) and is also sexually transmitted from male ticks to females via seminal fluid (Plowright et al., 1974). The possible occurrence of vertical and sexual transmission of the other invertebrate iridescent viruses in their natural hosts should be examined. There is also one iridescent virus which is probably transmitted vertically through the sperm. This virus (IV-27) of the marine worm *Nereis discolor* infects only the maturing spermatocytes (Devauchelle, 1977; Devauchelle and Durchon, 1973b).

(Figure 3 cont'd) of S. Hembree and the American Mosquito Control Association. C, cuticle; MV, microvilli, P, peritrophic membrane; V, virions.

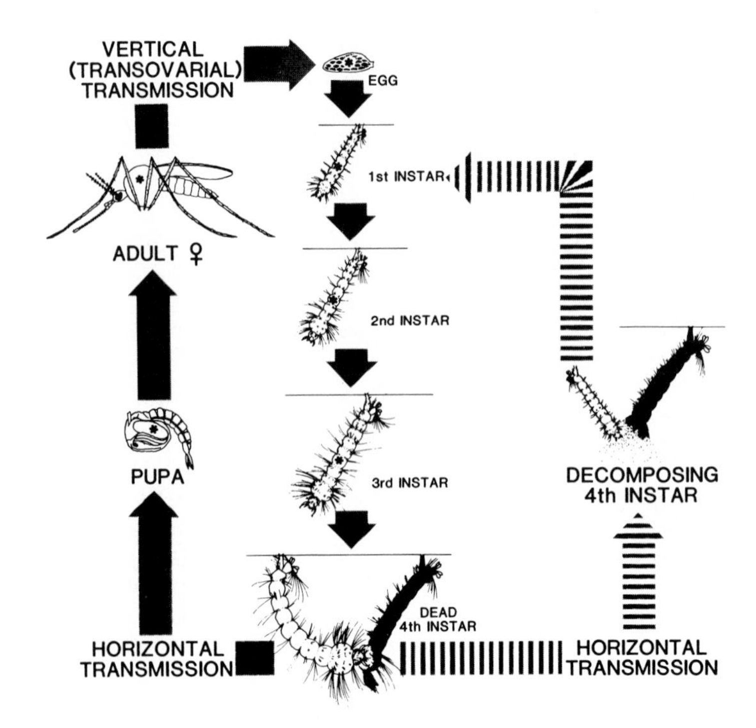

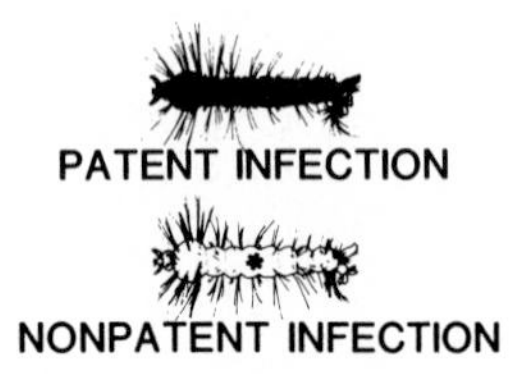

FIGURE 4. Epizootiological cycles of iridescent virus type 3. Solid arrows indicate the major cycle; broken arrows indicate a possible minor cycle.

X. POTENTIAL FOR BIOLOGICAL CONTROL

In field trials where IV-1 was introduced into plots via live and dead *Tipula* larvae, in a bran bait, and as a spray, the resulting incidence of infection varied between 1 and 17% of the tipulid population (Carter, 1978). Attempts to transmit IV-9 to field populations of *Wiseana* sp. were completely unsuccessful (Moore et al., 1974b).

Due to their low infectivity per os and relatively short persistence in the field, the iridescent viruses do not at present appear to have much potential for biological control applications. If these problems could be overcome, the insect iridescent viruses appear to be safe to most animals other than insects, and it is likely that mass production techniques could be developed.

REFERENCES

Anderson, J. F. (1970). An iridescent virus infecting the mosquito *Aedes stimulans*. *J. Invertebr. Pathol. 15*, 219-224.

Avery, S. W., and Bauer, L. (1984). Iridescent virus from *Prosimulium* collected in Maine. *J. Invertebr. Pathol.*, in press.

Bailey, L., and Ball, B. V. (1978). *Apis* iridescent virus and "clustering disease" of *Apis cerana*. *J. Invertebr. Pathol. 31*, 368-371.

Bailey, L., Ball, B. V., and Woods, R. D. (1976). An iridovirus from bees. *J. Gen. Virol. 31*, 459-461.

Barry, S., and Devauchelle, G. (1979). Study on structural polypeptides of *Chilo* iridescent virus. (Iridovirus type 6). *Can. J. Microbiol. 25*, 841-849.

Batson, B. S., Johnston, M. R. L., Arnold, M. K., and Kelly, D. G. (1976). An iridescent virus from *Simulium* sp. (Diptera: Simuliidae) in Wales. *J. Invertebr. Pathol. 27*, 133-135.

Bellett, A. J. D. (1968). The iridescent virus group. *Advan. Virus Res. 13*, 225-246.

Bellett, A. J. D., and Inman, R. B. (1967). Some properties of deoxyribonucleic acid preparations from *Chilo*, *Sericesthis*, and *Tipula* iridescent viruses. *J. Mol. Biol. 25*, 425-432.

Bellett, A. J. D., and Mercer, E. H. (1964). The multiplication of *Sericesthis* iridescent virus in cell cultures from *Antherea eucalypti* Scott. I. Qualitative experiments. *Virology 24*, 645-653.

Bird, F. T. (1961). The development of *Tipula* iridescent virus in the crane fly, *Tipula paludosa* Meig., and the wax moth, *Galleria mellonella* L. *Can. J. Microbiol. 7*, 827-830.

Bird, F. T. (1962). On the development of the *Tipula* iridescent virus particle. *Can. J. Microbiol. 8*, 533-534.

Black, P. N., Blair, C. D., Butcher, A., Capinera, J. L., and Happ, G. M. (1981). Biochemistry and ultrastructure of iridescent virus type 29. *J. Invertebr. Pathol. 38*, 12-21.

Brown, D. A., Lescott, T., Harrap, K. A., and Kelly, D. C. (1977). The replication and titration of iridescent virus type 22 in *Spodoptera frugiperda* cells. *J. Gen. Virol. 38*, 175-178.

Buchatsky, L. P. (1977). An iridovirus from larvae of *Culiseta annulata* and *Culex territans*. *Acta Virol. 21*, 85-86.

Carey, G. P., Lescott, T., Robertson, J. S., Spencer, L. K., and Kelly, D. C. (1978). Three African isolates of small iridescent viruses: Type 21 from *Heliothis armigera* (Lepidoptera: Noctuidae), Type 23 from *Heteronychus arator* (Coleoptera: Scarabaeidae), Type 28 from *Lethocerus columbiae* (Hemiptera Heteroptera: Belostomatidae). *Virology 85*, 307-309.

Carter, J. B. (1973a). The mode of transmission of *Tipula* iridescent virus: I, Source of infection. *J. Invertebr. Pathol. 21*, 123-130.

Carter, J. B. (1973b). The mode of transmission of *Tipula* iridescent virus: II, Route of infection. *J. Invertebr. Pathol. 21*, 136-143.

Carter, J. B. (1976). A survey of microbial, insect and nematode parasites of Tipulidae (Diptera) larvae in north-east England. *J. Appl. Ecol. 13*, 103-122.

Carter, J. B. (1978). Field trials with *Tipula* iridescent virus against *Tipula* spp. larvae in grassland. *Entomophaga 23*, 169-174.

Cerutti, M., and Devauchelle, G. (1979). Cell fusion induced by an invertebrate virus. *Arch. Virol. 61*, 149-155.

Cerutti, M., and Devauchelle, G. (1980). Inhibition of macromolecular synthesis in cells infected with an invertebrate virus. *Arch. Virol. 63*, 297-303.

Chapman, H. C., Clark, T. B., Woodard, D. B., and Kellen, W. R. (1966). Additional mosquito hosts of the mosquito iridescent virus. *J. Invertebr. Pathol. 8*, 545-546.

Chapman, H. C., Petersen, J. J., Woodard, D. B., and Clark, T. B. (1968). New records of parasites of Ceratopogonidae. *Mosquito News 28*, 122-123.

Chapman, H. C., Clark, T. B., Petersen, J. J., and Woodard, D. B. (1969). A two-year survey of pathogens and parasites of Culicidae, Chaoboridae, and Ceratopogonidae in Louisiana. *Proc. 56th Annu. Meet. New Jersey Mosquito Extermination Assoc.*, pp. 203-212.

Chapman, H. C., Clark, T. B., Anthony, D. W., Glenn, F. E., Jr. (1971). An iridescent virus from larvae of *Corethrella brakeleyi* (Diptera: Chaoboridae) in Louisiana. *J. Invertebr. Pathol. 18*, 284-286.

Cole, A., and Morris, T. J. (1980). A new *Iridovirus* of two species of terrestrial isopods, *Armadillidium vulgare* and *Porcellio scaber*. *Intervirology 14*, 21-30.

Cunningham, J. C., and Tinsley, T. W. (1968). A serological comparison of some iridescent non-occluded insect viruses. *J. Gen. Virol. 3*, 1-8.

Day, M. F., and Gilbert, N. (1967). The number of particles of *Sericesthis* iridescent virus required to produce infections of *Galleria* larvae. *Austr. J. Biol. Sci. 20*, 691-693.

Day, M. F., and Mercer, E. H. (1964). Properties of an iridescent virus from the beetle *Sericesthis priunosa*. *Austr. J. Biol. Sci. 17*, 892-902.

DeBlois, R. W., Uzgiris, E. E., Cluxton, D. H., and Mazzone, H. M. (1978). Comparative measurements of size and polydispersity of several insect viruses. *Anal. Biochem. 90*, 273-288.

Devauchelle, G. (1977). Ultrastructural characterization of an iridovirus from the marine worm *Nereis diversicolor* (O. F. Muller). Architecture of the virion and virus morphogenesis. *Virology 81*, 237-246.

Devauchelle, G., and Durchon, M. (1973a). Cytopathogenic effects of iridovirus in *Neireis diversicolor* (Annelida: Polychaeta). *J. Microsc. 17*, 44A.

Devauchelle, G., and Durchon, M. (1973b). Sur la présence d'un virus, du type iridovirus dans les cellules mâles de *Nereis diversicolor* (O. F. Muller). *C. R. Acad. Sci. D277*, 463-466.

Elliott, R. M., Lescott, T., and Kelly, D. C. (1977). Serological relationships of an iridescent virus (type 25) recently isolated from *Tipula* sp. with two other iridescent viruses (types 2 and 22). *Virology 81*, 309-316.

Engelmann, F. (1979). Insect vitellogenin: Identification, biosynthesis, and role in vitellogenesis. *Advan. Insect Physiol. 14*, 49-108.

Federici, B. A. (1980). Isolation of an iridovirus from two terrestrial isopods, the pill bug, *Armadillidium vulgare*, and the sow bug, *Porcellio dilatatus*. *J. Invertebr. Pathol. 36*, 373-381.

Federici, B. A., and Hazard, E. I. (1975). Iridovirus and cytoplasmic polyhedrosis virus diseases in the fresh water daphnid *Simocephalus expinosus*. *Nature (London) 254*, 327-328.

Fine, P. E. M. (1975). Vectors and vertical transmission: An epidemiologic perspective. *Ann. N.Y. Acad. Sci. 266*, 173-194.

Fowler, M., and Robertson, J. S. (1972). Iridescent virus infection in field populations of *Wiseana cervinata* (Lepidoptera: Hepialidae) and *Witlesia* sp. (Lepidoptera: Pyralidae) in New Zealand. *J. Invertebr. Pathol. 19*, 154-155.

Fukaya, D. R., and Nasu, S. (1966). A *Chilo* iridescent virus from the rice stem borer, *Chilo suppressalis* Walker (Lepidoptera: Pyralidae). *Appl. Entomol. Zool. 1*, 69-72.

Fukuda, T. (1971). *Per os* transmission of *Chilo* iridescent virus to mosquitoes. *J. Invertebr. Pathol. 18*, 152-153.

Fukuda, T., and Clark, T. B. (1975). Transmission of the mosquito iridescent virus (RMIV) by adult mosquitoes of *Aedes taeniorhynchus* to their progeny. *J. Invertebr. Pathol. 25*, 275-276.

Galfre, G., and Milstein, C. (1981). Preparation of monoclonal antibodies: strategies and procedures. *Methods Enzymol. B73*, 3-46.

Glitz, D. G., Hills, G. J., and Rivers, C. F. (1968). A comparison of the *Tipula* and *Sericesthis* iridescent viruses. *J. Gen. Virol. 3*, 209-220.

Goorha, R., and Granoff, A. (1979). Icosahedral cytoplasmic deoxyriboviruses. *Comp. Virol. 14*, 347-399.

Guerillon, J., Barray, S., and Devauchelle, G. (1982). Crossed immunoelectrophoretic characterization of *Chilo* iridescent virus surface antigens. *Arch. Virol. 73*, 161-170.

Hall, D. W., and Anthony, D. W. (1971). Pathology of a mosquito iridescent virus (MIV) infecting *Aedes taeniorhynchus*. *J. Invertebr. Pathol. 18*, 61-69.

Hall, D. W., and Anthony, D. W. (1976). An "R" type *Iridovirus* from *Aedes vexans* (Meigen). *Mosquito News 36*, 536-537.

Hall, D. W., and Lowe, R. E. (1971). A new distribution record for the mosquito iridescent virus (MIV). *Mosquito News 31*, 448-449.

Hall, D. W., and Lowe, R. E. (1972). Physical and serological comparisons of "R" and "T" strains of mosquito iridescent virus from *Aedes taeniorhynchus*. *J. Invertebr. Pathol. 19*, 317-324.

Hasan, S., Vago, C., and Kuhl, G. (1971). Infection of *Aedes detritus* Hal. with mosquito iridescent virus. *Bull. WHO 45*, 268-269.

Hembree, S. C. (1979). Non-participation of male *Aedes taeniorhynchus* (Wiedemann) in vertical transmission of regular mosquito iridescent virus. *Mosquito News 39*, 672-673.

Hembree, S. C., and Anthony, D. W. (1980). Possible site of entry of the regular mosquito iridescent virus (RMIV) in *Aedes taeniorhynchus* larvae. *Mosquito News 40*, 449-451.

Hibbin, J. A., and Kelly, D. C. (1981). Iridescent virus type 22 DNA. *Arch. Virol. 68*, 9-18.

Hukuhara, T. (1964). Induction of an epidermal tumour in *Bombyx mori* (L.) with *Tipula* iridescent virus. *J. Insect Pathol. 6*, 246-249.

Hukuhara, T., and Hashimoto, Y. (1966). Development of *Tipula* iridescent virus in the silkworm, *Bombyx mori* L. (Lepidoptera: Bombycidae). *Appl. Entomol. Zool. 1*, 166-172.

Jensen, D. D., Hukuhara, T., and Tanada, Y. (1972). Lethality of *Chilo* iridescent virus to *Colladonus montanus* leafhoppers. *J. Invertebr. Pathol. 19*, 276-278.

Kalmakoff, J., and Tremaine, J. H. (1968). Physiochemical properties of *Tipula* iridescent virus. *J. Virol. 2*, 738-744.

Kalmakoff, J. S., Moore, S. G., and Pottinger, R. P. (1972). An iridescent virus from the grass grub *Costelytra zealandica*: Serological study. *J. Invertebr. Pathol. 20*, 70-76.

Kelly, D. C. (1981). Non-occluded viruses. *In* "Pathogenesis of Invertebrate Microbial Diseases" (E. D. Davidson, ed.), pp. 39-60. Allanheld, Osmun Publishers, Totowa.

Kelly, D. C., and Avery, R. J. (1974). The DNA content of four small iridescent viruses: Genome size, redundancy, and homology determined by renaturation kinetics. *Virology 57*, 425-435.

Kelly, D. C., and Robertson, J. S. (1973). Icosahedral cytoplasmic deoxyriboviruses. *J. Gen. Virol. Suppl. 20*, 17-41.

Kelly, D. C., and Tinsley, T. W. (1972). Proteins of iridescent virus types 2 and 6. *J. Invertebr. Pathol. 19*, 273-275.

Kelly, D. C., and Tinsley, T. W. (1973). Ribonucleic acid polymerase activity associated with particles of iridescent virus types 2 and 6. *J. Invertebr. Pathol. 22*, 199-202.

Kelly, D. C., and Tinsley, T. W. (1974a). Iridescent virus replication: a microscope study of *Aedes aegypti* and *Antherea eucalypti* cells in culture infected with iridescent virus types 2 and 6. *Microbios 9*, 75-93.

Kelly, D. C., and Tinsley, T. W. (1974b). Iridescent virus replication: patterns of nucleic acid synthesis in insect cells infected with iridescent virus types 2 and 6. *J. Invertebr. Pathol. 24*, 169-178.

Kelly, D. C., and Vance, D. E. (1973). The lipid content of two iridescent viruses. *J. Gen. Virol. 21*, 417-423.

Kelly, D. C., Edwards, M. L., and Robertson, J. S. (1978). The use of enzyme-linked immunosorbent assay to detect, and discriminate between, small iridescent viruses. *Ann. Appl. Biol. 90*, 369-374.

Kelly, D. C., Ayers, M. D., Lescott, T., Robertson, J. S., and Happ, G. M. (1979). A small iridescent virus (type 29) isolated from *Tenebrio molitor*: A comparison of its proteins and antigens with six other iridescent viruses. *J. Gen. Virol. 42*, 95-105.

Kelly, D. C., Elliott, R. M., and Blair, G. E. (1980). Phosphorylation of iridescent virus polypeptides *in vitro*. *J. Gen. Virol. 48*, 205-212.

Klug, A., Franklin, R. E., and Humphreys-Owen, S. P. F. (1959). The crystal structure of *Tipula* iridescent virus as determined by Bragg reflection of visible light. *Biochim. Biophys. Acta 32*, 203-219.

Kolata, G. (1983). Is tyrosine the key to growth control? *Science 219*, 377-378.

Krell, P., and Lee, P. E. (1974). Polypeptides in Tipula iridescent virus (TIV) and in TIV-infected hemocytes of *Galleria mellonella* (L.) larvae. *Virology 60*, 315-326.

Krøll, J. (1981). Production of specific antisera by immunization with precipitin lines. *Methods Enzymol. B73*. 52-57.

Linley, J. R., and Nielsen, H. T. (1968a). Transmission of a mosquito iridescent virus in *Aedes taeniorhynchus*. I. Laboratory experiments. *J. Invertebr. Pathol. 12*, 7-16.

Linley, J. R., and Nielsen, H. J. (1968b). Transmission of a mosquito iridescent virus in *Aedes taeniorhynchus*. II. Experiments related to transmission in nature. *J. Invertebr. Pathol. 12*, 17-24.

Longworth, J. F., Scotti, P. D., and Archibald, R. D. (1979). An iridovirus from the black beetle, *Heteronychus arator* (Coleoptera: Scarabaeidae). *New Zeal. J. Zool. 6*, 637-639.

Manyakov, V. F. (1977). Fine structure of the iridescent virus type I capsid. *J. Gen. Virol. 36*, 73-79.

Marx, J. L. (1981). Tumor viruses and the kinase connection. *Science 211*, 1336-1338.

Mathieson, W. B., and Lee, P. E. (1981). Cytology and autoradiography of *Tipula* iridescent virus infection of insect suspension cell cultures. *J. Ultrastr. Res. 74*, 59-68.

Matta, J. F. (1970). The characterization of a mosquito iridescent virus (MIV). II. Physiochemical characterization. *J. Invertebr. Pathol. 16*, 157-164.

Matta, J. F., and Lowe, R. E. (1970). The characterization of a mosquito iridescent virus (MIV). I. Biological characteristics, infectivity, and pathology. *J. Invertebr. Pathol. 16*, 38-41.

Matthews, R. E. F. (1982). Classification and nomenclature of viruses. *Intervirology 17*, 1-200.

Mitsuhashi, J. (1966). Appearance of iridescence in the tissues of the rice stem borer larva, *Chilo suppressalis* Walker, infected with *Chilo* iridescent virus (Lepidoptera: Pyralidae). *Appl. Entomol. Zool. 1*, 130-137.

Mitsuhashi, J. (1967). Infection of leafhopper and its tissues cultivated *in vitro* with *Chilo* iridescent virus. *J. Invertebr. Pathol. 9*, 432-434.

Monnier, C., and Devauchelle, G. (1976). Enzyme activities associated with an invertebrate iridovirus: nucleotide phosphohydrolase activity associated with iridescent virus type 6 (CIV). *J. Virol. 19*, 180-186.

Monnier, C., and Devauchelle, G. (1980). Enzyme activities associated with an invertebrate iridovirus: protein kinase activity associated with iridescent virus type 6 (*Chilo* iridescent virus). *J. Virol. 35*, 444-450.

Moore, S. G., Kalmakoff, J., and Miles, J. A. R. (1974a). An iridescent virus and a rickettsia from the grass grub

Costelytra zealandica (Coleoptera: Scarabaeidae). *New Zeal. J. Zool. 1*, 205-210.
Moore, S. G., Kalmakoff, J., and Miles, J. A. R. (1974b). Virus diseases of porina (*Wiseana* spp.; Lepidoptera: Hepialidae). 2. Transmission trials. *New Zeal. J. Zool. 1*, 85-95.
Ohba, M. (1975). Studies on the multiplication of *Chilo* iridescent virus. 3. Multiplication of CIV in the silkworm, *Bombyx mori* L. and field insects. *Sci. Bull. Fac. Agr., Kyushu Univ. 30*, 71-81.
Ohba, M., and Aizawa, K. (1979). *In vivo* insect hemocyte destruction by UV-irradiated *Chilo* iridescent virus. *J. Invertebr. Pathol. 34*, 32-40.
Ohba, M., and Aizawa, K. (1981). Lethal toxicity of arthropod iridoviruses to an amphibian, *Rana limnocharis*. *Arch. Virol. 68*, 153-156.
Ohba, M., and Aizawa, K. (1982). Mammalian toxicity of an insect iridovirus. *Acta Virol. 26*, 165-168.
Ono, M., and Fukaya, M. (1969). The juvenile-hormone-like effect of *Chilo* iridescent virus (CIV) on the metamorphosis of the silkworm, *Bombyx mori* L. *Appl. Entomol. Zool. 9*, 211-212.
Ono, M., Yagi, S., and Fukaya, M. (1972). The infectivity of *Chilo* iridescent virus to the silkworm, *Bombyx mori* L. *Bull. Sericult. Exp. Sta. Tokyo Univ. Educ. 25*, 77-102.
Plowright, W., Perry, C. T., and Pierce, M. A. (1970). Transovarial infection with African swine fever virus in the argasid tick, *Ornithodoros moubata porcinus*, Walton. *Res. Vet. Sci. 11*, 582-584.
Plowright, W., Perry, C. T., and Greig, A. (1974). Sexual transmission of African swine fever virus in the tick *Ornithodoros moubata porcinus*, Walton. *Res. Vet. Sci. 17*, 106-113.
Poinar, G. O. (1981). *Thaumamermis cosgrovei* n. gen., n. sp. (Mermithidae: Nematoda) parasitizing terrestrial isopods (Isopoda: Oniscoidea). *Syst. Parasitol. 2*, 261-266.
Poinar, G. O., Hess, R. T., and Cole, A. (1980). Replication of an iridovirus in a nematode (Mermithidae). *Intervirology 14*, 316-320.
Popelkova, Y. (1982). *Coelomomyces* from *Aedes cinereus* and a mosquito iridescent virus of *Aedes contans* in Sweden. *J. Invertebr. Pathol. 40*, 148-149.
Ricou, G. (1975). Production de *Tipula paludosa* Meig. en prairie en function de l'humiditié du sol. *Rev. Ecol. Biol. Sol. 12*, 69-89.
Rivers, C. F. (1966). The natural and artificial dispersion of pathogens. *Proc. Intern. Colloq. Insect Pathol. Biol. Control, Wageningen*, pp. 252-263.

Runnger, D., Rastelli, M., Braendle, E., and Malsberger, R. G. (1971). A virus-like particle associated with lesions in the muscle of *Octopus vulgaris*. *J. Invertebr. Pathol. 17*, 72-80.

Singh, Y. (1979). Iridescent virus in the Indian honeybee *Apis cerana indica*. *Amer. Bee J. 119*, 398.

Smith, K. M. (1958a). A study of the early stages of infection with *Tipula* iridescent virus. *Parasitology 48*, 459-462.

Smith, K. M. (1958b). Early stages of infection with the *Tipula* iridescent virus. *Nature (London) 181*, 966-967.

Smith, K. M. (1976). "Virus-Insect Relationships." Longman, London.

Stadelbacher, E. A., Adams, J. R., Faust, R. M., and Tompkins, G. J. (1978). An iridescent virus of the bollworm *Heliothis zea* (Lepidoptera: Noctuidae). *J. Invertebr. Pathol. 32*, 71-76.

Steiger, U., Lamparter, H. E., Sandri, C., and Akert, K. (1969). Virus-ähnliche Partikel im Zytoplasma von Nerven- und Gliazellen der Waldameise. *Arch. Virusforsch. 26*, 271-282.

Steinhaus, E. A., and Leutenegger, R. (1963). Icosahedral virus from a scarab (*Sericesthis*). *J. Insect Pathol. 5*, 266-270.

Stoltz, D. B. (1971). The structure of icosahedral cytoplasmic deoxyriboviruses. *J. Ultrastr. Res. 37*, 219-239.

Stoltz, D. B. (1973). The structure of icosahedral cytoplasmic deoxyriboviruses. II. An alternative model. *J. Ultrastr. Res. 43*, 58-74.

Stoltz, D. B., and Summers, M. D. (1971). Pathway of infection of mosquito iridescent virus. I. Preliminary observations on the fate of ingested virus. *J. Virol. 8*, 900-909.

Stoltz, D. B., Hilsenhoff, W. L., and Stich, H. F. (1968). A virus disease in *Chironomus plumosus*. *J. Invertebr. Pathol. 12*, 118-128.

Takaoka, H. (1980). Pathogens of blackfly larvae in Guatemala and their influence on natural populations of 3 species of onchocerciasis vectors. *Amer. J. Trop. Med. Hyg. 29*, 467-472.

Thomas, R. S. (1961). The chemical composition and particle weight of the *Tipula* iridescent virus. *Virology 14*, 240-252.

Thomas, D., and Gouranton, J. (1978). Development of an iridovirus in *Tenebrio molitor*. *J. Invertebr. Pathol. 32*, 114-116.

Tinsley, T., and Harrap, J. A. (1978). Viruses of invertebrates. *Compr. Virol. 12*, 1-101.

Tinsley, T. W., and Kelly, D. C. (1970). An interim nomenclature system for the iridescent group of insect viruses. *J. Invertebr. Pathol. 16*, 470-472.

Tinsley, T. W., Robertson, J. S., Rivers, C. F., and Service, M. W. (1971). An iridescent virus of *Aedes cantans* in Great Britain. *J. Invertebr. Pathol. 18*, 427-428.

Tojo, S., and Kodama, T. (1968). Purification and some properties of *Chilo* iridescent virus. *J. Invertebr. Pathol. 12*, 66-72.

Torybaev, Kh. K. (1970). Discovery of *orange iridescent virus* of mosquito in the larvae of *Aedes caspius caspius* in the southeastern part of Kazakhstan. *Vestn. Acad. Nauk Kaz. SSR 7*, 68-69.

Vago, C., Roux, J. A., Duthoit, J. L., and Dedet, J. P. (1969). Infection spontanée à virus irisant dan une population d'*Aedes detritus* (Hal., 1833) des environs de Tunis. *Ann. Parasitol. 44*, 667-676.

Wagner, G. W., and Paschke, J. D. (1977). A comparison of the DNA of the "R" and "T" strains of mosquito iridescent virus. *Virology 81*, 298-308.

Wagner, G. W., Paschke, J. D., Campbell, W. R., and Webb, S. R. (1973). Biochemical and biophysical properties of two strains of mosquito iridescent virus. *Virology 52*, 72-80.

Wagner, G. W., Paschke, J. D., Campbell, W. R., and Webb, S. R. (1974a). Proteins of two strains of mosquito iridescent virus. *Intervirology 3*, 97-105.

Wagner, G. W., Webb, S. R., Paschke, J. D., and Campbell, W. R. (1974b). A picornavirus isolated from *Aedes taeniorhynchus* and its interaction with mosquito iridescent virus. *J. Invertebr. Pathol. 24*, 380-382.

Webb, S. R., Paschke, J. D., Wagner, G. W., and Campbell, W. R. (1976). Pathology of mosquito iridescent virus of *Aedes taeniorhynchus* in cell cultures of *Aedes aegypti*. *J. Invertebr. Pathol. 27*, 27-40.

Weiser, J. (1965). A new virus infection of mosquito larvae. *Bull. WHO 33*, 586-588.

Weiser, J. (1968). Iridescent virus from the black fly *Simulium ornatum* Meigen in Czechoslovakia. *J. Invertebr. Pathol. 12*, 36-39.

Wigglesworth, V. B. (1972). "The Principles of Insect Physiology," 7th ed. Halsted Press, New York.

Williams, R. C., and Smith, K. M. (1958). The polyhedral form of the *Tipula* iridescent virus. *Biochim. Biophys. Acta 28*, 464-469.

Willison, J. H. M., and Cocking, E. C. (1972). Frozen-fractured viruses: a study of virus structure using freeze-etching. *J. Microsc. 95*, 397-411.

Woodard, D. B., and Chapman, H. C. (1968). Laboratory studies with the mosquito iridescent virus (MIV). *J. Invertebr. Pathol. 11*, 296-301.

Wrigley, N. G. (1969). An electron microscope study of the structure of *Sericesthis* iridescent virus. *J. Gen. Virol. 5*, 123-134.

Wrigley, N. G. (1970). An electron microscope study of the structure of *Tipula* iridescent virus. *J. Gen. Virol. 6*, 169-173.

Xeros, N. (1954). A second virus disease of the leather jacket, *Tipula palusoda. Nature (London) 174*, 562-563.

Xeros, N. (1964). Development of the *Tipula* iridescent virus. *J. Insect Pathol. 6*, 261-283.

Yule, G. B., and Lee, P. E. (1973). A cytological and immunological study of *Tipula* iridescent virus-infected *Galleria mellonella* larval hemocytes. *Virology 51*, 409-423.

Zola, H., and Brooks, D. (1982). Techniques for the production and characterization of monoclonal hybridoma antibodies. *In* "Monoclonal Hybridoma Antibodies: Techniques and Applications" (J. G. R. Hurrell, ed.), pp. 1-57. CRC Press, Inc., Boca Raton, Florida.

PATHOLOGY ASSOCIATED WITH DENSOVIRUSES

SHIGEMI KAWASE

Faculty of Agriculture
Nagoya University
Nagoya 464 Japan

I. INTRODUCTION

Viruses containing linear single-stranded DNA belong to Parvoviridae; the genus *Densovirus* is in this family. Members of *Densovirus* are commonly called densonucleosis virus (DNV). The first DNV was isolated from larvae of the greater wax moth, *Galleria mellonella*, by Meynadier et al. (1964), who reported that this virus caused a fatal disease in this insect. Subsequently, similar viruses or denso-like viruses were isolated from Lepidoptera, Diptera, Orthoptera, Blattariae, and Odonata, i.e., from *Junonia coenia (Aglais urticae)* (Rivers and Longworth, 1972), *Euxoa auxiliaris* (Sutter, 1973), *Aedes aegypti* (Lebedeva et al., 1973, *Acheta domestica* (Meynadier et al., 1977c, *Diatraea saccharalis* (Meynadier et al., 1977b), *Sibine fusca* (Meynadier et al., 1977a), *Simulium vittatum* (Federici, 1976), *Periplaneta fuliginosa* (Suto, 1979); *Leucorrhinia dubia* (Charpentier, 1979), *Agraulis vanillae* (Kelly et al., 1980a), and *Pieris rapae* (Sun et al., 1981).

In 1968, a disease of the silkworm, *Bombyx mori*, prevalent in sericultural farms around the suburbs of Ina city, Japan, caused great economic damage. The disease was initially believed to be caused by the infectious flacherie virus (IFV),

ISBN 0-12-470295-3

a small spherical RNA virus (diameter: 26 nm) (Kawase et al., 1974), because of similarity of symptoms in the infected larvae. However, Shimizu (1975) found from histopathological study of infected larvae that the "Ina isolate virus" (Ina isolate) was not an IFV but a heretofore undescribed virus in the silkworm. Several investigators studied the characteristics of this new virus and its disease in some detail. They concluded that the virus differed from IFV, and was similar in cytopathological, chemical, and physical characteristics to a DNV described from the greater wax moth, *G. mellonella* (Watanabe et al., 1976; Kawase and Kang, 1976; Maeda and Watanabe, 1978; Maeda et al., 1977; Kang et al., 1978). They proposed that the virus be named *Bombyx* densonucleosis virus (*Bombyx* DNV). Similar denso-like viruses were isolated in the silkworm by Matsui (1973) and Furuta (1973). Recently, the virus isolated by Furuta was found to be identical or closely similar to the Ina isolate (Kawase et al., 1982).

More recently, a new type of DNV was discovered in diseased silkworms in mainland China (Iwashita and Chun, 1982). Although, at present only little is known about the new DNV (China isolate), the histopathological features of the infected cells clearly differ from those of the Ina isolate. Other new types of DNV were isolated from the silkworms at Saku city (Saku isolate) (H. Watanabe, personal communication) and at Yamanashi Prefecture (Yamanashi isolate) (H. Seki and Y. Iwashita, personal communication) in Japan; they cause a fatal disease in silkworm strains that are resistant to the Ina isolate. In order to avoid confusion in names, the author has designated the ina isolate as the *Bombyx* DNV.

Of the various DNVs known, only *Galleria* and *Bombyx* DNVs have been investigated in detail. An earlier review on the DNV from *G. mellonella* by Kurstak (1972) has covered in some detail several of the initial studies.

II. CLASSIFICATION

The International Committee on the Nomenclature of Viruses in 1982 (Matthews, 1982) established three genera in the family Parvoviridae: (1) *Parvovirus*, (2) *Dependovirus*, and (3) *Densovirus*. *Dependovirus*, the common name, is an adeno-associated virus (AAV), which multiplies only in the presence of a helper virus such as adenovirus. *Parvovirus* is characterized by the presence of only the minus strand of DNA in the virion, but the second and third genera have complementary (plus and minus) DNA strands that are separately encapsidated.

The common name of *Densovirus*. densonucleosis virus, was first used by Kurstak and Vago (1967) from the characteristic appearance of the nuclei in infected larvae. The name was subsequently shortened to "densovirus." The other two genera have not yet been discovered in insects.

III. HISTOPATHOLOGICAL SYMPTOMS

In *Galleria* DNV, the virus multiplies in almost all of the insect tissues with the exception of the midgut. The susceptible tissues are fat body, nerve cells, silk gland, muscular membrane, Malpighian tubule, gonads, and molting gland. The histopathological aspects of DNV infection are characteristic, i.e., the main lesions occur in the nuclei of infected cells. The nuclei become greatly hypertrophied very rapidly and are eosinophilic. A voluminous dense body appears in each infected nucleus. The tissues are destroyed in 4 to 6 days after viral inoculation (Amargier et al., 1965). After staining with acridine orange at pH 3.8, a greenish yellow fluorescent area is visible in the nucleus using fluorescent microscopy, indicating the presence of double-stranded (ds) nucleic acids. This area becomes apparent 13-15 hr after infection (Vago et al., 1966; Kurstak et al., 1968a). The fluorescence in the nucleus later changes in color from green to red suggesting the accumulation of single-stranded (ss) nucleic acids (Kurstak et al., 1968b).

When observations of sections from the larva of the mosquito, *A. aegypti* infected with *Aedes* DNV are made using the light microscope, the cells of the hypodermis, imaginal disk, and Malpighian tubule are shown to be affected. However, the most obvious pathological changes occur in cells of the fat body. Infected nuclei are much larger than healthy ones, and their internal structures are not discernible. When stained with Feulgen, the infected nuclei acquire a bluish purple color (Lebedeva et al., 1973).

In the case of the lepidopterous insect, *S. fusca*, a notable modification caused by the *Sibine* DNV infection develops in the digestive tract (Meynadier et al., 1977a). In healthy larvae, the intestinal content is visible through the thin digestive wall, but in diseased larvae, the wall becomes opaque, whitish, and thickened. This thickening of the wall is the most characteristic sign that the disease is at an advanced stage. In certain larvae, the nucleus increases in density without a distinct hypertrophy and the cytoplasm vacuolates or retracts, surrounding the nucleus in a thin layer. Lesions are also observed in infected cells of hypodermal tissue, tracheal cell, muscle, and Malpighian tubule.

The cockroach, *P. fuliginosa*, infected with the *Periplaneta* DNV, displays the following signs and symptoms. Prior to death, the hind legs become paralyzed, with movements becoming uncoordinated. The abdomen is swollen with enlarged milky-white colored fat body which is in contrast to the brownish white tissues observed in an uninfected cockroach. Over one-half of the dead cockroaches develop ulcers in the hindgut (Suto et al., 1979).

In the dragonfly, *Leucorrhnia dubia*, infected with a denso-like virus, the nymphs become sluggish and flaccid, but there is no other external sign of the disease (Charpentier, 1979).

The larva of the butterfly, *A. vanillae*, when infected with the *Agraulis* DNV, is flaccid and discolored after death. It is killed mainly at the prepupal or pupal stage and the dead chrysalis turns light brown caused by secondary bacterial invasion. If the insect survives, the prepupae frequently fails to shed its larval skin or the adult fails to eclose. Adults that eclose have deformed wings. The disease appears to affect the entire insect although epidermis is uneffected (Kelly et al., 1980a).

In the silkworm larva infected with the *Bombyx* DNV, the histopathological changes are very different from those of other DNV infections. When silkworm larvae are perorally infected with the DNV, they usually die after 7 days; body flaccidity is the major sign. The alimentary canal of the diseased larva becomes pale yellow in color with little internal contents. This sign is similar to that of IFV infection, but the midgut histopathology differs between the two viruses. In DNV infection, the goblet cells are relatively intact, but the degraded columnar cells with hypertrophied nuclei are discharged into the gut lumen (Shimizu, 1975; Watanabe et al., 1976). In the case of the midgut infected with IFV, the goblet cells are first degenerated and discharged into the gut lumen (Iwashita, 1965).

Bombyx DNV multiplies only in the nuclei of columnar cells of the midgut epithelium, and no histopathological changes are observed in other tissues (Watanabe et al., 1976). When the progress of the disease is followed by fluorescent microscopy after staining with acridine orange, the greenish yellow areas in the nuclei of columnar cells do not change and are almost the same as those of healthy cells, i.e., there are no marked color changes of fluorescence in the nuclei, which is different from the case of *G. mellonella* (Nakagaki and Kawase, unpublished).

In an isolate of the silkworm DNV discovered in China, the histopathological changes also occur in the columnar midgut cells (Iwashita and Chun, 1982). The infected nucleus hypertrophies as the virus multiplies. At the final stage of infection, the infected nucleus is more than 2.5 times larger than normal. The entire nuclear area changes into a dense, homogeneous structure which stains strongly with methyl green and Feulgen reaction. When the midgut is stained with acridine orange, the orange yellow or red intranuclear inclusion occupies the entire nucleus, which suggests the presence of viruses with ssDNA. These histopathological features clearly differ from those of *Bombyx* DNV described above.

Histopathological aspects of midgut cells infected with both Saku and Yamanashi isolates of silkworm DNVs are fairly similar to those of the China isolate, and the columnar cells are rarely discharged from the epithelium into the gut lumen, which differ from *Bombyx* DNV (H. Watanabe, personal communication; H. Seki and Y. Iwashita, personal communication).

IV. PROPERTIES OF VIRIONS

A. *MORPHOLOGY*

1. *Size*

The particle sizes of several DNVs were determined by measuring the purified virions, which had been negatively stained with phosphotungstate (PTA), and by measuring virions in ultrathin sections. Although they all seemed to have diameters of the order of 19 to 24 nm, some variation existed. Vago et al. (1964; 1966) first reported that the semipurified *Galleria* DNV stained with 2% PTA had a diameter of 23 nm, but those in sections of insect tissues were 19-20 nm. The sizes of DNVs reported by several investigators were approximately 20 nm (Longworth et al., 1968; Kurstak and Côté, 1969). Thereafter, Tijssen et al. (1977) reported that the DNV from *G. mellonella* was heterogenous and consisted of two types with mean diameters of 24 and 21 nm for types I and II, respectively. According to their report, the mixture of these particles gave a two-peak histogram.

The sizes of other DNVs are as follows: *J. coenia*, 20-22 nm (Rivers and Longworth, 1972); *A. aegypti*, 20 nm (Lebedeva et al., 1973); *A. domestica*, 22 nm (Meynadier et al., 1977c); *D. saccharalis*, 22 nm (Meynadier et al., 1977b); *S. fusca*, 20-22 nm (Meynadier et al., 1977a); *P. fuliginosa*, 19-21 nm (Suto et al., 1979); and *A. vanillae*, 20 nm (Kelly et al., 1980a) (Fig. 1).

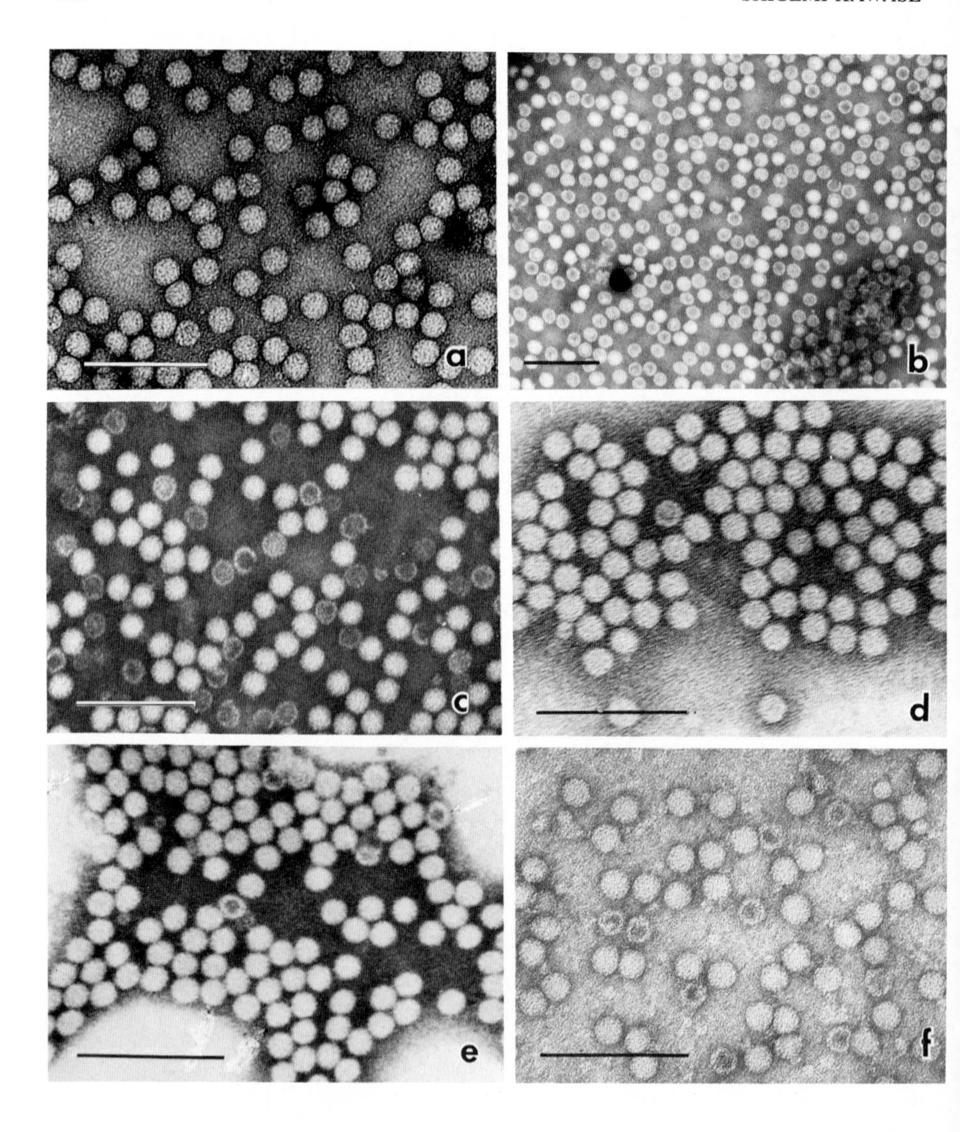

FIGURE 1. Several negatively-stained DNVs. (a) Galleria DNV (courtesy of Dr. P. Tijssen); (b) Diatraea DNV and (c) Acheta DNV (both courtesy of Dr. G. Meynadier); (d) Bombyx DNV (Nakagaki and Kawase, original photograph); (e) Periplaneta DNV (courtesy of Dr. C. Suto); and (f) China isolate of silkworm DNV (courtesy of Dr. Y. Iwashita). Bar indicates 100 nm.

The size of the *Bombyx* DNV was measured by negatively stained PTA preparations. The initially reported diameter was about 20 nm (Watanabe et al., 1976; Kawase and Kang, 1976). The following values were subsequently given: 21 nm (Maeda et al., 1977); 20 ± 1 nm (Kang et al., 1978). Nakagaki and Kawase (1980a) obtained a value of 22 ± 0.5 nm based on tobacco mosaic virus particles as an intenal standard. These variations may be due either to technical error, or to buffer and staining effects. The presence of different sized particles of *Bombyx* DNV has not been reported, but the occurrence of particles with two different densities have been observed in *Periplaneta* DNV (1.41 and 1.44 gm/ml) and in *Bombyx* DNV (1.40 and 1.45 gm/ml) (Suto, 1979; Nakagaki and Kawase, 1980a).

Sizes of other DNV isolates from the silkworm are as follows: DNV isolated by Matsui, 20 nm (Matsui and Watanabe, 1974); DNV isolated by Furuta, 22 nm (Kawase et al., 1982); China isolate, 21-23 nm (Iwashita and Chun, 1982); Saku isolate, 20 nm (H. Watanabe, personal communication); and Yamanashi isolate, 20 nm (H. Seki and Y. Iwashita, personal communication).

2. *Structure*

Because of the small size, the determination of the fine structure of DNV is very difficult to resolve with any degree of certainty. There are several reports on the number of capsomeres in Parvoviridae. For example, in Kilham rat virus (KRV) H-1, H-3, HT, and HB, the numbers are 32, and in AAV, 12 or 32 (Hoggan, 1971). The structure of the capsid of *Galleria* DNV has been investigated in detail by Kurstak and Côté (1969). They report that the DNV exhibits icosahedral symmetry. In highly magnified micrographs, the capsomeres measure between 2.0 to 3.5 nm with a 1.5 nm central hole. The number of capsomeres of the DNV reported initially by Kurstak and Garzon (1971) was 42. Subsequently, the number was corrected to 12 (Kurstak et al., 1977; Tijssen and Kurstak, 1979b).

In *Galleria* DNV, two types (DNV-I and II) have been isolated, with both types having similar isoelectric points. Although the viral proteins found in each type are identical, their stoichiometry differs. Localization of the structural protein, by labeling with periodate-oxidized glycoprotein, supports the hypothesis that sixty molecules of molecular weight (MW) 49,000 (p49) aggregates into a dodecahedron (12 pentamers), whereas two other proteins (p59 and p69), which may have a stabilizing function, are localized on the outer surface of the p49 dodecahedron (Tijssen and Kurstak, 1979b).

Capsid structure of the *Bombyx* DNV was investigated in highly magnified micrographs of greatly purified virions, using a rotation technique devised by Nakagaki and Kawase (1982) (Fig. 2). They reported that the virion had two-, three-, and fivefold axes of symmetry. They also found good agreement between the shape of the virus particle and the twelve capsomere model. Globular structures, which were compatible with the theoretical size of such capsomeres, were found in some virus preparations. Therefore, they suggested

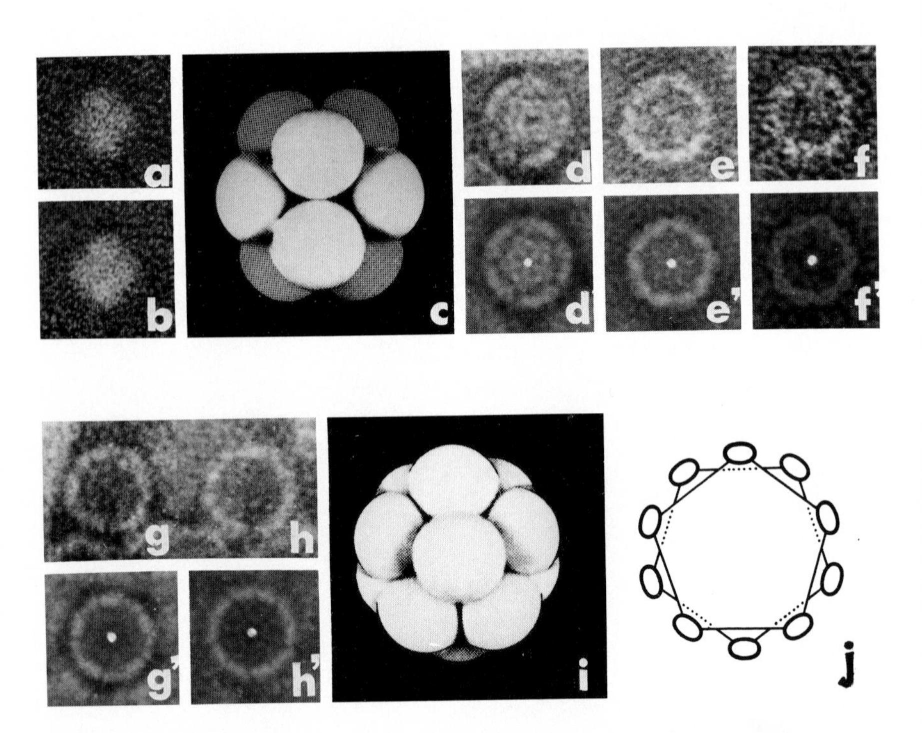

FIGURE 2. Bombyx DNV particles and the 12 subunits models. (a,b) Two selected DNV particles, probably oriented with one of its twofold axes of symmetry perpendicular to the plane of the micrograph. (c) The 12 subunits model, oriented with a twofold axis perpendicular to the plane of the photograph. (d-h) Five selected DNV particles, probably oriented with one of its fivefold axes of symmetry perpendicular to the plane of the micrograph. (d'-h') Particles enhanced by use of rotation technique (rotation × 5). Ten clear peripheral subunits are resolved. (i) The 12 subunits model oriented with a fivefold axis perpendicular to the plane of the photograph. (j) Drawing to illustrate the arrangement of periphery subunits: the upper set of 5 subunits and the lower set of 5 subunits. (Nakagaki and Kawase, 1982).

that the surface of the *Bombyx* DNV particle had a form of an icosahedron derived from 12 capsomeres. Protein analysis of the *Bombyx* DNV (which will be given later) showed that the virion had sixty molecules of VP1 (MW: 50,000) which appeared to be the repeating units for the capsid structure. On the basis of the electron micrographs, they concluded that each capsomere was a pentamer of VP1; presumably, the 12 capsomeres were arranged as an icosahedron.

B. PHYSICOCHEMICAL PROPERTIES

Parvoviruses contain DNA and protein, but no essential lipid, since they are resistant to ether and chloroform. They are also resistant to high temperature (Hoggan, 1971). In addition to DNA and protein, small quantities of polyamines have been detected in some DNVs (Kelly and Elliott, 1977; Bando et al., 1983). *Galleria* and *Periplaneta* DNVs, and silkworm DNV isolated by Furuta were stable at 56°-65°C (Boemare et al., 1970; Suto, 1979; Furuta, 1977). *Galleria* and *Periplaneta* DNVs are not inactivated after treatment with lipid solvents (Boemare et al., 1970; Suto, 1979).

1. Nucleic Acid

As with all parvoviruses, the DNV genome is a small ssDNA molecule. Truffaurt et al. (1967) first showed that *Galleria* DNV contained DNA. Subsequently, Kurstak et al. (1968b) and Longworth et al. (1968) confirmed the occurrence of DNA in the virion by staining with acridine orange, digestion with DNase, and diphenylamine reaction. The DNA content in the virion was about 37% (Longworth et al., 1968). The DNA initially extracted by Truffaurt et al. (1967) showed typical characteristics of double strands. However, it was subsequently demonstrated by Barwise and Walker (1970) through the effect of formaldehyde on the ultraviolet absorption spectrum of the virion, that the DNV possessed ssDNA, since the addition of formaldehyde to the virion produced an increase in absorption over a 24-hr period, and a shift in absorption maximum from 261 to 265 nm. These workers suggested that the DNV contained two separate complementary strands in different particles of DNV. When extracted under a solution of high ionic strength, the two complementary single strands were joined together by complementary base pairing and became one dsDNA. This phenomenon had been already shown for AAV, which contained complementary ssDNA molecules that were separately encapsidated (Berns and Rose, 1969). However, it should be noted that AAV is a satellite virus that depends on the adenovirus for its replication; DNV is an autonomously replicating virus.

Based on the MW of complete and empty virions, measured by sedimentation equilibrium as 5.7×10^6 and 3.5×10^6, respectively, the MW of the ssDNA in *Galleria* DNV has been calculated to be 2.2×10^6 (Barwise and Walker, 1970). The following values have been determined from sedimentation data: 4×10^6 from the dsDNA (Kurstak, 1972), and from direct measurement of ssDNA using electron microscopy as 1.6×10^6 (Tijssen et al., 1976). Two types of *Galleria* DNV, DNV-I (1.40 g/ml) and II (1.44 g/ml), contained ssDNA with similar physicochemical properties (Tijssen et al., 1977). Subsequently, Kelly et al. (1977) revealed that in both *Galleria* and *Junonia* DNVs existed as ssDNA's with limited secondary structure within the particle. This was assessed by spectral changes induced with formaldehyde, by melting profiles, and by circular dichroism studies. They confirmed that the ssDNA had an apparent MW of 1.9 to 2.2×10^6, determined by difference in the MW of virus particles and "top component" particles, based on naturally occurring empty particles and by the percentage of nucleic acids. They also showed that ssDNA was extracted from virus particles in a low salt buffer, and under appropriate conditions of high salt and elevated temperature, dsDNA was obtained. Similar phenomena were observed in the DNA extracted from both *Periplaneta* and *Bombyx* DNVs (Suto, 1979; Nakagaki and Kawase, 1980a). The linear dsDNA extracted from both *Galleria* and *Junonia* DNVs had MW's of about 3.9 to 4.1×10^6, as determined by neutral sedimentation and electron microscopy; a similar genome size was obtained by reassociation kinetics (Kelly et al., 1977). Their study showed that about 87% of the DNAs was homologous between *Galleria* and *Junonia* DNVs. Kelly et al. (1977) considered that the genome size, determined by reassociation kinetics, was equivalent to the physical size of the dsDNA extracted from the virus particles. Furthermore, they revealed that DNV is not a multicomponent virus, i.e., it is not a virus with a large genome divided into physically identical small molecules packaged in equivalent amounts in different particles. This has special importance in relation to the coding capacity of the virus, which is a protein of MW 1.76×10^5, if the entire genome is transcribed and translated. As will be stated later, the combined MW's of the four polypeptides of *Galleria* DNV range from 2.2 to 2.6×10^5 (MacLeod et al., 1971; Kelly et al., 1977). Therefore, all the DNV structural polypeptides are unlikely to represent primary gene products (Kelly et al., 1977).

Kelly and Bud (1978) reported subsequently that dsDNA from both *Galleria* and *Junonia* created by the annealing of complementary strands of ssDNA on release from virus particles could exist primarily not only as linear monomers, but also as circular monomers and concatamers, although predominantly as linear monomers (Fig. 3). By electron microscopy and agarose gel electrophoresis, they concluded that a limited nucleotide sequence permulation, which was probably nonrandom, comprised about 2.7% of the genome (160 base pair), which was considered to be the structural feature causing circulization and concatamerization. Furthermore, they stated that evidence of an inverted terminal repetition, observed by circulization of ssDNA, was obtained by electron microscopy and S_1 nuclease digestion. Estimates of the size of the terminal repeats varied from 60 to 380 base pairs. Restriction endonuclease (REN) analyses of DNAs from *Galleria* and *Junonia* DNVs also indicated a close genetic relationship (Kelly unpublished, cited in Harrap and Payne, 1979). *Hha*II produced identical fragment profiles for the two viral DNVs, while *Hind*III and *Hpa*II cleavages produced minor differences in the sizes of DNA fragments.

In *Bombyx* and *Periplaneta* DNVs, as well as in *Galleria* DNV, similar properties of the DNA extracted from the virion had been observed (Kawase and Kang, 1976; Kang et al., 1978; Suto, 1979; Nakagaki and Kawase, 1980a). When the DNA was extracted in low salt buffer, it possessed properties typical of single-stranded molecule. Under conditions of an appropriate high salt and an elevated temperature, dsDNA was extracted. This phenomenon was also confirmed by acridine orange staining of the extracted DNA and of the virion, respectively. The DNA content in the virion of *Bombyx* DNV was 28 ± 2% (Nakagaki and Kawase, 1980a), and the base ratio was G:A:C:T = 22.1:30.0:20.6:27.2 (Kawase and Kang, 1976).

An electron micrograph of the dsDNA of *Bombyx* DNV revealed that the DNA was composed of linear molecules with

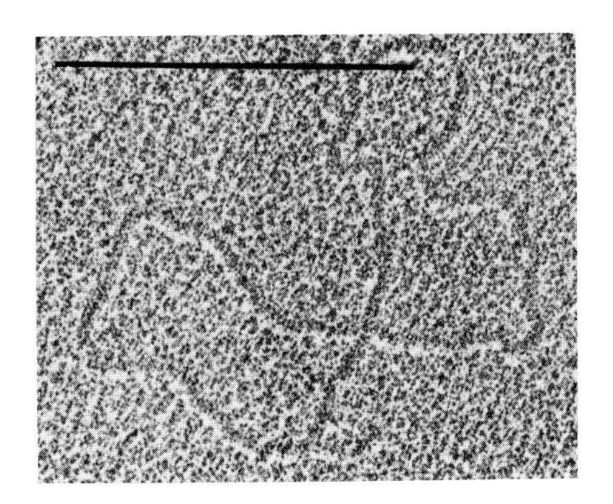

FIGURE 3. Electron micrograph of galleria DNV dsDNA (circular monomer) (Kelly and Bud, 1978). Bar indicates 500 nm.

an average length of 1.7 μm and other less well-defined structures (Nakagaki and Kawase, 1980a) (Fig. 4). The ds molecule had a MW of about 3.4×10^6, which was determined by both electron microscopy and agarose gel electrophoresis. When the dsDNA was alkali-denatured and examined under the electron microscope, the linear single-stranded molecule with appropriate length of 1.7 μm was observed. This indicated that the MW of the linear single-stranded molecule in the virion was 1.7×10^6 (Nakagaki and Kawase, 1980a).

Although no circular DNA molecule was observed in *Bombyx* DNV preparations, molecules with a length of 3.4 which were assumed to be linear duplex dimers, were rarely observed by electron microscopy (Nakagaki and Kawase, unpublished). This fact revealed that a cohesive end was also present in the DNA molecules of *Bombyx* DNV, through which two molecules were held together by short complementary ss regions, as in the case of AAV DNA (Berns and Hausworth, 1979).

2. *Protein*

Several vertebrate parvoviruses are reported to have three major polypeptides with MW's in the range of 52,000 (VP3) to 92,000 (VP1) with either VP1 or VP2 (MW: 62,000-79,000) being the major component (Kurstak et al., 1977). In contrast, when examined on 10% polyacrylamide gels, all invertebrate DNVs reported to date have four polypeptides with MW's 43,000-110,000 (Table I), although both MacLeod et al. (1971) and Kurstak (1972) identified only three polypeptides on 5% gels.

Tijssen et al. (1976) reported that four polypeptides were present in *Galleria* DNV as shown in Table I, and they suggested that the polypeptide of 98,000 (p98), which was corrected as 92,000 by Tijssen et al. (1982), may be a dimer of a polypeptide of 49,000 (p49). However, Tijssen and Kurstak (1981), using peptide mapping and enzyme-linked immunoadsorbent assay

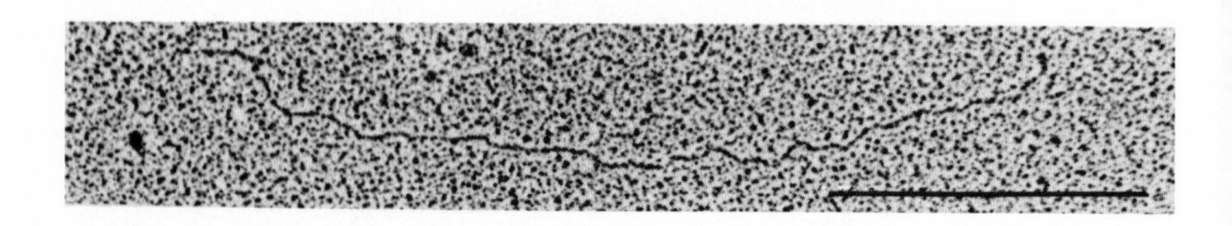

FIGURE 4. Electron micrograph of Bombyx DNV dsDNA (linear monomer) (Nakagaki and Kawase, 1980a). Bar indicates 500 nm.

TABLE I. *Densovirus Structural Proteins*[a,b]

	Galleria			*Junonia*	*Bombyx*	*Periplaneta*	*Agraulis*
	(1)[b]	(2)	(3)	(2)	(4)	(5)	(6)
VP1	49,000	42,600	49,000	41,900	50,000	48,000	43,000
VP2	58,500	61,100	58,500	58,700	57,000	52,000	62,000
VP3	69,000	70,900	67,000	70,800	70,000	61,000	72,000
VP4	98,000	107,300	92,000	109,600	77,000	76,000	110,000

[a]Underlined figures indicate the main protein for each virus.

[b](1) Tijssen *et al.* (1976); (2) Kelly *et al.* (1980b); (3) Tijssen *et al.* (1982); (4) Nakagaki and Kawase (1980b); (5) Suto (1979); (6) Kelly *et al.* (1980a).

(ELISA) technique, concluded that p98 was a monomer and not a dimer of p49, since the supplementary fragments of p59 and p69, in addition to the p49 fragments, were found in the p98 digest. The summed MW's of four polypeptides of *Galleria* and *Junonia* DNVs were approximately 2.6 × 10^5 (Kelly et al., 1977; 1980b).

According to Tijssen et al. (1976), the mass of the main protein of *Galleria* DNV was equivalent to about sixty protein molecules per virion. This is useful supporting evidence for the sixty subunits model first suggested by Barwise (1969). Furthermore, Tijssen and Kurstak (1979b) reported that the faster sedimenting component of *Galleria* DNV, which they termed DNV II, had less protein per virion than the major component, and that polypeptides of p59 and p69 were present in about one-half the concentration. If the sixty subunit model of twelve pentamers is assumed to be correct, and about one-half of each of these two polypeptides is lost in the formation of DNV II, although DNV II still remains intact, a surface location, perhaps at the center of each pentamer, may be suggested.

Tijssen et al. (1982) also demonstrated important stoichiometric differences in the DNV preparations from different origins. For example, the relative quantity of p49 in the preparation from naturally infected larvae at a younger stage was one-half of that of older infected larvae. They suggest the variation is very likely due to an age-related alteration of the synthesizing and processing mechanisms of proteins.

According to the data of Moore and Kelly (1980) who used ^{125}I-labeled proteins of *Galleria* and *Junonia* DNVs, the peptide maps obtained by limited proteolysis of isolated proteins of both types indicated a common origin of the virus proteins and a homology between the different viruses. The structure of *Junonia* DNV and its homologous top component were compared by proteolysis using several proteases and the bifunctional cross-linking reagents, dimethyl suberimidate (DMS) and dimethyl malonimidate. Similar susceptibilities of both components with protease were obtained. Only the top components were accessible to the action of the cross-linking reagent DMS. The lowest MW major structural polypeptide was most resistant to the action of the protease and DMS. Tijssen and Kurstak (1979a; 1981) carried out an experiment on peptide mapping after dansylation using SDS-PAGE. Furthermore, they conducted a comparative analysis of the viral proteins using ELISA. Their data showed that *Galleria* DNV had four unique structural proteins, all with extensive sequence homologies. All of the structural proteins contained intraprotein, but no interprotein, disulfide linkages.

In the case of *Periplaneta* DNV, four polypeptides with MW of 48,000 to 76,000 were detected on 10% gel system, and about 60% of the total viral protein was composed of the major polypeptide VP2 (MW: 52,000) (Suto, 1979).

The structural proteins of *Bombyx* DNV were investigated in detail using SDS-PAGE by Nakagaki and Kawase (1980b). Four structural proteins were found as in other DNVs. The MW was estimated from the relative mobility in various gel concentrations and the retardation coefficient. The major protein (VP1), accounting for 65% of the total virion protein, had a MW of about 50,000. The *Bombyx* DNV particle contained about sixty molecules of VP1, which is believed to be capsid protein. In *Galleria* DNV, the estimates of the MW's of viral proteins varied considerably with a change in gel concentration (Tijssen et al., 1976). However, in *Bombyx* DNV, the variations reported by MacLeod et al. (1971) and Tijssen et al. (1976) did not occur when the electrophoresis was conducted at different gel concentrations (Nakagaki and Kawase, 1980b).

Bombyx DNV possesses four structural proteins with a total MW of about 2.54×10^5. If the total MW represents the products of individual genes, this would exceed the coding capacity of the DNA (which is about 1.6×10^5, assuming the unlikely situation that the total genome is transcribed and translated). Since these phenomena are also observed with many other parvoviruses, a possibility exists of either an incorporation of host proteins into the virion or the presence of overlapping genes in the viral DNA. For this reason, the peptide mapping technique has been applied, and amino acid compositions of the four structural proteins have been investigated (H. Bando and S. Kawase, unpublished).

Table II shows the amino acid compositions of the four structural proteins of *Bombyx* DNV. The data indicate some differences, as well as similarities, among the four structural proteins. These proteins contained more acidic than basic amino acids and the relative concentrations of each amino acid were very similar. However, it should be noted that VP2 and VP4 showed distinctly higher serine concentration than VP1 and VP3, and VP2 (MW: 57,000) contained a larger amount of serine than VP3 (MW: 70,000).

The peptide map of VP1 almost coincided with that of VP2 (Fig. 5). All of the fragments of VP3 obtained after the digestion with *Staphylococcus aureus* V8 protease were also found in VP4, whereas the digests of VP3 and VP4 contained at least three extra bands as compared to VP1 or VP2. On the other hand, at least four extra bands were present in the maps of VP1 and VP2 as compared to other proteins of higher MW. Moreover, one unique band was identified in VP2 (H. Bando and S. Kawase, unpublished). The strikingly similar peptide patterns of the structural proteins suggested that VP3 and VP4 might have extensive sequence homology.

TABLE II. The Amino Acid Composition of the Four Structural Proteins of Bombyx DNV[a]

Amino acids	VP1		VP2		VP3		VP4	
	M%	Res	M%	Res	M%	Res	M%	Res
Asp	14.05	70.3	14.35	81.8	14.16	99.1	13.66	105.2
Thr	8.26	36.5	8.29	47.3	8.05	56.4	7.61	58.6
Ser	7.97	39.9	10.28	58.6	6.11	42.8	10.01	77.1
Glu	10.36	51.8	9.66	55.1	12.65	88.6	11.65	89.7
Gly	8.27	41.3	8.64	49.3	8.22	57.5	8.41	64.7
Ala	10.31	56.5	9.91	56.4	8.36	58.5	8.16	62.8
Val	7.18	35.9	6.72	38.3	7.43	52.0	6.93	53.4
Met	1.18	6.0	1.64	9.1	1.62	11.4	1.53	11.7
Ile	6.79	34.0	6.56	37.4	6.74	47.2	6.34	48.8
Leu	7.54	37.7	7.68	43.8	8.64	60.5	8.23	63.4
Tyr	4.39	22.0	4.12	23.4	3.33	23.3	3.30	25.4
Phe	3.86	19.5	3.38	19.4	4.54	31.8	4.14	31.9
Lys	3.08	15.5	2.72	15.4	3.09	21.6	2.77	21.3
His	1.08	5.5	1.01	5.8	1.01	7.0	1.06	8.1
Arg	5.69	28.4	5.02	28.5	6.32	44.2	6.22	47.9

[a]M%, Percentage (mol %) of total recovered amino acids; Res, number of residues calculated from Mol. Wt. of each structural protein. From H. Bando and S. Kawase, unpublished data.

Moreover, two smaller proteins, VP1 and VP2, might be derived, at least partially, from a common DNA sequence. These data suggest that the derivation of the structural proteins of *Bombyx* DNV is similar but more complex than those of other parvoviruses, including *Galleria* DNV.

TABLE III. Polyamine Content of DNV Particles

Polyamine	Polyamine content (μg/mg of viral protein)		
	Galleria (1)[a]	Aglais (1)	Bombyx (2)
Spermine	0.64 ± 0.02	0.62 ± 0.02	3.16 ± 0.30
Spermidine	20.08 ± 1.00	20.10 ± 1.00	4.37 ± 0.50
Putrescine	1.67 ± 0.07	1.67 ± 0.07	0.74 ± 0.03

[a](1) Kelly and Elliott (1977); (2) Bando et al. (1983).

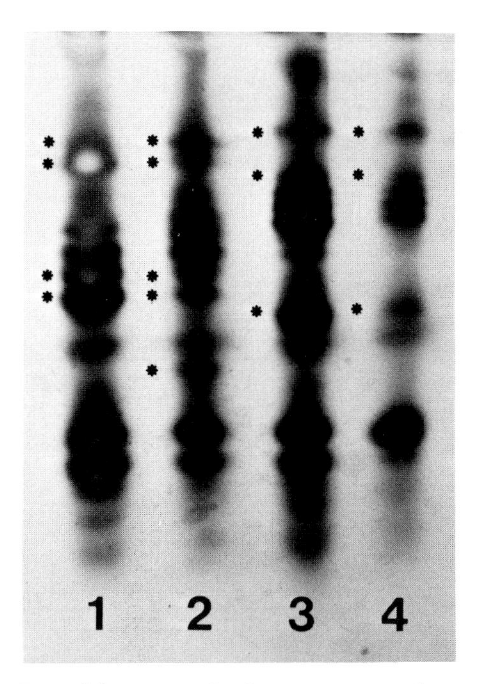

FIGURE 5. Peptide map of the structural proteins of Bombyx DNV generated by Staphylococcus aureus V8 protease. Lanes 1-4 were loaded with the gel slices containing each structural protein, VP1, VP2, VP3, and VP4, respectively. The dots indicate the peptide bands which were not common for all the structural proteins (Bando and Kawase, unpublished data).

3. *Polyamine*

The naturally occurring polyamines, spermine, spermidine, and putrescine, and related polyamines occur in many viruses as well as in biological materials. Polyamines were reported in *Galleria* and *Aglais* DNVs by Kelly and Elliott (1977). Recently, quantitative analyses of the polyamine content in *Bombyx* DNV were carried out by Bando et al. (1983). Table III shows the polyamine contents in *Galleria*, *Aglais*, and *Bombyx* DNVs.

Both *Galleria* and *Aglais* DNVs contain spermine, spermidine and putrescine, in their virions, and the relative amounts of the three polyamines from both viruses are virtually identical, despite the fact that they originate in different hosts. However, in the case of *Bombyx* DNV, the contents of spermine, spermidine, and putrescine clearly differ from other DNVs. In *Bombyx* DNV, 68 molecules of spermine, 130 molecules of spermidine, and 37 molecules of putrescine are contained in each particle, and the sum of these polyamines comprise about 0.6% of the virus particle by weight. Since the MW of the virus DNA is 1.7×10^6 (Nakagaki and Kawase, 1980a), there are about 5600 phosphate charges, assuming one charge per nucleotide. On the basis of spermine with four nitrogen atoms, spermidine with three, and putrescine with two, and provided that all of the polyamines are found in the viral DNA, there are sufficient polyamines present to neutralize a maximum of about 13% of viral DNA phosphate groups.

C. IMMUNOLOGICAL CHARACTERIZATION AND HEMAGGLUTINATION

1. Serological Relationship

Galleria and *Junonia* DNVs are reported to be very closely related serologically, even though these viruses differ widely in host range and in histopathology in their respective hosts (Rivers and Longworth, 1972). *Agraulis* DNV isolated from *A. vanillae* is related to both *Galleria* and *Junonia* DNVs (Kelly et al., 1980a), but *Bombyx* DNV is not related to *Galleria* DNV (H. Watanabe, personal communication). A virus, which multiplies in the midgut of the mulberry pyralid, *Glyphodes pyloalis*, is not distinguishable serologically from *Bombyx* DNV (Watanabe and Shimizu, 1980; Watanabe, 1981). *Galleria* DNV is not related to AAV-1, latent rat virus, MVM or bovine parvovirus (Hoggan, 1971). In partial identity patterns, the *Junonia* DNV produces a larger spur than that of *Galleria* DNV, suggesting that, while each virus shares common antigens, *Junonia* DNV has more antigens which differ from those of *Galleria* DNV (Longworth, 1978).

Among several isolates of silkworm DNVs, the DNVs isolated by Furuta and by Matsui are closely related serologically to *Bombyx* DNV (Furuta, 1975; Matsui et al., 1977; Kawase et al., 1982). No serological relationships occur between *Bombyx* DNV and the China isolate, and between *Bombyx* DNV and the Saku isolate (Y. Iwashita, personal communication; H. Watanabe, personal communication).

2. Hemagglutination

Some parvoviruses agglutinate erythrocytes of rat, human (type O), and guinea pig. RV, MVM, H-1, and Haden viruses agglutinate these erythrocytes, but *Galleria* DNV and AAV-1 and -3 do not (Hoggan, 1971). Boemare et al. (1970) and Kurstak (1972) report that *Galleria* DNV does not agglutinate the erythrocytes from monkey, horse, cow, duck, goat, hamster, sheep, pig, mouse, human, guinea pig, and baboon. *Bombyx* DNV also does not agglutinate erythrocytes from guinea pig and sheep (M. Nakagaki and S. Kawase, unpublished). *Periplaneta* DNV does not agglutinate erythrocytes from man (O and B), savannah monkey, white rat, guinea pig, mouse, and frog (Suto et al., 1979).

V. REPLICATION AND MORPHOGENESIS

A. *HOST RANGE*

Densovirus was originally isolated from the larvae of the lepidopterous *G. mellonella* (Meynadier et al., 1964). It kills the larvae in 4-6 days at 28°C (Vago et al., 1964). The host range of *Galleria* DNV is restricted to *G. mellonella*. In contrast, *Junonia* DNV infects *Mamestra brassicae*, *Bombyx mori*, and *Lymantria dispar*, but not *G. mellonella* (Rivers and Longworth, 1972; Longworth, 1978). Interestingly, *Galleria* DNV can be transmitted from an infected *G. mellonella* larva to a healthy larva via the ovipositor of the parasite, *Nemeritis vanescens* (Kurstak and Vago, 1967).

Aedes DNV was infectious for different species of mosquitoes: *Aedes vexans*, *A. geniculatus*, *A. caspius dorsalis*, *A. cantans*, *A. albopictus*, *Culex pipiens pipiens*, *C. pipiens molestus*, and *Culiseta annulata* (Lebedinets et al., 1978). The infectivity of the DNV of *Periplaneta fuliginosa* against other species of *Periplaneta* has been investigated. Adults of *P. australasiae* and *P. brunnea* and the nymphs of *P. japonica* are susceptible (Suto et al., 1979). In the case of *Bombyx* DNV, there is no information concerning its cross-infection to other insects except between *Bombyx mori* and *Glyphodes pyoalis*, i.e., the DNV of *Glyphodes* infects DNV-susceptible strains of silkworm (Watanabe, 1981). Peroral inoculation tests of *Bombyx* DNV to various silkworm strains reveal that most strains are nonsusceptible, while some strains are highly susceptible to DNV infection (Watanabe and Maeda, 1978). Similar results have been obtained with another DNV isolated by Furuta (Furuta, 1977).

In order to determine the mode of inheritance of nonsusceptibility in silkworm strains, the infectivity tests have been conducted with susceptible (S) and nonsusceptible (N) parent strains, their reciprocal F_1 hybrids, and the back-crossed hybrids to either of the parents. The -log IC_{50} values of reciprocal F_1 hybrids and those of their back-crossed hybrid to the S strain are nearly the same as that of the S strain, while the -log IC_{50} value of F_2 hybrid is a little less; the value of the back-crossed hybrid to the N strain is significantly less than that of the S strain. When high concentrations of DNV that result in 100% infection in S strain are administered to the F_2 hybrid and to the back-crossed hybrid to the N strain, the susceptible and nonsusceptible larvae segregate at a 3:1 and a 1:1 ratio, respectively. These results indicate that nonsusceptibility to DNV infection is controlled genetically by a recessive gene which is not sex linked. Therefore, the practical rearing of the silkworm strain which contains homozygously the nonsusceptible recessive gene is one of the best ways to protect the silkworm against epizootics of densonucleosis in sericultural farms. The mechanism by which the nonsusceptibility gene causes resistance to DNV in the silkworm is unknown (Watanabe and Maeda, 1978).

B. REPLICATION IN MULTIPLE INFECTION AND AT HIGH TEMPERATURE

1. Multiple Infection

Kurstak and Garzon (1975) and Odier (1975) showed a selective inhibition in the production of *Galleria* DNV in cells of the fat body if the DNV inoculation was preceded at least 16 hr by the inoculation of a nuclear polyhedrosis virus (NPV). The silk glands, however, which are not susceptible to infection by NPV, developed typical DNV cytopathogenicity. The simultaneous inoculation of DNV and NPV produced a characteristic DNV in all cells. Odier (1975) observed the two viruses in the same nucleus of *G. mellonella*, i.e., the virogenic stroma of DNV and bacilloformic NPV virions at different stages of development. The initial infection of DNV suppressed most of the development of NPV in the nucleus. In another study by Kurstak and Garzon (1975), single cells of *G. mellonella* were observed to be coinfected with *Tipula* iridescent virus (TIV) and DNV. Under these conditions, the simultaneous production of both viruses was observed, without any apparent mutual or antagonistic effect other than exhausting the cell by metabolic competition.

Cytopathological studies were carried out by Matsui and Watanabe (1974) on the viral multiplication in the coinfection of silkworm DNV isolated by Matsui (SFV) and of IFV in the silkworm midgut. Their optical microscope observations revealed that the infected nuclei of columnar cells became swollen and contained a few inclusions which stained with Feulgen reaction. Using electron microscopy, they reported that the virogenic stromata, in which SFV particles were produced, were formed in the nuclei of columnar cells. They observed that SFV and IFV frequently coinfected the same columnar cell, the former of about 19 nm in diameter in the nucleus and the latter of about 25 nm in the cytoplasm.

2. *High Temperature*

Multiplication of *Bombyx* DNV was greatly reduced when the infected larvae of the silkworm, which had been reared at 25° to 28°C, were submitted to a high temperature of 37°C (Watanabe and Maeda, 1979). Autoradiographic results with [^{3}H]thymidine and [^{3}H]tyrosine revealed that the synthesis of both viral DNA and protein was greatly reduced in the infected larvae maintained at 37°C. Fluorescent antibody studies also confirmed that the synthesis of DNV antigen in the larva was inhibited at 37°C. These results indicated that high temperature reduced the activity of the enzymes concerned with viral DNA and protein syntheses.

C. *METABOLISM IN THE TISSUES INFECTED WITH DNV*

The hemolymph and cell metabolism of various tissues are affected by virus infection. In *G. mellonella* infected with *Galleria* DNV, electropherograms of the hemolymph from the infected larvae showed a significant decrease of several fractions of hemolymph protein compared with those of the healthy larvae (Weiser and Lysenko, 1972). In the case of *Bombyx* DNV, the target organ is the midgut. The changes in protein, nucleic acid, and polyamine concentrations in the midgut during the course of the DNV infection were investigated by H. Bando and S. Kawase (unpublished), to establish the dynamic changes in the metabolism of the midgut infected with DNV. By using a single-radial immunodiffusion technique, they found that *Bombyx* DNV multiplication, without apparent damage, was almost completed by the 6th day after the oral inoculation of the virus. No marked differences were observed between the infected and the healthy midgut epithelia based on concentrations of DNA, RNA, and protein, except for the rather high concentration of DNA in the infected midgut during the last 4 days. Three kinds of polyamines, spermine, spermidine, and putrescine were found in the midguts. The spermine:spermidine ratio in the infected midgut during the high viral multiplication period was lower than that of the healthy midgut.

D. CYTOPATHOLOGY

1. Electron Microscopy

Vago et al. (1966) have reported that, in ultrathin sections the Feulgen-positive dense inclusions which appear in the nuclei of the fat body of *G. mellonella* after infection with the DNV, are composed of particles of 19-20 nm in size. According to Garzon and Kurstak (1976), the first ultrastructural changes in *Galleria* DNV infection are an increase in the number of free ribosomes and the formation of "microbody-like" structures arising from the accumulation of small, round bodies of 17 to 20 nm inside of vesicles. In the nucleus, the nucleolus undergoes hypertrophy which is accompanied by a segregation of its fibrillar and granular components. The development of the granular portion coincides with the synthesis of dsDNA of the replicative form within the virogenic stroma. As the infection progresses, the granular portion of the nucleolus regresses in favor of the fibrillar portion. The DNV seems to stimulate the formation of intranuclear bodies associated with the virogenic stroma. The virions are assembled inside the virogenic stroma. The replication of *Galleria* DNV shows certain similarities with those of other parvoviruses, particularly AAV and H-1 (Garzon and Kurstak, 1976).

In *Aedes* DNV, the first ultrastructural changes are observed in the cytoplasm of virus-infected cells. They consist of the formation of paracrystalline structures containing particles of 18-20 nm diameter. Virogenic stromata and paracrystalline virion arrays are found in the nuclei of virus-infected cells (Buchatsky and Raikova, 1979). A similar crystalline array of virus particles was observed in the cytoplasm of the fat body, muscle, and pericardial cells of *P. fuliginosa* infected with *Periplaneta* DNV (Suto et al., 1979).

In the case of *Sibine* DNV, cell alterations, mainly at nuclear level, have been found in thin sections of the thickened intestinal wall (Meynadier et al., 1977a). At a less advanced stage of infection, the structure of the nucleus is already altered, and the nuclear substance is divided into opaque masses. The virogenic mass continues to grow with small cavities around which virus particles are formed. The mitochondria hypertrophy and burst, and their contents released leaving vacuoles in their place.

The first ultrastructural change that occurs in the silkworm infected with *Bombyx* DNV is the appearance of electron-dense bodies in the nuclei of the midgut cell. Thereafter, the bodies occupy most of the nucleus, suggesting that these are virogenic stromata. No virogenic stroma is observed in the cytoplasm (Watanabe et al., 1976) (Fig. 6). The ultrastructural changes of midgut cells infected with the China isolate of silkworm DNV occur in both the cytoplasm and nucleus (Iwashita and Chun, 1982). In the nucleus at an early stage of infection, the chromatin disperses and many nucleolei appear. The nucleus hypertrophies greatly, and is gradually filled with a network of electron-dense granules, which is probably the precursor of the virogenic stroma. In this replication, two different patterns occur in the infected nuclei. In one of them, the virions replicate in linear array and, in the other, the virions first disperse as patches, gradually condense, and finally form large masses.

2. *Autoradiography*

The site of synthesis of the viral nucleic acids has been studied by autoradiography with [^{3}H]thymidine in both *Galleria* and *Bombyx* DNVs (Kurstak, 1970; Kurstak et al., 1970; Morris, 1970; Watanabe et al., 1976). In *Galleria* DNV, the autoradiograms examined by electron microscopy exhibit very strong in-

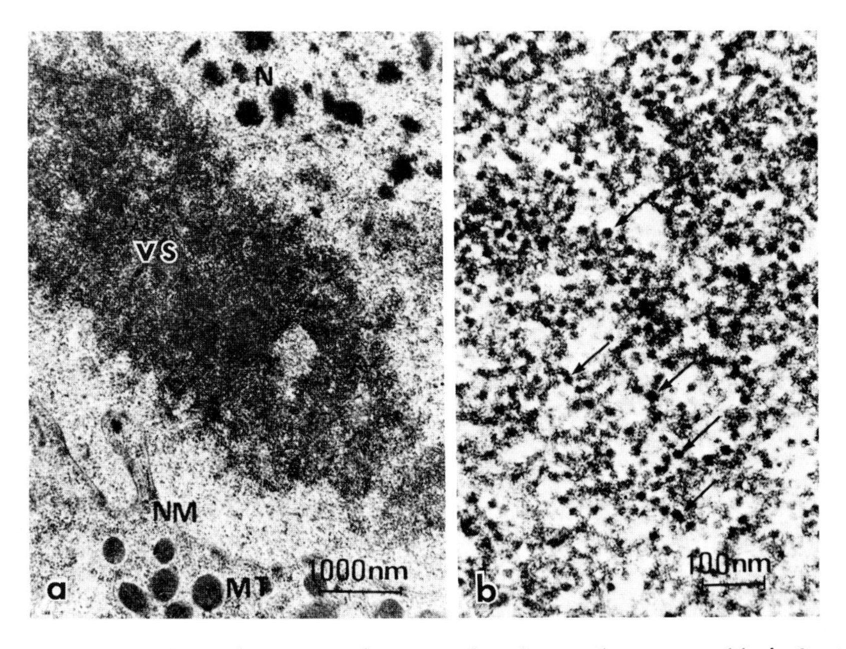

FIGURE 6. Electron micrograph of a columnar cell infected with Bombyx DNV. (a) Virogenic stroma in the nucleus. (b) High magnification of an area of virogenic stroma. VS, virogenic stroma; N, nucleus; NM, nuclear membrane; MT, mitochondrion. Arrows indicate the virions (Watanabe et al., 1976).

corporation of the isotope in the nuclei of infected cells (Kurstak, 1972). According to Kurstak (1970), the first silver grains are noticeable within 1-2 hr after oral inoculation of the virus. After 2 hr, the synthesis of the viral DNA increases up to 6 hr. However, when the virions appear in the nucleus, the nuclear DNA synthesis decreases and terminates at approximately 12 hr.

Changes in several amino acids in *G. mellonella* infected with the DNV have also been studied by autoradiography to determine the effect of the virion on protein metabolism (Morris, 1971). During the DNV infection cycle, the enlargement of the nucleus is accompanied by an increase of amino acids in the nucleus and a simultaneous decrease in the cytoplasm. Total cell protein increases at the intermediate stage of infection and then decreases. The rate of change in the nuclear protein content is inversely proportional to the rate of change in the cytoplasm.

In the case of *Bombyx* DNV, the DNA synthesis occurs predominantly in infected nuclei of columnar cells of the silkworm midgut, indicating that the virus multiplies in the nucleus (Watanabe and Maeda, 1979) (Fig. 7).

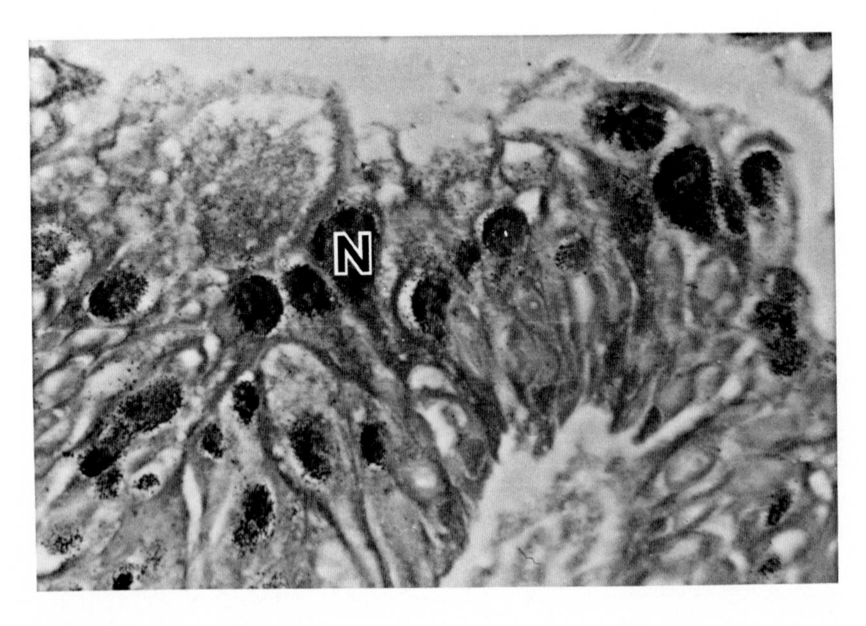

FIGURE 7. Radioautograph of DNV-infected midgut epithelium of silkworm showing incorporporation of [^{3}H]thymidine. N, nucleus (Watanabe and Maeda, 1979).

3. *Fluorescent Antibody and Immunoperoxidase Techniques*

Kurstak and Stanislawski-Birencwaijg (1968) have observed the localization of *Galleria* DNV infected cells, both *in vitro* and *in vivo*, using fluorescent antibody technique. The first fluorescence appears around 4 and 7 hr after virus inoculation. It is exclusively cytoplasmic and corresponds to the production of early antigens. The second pattern, which is intranuclear, corresponding to the production of structural antigens of *Galleria* DNV, begins close to the nuclear membrane in the form of a fluorescent halo. To confirm the results obtained by the immunofluorescence technique, the immunoperoxidase technique has been used to localize the DNV antigen (Kurstak et al., 1969b; 1970; Kurstak, 1972). In this technique, specific antibodies to *Galleria* DNV are labeled with peroxidase enzyme and are used to localize the DNV antigen in infected cells. Early proteins are observed in the cytoplasm 4 hr after viral inoculation at 32°C. At approximately 6-8 hr, the intracytoplasmic localization is replaced by an intranuclear one, first close to the nuclear membrane and then spreading throughout the nucleus. With this technique, the first viral antigens are detected in the cytoplasm 3 hr after oral inoculation of the virus.

The fluorescent antibody technique has also been applied with *Bombyx* DNV and the DNV isolated by Furuta (Maeda and Watanabe, 1978; Sato and Inoue, 1978). Of the various tissues in the infected silkworm larvae, only the midgut epithelium shows specific fluorescence. This fluorescence is restricted to the nuclei of midgut columnar cells, which shows that this is the site of virus multiplication. Fluorescence appears sporadically in a few columnar cells at 24 hr after virus inoculation and becomes more intense, the number of fluorescing cells increasing with time. At 72 hr, most of the columnar cells near the cardiac valve do not show any fluorescence even at late stages of infection (Fig. 8).

E. *VIRUS MULTIPLICATION IN CULTURED CELLS AND ORGANS*

Galleria DNV infects primary cultures of ovarian cells of *G. mellonella* and *Bombyx mori* (Vago et al., 1966). Kurstak et al. (1969c) have reported that *Galleria* DNV replicates in mouse L cells, but extensive attempts to infect mice have failed (Kurstak et al., 1977). In the mouse L cell line, 3-4 days after the viral inoculation at 37°C, some basophillic, Feulgen-positive intranuclear inclusions appear in the cells. The inclusions show yellow-orange fluorescence after staining with acridine orange, and the fluorescence is similar to that

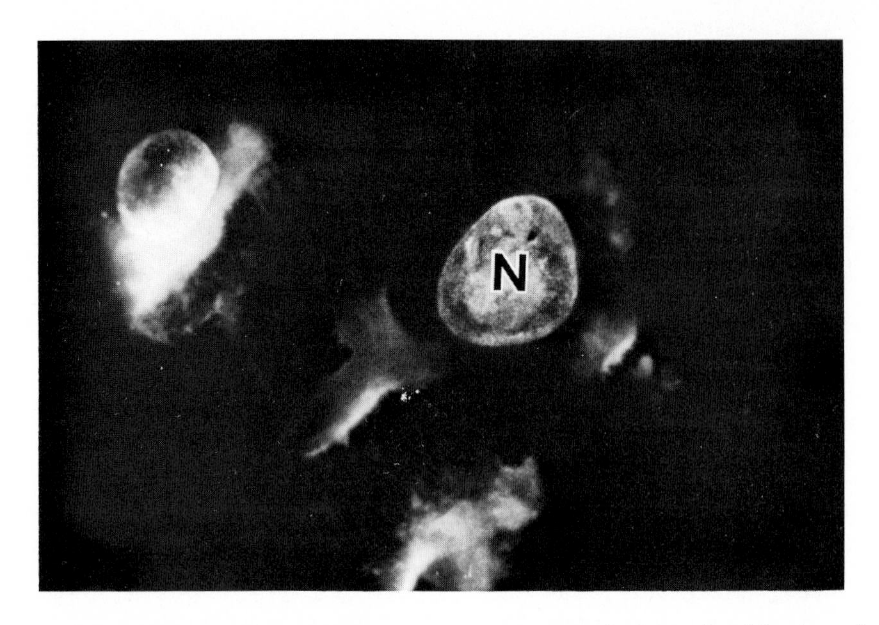

FIGURE 8. Fluoresced nucleus of DNV-infected midgut epithelium of silkworm stained with fluorescent antibody. N, nucleus (Maeda and Watanabe, 1978).

observed in the nucleus of *G. mellonella* cells (Kurstak et al., 1969a; 1969c). The virions, which were isolated from the L cells infected 6 days earlier using successive freezing and thawing, reproduce the densonucleosis in *G. mellonella* larva. The DNV antigens have also been detected by immunofluorescence in the cytoplasm and nuclei of infected L cells (Kurstak et al., 1969b). Interestingly, a portion of the infected L cells acquires the properties of transformed cells (Kurstak et al., 1969a). By day 3 or 4 of infection at 37°C, the uniform and oriented growth of L cells is disrupted, and by day 5-6, readily recognizable foci of round cells are observed throughout the entire culture. On day 7-8, these foci show the morphological characteristics of transformed cells. However, according to Longworth (1978), other workers have not found it possible to infect either vertebrate or invertebrate cells with *Galleria* DNV, despite attempts to coinfect with a baculovirus or an iridescent virus.

The development of *Galleria* DNV has been observed in ovaries, ovarioles of chrysalids, and in silk glands, fat bodies, and dorsal vessels of larvae of *G. mellonella* maintained in organotypic culture on a gell-type medium (Quiot et al., 1970). Cultures are infected with the hemolymph of diseased larvae or with the purified virus and suspended in a balanced salt solution of the culture medium. Nuclear lesions have been observed on these organs in culture.

There is no information on the multiplication of other DNV in cultured cells or organs. However, it is recognized by fluorescent antibody technique that the *Bombyx* DNV multiplies in the cultured embryos of susceptible strains of silkworm, but not in those of resistant strains (Y. Furuta, personal communication).

VI. EPIZOOTIOLOGY

There are few investigations concerning the epizootiology of DNVs. Lavie et al. (1965) and Giauffret (1966) applied *Galleria* DNV as a microbial insecticide for the elimination of *G. mellonella* from bee hives. Since this virus multiplies in cultured mammalian cells, and, according to Kurstak and Onji (1972), rat embryo fibroblasts are transformed *in vitro* in the presence of *Galleria* DNV, the massive use of this virus as a viral insecticide should be avoided until its pathogenicity to vertebrates has been investigated in greater detail.

Watanabe and Shimizu (1980) made epizootiological investigations on the occurrence of *Bombyx* DNV in sericultural farms in two separate districts of Japan. In Saitama and Nagano Prefectures, an enzootic of densonucleosis was only noted at a few farms in the Nagano Prefecture. The enzootic was mainly due to the rearing of silkworm strains which were nonsusceptible or highly resistant to DNV infection. However, DNV was detected generally in the dusts of every farm in the Nagano Prefecture, even in farms where nonsusceptible strains were reared and had no occurrence of densonucleosis. They also established that the *Bombyx* DNV detected in the dust originated from mulberry leaves contaminated with the virus which had multiplied chronically in the mulberry pyralid, *Glyphodes pyloalis*, infesting the mulberry plantations. They reported that the virus obtained from the midgut of the mulberry pyralid was indistinguishable serologically from *Bombyx* DNV, and was also pathogenic to susceptible silkworm larvae. In Saitama Prefecture, on the other hand, the DNV was not detected in the dust of any farm, or did larvae of the mulberry pyralid contain DNV in their midguts.

The above results suggest that the epizootic of *Bombyx* densonucleosis in sericultural farms is caused by two major factors: (1) the rearing of silkworm strains susceptible to *Bombyx* DNV, and (2) the infestation of mulberry pyralid infected with DNV in the mulberry plantations (Watanabe, 1981).

The results also suggest that the virus isolated as *Bombyx* DNV is originally derived from the virus of the mulberry pyralid. However, comparative studies on the chemical and physical nature of both viruses are required to obtain further information on the genetical variations and the micro-evolution of these insect viruses (Watanabe, 1981).

VII. CONCLUSIONS

Densoviruses have been discovered during the last two decades, and presumably numerous other new parvoviruses will be discovered from various host insects in the near future. At present, only *Galleria* and *Bombyx* DNVs have been investigated in some detail. More information on other DNVs is required to resolve the properties of the virion and the mechanism of viral replication. The mode of replication of DNV from a chemical and a molecular biological point of view must be studied to control the densonucleosis in sericulture or to use the DNV for microbial control.

The mode of replication of DNV DNA is still obscure. According to Kurstak et al. (1977), the replication of single-stranded viral DNA should include three stages from a theoretical point of view: (1) conversion of viral DNA to a replicative form (RF), (2) multiplication of the RF, and (3) the synthesis of the progeny single strands. The first stage probably relies on the host's replicative system and, thus, would not require virus-coded proteins. To resolve these problems, we must first establish suitable cell line systems for DNV multiplication.

ACKNOWLEDGMENTS

The author wishes to thank Dr. Y. Tanada, University of California, who has kindly revised this manuscript.

REFERENCES

Amargier, A., Vago, C., and Meynadier, G. (1965). Étude histopathologique d'un nouveau type de virose mis en évidence chez le lépidoptère *Galleria mellonella*. *Arch. Ges. Virusforsch. 15*, 659-667.

Bando, H., Nakagaki, M., and Kawase, S. (1983). Polyamines in densonucleosis virus from the silkworm, *Bombyx mori*. *J. Invert. Pathol. 42*, 264-266.

Barwise, A. H. (1969). Thesis, University of Oxford. (Cited in Kelly and Bud, 1978.)

Barwise, A. H., and Walker, I. O. (1970). Studies on the DNA of a virus from *Galleria mellonella*. *FEBS Lett. 6*, 13-16.

Berns, K. I., and Hausworth, W. W. (1979). Adeno-associated viruses. *Advan. Virus Res. 25*, 407-449.

Berns, K. I., and Rose, J. A. (1969). Evidence for a single-stranded adenovirus-associated virus genome: Isolation and separation of complementary single strands. *J. Virol. 5*, 693-699.

Boemare, N., Croizier, G., and Veyrunes, J. C. (1970). Contribution à la connaissance des propriétés du virus de la densonucléose. *Entomophaga 15*, 327-332.

Buchatsky, L. P., and Raikova, A. P. (1979). Electron microscope study of mosquito densonucleosis virus maturation. *Acta Virol. 23*, 170-172.

Charpentier, R. (1979). A nonoccluded virus in nymphs of the dragonfly *Leucirrhinia dubia* (Odonata, Anisotera). *J. Invert. Pathol. 34*, 95-98.

Federici, B. A. (1976). Pathology and histochemistry of a densonucleosis virus in larvae of the blackfly, *Simulium vittatum*. *Proc. Intern. Colloq. Invert. Pathol., Kingston, Ontario*, pp. 341-342.

Furuta, Y. (1973). Studies on the previously unreported virus infecting the silkworm, *Bombyx mori*. *J. Sericult. Sci. Japan 42*, 443-453.

Furuta, Y. (1975). Studies on the previously unreported flacherie virus infecting the silkworm, *Bombyx mori*. III. Serological difference between the small flacherie virus and the infectious one. *J. Sericult. Sci. Japan 44*, 375-380.

Furuta, Y. (1977). Studies on the small flacherie virus of the silkworm, *Bombyx mori* L. I. Difference between SFV and IFV in infectivity and the sensitivity to formaldehyde and heat. *J. Sericult. Sci. Japan 46*, 297-300.

Garzon, S., and Kurstak, E. (1976). Ultrastructural studies on the morphogenesis of the densonucleosis virus (parvovirus). *Virology 70*, 517-531.

Giauffret, A. (1966). *Bull. Apicol. 9*, 35. (Cited in Kurstak, 1972.)

Harrap, K. A., and Payne, C. C. (1979). The structural properties and identification of insect viruses. *Advan. Virus Res.* *25*, 273-301.

Hoggan, M. D. (1971). Small DNA viruses. *In* "Comparative Virology" (K. Maramorosch and E. Kurstak, eds.), pp. 43-79. Academic Press, New York.

Iwashita, Y. (1965). Histo- and cyto-pathological studies on the midgut epithelium of the silkworm larvae infected with infectious flacherie. *J. Sericult. Sci. Japan 34*, 263-273.

Iwashita, Y., and Chun, C. Y. (1982). The development of a densonucleosis virus isolated from silkworm larvae, *Bombyx mori*, of China. *In* "Ultrastructure and Functioning of Insect Cells" (H. Akai, R. C. King, and S. Morohoshi, eds.), pp. 161-164. Soc. Insect Cells, Japan, Tokyo.

Kang, S. K., Nakagaki, M., Shimizu, T., and Kawase, S. (1978). Purification of Ina-flacherie virus and some properties of its nucleic acid. *J. Sericult. Sci. Japan 47*, 39-46.

Kawase, S., and Kang, S. K. (1976). On the nucleic acid of a newly isolated virus from the flacherie-diseased silkworm larvae. *J. Sericult. Sci. Japan 45*, 87-88.

Kawase, S., Suto, C., Ayuzawa, C., and Inoue, H. (1974). Chemical properties of infectious flacherie virus of the silkworm, *Bombyx mori* L. (Lepidoptera: Bombycidae). *Appl. Entomol. Zool. 9*, 100-101.

Kawase, S., Nakagaki, M., Bando, H., and Furuta, Y. (1982). Studies on the chemical properties of "flacherie virus isolated by Furuta" in the silkworm, *Bombyx mori*. *J. Sericult. Sci. Japan 51*, 58-65.

Kelly, D. C., and Bud, H. M. (1978). Densonucleosis virus DNA: Analysis of fine structure by electron microscopy and agarose gel electrophoresis. *J. Gen. Virol. 40*, 33-43.

Kelly, D. C., and Elliott, R. M. (1977). Polyamines contained by two densonucleosis viruses. *J. Virol. 21*, 408-410.

Kelly, D. C., Barwise, A. H., and Walker, I. O. (1977). DNA contained by two densonucleosis viruses. *J. Virol. 21*, 396-407.

Kelly, D. C., Ayres, M. D., Spencer, L. K., and Rivers, C. F. (1980a). Densonucleosis virus 3: A recent insect parvovirus isolated from *Agraulis vanillae* (Lepidoptera: Nymphalidae). *Microbiologica 3*, 455-460.

Kelly, D. C., Moore, N. F., Spilling, C. R., Barwise, A. H., and Walker, I. O. (1980b). Densonucleosis virus structural proteins. *J. Virol. 36*, 224-235.

Kurstak, E. (1970). Étude par l'autographie en haute résolution de la synthèse de l'acide désoxyribonucléique du virus de la densonucléose. *Rev. Can. Biol. 29*, 207-211.

Kurstak, E. (1972). Small densonucleosis virus (DNV). *Advan. Virus Res. 17*, 207-241.

Kurstak, E., and Côté, J. R. (1969). Proposition de classification du virus de la densonucléose (VDN) basée sur l'étude de la structure moléculaire et des propriétés physicochimiques. *C.R. Acad. Sci. 268*, 616-619.
Kurstak, E., and Garzon, S. (1971). *Proc. Can. Fed. Biol. Soc. 14*, 630. (Cited in Kurstak, 1972.)
Kurstak, E., and Garzon, S. (1975). Multiple infection of invertebrate cells by viruses. *Ann. N.Y. Acad. Sci. 266*, 232-240.
Kurstak, E., and Onji, P. A. (1972). *Proc. Can. Fed. Biol. Soc. 15*, 686. (Cited in Kurstak, 1972.)
Kurstak, E., and Stanislawski-Birencwajg, M. (1968). Localisation du matériel antigénique du virus de la densonucléose au cours de l'infection chez *Galleria mellonella* L. *Can. J. Microbiol. 14*, 1350-1352.
Kurstak, E., and Vago, C. (1967). Transmission du virus de la densonucléose par le parasitisme d'un Hyménoptère. *Rev. Can. Biol. 26*, 311-316.
Kurstak, E., Garzon, S., Goring, I., and Côté, J. R. (1968a). Parasitisme des cellules intestinales de *Galleria mellonella* L. par le virus de la densonucléose. *Rev. Can. Biol. 27*, 261-265.
Kurstak, E., Goring, I., Garzon, S., and Côté, J. R. (1968b). Etude de la densonucléose de *Galleria mellonella* L. (Lepidoptera) par les techniques de fluorescence. *Natur. Can. 95*, 773-783.
Kurstak, E., Belloncik, S., and Brailovsky, C. (1969a). Transformation de cellules L de souris par un virus d'invertébrés: Le virus de la densonucléose (VDN). *C.R. Acad. Sci. 269*, 1716-1719.
Kurstak, E., Côté, J. R., and Belloncik, S. (1969b). Étude de la synthèse et de la localisation des antigènes du virus de la densonucléose (VDN) a l'acide d'anticorps conjugués à l'enzyme peroxydase. *C.R. Acad. Sci. 268*, 2309-2312.
Kurstak, E., Côté, J. R., Belloncik, S., Garzon, S., and Trudel, M. (1969c). Infection des cellules L de la sources par le virus de la densonucléose. *Rev. Can. Biol. 28*, 139-141.
Kurstak, E., Belloncik, S., and Garzon, S. (1970). Immunoperoxydase ultrastructurale: Localisation d'antigènes de virus de la densonucléose (VDN) basée sur les propriétés structurales de la peroxydase. *C.R. Acad. Sci. 271*, 2426-2429.
Kurstak, E., Tijssen, P., and Garzon, S. (1977). Densonucleosis viruses (Parvoviridae). *In* "Atlas of Invertebrate and Plant Viruses" (K. Maramorosch, ed.), pp. 67-91. Academic Press, New York.

Lavie, P., Fresnaye, J., and Vago, C. (1965). L'action de la virose a noyaux denses sur les larves de *Galleria mellonella* dans les suches. *Ann. Abeille 8*, 321-323.
Lebedeva, O. P., Kuznetsova, M. A., Zelenko, A. P., and Gudz-Gorban, A. P. (1973). Investigation of a virus disease of the densonucleosis type in a laboratory culture of *Aedes aegypti*. *Acta Virol. 17*, 253-256.
Lebedinets, N. N., Tsarichkova, D. B., Karpenko, L. V., Kononko, A. G., and Buchatskij, L. P. (1978). Studies of the *Aedes aegypti* L. densonucleosis virus effect on pre-imaginal stages of different species of blood-sucking mosquitoes. *Mikrobiol. Zh. 40*, 352-356.
Longworth, J. F. (1978). Small isometric viruses of invertebrates. *Advan. Virus Res. 23*, 103-157.
Longworth, J. F., Tinsley, T. W., Barwise, A. H., and Walker, I. O. (1968). Purification of a non-occluded virus of *Galleria mellonella*. *J. Gen. Virol. 3*, 167-174.
MacLeod, R., Longworth, J. F., and Tinsley, T. W. (1971). *Paper 4th Ann. Meet. Soc. Invert. Pathol.* (Cited in Tinsley and Longworth, 1973; Longworth, 1978.)
Maeda, S., and Watanabe, H. (1978). Immunofluorescence observation of the infection of densonucleosis virus in the silkworm, *Bombyx mori*. *Japan. J. Appl. Entomol. Zool. 22*, 98-101.
Maeda, S., Watanabe, H., and Matsui, M. (1977). Purification of an Ina-isolate virus of the silkworm, *Bombyx mori*. *J. Sericult. Sci. Japan. 46*, 313-317.
Matsui, M. (1973). Smaller virus particles than the flacherie virus found in the diseased silkworm larvae, *Bombyx mori* L. *Japan. J. Appl. Entomol. Zool. 17*, 113-115.
Matsui, M., and Watanabe, H. (1974). Cytopathological studies on the multiplication of a smaller flacherie virus (SFV) in the midgut epithelium of silkworm, *Bombyx mori* L. *Japan. J. Appl. Entomol. Zool. 18*, 133-138.
Matsui, M., Maeda, S., and Watanabe, H. (1977). Some properties of a small flacherie virus of the silkworm, *Bombyx mori* L. *Japan. J. Appl. Entomol. Zool. 21*, 79-84.
Matthews, R. E. F. (1982). Classification and nomenclature of viruses. *Intervirology 17*, 72-75.
Meynadier, G., Vago, C., Plantevin, G., and Atger, P. (1964). Virose d'un type habituel chez le Lépidoptère, *Galleria mellonella* L. *Rev. Zool. Agr. Appl. 63*, 207-209.
Meynadier, G., Amargier, A., and Genty, Ph. (1977a). Une virose de type densonucléose chez le lépidoptère *Sibine fusca* Stoll. *Oléagineux 32*, 357-361.
Meynadier, G., Galichet, P. F., Veyrunes, J. C., and Amargier, A. (1977b). Mise en évidence d'un densonucléose chez *Diatraea saccharalis* (Lep.: Pyralidae). *Entomophaga 22*, 115-120.

Meynadier, G., Matz, G., Veyrunes, J. C., and Bres, N. (1977c). Virose de type densonucléose chez les Orthoptères. *Ann. Soc. Entomol. Fr.* [*N.S.*] *13*, 487-493.

Moore, N. F., and Kelly, D. C. (1980). Interrelationships of the proteins of two insect parvoviruses (densonucleosis virus type 1 and 2). *Intervirology 14*, 160-166.

Morris, O. N. (1970). Metabolic changes in diseased insects. III. Nucleic acid metabolism in Lepidoptera infected by densonucleosis and *Tipula* iridescent viruses. *J. Invert. Pathol. 16*, 180-186.

Morris, O. N. (1971). Metabolic changes in diseased insects. IV. Radioautographic studies on protein changes in nuclear polyhedrosis, densonucleosis and *Tipula* iridescent virus infection. *J. Invert. Pathol. 18*, 191-206.

Nakagaki, M., and Kawase, S. (1980a). DNA of a new parvo-like virus isolated from the silkworm, *Bombyx mori*. *J. Invert. Pathol. 35*, 124-133.

Nakagaki, M., and Kawase, S. (1980b). Structural proteins of densonucleosis virus isolated from the silkworm, *Bombyx mori*, infected with the flacherie virus. *J. Invert. Pathol. 36*, 166-171.

Nakagaki, M., and Kawase, S. (1982). Capsid structure of the densonucleosis virus of the silkworm, *Bombyx mori*. *J. Sericult. Sci. Japan 51*, 420-424.

Odier, F. (1975). Les complexes de viroses: Entités pathologiques transmissibles chez les invertébrés. *C.R. Acad. Sci. 280*, 2277-2280.

Quiot, J. M., Vago, C., Luciani, J., and Amargier, A. (1970). Dévelopment du virus de la densonucléose en culture organotypique de lépidopterès. *Bull. Soc. Zool. Fr. 95*, 341-348.

Rivers, C. F., and Longworth, J. F. (1972). A non-occluded virus of *Junonia coenia* (Nymphalidae: Lepidoptera). *J. Invert. Pathol. 20*, 369-370.

Sato, F., and Inoue, H. (1978). Double infection of viruses in the midgut of the silkworm, *Bombyx mori*. *Bull. Sericult. Exp. Sta. 27*, 427-444.

Shimizu, T. (1975). Pathogenicity of an infectious flacherie virus of the silkworm, *Bombyx mori*, obtained from sericultural farms in the suburbs of Ina city. *J. Sericult. Sci. Japan 44*, 45-48.

Sun, F. L., Ma, G. H., and Chen, M. S. (1981). A new insect virus of *Pieris rapae* L. I. Isolation and characterization of the virus. *Acta Microbiol. Sin. 21*, 41-44.

Suto, C. (1979). Characterization of a virus newly isolated from the smoky-brown cockroach, *Periplaneta fuliginosa* (Serville). *Nagoya J. Med. Sci. 42*, 13-25.

Suto, C., Kawamoto, F., and Kumada, N. (1979). A new virus isolated from the cockroach, *Periplaneta fuliginosa* (Serville). *Microbiol. Immunol. 23*, 207-211.

Sutter, G. R. (1973). A nonoccluded virus of the army cutworm. *J. Invert. Pathol. 21*, 62-70.

Tijssen, P., and Kurstak, E. (1979a). A simple and sensitive method for the purification and peptide mapping of proteins solubilized from densonucleosis virus with sodium dodecyl sulfate. *Anal. Biochem. 99*, 97-104.

Tijssen, P., and Kurstak, E. (1979b). Studies on the structure of the two infectious types densonucleosis virus. *Intervirology 11*, 261-267.

Tijssen, P., and Kurstak, E. (1981). Biochemical, biophysical, and biological properties of densonucleosis virus (parvovirus). III. Common sequences of structural proteins. *J. Virol. 37*, 17-23.

Tijssen, P., van den Hurk, J., and Kurstak, E. (1976). Biochemical, biophysical, and biological properties of densonucleosis virus. I. Structural proteins. *J. Virol. 17*, 686-691.

Tijssen, P., Tijssen-van der Slikke, T., and Kurstak, E. (1977). Biochemical, biophysical, and biological properties of densonucleosis virus (parvovirus). II. Two types of infectious virions. *J. Virol. 21*, 225-232.

Tijssen, P., Kurstak, E., Su, T. M., and Garzon, S. (1982). Densonucleosis viruses: Unique pathogens of insects. *Proc. 3rd Intern. Colloq. Invert. Pathol., Brighton*, pp. 148-153.

Tinsley, T. W., and Longworth, J. F. (1973). Parvoviruses. *J. Gen. Virol. Suppl. 20*, 7-15.

Truffaut, N., Berger, G., Niveleau, A., May, P., Bergoin, M., and Vago, C. (1967). Recherches sur l'acid nucléique du virus de la densonucléose du lépidoptère *G. mellonella* . *Arch. Ges. Virusforsch. 21*, 469-474.

Vago, C., Meynadier, G., and Duthoit, J. L. (1964). Étude d'un noveau type de maladie à virus chez les Lépidoptères. *Ann. Epiphyt. 15*, 475-479.

Vago, C., Duthoit, J. L., and Delahaye, F. (1966). Les lésions nucleaires de la virose à noyaux denses du Lépidoptère *Galleria mellonella*. *Arch. Ges. Virusforsch. 18*, 344-349.

Watanabe, H. (1981). Characteristics of densonucleosis in the silkworm, *Bombyx mori*. *Japan Agricult. Res. Quart. 15*, 133-136.

Watanabe, H., and Maeda, S. (1978). Genetic resistance to peroral infection with a densonucleosis virus in the silkworm, *Bombyx mori*. *J. Sericult. Sci. Japan 47*, 209-214.

Watanabe, H., and Maeda, S. (1979). Multiplication of a densonucleosis virus in the silkworm, *Bombyx mori* L., reared at high temperature. *Japan. J. Appl. Entomol. Zool. 23*, 151-155.

Watanabe, H., and Shimizu, T. (1980). Epizootiological studies on the occurrence of densonucleosis in the silkworm, *Bombyx mori*, reared at sericultural farms. *J. Sericult. Sci. Japan 49*, 485-492.

Watanabe, H., Maeda, S., Matsui, M., and Shimizu, T. (1976). Histopathology of the midgut epithelium of the silkworm, *Bombyx mori*, infected with a newly-isolated virus from the flacherie diseased larvae. *J. Sericult. Sci. Japan 45*, 29-34.

Weiser, J., and Lysenko, O. (1972). Protein changes in the haemolymph of *Galleria mellonella* (L) larvae infected with virus and protozoan pathogens. *Acta Entomol. Bohemoslov. 69*, 97-100.

PATHOLOGY ASSOCIATED WITH SMALL RNA VIRUSES OF INSECTS

NORMAN F. MOORE

Institute of Virology
Oxford, United Kingdom

I. INTRODUCTION

Undoubtedly a great many insects are infected with small RNA viruses. However, relatively few diseases have been attributed to these viruses and this is due to the problems of detecting viruses with diameters less than 40 nm. Electron microscopy, probably in combination with ultracentrifugation, is required. When a small RNA virus has been identified, the virus can be purified and antiserum raised against it to assist in future identification. It is difficult to identify these viruses in decaying material and most isolations have been from field-collected dead insects. Even under carefully controlled laboratory conditions these virions are normally seen in the infected cell by electron microscopy only after crystalline arrays have formed. Hence, the pathology at a cellular level is an area on which little information is available. At the level of the whole insect most information is available on the pathology caused by viruses of bees and *Drosophila melanogaster* flies. Bees are of obvious economic importance. *Drosophila* flies are widely used in the study of genetics and hence many laboratories have insect populations which are closely monitored.

There are three groups of small RNA viruses whose effect on insects has been examined: the insect picornaviruses,

ISBN 0-12-470295-3

Nodamura-like viruses and the *Nudaurelia* β group of viruses. Several viruses from bees and *Drosophila* flies do not belong to these distinct families and the pathology of these will be considered separately. Finally, the pathology of a miscellaneous group of "other viruses" will be briefly considered.

II. PICORNAVIRUSES

Three viruses of insects have been classified as members of the *Picornaviridae* but have not yet been assigned to genera (see Matthews, 1982, p. 131). These are *Gonometa* virus, *Drosophila* C virus, and Cricket paralysis virus. In 1965, a large area of *Pinus patula* in the Kigezi district of Uganda was defoliated by a large number of *Gonometa podocarpi* larvae. Many dead and moribund larvae were found on and beneath trees. The infected larvae were often found hanging in a limp state by the last pair or two pairs of prolegs (Harrap et al., 1966). Extraction of larvae and pupae produced a high yield of virus which was later shown to have many of the physical characteristics of a picornavirus (Longworth et al., 1973). The virus developed in the cytoplasm of gut and fat body cells. The cytoplasm of columnar cells was filled with viruses, and mitochondria and cell membranes showed pronounced degeneration (Longworth et al., 1973). Jousset et al. (1972) found viruses in laboratory populations of *Drosophila melanogaster* when dead flies were extracted. On injection of the extracts into uninfected *D.melanogaster*, mortality occurred within 3 days. Virions were apparent in tracheal cells, cells around the cerebral ganglion, and in the cytoplasm of cells surrounding the intestine and Malpighian tubules. There have now been several isolates of *Drosophila* C virus (DCV) from natural populations of *D.melanogaster* from Morocco, France, and the French Antilles (Plus et al., 1975, 1978; Moore et al., 1982). Jousset and Plus (1975) studied the effect of infecting virus-free *D.melanogaster* with DCV by exposing eggs to a virus suspension and by feeding virus to adults. Infection by feeding led to a reduction in the lifespan of adults and exposure of eggs to the virus led to a significant decrease in the number of adult flies. Plus et al. (1978) demonstrated that five different isolates of DCV were serologically indistinguishable and had similar polypeptides, but that some of the isolates could be distinguished by their different pathogenicity for *Drosophila melanogaster* or *Galleria mellonella*. For example, with the Charolles isolate of DCV there was definite growth in *G.mellonella*, but no mortality. However, a Moroccan isolate (DCV_0) caused 30% mortality over the same time period.

Cricket paralysis virus (CrPV) was originally identified in two species of the Australian field cricket, *Teleogryllus oceanicus* and *T. commodus*. Early instar nymphs became uncoordinated and died following paralysis of their hind legs. When the virus extract was injected into young crickets the same symptoms became evident and mortality occurred within 3 days (Reinganum et al., 1970; Reinganum, 1973). In natural populations a low level of the virus was detectable without apparent signs of disease (Reinganum et al., 1970). In a search for pathogenesis of the black field cricket, *T. commodus*, in 232 sites in the Western District of Victoria, Australia, Reinganum et al. (1981) demonstrated the presence of CrPV in 42.7% of the sites.

The host ranges of DCV and CrPV have been investigated (Jousset, 1976; Reinganum, 1975; Plus et al., 1978) and this has been recently reviewed (Brun and Plus, 1980; Scotti et al., 1981). An interesting investigation by Reinganum (1975) may give some indication to why CrPV is apparently found in a large range of insect species. Reinganum (1975) found larvae of the emperor gum moth (*Antheraea eucalypti*) hanging from the food plants by one or more pairs of their thoracic legs. They were flaccid with scattered brown spots. Of three types of viruses isolated from the diseased larvae, one was indistinguishable from CrPV. CrPV from crickets and from *Antherea* were cross-infectious and the crickets could obtain lethal doses of virus by feeding on infected gum moths. Reinganum (1975) demonstrated that CrPV could infect several species of Orthoptera and Lepidoptera and suggested that the virus may have originated in the Lepidoptera. Hence, its appearance in the cricket may be attributable to the end of a food chain.

III. NODAVIRUSES

Nodamura virus, the type member of the *Nodaviridae*, was isolated from a single pool of mosquitoes (*Culex tritaeniorhynchus*) collected at the village of Nodamura near Tokyo, Japan. It was at first considered to be an arbovirus, although it was unusual in that its infectivity was not destroyed by diethyl ether and chloroform (Scherer and Harlbut, 1967; Scherer et al., 1968). It has since been demonstrated that Nodamura virus is the first member of a new group of small RNA viruses, the *Nodaviridae*, whose members contain two segments of RNA, a single major capsid protein, and probably no lipid (Newman and Brown, 1973, 1977). The virus is transmissible to suckling mice by infected *Aedes aegypti* mosquitoes. In mice bilateral extensor paralysis of hindlimbs occurs and viremia is apparent 3-6 days after inoculation. A high percentage of infected suckling mice die (Scherer and Hurlbut, 1967). Bailey and Scott (1973) examined the pathogenicity of Nodamura virus for insects. When homogenized and clarified brain tissue from mice killed by Nodamura virus was injected into adult bees paralysis of the two anterior pairs of legs preceded mortality by a few hours. *Galleria mellonella* injected with Nodamura virus also died within 7 to 14 days at 30°C. Nodamura virus also multiplied in cell cultures of *Aedes albopictus*, *Aedes aegypti*, and BHK without apparent cytopathic effect (Bailey et al., 1975).

Tesh (1980) examined the pathogenicity of Nodamura virus for three species of mosquitoes using several different routes of infection. The virus when injected into the thoraces of adult *Aedes albopictus* and *Toxorhynchites amboinensis* mosquitoes caused loss of balance, inability to fly, and, finally, paralysis followed by death. However, *Culex quinquefasciatus* adults infected by a similar route were alive and apparently uninfected by the 10th day, although the two other species had died between the 6th and 10th day after inoculation. Infectivity studies demonstrated that the virus replicated poorly (or not at all) in the surviving species. When adult *A.albopictus* were orally infected with the virus no mortality resulted and a similar result was obtained when larvae were infected by immersion in virus suspension. Examination of the "brains" of infected insects using a fluorescent antibody technique indicated similar amounts of virus antigen using all infection routes. This evidence appears to suggest that the more natural route of infection, i.e., orally, results in virus multiplication but no mortality.

Two other members of the *Nodaviridae* have been isolated in New Zealand. The first of these was the virus isolated from the black beetle, *Heteronychus arator* (F.) (Coleoptera: Scarabaeidae) (Longworth and Archibald, 1975; Longworth and Carey, 1976) and the second from grass grubs *Costelytra zealandica* (White) (Dearing et al., 1980; Scotti et al., 1983). Substantial mortality was recorded in high density populations of *H.arator* in kikuyu pasture near Helensville in New Zealand in 1974, and aqueous extracts were made of each of 33 larvae, 30 pupae, and 30 adults which were dead on collection or soon after. Twenty-five percent of the 3rd instar larvae and 7% of the adults contained a 30 nm virus which developed in the cytoplasm of gut and fat body cells (Longworth and Archibald, 1975). While the virus was not infective to *H.arator* adults when injected into the body cavity it was transmitted to *Pseudaletia separata* (Lepidoptera: Noctuidae) and *Galleria mellonella* (Lepidoptera: Galleriidae) by injection into the body cavity. The *G.mellonella* larvae became inactive and flaccid, and 50% had died by 20 days. However, many survived in a paralyzed state for up to 37 days postinfection before mortality occurred. The second virus, designated Flock House virus, was isolated from grass grubs using a continuous line of *Drosophila melanogaster* cells (Dearing et al., 1980). The grubs were alive when collected from a site near Bulls in the North Island of New Zealand, but they appeared to be "retarded" in their development (Scotti et al., 1983). The final member of this group of small RNA viruses is Arkansas bee virus. This virus was found as an inapparent infection of honeybees in Arkansas and apparently shortens the lives of adult bees when injected (Bailey and Woods, 1974).

IV. NUDAURELIA β VIRUS GROUP

The first member of the *Nudaurelia* β virus group (see Matthews, 1982, p. 135) was identified and isolated by Grace and Mercer (1965). They found many laboratory-reared larvae of the emperor gum moth, *Antheraea eucalypti* Scott (Saturniidae), became infected at all instars and died. Infected larvae fell from the foliage and lay in a moribund state or hung from the foliage by the back pair of prolegs. Brzostowski and Grace (1970) also found that larvae of *A.helena* were equally susceptible to the virus of *A.eucalypti*. A similar disease was recognized in the pine emperor moth, *Nudaurelia cytherea capensis* Stoll, which is closely related to the emperor gum moth. Infected larvae either hung from the branches by their prolegs or dropped to the ground (Hendry et al., 1968; Juckes, 1970; Tripconey, 1970). The body contents liquefied and the larvae assumed a darker than normal coloration. The predominant virus present in infected larvae was designated *Nudaurelia* β virus and it was serologically related to the virus isolated from the gum moth, as well as to a number of more recent isolates from other insect species (Juckes et al., 1973; Reinganum et al., 1978). The biophysical properties of *Nudaurelia* β virus were extensively studied and it has become the type member of this relatively new group of insect viruses (Finch et al., 1974; Struthers and Hendry, 1974; Reinganum et al., 1978). Two other viruses isolated from species of Lepidoptera: Limacodiidae, *Darna (Orthocraspida) trima* Moore from Southeast Asia, which is a severe pest of coconut and oil palms in Malaysia, and *Thosea asigna* (Moore), which is also a pest of oil palms in Malaysia, appear to cause similar fatal diseases in their host larvae (Reinganum et al., 1978). However, many of the virus preparations contain more than one size of particle when initially purified from field-collected material. The diameters of members of the *Nudaurelia* β group have been estimated to be between 32-39 nm, and the *A. eucalypti* virus preparation contained 14 nm particles (Brzostowski and Grace, 1970). The preparations from *N. cytherea* contained four additional types of viruses (Juckes, 1970, 1979) and purification of viruses from *Darna trima* demonstrated the presence of a possible picornavirus in addition to the *Nudaurelia* β-like virus (Reinganum et al., 1978). Hence the diseases detected in the larvae may be attributable to more than one type of virus.

Several additional members of the *Nudaurelia* β group have been identified mainly from dead larvae (Greenwood and Moore, 1982). Hess et al. (1977) found a new icosahedral insect virus in mixed infections with the nuclear polyhedrosis virus (NPV) isolated from the alfalfa looper, *Autographa californica*, and later work by Morris et al. (1979) demonstrated that the new isolate was very similar to *Nudaurelia* β virus. Vail et al. (1983a,b) studied the effect on insects of this virus by itself and when coinfected with baculovirus. They demonstrated that infected larvae were severely stunted, weighing considerably less than control larvae. While stunting occurred with extremely low amounts of virus (0.1 ng), doses 100 times lower did not cause the loss of weight although infection could be detected by enzyme-linked immunosorbent assay (ELISA). If the virus was present at very low levels in the population it would possibly not be detected (Vail et al., 1973a). When the RNA virus was mixed with *Autographa californica* NPV little effect was found on the bioassay of the latter virus, although larval weight loss occurred.

V. VIRUSES OF *DROSOPHILA*

Several small RNA viruses have been isolated from *Drosophila* flies which have different physical properties than the three groups of viruses previously described. *Drosophila* P virus, originally identified by Plus and Duthoit (1969), is present in many *Drosophila* populations and in natural populations appears to have little effect on the flies. However, on injection female sterility resulted and the lifespan of both sexes was reduced (David and Plus, 1971). Ovaries and Malgiphian tubules, but not muscle and brain, were found to contain a large number of virions when *Drosophila* flies were injected with P virus (Teninges and Plus, 1972). Before death P-infected flies became turgid, and as the Malgiphian tubules are responsible for water exchange in insects, Teninges and Plus (1972) suggested that the virus multiplication was causing a malfunction in these organs. *Drosophila* A virus was originally identified by Plus et al. (1975) and it also has a low pathogenicity for flies. The *Drosophila* A and P viruses belong to different serotypes than DCV, and have different buoyant densities and protein compositions to that normally found with mammalian picornaviruses (Plus et al., 1976). While *Drosophila* A and P virus multiply in the gut and Malpighian tubules and have low pathogenicity, DCV infects the tracheal system, particularly the cells surrounding the cerebral ganglion, and it is very pathogenic (Brun and Plus, 1980) Another virus belonging to the same serotype as *Drosophila* P virus was found by Jousset (1970) in *Drosophila immigrans*. This virus was injected into *D.melanogaster* and induced CO_2 sensitivity in males (Jousset, 1972). This effect has only previously been recorded with the hereditary *Drosophila* rhabdovirus, Sigma.

VI. VIRUSES OF THE HONEY BEE *Apis mellifera L.*

Much of the work on small RNA viruses of honey bees has already been reviewed by Bailey (1976) and Bailey and Woods (1977) and the pathology caused by a relatively large number of viruses has been thoroughly documented. These viruses include sacbrood, chronic bee paralysis and its associate acute bee paralysis, Arkansas, X, slow bee paralysis, black queen cell, and Kasmir bee. Bailey et al. (1981) have also described the seasonal appearance of the diseases caused by many of these viruses. Several of the viruses may belong to the previously described established groups, but at present insufficient evidence is available on their physicochemical properties.

VII. OTHER VIRUSES

For many years a disease of unknown origin (Flacherie) was described in the silkworm (*Bombyx mori*) which led to a serious loss of cocoons. Young larvae were most susceptible but by fourth instar the larvae were more resistant and by the fifth there was a marked resistance to infection. A virus was identified as the causative agent (Aizawa et al., 1964; Ayuzawa, 1972). The virus appears to replicate in the cytoplasm of midgut cells. Several types of virus have been found associated with the disease (Himeno et al., 1979, 1980) and at least one of these appears to be a picornavirus (Hashimoto and Kawase, 1983).

More recently another probable picornavirus has been purified from the aphid *Rhopalosiphum padi* (D'Arcy et al., 1981a) and was shown to have a significant effect on the longevity of infected aphids (D'Arcy et al., 1981b). Longevity of offspring from infected adults was also significantly reduced.

There have been numerous records of virus-like particles in infected insects and several of these had sizes considerably less than those recognized for virus particles. Only one of these isolates has been characterized in detail--the 13 nm virus isolated from the Grasshopper [*Melanoplus bivittatus* (Say) (Acridiidae)]. The virus was called "crystalline-array virus" (CAV) and was isolated from nymphs of the grasshopper. The symptoms of the disease were loss of appetite and vigor, and lack of coordination before death (Jutila et al., 1970). The virus was transmissible to several other grasshopper species with some differences in susceptibility being noted. Replication took place in the cytoplasm of tracheal matrix cells, muscle, and pericardial cells.

VIII. DISCUSSION

The small RNA viruses of insects have been described in detail by Longworth (1978) and Moore and Tinsley (1982). While considerably more information is becoming available on their replicative events (see accompanying chapter), relatively little is known about the pathology with the exception of the viruses of bees and *Drosophila*. The pathology caused by the cytoplasmic and nuclear polyhedrosis and entomopox viruses was studied in greater detail because it was possible to easily recognize large polyhedral structures in the light microscope and relate these structures to the diseased state of the insect. In only a few instances has a purified small RNA virus been injected into a colony of susceptible insects to study the pathology of the resulting disease. It would, of course, be difficult to recreate the natural distribution of the virus found in the field in overcrowded and probably stressed insects in a laboratory. Many of the studies have involved injecting insects with the virus. This is an artificial route of infection unless the insect is parasitized by another insect species which carries the virus. Injection seems to lead to a much greater mortality than the natural orally or even sexually transmitted disease.

ACKNOWLEDGMENTS

The author wishes to thank Gail Davies and Steven Eley for assisting in the preparation of this paper.

REFERENCES

Aizawa, K., Furuta, Y., Kurata, K., and Sato, F. (1964). On the etiologic agent of the infectious flacherie of the silkworm, *Bombyx mori* (Linnaeus). *Bull. Sericult. Exp. Sta., Tokyo 19*, 223-240.

Ayuzawa, C. (1972). Studies on the infectious flacherie of the silkworm, *Bombyx mori* L. I. Purification of the virus and some properties. *J. Sericult. Sci. Japan 41*, 338-344.

Bailey, L. (1976). Viruses attacking the honey bee. *Advan. Virus Res. 20*, 271-304.

Bailey, L. and Scott, H. A. (1973). The pathogenicity of Nodamura virus for insects. *Nature (London) 241*, 545.

Bailey, L. and Woods, R. D. (1974). Three previously undescribed viruses from the honey bee. *J. Gen. Virol. 25*, 175-186.

Bailey, L. and Woods, R. D. (1977). Bee viruses. *In* "The Atlas of Insect and Plant Viruses" (K. Maramorosch, ed.), pp. 141-156. Academic Press, New York.

Bailey, L., Newman, J. F. E., and Porterfield, J. S. (1975). The multiplication of Nodamura virus in insect and mammalian cell cultures. *J. Gen. Virol. 26*, 15-20.

Bailey, L., Ball, B. V., and Perry, J. N. (1981). The prevalence of viruses of honey bees in Britain. *Ann. Appl. Biol. 97*, 109-118.

Brun, N. and Plus, N. (1980). The viruses of *Drosophila*. *In* "The Genetics and Biology of *Drosophila*" (M. Ashburner and T. R. Wright, eds.), Vol. 2D, pp. 625-702. Academic Press, New York.

Brzostowski, H. W. and Grace, T. D. C. (1970). Observations on the infectivity of RNA from *Antheraea* virus (AV). *J. Invertebr. Pathol. 16*, 277-279.

D'Arcy, C. J., Burnett, P. A., Hewings, A. D., and Goodman, R. M. (1981a). Purification and characterisation of a virus from the aphid *Rhopalosiphum padi*. *Virology 112*, 346-349.

D'Arcy, C. J., Burnett, P. A., and Hewings, A. D. (1981b). Detection, biological effects, and transmission of a virus from the aphid *Rhopalosiphum padi*. *Virology 114*, 268-272.

David, J. and Plus, N. (1971). Le virus P de la Drosophile: comparison de la longévité et de la fécondité des mouches infectées par injection ou par contamination naturelle. *Ann. Inst. Pasteur (Paris) 120*, 107-119.

Dearing, S. C., Scotti, P. D., Wigley, P. J., and Dhana, S. D. (1980). A small RNA virus isolated from the grass grub, *Costelytra zealandica* (Coleoptera: Scarabaeidae). *New Zeal. J. Zool. 7*, 267-269.

Finch, J. T., Crowther, R. A., Hendry, D. A., and Struthers, J. K. (1974). The structure of *Nudaurelia capensis* β virus: the first example of a capsid with isosahedral surface symmetry T=4. *J. Gen. Virol. 24*, 191-200.

Grace, T. D. C. and Mercer, E. H. (1965). A new virus of the saturniid *Antheraea eucalypti* Scott. *J. Invertebr. Pathol. 7*, 241-244.

Greenwood, L. K. and Moore, N. F. (1982). The *Nudaurelia* β group of small RNA-containing viruses of insects: serological identification of several new isolates. *J. Invertebr. Pathol. 39*, 407-409.

Harrap, K. A., Longworth, J. F., Tinsley, T. W., and Brown, K. W. (1966). A non-inclusion virus of *Gonometa podocarpi* (Lepidoptera: Lasiocampidae). *J. Invertebr. Pathol. 8*, 270-272.

Hashimoto, Y. and Kawase, S. (1983). Characteristics of structural proteins of infectious Flacherie virus from the silkworm, *Bombyx mori*. *J. Invertebr. Pathol. 41*, 68-76.

Hendry, D. A., Bekker, M. F., and Van Regenmortel, M. H. V. (1968). A non-inclusion virus of the pine emperor moth, *Nudaurelia cyntherea capensis* Stoll. *S. Afr. Med. J. 42*, 117-118.

Hess, R. T., Summers, M. D., Falcon, L. A., and Stoltz, D. B. (1977). A new icosahedral insect virus: apparent mixed nuclear infection with the beculovirus of *Autographa californica*. *IRCS Med. Sci. 5*, 562.

Himeno, M., Maeda, M., Aoki, H., and Komano, T. (1979). Biochemical and biological properties of flacherie virus of the silkworm. *J. Invertebr. Pathol. 33*, 348-357.

Himeno, M., Asada, Y., and Komano, T. (1980). Polypeptide composition of flacherie viruses of the silkworm. *J. Invertebr. Pathol. 36*, 172-179.

Jousset, F.-X. (1970). Un virus extrait de *Drosophila immigrans* provoquant un symptôme de sensibilite au CO_2 chez les males de *Drosophila melanogaster*. *C.R. Acad. Sci. 271*, 1141-1144.

Jousset, F.-X. (1972). Le virus Iota de *Drosophila immigrans* étudié chez *D. melanogaster*: symptôme de la sensibilite au CO_2, descriptions des anomalies provoquées chez l'hôte. *Ann. Inst. Pasteur 121*, 275-288.

Jousset, F.-X. (1976). Étude expérimentale du spectra d'hôtes du virus C de *Drosophila melanogaster* chez quelques Diptères et Lépidoptères. *Ann. Inst. Pasteur 127A*, 529-544.

Jousset, F.-X. and Plus, N. (1975). Étude de la transmission horizontale et de la transmission verticale des picornavirus de *Drosophila melanogaster* et de *Drosophila immigrans*. *Ann. Microbiol. 126B*, 231-249.
Jousset, F.-X., Plus, N., Croizier, G., and Thomas, M. (1972). Existence chez *Drosophila* de deux groupes de picornavirus de propriétés sérologiques et biologiques différentes. *C.R. Acad. Sci. D275*, 3043-3046.
Juckes, I. R. M. (1970). Viruses of the pine emperor moth. *Bull. S. Afr. Soc. Plant Pathol. Microbiol. 4*, 18.
Juckes, I. R. M. (1979). Comparison of some biophysical properties of the *Nudaurelia* β and ε viruses. *J. Gen. Virol. 42*, 89-94.
Juckes, I. R. M., Longworth, J. F., and Reinganum, C. (1973). A serological comparison of some non-occluded insect viruses. *J. Invertebr. Pathol. 21*, 119-120.
Jutila, J. W., Henry, J. E., Anacker, R. L., and Brown, W. R. (1970). Some properties of a crystalline-array virus (CAV) isolated from the grasshopper *Melanoplus bivittatus* (Say) (Orthoptera: Acridiidae). *J. Invertebr. Pathol. 15*, 225-231.
Longworth, J. F. (1978). Small isometric viruses of invertebrates. *Advan. Virus Res. 23*, 103-157.
Longworth, J. F. and Archibald, R. D. (1975). A virus of black beetle, *Heteronychus arator* (F.) (Coleoptera: Scarabaeidae). *New Zeal. J. Zool. 2*, 233-236.
Longworth, J. F. and Carey, G. P. (1976). A small RNA virus with a divided genome from *Heteronychus arator* (F.) (Coleoptera: Scarabaeidae). *J. Gen. Virol. 33*, 31-40.
Longworth, J. F., Payne, C. C., and Macleod, R. (1973). Studies on a virus isolated from *Gonometa podocarpi* (Lepidoptera: Lasiocampidae). *J. Gen. Virol. 18*, 119-125.
Matthews, R. E. F. (1982). Classification and nomenclature of viruses. Fourth Report of the International Committee on Taxonomy of Viruses. *Intervirology 17(1-3)*, 1-199.
Moore, N. F. and Tinsley, T. W. (1982). The small RNA-viruses of insects. *Arch. Virol. 72*, 229-245.
Moore, N. F., Pullin, J. S. K., Crump, W. A. L., and Plus, N. (1982). The proteins expressed by different isolates of *Drosophila* C virus. *Arch. Virol. 74*, 21-30.
Morris, T. J., Hess, R. T., and Pinnock, D. E. (1979). Physiochemical characterization of a small RNA virus associated with baculovirus infection in *Trichoplusia ni*. *Intervirology 11*, 238-247.
Newman, J. F. E. and Brown, F. (1973). Evidence for a divided genome in Nodamura virus, an arthropod-borne picornavirus. *J. Gen. Virol. 21*, 371-384.

Newman, J. F. E. and Brown, F. (1977). Further physiochemical characterization of Nodamura virus. Evidence that the divided genome occurs in a single component. *J. Gen. Virol. 38*, 83-95.

Plus, N. and Duthoit, J. L. (1969). Un nouveau virus de *Drosophila melanogaster*, le virus P. *C.R. Acad. Sci. D268*, 2313-2315.

Plus, N., Croizier, G., Jousset, F.-X., and David, J. (1975). Picornaviruses of laboratory and wild *Drosophila melanogaster*: geographical distribution and serotypic composition. *Ann. Microbiol. 126A*, 107-117.

Plus, N., Croizier, G., Veyrunes, J. C., and David, J. (1976). A comparison of buoyant density and polypeptides of *Drosophila* P, C and A viruses. *Intervirology 7*, 346-350.

Plus, N., Croizier, G., Reinganum, C., and Scotti, P. D. (1978). Cricket paralysis virus and *Drosophila* C virus: serological analysis and comparison of capsid polypeptides and host range. *J. Invertebr. Pathol. 31*, 296-302.

Reinganum, C. (1973). Studies on a non-occluded virus of the field cricket *Teleogryllus* spp. Master's thesis, Monash University, Melbourne.

Reinganum, C. (1975). The isolation of cricket paralysis virus from the emperor gum moth, *Antheraea eucalypti* Scott, and its infectivity towards a range of insect species. *Intervirology 5*, 97-102.

Reinganum, C., O'Loughlin, G. T., and Hogan, T. W. (1970). A non-occluded virus of the field crickets *Teleogryllus oceanicus* and *T.commodus* (Orthoptera: Gryllidae). *J. Invertebr. Pathol. 16*, 214-220.

Reinganum, C., Robertson, J. S., and Tinsley, T. W. (1978). A new group of RNA viruses from insects. *J. Gen. Virol. 40*, 195-202.

Reinganum, C., Gagen, S. J., Sexton, S. B., and Vellacott, H. P. (1981). A survey for pathogens of the black field cricket, *Teleogryllus commodus*, in the western district of Victoria, Australia. *J. Invertebr. Pathol. 38*, 153-160.

Scherer, W. F. and Hurlbut, H. S. (1967). Nodamura virus from Japan: a new and unusual arbovirus resistant to diethyl ether and chloroform. *Amer. J. Epidemiol. 86*, 271-285.

Scherer, W. F., Verna, J. E., and Richter, G. W. (1968). Nodamura virus, an ether- and chloroform-resistant arbovirus from Japan. *Amer. J. Tropical Med. Hyg. 17*, 120-127.

Scotti, P. D., Longworth, J. F., Plus, N., Croizier, G., and Reinganum, C. (1981). The biology and ecology of strains of an insect small RNA virus complex. *Advan. Virus Res. 26*, 117-143.

Scotti, P. D., Dearing, S., and Mossop, D. W. (1983). Flock House virus: A Nodavirus isolated from *Costelytra zealandica* (White) (Coleoptera: Scarabaeidae). *Arch. Virol. 75*, 181-189

Struthers, J. K. and Hendry, D. A. (1974). Studies of the protein and nucleic acid components of *Nudaurelia capensis* β virus. *J. Gen. Virol. 22*, 355-362.

Teninges, D. and Plus, N. (1972). P virus of *Drosophila melanogaster*, as a new picornavirus. *J. Gen. Virol. 16*, 103-109.

Tesh, R. B. (1980). Infectivity and pathogenicity of Nodamura virus for mosquitoes. *J. Gen. Virol. 48*, 177-182.

Tripconey, D. (1970). Studies on a nonoccluded virus of the pine tree emperor moth. *J. Invertebr. Pathol. 15*, 268-275.

Vail, P. V., Morris, T. J., and Collier, S. S. (1983a). An RNA virus in *Autographa californica* nuclear polyhedrosis virus preparations: gross pathology and infectivity. *J. Invertebr. Pathol. 41*, 179-183.

Vail, P. V., Morris, T. J., and Collier, S. S. (1983b). An RNA virus in *Autographa californica* nuclear polyhedrosis virus preparations: incidence and influence on baculovirus activity. *J. Invertebr. Pathol. 41*, 171-178.

III.

Ecology and Environmental Biology

NATURAL DISPERSAL OF BACULOVIRUSES IN THE ENVIRONMENT

D. L. HOSTETTER

USDA-ARS
Biological Control of Insects Research Laboratory
Columbia, Missouri

M. R. BELL

USDA-ARS
Western Cotton Research Laboratory
Phoenix, Arizona

I. INTRODUCTION

A. *BACULOVIRUSES*

The occurrence of worldwide decreases in insect population levels have been reported to occur spontaneously. Effects have been made to isolate and study the causative agents in order to develop biological control methods that are compatible with the environment. Among the various methods, viruses have emerged as effective and safe control agents due to their self-limiting pathogenicity and their ability to persist within the host population. Payne and Kelly (1981) described six main groups of viruses responsible for disease in insects and mites: baculoviruses, cytoplasmic polyhedrosis viruses, entomopox viruses, iridescent viruses, densonucleosis viruses, and a variety of small RNA viruses. This chapter will concentrate on Baculoviridae and their natural dispersion within the environment.

The Baculoviridae include the nuclear polyhedrosis viruses (NPVs) (Subgroup a), the granulosis viruses (GVs) (Subgroup b), and the "*Oryctes*-like" viruses (Subgroup c). Baculoviruses have been isolated from Lepidoptera, Hymenoptera, Diptera, Neuroptera, Coleoptera, Tricoptera, as well as Crustacea and mites (Fenner, 1976). Both GV and NPV inclusion bodies (IB's) can be seen in wet mounts of infected insect tissue or tissue extracts

VIRAL INSECTICIDES
FOR BIOLOGICAL CONTROL

ISBN 0-12-470295-3

by phase-contrast and dark-field microscopy. The *Oryctes* virus can only be observed by electron microscopy. Differences among the three subgroups of baculoviruses occur during the embedding of virions in the IBs. NPVs may contain more than 100 virions/IB while GVs contain one, or rarely two, and *Oryctes* viruses have not been seen within IBs, either *in vivo* (Huger, 1966a; Payne, 1974) or *in vitro* (Kelly, 1976). Definitive identifications made through biochemical and serological methods are specific diagnostic tools for these viruses.

Larvae infected with NPV or GV undergo distinct changes in their appearance and behavior as infection progresses. In general, their movements become restricted, their integument becomes flacid and changes color, and they cease feeding. Larvae infected with NPV may climb to the higher portions of the host plant and assume an inverted hanging position prior to death. The body wall ruptures shortly after death thereby releasing massive numbers of IBs. NPVs and GVs infecting Lepidoptera are generally found in the cell nucleus and/or cytoplasm of most tissues, while in the order of Hymenoptera the NPVs replicate only in the midgut epithelium. The *Oryctes* virus replicates in midgut epithelium cells of adult rhinoceros beetles, is shed into the gut lumen, and is voided in the feces.

B. NATURAL DISSEMINATION

The baculoviruses are disseminated within insect populations vertically (Generation to generation), and horizontally (from one susceptible insect host to another) by both abiotic and biotic agents (Cunningham and Entwistle, 1981; Federici, 1980; Tanada, 1976). Vertical transmission of viruses occurs primarily at the time of oviposition. Transovum transmission is due to the direct contamination of the egg surfaces or foliage in the vicinity of the neonate larvae by infected or surface-contaminated adults. Transovarian transmission, a mode of transovum transmission, refers to the passage of infective pathogens within the confines of the ova. Transstadial transmission is the continued passage of the pathogen throughout part or all the host's life cycle. Concrete evidence of transovarian transmission is still limited. Bird (1961) suggested that sawfly NPV was transmitted within the egg, while Neilson and Elgee (1968) found that diseased and/or contaminated adults were responsible for virus dissemination onto foliage. They concluded that transovarian and even transovum transmission rarely, if ever, occurred. Extensive tests with the codling moth, *Cydia* (=*Laspeyresia*) *pomonella* (L.) GV failed to demonstrate transovarian transmission (Etzel and Falcon, 1976).

Examples of abiotic dispersal include rain-splash (often very localized) and wind-carried pathogens in the form of dust or aerosols. The significance of this type of dispersal has not been resolved in laboratory or field studies.

Agents responsible for biotic horizontal passage include infected adults acting as specific agents and casual agents such as parasites, predators, and sacrophages. Since viruses can survive in the alimentary tracts of birds, mammals, and a wide variety of predatory invertebrates, their dispersal is greatly enhanced (Bird, 1955).

II. INDIRECT DISPERSAL BY AGRONOMIC PRACTICES

A. *THE SOIL RESERVOIR*

Results of an extensive series of tests conducted with the NPV of the cabbage looper, *Trichoplusia ni* (Hübner), indicate that the levels of NPV in the soil increase as epizootics progress (Jaques, 1964, 1967, 1970, 1974a; Jaques and Harcourt, 1971). Compounding these results with persistence, data has established that soil is the principle reservoir of insect viruses infecting a wide variety of crops and natural habitats.

Liquefied cadavers are transported by gravity and rain to the soil. Once in the soil, the virus persists in the upper layers where it is fairly accessible and capable of infecting larvae. Jaques (1969) noted that the activity of *T. ni* NPV applied to the soil surface did not decline significantly over a period of 231 weeks. He found that little virus was detectable at depths below 7.5 cm, indicating that *T. ni* NPV did not readily leach into the soil. In addition, Jaques (1975) found that about 15% of the original viral activity remained in soil samples taken from a depth of 2.5 cm 318 weeks following a single application of virus onto the soil. Soil in these plots was not planted or disturbed during this period. Jaques presented data to support his hypothesis that the virus from the soil reaches the leaves by a variety of means. He speculated that the virus was deposited onto the foliage by wind, rain, during cultivation, and by insects or other animals. Moreover, the concentration of viruses in the surface soil was relatively constant during the winter months but decreased in the spring due to irrigation and cultivation.

Thomas et al. (1972) reported that low levels of *T. ni* NPV were present in all field plots before treatment, even though no cole crops had been planted in the field for 9 years. The *T. ni* NPV accumulated in the upper 1 cm of the soil and viral residues persisted in the soil throughout the winter. Considerable amounts (2.6×10^{10} IB/acre) of active virus were discovered in the spring prior to cultivation. They also suggested that the virus residues in the upper portion of the soil served as a reservoir for future contamination of the plants. Heimpel et al. (1973), in studies with *T. ni* NPV on cabbage fields in Maryland, noted that the IBs survived in the soil for at least 9-10 years and were subsequently distributed by wind-blown and rain-splashed soil onto young plants.

Jaques (1976) found that 68 and 45% of the *T. ni* NPV residues were still detectable after 1 and 2 years, respectively, posttreatment. GV residues of the imported cabbageworm, *Artogeia* (=*Pieris*) *rapae*, were also found to be 18 and 19%, respectively, in these soil samples. Jaques concluded that the IBs of *T. ni* NPV and *A. rapae* GV remain active in the soil for long periods of time, and that the *A. rapae* GV was significantly more stable in soil than the *T. ni* NPV. He further suggested that the IBs adhere to soil particles so that little downward movement or leaching occurred. The author observed little lateral movement in the subsurface water and noted that the movement of viruses in runoff water and streams has not been investigated.

Jaques and Harcourt (1971) observed the effects of soil pH on the viruses of the cabbage looper and the imported cabbageworm. Soil pH as low as 5.6 had no effect upon these viruses. The *A. rapae* GV was more frequently found in alkaline soils (pH 5.6-6). Thomas et al. (1973) noted that the *T. ni* NPV was inactivated more rapidly in loamy sand soil with a pH of 4.83 to 7.17. They suggested that lime be added to enhance the persistence of virus in soils of this type.

Similar studies conducted by Crawford and Kalmakoff (1977) also established soil as the primary reservoir for three viruses infecting a lepidopteran soil-dwelling pasture pest (*Wiseana* sp.) in New Zealand. Entwistle and Adams (1977) demonstrated with an NPV of the European spruce sawfly, *Diprion* (=*Gilpinia*) *hercyniae*, that soil was the major reservoir of the virus in addition to providing protection for the overwintering stages of the sawfly.

In studies with the soybean looper, *Pseudoplusia includens* NPV, Young and Yearian (1979) demonstrated that the soil beneath soybeans was the primary reservoir of this virus. McLeod et al. (1982) reported that the *P. includens* NPV persisted in soil through harvest, 2 weeks postharvest, and throughout the fall of the year. Virus concentration decreased in November and rose to 1.13×10^{14} IBs/ha in May after which it began to decline. Soil sampled 2 weeks posttreatment contained 194.8×10^{12} IBs/ha; at 3 weeks posttreatment it decreased to 35.5×10^{12} IBs/ha. Nontreated fields remained below 3.0×10^{12} IBs/ha. Virus concentration in the soil samples increased about 1 month after treatment, possibly due to maturing plants shedding their leaves which had been contaminated with virus. The maximum NPV concentration occurred 7 days after application, primarily due to the release of additional NPV from larval cadavers.

Extensive studies with baculoviruses in forest environments have established not only soil but the floor litter to be major reservoirs of viral pathogens. Hukuhara and Namura (1972) observed that fall webworm (*Hyphantria cunea*) cadavers dropped from trees to the ground, disintegrated, and released IBs into the soil. They also noted that *H. cunea* NPV could be collected within the plant litter and soil from up to 15 m from these large trees. Concentrations were highest at a depth of 0.01 cm and at 4-5 cm; they were lower at 10-11 cm and practically nil below 30 cm in the forest soil.

Thompson and Scott (1979) working with Douglas fir tussock moth, *Orgyia pseudotsugata*, established that the NPV produced in cadavers is eventually released into the soil. However, once in the dust layer of the forest floor, little vertical distribution occurred. The majority of the virus was inactivated prior to reaching the soil. The IBs reaching the soil became relatively stable once they were incorporated into the dust layer, remaining active for at least 1 year. Soil samples from an endemic area were air-dried and dusted onto fir foliage. About 32% of the samples were positive for NPV, which demonstrated that the Douglas fir tussock moth NPV can be transported via dust. Ideally, the tussock moth NPV could be reintroduced into the forest canopy through a variety of dust-borne dispersions. However, except in unusual circumstances, such as road construction or livestock drives, widespread foliage contamination by dust-borne virus would be very unlikely within the forest ecosystem.

Further investigations on tussock moth NPV were conducted by Thompson et al. (1981) in a California white fir management area. By sampling undisturbed litter from the forest floor beneath large trees, they found that viable NPV occurred in sufficient amounts (45 IB/cm^3) to cause infection 41 years after the last recorded epizootic which occurred in 1937 and 1938. The authors estimated that approximately 50-75% of the activity was lost in the first 10 years and that >99% was lost by the 41st year. They concluded that in the absence of direct tussock moth control measures, the NPV may remain a natural component of the forest ecosystem for long periods of time. When direct control measures are deemed necessary, they advised the application of a moderate dose of NPV to achieve population control which would have a minimal effect upon the ecosystem.

Granulosis viruses have also been shown to be very stable within the soil. David and Gardiner (1967) demonstrated that the *Pieris brassicae* GV lost little or no activity after 2 years in garden soil and sand. They hypothesized that the

P. brassicae GV is unlikely to be inactivated by organisms or other conditions in normal soil, and, therefore, can easily persist from year to year in the environment.

Hukuhara et al. (1978) conducted a series of tests with the fall webworm GV in Japan and determined that the soil was the reservoir of this GV which persisted in an enzootic state within the test plots. Soil bioassays were positive (although limited) up to 60 m from the test plots. The GV was presumably dispersed by both abiotic and biotic agents, such as wind, predators, or infected adults. This GV, although present in the soil, was not evenly distributed in the environment. Similar studies with *Artogeia rapae* (Jaques, 1974b), the potato moth, *Phthorimaea operculella* (Zell.) (Reed, 1971), and the gray grain moth, *Apamea anceps* Schiff (Shekhurina, 1980), all showed that the soil was the main reservoir of these granulosis viruses.

Tanada (1971), in a review of the persistence of entomopathogens in the ecosystem, noted that the primary biotic element of viruses was the susceptible host and that the primary abiotic element was the soil. Tinsley (1979) reviewed the potential of pathogenic insect viruses and suggested that the spread of viruses most probably required a high host density and that the persistence depended on the retention of infective virus on foliage, in the soil, in cadavers, and in infected overwintering hosts. These demonstrations of persistence and natural occurrence of viruses in cropping and forest systems indicate that viruses have been present in nature and on harvested crops for many years.

B. *DISPERSAL BY CULTIVATION*

Cultivation is an integral element in the dispersal of viruses onto the foliage of crops. Cultivation practices such as plowing, disking, tilling, seeding, and transplanting disturb the upper 2-3 inches of soil where most of the viruses have accumulated which provides an excellent opportunity for their dust-borne application. Low-growing crops, such as cole crops that are in close proximity to the soil, may be more available for viral contamination by cultivation dust. Foliage of crops, such as cotton and soybeans, are not as close to the soil surface, and the probability of viral dissemination to the susceptible insects is less likely, although possible.

Upper surfaces of the soil may become heavily contaminated with the IBs of nuclear polyhedrosis or granulosis viruses following an epizootic (Jaques, 1969, 1970, 1974a; Jaques and Harcourt, 1971). If the soil remains undisturbed after harvesting, the virus would be readily available for dissemination. However, standard cultivation practices such as cotton plow-downs or secondary crop rotations dictate immediate soil preparation. Seasonal variations in soil-virus concentrations fluctuate depending upon cultivation practices. Postharvest plowing turns the soil 8-10 inches and the virus is temporarily lost to natural dispersal. During this time the virus lies protected until the pre-plant plowing when it is returned to the surface and again becomes available for dispersal. Any activity such as cultivation, vehicular application of fertilizers and pesticides, or animal movement can mechanically translocate soil particles containing IBs to plant foliage and susceptible larvae. As the incidence of disease increases, the IBs return to the soil, and the virus completes its cycle. It is particularly important that viruses and susceptible hosts remain in close proximity to naturally perpetuate the virus.

C. DISPERSAL BY IRRIGATION WATER

Irrigation water is considered to be an important dispersal agent for the virus of the alfalfa butterfly, *Colias eurytheme*, in California (Thompson and Steinhaus, 1950). It was suggested that the irrigation water deposited virus onto the short alfalfa, thereby assisting in developing epizootics. Moreover, these authors suggested that irrigation water may be one of the primary methods by which this virus was introduced to uninfected fields. This was an important observation which could account for the uniform spread of this virus into this particular ecosystem. Irrigation water could be implicated in other crops but probably not to the extent of irrigated alfalfa. Cotton, soybeans, and other row crops that are routinely flood-irrigated may be innoculated in this manner. However, the proximity of row crops to the soil may preclude the active transport of the virus onto the plant foliage in areas where insect infestation occurs.

D. DISPERSAL BY MARKETING

Heimpel et al. (1973) conducted one of the few tests where the end product, cabbage heads, were assayed for the presence of *T. ni* NPV at the time of marketing. They determined that during the Maryland growing season, the amount of viable *T. ni* NPV was ca. 7×10^6 IBs/in.^{2} during natural epizootic peaks. These findings clearly demonstrated that vegetable crops marketed in a relatively natural state, such as cabbage, lettuce, and tomatoes, can conceivably harbor large quantities of insect viruses when produced in an area where an epizootic has occurred. Jaques (1974a,b) also reported the presence of infective *T. ni* NPV and *A. rapae* GV on cabbage foliage at harvest time. It is also conceivable that many forage crops (e.g., alfalfa) harvested and often transported great distances, could shed infected IBs leading to the dispersal of virus within the environment.

III. DISPERSAL VIA METEROLOGICAL FACTORS

A. *WIND*

The wind may be one of the primary movers of viruses from the soil to foliage, where susceptible larvae come into contact with the IBs and begin new foci of infection. Transfer of virus by wind is difficult to prove in field situations. The wind is obviously capable of transporting viruses contained on soil particles or in liquefied remains of larvae, although little information pertaining to this means of distribution is available. The formation of aerosols during thunder storms could transport viruses over considerable distances, from plant to plant, or merely throughout a single plant. Wind is also important in the movement of diseased insects, noninfected carriers, and other air-borne vectors. Early studies by Thompson and Steinhaus (1950) implied that wind may account for some localized transmission of the disease, but the data were not conclusive. They felt that wind played an important role in infecting alfalfa fields, particularly in some newly established fields. David and Gardiner (1967) discussed the importance of wind in transferring virus from the soil to foliage where it became accessible to susceptible larvae. In 1979, Young and Yearian, working with the NPV of the soybean looper, noted that the wind played a major role in the movement of these viruses within the soybean ecosystem. However, they stipulated that the distance of the soybean foliage from the ground may prevent sufficient viral exposure to initiate an epizootic. The wind also moves the virus and/or infected foliage back to the soil which acts as its primary reservoir.

Wind can certainly be considered a primary dispersal agent when infected flying adults are carried great distances by air currents. Man-made air currents within enclosed areas, such as greenhouses, have been associated with the transmission of lepidopteran viruses by coprophagus and necrophagus mites (Szalay-Marzso and Vago, 1975). Infective IBs were found within the folds of the mite integument following their feeding upon diseased cadavers. When the mites molt, their cast skins are distributed within the enclosed greenhouse area by air currents, dispersing virus to new locations where ingestion can occur by susceptible larvae. It is also reasonable to assume that insect predators and parasites which have fed on infected larvae could also be distributed over long distances by the wind, thereby dispersing the virus to new populations in new locations.

Stairs (1965) noted that adults of the forest tent caterpillar infected with an NPV were transmitted up to 30 miles in the direction of the prevailing winds. He reported that the virus spread throughout a population area limited to 4 acres, and then spread throughout 250,000 acres in a single generation. Magnoler (1974) examined wind and its effect on *Calisoma syncophanta*, a carabid predator of gypsy moth larvae. He suggested that wind was an important element in the widespread epizootic of the gypsy moth NPV in the Northeast United States. Smith (1976) observed that falling leaves were heavily contaminated with baculoviruses prior to their deposition on the soil beneath trees containing infected larvae. Wind dispersed contaminated foliage up to 30 m from the trees. Theoretically, violent summer storms with high winds and rain could have dramatic effects on the dispersal of viruses. The importance of wind is implied; nevertheless, the experimental data is almost nonexistent.

B. *RAIN*

Results of extensive investigations on the dispersal potentials of viruses by rain within forest environments have been widely published. Ossowski (1960) in a study on the biological control of the wattle bagworm, *Cryptothelea (=Kotochalia) junodi* (Heyl), reported that the bags containing infected larvae remained in the trees for several years and disintegrated very slowly. He observed that the rain did not wash the NPV off the foliage, but implied that it aided in the distribution of the NPV throughout the rough hairy foliage of the trees. Rain was also the chief agent in dispersing sawfly viruses in Canadian forests (Bird, 1961). In further studies with sawflies and the epizootics involved with sawfly populations, Bird established that the disease was spread faster down a tree than upward. Significantly, an infected colony at the top of a tree resulted in rapid mortality throughout the tree while almost no spread occurred upward from an isolated colony at the base of the tree. The rapid downward spread of the disease within the tree was attributed to rain. Bird noted that it seemed difficult to remove the virus by washing with water; however, diseased cadavers were loosened from the foliage by rain that rinsed into the soil. Smirnoff (1962) working with the NPV of *Neodiprion swainei*, a defoliator of eastern Canadian forests, implied the importance of wind, rain, animals, parasites, and predators as important factors in the natural dispersion of this virus.

Orlovskaya (1968), examining the geographical distribution and manifestation of viral diseases in the USSR, noted that about 27 baculoviruses infected forest insects in the USSR and that they were widely distributed. She suggested that rainfall probably contributed to the dispersal and translocation of viruses infecting the lacky (*Malacosoma neustria*) and the brown tail (*Euproctis chrysorrhoea*) moths. The author observed that during very hot weather, when epizootics were in progress, uninfected larvae "drank" the IB-laden fluid exuding from dead larvae. Rainfall and the ingestion of cadaver fluids was strongly implicated in the dispersion of baculoviruses among forest insects in the USSR. Doane (1970) studying pathogens of the gypsy moth and their role in the development of an epizootic within a population, suggested that heavy rains tended to soften infected cadavers and to wash them from the leaves into the understory, thereby contaminating additional foliage and dispersing virus within the canopy. He noted the location of the infected cadavers and suggested that those in the upper portions of the trees tended to facilitate viral spread. Yendol and Hamlen (1973), in a dissertation dealing with the ecology of entomogenous viruses and fungi, discussed the importance of rain in washing virus through the canopy and into the soil. They observed *Neodiprion sertifer* NPV cadavers at high foci within the tree canopy, and the dispersal of IBs to lower parts of the crown by rain which resulted in new foci of infection. These reports all implicate the soil as the primary reservoir for baculoviruses and rain as one of the abiotic transport agents in the dispersal and transfer of IBs to the soil. The authors also mention that contaminated soil is transferred via rain-splash and wind to foliage and that these abiotic mechanisms are important for dispersal of the virus to foliage close to the ground. They doubted that wind and rain-splash would be important for foliage more than a few feet above the ground. Smith (1976) referred to the rain and wind as factors in the dispersal of baculoviruses, although he minimized the importance of rain, stating that it did not seem to be an important factor.

Magnoler (1974) reported that the highest rate of NPV infection in gypsy moth larvae occurred after rainy days, indicating that the washing activity of rain probably distributed the infectious material on the foliage. Lautenschlager et al. (1980) in describing the incidence of natural occurrence of the NPV of the gypsy moth, noted the importance of rain and wind in the dispersion of this virus within the ecosystem. They stated that the movement of an NPV in the wild very likely results from a combination of all of the abiotic and biotic mechanisms, with the primary sources of movement varying with the biological and physical conditions within the ecosystem. Thompson and Steinhaus (1950) discussed the multifaceted importance of rain in the dispersion of the NPV of the alfalfa butterfly, *Colias eurytheme*, in California alfalfa fields. David and Gardiner (1967), in a study of persistence of a GV virus of *Pieris brassicae* in soil and sand, noted that liquefied larvae were transported by gravity and by rain to the soil, and that most of the virus material reached the soil, which is the primary reservoir. Once this GV is in the soil, it is not readily removed by trickling water, which would be analogous to rain. An additional consideration is the pH of the microenvironment in which the rain is falling and its effect upon the virus. The rain may be far from neutral pH, which could dramatically change the pH of the soil. Elmore and Howland (1964), in a study of natural versus artificial dissemination of *T. ni* NPV, suggested that rain may have increased virus dispersion within the field. They concluded that the *T. ni* NPV was poorly dispersed by natural means. Rain does not readily remove virus from foliage. However, rain is implicated in the spread of virus on foliage of trees and other plants (Bird 1961; Ignoffo et al., 1965; Burgerjon and Grison, 1965). Heimpel et al. (1973) discussed the importance of rain for the transfer of baculoviruses from the soil to the lower foliage of cole crops (particularly cabbage) and attributed the initiation of epizootics in Maryland to this phenomenon. Jaques (1976) also noted the importance of rain in the spread of baculoviruses onto cole crops. He attributed the spread, dispersal, and translocation of NPV over cabbages and onto the soil surfaces to rain. In 1971, Tanada discussed the role of rain in the distribution of the NPV of sawflies and cabbage looper viruses. This was followed by a review in 1973 on the importance of rain in disseminating viruses onto plants and within the environment. Undoubtedly, wind and rain are important factors in the dispersal and translocation of baculoviruses within the environment and perform a very necessary function in their maintenance within ecosystems.

C. NATURAL WATER MOVEMENT

Very little has been done on the movement of baculoviruses through natural water systems. The IBs of the baculoviruses can readily persist in aqueous environments, such as lakes and streams, waterways, and irrigation systems, all of which are abiotic dispersal sources. Other animals could feed upon infected individuals or cadavers within the aquatic environment, which could then be transported through fecal contamination into the stream or by direct mechanical transport to other areas adjoining the stream where new foci of infection could occur. In the opinion of Tinsley (1979), viruses retain infectivity and are dispersed in fast-flowing streams. Although very little data are available, all indications lead to the assumption that natural waterways (streams, lakes, creeks, etc.) could very readily disperse insect baculoviruses.

IV. DISPERSAL VIA TERRESTRIAL AND AQUATIC ANIMALS

A. MAMMALIA

1. Man

Among mammals, man is the primary unnatural dispersing agent of baculoviruses in the environment. Farmers may transport and disperse viruses by cultivation, scouting, and sampling procedures, and then by harvesting and marketing crops from fields where baculoviruses occur. Thus, if we consider activities such as cultivation, vehicular movement, and equipment use during work and recreation on farms, in forests, and in other ecosystems where baculoviruses abound, we can easily visualize how man's activities can result in the dispersal of viruses.

2. Domestic Animals

The role of man's domestic animals in the dispersal of baculoviruses has not received much consideration in the overall scope of virus transmittal in the environment. One interesting example was cited by Crawford and Kalmakoff (1977) working with *Wiseana* spp., a soil-dwelling Lepidopteran of New Zealand. Three pest species occur, *Wiseana cervinata*, *W. umbraculata*, and *W. signata*; the most predominant is *W. cervinata*. They are serious pasture pests and cause between 10 and 30 million dollars damage per year in New Zealand. These Lepidopterans are infected by three viruses: a NPV, a GV, and an entomopox. Livestock movement within the pastures is thought to spread the virus over the soil surface. This virus maintains itself from year to year in the soil, perpetually infecting these insects in early instars. Moore et al. (1973) found NPV to be the most prevalent and most lethal of the three viral diseases. In a survey of 43 New Zealand pastures, 87% of the pastures over 5 years old contained virus-infected larvae. In younger pastures, the incidence of infection was <10%. They hypothesized that the virus accumulated in older pastures and that cultivation had removed the viruses from the larval microhabitat of younger pastures. The movement of sheep through pastures appeared to be a principle dispersal factor. When stock movement was maximized in pastures during the winter months, a viral epizootic occurred in the *Wiseana* population the following spring.

Wiseana spp. larvae maintain solitary habitats. The larvae live their entire subterranean larval stage in one tunnel within a defined territory. Virus transmission through larval interactions or contact is highly unlikely. Therefore, the movement of livestock is the logical dispersal factor between these isolated larval habitats. The combination of recent cultivation and stock movement would predispose a field to a *Wiseana* viral infection. Other domestic animals within an endemic area could conceivably aid in the dispersal of virus throughout that area. Cattle, horses, or any large animals moving through an area disturbing the soil, foliage, and other habitats where larvae cadavers or infective IBs are found could lead to the dispersal of these viruses.

3. *Wild Animals*

Dispersal of viruses by wild animals is quite well documented, particularly in forest ecosystems of the gypsy moth and the sawflies. Since soil is the primary reservoir for the gypsy moth NPV and the NPV remains infectious in forest litter for at least 1 year, it is very probable that NPV carried externally on feet, hair, and on other parts of the mammalian body can be washed or shaken loose at a later time (although no field data are available to support this assumption). Another significant consideration is that in most northeastern forests mammals are more numerous than birds and are present year round to aid in the passive transport of NPV.

Lautenschlager et al. (1980) studied the natural occurrence of gypsy moth NPV in wild birds and mammals. They captured and examined 3 species of birds and 5 species of mammals in plots with naturally occurring NPV. Of the 5 mammals, the white-footed mouse, *Peromyscus leucopus* (Rafinesque), red-backed vole, *Celethrionomys gapperi* (Vigers), raccoon, *Procyon lotor*, and chipmunk, *Tamias striatus* contained infectious NPV in their alimentary tracts. It is interesting to note that the masked shrew, *Sorex cinereus*, also captured in these plots, did not contain NPV. An explanation for the absence of infectious NPV may be related to the very rapid metabolism of the masked shrew, since ingested material passes their alimentary tracts very swiftly. These studies clearly demonstrate that small mammals found in the litter and understories of the forest ecosystem can passively transport infectious gypsy moth NPV. Animals used in this study were collected at a time when NPV-infected gypsy moth larvae were not available for food. Transport must then have resulted from the ingestion of cadaver-contaminated foliage, seeds, and berries, which might explain why a herbivorous animal such as the red-backed vole showed a high incidence of infection and had large amounts of NPVs in its alimentary tract. In another study, Lautenschlager and Podgwaite (1979) concluded that both mammals and birds have features which contribute to their ability to passively transport significant amounts of NPV within the environment. All tested animals, whether fed NPV-infected foliage or gypsy moth larvae, passed intact, infectious IBs through their alimentary tract. The white-footed mouse, considered an important gypsy moth predator, actually preferred the infected larvae over the healthy ones. The opossum, *Didelphis marsupialis*, a more primitive marsupial, could also be an important source of NPV dispersal due to its low metabolic rate and long alimentary tract. Lautenschlager and Podgwaite examined the alimentary tracts from 169 mammals collected within NPV epizootic areas and found that about 75% of the alimentary tracts contained enough NPV to cause gypsy moth mortality. From studies such as these, the importance of mammals in the dispersion of baculoviruses can be appreciated.

B. AVES

Franz et al. (1955) were one of the first workers to report that baculoviruses could pass successfully through the alimentary tract of birds. While working with the NPV of the pine sawfly, *Neodiprion sertifer*, they observed this virus pass through the alimentary tracts of the robin, *Erithacus rubecula* (L.) and a predatory bug, *Rhinocoris annulatus* (L.). The authors suggested that animals of this type, particularly birds, should be considered vectors and as well as distributors of pine sawfly NPV in the environment. Bird (1955) reported that birds acted as predators on *N. sertifer* populations and were responsible for the dispersal of NPV to new localities and populations. He noted that the catbird (*Dunatella carolinensis*), and cedar wax wing (*Bombycilla cedrorum*), were particularly influential in transmitting these viruses through their droppings after eating the infected larvae. In a South African study on the biological control of the wattle bagworm, *Cryptothelea (=Kotochalia) junodi* (Heyl), Ossowski (1960) observed that bags of infected larvae remained in trees for several years. He observed many biotic and abiotic dispersal factors of the wattle bagworm virus, but emphasized that birds feeding on infected bagworms transferred virus through feces over long distances.

In an article on parasites and predators as vectors of insect viruses, Smirnoff (1975) noted that the major bird predators of the jack pine sawfly, *N. swainei*, were the slate-colored Junco, the hermit thrush, the myrtle warbler, and the white-throated sparrow. While studying an epizootic in Wales, Entwistle et al. (1977a) investigated 16 species of arboreal birds in 6 families comprising 80% of the bird population. They found positive virus in feces during the months of September and October; 90% of the droppings collected were infective. Droppings collected outside of the epizootic area, and from birds netted in January, were also infective which indicated that birds dispersed virus all year. They observed that the defecation of virus was the greatest approximately 1 hr after ingestion, and continued up to 3 days, depending upon the bird species. A prolonged retention in the alimentary tract would increase the opportunity for wide dispersal of virus. Entwistle et al. (1977b) noted that it was possible for birds, via infective feces, to disperse the European spruce sawfly *Diprion* (=*Gilpinia*) *hercyniae* NPV up to 7 km. He observed that birds served as dispersal agents during the winter, and that they routinely passed feces containing infective NPV throughout the months when larvae were not available (primarily the months of November-June). It was theorized that the birds may have fed upon hanging cadavers, which remained from the summer epizootic, liberating and mobilizing the viral innoculum pool. An estimated density of 5 birds/ha could potentially result in 156 droppings/tree/year. Further calculations indicated up to 14 infectious droppings/tree could potentially be deposited during the 4-month larval season (July-October). Theoretically, the occurrence of droppings every 3.5 minutes/ 10 hr/day would be enough to infect 2000 trees/ha. Entwistle and his associates concluded that birds may very well be the single most important carrier of viral innoculum over long distances, and that they substantially contributed to more local dispersal in the immediate range within their ecosystem.

In a similar study of the European spruce sawfly in Wales, Buse (1977) observed that adult sawflies, parasites, predators, scavengers, insect frass, rain, and particularly birds, were the major agents of virus dispersal within the ecosystem. Bird droppings were collected from 23 birds of 12 species along with the alimentary tracts from 19 birds killed during the larval season. No droppings or gut samples contained baculovirus. This may be attributed to their solitary behavior and because larvae were only a very small portion of the available food source at the time. Buse also noted that bird populations were generally static during the sawfly larval season and further suggested that virus transmission by birds was likely to be related to the availability of infected sawfly larvae and alternative food sources. Viable NPV were recorded from bird feces collected from areas with large populations of infected sawfly larvae. In the field, however, it was difficult to show that droppings were deposited in a suitable form, in large enough quantities, and in appropriate locations for larvae to become infected. Buse indicated that the sawfly adult was the most important viral dispersal agent. Entwistle et al. (1978) noted that the peak passage of European spruce sawfly polyhedra through caged birds occurred in <1 hr. They observed that the feces of all birds tested remained infective through the 2nd day and those of 6 birds through the 3rd day postingestion. One bird remained infective 7 days after ingesting the sawfly NPV. Commenting on the comparatively long retention and passage of infective virus, they suggested that birds may be active in the short- and long-distance transport of baculoviruses. Small birds pass droppings at a rapid rate and may be primarily responsible for transporting viruses over short distances within the forest ecosystem. The IBs of the baculoviruses pass unchanged through the alimentary tract of all bird species examined. The passerine birds generally pass food through the alimentary tract in about 1.5 hours; they can pass NPV in droppings for up to 3 days following ingestion. Entwistle et al. (1978) is convinced that birds are very important in dispersing these viruses throughout the forest ecosystem.

Elmore and Howland (1964), studying the natural and artificial dissemination of *T. ni* NPV, noted that it was poorly dispersed by natural means, although birds were mentioned as important carriers. These authors were particularly interested in blackbirds feeding on diseased *T. ni* larvae in cauliflower fields. Hostetter and Biever (1970), while working with an epizootic of the *T. ni* NPV in St. Louis county, Missouri, reported that English sparrows, *Passer domesticus*, fed extensively on infected *T. ni* larvae. The infected larvae, which lost their protective green color due to the infection with the NPV, combined with the modification of their behavior (e.g., migration toward the outer, unprotected areas of the cabbage plant) became easy prey for the birds. The English sparrows were observed methodically hopping down the rows of cabbage searching for and rapidly devouring the infected larvae. Bird droppings collected during these observations were found to be positive for the NPV of the cabbage looper. The authors suggested that birds were a source of viral transfer between fields of cabbage on an intra-farm basis and would certainly enter into the dissemination of virus within areas where cabbage and other cole crops were being produced.

Polson and Gitay (1972) investigated the cattle egret, *Bubulcus ibis*, known to migrate over Europe and Asia. The feces contained infective GV of *Heliothis armigera* and the authors speculated this would be an excellent dispersal source for GV virus throughout the egret's migration range. Reed (1971) investigated some of the factors that affected the status of GV as a control agent for the potato moth, (*Phthorimaea operculella*), isolated its GV from bird feces. He reported that healthy larvae were translucent, green, or gray in color, and that infected larvae became opaque and whitish in color.

The diseased larvae were readily visible and remained longer on the foliage after leaving the leaf mine, predisposing them to predators. Large numbers of healthy and diseased larvae are often found moving over the ground in search of new food. Migrating larvae are easy targets for birds in search of prey. The white-fronted chat (*Epithianura albifrons*) was observed to prey almost exclusively upon *P. operculella* larvae. Reed found an average of 2 to 4 virus-infected droppings/plant over a plot of approximately 5000 plants, which is a considerable source of virus within a plot area. Reed suggested that predator birds and volunteer potato plants should be considered important agents in the continuation of the potato moth GV cycle. Although documentation is not available for other birds, either domestic or wild, they should be considered very active viral dispersal and transport vectors. Dispersal of virus may be particularly important during the reproduction cycle of wild birds, when younger birds are in a more insectivorous stage of their life. During this period they are apt to feed on diseased larvae that would be readily available and perhaps easier prey due to infection. Some of the gallanaceous birds (i.e., quail and pheasants), which are often found in close conjunction with the cultivated crops of man, may have ready access to such larvae. Domestic birds, such as, chickens and perhaps ducks, and geese that may be found in areas where cultivated crops are being produced, should also be considered as dispersal vectors for some baculoviruses.

The expression of a full epizootic requires several generations of insects which necessitates persistence between generations. The initial spread and the continuous growth of epicenters and their overlap generates the epizootic. Obviously, birds play an important part in generating new foci of infection which can become epicenters of infection.

C. PISCES AND AQUATIC ANIMALS

Smirnoff (1975), in one of the few references to reptiles, reported that the red-spotted newt and the blue-spotted salamander were predators of the jack pine sawfly and indicated that they may serve as dispersal agents of baculoviruses.

Savan et al. (1979) conducted a laboratory study on the passage of spruce budworm NPV through rainbow trout, *Salmo gairdneri*, and white suckers, *Castostomus commersoni*. Fish, in forest ecosystems where virus-infected larvae may fall into streams or lakes, could ingest and transport IBs throughout the aquatic environment. The probability of removal or extraction from that aquatic environment is not known and is probably quite unlikely. However, fish may be active transporters of baculoviruses, in addition to other members of the food chain in an aquatic environment. It is conceivable that these animals could disperse baculoviruses over considerable distances throughout the ecosystem.

V. DISPERSAL VIA INSECTS

A. PRIMARY HOST SPECIES

Many investigations concerning transmission of viruses within insects and insect communities have been conducted in the forest ecosystem. Several of the early studies implicated host insects in ovarial transmission (Thompson and Steinhaus, 1950; Bird, 1953, 1955; Bird and Elgee, 1957; Bird and Burk, 1961). Concrete evidence of transovarian transmission is still limited, although some of the early literature suggested, or at least took for granted, that these viruses were readily transmitted transovarially from infected females to the progeny within the egg. More recent studies, however, seem to cast considerable doubt on true transovarial transmission. Most of these later investigations suggest that infection occurs as the egg passes through the contaminated genitalia during deposition of the egg.

The importance of ovarial transmission in the development of epizootics among sawfly populations, and the significance of viral transport by eggs, either externally as contaminant or transovarially, was discussed by Bird (1961). He examined the importance of ovarial transmission in the initiation and development of epizootics within sawfly populations within the forest ecosystem. Working primarily with the European spruce sawfly *Gilpinia* (=*Diprion*) *hercynaiea*, the European pine sawfly (*Neodiprion sertifer*), the red-headed pine sawfly (*N. leconti*), and the jack pine sawfly (*N. swainei*). This was a significant study and encompassed many abiotic and biotic dispersal agents. There was very strong evidence of ovarial infection with *Gilpinia* prior to oviposition due to the consistent frequency of infection among progeny and the consistency in the appearance of the disease in late second and third instars. Bird suggested that this transmission may have occurred through the external contamination of the chorion at the time of oviposition. The author found that larvae fed a heavy dose of virus upon hatching, died at about the same stage as those suspected of ovarial infection and concluded that this was reasonable evidence supporting transoverian transovarian of this particular virus. Bird presented further evidence of ovarial transmission when epizootics were generated from the infection of <10% of the egg clusters and spread to other sawfly colonies.

He discussed how ovarial infection was important as a means of pathogen overwintering and dispersal, depending on the flight habits of the adults. Thus, a focus of infection may then begin wherever an infected female flies, or is carried by air currents. Because eggs are laid singly, many foci of infection may be initiated, which may explain the phenomenal spread of disease in low population levels (Bird and Elgee, 1957; Bird and Burk, 1961).

The *G. hercyanae* virus is disseminated under many different weather conditions and consistent results are usually obtained. The development of epizootics is dependent on sufficient numbers of hosts for propagation and transmission. The theoretical mechanism for adult transmission and dispersal of the NPV occurs when a very high proportion of the final instar larvae ingest sufficient virus within a very critical time period prior to pupation (Bird, 1953, 1961). Pupation occurs in the litter of the forest floor and the resulting cocoons may enter diapause from 1 to 6 winters. The infective virus remains in the lumen of the gut. Upon emergence from the cocoon, the adult is infective and thereby becomes a dispersal agent. Geographically this can be related to dispersal flights of at least 2 km and probably further (Entwistle, 1976). Bird found that *N. sertifer* had only one generation per year, laid a single cluster of eggs, and that the larvae fed gregariously. Usually the entire colony of about 60 infected individuals will die at an early age producing a single focus of infection.

Smirnoff (1960) observed the migration of larvae of *N. swainei* in Canada. He noted that the larvae migrated enmasse when all the foliage was devoured on a particular tree and that these larvae could successfully migrate at least 200 yards in the search for new food sources. The infected fourth and fifth instar larvae survived at least 20 days before dying and adult females transmitted NPV. This suggested that larval migration was a means of viral dispersal over considerable distances within the pine forest. In an additional study, Smirnoff (1962) investigated the transovum transmission of the NPV of *N. swainei*. He subjected older larvae to a low concentration of virus. Almost 60% of the exposed larvae retained the faculty of transovum transmission which resulted in about 90% mortality among the larvae that hatched from the eggs of the female carrier. Observations over a 4-year period indicated that this virus was carried by the adult sawflies, and that it did not affect their fecundity, the incubation period, or the natural mortality of their eggs. Transmission of this NPV from the parent to the progeny through the eggs occurred each year.

Smith (1976), in referring to the NPV of sawflies and the GV of *P. brassicae*, noted that transovarian and transovum transmission were probably the most effective methods of dispersal and transmission of viruses among insects. The host insect was a major dispersing vehicle of the virus and could be partially influenced by certain meterological conditions (e.g., wind).

Buse (1977) conducted a series of studies with the European spruce sawfly and its NPV in mid-Wales of the U.K. He considered alternate means of dispersion within the population, but concluded that it was probable that the adult sawflies were the most important means of dispersal of this NPV in the ecosystem. In another study dealing with the sawfly NPV, Neilson and Elgee (1968) showed that this particular NPV was transmitted from one generation to the next by foliage contamination and by externally contaminated or diseased adults. They suggested that this accounted for the persistence of NPV in low population densities. The detection of the NPV at the limits of the sawflies distribution areas further indicated adult transmission and dispersal of this NPV. Cunningham and Entwistle (1981), studying the control of sawflies by baculoviruses, emphasized the importance of vertical passage in spreading the virus throughout the forest ecosystem. They presented evidence to indicate that most of the transmission occurred from the adult *G. hercyniae* to the progeny. The behavior of the sawflies led them to believe that the primary method of viral dispersal was by infected adults contaminating the oviposition area, thereby providing a source of innoculum for hatching larvae. Sawfly eggs hatch by bursting. The neonates do not feed upon the egg chorion or on the spruce foliage on which the eggs have been deposited but become infected passively via contamination.

Tanada (1976) noted that the site of infection with baculoviruses considerably influenced subsequent dissemination, particularly with insects such as sawflies where the midgut epithelium is the primary site of infection. When the infection is confined to the midgut epithelium considerably more virus can be shed and disseminated than when the insect is systemically infected. These insects live longer, producing virus in the gut lumen which is then excreted in the feces as the adult insects move throughout the environment. Tanada suggested that the effectiveness of sawfly virus results from the midgut infection sites.

In a discussion pertaining to disease prevalence and epizootics in insect populations, Federici (1980) referred to a NPV epizootic of the European spruce sawfly, which had been an economically important pest during the 1930s. The virus of this insect was thought to have been introduced into Canada via parasites during the mid-1930s. The NPV spread through a sawfly population which exceeded thousands per square kilometer within a few years. By 1942, populations were below the economic threshholds, which were calculated to be 1.5 larvae and 1.3 cocoons/ft^2 of forest floor.

Similar results have been obtained on disease-free populations near Ontario and in Wales (Bird and Burk, 1961; Entwistle, 1974). This particular NPV replicates in the midgut epithelium which results in the routine voiding of NPV in the feces. It persists on the foliage for sufficient periods of time, which is probably one of the main means of dissemination of the NPV. Recurrent viral epizootics control the spruce sawfly population.

Nordin (1976), working with the fall webworm, *Hyphantria cunea* (Drury), and the transovum transmission of its NPV, noted that transmission to the progeny by contaminated females and males resulted in approximately 94% mortality. He also concluded that the NPV was transmitted from contaminated males to females via mating and from the female to the progeny via egg contamination. Scanning electron microscopy revealed IBs on egg surfaces which was supportive evidence for contamination at the time of oviposition.

Hukuhara et al. (1978) found the incidence of GV infection among fall webworm larvae ranged from 1.2 to 3.9%. The GV persisted in an enzootic state in the study plot, although only at a low incidence. Positive soil bioassays 12-60 m from the plot indicated the soil was the reservoir for this virus. The authors suggested that this GV was dispersed by physical or biological agents, such as, the wind, predators, and infected adults and concluded that the GV was not a major mortality factor.

Suzuki and Kunimi (1981) studied the dispersal and survival rate of the fall webworm, using NPV as an adult marker to examine the transovum transmission of the virus and the gregarious habits of the larvae. The maximum and minimum distances traversed by females ranged from 1.25 to 188 m in mulberry trees in Japan and was slightly influenced by the wind. NPV is physically transmitted from the adult to the progeny, probably on the abdominal hairs of females which adhere to the surface of the egg mass during the act of oviposition (females of this species deposit a single egg mass). Ninety percent of the egg masses and colonial webs were recovered within a radius of about 37 m; approximately 96% were infected with the NPV. The NPV dissemination rate was quite significant since no natural populations existed in the study area during this test; moreover, flight activity was apparently low for this species. Kunimi (1982), in another transovum transmission study with the fall webworm NPV, smeared a NPV paste or a suspension of virus dust in vaseline onto the abdominal areas of female moths. Thorough surface contamination during the act of oviposition caused between 90 and 100% mortality in larvae

10 days postemergence, indicating that this method of transmission was probable.

Morris (1963), referring to the gypsy moth and tussock moth, noted that transovum transmission does occur and can be very significant. Further, the development or collapse of a population can be predicted based on the percentage of egg masses whose hatching larvae succumb to virus. Other significant references to transovum transmission include those by Knipling (1960), Martignoni and Milstead (1962), and Nordin (1976). Transmission of the gypsy moth NPV occurred on the surface of the eggs and larvae became infected as they hatched, thus providing vertical transmission from generation to generation (Doane, 1969, 1970).

Magnoler (1974) investigated the field dissemination of NPV in gypsy moth and found wind-borne dispersion of infected 1st instar larvae. Migration of larvae may have been the primary factor in establishing foci of infection within the infestation area. Moreover, he also noted that the migration of larvae and predation were also very important in the dispersal of this virus within the forest ecosystem. The larvae became infected following contact with dead larvae on foliage or through aggregation of infected larvae in favorable temperature zones on the tree. Throughout the study, the migratory activities of early and late instar larvae aided in the dispersal of the NPV and spread of the disease.

Doane (1976), dealing with the ecology of pathogens of the gypsy moth, noted that neonate larvae are infected with NPV via the egg surface and the scales on the egg mass contaminated by females during oviposition on the trees. He suggested that neonates constituted the primary innoculum for horizontal transmission. The fifth instar larvae deposit a silken matting or a "debris" mat which is thought to be the source of future contamination. Debris mats are often used by the larvae of following generations for shelter, where they could be exposed to NPV.

Thompson and Steinhaus (1950) noted that transmission of the alfalfa caterpillar NPV occurs via mechanical factors and listed the contaminated adult, parasites, and carniverous insects as being the main biotic dispersal agents of this particular virus. They suggested transmission of this virus occurred through contaminated eggs. The parasite, *Cotesia (=Apanteles) medicaginis* Muesebeck, was also suspected of being instrumental in establishing foci of infection, because it can transmit virus from infected larvae to healthy larvae through stinging. The virus may also be transmitted by the parasite's contaminated body. In new alfalfa fields, the virus was introduced via a series of biotic and abiotic factors.

In a study on natural versus artificial dissemination of the NPV of cabbage loopers, Elmore and Howland (1964) determined that moths from a diseased field population did not transmit *T. ni* virus to their progeny. They also concluded that the *T. ni* NPV was poorly dispersed by natural means. Vail and Hall (1969) determined that transovarian transmission of *T. ni* NPV did not occur. In a somewhat related study, Jaques (1976) observed that larval behavior of some insects, once infected with their respective viruses, can aid in the distribution of the virus, i.e., through regurgitation of food from the gut, through the feces, through cannibalism, and contact with one another in the environment. Jaques also mentioned that baculoviruses could be transmitted either in or on the eggs, but that there was no evidence that the *T. ni* NPV, *A. rapae* GV, or *P. brassicae* GV were transmitted in this way with any significant frequency. David and Taylor (1976) were unable to substantiate transovarian transmission of *P. brassicae* GV after feeding GV or injecting free virions into the adults. Tatchell (1981) found, via artificial contamination, that *P. brassicae* GV was readily transmitted to progeny.

Hamm and Young (1974) fed IBs to adult female bollworms and observed passage of NPV through the alimentary tract in sufficient quantities to contaminate eggs laid during oviposition. They suggested that this could also occur in natural populations. Transovum transmission was alluded to by the use of scanning electron micrographs which revealed IBs on the egg surfaces. IBs were also found in the lumen of the gut and in the feces of adult moths, which is particularly important with species such as *Heliothis* whose larvae devour the eggshells at the time of hatching. Federici (1978) referred to the possibility of transovum transmission for clover cutworm, *Scotogramma trifolli*, NPV, and suggested that during NPV epizootics the larval survivors may become carriers as adults. Shekhurina (1980) noted that the GV of the gray grain moth, *Apamea anceps* (Schiff), spreads primarily via food ingestion, transovarially, and through parasites. The soil acts as the GV reservoir and this GV, preserved in a latent form, is transmitted onto the ovum. This virus is always present in pest populations and epizootics arise periodically. The gray grain moth migrates from highly concentrated population areas and in doing so disperses the virus. GV infects up to 80% of the gray grain moths overwintering as larvae, causing their death in the spring.

Most researchers agree that transovum contamination of the eggs occurs at the time of oviposition, but that true transovarian transmission rarely takes place. *Spodoptera littoralis* (Boisd.) adults exposed to cotton foliage which had been treated with the NPV transmitted NPV to their progeny. The females were able to transmit low levels to progeny through oviposition on untreated plants. In a similar test, when the adults were exposed to a sugar-virus feeding solution, the NPV was also successfully transmitted to their progeny (Elnager et al., 1982). The authors concluded that flying female moths may contribute to the field dissemination of this virus over considerable distances, and noted that females seemed to be better carriers than males.

Tanada (1971), dealing with the persistence of entomopathogens in the ecosystem, noted that these pathogens, particularly viruses, may be carried on the external body surfaces of the host insect as well as nonsusceptible insects and other animals within that ecosystem. He further concluded that persistence of insect viruses occurs mainly in the primary or alternate host and within the soil. When baculoviruses occur within the host, either transovum or transovarial transmission is very likely.

B. PREDATORY INSECTS

Bird (1961) investigated the transmission of viruses by parasites, predators, scavengers, and birds, and presented several interesting findings. In particular, he noted that the NPV within *N. sertifer* populations spread from tree to tree relatively slowly and that this species had few parasites. However, in the *N. leconti*, a native North American species, the transmission of virus was very rapid when compared to *N. sertifer*. Further, he suggested that this was probably due to a well-established natural enemy complex which aided greatly in the dispersion and spread of this NPV in the *N. leconti*. Bird (1961) concluded that a native virus in a native host probably spreads faster due to the complex of natural enemies, particularly, parasites.

Smirnoff (1975) reported that the major invertebrate predators of *N. swainei* were Coleoptera, ants, wasps, and an unidentified species of spider. He also observed that the wasp, *Vespa rufa consobrina*, built nests in numbers directly proportional to the *N. swainei* population. NPV diseased larvae were easier prey for wasps; after capturing and partially devouring the larvae, the wasps often carried them great distances. At least 3 species of Hemiptera fed on the infected larvae. Ants were active on all larval instars and adults and some Elaterids also attacked cocoons in the forest litter. Smirnoff (1959) reported a hemipteran, *Pilophorus uhler* (Knight), was a predator of neonatal and some second stage larvae of *N. swainei*. Burgess and Collins (1915) reported that *Calosoma syncophanta*, the predatory beetle of the gypsy moth, were recovered as far as 20 miles beyond the areas of release suggesting that they are highly mobile. These predators, when they disperse from collapsed gypsy moth populations, may establish in new areas where food is available and thereby initiate new foci of infection (Doane and Schaefer, 1971). Capinera and Barbosa (1975) found that there were sufficient IBs in the feces of all field-collected *C. sycophanta* adults to infect third-stage gypsy moth larvae. The number of IBs in fecal samples of these beetles averaged 3.1×10^9 IBs/ml. The highest count was 12×10^9 IBs/ml. The authors discussed the possibility that this predator transmitted the gypsy moth NPV between local foci of gypsy moth populations.

The importance of parasites and predators as vectors of insect diseases has been well documented (Kaya, 1982). Parasites can inoculate directly onto or into susceptible hosts and they may or may not be infected by the pathogen. These parasites and predators can also contaminate the substrate of the host insects, which then ingest the virus and become diseased. Tachinids and sarcophagid flies have also been implicated in the dispersal of NPVs and GVs, at least within insect populations. These viruses have not been reported to infect the developing parasite larvae within the host.

C. PARASITOIDS

Stairs (1966) found that NPV was transmitted from generation to generation by infected *Malacosoma dissteria* adults. He also observed that adult flies, *Sarcophaga aldrichi* Parker, were attracted to dead and diseased larvae. (The sarcophagid flies are very strong flyers and behave much like oversized houseflies, thoroughly contaminating the area in which they are found.) He suggested that the foliage became contaminated where healthy larvae were feeding, or even that NPV was carried from larvae to larvae by this Sarcophagid fly, and determined that the Sarcophagid remained infectious for at least 24 hr. Moreover, in field studies, it was demonstrated that the intensity of virus epizootics varied directly with the prevalence of adult parasites. From these studies, Stairs concluded that ovarial transmission appeared to be the most important means of survival for this virus in an insect population and that the tent caterpillar had a relatively high incidence of transmission of this virus via the egg. The most severe NPV epizootics in the tent caterpillar populations probably resulted when both the incidence of egg transmission and the population of the sarcophagid were high.

Beegle and Oatman (1975) determined that 50% of the female *Hyposotor exiguae* parasites from virus-infected cabbage loopers transmitted enough virus to other hosts to infect them. *Hyposotor exiguae* did not discriminate against NPV-infected or noninfected hosts. The authors suggested that the virus was carried and transmitted by the ovipositor and not on other areas of the parasites (about 90% of the male parasites developing within infected hosts transferred infective virus to approximately 21% of the uninfected hosts exposed to them) which indicated external contamination of the male parasites. This series of tests also demonstrated that parasites developing in infected hosts were more effective in spreading virus to healthy hosts than were parasites which were initially free of the virus and become contaminated by ovipositing in infected hosts.

Levin et al. (1979) and Irabagon and Brooks (1974), working with two parasites, *Cotesia* (=*Apanteles*) *glomerata* (L.) with its host *Artogeia (=Pieris)*, and *Campoletis sonorensis* with its host *Heliothis virescens*, reported that both of these parasites could transmit virus to their host. Inclusion bodies were not detected within the cells of the parasite. However, numerous IBs were found in midgut lumens of developing parasitic larvae. Ingested IBs were voided in the myconium and the resulting adult parasites were free of virus infection. They found that females developing in NPV-infected hosts transmitted the NPV to healthy *H. virescens* larvae. Transmission during this study with *C. sonorensis* and *H. virescens* was considered to have occurred during the act of oviposition through a contaminated ovipositor.

D. OTHER ARTHROPODS

Mites of the species *Tyrophagus putrescentiae* Schrank feed on molds which are found primarily in storehouses or warehouses. These mites were allowed to feed on cadavers of codling moth larvae killed by the GV. The mites, through external contamination, were expected to transmit this virus to healthy codling moth larvae, which was confirmed through electron microscopy, i.e., numerous IBs were found in the intestinal cavities of mites and on the legs and in folds of integument after they had fed on the infected cadavers (Szalay-Marzso and Vago, 1975). The authors concluded that coprophagus, mycophagus, or necrophagus mites may readily disseminate baculoviruses naturally, which is important from an epidemiological point of view. Thus, mites are an example of arachnids that may act as vectors of insect viruses.

VI. ORYCTES-LIKE VIRUSES

The palm rhinoceros beetle, *Oryctes rhinoceros*, occurs throughout southeast Asia and a number of South Pacific countries. Adult beetles fly to the central crown of palm trees, crawl down the axil of young fronds, and bore into the unopened fronds through the heart of the palm. Palms may be killed by repeated or severe beetle attacks which destroy the apical meristem. Coconut palms of all ages and young oil palms are the main economic palms attacked. Eggs are laid, larvae develop and pupate in the tops of dead palms, decaying trunks of palms and other wood, and in heaps of compost, sawdust, manure or other decomposing vegetable matter.

The *Oryctes* baculovirus was first discovered in rhinoceros beetle larvae in Malaysia (Huger, 1966b), and has since been isolated from beetles found throughout the Philippines and in Indonesia on the islands of Sumatra and west Kalimantan (Zelazny, 1977). The virus was first observed in the nuclei of cells extracted from larval fat bodies (Huger, 1966a), but was later observed in nuclei of midgut epithelium of larvae and adults (Huger, 1973; Payne, 1974), in the wall of the ovarian sheath, and in the inner wall of spermatheca (Monsarrat et al., 1973).

Oryctes rhinoceros larvae, infected with *Oryctes* virus, undergo extensive changes as the infection progresses. The abdomen becomes turgid and glassy, the fat body disintegrates, and the amount of hemolymph increases giving larva a translucent appearance when viewed against light. Internal turgor may increase causing a prolapse or extroversion of the rectum. In the terminal phase, chalky white bodies may appear under the abdominal integument (Huger, 1966a).

Most viral multiplication occurs in the midgut epithelium (Payne, 1974). Virus multiplies in the nuclei of midgut epithelial cells of adult beetles. New cells are produced in regenerative crypts, become infected, nuclei hypertrophy, and the gut lumen eventually fills with disintegrating cells and virus particles (Huger, 1973; Monty, 1974). Infected adults defecate virus into the environment; thus, the adults are virus reservoirs (Huger, 1973). It is estimated that up to 0.3 mg virus/day may be produced per infected adult (Monsarrat and Veyrunes, 1976). Adults are the natural vectors of the disease through defecation and are primarily responsible for its spread and transmission. Diseased beetles generally show no external symptoms and the virus is not normally passed from the larval to the adult stage (Zelazny, 1973a).

Oryctes virus is transmitted most frequently during mating and/or, via oral contact with feces from an infected partner. It is also transmitted through contact with feces when both healthy and infected beetles feed in the palms and when adults visit larval breeding sites where they encounter dead and dying larvae.

Zelazny (1973b) found that breeding sites were visited less frequently by infected females than healthy ones. Mated females collected from palms were more often infected than ovipositing females collected from breeding sites. Significantly more males than females were found to be infected and infections in male adults increased with age.

The *Oryctes* virus was introduced in Western Samoa in 1967. It became established in 1 year and subsequently spread to other parts of the islands (Marschall, 1970). Studies in 1970-1971 showed that only about 3% of larvae and 35% of the adults in the wild were infected while an average of 7.3% of the breeding sites contained infected individuals (Zelazny, 1973b). *Oryctes* virus was introduced into Fiji in 1970 and established in *Oryctes* populations by 1974, significantly reducing palm damage.

The *Oryctes* virus was introduced into breeding sites on Wallis Island over a 10-month period (September, 1970-June, 1971). In less than 2 months after initiation, the virus had spread throughout the entire island (Hammes, 1971). On Tongatapu, the *Oryctes* virus was released over a 3-month period (November 1970-January 1971) and its subsequent rate of dispersal in the epizootic phase was 3 km/month (Young, 1974). After introduction into African Samoa (January-April, 1972) the virus spread through the population at a rate of 0.8 to 1.6 km/month (Swan, 1972). The *Oryctes* virus inactivates rapidly and can persist in an area only if adequate beetle populations are available to propagate and transmit it. The utilization of this virus in future control programs directed at curtailing the serious damage inflicted by rhinoceros beetles should be encouraged.

REFERENCES

Beegle, C. C., and Oatman, E. R. (1975). Effect of a nuclear polyhedrosis virus on the relationship between *Trichoplusia ni* (Lepidoptera: Noctuidae) and the parasite *Hyposotor exiguae* (Lepidoptera: Ichneumonidae). *J. Invertebr. Pathol. 25*, 59-71.

Bird, F. T. (1953). The effect of metamorphosis on the multiplication of an insect virus. *Can. J. Zool. 31*, 300-303.

Bird, F. T. (1955). Virus diseases of sawflies. *Can. Entomol. 87*, 124-127.

Bird, F. T. (1961). Transmission of some insect viruses with particular reference to ovarial transmission and its importance in the development of epizootics. *J. Insect Pathol. 3*, 352-380.

Bird, F. T., and Burk, J. M. (1961). Artificially disseminated virus as a factor controlling the European Spruce sawfly,

Diprion hercyniae (Htg.) in the absence of introduced parasites. *Can. Entomol. 93*, 228-238.
Bird, F. T., and Elgee, D. E. (1957). Virus disease and introduced parasites as factors controlling the European Spruce sawfly, *Diprion hercyniae* (Htg.) in central New Brunswick. *Can. Entomol. 89*, 371-378.
Burgerjon, A., and Grison, P. (1965). Adhesiveness of preparations of *Smithiavirus pityocampae* Vago on pine foliage. *J. Invertebr. Pathol. 7*, 281-284.
Burgess, A. F., and Collins, C. W. (1915). The *Calosoma* beetle (*Calosoma sycophanta*) in New England. *USDA Bull. 251*, 1-40.
Buse, A. (1977). The importance of birds in the dispersal of nuclear polyhedrosis virus of the European Spruce sawfly. *Gilpinia hercyniae* (Hymenoptera: Diprionidae) in Mid-Wales. *Entomol. Exp. Appl. 22*, 191-199.
Capinera, J. L., and Barbosa, P. (1975). Transmission of nuclear polyhedrosis virus to gypsy moth larvae by *Calosoma sycophanta*. *Ann. Entomol. Soc. Amer. 68*, 593-594.
Crawford, A. M., and Kalmakoff, J. (1977). A host-virus interaction in a pasture habitat. *J. Invertebr. Pathol. 29*, 81-87.
Cunningham, J. C., and Entwistle, P. F. (1981). Control of sawflies by baculoviruses. "Microbial Control of Pests and Plant Diseases, 1970-1980," ed. H. D. Burges, 949 pp. Academic Press, London.
David, W. A. L., and Gardiner, B. O. C. (1967). The persistence of a granulosis virus of *Pieris brassicae* in soil and in sand. *J. Invertebr. Pathol. 9*, 342-347.
David, W. A. L., and Taylor, C. E. (1976). Transmission of a granulosis virus in the eggs of a virus-free stock of *Pieris brassicae*. *J. Invertebr. Pathol. 27*, 71-75.
Doane, C. C. (1969). Transovum transmission of an NPV in the gypsy moth and the inducement of virus susceptibility. *J. Invertebr. Pathol. 14*, 199-210.
Doane, C. C. (1970). Primary pathogens and their role in the development of an epizootic in the gypsy moth. *J. Invertebr. Pathol. 15*, 21-23.
Doane, C. C. (1976). Ecology of pathogens of the gypsy moth. *In* "Perspectives in Forest Entomology" (J. F. Anderson and H. K. Kaya, eds.), pp. 285-293. Academic Press, New York.
Doane, C. C., and Schaefer, P. W. (1971). Aerial application of insecticides for control of the gypsy moth. *Conn. Agri. Exp. Sta. Bull. (New Haven) 724*, 23 pp.
Elmore, J. C., and Howland, A. F. (1964). Natural versus artificial dissemination of nuclear polyhedrosis virus by contaminated adult cabbage loopers. *J. Invertebr. Pathol. 6*, 430-438.

Elnagar, S., Tawfik, M. F. S., and Abdelrahman, T. (1982). Transmission of the nuclear polyhedrosis virus (NPV) disease of *Spodoptera littoralis* (Boisd.) via exposure of adults to the virus. *Z. Angew. Entomol. 94*, 152-156.

Entwistle, P. F. (1974). New perspectives in pest control with pathogenic viruses. *Land, Oxford 1 1*, 84-88.

Entwistle, P. F. (1976). The development of an epizootic of a nuclear polyhedrosis virus disease in European Spruce sawfly *Gilpinia hercuniae. Proc. 1st Intern. Colloq. Invertebr. Pathol. and the IX Ann. Mtg. Soc. Invertebr. Pathol.*, Queens Univ., Kingston, Canada, 461 pp.

Entwistle, P. F., and Adams, P. H. W. (1977). Prolonged retention of infectivity in the nuclear polyhedrosis virus of *Gilpinia hercyniae* (Hymenoptera: Diprionidae) on foliage of spruce species. *J. Invertebr. Pathol. 29*, 392-394.

Entwistle, P. F., Adams, P. H. W., and Evans, H. F. (1977a). Epizootiology of a nuclear polyhedrosis virus in European Spruce sawfly (*Gilpinia hercyniae*). The status of birds as dispersal agents during the larval season. *J. Invertebr. Pathol. 29*, 354-360.

Entwistle, P. F., Adams, P. H. W., and Evans, H. F. (1977b). Epizootiology of a nuclear polyhedrosis virus in European Spruce sawfly (*Gilpinia hercyniae*). Birds as dispersal agents of the virus during winter. *J. Invertebr. Pathol. 30*, 15-19.

Entwistle, P. F., Adams, P. H. W., and Evans, H. F. (1978). Epizootiology of a nuclear polyhedrosis virus in European Spruce sawfly (*Gilpinia hercyniae*): The rate of passage of infective virus through the gut of birds during cage tests. *J. Invertebr. Pathol. 31*, 307-312.

Etzel, L. K., and Falcon, L. A. (1976). Studies of transovum and transstadial transmission of a granulosis virus of the codling moth. *J. Invertebr. Pathol. 27*, 13-26.

Federici, B. A. (1978). Baculovirus epizootic in a viral population of the clover cutworm, *Scotogramma trifoilii*, in southern California. *Environ. Entomol. 7*, 423-427.

Federici, B. A. (1980). Disease prevalence and epizootics in insect populations. *In* "Characterization, Production, and Utilization of Entomopathogenic Viruses" (C. M. Ignoffo et al., eds.), Proc. 2nd Conf. Proj. V, Microbiological Control of Insect Pests of the US/USSR Joint Working Group. Clearwater Beach, Florida, Jan. 7-10, 1980.

Fenner, F. (1976). Classification and nomenclature of viruses. Second report of the International Committee on Taxonomy of Viruses. *Intervirology 7*, 1-115.

Franz, J. M., Krieg, A., and Langenbuch, R. (1955). Untersuchungen uber den Einflub der passage durch den Darm von Raubinsekten und Vogeln auf die Infektiositat insekten-Pathogener Viren. *Z. Pflanzzenkr. Pflanzenpathol. Pflanzenschutz. 62*, 721-726.

Gitay, H., and Polson, A. (1971). Isolation of granulosis virus from *Heliothis armigera* and its persistence in avian faeces. *J. Invertebr. Pathol. 17*, 288-290.

Hamm, J. J., and Young, J. R. (1974). Mode of transmission of nuclear polyhedrosis virus to progeny of adult *Heliothis zea*. *J. Invertebr. Pathol. 24*, 70-81.

Hammes, C. (1971). Multiplication et introduction d'un virus d' *Oryctes rhinoceros* a l'ile Wallis. *C. R. Acad. Sci. 273*, 1048-1050.

Heimpel, A. M., Thomas, E. O., Adams, J. R., and Smith, L. J. (1973). The presence of nuclear polyhedrosis viruses of *Trichoplusia ni* on cabbage from the market shelf. *Environ. Entomol. 2*, 72-75.

Hostetter, D. L., and Biever, K. D. (1970). The recovery of virulent nuclear polyhedrosis virus of the cabbage looper, *Trichoplusia ni*, from the feces of birds. *J. Invertebr. Pathol. 15*, 173-176.

Huger, A. L. (1966a). A virus disease of the Indian rhinoceros beetle, *Oryctes rhinoceros* (L.) caused by a new type of insect virus *Rhabdionvirus oryctes* gen. n., sp. n. *J. Invertebr. Pathol. 8*, 38-51.

Huger, A. L. (1966b). Untersuchungen uber mikrobielle begvenzungsfaktoren von populationen des indischen nashornkafers. *Oryctes rhinoceros* (L.) in *So-Asien* und in der Sudsee. *Z. Angew. Entomol. 58*, 89-95.

Huger, A. L. (1973). Grundlagen zur biologischen bekafers, *Oracytes rhinoceros* (L.), in So-Asien und in der sudsee. *Z. Angew. Entomol. 72*, 303-319.

Hukuhara, T., and Namura, H. (1972). Distribution of a nuclear polyhedrosis virus of the fall webworm, *Hyphantria cunea*, in soil. *J. Invertebr. Pathol. 19*, 308-316.

Hukuhara, T., Kitajima, K., and Tamura, M. (1978). Introduction of a granulosis virus of the fall webworm, *Hyphantria cunea* (Lepidoptera: Arctiidae), into Japan. *Appl. Entomol. Zool. 13*, 29-33.

Ignoffo, C. M., Chapman, A. J., and Martin, D. F. (1965). The nuclear polyhedrosis virus of *Heliothis zea* (Boddie) and *Heliothis virescens* (F.). III. Effectiveness of the virus against field populations of *Heliothis* on cotton, corn, and grain sorghum. *J. Invertebr. Pathol. 7*, 227-235.

Irabagon, T. A., and Brooks, W. M. (1974). Interaction of *Campoletis sonorensis* and a nuclear polyhedrosis virus in larvae of *Heliothis virescens*. *J. Econ. Entomol. 67*, 229-231.

Jaques, R. P. (1964). The persistence of a nuclear polyhedrosis virus in soil. *J. Insect. Pathol. 6*, 251-254.

Jaques, R. P. (1967). The persistence of a nuclear polyhedrosis virus in the habitat of the host insect, *Trichoplusia ni*. II. Polyhedra in the soil. *Can. Entomol. 99*, 820-829.

Jaques, R. P. (1969). Leaching of the nuclear polyhedrosis virus of the *Trichoplusia ni* from soil. *J. Invertebr. Pathol. 13*, 256-263.

Jaques, R. P. (1970). Natural occurrence of viruses of the cabbage looper in field plots. *Can. Entomol. 102*, 36-41.

Jaques, R. P. (1973). Methods and effectiveness of distribution of microbial insecticides. *Ann. N.Y. Acad. Sci. 217*, 109-119.

Jaques, R. P. (1974a). Occurrence and accumulation of viruses of *Trichoplusia ni* in treated field plots. *J. Invertebr. Pathol. 23*, 140-152.

Jaques, R. P. (1974b). Occurrence and accumulation of the granulosis virus *Pieris rapae* in treated field plots. *J. Invertebr. Pathol. 23*, 351-359.

Jaques, R. P. (1975). Persistence, accumulation, and denaturation of nuclear polyhedrosis and granulosis viruses. *In* "Baculoviruses for Insect Pest Control: Safety Considerations" (M. Summers et al., eds.), pp. 90-101. *Amer. Soc. Microbiol.* Washington, 186 pp.

Jaques, R. P. (1976). Epizootiology of invertebrate pathogens: Baculovirus diseases of pests of cole crops. *Proc. 1st Intern. Colloq. Invertebr. Pathol. and IX Annual Meeting Soc. Invertebr. Pathol., Queens University, Kingston, Canada*, 461 pp.

Jaques, R. P., and Harcourt, D. G. (1971). Viruses of *Trichoplusia ni* (Lepidoptera: Noctuidae) and *Pieris rapae* (L.: Pieridae) in soil in fields of crucifers in southern Ontario. *Can. Entomol. 103*, 1285-1290.

Kaya, H. K. (1982). Parasites and predators as vectors of insect diseases. *Proc. 3rd Intern. Colloq. Invertebr. Pathol., Brighton, U.K., Sept. 6-10, 1982*, pp. 39-44.

Kelly, D. C. (1976). "Oryctes" virus replication: Electron microscopic observations on infected moth and mosquito cells. *Virology 69*, 596-606.

Knipling, E. F. (1960). Use of insects for their own destruction. *J. Econ. Entomol. 53*, 415-420.

Kunimi, Y. (1982). Transovum transmission of a nuclear polyhedrosis of the fall webworm, *Hyphantria cunea* Drury. *Appl. Entomol. Zool. 17*, 410-417.

Lautenschlager, R. A., and Podgwaite, J. D. (1979). Passage of nucleopolyhedrosis virus by avian and mammalian predators of the gypsy moth, *Lymantria dispar*. *Environ. Entomol. 8*, 210-214.

Lautenschlager, R. A., Podgwaite, J. D., and Watson, D. E. (1980). Natural occurrence of the nucleopolyhedrosis virus of the gypsy moth, *Lymantria dispar* (Lepidoptera: Lymantriidae) in wild birds and mammals. *Entomophaga 25*, 261-267.

Levin, D. B., Laing, J. E., and Jaques, R. P. (1979). Transmission of granulosis virus by *Apanteles glomeratus* to its host *Pieris rapae*. *J. Invertebr. Pathol. 34*, 317-318.

McLeod, P. J., Young, S. Y., and Yearian, W. C. (1982). Application of baculovirus of *Psudoplusia includens* to soybean: Efficacy and seasonal persistence. *Environ. Entomol. 11*, 412-416.

Magnoler, A. (1974). Field dissemination of a nucleopolyhedrosis virus against the gypsy moth, *Lymantria dispar* L. *Z. Pflanzenkr. Pflanzenpathol. Pflanzenschutz. 81*, 497-511.

Marschall, K. J. (1970). Introduction of a new virus disease of the coconut rhinoceros beetle in western Samoa. *Nature (London) 225*, 228-229.

Martignoni, M. E., and Milstead, J. E. (1962). Trans-ovum transmission of the nuclear polyhedrosis virus of *Colias eurytheme* Boisdural through contamination of the female genitalia. *J. Insect Pathol. 4*, 113-121.

Monsarrat, P., and Veyrunes, J. C. (1976). Evidence of *Oryctes* virus in adult feces and new data for virus characterization. *J. Invertebr. Pathol. 27*, 387-389.

Monsarrat, P., Veyrunes, J., Meynadier, G., Croizier, G., and Vago, C. (1973). Purification ef e'tude structurale du virus du Coleoptere *Oryctes rhinoceros*. *L. C. R. Acad. Sci. D277*, 1413-1415.

Monty, J. (1974). Teratological effects of the virus *Rhabdionvirus oryctes* on *Oryctes rhinoceros* (L.) (Coleoptera: Dynastidae). *Bull. Entomol. Res. 64*, 633-636.

Moore, S. G., Kalmakoff, J., and Miles, J. A. R. (1973). Virus diseases of Porina (*Wiseana* spp. Lepidoptera: Hepialidae). II. Transmission trials. *N.Z. J. Zool. 1*, 85-95.

Morris, O. N. (1963). The natural and artificial control of the Douglas Fir tussock moth, *Orygyia pseudotsugata* McDunough, by a nuclear polyhedrosis virus. *J. Insect Pathol. 5*, 401-414.

Neilson, M. M., and Elgee, D. E. (1968). The method and role of vertical transmission of a nucleopolyhedrosis virus in the European Spruce sawfly, *Diprion hercyniae*. *J. Invertebr. Pathol. 12*, 132-139.

Nordin, G. L. (1976). Transovum transmission of a nuclear polyhedrosis virus of fall webworm, *Hyphantria cunea*. *J. Kansas Entomol. Soc. 49*, 589-594.

Orlovskaya, Y. E. (1968). Geographical distribution and manifestation of viral diseases in dendrophilous insect pests in the USSR. *Entomol. Rev. 47*, 455-463.

Ossowski, L. L. J. (1960). The biological control of the wattle bagworm, *Kotochalia junodi* (Heyl) by a virus. *Ann. Appl. Biol. 48*, 399-313.

Payne, C. C. (1974). The isolation and characterization of a virus from *Oryctes rhinoceros*. *J. Gen. Virol. 25*, 105-116.

Payne, C. C., and Kelly, D. C. (1981). Identification of insect and mite viruses. "Microbial Control of Pests and Plant Diseases, 1970-1980," ed. H. D. Burges, 949 pp. Academic Press, London.

Polson, A., and Gitay, H. (1972). A possible role of the cattle egret in the dissemination of the granulosis virus of the bollworm. *Ostrich 43*, 231-232.

Reed, E. M. (1971). Factors affecting the status of a virus as a control agent for the potato moth (*Phthorimaea operculella*) (Zell.) (Lepidoptera: Gelechiidae). *Bull. Entomol. Res. 61*, 207-222.

Savan, M., Budd, J., Reno, P. W., and Darley, S. (1979). A study of two species of fish innoculated with spruce budworm nuclear polyhedrosis virus. *J. Wildlife Dis. 15*, 331-334.

Shekhurina, T. A. (1980). Virus epizootics in populations of the gray grain moth (*Apamea anceps* Schiff.) and prognosis of their development. *In* "Characterization, Production, and Utilization of Entomopathogenic Viruses" (C. M. Ignoffo et al., eds.). Proc. 2nd Conference, Project V, Microbiological Control of Insect Pests of the US/USSR Joint Working Group, Clearwater Beach, Florida.

Smirnoff, W. A. (1959). Predators of *Neodiprion swainei* Midd (Hymenoptera: Tenthredinidae) larval vectors of virus diseases. *Can. Entomol. 91*, 246-249.

Smirnoff, W. A. (1960). Observations on the migration of larvae of *Neodiprion swainei*. *Can Entomol. 152*, 957-958.

Smirnoff, W. A. (1962). Transovum transmission of virus of *Neodiprion swainei* Middleton (Hymenoptera: Tenthredinidae). *J. Insect Pathol. 4*, 192-200.

Smirnoff, W. A. (1975). Parasites and predators as vectors of insect viruses. *In* "Baculoviruses for Insect Control: Safety Considerations" (M. Summers et al., eds.), pp. 131-133. *Amer. Soc. Microbiol.* Washington, 186 pp.

Smith, K. M. (1976). "Virus-insect relationships." Longman Group, London. 291 pp.

Stairs, G. R. (1965). The study of virus epizootics in insect populations. Insect Path. Microb. Contr. VII-Vol. 3, pp. 275-279. North Holland, Amsterdam.

Stairs, G. R. (1966). Transmission of virus in tent caterpillar populations. *Can. Entomol. 98*, 1100-1104.

Suzuki, N., and Kunimi, Y. (1981). Dispersal and survival rate of adult females of the fall webworm, *Hyphantria cunea* Drury (Lepidoptera: Arctiidae), using the nuclear polyhedrosis virus as a marker. *Appl. Entomol. Zool. 16*, 374-385.

Swan, D. I. (1972). UN/SPC Rhinoceros Beetle Project, Annu. Rept., pp. 166-169. S. Pacific Commission, Noumea.
Szalay-Marzso, L., and Vago, C. (1975). Transmission of baculoviruses by mites. Study of granulosis virus of codling moth (*Laspeyresia pomonella* L.) *Acta. Phytopathol. Acad. Sci. Hung. 10*, 113-122.
Tanada, Y. (1971). Persistence of entomopathogenous viruses in the insect ecosystem. "Entomological Essays to Commemorate the Retirement of Professor K. Yasumatsu," S. Asahina et al., eds., pp. 367-379. Hokuryukan Publ. Co. Ltd., Tokyo. 389 pp.
Tanada, Y. (1973). Environmental factors external to the host. *In* "Regulation of Insect Populations by Microorganisms" (L. A. Bulla, Jr., ed.), pp. 120-130. (*Ann. N.Y. Acad. Sci. 217*, 243 pp.) Publ. NY Acad. of Sciences.
Tanada, Y. (1976). Ecology of insect viruses. *In* "Perspectives in Forest Entomology" (J. F. Anderson and H. K. Kaya, eds.), 428 pp. Academic Press, New York.
Tatchell, G. M. (1981). The transmission of a granulosis virus following the contamination of *Pieris brassicae* adults. *J. Invertebr. Pathol. 37*, 210-213.
Thomas, E. D., Reichelderfer, C. F., and Heimpel, A. M. (1972). Accumulation and persistence of a nuclear polyhedrosis virus of the cabbage looper in the field. *J. Invertebr. Pathol. 20*, 157-164.
Thomas, E. D., Reichelderfer, C. F., and Heimpel, A. M. (1973). The effect of soil pH on the persistence of cabbage looper nuclear polyhedrosis virus in soil. *J. Invertebr. Pathol. 21*, 21-25.
Thompson, C. G., and Scott, D. W. (1979). Production and persistence of the nuclear polyhedrosis virus of the Douglas-fir tussock moth, *Orgyia pseudotsugata* (Lepidoptera: Lymantriidae), in the forest ecosystem. *J. Invertebr. Pathol. 33*, 57-65.
Thompson, C. G., and Steinhaus, E. A. (1950). Further tests using a polyhedrosis virus to control the alfalfa caterpillar. *Hilgardia 19*, 412-429.
Thompson, C. G., Scott, D. W., and Wickman, B. E. (1981). Long-term persistence of the nuclear polyhedrosis virus of the Douglas-fir tussock moth, *Orygyia pseudotsugata* (Lepidoptera: Lymantriidae), in forest soil. *Environ. Entomol. 10*, 254-255.
Tinsley, T. W. (1979). The potential of insect pathogenic viruses as pesticidal agents. *Annu. Rev. Entomol. 24*, 63-87.
Vail, P. V., and Hall, I. M. (1969). The influence of infections of nuclear polyhedrosis virus on adult cabbage loopers and their progeny. *J. Invertebr. Pathol. 13*, 358-370.
Yendol, W. C., and Hamlen, R. A. (1973). Ecology of entomogenous viruses and fungi. *In* "Regulation of Insect

Populations of Microorganisms" (L. A. Bulla, Jr., ed.), pp. 18-30. (*Ann. N.Y. Acad. Sci. 217*, 243 pp.)

Young, E. C. (1974). The epizootiology of two pathogens of the coconut palm rhinoceros beetle. *J. Invertebr. Pathol. 24*, 82-92.

Young, S. Y., and Yearian, W. C. (1979). Soil application of *Pseudoplusia* NPV: Persistence and incidence of infection in soybean looper caged on soybean. *Environ. Entomol. 8*, 860-864.

Zelazny, B. (1973a). Studies on *Rhabdionvirus oryctes*. II. Effect on adults of *Oryctes rhinoceros*. *J. Invertebr. Pathol. 22*, 122-126.

Zelazny, B. (1973b). Studies on *Rhabdionvirus oryctes*. III. Incidence in the *Oryctes rhinoceros* population of western Samoa. *J. Invertebr. Pathol. 22*, 359-363.

Zelazny, B. (1977). *Oryctes rhinoceros* populations and behavior influenced by a baculovirus. *J. Invertebr. Pathol. 29*, 210-215.

STABILITY OF INSECT VIRUSES IN THE ENVIRONMENT

ROBERT P. JAQUES

Research Station, Agriculture Canada
Harrow, Ontario, Canada

I. INTRODUCTION

The effectiveness of a naturally occurring or introduced pathogen in the regulation of a pest insect is dependent primarily on the ability of the pathogen to kill, incapacitate or otherwise harm the pest and on the probability of the pest contracting the pathogen under conditions favoring infection. The latter parameter is affected significantly, in turn, by the ability of the pathogen to be present and remain active on or in a substrate in a location from which infection of the target host insect may occur.

Stability of a pathogen generally refers to maintenance of activity and relates mainly to characteristics of the pathogen including tolerance or resistance to environmental factors and conditions. Persistence, on the other hand, designates the ability to remain present and, for the purpose of this discussion, present in the active state. Persistence, therefore, has a wider connotation, including effects of characteristics of the host, the habitat, and related biotic and abiotic factors, as well as the characteristics of the pathogen.

Stability and persistence in the habitat of the pest insect are particularly important for entomoviruses because viruses are obligate pathogens and they cannot multiply outside

ISBN 0-12-470295-3

TABLE I. Persistence of Representative Entomopathogens in the Field Habitat Following Introduction

Pathogen	Preparation[b]	Crop or substrate
VIRUSES[a]		
Choristoneura fumiferana NPV	PIB, suspension	White spruce foliage
Heliothis NPV	PIB, suspension	Cotton leaves
	PIB, suspension	Cotton leaves
	PIB, suspension	Cotton leaves
	PIB, suspension (Virion-H)[c]	Cotton leaves
	PIB, suspension (Virion-H)[c]	Soybean leaves
Lymantria dispar NPV	PIB, suspension	Bark
Malacosoma disstria NPV	PIB, suspension	Sweet gum foliage
Orgyia pseudotsugata NPV	Debris from epizootic	Soil
Trichoplusia ni NPV	PIB, suspension	Soil
	PIB, suspension	Soil
	PIB, suspension	Cabbage leaves
Cydia pomonella GV	GIB, suspension	Apple
Pieris brassicae GV	GIB, suspension	Soil or sand
	GIB, suspension	Cabbage leaves
Pieris rapae GV	GIB, suspension	Soil Cabbage leaves
Melolontha melolontha EPV	Suspension	Soil
Panonychus citri virus	Suspension	Citrus foliage
BACTERIA		
Bacillus thuringiensis	Suspension (Thuricide)	Soil
	Suspension (Thuricide)	Soybean leaves
	Suspension (Dipel)	Cotton leaves

Estimated percentage inactivation	Reference
100% in 1 day	Morris and Moore, 1975
>75% in 12 hours	Ignoffo and Batzer, 1971
90% in 24 hours	Ignoffo and Batzer, 1971
>95% in 3 days	Bullock, 1967
75% in 3 days	Ignoffo et al., 1972
50% in 2 days	Smith and Hostetter, 1982, Ignoffo et al., 1974a
<10% during winter	Podgwaite et al., 1979
100% in 10 hours	Broome et al., 1974
Activity after 40 years	Thompson et al., 1981
<10% in 4 years	Jaques, 1969
75% in 5 years	Jaques, 1967b
50% in 2 days	Jaques, 1967a, 1972b
50% in 4 days	Laing and Jaques, unpublished
<10% in 2 years	David and Gardiner, 1967a
>50% in 3 hours	David et al., 1968
20% in 7 months	Jaques, 1974b
75% in 2 days	Jaques, 1972b, 1974b
25% in 4 years	Hurpin and Robert, 1976
>75% in 2 to 6 hours	Reed et al., 1973
<10% in 3 months	Saleh et al., 1970
50% in 24 hours	Ignoffo et al., 1974a
50% in 1.5 to 2 days	Beegle et al., 1981

Table I (Continued)

Pathogen	Preparation[b]	Crop or substrate
MICROSPORIDIA		
Vairimorpha necatrix	*Suspension of spores*	*Soil*
	Suspension of spores	*Cabbage leaves*
	Suspension of spores	*Cotton leaves*
FUNGI		
Beauveria bassiana	*Suspension of spores*	*Soil*

[a]NPV, GV, and EPV refer to nuclear polyhedrosis virus, granulosis virus, and entomopox virus, respectively.

[b]PIB and GIB refer to polyhedral inclusion bodies and granular inclusion bodies, respectively.

[c]Virion-H, Thuricide, and Dipel are registered formulations of the designated entomopathogen.

living tissue. In addition, most entomoviruses are specific for a species of insects and/or for a group of related insects. By comparison, many other entomopathogens are facultative pathogens that can multiply saprophytically as well as in live tissue. Furthermore, many entomopathogens have a wide host range (e.g., the fungus *Beauveria bassisana* and the microsporidium *Vairimorpha necatrix*) and can survive in alternate hosts in the absence of the target host insect. Therefore, whereas environmental stability and persistence have substantial effect on the effectiveness of all types of pathogens, the capacity to remain active and persist in a suitable location to contact the target insect is especially important for entomoviruses and is a major concern in the development of entomoviruses as microbial insecticides.

Several biotic and abiotic environmental factors and conditions affect persistence and stability of entomoviruses. It is apparent, however, that the relative importance of the effect of the factors on the presence and activity of viruses may be different for naturally occurring viruses than for applied or introduced viruses and for different host-virus systems.

Estimated percentage inactivation	*Reference*
>50% in 3 months	*Chu and Jaques, 1981*
>25% in 2 days	*Chu and Jaques, 1981*
50% in 0.6 days	*Fuxa and Brooks, 1978*
50% in 14 days	*Lingg and Donaldson, 1981*

II. PERSISTENCE AND STABILITY

A. *PERSISTENCE OF ENTOMOVIRUSES AND OTHER ENTOMOPATHOGENS*

The period that entomoviruses and other entomopathogens remain active in the habitat depends largely on the substrate in or on which the virus or pathogen is located, the environmental factors acting on the pathogen, and on the form of virus in the habitat. For example, entomoviruses that are occluded in inclusion bodies, notably the baculoviruses [the nuclear polyhedrosis viruses (NPV) and the granulosis viruses (GV)] and the cytoplasmic polyhedrosis viruses (CPV), appear to retain activity for periods similar to spore-forming bacteria [e.g., *Bacillus thuringiensis* (*B.t.*)] or microsporidia (e.g., *V. necatrix*) (Table I). The baculoviruses, *B.t.*, and microsporida retain activity for relatively short periods on foliage exposed to sunlight with a half-life of activity of 2 days or less (Beegle et al., 1981; Bullock, 1967; Fuxa and Brooks, 1978; Jaques, 1975), a shorter period than for many fungi (Roberts and Campbell, 1977) (Table I). On the other hand, the viruses, *B.t.*, and spores of *Bacillus popilliae* have retained activity for long periods in soil (Jaques, 1977b; Ladd and McCabe, 1967; Saleh et al., 1970; Thompson et al., 1981).

B. *PERSISTENCE OF TYPES OF VIRUSES*

The proteinaceous inclusion body in which virions of some entomoviruses, notably the polyhedrosis and granulosis viruses, are embedded appears to protect the virions somewhat from unfavorable environmental conditions. Therefore, occluded viruses remain active in the field habitat for several days following application whereas nonoccluded viruses remain active for shorter periods (Table I). Similarly, virions of occluded types of viruses that are not occluded in an inclusion body for various reasons may be quite unstable when exposed to environmental conditions. Virions released from inclusion bodies by dissolution of the inclusion body protein by exposure to alkali (Section IV, D, 1) have been found to be generally less stable than those occluded in the inclusion body. Shapiro and Ignoffo (1969) found that virions of the NPV of *Heliothis* spp. (HNPV) released from polyhedral inclusion bodies (PIBs) by treatment with alkali were 50% inactivated after 30, 60, and 120 days in an aqueous suspension at 50°, 37°, and 5°C compared to the much greater stability of virions in intact polyhedral bodies. Similarly, Watanabe (1951) reported that virions of the NPV of *Bombyx mori* (BmNPV) were practically inactive after storage for 65 days at 4°C; Steinhaus (1960) reported that virions in intact polyhedral bodies of the virus lost less than 50% activity in 20 years.

Free virions not occluded in inclusion bodies in hemolymph of infected insects are also generally less stable than occluded virions. For example, Stairs and Milligan (1979) calculated that free virions of *Galleria mellonella* NPV (GmNPV) in hemolymph of diseased larvae of the wax moth, *Galleria mellonella*, were sevenfold more susceptible to heat irradiation than were virions of the NPV of the Douglas fir tussock moth occluded in inclusion bodies in tests by Martignoni and Iwai (1977). On the other hand, virions of GmNPV released from polyhedral bodies by treatment with alkali were 3.5-fold more resistant to heat than were nonoccluded virions in hemolymph (Stairs and Milligan, 1980).

The nonoccluded virus of *Panonychus citri* appeared to be less stable than occluded viruses. Reed (1974) noted that the virus in intact mite cadavers was inactivated in 6 hours at 46°C, a temperature that would not inactivate occluded viruses in this time (Section IV, B). Similarly, the *P. citri* virus appeared to be less stable on foliage being inactivated in 2-6 hr after application (Reed et al., 1973) compared to occluded viruses that have a 1- to 3-day halflife in the field (Jaques, 1975).

The possibility of increased sensitivity to environmental factors resulting in reduced effectiveness is of concern when entomoviruses are propagated *in vitro* in cell culture because of the probability that such preparations would contain a higher proportion of nonoccluded virions than do preparations of entomoviruses propagated *in vivo*. Although field persistence of viruses propagated by the two methods was not compared, field plot tests to evaluate effectiveness of applications of the NPV of *Trichoplusia ni* (TnNPV) (Jaques, 1977a) and of the NPV of *Autographa californica* (AcNPV) (Ignoffo et al., 1974b) for control of the cabbage looper on cabbage indicated that effectiveness of *in vitro* and *in vivo* preparations were similar. Likewise, preparations of GmNPV produced *in vivo* or *in vitro* were equally infective against the greater wax moth, *G. mellonella*, in laboratory and field tests by Dougherty and co-workers (1982), further indicating similar environmental persistence.

C. *SIGNIFICANCE OF FEEDING HABITS OF HOST INSECTS*

The life cycle of the host target insect, particularly its feeding habits, significantly influences the impact of environmental stability on effectiveness of an entomovirus, especially of an introduced virus. The usual route of entry of entomoviruses into the host insect is through the digestive tract with the virus being ingested with food. In some cases, however, infection may be initiated by transfer of virus in the egg, by oviposition by parasites, by wounding by another infected host, or by other means. Because the majority of infection is by ingestion, persistence or stability of the virus on the food of the host insect or on a substrate from which food could be contaminated is considered to have substantial influence on effectiveness of the applied virus or the extent to which an epizootic develops by natural dissemination. Likewise, because most infection is by ingestion of the virus, infection is largely limited to the larval stage of the host. In addition, partly because of this route of infection, entomoviral diseases are more common among foliage-eating insects, especially those of the order Lepidoptera, than among insects with other feeding habits (Martignoni and Iwai, 1981).

Inactivation of deposits of the virus may be more detrimental to effectiveness of the virus against pest species that feed on an exposed substrate for a comparatively short period of time than against species that feed on exposed foliage for longer periods. In regard to naturally induced epizootics, it is apparent that insects like the cabbage looper, *T. ni*, or the European spruce sawfly, *Diprion* (=*Gilpinia*) *hercyniae*, that feed on exposed foliage for all or nearly all of their larval feeding period would have a greater opportunity to feed, at some time during their life cycle, on foliage that is contaminated by NPV from a cadaver than would, for example, a larva of the codling moth, *Cydia* (=*Laspeyresia*) *pomonella* that feeds on exposed apple leaves or fruit for only a very short time before burrowing into the apple. Similarly, feeding habits influence the impact of stability on effectiveness of applied viruses. AcNPV or TnNPV are applied several times in programs for control of *T. ni* (Jaques, 1972a, 1973a,b, 1977a). Larvae that escape or otherwise survive an application of the virus, perhaps because they did not emerge from the egg until the residue of the spray was inactivated, could be killed by a subsequent application even if the applications were spaced 7 to 10 days apart. On the other hand, a larva of the codling moth that escaped an application of *C. pomonella* granulosis virus (CpGV) to apple trees because the deposit of virus from the spray was inactivated would have burrowed into the apple when a subsequent spray was applied and would not contact the virus. Thus, because of the relatively short period of activity of CpGV on apple leaves or fruit (Table I), the virus must be applied frequently (Huber and Dickler, 1977; Jaques et al., 1977, 1981) to maintain a deposit of lethal concentration of the virus. Likewise, some insects like leafrollers that feed on foliage protected from deposits of applied viruses may be difficult to control with entomoviruses.

D. *PERSISTENCE, ESTABLISHMENT, AND CARRY-OVER OF EFFECTIVENESS*

The long-term effect of entomoviruses resulting from persistence in the environment significantly influences their effectiveness as naturally occurring pathogens and as introduced or applied pathogens of insects. Persistence in the habitat and/or environment of the host insect by establishment and self-perpetuation in the population of the host insect is essential for development of natural epizootics. In addition, persistence in the population increases effectiveness of application or introduction of entomoviruses enhancing their competitive status relative to other types of pathogens and to chemical insecticides.

Viruses have become established as natural mortality agents in populations of many pest species of insects, in some cases contributing substantially to suppression of the pests. For example, epizootics and enzootics of NPV in populations of the European spruce sawfly, *D. hercyniae*, in New Brunswick, Canada in the 1930s resulting from an introduction of the virus and subsequent establishment of the virus, have contributed substantially to suppression of the pest (Bird and Burk, 1961;

Bird and Elgee, 1957). Likewise, *Lymantria dispar* NPV (LdNPV), became established in populations of the host in the early 1900's in the United States (Lewis, 1981) but although severe outbreaks of the disease occurred, the suppression of the pest was not sufficient to maintain populations at economically acceptable densities. The persistence of a virus in the habitat would be expected to have greater impact on pests of forests than on pests of annual agricultural crops because of the greater stability of the forest habitat. Nevertheless, it is noteworthy that TnNPV and the GV of *Pieris rapae* (PrGV) persisted and accumulated in soil in sufficient concentrations to give effective late-season protection of the crop in plots repeatedly cropped to cole crops but not treated to control the host larvae (Jaques, 1971a). In addition, these viruses accumulated in soil in fields of cole crops in which standard production practices were followed (Jaques and Harcourt, 1971).

The persistence of applied entomoviruses in the habitat and/or host population following application to give a carry-over in effectiveness is a significant consideration in assessing the potential of entomoviruses as components of integrated management systems to control pest insects. TnNPV is spread within the population within a growing season by susceptible individuals feeding on foliage contaminated by virus released from cadavers of larvae killed by an application of the virus, by feeding on foliage contaminated in some other way with the virus or, more rarely, by feeding on cadavers of virus-killed insects. This gives some carry-over effect of application (Jaques, 1974b, 1975). Carry-over effect of application of the NPV of the spruce budworm, *Choristoneura fumiferana*, (CfNPV) was noted by several workers (Cunningham, 1978; Cunningham et al., 1975; Morris, 1977, 1980). The 3-year carry-over effect of application of CfNPV-fenitrothion is particularly noteworthy. Likewise, the establishment and dissemination of the NPV of the gypsy moth, *L. dispar* (Lewis, 1981; Magnoler, 1974; Yendol et al., 1977), greatly enhances the usefulness of the virus in control of the pest. The effect of application of the NPV of the European pine sawfly, *Neodiprion sertifer* (Bird, 1961; Mohamed et al., 1982) was extended into subsequent generations of the host, being disseminated quite widely from the points of introduction. The rapid dissemination of *Oryctes rhinoceros* NPV in populations of the host and the establishment of the virus in the habitat following release results in a long-term effect, significantly enhancing the usefulness of the virus (Bedford, 1981; Young and Longworth, 1981). On the other hand, spores and endotoxin-containing crystals of the bacterium *B. thuringiensis* applied, for example, to cabbage for control of larvae of *T. ni* and *P. rapae* on cole crops are used in a manner similar to a chemical insecticide. There is no carry-over effect because,

TABLE II. Loss of Activity of Some Representative Entomoviruses in Storage

Virus	*Preparation stored*[a]	*Storage temperature (°C)*
Bombyx mori NPV	*PIB, powder in ampules*	4
	Alkali-freed virions, suspension	4
	Virions in hemolymph, lyophilized	4
Diprion hercyniae NPV	*Cadavers*	4.5
		21
Heliothis NPV	*PIB, dried powder*	2
	Alkali-freed virions, suspension	5
Lambdina f. fiscellaria NPV	*PIB, suspension*	4
Trichoplusia ni NPV	*PIB, suspension*	4
Pieris brassicae GV	*GIB, dried powder*	0
	GIB, lyophilized	-20
		3
Pieris rapae GV	*GIB, suspension*	4
Panonychus citri	*Nonoccluded virus, dried*	4
	In cadavers	22-27

[a]*PIB and GIB refer to polyhedral inclusion bodies and granular inclusion bodies, respectively.*

although the spores of *B.t.* introduced into the field habitat can multiply saprophytically or in alternate hosts, multiplication is minimal with an insufficient concentration of spores and/or crystals to reinfect host insects.

The relationship of carry-over effect to persistence is discussed further in Section VI.

Period of storage	Estimated percentage loss of activity	Reference
20 years	<50	Steinhaus, 1960
65 days	>75	Watanabe, 1951
12 months	<10	Vaughn, 1972
5 years	<5	Neilson and Elgee, 1960
5 years	<5	
2 years	40	Dulmage and Bergerjon, 1977
120 days	50	Shapiro and Ignoffo, 1969
6 years	>90	Cunningham, 1970
4 years	<10	Jaques, 1977b
4 years	<10	David and Gardiner, 1967b
7 days	0	David et al., 1971c
7 days	<10	
4 years	<10	Jaques, 1977b
6.5 years	<10	Shaw et al., 1972
14 days	<10	Tashiro et al., 1970

E. STORABILITY OF ENTOMOVIRUSES

The maintenance of activity of entomoviruses during long-term storage indicates the degree of stability of the virus to be expected when the entomoviruses are exposed to environmental factors in the field or forest. In addition, storability indicates shelf-life of the formulated product, a major concern in commercial utilization of microbial agents. Storage data show that the baculoviruses and other entomoviruses retain activity for long periods when stored in inclusion bodies as dried powders or suspensions in the dark, frozen or refrigerated or, in some cases, at room temperature (Table II). For example, BmNPV in glass ampules retained much of its original activity for 20 years (Steinhaus, 1960). A dried powder preparation of *Pieris brassicae* GV (PbGV) had lost little activity after being stored for 4 years at 4°C (David and Gardiner, 1967b). Likewise, the viruses are quite stable if frozen, retaining activity for long periods. David and Gardiner (1967b) found that deep-freezing had little effect on activity of PbGV and Tanada (1953) reported that PrGV was hardly affected by storing at -32°C for 36 days.

Dulmage and Burgerjon (1977) compared activity of dry preparations of HNPV and *Mamestra brassicae* NPV (MbNPV) that had been purified by an acetone-lactose coprecipitation procedure and stored in the dark at 22°C, in a refrigerator at 4°C, or in a freezer at approximately -18°C. They found that the frozen sample of HNPV was stable over the 2-year experiment whereas the samples stored at room temperature or in the refrigerator lost 40 to 20% of activity, respectively. The LC_{50}'s for MbNPV frozen or stored at 5°C were similar, being approximately 10 and 40 PIB/mm^2 of diet after 14 and 22 months, respectively; LC_{50} values for MbNPV stored at room temperature (20°C) were 25 and 200 after 14 and 22 months, respectively, indicating a greater loss of activity at the higher temperature.

Virus particles, or virions, not occluded in inclusion body protein are generally less stable than are virions embedded in intact inclusion bodies. Most nonoccluded virions retain activity for several hours or days compared to the retention of activity of viruses in intact inclusion bodies for several months or years. For example, Shapiro and Ignoffo (1969) showed that 50% of activity of virions of HNPV released from polyhedral inclusion bodies (PIB) was lost in 30 days at 50°C, 60 days at 37°C, or 120 days at 5°C. Likewise, Watanabe (1951) noted that free virions of BmNPV were practically inactivated in 65 days at 4°C. On the other hand, Aizawa (1963) reported that exposure of free virions to 50°C for 10 minutes caused nearly complete inactivation, although virions stored for 30 days at 5° or -25°C lost little activity. Similarly, Vaughn (1972) found that virions of BmNPV lost little activity when stored frozen under liquid nitrogen. It is interesting that the nonoccluded virus of the citrus red mite, *P. citri*, is quite stable; the virus retained original activity for 5 weeks at 25°C (Gilmore and Munger, 1963) and lost less than 10% activity in 14 days at 22° to 27°C (Tashiro et al., 1970) or in 6.5 years at 4°C (Shaw et al., 1972).

It is evident that the stability of occluded viruses is related to the protection afforded the virions by the inclusion body protein. It is interesting in this regard that inclusion body protein, although readily dissolved by weak alkalis (Section IV, D), was not decomposed by proteolytic microorganisms found in the soil (Jaques and Huston, 1969), perhaps an important factor in long-term persistence of occluded viruses in soil (Jaques, 1975). It is also interesting that holes and cracks appeared in polyhedral bodies of *Lambdina fiscellaria fiscellaria* NPV (Cunningham, 1970) and of TnNPV (Jaques, unpublished) stored for long periods in aqueous suspension, suggesting a gradual decomposition or erosion of the polyhedral protein.

Because of the long-term retention of activity of entomoviruses that are deep-frozen, this type of storage is used routinely. Deep-freezing is particularly useful for storage of unstable viruses like *O. rhinoceros* NPV that remain active for only short periods in other storage conditions (Bedford, 1981). In regard to storage by deep-freezing, it is noteworthy that repeated freezing and thawing of suspensions of PbGV 10 times in 12 days did not reduce appreciably the activity of virus in tests by David and co-workers (1971a) indicating that frozen supplies in storage could be thawed from time to time, if necessary, for utilization of a portion of the stored material without reducing activity.

III. STABILITY OF ENTOMOVIRUSES IN THE HABITAT

The persistence and stability of the virus on or in various substrates in the field habitat differ substantially not only because of differences in effects of substrates on the viruses but also because of exposure of the substrates to different environmental factors. Because the majority of entomoviruses of current interest in pest management contact the target pest insect by ingestion of the virus as a contaminant of food and because most of these entomoviruses feed on leaves, persistence on foliage is of major significance. Likewise, sources of virus that could contaminate foliage areof major concern.

A. STABILITY ON FOLIAGE

Deposits of viruses on foliage are exposed to several environmental factors that may contribute to inactivation. Exposure to sunlight, especially the ultraviolet portion of sunlight, appears, however, to have the major impact on activity with temperature, humidity, surface moisture, and chemicals having less impact (Jaques, 1977b; Tanada, 1971; Yendol and Hamlen, 1973). Deposits of viruses on foliage of plants retain

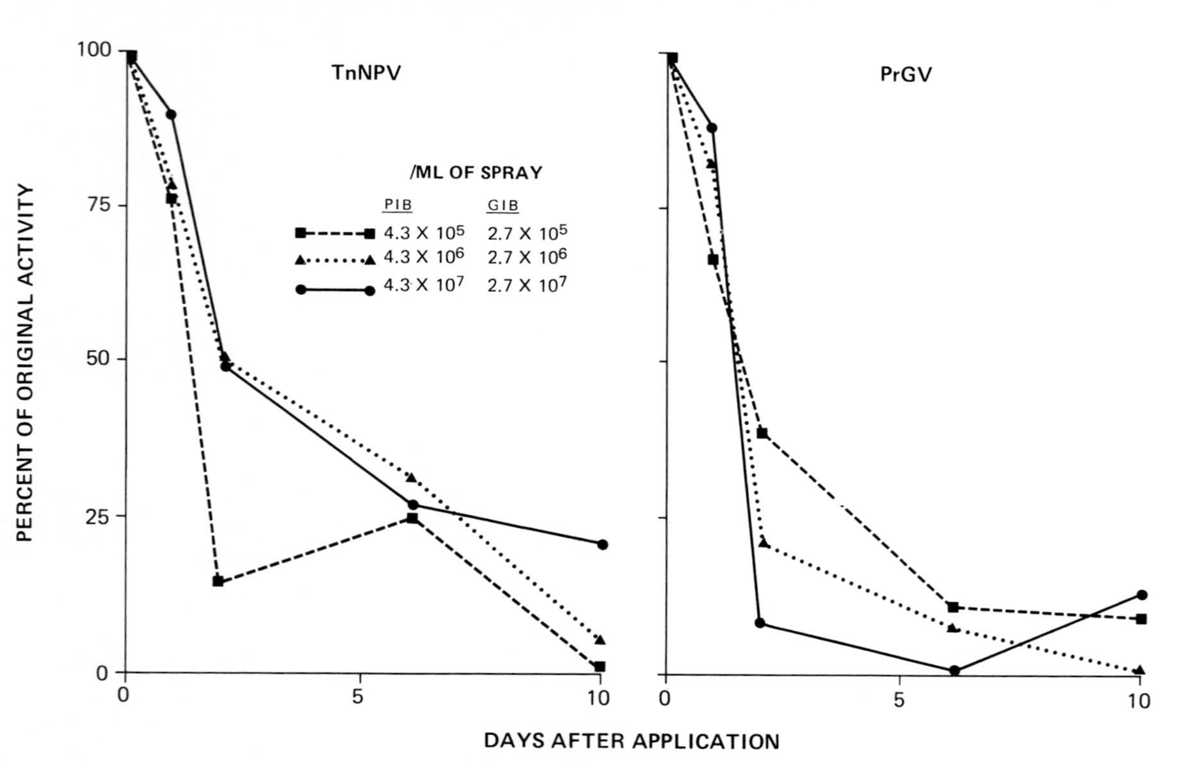

FIGURE 1. The activity of the nuclear polyhedrosis virus of Trichoplusia ni (TnNPV) and the granulosis virus of Pieris rapae (PrGV) after application of suspensions of virus inclusion bodies to leaves of cabbage plants in field plots. (After Jaques, 1975.)

activity for relatively short periods, often being 50% inactivated in 2 or 3 days and usually being nearly completely inactivated within 2 weeks (Table I).

The decline in activity of deposits of TnNPV and PrGV on foliage of cabbage plants in studies by Jaques (1975) (Fig. 1) is typical of other viruses. In this and other studies (Jaques, 1967a, 1971b, 1972b), Jaques reported that deposits of TnNPV or PrGV retained 50% or less of original activity 2 days after application to foliage of cabbage plants in field plots and retained less than 35% of original activity on the 5th day. The decline of PrGV appeared to be slightly more rapid than that of TnNPV. Likewise, David et al. (1968) found that deposits of purified suspensions of PbGV were inactivated rapidly being 75% inactivated within 3 hours; total inactivation occurred in 12-19 hours. Bullock (1967) observed that HNPV retained little activity 2 days after application to cotton. A similar rate of inactivation of HNPV was observed on cotton (Ignoffo and Batzer, 1971; Roome and Daoust, 1976; Young and Yearian, 1974), on corn silks (Ignoffo et al., 1973), and on soybean foliage (Ignoffo et al., 1974a). Ignoffo et al. (1972) found that persistence of HNPV on cotton differed slightly among 6 test locations; mean data for the locations indicated that the virus was approximately 75 and 95% inactivated in 3 and 6 days, respectively. Deposits of *Epiphyas postvittana* NPV (MacCollom and Reed, 1971) and CpGV (Laing and Jaques, unpublished) on leaves and fruit of apple were approximately 50% inactivated in 2 days after application. *Malacosoma disstria* NPV on leaves of gum trees was less persistent, being completely inactivated in 10 hours after application (Broome et al., 1974). Similarly, the nonoccluded virus of the citrus red mite, *P. citri*, survived 6 to 24 hr following application to foliage of citrus (Reed et al., 1973).

Although viruses deposited on foliage are usually inactivated quite readily by exposure to sunlight, in some instances significant residues of active viruses may persist on leaves. For example, substantial concentrations of TnNPV have been found on wrapper leaves and heads of cabbage harvested from plots that were either treated or not treated with the virus (Heimpel et al., 1973; Jaques, 1974a,b; Thomas et al., 1974). This concentration may be attributed to repeated contamination of foliage by virus produced in epizootics of diseases, to contamination from virus persisting in soil, or to virus that was deposited in locations in the cabbage plant not exposed to sunlight (Jaques, 1975).

There is good evidence that the persistence of viruses on leaf surfaces in the field habitat is influenced, to some extent at least, by the leaf. Activity of residues of TnNPV or PrGV on leaves of cabbage plants retained dry in the dark at room temperature was reduced somewhat (Jaques, 1972b)

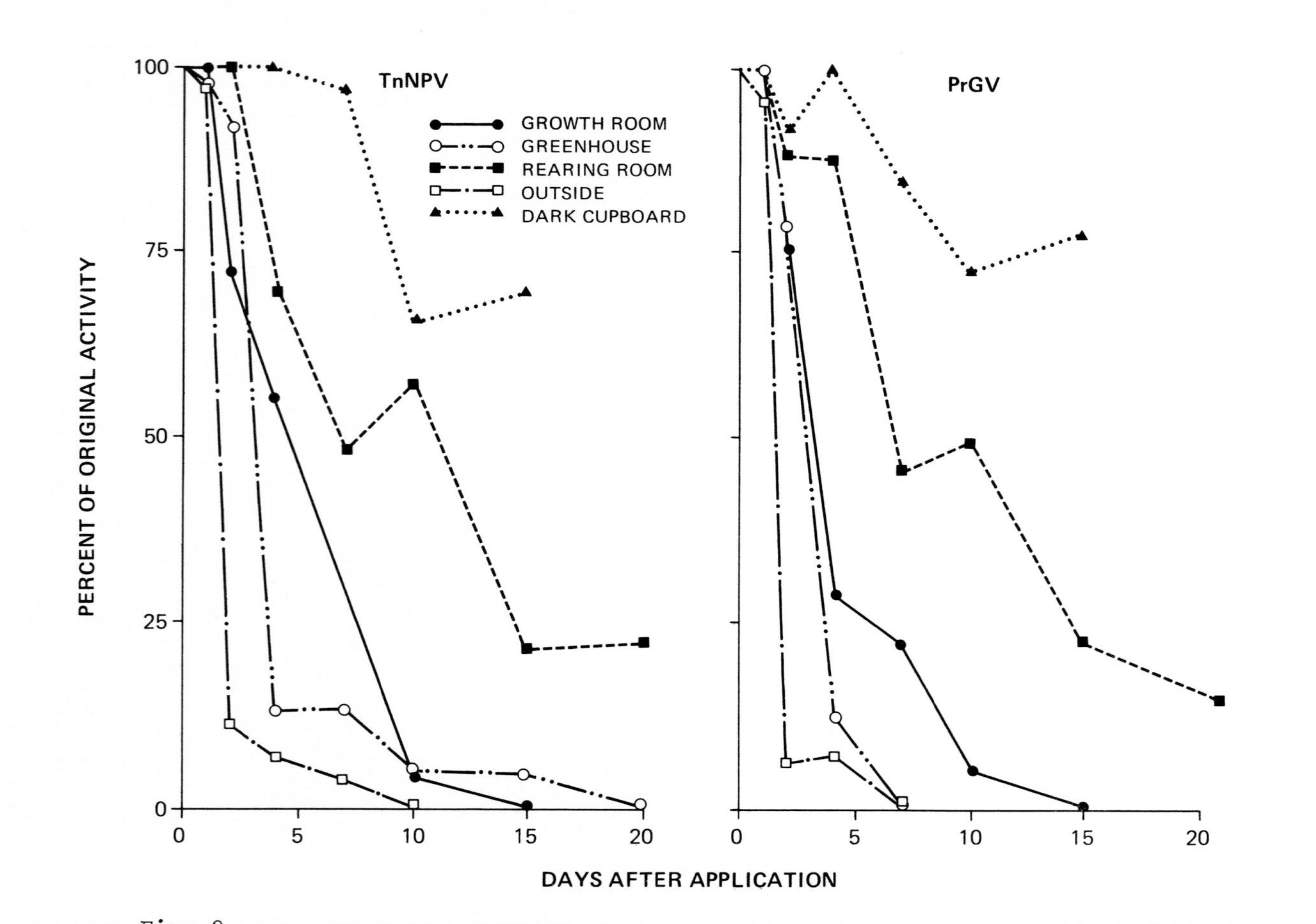
TnNPV
PrGV
GROWTH ROOM
GREENHOUSE
REARING ROOM
OUTSIDE
DARK CUPBOARD
PERCENT OF ORIGINAL ACTIVITY
100
75
50
25
0
0
5
10
15
20
DAYS AFTER APPLICATION

Fig. 2

(Fig. 2) suggesting that substances of plant origin contribute to inactivation. It has been found, for example, that components, notably terpene compounds, of leaves of some coniferous and deciduous trees inhibited activity of viruses or *B.t.* (Morris, 1972; Smirnoff, 1968) perhaps explaining the differences observed in persistence of activity of NPV and *B.t.* applied to different trees (Jaques and Morris, 1981).

The effect of the plant on stability of deposits of entomoviruses was indicated by the period of activity of deposits of HNPV on different crop plants. HNPV applied to cotton at a rate of 200 LE/ha (= 1.2×10^{12} PIB/ha) in Botswana was about 50% inactivated in 2 days after application whereas deposits on sorghum retained activity much longer, being 50% inactivated in about 10 days (Romme and Daoust, 1976). Likewise, deposits of HNPV on cotton leaves were not as stable as were deposits on soybean leaves (Young et al., 1977) suggesting differences in exudates from leaves. Andrews and Sikorowski (1973) noted that deposits of HNPV on cotton leaves were inactivated during the night when dew caused the deposits to become wet. They found that the dew on the cotton leaves was alkaline (pH 8.2-9.2) due to basic ions and that polyhedral bodies of HNPV immersed in the dew on leaves swelled as they do in weakly alkaline solutions (Section IV, D, 1). Laboratory studies (Young et al., 1977) showed that cotton dew was normally more alkaline than dew found on soybean leaves. Furthermore, it was found (McLeod et al., 1977) that the alkalinity of dew on cotton plants differed; the dew on leaves of some plants was less alkaline (pH 7.4-8.8) than others (pH 9.3). Falcon (1971) also reported that substances on cotton leaves may adversely affect persistence of HNPV. Various chemicals (bicarbonates, carbonates, sodium, potassium, and calcium) found in dew or on leaves could have caused the observed effect on activity of the virus. While composition of dew is undoubtedly substantially influenced by deposits from the air, these observations suggest that substances produced by leaves may influence activity of viruses.

FIGURE 2. The activity of deposits of the nuclear polyhedrosis virus of Trichoplusia ni (TnNPV) and the granulosis virus of Pieris rapae (PrGV) on leaves of collard plants retained in five locations. Plants were treated on May 31 with a suspension containing 3.6 × 10^5 polyhedral inclusion bodies of TnNPV and 5 × 10^5 granular inclusion bodies of PrGV/ml. (After Jaques, 1972b.)

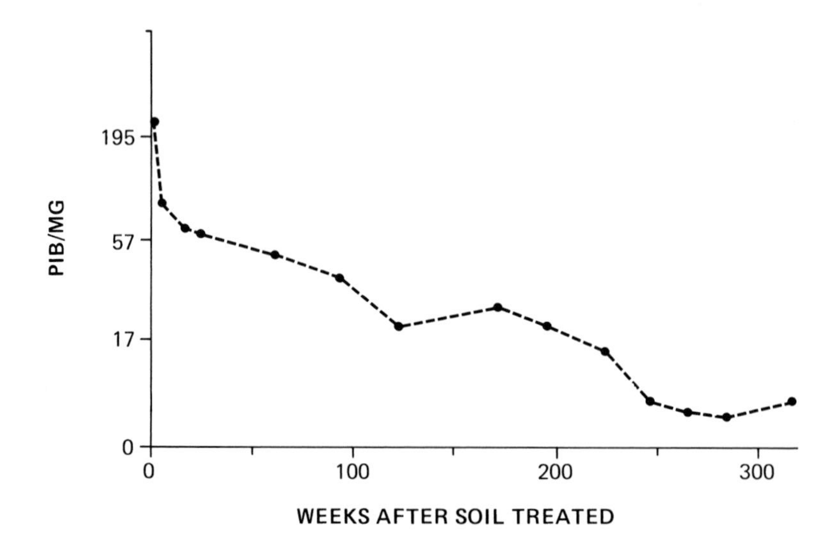

FIGURE 3. The activity of the nuclear polyhedrosis virus of Trichoplusia ni in soil in plots after surface application of 3.3 × 10^{10} polyhedral inclusion bodies/m^2. (After Jaques, 1975.)

B. STABILITY IN SOIL AND DEBRIS

The prolonged persistence of entomoviruses in soil and debris was suggested over 50 years ago by early workers (Komarek and Breindl, 1924; Prell, 1926). More recently, Steinhaus (1948) and Thompson and Steinhaus (1950) presented evidence that leaf debris and soil were infectious for *Colias eurytheme* larvae. Studies by Jaques (1964, 1967b) showed that TnNPV applied to the surface of field plot soil retained about 15% of original activity 318 weeks after application (Fig. 3). This study was especially credible because the host insect, *T. ni*, did not occur in the experimental area during the sampling period and, therefore, the NPV detected was the virus originally applied to the soil. He also demonstrated (Jaques, 1967b) that leaves of plants grown in the treated soil were contaminated with infectious virus probably due to virus-laden soil being splashed by rain onto the leaves or blown onto leaves by wind indicating that virus persisting in soil could serve as a reservoir to initiate epizootics. Subsequent studies (Jaques, 1974b) showed that PrGV may persist in field soil for similarly long periods following application. David (1965) reported that *P. brassicae* GV persisted in soil for at least 14 months. Later, David and

Gardiner (1967a) reported that this virus lost little activity when stored in soil or sand for 2 years. Also NPV of *Hyphantria cunea* lost little activity in 8 months in soil at room temperature (Hukahara and Namura, 1972). Indeed, Hukahara and Namura (1971) observed inclusion bodies of the virus absorbed onto soil particles. Young and Yearian (1979) applied *Pseudoplusia includens* NPV to soil in field plots at the time that soybeans were planted and found that the virus persisted until the following season after an initial rapid decline in concentration in soil immediately following application. Similar persistence of HNPV in soil in fields of sorghum was noted by Roome and Daoust (1976). Likewise, Mohamed et al. (1982) recovered the *N. sertifer* NPV from soil 21 months after application to the forest. Early studies on persistence of *Orgyia pseudotsugata* NPV in the forest ecosystem (Thompson and Scott, 1979) showed that the virus persisted for at least 11 years in the soil and debris on the forest floor. Later work by Thompson and co-workers (1981) suggested that the virus had persisted in the soil and debris for at least 40 years.

Hurpin and Robert (1976) found that the EPV of *M. melolontha* (MmEPV), like baculoviruses, remained active for long periods in soil. In their study, MmEPV in soil in pots retained in the laboratory or outside in the field lost little of its original activity in 4 years. Subsequently, they (Hurpin and Robert, 1977) reported that MmEPV applied to soil in field plots persisted in the soil longer than did *Bacillus popilliae* var. *melolontha* and *Rickettsiella melolontha*, which lost much of their activity in 2 years. Persistence of a pathogen for at least 2 years is required to initiate a new epizootic because of the cycles of activity of this host insect.

Viruses produced in epizootics of some diseases reach the soil either directly by cadavers falling to and decomposing on the soil surface or by foliage or other substrates contaminated by virus from decomposed cadavers reaching the soil. Because the viruses remain active in the soil, they accumulate and apparently serve as a reservoir for initiation of epizootics in subsequent generations. Jaques (1970a,b, 1974a,b, 1975, 1977b) showed that concentrations of TNNPV and PRGV in soil in field plots varied over a 5-year period, increasing in the autumn when concentrations on foliage were high, decreasing slightly during winter, decreasing substantially with tillage of the plots in spring, and again increasing as epizootics of the diseases developed in populations of the host insect (Fig. 4). It is noteworthy that the concentrations in soil in treated and nontreated plots were similar after the initial year of the studies. Likewise, concentrations of deposits of TnNPV and PrGV on cabbage foliage and on the cabbage heads were similar at harvest in plots treated or not treated with the viruses. Indeed, concentrations were often lower on plants in plots treated with the virus presumably because few larvae survived to produce virus. Probably for similar reasons, the concentration of the viruses was lower in plots treated with full dosages of chemical insecticides. Similarly, Thomas et al. (1972, 1974) found that concentrations of residues of TnNPV on leaves, on heads of cabbage, and in soil were similar in virus-treated and nontreated plots. Podgwaite and co-workers (1979) found that LdNPV, like TnNPV, accumulated in soil and debris. It is significant that concentrations of LdNPV were similar in virus-treated and nontreated plots. Young and Yearian (1979) noted seasonal fluctuations in concentrations of *P. includens* NPV in soil and on foliage following application of the virus to soybean. In a subsequent study, McLeod et al. (1982) found that the concentration of the virus in soil was insufficient to initiate an epizootic in populations in the second year after application. On the other hand, Zether (1980) noted that the GV of *Agrotis segetum* applied to soil persisted from one year to the next causing substantial mortality of the host insect in the following year affording considerable protection to root crops.

Observations on plots of cabbage not treated with a virus or other pesticide indicated that natural accumulation of viruses resulting from epizootics of NPV and GV disease in populations of *T. ni* and *P. rapae* were sufficient to control the pests (Jaques, 1970b, 1971a). Similarly, Crawford and Kalmakoff (1977) found that *Wiseana* spp., a soil-dwelling group of pests infesting pastures in New Zealand, could be regulated to economically acceptable populations by epizootics of diseases caused by an NPV and, to a less extent, an EPV and GV, that persisted from year in the soil. Tanada and Omi (1974) found several viruses of *C. eurytheme*, *Spodoptera exigua*, and *A. californica* in soil in California fields and concluded that the viruses had accumulated from epizootics of the disease. Likewise, comparisons of the residues of the viruses of *T. ni* and *P. rapae* occurring naturally in soil and on leaves of crucifer crops in fields showed that the residues in soil were related qualitatively and quantitatively to residues of the viruses on leaves of the plants and to mortality by viruses in the host populations (Jaques, 1970b; Jaques and Harcourt, 1971).

Hukahara (1973) related concentrations of *H. cunea* NPV in soil to numbers of inclusion bodies produced per cadaver of the host insect, *H. cunea*, to numbers of larvae killed. In addition, he noted that the pattern of distribution of leaves falling from trees infested with NPV-infected host insects fitted the distribution of the virus in soil. This demonstrated that the virus from cadavers deposited on leaves

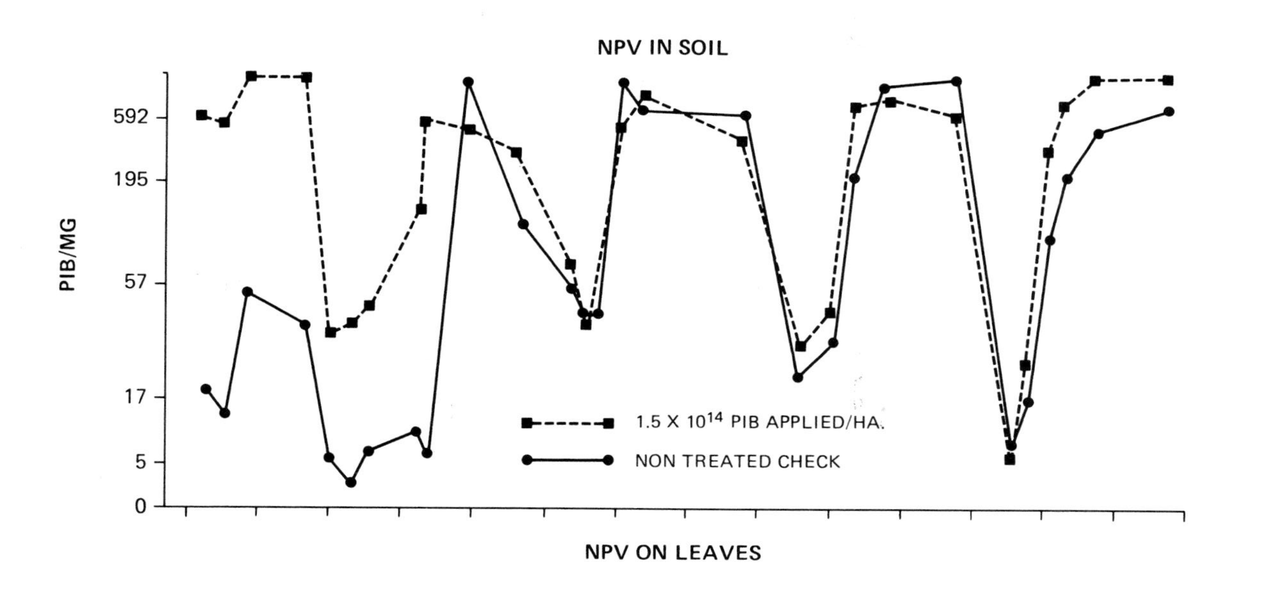
NPV IN SOIL
PIB/MG
592
195
57
17
5
0
1.5 X 10^14 PIB APPLIED/HA.
NON TREATED CHECK
NPV ON LEAVES

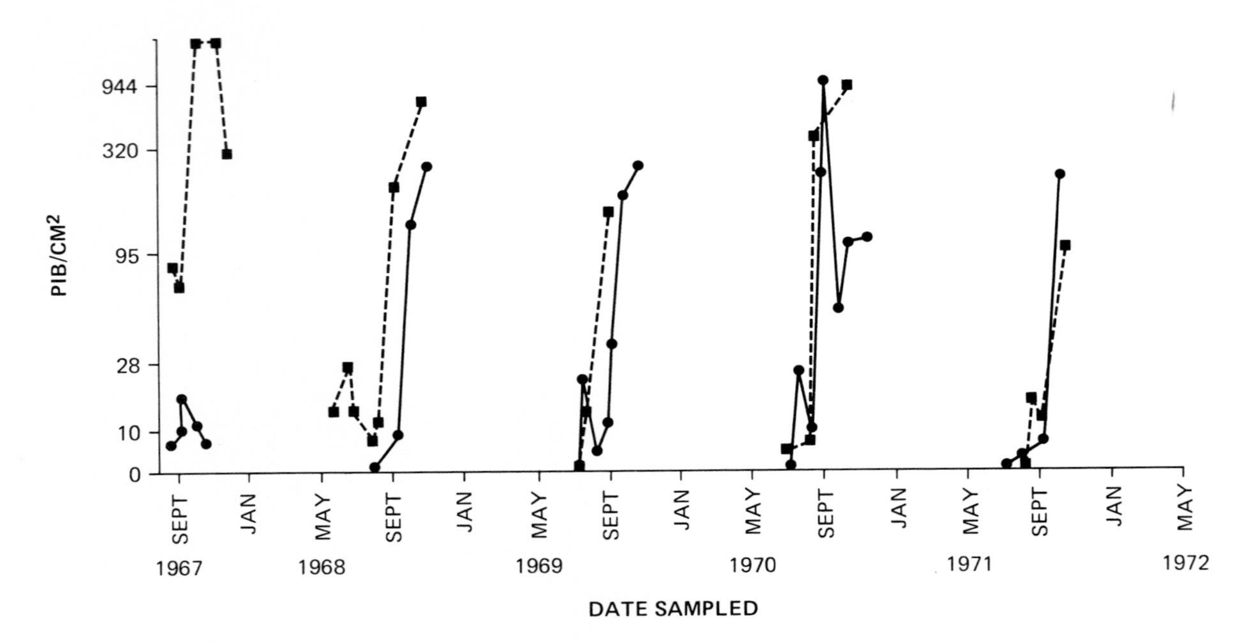

FIGURE 4. Concentrations of the nuclear polyhedrosis virus of Trichoplusia ni on leaves of cabbage plants and in soil in nontreated plots of cabbage and in plots in which soil was treated on July 19 and August 16, 1967 with 7.5 × 10^{13} polyhedral inclusion bodies/ha. (After Jaques, 1974b.)

reached the soil as the leaves fell. The virus was found in soil up to 15 m from the point of introduction (Hukahara and Namura, 1972) indicating the extent of dissemination by this means. Thompson and Scott (1979) demonstrated that viruses from killed *O. pseudotsugata* larvae, especially late-instar larvae, contaminated the duff layer in the forest and that the virus was reintroduced into the forest canopy by contamination of foliage by virus-laden dust. It is probable that epizootics of NPV and GV in populations of the fall armyworm, (*Spodoptera frugiperda*) (Fuxa, 1982), were initiated by virus persisting in soil. Kaupp (1980) reported on a technique for measuring the quantity of *N. sertifer* NPV reaching the soil from epizootics of the disease in populations of *N. sertifer* in pine forests.

The greater persistence of entomoviruses in soil compared to persistence on foliage is implicated in intergeneration dissemination of naturally occurring pathogens. Furthermore, the feature may be exploited in application or introduction of viruses for control of insect pests in integrated pest management systems. This is discussed later in this chapter (Section VI).

The stability of entomoviruses, especially those occluded in inclusion bodies, in soil is affected by various environmental factors. As mentioned previously, stability is influenced by the decomposition of the inclusion body. It is noteworthy, therefore, that inclusion body protein is not readily decomposed by proteolytic microorganisms commonly found in soil (Jaques and Huston, 1969). Probably because of dissolution of the inclusion body protein by strong acids and alkalis (Section IV, D, 1) the pH reaction of the soil influenced stability. Thomas et al. (1973) found that TnNPV was inactivated more quickly in soils of low pH than in soils that were near neutral. Laboratory studies by Jaques (unpublished) suggested that activity of the virus was not affected by soil with a pH 5.5-9.0, the range suitable for most crops, but was less stable in highly acid or alkaline soils.

Persistence of inclusion bodies near the surface of soil is enhanced by the resistance to leaching through soil by rain. The strong absorption of inclusion bodies onto soil particles was clearly demonstrated by laboratory and field studies on persistence of *B. mori* CPV and *H. cunea* NPV and GV by Hukahara and co-workers (Hukahara, 1972; Hukahara and Namura, 1971; Hukahara and Wada, 1972). In a field plot study, Jaques (1969) treated the surface of soil with TnNPV and found that most viral activity remained in the top 0-2.5 cm of the undisturbed soil during the 223-week period of observation. Little viral activity was detected at the 10-12.5 cm depth. In laboratory tests, TnNPV (Jaques, 1969), *P. brassicae* GV (David and Gardiner, 1967a), or *B. mori* CPV (Hukahara and Wada, 1972) applied to top of columns of soil did not move downward appreciably in the columns when substantial quantities of water were passed through. Hukahara and Wada (1972), found, for example, that *B. mori* CPV remained in the upper 4 cm of treated columns of soil through which 11.2 m of water had been passed. Likewise, in laboratory studies by Jaques (1969), TnNPV remained near the top of columns of soil through which 95.5 or 67.8 m of water was passed in 30 or 20 days, respectively.

C. STABILITY IN WATER

There is very little information on persistence of entomoviruses in water under natural conditions. It is known that viruses that are occluded in inclusion bodies remain active when stored as aqueous suspensions (Table 1) but little is known of persistence of viruses in water in moving or stagnant bodies of water in the natural environment. This lack of information is partly the result of limited interest in use of viruses to control insects in aquatic habitats. (Chapman, 1978). A discussion of stability of entomoviruses in fresh water and marine habitats (Tanada, 1975) concerned only factors thay may influence stability of viruses.

IV. ABIOTIC FACTORS AFFECTING ENVIRONMENTAL STABILITY OF ENTOMOVIRUSES

A. INACTIVATION BY SUNLIGHT AND OTHER IRRADIATION

1. Sunlight

Numerous studies have demonstrated the rapid decline in activity of deposits of entomoviruses on foliage of various crops and other surfaces exposed to sunlight suggesting the major role of solar irradiation in inactivation of foliar deposits of entomoviruses in the field habitat (Section III, A) (Table I) (Jaques, 1975, 1977b; Tanada, 1971; Yendol and Hamlen, 1973). In an attempt to demonstrate this role, Jaques (1967a, 1972b) compared the activity of deposits of known dosages of TnNPV and PrGV on leaves of potted collard plants retained in different locations. In one such test (Fig. 2) (Jaques, 1972b), deposits of TnNPV and PrGV on leaves of collard plants lost about 90% of original activity in 2 days on plants kept outside exposed to sunlight. Deposits on plants retained in a greenhouse were inactivated less quickly than those exposed to sunlight but more quickly than were those exposed only to artificial light in a lighted growth chamber. By contrast, the deposits on plants retained in a dark cupboard or in a dark rearing room or refrigerator lost 25% or less activity in 8 or 15 days.

The inactivation of virus on foliage exposed to direct sunlight may be more rapid than demonstrated by Jaques in his test. For example, *M. disstria* NPV was 100% inactivated by a 10-hour exposure on sweet gum foliage (Broome et al., 1974) and HNPV on glass plates retained only 31% of original activity after a 4-hour exposure (Ignoffo and Batzer, 1971). Likewise, TnNPV deposits on filter disks were more than 75% inactivated by a 1-hour exposure to sunlight (Cantwell, 1967) and were found to be completely inactivated in 3 days on cotton leaves exposed to sunlight (Bullock, 1967). Morris (1971) exposed suspensions of the NPV of *Lamdina fiscellaria somniaria* to sunlight and noted 50% inactivation in 10 hours. Smirnoff (1972) demonstrated that deposits of the NPV of *Neodiprion swainei* were inactivated much more rapidly during periods that the deposits were exposed to direct sunlight than during periods in the shade, which verifies the role of sunlight in inactivating viruses.

2. *Ultraviolet Irradiation and Sunlight*

Radiant energy in the ultraviolet wavelengths is considered to be the most germicidally effective portion. This is light with a wavelength approximating 253.7 nm. In regard to viricidal effect it is quite significant that effective wavelengths resemble the absorption of nucleic acids with a peak near 260 nm and substantial decrease at 300 nm. The absorption of ultraviolet light by atmospheric ozone reduces or prevents light of a wavelength less than 290 nm from reaching the earth's surface. This suggests that inactivation of entomoviruses by natural sunlight may be largely attributable to light of wavelengths exceeding 290 nm (i.e., in the near uv range).

Laboratory tests have described inactivation of viruses exposed to specific wavelengths of uv and to sunlight. Ignoffo and Batzer (1971) found that uv (253.7 nm) inactivated HNPV at approximately the same rate as did a similar intensity of sunlight. David (1969) compared the inactivation of *P. brassicae* GV by monochromatic light of various wavelengths and found that the inactivating effect of uv decreased progressively as the wavelength was increased. Although uv of 253.7 nm was most effective, uv of the shorter wavelength present in sunlight reaching the earth (291.5-320 nm) nevertheless caused inactivation of the virus. Similarly, Bullock et al. (1970) found that exposure of suspensions of HNPV to ultraviolet light (253.7 nm) reduced activity more than did exposure to a similar quantity of uv of longer wavelength (307.5 nm) and exposure to a similar dosage of uv of a somewhat longer wavelength (364 nm) or a broad-band mixture of visible and infrared light did not affect activity. Far uv (253.7 nm) had more effect on CpGV and *M. brassicae* NPV (MbNPV), than did near uv (285-380 nm) in tests by Krieg and co-workers (Krieg et al., 1981). In an attempt to simulate sunlight, Morris (1971) exposed *L. f. sominaria* NPV to a lamp emitting long-wave uv (366 nm). Long periods of exposure caused only limited reduction in activity.

The viricidal effect of the short-wave uv was demonstrated by the rapid inactivation of TnNPV exposed as foliar deposits or suspensions to 253.7 nm-UV by Jaques (1967a, 1968, 1971b) and the inactivation of dry deposits of HNPV (McLeod et al., 1977). Virions of NPV from hemolymph of *G. mellonella* were inactivated by exposure to near uv (300-380 nm) as well as by far uv (253.7 nm) in tests by Witt and Stairs (1975). They considered that the differences in dosage required for inactivation and the pattern of response indicated that the mode of action of the types of radiation were distinctive. A comparison of inactivation of naked and occluded virions was not made.

The inactivation responses of suspensions of HNPV and TnNPV exposed to a uv source differed in studies by Gudauskas and Canerday (1968). Whereas the relationship of inactivation to time of exposure to uv was similar for the two viruses the relationship of inactivation to distance from the uv source was nearly linear for HNPV but sigmoid for TnNPV suggesting differences between the viruses.

It is noteworthy that far uv was more active against all pathogens than was near uv. The studies are significant in that they did indicate that near uv, the wavelength of uv predominating in sunlight reaching the earth's surface, inactivated entomoviruses but not as readily as uv of a shorter wavelength. Comparisons of susceptibility of entomopathogens to inactivation by ultraviolet radiation vary in ranking among investigators probably because of differing test conditions and pathogens. Ignoffo and co-workers (1977) reported a relative ranking in stability when exposed to uv as *B. thuringiensis* > *Nomuraea rileyi* (conidia) > EPV > NPV = CPV > *V. necatrix* > GV. On the other hand, Krieg and co-workers (1980) found that *B. thuringiensis* was more sensitive than *M. brassicae* NPV to near uv (285-380 nm). Later they (Krieg et al., 1981) reported a ranking in stability to near uv as *B. bassiana* conidia ≥ *M. brassicae* NPV > *B. thuringiensis* spores > *C. pomonella* granules > *B. thuringiensis* vegetative cells. Their ranking of stability when exposed to far uv (253.7 nm) was *M. brassicae* NPV > *B. bassiana* conidia > *C. pomonella* GV

> *B. thuringiensis* spores > *B. thuringiensis* vegetative cells.

Work by Smirnoff (1972) showed that intensity of the uv portion of sunlight, as well as its wavelength, affected the rate of inactivation of *N. swainei* NPV. He suggested that entomoviruses should be applied in the evening rather than in the morning to increase the time of availability to the host before the deposits of virus would be exposed to sunlight of sufficient intensity to cause inactivation. In this regard, Shaw and co-workers (1968) noted that virus of *P. citri* applied to California orange groves after sundown was more effective than virus applied in daylight.

3. *Protection of Entomoviruses Against Solar Radiation*

Purity of suspensions influences effectiveness of some entomoviruses applied to foliage presumably partly due to an influence on stability of the viral deposit. For example, early work by Ignoffo (1964) indicated that impure suspensions of TnNPV were more effective than were pure ones. Subsequent findings that foliar deposits of impure suspensions were not inactivated as quickly as were those of purified suspensions when exposed to sunlight suggested that greater activity of impure suspensions was due at least in part to protection of the virus by substances in the impure suspensions. David et al. (1968, 1971c) noted that purified suspensions of *P. brassicae* GV deposited on leaves exposed to sunlight were inactivated more rapidly than were the deposits from impure suspensions used in previous experiments (David and Gardiner, 1966). Tests on inactivation of purified and impure TnNPV exposed to a UV source as suspensions and deposits on leaves showed that pure suspensions and deposits were inactivated more than were those of impure virus (Jaques, 1971b, 1972b). In addition, deposits of impure suspensions of TnNPV on leaves of collard plants were not inactivated as quickly as deposits of purified suspensions when exposed to sunlight.

The consideration that the dark color of the impure suspensions or the proteinaceous debris in the suspensions, or both, contributed to the extension of activity by affording the virus protection against ultraviolet radiation prompted substantial interest in the development of protectant additives to prolong activity of residues (Table III). The importance of the proteinaceous component was suggested by the finding that incorporation of hemolymph of the host insect in suspensions prolonged activity of capsules of *P. brassicae* GV exposed on glass slides to uv (David et al., 1971b). Likewise, Jaques (1971b, 1972b) found that incorporation of yeast extract, hydrolysates of some proteins, and other proteinaceous

TABLE III. *Some Representative Uses of Protectant Additives in Applications of Entomoviruses*

Virus	Crop	Application	Protectant additive	Reference
Autographa californica NPV	Lettuce	Spray	Shade[a]	Vail et al., 1980
	Cabbage	Spray	Carbon	Hostetter et al., 1979
	Cabbage	Spray	Skim milk	Sears et al., 1983
				Jaques, unpublished
Choristoneura fumiferana NPV	White spruce	Spray	Shade	Morris and Moore, 1975
Cydia pomonella GV	Apple	Spray	Skim milk	Huber and Dickler, 1977
			Skim milk	Jaques et al., 1977, 1981
			2-hydroxy-4-methoxy-benzophenone	Kreig et al., 1980
Heliothis NPV	Cotton	Spray	Carbon	Ignoffo et al., 1972
			Shade-polyvinyl alcohol	Smith et al., 1978
		Bait	Cotton seed oil	Bell and Kanavel, 1978
			Cotton seed meal	Smith et al., 1980
Lymantria dispar NPV	Deciduous forest	Spray (aerial)	Shade	Lewis, 1981; Wollam et al., 1978
Orgyia pseudotsugata NPV	Fir	Spray	Molasses	Ilnytzky et al., 1977
Neodiprion sertifer NPV	Pine	Spray	Shade	Mohamed et al., 1982
Pieris rapae GV	Cabbage	Spray	Carbon	Hostetter et al., 1973
			Skim milk	Sears et al., 1983
				Jaques, unpublished
Trichoplusia ni NPV	Cabbage	Spray	San 285[b]	Jaques, 1977a

[a]Shade is a protectant developed by International Minerals and Chemical Company and now produced by Sandoz Inc.

[b]San 285 is an experimental protectant produced by Sandoz Inc.

materials, including casein, extended activity of TnNPV or PrGV exposed to uv and sunlight. In addition, dark coloring or dark particles (India ink) prolonged activity, especially if combined with a proteinaceous material. For example, deposits of TnNPV applied alone on collard leaves retained in a growth room were inactivated in 5 days but deposits of the virus applied in brewer's yeast-charcoal mixtures retained over 80% of original activity 15 days after application (Fig. 5). In field tests the addition of skim milk-charcoal or egg albumen-charcoal mixtures to sprays of TnNPV or PrGV resulted in a 2.5- to 3-fold increase in the period over which 50% of original activity was retained on cabbage plants.

Ignoffo and Batzer (1971) found that shielding HNPV by the addition of carbon to spray mixtures protected the virus against inactivation by sunlight as well as did prior encapsulation of the virus with carbon. Bull et al. (1976) reported that encapsulation with carbon or titanium was an effective protectant mechanism but they did not include nonencapsulated virus and carbon in the test for comparison. In their studies HNPV protected by encapsulation or shielding with carbon retained 84 to 92% of original activity after a 4-hr exposure to direct sunlight compared to 30% original activity retained by unprotected HNPV.

Several materials have been incorporated into experimental formulations of entomoviruses in laboratory and field tests to assess their ability to protect viruses against inactivation by solar irradiation (Table III). At least one protectant material (Shade) has been formulated commercially and marketed for use with microbial insecticides. Couch and Ignoffo (1981) reviewed a wide range of additives and adjuvants with potential for use with entomoviruses and other pathogens and noted that although use of the additives usually improved persistence there was little evidence that effectiveness in protecting the crop in the field was enhanced substantially. For example, carbon and an experimental protectant adjuvant were included in sprays of AcNPV (Hostetter et al., 1979) and of PrGV (Hostetter et al., 1973) on cabbage with seemingly little influence on effectiveness. Likewise, inclusion of a protectant in a suspension of AcNPV sprayed on lettuce in California apparently had little benefit on effectiveness (Vail et al., 1980). Skim milk was used by Jaques and co-workers as a protectant in plot tests to assess the efficacy of TnNPV, AcNPV, and PrGV on cabbage (Jaques, 1977a; Jaques and Laing, 1978) and of CpGV on apple (Jaques et al., 1977, 1981) but the protectant effect of the adjuvant was not determined. Ignoffo and co-workers (1972) compared persistence of protected and nonprotected deposits of HNPV on cotton leaves at six locations in the cotton-growing belt of the United States. Three days after application the average

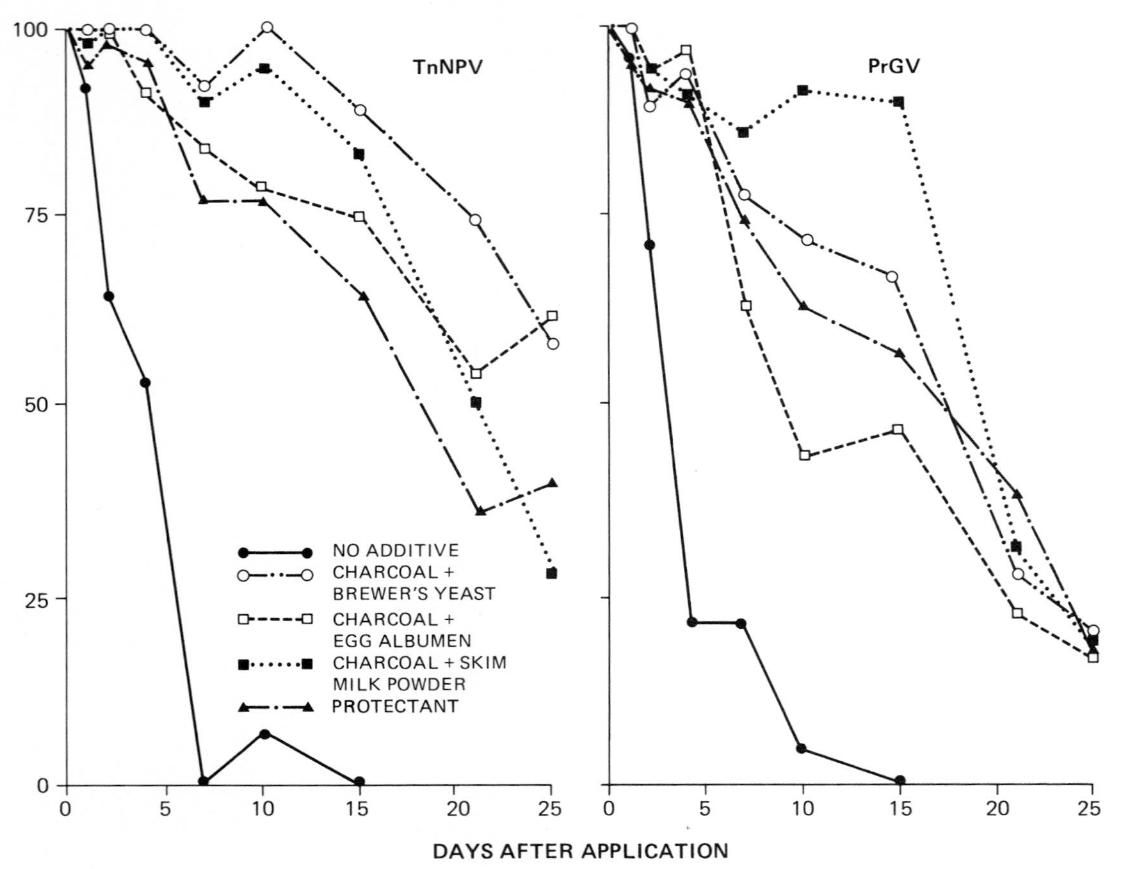
PERCENT OF ORIGINAL ACTIVITY
100
75
50
25
0
TnNPV
PrGV
NO ADDITIVE
CHARCOAL + BREWER'S YEAST
CHARCOAL + EGG ALBUMEN
CHARCOAL + SKIM MILK POWDER
PROTECTANT
0
5
10
15
20
25
DAYS AFTER APPLICATION

Fig. 5

activities of deposits at the six locations that were not protected, or protected with carbon or an experimental protectant were reduced by 77, 44, and 73%, respectively, indicating that the virus was protected by the carbon additive but not by the commercial experimental protectant. On the other hand, persistence of HNPV on glass surfaces was greatly enhanced by inclusion of this commercial protectant in a spray formulation (Ignoffo et al., 1976a); the adjuvant not only afforded protection from sunlight but also stimulated feeding by test larvae. Laboratory tests with HNPV on soybean foliage (Smith et al., 1978) showed that effectiveness of the protectant Shade as an extender of activity was enhanced by formulation with polyvinyl alcohol and an algin material (Keltose). It is noteworthy that Ignoffo et al. (1976b) found that a 10-45% increase in persistence of new formulations of HNPV was accompanied by a 15-115% increase in cotton yields.

Krieg and co-workers (Krieg et al., 1980) assessed the protective efficiency of several uv-absorbing compounds. Those that protected *M. brassicae* NPV best in the laboratory were India ink, 2-hydroxy-4-methylbenzophenone and 3-(4-methylbenzyliden)camphor. In addition, the benzophenone derivative enhanced the persistence of CpGV twofold; it was the best of the materials tested in an orchard trial by these authors. MacCollom and Reed (1971), on the other hand, found that addition of charcoal or other additives did not enhance appreciably persistence of activity of *E. postvittana* NPV on foliage of apple. Reed and co-workers (1973) tested several materials to protect *P. citri* virus and found that materials high in carbohydrate extended activity of the virus from 6 to 12 hr to 144 hr. Morris and Moore (1975) found that *C. fumiferana* NPV was 100% inactivated, if not protected but CfNPV protected by Shade was only 50 and 70% inactivated in 1 and 3 days, respectively, after application to white spruce.

Viruses formulated as bait are not only more attractive to target insects but also appear to be less susceptible to inactivation by sunlight (Fig. 6), especially if the bait formulations contain a protectant. For example, Stacey and co-workers (Stacey et al., 1977) noted that addition of Shade to baits of HNPV improved their effectiveness. A cottonseed meal bait formulation of HNPV was the best treatment for bollworms in labora-

FIGURE 5. The activity of deposits of the nuclear polyhedrosis virus of Trichoplusia ni (TnNPV) and the granulosis virus of Pieris rapae (PrGV) on leaves of collard plants sprayed with the viruses and protectant additives and retained in a growth room. Suspensions contained 3.6×10^5 polyhedra of T. ni NPV and 5×10^5 granules of P. rapae GV/ml and additives at 1% concentration (w/v). (After Jaques, 1972b.)

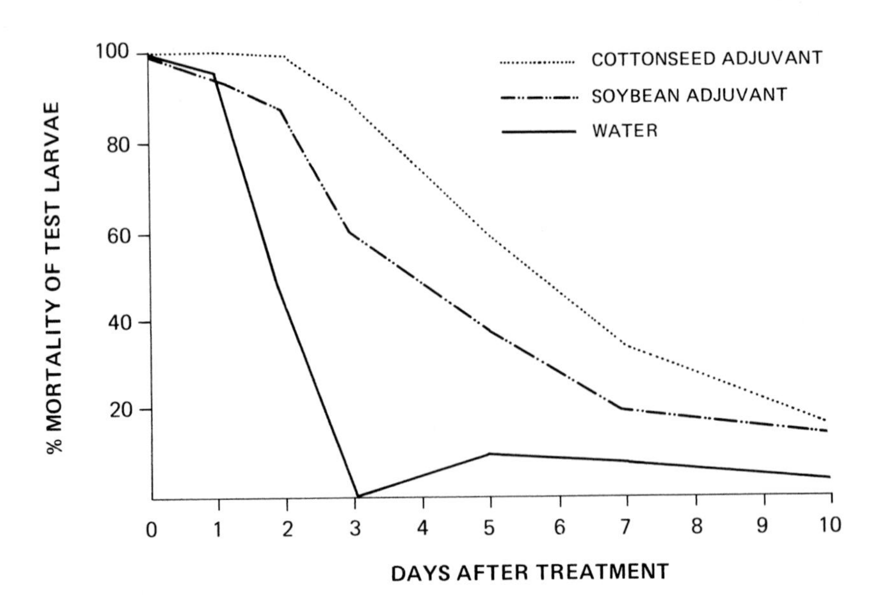

FIGURE 6. Mortality of Heliothis zea larvae fed soybean leaves on indicated days after treatment with suspensions of the nuclear polyhedrosis virus of Heliothis spp. containing a cottonseed or soybean adjuvant or no adjuvant. (Modified from Smith and Hostetter, 1982.)

tory tests on soybeans, causing higher mortality of *H. zea* than did sprays of HNPV formulated with carbon or other protectants (Smith et al., 1980). Later studies (Smith et al., 1982) indicated that citrus pulp, cottonseed, and soybean adjuvant formulations of HNPV were equally as effective against *H. zea* as were formulations that included a uv protectant. Smith and Hostetter (1982) showed that addition of aqueous cottonseed or soybean adjuvants to sprays of HNPV applied with coarse nozzles increased persistence of the virus on soybean leaves by 3.3- and 2.2-fold, respectively, greatly enhancing effectiveness of the virus (Fig. 6). The extension of activity was particularly notable on day 3 following application. Similarly, the enhanced effectiveness of other bait formulations of HNPV applied to cotton and other crops (Andrews et al., 1975; Bell and Kanavel, 1978; McLaughlin et al., 1971) may have been attributable, at least in part, to protection of the virus from inactivation by sunlight. For example, it is conceivable that the increased effectiveness of formulations of HNPV containing cottonseed oil, cottonseed flour, and sucrose (Bell and Kanavel, 1978) was partially attributable to protection of the virus from sunlight. Likewise, AcNPV and *Rachiplusia ou* NPV (RoNPV) formulated as baits in wheat bran were more effective than foliar sprays against the black cutworm, *Agrotis ipsilon*, in field and greenhouse tests, presumably because of increased attractiveness and persistence of the bait formulation compared to aqueous suspension formulations (Johnson and Lewis, 1982).

4. *uv-Resistant Strains of Entomoviruses*

Selection of uv-resistant or tolerant strains of an entomovirus may be a feasible procedure for prolonging activity of viral deposits on foliage. Selection of CpGV for resistance to inactivation by sunlight yielded a strain that was 5.6 times more resistant to inactivation by uv than was the wild strain (Brassel and Benz, 1979). In another study a selected uv-tolerant strain of this virus persisted about 2 times as long as the unselected strain of the virus and remained active as long as did the unselected strain protected by a protectant spray adjuvant (Krieg et al., 1980). Likewise, Witt and Hink (1979) selected a strain of AcNPV with greater resistance to inactivation by near uv radiation but with no increased resistance to far uv. The uv-resistant virus apparently was less effective against the host, however.

5. *Other Irradiation*

Forms of irradiation other than ultraviolet light and a part of the visible spectrum appear to have little effect on entomoviruses. David et al. (1971a) showed that short exposures of less than 1 hr to infrared light did not appreciably reduce the activity of dried deposits of PbGV. Prolonged exposure (2, 7, or 14 days) of the virus to infrared light did inactivate it, however. Substantial dosages of γ-radiation were required to inactivate TnNPV (Jaques, 1968) and *L. f. somniaria* NPV (Morris, 1971). Likewise, doses of γ-irradiation of up to 800,000 rads had little effect on the activity of *Malacosoma americanum* NPV in tests by Smirnoff (1967).

B. *TEMPERATURE*

Entomoviruses, especially those in inclusion bodies, appear to be quite stable at temperatures prevailing in habitats in which agricultural crops and forests can flourish. Because viruses are, for the most part, crystalline structures being largely devoid of life processes and enzymes, they are affected less by low temperature than are other types of entomopathogens. On the other hand, being constructed largely of nucleoprotein they are somewhat susceptible to denaturation by exposure to

TABLE IV. *Loss of Activity of Representative Entomoviruses Exposed to Various Temperature Conditions*

Virus	Preparation exposed	Period of exposure	Temperature (°C)	Estimated percentage loss of activity	Reference
Heliothis NPV	PIB, suspension	10 min	60	<5	Gudauskas and Canerday, 1968
		10 min	70	<10	
		10 min	80	>95	
	PIB, suspension	2 hours	38	<5	Stuermer and Bullock, 1968
		2 hours	60	<5	
		1 hour	71	75	
		15 min	71	50	
		15 min	82	65	
	PIB, dried powder	2 years	-18	>5	Dulmage and Burgerjon, 1977
		2 years	2	40	
		2 years	22	75	
	Virions, alkali-freed	100 days	5	10	Shapiro and Ignoffo, 1969
		100 days	37	90	
		50 days	50	>90	
Lamdina f. somniaria NPV	PIB, suspension	200 hours	45	<5	Morris, 1971
Spodoptera littoralis NPV	PIB, suspension	10 min	90	>95	Harpaz and Raccah, 1978
Trichoplusia ni NPV	PIB, suspension	10 min	70	<10	Gudauskas and Canerday, 1968
		10 min	75	<10	
		10 min	80	75	
		10 min	90	>95	

Oryctes rhinoceros NPV	PIB, sawdust	7 days	26	>99	Zelazny, 1972
		1 month		100	
		10 min	60	80	Bedford, 1981
		10 min	70	>95	
Pieris brassicae GV	GIB, suspension	10 days	40	40	David and Gardiner, 1967b
		20 days	40	75	
		5 days	50	<90	
		10 days	50	100	
Colias eurytheme CPV	PIB, suspension	10 min	60	10	Tanada and Chang, 1968
			70	<10	
			85	100	

high temperatures. The prolonged persistence of viruses in soil and other substrates of the habitat protected from ultraviolet light and deleterious chemicals (Jaques, 1975, 1977b; Tanada, 1973, 1975; Yendol and Hamlen, 1973) attest to the relatively minor effect of temperatures encountered in many habitats of importance to agricultural and forestry production.

Most entomoviruses are quite stable at low temperatures (Tables II and IV) (Section II, E). Data show that viruses, especially those in intact inclusion bodies, stored in cadavers, as dry powders, or in suspensions in the dark at 0° to 4°C retain much of their original activity for several years. On the other hand, naked virions, especially those released from inclusion bodies, were unstable even at low temperatures (Aizawa, 1963). Similarly, deep-freezing has been shown to have little effect on activity of viruses. On the other hand, nonoccluded virions are less stable even at low temperatures.

It is noteworthy that repeated freezing and thawing of suspensions of *P. brassicae* GV 10 times in 12 days did not reduce appreciably the activity of the virus (David et al., 1971a) indicating that repeated freezing and thawing of a virus in surface soil or debris during winter months in the field or forest would not inactivate the virus. Evans et al. (1980) found, however, that *M. brassicae* NPV stored in frozen samples of soil (-20°C) lost activity during the initial 5 days of storage but retained activity for the subsequent 25 days of the test. Studies on persistence of viruses in soil in field plots have not indicated a substantial loss of activity during winter (Section III, B).

Entomoviruses, like other viruses, are inactivated by relatively short exposures to high temperatures (Fig. 7). Representative experimental data summarized in Table IV and elsewhere (Bergold, 1958; Jaques, 1977b; Martignoni and Iwai, 1977) demonstrated that most entomoviruses were completely or nearly completely inactivated by exposure as suspensions or dried powders to temperatures of 70° to 80°C for 10 minutes. The thermal inactivation data for HNPV and TnNPV reported by Gudauskas and Canerday (1968), reproduced in Fig. 7, are considered to be typical of several viruses. These viruses and others, that are inactivated by short exposure to 70° to 80°C, have been found to be quite tolerant to exposures to slightly lower temperatures such as 50° or 60°C (Table IV). David and Gardiner (1967b) found that PbGV was inactivated by a 10-minute exposure to 70° but not by a 10-minute exposure to 65°C or lower. Although exposure to 60°C for 1 day inactivated PbGV in these tests, some activity remained after exposure to 50°C for 5 days and a substantial proportion survived for 10 days at 40°C. Likewise, suspensions of *L. f. somniaria* NPV were not inactivated appreciably by exposure to 45°C for 200 hours (Morris, 1971). Martignoni and Iwai (1977) found that two

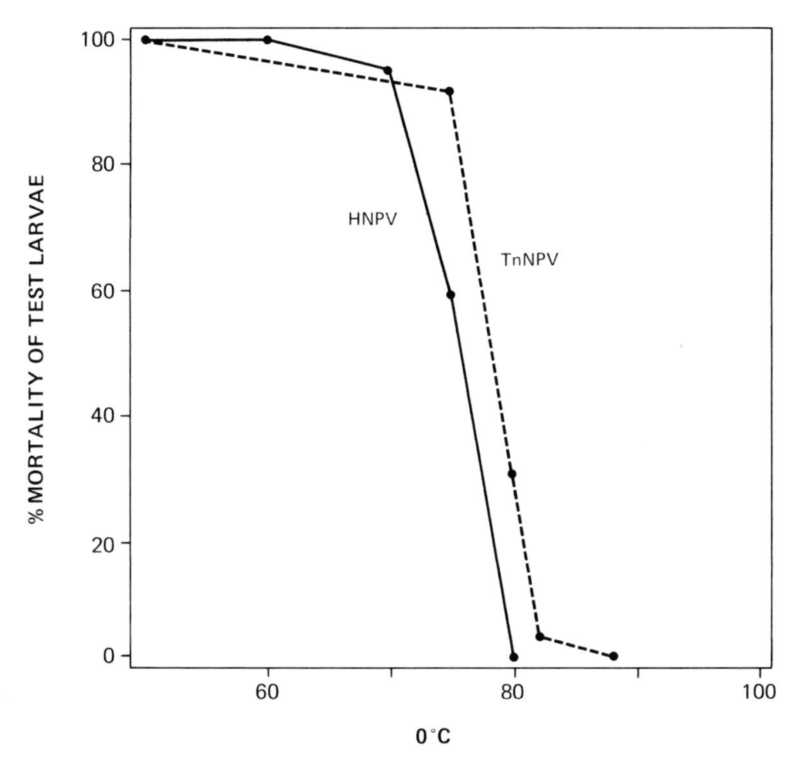

FIGURE 7. Activity of suspensions of inclusion bodies of the nuclear polyhedrosis virus of Heliothis spp. (HNPV) and of Trichoplusia ni (TnNPV) after 10-minute exposures to indicated temperature. (After Gudauskas and Canerday, 1968.)

NPV's of *O. pseudotsugata* differed in thermostability as well as in other characteristics. The single-embedded type was less stable than the multiple-embedded type being inactivated by exposure to 60° and 65°C for 4 or 0.5 hr, respectively, whereas thermal inactivation of the multiple-embedded type required exposure to the respective temperatures for 14 and 1 hr. It is interesting that the densonucleosis virus of *G. mellonella* was found by Boemare et al. (1970) to be partially thermoresistant, not being inactivated to a significant extent by exposure to 60°C for up to 60 min. Likewise, the thermostability of *Spodoptera littoralis* NPV and *Spodoptera* (= *Prodenia*) *litura* NPV are noteworthy; Harpaz and Raccah (1978) reported that *S. littoralis* NPV retained some activity after a 10-minute exposure to 90°C and earlier Pawar and Ramakrishnan (1971) reported similar thermostability of *S. litura* NPV.

It is significant that TnNPV and PrGV added to sauerkraut before fermentation or after fermentation were nearly 100% inactivated by pasteurization in which the sauerkraut in cans was heated to 82°C and subsequently cooled indicating that residues of the virus on cabbage would retain little activity after exposure to normal processing procedures (Jaques and Laing, 1976). Substantial concentrations of TnNPV were found on heads of cabbage harvested from plots treated or not treated with the virus (Heimpel et al., 1973; Jaques, 1974b, 1975; Thomas et al., 1974).

Studies on stability of insect viruses exposed to temperatures in the range expected to be encountered in the field habitat are particularly pertinent to effectiveness of viral insecticides. Air temperatures exceeding 40°C (104°F) for long periods may occur in forest or crop production areas but probably not in streams and on leaf surfaces even if the temperature of the air is higher. Temperature of soil, particularly at the surface, exceeding 50°C (123°F) would not be uncommon, especially if the soil were exposed and not covered by a plant canopy or otherwise protected from the sunlight. The data reported by McLeod et al. (1977) and Hunter et al. (1973) are, therefore, of particular interest to this discussion because of the temperatures to which the viruses were exposed in their studies. McLeod and co-workers showed that the LD_{50} values of deposits of HNPV exposed for 12-, 24-, or 48-hr periods to 15°C were 2, 10, and 48 PIB/mm^2 of diet, to 30°C, 5, 25, and 124 PIB/mm^2, and to 45°C, 22, 112, and 565 PIB/mm^2, respectively, indicating substantial decreases in activity of the virus exposed for a long period to 15°C, a common field temperature in spring and autumn, especially at night. A temperature of 30°C, likely to be encountered very frequently during the growing season in most crop-growing areas, was quite detrimental to activity of the virus while even a short exposure (12 hr) to 45°C (113°F), a temperature not uncommon in field habitats, reduced activity considerably. *Cadra cautella* NPV formulated in bran was more stable (Hunter et al., 1973) than were dried deposits of HNPV studied by McLeod et al. *Cadra cautella* NPV retained at 32°C or lower lost little activity in 28 days but the virus was 80% inactivated when retained at 42°C for the period. This virus withstood exposure for 7 days at 42°C and was only partially inactivated in 14 days indicating that the virus could be expected to withstand substantial exposure to this high temperature.

The activity of the granulosis virus of *Plodia interpunctella* did not decrease appreciably in a year in storage bins of wheat in which temperatures ranged from -19° to 48°C (-2° to 118°F) (Kinsinger and McGaughey, 1976). However, the virus was about 90% inactivated by a 42-week storage period at a constant temperature of 42°C. Spores of *B.t.*, on the other hand, were about 15% less active after the storage period at the higher temperature.

Reed (1974) noted that the nonoccluded virus of the citrus red mite, *P. citri*, retained activity when exposed in intact host cadavers to 40.5°C for 24 hr or to 38°C for 28 days but was inactivated by exposure to 46°C for 6 hours or 60°C for 1 hour. Aqueous suspensions of *P. citri* virus retained activity for 5 weeks at 25°C but for only 6 hours at 40°C and the virus was inactivated in 2 hours at 50°C (Gilmore and Munger, 1963) indicating that the virus may remain active for a relatively short period in California citrus groves in the summer.

Free virions of GmNPV, that is, virions in the hemolymph of infected hosts, were inactivated by exposure to temperatures about 38°C (Stairs and Milligan, 1979). Virions of GmNPV released from polyhedral inclusion bodies by exposure to alkali were found, however, to be 3.5 times more thermostable than were the nonoccluded virions (Stairs and Milligan, 1980). These data suggest that free virions may be inactivated by temperatures that frequently occur in some field habitats. In addition, the low tolerance of free virions to these temperatures may influence accumulation of viruses in the habitat by inhibiting or retarding development of disease in host insects. Exposure of host larvae to such temperatures following ingestion of viruses has been found to retard development of disease. For example, Hunter and Hartsell (1971) noted that larvae of *P. interpunctella* fed GV and held at 37°C did not exhibit symptoms to indicate development of the disease and Kobayashi et al. (1981) found that pupae of *B. mori* inoculated with NPV and incubated at 35°C survived longer than those reared at low temperatures following inoculation. Similar observations were made by Carter (1975) and Thompson (1959).

C. HUMIDITY, WATER, AND RAIN

Relative humidity of the air generally has much less effect on viruses than on other entomopathogens, especially the entomofungi. David et al. (1971c), for example, found that relative humidity had no effect on stability of deposits of PbGV on glass surfaces. On the other hand, the virus in dry deposits was not as stable as was the virus retained in aqueous suspensions. Surface moisture as dew was found to reduce the stability of HNPV on cotton foliage due apparently to substances dissolved in the dew which altered the pH of the dew rather than resulting from a direct effect of surface moisture (Andrews and Sikorowski, 1973) (Section III, A and IV, D, 1). In addition, surface moisture may influence susceptibility of deposits of virus to inactivation by ultraviolet light.

Wetted deposits of PbGV were more readily inactivated by uv light (David, 1969) than were dry deposits and wetted deposits of TnNPV were more readily inactivated by exposure to uv than were dry deposits of the virus (Jaques, 1967a). In addition, Jaques found that deposits of TnNPV on plants exposed to sunlight and kept wet under a sprinkler did not retain activity as long as did dry deposits or deposits on leaves exposed to constant flowing water in a photoprint washer in subdued light in the laboratory. Furthermore, suspensions of TnNPV were more readily inactivated by exposure to uv light than were dry deposits. These findings suggest that surface moisture, like that resulting from dew or following rain, may favor the inactivation of deposits of entomoviruses exposed to sunlight.

Deposits of virus appear to adhere quite well to leaf surfaces and, therefore, the washing of virus from leaves by rain does not appear to be a major factor affecting persistence of foliar deposits (Tanada, 1971). For example, Burgerjon and Grison (1965) exposed deposits of polyhedra of *Thaumetopoea pityocampa* CPV on pine foliage to the equivalent of 50 mm of simulated rain and noted little loss of activity. Similarly, the activity of pure and crude preparations of PbGV deposited on leaves of cabbage was not reduced substantially by exposure to 250 cm of simulated rain in a 5 hr period (David and Gardiner, 1966). Exposure to natural rainfall was found to have little effect on persistence of HNPV on cotton (Bullock, 1967; Ignoffo et al., 1965). David and Gardiner (1966) did not remove deposits of purified *P. brassicae* GV from cabbage leaves by scrubbing. Similarly, Jaques (1967a) found that deposits of TnNPV on cabbage leaves placed in a photoprint washer lost only 50% of their original activity in 4 days.

Rain, although not a significant contributor to inactivation of viruses deposited on foliage, has a very important role in dissemination of some viruses in the habitat (Tanada, 1971), thus enhancing persistence of the virus in the field habitat. For example, Bird (1961) reported that rain appeared to be a principal agent responsible for spreading NPV's of three sawflies throughout conifer trees, thus substantially aiding development of epizootics of the diseases (Cunningham and Entwistle, 1981). Similarly, there is good evidence that *L. dispar* NPV which persists in soil, cadavers, debris, and egg masses of the host is disseminated by rain (Doane, 1975; Podgwaite et al., 1979). Likewise, TnNPV persisting in soil apparently was splashed onto leaves of cabbage plants grown in the soil and was ingested by susceptible hosts (Jaques, 1967b, 1974a,b, 1975).

D. *CHEMICALS*

The inclusion body protein in which some entomoviruses, including the baculoviruses, which are of particular interest for development as microbial insecticides, are embedded provides the virus particle with some degree of protection against adverse effects of chemicals in the habitat just as the protein protects the virus against other adverse environmental factors (Bergold, 1958; Stairs, 1968). Generally, therefore, this type of entomoviruses is affected less than nonoccluded viruses by chemicals. Likewise, viruses are, as a group, affected less adversely by chemical components of the habitat than are other types of pathogens (Benz, 1971; Jaques and Morris, 1981).

1. *Acids, Alkalis, and pH*

Acids and alkalis may affect stability of inclusion type viruses by dissolving or otherwise denaturing the inclusion body protein or by less direct effects on activity.

The effects of acids and alkalis on entomoviruses are important in stability of entomoviruses in the habitat because concentrations of alkalis and acids sufficient to cause inactivation of entomoviruses are frequently encountered. Furthermore, the alkalinity of formulations is important to storage life. Early studies showed that the activity of HNPV was not affected significantly by suspension for 24 hr in buffers at pH 4, 7, and 9 but was 99% inactivated by exposure to 0.02 *M* $Ca(OH)_2$ (pH 12.4) (Ignoffo and Garcia, 1966). Likewise, a highly acid reaction (pH 1.2) inactivated the virus. HNPV was about 50 and 90% inactivated by a 30-min and 24-hour exposure, respectively, to buffers of pH 2 or 12 but was not inactivated appreciably in buffers of pH 5, 7, and 9 in studies by Gudauskas and Canerday (1968). Shapiro and Ignoffo (1969) showed that virions of HNPV released from inclusion bodies by exposure for 3 hours to 0.007 *M* Na_2CO_3 + 0.017 *M* NaCl were 99.9% less active than were occluded virions when fed to neonate *H. zea*. Inactivation of the *P. citri* nonoccluded virus was proportional to the concentration of salt in the suspending solution, to time, and to temperature (Chambers, 1968) with even slight alkaline deviations from neutral being detrimental to activity, especially at higher temperature. Other studies showed that buffering suspensions to pH 6.0 significantly enhanced infectivity of *P. citri* virus (Shaw et al., 1968). The interaction of temperature and pH on virions was also noted by Hayashi et al. (1970) who found that virions of *B. mori* CPV in suspension at pH 11 were not inactivated at room temperature (21°) but were inactivated at 40°C.

Microhabitats that are strongly acid or strongly alkali may affect the activity of the virus directly or, in the case of viruses that form inclusion bodies, adversely affect the virus by dissolving or otherwise denaturing the inclusion body protein, thus exposing the virus particle directly to environmental factors. The inclusion body protein of most occluded insect viruses is dissolved in the laboratory by suspension for a few minutes in an alkaline buffer, often 0.01 *M* Na_2CO_3 + NaCl, (e.g., Bergold, 1958; Kawarabata et al., 1980; Shapiro and Ignoffo, 1969; Stairs, 1968; Yamamoto and Tanada, 1978). Such dissolution appears to closely resemble dissolution in the gut of the host larva (Summers, 1971).

Prolonged exposures to solutions containing lower concentrations of alkali, such as 0.005 *M* Na_2CO_3, caused swelling of the inclusion bodies of HNPV but not dissolution of the inclusion body protein (Andrews and Sikorowski, 1973). Exposure to weakly alkaline solutions (pH 9-10) dissolved polyhedral inclusion bodies of HNPV if the exposure was long enough and if the ionic content of the buffer was sufficient (Gudauskas and Canerday, 1968; Ignoffo and Garcia, 1966). In this regard it is noteworthy that the dew on cotton contained more salts than did dew on soybean leaves indicating that higher ionic concentration as well as higher pH of cotton dew compared to soybean dew or tomato dew may have contributed to inactivation of HNPV on cotton foliage at night, which was observed by Andrews and Sikorowski (1973) (Section III, A). They found that the alkaline dew on leaves of cotton (pH 8.2-9.2) contained several cations, especially potassium, calcium, and sodium, and substantial concentrations of carbonate. Immersion of HNPV in leaf washings (pH 9.8) caused swelling similar to that noted when PIBs were suspended in 0.005 *M* Na_2CO_3 + 0.005 *M* NaCl adjusted to pH 10. Young et al. (1977) showed that cotton dew normally was more alkaline (pH 8.8) and contained a higher concentration of ions than did dew on soybean (pH 7.8). HNPV held in either dew retained activity, but if the suspension was dried and resuspended, the virus in the cotton dew was inactivated whereas that in soybean dew was not. McLeod et al. (1977) noted that HNPV was stable in moderately alkaline cotton dew (pH 7.4 or 8.8) but was inactivated in more alkaline dew (pH 9.3) on cotton.

Virions released from inclusion bodies by low concentrations of alkali may retain at least partial activity (Bergold, 1958). For example, virions of *Mythimna* (=*Pseudaletia*) *unipuncta* GV released from the capsules by suspension in 0.02 *N* NaOH retained a substantial proportion of original activity (Yamamoto and Tanada, 1979). Virions of *G. mellonella* NPV released from polyhedral inclusion bodies (PIBs) were 10^4 less infective than were virions fed in intact polyhedra indicating the severe effect of treatment with alkali (Stairs, 1980). The infectivity of virions released from inclusion bodies differs substantially from that of nonoccluded forms of the virus. A comparative study by Volkman and Summers (1977) showed that alkali-liberated virions of ACNPV were up to 50 times less infective than were virions from cell cultures which were not occluded. The latter existed singly whereas the former were predominantly in bundles accounting for a part of the difference.

Dissolution of occluded viruses by the alkaline gut juices of susceptible insects is essential for infection. In addition, susceptibility to inactivation by gut juices may contribute to specificity of virions. It is interesting that inclusion body protein of *S. littoralis* NPV was not dissolved when exposed to contents of the larval midgut at the normal pH of 8.5-9.5 but was dissolved if the pH was increased to 10.5 (Harpaz and Raccah, 1978) suggesting involvement of other factors. On the other hand, inclusion bodies of BmNPV were readily dissolved in 2 minutes when exposed to gut juices from the host larvae (Kawarabata et al., 1980). The dissolving activity of gut juices was influenced by rearing conditions of the larvae, for example, decreasing after a 48-hr starvation period. It is noteworthy that virions of BmNPV released by exposure to gut juices were slightly less infective than were virions released by exposure to an alkaline buffer (0.08 *M* Na_2CO_3 + 0.025 *M* NaCl). Likewise, virions of AcNPV released from polyhedral bodies by exposure to the gut juice from larvae of *Estigmene acrea* (pH 10.4-10.6) for up to 8 minutes retained infectivity for a *T. ni* cell line (Vail et al., 1979). Longer exposure reduced activity.

The effect of gut contents on viral activity is a consideration in dissemination of viruses by birds and other animals (Section V, E). In addition, the inactivation of entomoviruses by gut juice may have implications in safety to humans and mammals (Section IV, D, 4).

The pH of soil has been found to affect stability of viruses. Thomas et al. (1973) reported that TnNPV was inactivated more quickly in soil of low pH (pH 4.8) than in soil near neutral in reaction. This virus persisted for long periods in a loam soil of pH 5.2-5.6 in field plots and in pots in studies by Jaques (1964, 1967b) but unpublished laboratory tests by Jaques showed the virus did not persist in highly alkaline or highly acid soil.

TnNPV and PrGV on cabbage were more than 50% inactivated during sauerkraut fermentation during which pH was reduced to 3.6 (Jaques and Laing, 1976). Because of the former virus was found to be highly resistant to proteolytic activity of bacteria (Jaques and Huston, 1969) it is assumed that loss of activity was due to the prolonged exposure to the highly acidic conditions rather than to direct microbial activity.

2. *The Effect of Pesticides and Adjuvants*

The compatibility of entomoviruses with pesticide chemicals, especially fungicides and insecticides, is important because in practical usage viruses may be frequently mixed with chemical pesticides either during application or on the crop, animal, or other substrate to which the virus is applied. In addition, it is equally important that naturally occurring entomoviruses should not be affected adversely by applied pesticides so that maximum benefit of naturally occurring pathogens can be exploited in integrated management of species of pest insects. Pesticides that inactivate an entomovirus or reduce its effect on the target insect could be considered to have an antagonistic effect on the virus. The combined effect of the chemical and the entomovirus may be beneficial to the activity of the chemical or of the entomovirus or of both; such joint action would be additive, supplementary, synergistic or potentiating. Relationships of viruses and pesticides have been reviewed recently by Benz (1971), Jaques and Morris (1981), and more recently in this treatise.

Pesticide chemicals generally have less effect on entomoviruses than on other entomopathogens. Antagonistic joint action of chemical pesticides and viruses have been reported but usually the effect on development of disease rather than a direct reduction of the activity of the virus per se was determined (Jaques and Morris, 1981). In studies by Ignoffo and Montoya (1966), methyl parathion was the only insecticide that reduced activity of HNPV against *H. zea* larvae in laboratory tests. Endrin, DDT, toxaphene, and carbaryl did not adversely affect the virus. An exposure time was established by diluting the insecticide--HNPV mixtures after a specified period so the mortality by the virus could be differentiated from that of the chemical. The tests with fungicidal materials may be more credible because the fungicides should have a minimum direct effect on the insect. The activity of LdNPV in laboratory tests appeared to be enhanced by mixture with boric acid or zinc sulfate (Jaques and Morris, 1981; Shapiro and Bell, 1982). The activity of *Lymantria monacha* NPV was increased by $CuSO_4$ whereas the fungicides captan, dinocap, sulfur, and zineb were antagonistic to CpGV (Jaques and Morris, 1981).

Most assessments of compatibility of pesticides and entomoviruses have evaluated the effectiveness of pesticide-virus mixtures against the insect. In such studies reviewed in this volume, the direct action of the pesticide on the virus was often not determined. The fact that additive or potentiating joint action did result would indicate, however, that mixing the materials in the spray tank and subsequently as deposits on the substrate treated did not adversely affect the activity of the virus and probably enhanced it. Conversely, an antagonistic joint action could suggest that the virus was inactivated. In this connection, therefore, it is noteworthy that Jaques (1972a, 1973a,b, 1977a) applied TnNPV and PrGV in mixtures with endosulfan and methomyl and obtained the expected mortality by the virus. Similarly, the enhanced effectiveness of mixtures in other field tests suggest compatibility of chemical insecticides and entomoviruses: e.g., mevinphos or naled with TnNPV (Creighton et al., 1970), chlordimeform and AcNPV against *T. ni* larvae (Jaques and Laing, 1978), chlordimeform and PrGV against *P. rapae* larvae (Jaques and Laing, 1978), and fenitrothion and *C. fumiferana* NPV against *C. fumiferana* (Morris, 1977; Morris et al., 1974). In addition, the enhanced effectiveness of some chemical-HNPV mixtures in tests by Pfrimmer (1979) indicated compatibility whereas others indicated antagonism. The joint action of most of the mixtures of insecticides and HNPV tested by Luttrell et al. (1979) against *Heliothis* spp. on cotton indicated compatibility. It is noteworthy that mixing entomoviruses with *B.t.* does not appear to have reduced activity of viruses in various studies (McVay et al., 1977; Bell and Romine, 1980; Jaques, 1977a; Johnson, 1982).

The accumulation of TnNPV and PrGV on leaves of cabbage plants treated with endosulfan or methomyl (Jaques, 1974a,b, 1975) also adds to circumstantial evidence that these insecticides did not directly adversely affect the activity of these viruses significantly.

The granulosis virus of *P. interpunctella*, the Indian meal moth, on stored grain was inactivated by fumigation with methyl bromide but was not inactivated by ethylene dichloride-CCl_4, CCl_4-carbon bisulfide, or phosphine (McGaughey, 1975). Ethylene oxide was found to be highly effective in inactivating AcNPV, TnNPV-MEV, and *Spodoptera frugiperda* GV in an insectary or glasshouse but *T. ni* CPV was less susceptible to inactivation (Tompkins and Cantwell, 1975).

Many adjuvants to formulations of entomoviruses have positive effects on efficacy with few having a direct effect on activity of the virus (Bell and Romine, 1980). Protectant-type additives have been discussed earlier (Section IV, A, 3). Several workers (Bell and Romine, 1980; Johnson, 1982; Luttrell et al., 1983) have noted that feeding stimulants (e.g., Coax and Gustol) did not directly affect activity of HNPV against *Heliothis* spp. on cotton but enhanced feeding and probably protected the virus to some extent. Early work by Ignoffo and co-workers (1976a) demonstrated that formulations containing a protectant-gustatory adjuvant enhanced stability of HNPV two- to elevenfold and increased feeding by *H. zea* larvae up to threefold. On the other hand, although most detergent-type spray additives do not directly affect activity of entomoviruses (Jaques, 1977b), Yamamoto and Tanada (1980) found, for example, that acylamines (cationic detergents), especially the hexylamines (6 carbons in acyl group) or the dodecylamines (12 carbons), enhanced infection of *P. unipuncta* larvae by NPV by a 10- to 100-fold. Similarly, boric acid and some borate salts enhanced the infection of *L. dispar* larvae by LdNPV by up to elevenfold (Shapiro and Bell, 1982). Sodium salts were equally as effective as the acid whereas potassium and ammonium salts had no effect.

3. *Disinfectants*

It is recognized that susceptibility of entomoviruses to inactivation by disinfectants and similar materials is of marginal relevance to this discussion of environmental stability. Nevertheless, a limited discussion is included here because laboratory conditions are often a part of the environment of the entomovirus, albeit an unnatural part.

Exposure to sodium hypochlorite or formalin inactivated a large proportion of viruses studied by various authors (Jaques, 1977b). Both compounds are especially useful for inactivation of viruses that externally contaminate eggs--formalin is used in diets and sodium hypochlorite is used extensively as a disinfectant for work areas and rearing containers. For example, dipping eggs of *T. ni* in 0.3% sodium hypochlorite or 10% formalin for 5 or 45 minutes, respectively, inactivated viral contaminants on most eggs without substantial loss of hatchability (Vail et al., 1968). Formalin is incorporated at low concentration (0.04%) into diets to inactivate viral contaminants to enhance disease-free rearing (David et al., 1969; Ignoffo and Garcia, 1968). David et al. (1972) showed that the antiviral activity of formalin in diet was partially as a fumigant. Formalin incorporated in the diet has been found to not only alter activity of the virus but may also reduce susceptibility to disease (Ignoffo and Garcia, 1968; Vail et al., 1968) increasing both LT_{50} and LD_{50} values causing discrepancies in bioassays.

4. *Natural Chemicals*

The effect of chemicals occurring naturally in the habitat on persistence of viruses has been only briefly studied. As discussed elsewhere (Section III, A), there is some evidence, for example, that exudates from certain leaves may inactivate virus. Smirnoff (1968) noted that some conifers release terpene compounds that could influence viral activity. In addition, natural deposits from the air on leaves may affect entomoviruses or leaves may produce salts that may influence viral stability; both factors may have contributed to the viral inactivating effect of cotton leaves noted by Andrews and Sikorowski (1973) (Section III, A and IV, D, 1). Likewise, soil contains many chemicals that, in addition to affecting viruses by the acid-alkali reaction, could influence persistence directly.

A natural chemical complex of considerable pertinence is the synergistic factor in the armyworm, *M. unipuncta*, studied by Tanada and co-workers. Synergism between a Hawaiian strain of granulosis virus and a nuclear polyhedrosis virus of *M. unipuncta* reported by Tanada (1959) was subsequently found to be associated with a factor, first identified as an enzyme (Tanada and Hara, 1975) and later as a phospholipid (Yamamoto and Tanada, 1978), which enhances infection of the host tissues by acting on the cell membrane of the microvilli of the midgut (Tanada et al., 1980).

The observation that juices in the gut of birds and mammals do not completely inactivate viruses, having seemingly minor effect on some viruses, is of significance in regard to persistence in the field environment because of the possible involvement of such animals in dissemination of some viruses. Infective NPV has been recovered after passage through the alimentary tract of birds (Bird, 1955; Hostetter and Biever, 1970) and mammals (Smirnoff and MacLeod, 1964; Ignoffo et al., 1971). HNPV was about 50% inactivated by a 1-hr exposure to gastric juice of rats (Ignoffo et al., 1971) or by a 40-min exposure to human gastric juice (pH 2.1) (Chauthani et al., 1968). It is interesting that exposure to human gastric juice adjusted to pH 7.9 had little effect on infectivity of the virus.

V. BIOTIC FACTORS AFFECTING PERSISTENCE IN THE HABITAT

Several biotic environmental factors influence the presence and activity of entomoviruses in the habitat of the host. In addition, dissemination of entomoviruses and/or their contact with a host individual are influenced by biotic environmental factors. Some of these factors pertain particularly to entomoviruses because they are obligate pathogens, whereas others pertain to other entomopathogens as well.

A. *PRESENCE OF HOST*

The presence of the host is particularly important to persistence of entomoviruses in the habitat because entomoviruses are obligate pathogens, which multiply only in living tissue. Most entomoviruses have been found to be relatively specific for a species of insect or a group of closely related insects, and, therefore, multiplication in the habitat requires the presence of the specific host.

The presence of the host insect is important to permit mechanisms of direct transmission such as infection by larval feeding on a cadaver of virus-killed host, ingestion of food contaminated by virus from the cadaver, by transmission through the egg, or by transmission of disease by a parasite. Likewise, presence of the host is necessary to propagate entomovirus to contaminate the habitat to serve as a reservoir for contamination of food to be consumed by healthy host individuals.

Some entomoviruses have been found to accumulate in substantial concentrations in the habitat and to persist at these high concentrations, especially in soil and debris, for long periods (Section III, B). Other entomoviruses accumulate to lesser concentrations and are less persistent. In addition, some viruses persisting in the habitat are transmitted readily within populations, whereas others are transmitted less readily, especially within sparse populations of the host. For example, Jaques (1974a,b, 1975) found that TnNPV and PrGV accumulated in high concentrations in soil as a result of epizootics of the diseases in the respective hosts, *T. ni* and *P. rapae*, on cole crops. In addition, the viruses produced in the epizootics persisted in the soil for long periods. More important, the virus persisting in the soil contaminated leaves of cabbage grown in the soil to initiate epizootics in subsequent generations. Indeed, Jaques (1971a; Jaques, unpublished) found that the residues of TnNPV and PrGV in soil accumulated to a sufficient concentration to regulate these pests to an economically acceptable population density on cabbage after cropping for several consecutive years without application of an insecticide or virus. Likewise, LdNPV that accumulated in the forest habitat initiated epizootics of disease in subsequent populations of *P. dispar* (Podgwaite et al., 1979). On the other hand, concentrations of *P. includens* NPV accumulating in soil from epizootics were insufficient to initiate epizootics (Young and Yearian, 1979; McLeod et al., 1982). Similarly, it is doubtful that significant quantities of virus accumulate now from the epizootics of NPV in the presently sparse populations of *D. hercyniae* in New Brunswick (Bird and Elgee, 1957). Some entomoviruses, such as CpGV (Huber and Dickler, 1977), do not accumulate in the habitat in detectable quantities because of the low yield of viruses from an individual host and because of the generally low populations of the host. Furthermore, as illustrated by *C. pomonella*, the biology of the host influences the probability of accumulated virus contributing to transmission of the disease.

The density of the population of host insects not only influences the quantity of virus accumulated in the habitat but affects the probability that a healthy individual will contract virus persisting in the habitat. For example, Hukahara (1973) related the concentration of *H. cunea* NPV in soil to numbers of host larvae killed. Jaques and Harcourt (1971) noted that TnNPV and PrGV were detected more frequently and in higher concentrations in soil sampled from fields in which infestations of the host species were severe than in fields in which infestations were sparse. Also, higher concentrations of TnNPV and PrGV accumulated on leaves of nontreated cabbage than on leaves of cabbage treated with a chemical insecticide lethal to the host insects (Jaques, 1974a,b).

Transmission of virus in or on the eggs from infected hosts would enhance development of epizootics in populations of host insects, especially in sparse populations and in populations of hosts that feed on poorly exposed food. This would increase accumulation of the virus in the habitat and increase persistence of the virus in the population. There is varying evidence for transmission in or on the egg. For example, the NPV of the armyworm, *Mythimna* (= *Pseudaletia*) *separata*, was found to be transmitted to neonates in or on the eggs from adults that had been fed the virus (Neelgund and Mathad, 1978). Likewise, Hamm and Young (1974) reported that the HNPV that passed through the digestive tract of adult *H. zea* contaminated the surface of eggs provided a source of virus to infect neonates. On the other hand, although capsules of GV were found on some orchard-collected eggs of *C. pomonella*, transovum transmission of the virus was not proved (Etzel and Falcon, 1976). It is interesting that the *O. rhinoceros* NPV is transmitted by contact of adults, especially during mating, with transmission being dependent on persistence of the virus in feces and not on the egg (Zelazny, 1976). Studies on the epizootiology of an entomopox virus of midges, *Chironomus* spp., indicated that transovum transmission was of substantial significance (Harkrider and Hall, 1978). Likewise, the CPV of *H. virescens* was found to be transferred from virus-infected parents to progeny (Sikorowski et al., 1973).

Lymantria dispar NPV persisting on the surface of the eggs and on hair surrounding the eggs in the egg mass where the virus is ingested by neonate larvae appears to play a significant role in initiation of natural epizootics of the disease in populations of this important pest of deciduous forests (Doane, 1969, 1970; Yendol et al., 1977). Debris mats composed of exuviae, larval and pupal cadavers, and pupal cases were the most concentrated source of virus to infect young larvae. The young larvae died in the first stadium usually in the upper part of the tree and larger larvae fed on the contaminated foliage which resulted in a continued epizootic.

B. ALTERNATE HOSTS

Most entomoviruses are considered to be specific for a species of insect (Ignoffo, 1968; Martignoni and Iwai, 1981). Some viruses, however, infect insects of several species; in most cases these species are closely related. The ability to infect alternate hosts was believed by Tanada and Omi (1974) to enhance persistence of viruses in alfalfa fields in California. They demonstrated substantial incidence of infection of several viruses in alternate hosts. Likewise, AcNPV infects several species of Noctuidae (Vail et al., 1971, 1978, 1982) including *A. californica*, *T. ni*, *H. zea*, and *H. virescens*. AcNPV and TnNPV (MEV) have been found to be reciprocally infective for their host (Vail et al., 1971). Also, AcNPV and *R. ou* NPV infected eight pluseine species (Harper, 1976). The viruses were found to be equally effective in preventing damage to corn seedlings by the black cutworm, *A. epsilon* (Johnson and Lewis, 1982). Larvae of the pea moth, *Cydia nigricana*, are susceptible to *C. pomonella* GV (Payne, 1981) but the niches occupied by these host larvae are quite different, reducing the probability of a substantial contribution to persistence.

The possibility of a wide host range of infectivity of entomoviruses was suggested by a study by Ignoffo and Garcia (1979) which showed that 7 of 38 viruses tested infected larvae of *A. ipsilon* at reasonable dosages. Likewise, the finding (Shapiro et al., 1982) that NPVs of three forest insect pests (*O. pseudotsugata* NPV, LdNPV, and *C. fumiferana* NPV) infected the larvae of the salt marsh caterpillar, *E. acrea*, also suggested that viruses may, indeed, persist in alternate hosts in the habitat of the primary host or in another habitat.

It is noteworthy that probability of infection and/or activity of some viruses may be increased by serial passage, enhancing the role of the alternate host in persistence of the virus. Wood et al. (1981) selected a mutant of AcNPV that killed *T. ni* larvae more quickly than did wild AcNPV and passaging AcNPV and TnNPV in *S. exiquae* increased virulence for *T. ni* (Tompkins et al., 1981). On the other hand, passage of AcNPV in larvae of *Ostrinia nubilalis*, the corn borer, did not increase activity of the virus in this alternate host. Similarly, serial passage of *C. fumiferana* NPV in alternate hosts did not affect infectivity but altered the pathology of the disease caused by the virus of *C. fumiferana* (Stairs et al., 1981). Infection of the soybean looper, *P. includens*, with the NPV of the velvet bean caterpillar, *Anticarsia gemmatalis*, was reported by Pavan and co-workers (1981) to have activated a latent virus in *P. includens* that after passage was infectious for both hosts. Similarly, studies by McKinley et al. (1981) on cross-infectivity of NPV of three *Spodoptera* and one *Heliothis* spp. suggested that activation of a virus native to the alternate host may be more common than is cross-infectivity.

C. *COMPETITION FOR HOST*

Competition with other pathogens for hosts influences persistence of the virus in the habitat by influencing multiplication of a virus. A virus may be directly antagonistic to another pathogen or, conversely, compatible or synergistic (Jaques and Morris, 1981) in regard to multiplication in a host insect or may compete by incubation period or other means.

The synergism of NPV and GV in *M. unipuncta* by virtue of a synergistic factor in the capsule of the GV (Tanada and Hara, 1975; Tanada et al., 1980) discussed earlier (Section IV, D, 4) illustrates a synergistic relationship. Undoubtedly the persistence of *M. unipuncta* NPV in the habitat is enhanced because this favorable relationship would favor increased propagation of the virus. On the other hand, antagonism between two GVs of *C. fumiferana* prevented normal development of the two viruses in the same cell or tissue (Bird, 1976). Similarly, development of NPV in *C. fumiferana* was interfered with by a CPV (Bird, 1969) and two NPVs of *M. unipuncta* competed and antagonized each other (Ritter and Tanada, 1978).

Pathogens that are not directly antagonistic may dominate over a second pathogen of a host by other mechanisms of competition. For example, the GV of *T. ni* (TnGV) accumulated and persisted in low concentrations in nontreated soil (Jaques, 1970b) and in soil treated with GV (Jaques, 1974b) apparently because GV developed more slowly than did TnNPV in *T. ni* larvae following infection, with the result that TnNPV predominated over TnGV causing TnGV to be crowded out of the habitat.

Applied pathogens may compete with entomoviruses to influence accumulation and persistence of the viruses in the habitat. Jaques (1974a,b) showed that application of formulations of *B.t.* reduced accumulation of TnNPV and PrGV on foliage and soil apparently because *B.t.* killed the host before the virus disease developed. Although applications of mixtures of *B.t.* and viruses in these and other studies have been generally more effective against the target insect than were the components used alone [e.g., *B.t.* + *Dendrolimus spectabilis* NPV

(Katagiri and Iwata, 1976), *B.t.* + TnNPV (McVay et al., 1977), *B.t.* + AcNPV (Jaques, 1977b), *B.t.* + *O. pseudotsugata* NPV (Stelzer et al., 1975)], the benefit of superior immediate control of the pest may be offset to some extent by reduced accumulation of the virus in the habitat resulting in reduced carry-over impact of the application. It is noteworthy that *V. necatrix* and HNPV were found to be antagonistic in their effect against larvae of *H. zea*, with the microsporidium often succeeding over the virus (Fuxa, 1979). On the other hand, in studies by Ferron and Hurpin (1974), *M. melolontha* NPV frequently dominated over *Beauveria tenella* in competition for the host, *M. melonotha*.

D. PARASITIC AND PREDACEOUS ARTHROPODS

Persistence of viruses in habitats and in populations has been generally enhanced by parasitic and predaceous arthropods. When viruses and parasitic arthropods have competed for a host, the virus has usually dominated unless the infection occurred after the parasite was quite well developed in the host. Levin et al. (1981, 1983) found that *Apanteles glomeratus* did not develop in larvae exposed to PrGV on or before 4 days after oviposition suggesting that the virus would compete favorably with this parasite. Similarly, adults of the parasite *Hyposoter exiguae* did not emerge from *T. ni* larvae that had been exposed to TnNPV more than 2 days before and up to 4 days after parasitization (Beegle and Oatman, 1975). In other laboratory studies survival of the parasite *Campoletis sonorensis* was substantially reduced by infection of *H. virescens* with HNPV within 2 days after oviposition (Irabagon and Brooks, 1974). On the other hand, Vail (1981) noted that the parasite *Voria ruralis* developed to maturity in *T. ni* larvae infected with NPV. Competition of entomoviruses with parasitic arthropods may be direct. For example, toxins produced by GV and NPV infections in *M. unipuncta*, *S. exiguae*, and *T. ni* inhibited growth and development of *Apanteles militaris* in embryonic and larval stages killing the parasite (Kaya and Tanada, 1972, 1973).

Parasitic arthropods may enhance persistence of an entomovirus in the habitat by aiding its dissemination to new host individuals. *Malacosoma disstria* NPV was believed to be transmitted within the forest to new *M. disstria* host individuals by *Sarcophaga aldrichii* (Stairs, 1966). Reardon and Podgwaite (1976) found a positive correlation between occurrence of LdNPV and parasitism by its parasite *Apanteles melanoscelus*. Intrahemocoelic inoculation of LdNPV by adults of *A. melonoscelus* that oviposited in noninfected *L. dispar* larvae after ovipositing in NPV-infected larvae was demonstrated by Raimo et al.

(1977). The parasite *H. exiguae* transmitted TnNPV from NPV-infected *T. ni* larvae to only a small proportion of healthy individuals in which it subsequently oviposited (Beegle and Oatman, 1975), but *C. sonorensis* adults transmitted HNPV to as many as 40% of a *H. virescens* larvae in which they oviposited after oviposition in NPV-infected larvae (Irabagon and Brooks, 1974). Levin et al. (1979, 1981, 1983) reported a higher frequency of transmission of PrGV by *A. glomeratus* with 81 to 95% of adult females transmitting the disease from GV-infected *P. rapae* larvae. Parasite adults that oviposited in GV-infected larvae transmitted GV to up to four healthy *P. rapae* larvae in which they subsequently oviposited. Their studies also showed that the frequencies with which the parasite oviposited in healthy and GV-treated larvae did not differ significantly. The finding that 84% of the parasite adults that emerged from GV-infected *P. rapae* transmitted the disease to healthy *P. rapae* larvae indicated the significance of transmission of this virus by the parasite in maintaining the virus in the field habitat. Versoi and Yendol (1982) reported, on the other hand, that *A. melanocelus* significantly preferred to oviposit in healthy *L. dispar* larvae over larvae infected with NPV, indicating competition for the host.

It is suggested that transmission of entomoviruses by parasitic arthropods may aid significantly in maintaining viruses of insects that feed alone on protected food, such boring insects and leafrollers, in the field habitat. An example is the possible role of the parasite *Agathis laticinctus* in maintaining the NPV of the eye-spotted bud moth, *Spilonota ocellana*, in populations of the host in apple orchards (Jaques and Stultz, 1966).

E. MAMMALS, BIRDS, AND OTHER SCAVENGERS

The ability of entomoviruses to maintain activity during passage through the digestive tract of scavenger insects, mammals, and birds indicates a further system of persistence of viruses in the habitat. For example, Hostetter (1971) recovered substantial quantities of active TNPV from the abdomen of Sarcophagid flies that had fed on liquefied cadavers of cabbage looper larvae killed by the virus.

Early studies by Bird (1955) and Franz et al. (1955) suggested that NPV contained in excrement from birds could aid in disseminating the disease. Similarly, Hostetter and Biever (1970) recovered virulent TnNPV from feces of birds. Entwistle et al. (1977a) noted that 16 species of birds in Wales dispersed infective NPV of *D. hercyniae*, the European spruce sawfly, in their feces. Active virus was found in 90%

of droppings of birds collected in September, with the proportion of infective droppings declining in May and June prior to the new larval period (Entwistle et al., 1977b). Data suggested that the birds carried NPV at least 6 km. Other studies (Entwistle et al., 1978) indicated that NPV-active excrement was passed up to 7 days after the birds had fed on NPV-infected larvae demonstrating the potential role of birds in dissemination of the virus by this means.

Infectious LdNPV was found in the alimentary tracts of 3 species of birds and 5 species of small mammals captured in a forest in which an epizootic of NPV occurred in populations of *L. dispar*, indicating that these animals could play a significant role in dissemination of this virus (Lautenschlager et al., 1980). Earlier studies had demonstrated the presence of LdNPV in feces samples from birds and in feces from small mammalian predators of *L. dispar* (Lautenschlager and Podgwaite, 1979; Lautenschlager et al., 1978, 1979). Smirnoff and MacLeod (1964) found, however, that NPVs of *N. swainei*, *G. mellonella*, and *Evannis tiliaria* (Harris) were inactivated by passage through the digestive tract of white mice, *Mus musculus*.

VI. SIGNIFICANCE OF PERSISTENCE IN USE OF ENTOMOVIRUSES IN INTEGRATED PEST MANAGEMENT

Persistence of entomoviruses has a substantial impact on their potential role in integrated pest management, affecting their effectiveness as introduced or applied regulatory agents as well as naturally occurring mortality agents. It is apparent that the rather rapid inactivation of entomoviruses exposed to sunlight reduces their effectiveness as natural mortality factors and, probably to a greater degree, their effectiveness as applied agents. The retention of activity of entomoviruses for long periods in soil, in cadavers of killed hosts, and in other protected locations in the habitat favors perpetuation of natural enzootics and epizootics and favors establishment of the viruses in the population following application and carry-over of effectiveness to succeeding generation of hosts. Likewise, persistence in excrement of birds and other predators of scavengers favors establishment as does transmission of the virus by parasitic insects and other means and development in alternate hosts.

It is apparent that the significance of persistence differs for different pathogen-host systems. For example, the rapid inactivation of foliar deposits of AcNPV, or TnNPV or HNPV applied for control of *T. ni* or *H. zea* (Section III, A) is a serious concern in the use of these viruses in management of these pests. Rapid inactivation on foliage may, however, be less limiting to overall effectiveness of these viruses than is a similar rate of inactivation of viruses of hosts that have a greater opportunity to escape contact with the virus. Larvae of *C. pomonella*, for example, feed for only a very short time on an exposed surface before burrowing into the apple. The long-term persistence of virus in soil and debris which may provide a inoculum to initiate epizootics or enzootics in succeeding generations of the host is important in control of such insects as *T. ni* or *P. rapae* on cabbage but may be more important for control of pests on alfalfa or of pests of forests which probably would not be treated annually with a virus. Likewise, dissemination of viruses, like LdNPV or PrGV, by parasites and *D. hercyniae* NPV by birds is important in maintaining natural epizootics and in providing carry-over impact in populations of these pests.

Carry-over effect from treatment is especially important in the use of viruses and other insecticides against insect pests of forests and forage crops partly because the acceptable damage threshold is higher than it is, for example, for an apple tree or cotton plant (Section II, D). In addition, the cost-benefit ratio in protection of forests, pastures or forage crops is such that annual applications may not be economically feasible. Therefore, the prolonged effectiveness of viruses against insects pests of such crops is of particular importance to the feasibility of their use. For example, LdNPV and NPVs of *N. sertifer* and other sawflies persisted in populations of the respective hosts following introduction and were disseminated considerable distances from introduction points greatly enhancing their usefulness. Similarly, the 3-year carry-over effect of *C. fumiferana* NPV-fenitrothion mixtures is particularly significant in assessing the economics of use of the virus. Likewise, *O. rhinoceros* NPV becomes well established in populations of the host resulting in its significant effectiveness as a control agent.

It is apparent that effectiveness of entomoviruses in protecting most crops, such as, cole crops, cotton or apple trees in which a high level of protection is required must be competitive with the effectiveness of alternate insecticides. This competitive status against some pests would be substantially enhanced by increasing the period of time over which lethal concentrations of the virus remain accessible to the target insect. Studies reviewed herein (Section III, A, 3) have demonstrated the potential for increasing the half-life of deposits of entomoviruses by adjuvants in formulations suggesting an area of further research. Likewise, adjuvants to enhance the attractiveness of entomoviruses to encourage feeding by target insects should be studied further.

Manipulation of the environment or microenvironment to offset deleterious environmental effects on viruses and to exploit positive features of environmental relationships of entomoviruses has been studied to only a limited extent (Allen et al., 1978; Jaques, 1978). For example, utilizing the finding that TnNPV and PrGV persisted for long periods in soil, Jaques (1970a, 1972a, 1973a) applied the virus to soil at the time when cabbage was transplanted and obtained acceptable control of *P. rapae* and moderate effectiveness against *T. ni*. Likewise, Ignoffo et al. (1980) treated seedlings with NPV before transplanting and obtained control of *T. ni*, exploiting the persistence of the virus in soil. Biever et al. (1982) applied the ability of a predaceous insect, *Podisus maculiventris*, to transmit TnNPV in populations of *T. ni* by releasing virus-contaminated predators in the field habitat to initiate epizootics of the disease. Ignoffo et al. (1978) released HNPV-infected *H. zea* larvae into populations obtaining a high mortality of the pest.

The early studies on selection of strains of entomoviruses that are more tolerant to adverse environmental factors (e.g., sunlight, Section III, A, 4) suggest the role that genetic manipulation may have in improving the environmental stability of entomoviruses.

REFERENCES

Aizawa, K. (1963). The nature of infections caused by nuclear-polyhedrosis viruses. *In* "Insect Pathology, An Advanced Treatise" (E. A. Steinhaus, ed.), Vol. I, pp. 381-412. Academic Press, New York.

Allen, G. E., Ignoffo, C. M., and Jaques, R. P. (1978) (eds.). "Microbial Control of Insect Pests: Future Strategies in Pest Management Systems" Proc. NSF-USDA-Univ. Florida Workshop, Jan., 1978. 290 pp.

Andrews, G. L., and Sikorowski, P. P. (1973). Effects of cotton leaf surface on the nuclear-polyhedrosis virus of *Heliothis zea* and *Heliothis virescens* (Lepidoptera: Noctuidae). *J. Invertebr. Pathol. 22*, 290-291.

Andrews, G. L., Harris, F. A., Sikorowski, P. P., and McLaughlin, R. E. (1975). Evaluation of *Heliothis* nuclear polyhedrosis virus in a cottonseed oil bait for control of *Heliothis virescens* and *H. zea* on cotton. *J. Econ. Entomol. 68*, 87-90.

Bedford, G. O. (1981). Control of the rhinoceros beetle by Baculovirus. *In* "Microbial Control of Pests and Plant

Diseases 1970-1980" (H. D. Burges, ed.), pp. 409-426. Academic Press, New York.
Beegle, C. C., and Oatman, E. R. (1975). Effect of nuclear polyhedrosis virus on the relationship between *Trichoplusia ni* Lepidoptera: Noctuidae) and the parasite, *Hyposoter exiguae* Hymenoptera: Ichneumonidae). *J. Invertebr. Pathol. 25*, 59-71.
Beegle, C. C., Dulmage, H. T., Wolfenbarger, D. A., and Martinez, E. (1981). Persistence of *Bacillus thuringiensis* Berliner insecticidal activity on cotton foliage. *Environ. Entomol. 10*, 400-401.
Bell, M. R., and Kanavel, R. F. (1978). Tobacco budworm: Development of a spray adjuvant to increase effectiveness of a nuclear polyhedrosis virus. *J. Econ. Entomol. 71*, 350-352.
Bell, M. R., and Romine, C. L. (1980). Tobacco budworm: Field evaluation of microbial control in cotton using *Bacillus thuringiensis* and a nuclear polyhedrosis virus with a feeding adjuvant. *J. Econ. Entomol. 73*, 427-430.
Benz, G. (1971). Synergism of microorganisms and chemical insecticides. *In* "Microbial Control of Insects and Mites." (H. D. Burges and N. W. Hussey, eds.), pp. 327-355. Academic Press, New York.
Bergold, G. H. (1958). Viruses of insects. *In* "Handbuch der Virusforschung" (R. Doerr and G. Hallauer, eds.), Vol. IV, 3, pp. 60-142. Springer, Vienna.
Biever, K. D., Andrews, P. L., and Andrews, P. A. (1982). Use of a predator, *Podisus maculiventris*, to distribute virus and initiate epizootics. *J. Econ. Entomol. 75*, 150-152.
Bird, F. T. (1955). Virus diseases of sawflies. *Can. Entomol. 87*, 124-127.
Bird, F. T. (1961). Transmission of some insect viruses with particular reference to ovarial transmission and its importance in the development of epizootics. *J. Insect Pathol. 3*, 352-380.
Bird, F. T. (1969). Infection and mortality of spruce budworm, *Choristoneura fumiferana*, and forest tent caterpillar, *Malacosoma disstria*, caused by nuclear and cytoplasmic polyhedrosis virus. *Can. Entomol. 101*, 1269-1285.
Bird, F. T. (1976). Effects of mixed infections of two strains of granulosis virus of the spruce budworm, *Choristoneura fumiferana* (Lepidoptera: Tortricidae), on the formation of viral inclusion bodies. *Can. Entomol. 108*, 865-871.
Bird, F. T., and Burk, J. M. (1961). Artificially dissemination virus as a factor controlling the European spruce sawfly. *Diprion hercyniae* (Htg.), in the absence of introduced parasites. *Can. Entomol. 93*, 228-238.

Bird, F. T., and Elgee, D. E. (1957). Virus disease and introduced parasites as factors controlling the European spruce sawfly, *Diprion hercyniae* (Htg.), in central New Brunswick. *Can. Entomol. 89*, 371-378.

Boemare, N., Croizier, G., and Veyrunes, J. C. (1970). Contribution a la connaissance des proprietes du virus de la densonucléose. *Entomophaga 15*, 327-332.

Brassel, J., and Benz, G. (1979). Selection of a strain of the granulosis virus of the codling moth with improved resistance against artificial ultraviolet radiation and sunlight. *J. Invertebr. Pathol. 33*, 358-363.

Broome, J. R., Sikorowski, P. P., and Neel, W. W. (1974). Effect of sunlight on the activity of nuclear polyhedrosis virus from *Malacosoma disstria*. *J. Econ. Entomol. 67*, 135-136.

Bull, D. L., Ridgway, R. L., House, V. S., and Pryor, N. W. (1976). Improved formulations of the *Heliothis* nuclear polyhedrosis virus. *J. Econ. Entomol. 69*, 731-736.

Bullock, H. R. (1967). Persistence of *Heliothis* nuclear-polyhedrosis virus on cotton foliage. *J. Invertebr. Pathol. 9*, 434-436.

Bullock, H. R., Hollingsworth, J. P., and Hartsack, A. W., Jr. (1970). Virulence of *Heliothis* nuclear polyhedrosis virus exposed to monochromatic ultraviolet irradiation. *J. Invertebr. Pathol. 16*, 419-422.

Burgerjon, A., and Grison, P. (1965). Adhesiveness of preparations of *Smithiavirus pityocampae* Vago on pine foliage. *J. Invertebr. Pathol. 7*, 281-284.

Cantwell, G. E. (1967). Inactivation of biological insecticides by irradiation. *J. Invertebr. Pathol. 9*, 138-140.

Carter, J. B. (1975). The effect of temperature upon *Tipula* irridescent virus infection in *Tipula oleracea*. *J. Invertebr. Pathol. 25*, 115-124.

Chambers, D. L. (1968). Effect of ionic concentration on the infectivity of a virus of the citrus red mite, *Panonychus citri*. *J. Invertebr. Pathol. 10*, 245-251.

Chapman, H. C. (1978). Insect pathogens in pest management systems for aquatic habitats. *In* "Microbial Control of Insect Pests: Future Strategies in Pest Management Systems" (G. E. Allen, C. M. Ignoffo, and R. P. Jaques, eds.), pp. 207-210. Proc. NSF-USDA-Univ. Florida Workshop, Jan., 1978. 290 pp.

Chauthani, A. R., Murphy, D., Claussen, D., and Rehnborg, C. S. (1968). The effect of human gastric juice on the pathogenicity of *Heliothis zea* nuclear-polyhedrosis virus. *J. Invertebr. Pathol. 12*, 145-147.

Chu, W. H., and Jaques, R. P. (1981). Factors affecting infectivity of *Vairimorpha necatrix* (Microsporidia: Nosematidae) in *Trichoplusia ni* (Lepidoptera: Noctuidae). *Can. Entomol.* *113*, 93-102.

Couch, T. L., and Ignoffo, C. M. (1981). Formulation of insect pathogens. *In* "Microbial Control of Pests and Plant Diseases 1970-1980." (H. D. Burges, ed.), pp. 621-635. Academic Press, New York.

Crawford, A. M., and Kalmakoff, J. (1977). A host-virus interaction in a pasture habitat. *Wiseana* spp. and its Baculoviruses. *J. Invertebr. Pathol.* *29*, 81-87.

Creighton, C. S., McFadden, T. L., and Bell, J. V. (1970). Pathogens and chemicals tested against caterpillars on cabbage. *Prod. Res. Rept.* *114*, 10 pp.

Cunningham, J. C. (1970). The effect of storage on the nuclear-polyhedrosis virus of the eastern hemlock looper, *Lambdina fiscellaria fiscellaria* (Lepidoptera: Geometridae). *J. Invertebr. Pathol.* *16*, 352-356.

Cunningham, J. C. (1978). The use of pathogens for spruce budworm control. *In* "Microbial Control of Insect Pests: Future Strategies in Pest Management Systems" (G. E. Allen, C. M. Ignoffo, and R. P. Jaques, eds.), pp. 249-260. Proc. NSF-USDA-U of Florida Workshop, Jan., 1978. 290 pp.

Cunningham, J. C., and Entwistle, P. F. (1981). Control of sawflies by Baculovirus. *In* "Microbial Control of Pests and Plant Diseases 1970-1980" (H. D. Burges, ed.), pp. 379-407. Academic Press, New York.

Cunningham, J. C., Kaupp, W. J., and McPhee, J. R. (1975). Aerial application of nuclear polyhedrosis virus against spruce budworm in Manitoulin Island, Ontario, in 1974, and a survey of the impact of the virus in 1975. *Can. Forest Serv. Inform. Rept.* lP-X-8. Sept., 1975. 30 pp.

David, W. A. L. (1965). The granulosis virus of *Pieris brassicae* L. in relation to natural limitation and biological control. *Ann. Appl. Biol.* *56*, 331-334.

David, W. A. L. (1969). The effect of ultraviolet radiation of known wavelengths on a granulosis virus of *Pieris brassicae*. *J. Invertebr. Pathol.* *14*, 336-342.

David, W. A. L., and Gardiner, B. O. C. (1966). Persistence of a granulosis virus of *Pieris brassicae* on cabbage leaves. *J. Invertebr. Pathol.* *8*, 180-183.

David, W. A. L., and Gardiner, B. O. C. (1967a). The persistence of a granulosis virus of *Pieris brassicae* in soil and in sand. *J. Invertebr. Pathol.* *9*, 342-347.

David, W. A. L., and Gardiner, B. O. C. (1967b). The effect of heat, cold and prolonged storage on a granulosis virus of *Pieris brassicae*. *J. Invertebr. Pathol.* *9*, 555-562.

David, W. A. L., Gardiner, B. O. C., and Woolner, M. (1968). The effects of sunlight on a purified granulosis virus of *Pieris brassicae* applied to cabbage leaves. *J. Invertebr. Pathol.* *11*, 496-501.

David, W. A. L., Ellaby, S. J., and Taylor, G. (1969). Formaldehyde as an antiviral agent against a granulosis virus of *Pieris brassicae*. *J. Invertebr. Pathol. 14*, 96-101.

David, W. A. L., Ellaby, S. J., and Gardiner, B. O. C. (1971a). Bioassaying an insect virus on leaves. 1. The influence of certain factors associated with the virus application. *J. Invertebr. Pathol. 17*, 158-163.

David, W. A. L., Ellaby, S. J., and Taylor, G. (1971b). The stabilizing effect of insect hemolymph on a granulosis virus held in darkness as dry films in intact capsules. *J. Invertebr. Pathol. 17*, 404-409.

David, W. A. L., Ellaby, S. J., and Taylor, G. (1971c). The stability of a purified granulosis virus of the European cabbageworm, *Pieris brassicae*, in dry deposits of intact capsules. *J. Invertebr. Pathol. 17*, 228-233.

David, W. A. L., Ellaby, S. J., and Taylor, G. (1972). The fumigant action of formaldehyde incorporated in a semisynthetic diet on the granulosis virus of *Pieris brassicae* and its evaporation from the diet. *J. Invertebr. Pathol. 19*, 76-82.

Doane, C. C. (1969). Trans-ovum transmission of nuclear-polyhedrosis virus in gypsy moth and the inducement of virus susceptibility. *J. Invertebr. Pathol. 14*, 199-210.

Doane, C. C. (1970). Primary pathogens and their role in the development of an epizootic in the gypsy moth. *J. Invertebr. Pathol. 15*, 21-23.

Doane, C. C. (1975). Infectious sources of nuclear polyhedrosis virus persisting in natural habitats of the gypsy moth. *Environ. Entomol. 4*, 392-394.

Dougherty, E. M., Cantwell, G. E., and Kuchinski, M. (1982). Biological control of the greater wax moth utilizing *in vivo*- and *in vitro*-propagated Baculovirus. *J. Econ. Entomol. 75*, 675-679.

Dulmage, H., and Burgerjon, A. (1977). Industrial and international standardization of microbial pesticides-II. Insect viruses. *Entomophaga 22*, 131-139.

Entwistle, P. F., Adams, P. H. W., and Evans, H. F. (1977a). Epizootiology of a nuclear-polyhedrosis virus in European spruce sawfly *Gilpinia hercyniae*: The status of birds as dispersal agents of the virus during the larval season. *J. Invertebr. Pathol. 29*, 354-360.

Entwistle, P. F., Adams, P. H. W., and Evans, H. F. (1977b). Epizootiology of a nuclear-polyhedrosis virus in European spruce sawfly, *Gilpinia hercyniae*: Birds as dispersal agents of the virus during winter. *J. Invertebr. Pathol. 30*, 15-19.

Entwistle, P. F., Adams, P. H. W., and Evans, H. F. (1978). Epizootiology of a nuclear polyhedrosis virus in European spruce sawfly *Gilpinia hercyniae*: The rate of passage of

infective virus through the gut of birds during cage tests. *J. Invertebr. Pathol. 31*, 307-312.
Etzel, L. K., and Falcon, L. A. (1976). Studies of transovum and transstadial transmission of a granulosis virus of the codling moth. *J. Invertebr. Pathol. 27*, 13-26.
Evans, H. F., Bishop, J. M., and Page, E. A. (1980). Methods for the quantitative assessment of nuclear-polyhedrosis virus in soil. *J. Invertebr. Pathol. 35*, 1-8.
Falcon, L. A. (1971). Microbial control as a tool in integrated control programs. *In* "Biological Control" (C. B. Huffaker, ed.), pp. 346-364. Plenum, New York.
Ferron, P., and Hurpin, B. (1974). Effets de la contamination simultanee ou successive par *Beauveria tenella* et par *Entomopoxvirus melolonthae* des larves de *Melolontha melolontha* (Col. Scarabaeidae). *Ann. Soc. Entomol. Fr.* [*N.S.*] *10*(3), 771-779.
Franz, J., Krieg, A., and Langenbuch, R. (1955). Untersuchungen uber den Einfluss der Passage durch den Darm von Raubinsekten and Vogeln auf die Infektiositat Insektenpathogener Viren. *Z. Pflanzenkr. Pflanzenpathol. Pflanzenschutz 62*, 721-726.
Fuxa, J. R. (1979). Interactions of the microsporidium *Vairimorpha necatrix* with a bacterium, virus, and fungus in *Heliothis zea*. *J. Invertebr. Pathol. 33*, 316-323.
Fuxa, J. R. (1982). Prevalence of viral infections in populations of fall armyworm, *Spodoptera frugiperda*, in southeastern Louisiana. *Environ. Entomol. 11*, 239-242.
Fuxa, J. R., and Brooks, W. M. (1978). Persistence of spores of *Vairimorpha necatrix* on tobacco, cotton, and soybean foliage. *J. Econ. Entomol. 71*, 169-172.
Gilmore, J. E., and Munger, F. (1963). Stability and transmissability of a virus-like pathogen of the citrus red mite. *J. Insect Pathol. 5*, 141-151.
Gudauskas, R. T., and Canerday, D. (1968). The effect of heat, buffer salt and H-ion concentration, and ultraviolet light on the infectivity of *Heliothis* and *Trichoplusia* nuclear-polyhedrosis viruses. *J. Invertebr. Pathol. 12*, 405-411.
Hamm, J. J., and Young, J. R. (1974). Mode of transmission of nuclear-polyhedrosis virus to progeny of adult *Heliothis zea*. *J. Invertebr. Pathol. 24*, 70-81.
Harkrider, J. R., and Hall, I. M. (1978). The dynamics of an entomopoxvirus in a field population of larval midges of the *Chironomus decorus* complex. *Environ. Entomol. 7*, 858-862.
Harper, J. D. (1976). Cross-infectivity of six Plusiine nuclear polyhedrosis virus isolates to Plusiine hosts. *J. Invertebr. Pathol. 27*, 275-277.
Harpaz, I., and Raccah, B. (1978). Nucleopolyhedrosis virus (NPV) of the Egyptian cottonworm, *Spodoptera littoralis*

(Lepidoptera, Noctuidae): Temperature and pH relations, host range, and synergism. *J. Invertebr. Pathol. 32*, 368-372.

Hayashi, Y., Kawarabata, T., and Bird, F. T. (1970). Isolation of a cytoplasmic-polyhedrosis virus of the silkworm, *Bombyx mori*. *J. Invertebr. Pathol. 16*, 378-384.

Heimpel, A. M., Thomas, E. D., Adams, J. R., and Smith, L. J. (1973). The presence of nuclear polyhedrosis viruses of *Trichoplusia ni* on cabbage from the market shelf. *Environ. Entomol. 2*, 72-75.

Hostetter, D. L. (1971). A virulent nuclear polyhedrosis virus of the cabbage looper *Trichoplusia ni*, recovered from the abdomens of sarcophagid flies. *J. Invertebr. Pathol. 17*, 130-131.

Hostetter, D. L., and Biever, K. D. (1970). The recovery of virulent nuclear-polyhedrosis virus of cabbage looper, *Trichoplusia ni*, from feces of birds. *J. Invertebr. Pathol. 15*, 173-176.

Hostetter, D. L., Pinnell, R. E., Greer, P. A., and Ignoffo, C. M. (1973). A granulosis virus of *Pieris rapae* as a microbial control agent on cabbage in Missouri. *Environ. Entomol. 2*, 1109-1112.

Hostetter, D. L., Biever, K. D., Heimpel, A. M., and Ignoffo, C. M. (1979). Efficacy of the nuclear polyhedrosis virus of the alfalfa looper against cabbage looper larvae on cabbage in Missouri. *J. Econ. Entomol. 72*, 371-373.

Huber, J., and Dickler, E. (1977). Codling moth granulosis virus: Its efficiency in the field in comparison with organophosphorus insecticides. *J. Econ. Entomol. 70*, 557-561.

Hukahara, T. (1972). Demonstration of polyhedra and capsules in soil with scanning electron microscope. *J. Invertebr. Pathol. 20*, 375-376.

Hukuhara, T. (1973). Further studies on the distribution of a nuclear-polyhedrosis virus of the fall webworm, *Hyphantria cunea*, in soil. *J. Invertebr. Pathol. 22*, 345-350.

Hukahara, T., and Namura, H. (1971). Microscopic demonstration of polyhedra in soil. *J. Invertebr. Pathol. 18*, 162-164.

Hukahara, T., and Namura, H. (1972). Distribution of a nuclear-polyhedrosis virus of the fall webworm, *Hyphantria cunea*, in soil. *J. Invertebr. Pathol. 19*, 308-316.

Hukahara, T., and Wada, H. (1972). Adsorption of polyhedra of a cytoplasmic virus by soil particles. *J. Invertebr. Pathol. 20*, 309-316.

Hunter, D. K., and Hartsell, P. L. (1971). Influence of temperature on Indian-meal moth larvae infected with a granulosis virus. *J. Invertebr. Pathol. 17*, 347-349.

Hunter, D. K., Hoffman, D. F., and Collier, S. J. (1973). Pathogenicity of a nuclear polyhedrosis virus of the almond moth, *Cadra cautella*. *J. Invertebr. Pathol. 21*, 282-286.

Hurpin, B., and Robert, P. H. (1976). Conservation dans le sol de trois germes pathogenes pour les larves de *Melolontha melolontha* [Col.: Scarabaeidae]. *Entomophaga 21*, 73-80.

Hurpin, B., and Robert, P. H. (1977). Effets en population naturelle de *Melolontha melolontha* (Col.: Scarabaeidae) d'une introduction de *Rickettsiella melolonthae* et de *Entomopoxvirus melolonthae*. *Entomophaga 22*, 85-91.

Ignoffo, C. M. (1964). Production and virulence of a nuclear-polyhedrosis virus from larvae of *Trichoplusia ni* (Hübner) reared on a semisynthetic diet. *J. Insect Pathol. 6*, 318-326.

Ignoffo, C. M. (1968). Specificity of insect viruses. *Bull. Entomol. Soc. Amer. 14*, 265-276.

Ignoffo, C. M., and Batzer, O. F. (1971). Microencapsulation and ultraviolet protectants to increase sunlight stability of an insect virus. *J. Econ. Entomol. 64*, 850-853.

Ignoffo, C. M., and Garcia, C. (1966). The relation of pH to the activity of inclusion bodies of a *Heliothis* nuclear polyhedrosis. *J. Invertebr. Pathol. 8*, 426-427.

Ignoffo, C. M., and Garcia, C. (1968). Formalin inactivation of nuclear-polyhedrosis virus. *J. Invertebr. Pathol. 10*, 430-432.

Ignoffo, C. M., and Garcia, C. (1979). Susceptibility of larvae of the black cutworm to species of entomopathogenic bacteria, fungi, protozoa, and viruses. *J. Econ. Entomol. 72*, 767-769.

Ignoffo, C. M., and Montoya, E. L. (1966). The effects of chemical insecticides and insecticidal adjuvants of a *Heliothis* nuclear-polyhedrosis virus. *J. Invertebr. Pathol. 8*, 409-412.

Ignoffo, C. M., Chapman, A. J., and Martin, D. F. (1965). The nuclear-polyhedrosis virus of *Heliothis zea* (Boddie) and *Heliothis virescens* (Fabricius). III. Effectiveness of the virus against field populations of *Heliothis* on cotton, corn, and grain sorghum. *J. Invertebr. Pathol. 7*, 227-235.

Ignoffo, C. M., Batzer, O. F., Barker, W. M., and Ebert, A. G. (1971). Fate of *Heliothis* nucleopolyhedrosis virus following oral administration to rats. *Proc. 4th Intern. Colloq. Insect. Pathol., College Park, Maryland*, pp. 357-362.

Ignoffo, C. M., Bradley, J. R., Jr., Gilliland, F. R., Jr., Harris, F. A., Falcon, L. A., Larson, L. V., McGarr, R. L., Sikorowski, P. W., Watson, T. F., and Yearian, W. C. (1972). Field studies on stability of the *Heliothis* nucleopolyhedrosis virus at various sites throughout the cotton belt. *Environ. Entomol. 1*, 388-390.

Ignoffo, C. M., Parker, F. D., Boening, O. P., Pinnell, R. E., and Hostetter, D. L. (1973). Field stability of the *Heliothis* nucleopolyhedrosis virus on corn silks. *Environ. Entomol. 2*, 302-303.

Ignoffo, C. M., Hostetter, D. L., and Pinnell, R. E. (1974a). Stability of *Bacillus thuringiensis* and *Baculovirus heliothis* on soybean foliage. *Environ. Entomol. 3*, 117-119.

Ignoffo, C. M., Hostetter, D. L., and Shapiro, M. (1974b). Efficacy of insect viruses propagated *in vivo* and *in vitro*. *J. Invertebr. Pathol. 24*, 184-187.

Ignoffo, C. M., Hostetter, D. L., and Smith, D. B. (1976a). Gustatory stimulant, sunlight protectant, evaporation retardant: three characteristics of a microbial insecticidal adjuvant. *J. Econ. Entomol. 69*, 207-210.

Ignoffo, C. M., Yearian, W. C., Young, S. Y., Hostetter, D. L., and Bull, D. L. (1976b). Laboratory and field persistence of new commercial formulations of the *Heliothis* nucleopolyhedrosis virus, *Baculovirus heliothis*. *J. Econ. Entomol. 69*, 233-236.

Ignoffo, C. M., Hostetter, D. L., Sikorowski, P. P., Sutter, G., and Brooks, W. M. (1977). Inactivation of representative species of entomopathogenic viruses, a bacterium, fungus, and protozoan by an ultraviolet light source. *Environ. Entomol. 6*, 411-415.

Ignoffo, C. M., Hostetter, D. L., Garcia, C., Biever, K. D., and Thomas, G. D. (1978). Autodissemination of entomopathogens: Release of living virus-infected larvae. *In* "Microbial Control of Insect Pests" (G. E. Allen, C. M. Ignoffo, and R. P. Jaques, eds.). pp. 69-71. Proc. NSF-USDA-U. of Florida Workshop, Jan., 1978. 290 pp.

Ignoffo, C. M., Garcia, C., Hostetter, D. L., and Pinnell, R. E. (1980). Transplanting: A method of introducing an insect virus into an ecosystem. *Environ. Entomol. 9*, 153-154.

Ilnytsky, S., McPhee, J. R., and Cunningham, J. C. (1977). Comparison of field propagated nuclear polyhedrosis virus from Douglas fir tussock moth with laboratory produced virus. *Can. Forest Serv. Bio-Mo. Res. Notes 33*, 5-6.

Irabagon, T. A., and Brooks, W. M. (1974). Interaction of *Campoletis sonorensis* and a nuclear polyhedrosis virus in larvae of *Heliothis virescens*. *J. Econ. Entomol. 67*, 229-231.

Jaques, R. P. (1964). The persistence of a nuclear-polyhedrosis virus in soil. *J. Insect Pathol. 6*, 251-254.

Jaques, R. P. (1967a). The persistence of a nuclear-polyhedrosis virus in the habitat of the host insect, *Trichoplusia ni*. I. Polyhedra deposited on foliage. *Can. Entomol. 99*, 785-794.

Jaques, R. P. (1967b). The persistence of a nuclear polyhedrosis virus in the habitat of the host insect, *Trichoplusia ni*. II. Polyhedra in soil. *Can. Entomol. 99*, 820-829.

Jaques, R. P. (1968). The inactivation of the nuclear-polyhedrosis virus of *Trichoplusia ni* by gamma and ultraviolet radiation. *Can. J. Microbiol. 14*, 1161-1163.

Jaques, R. P. (1969). Leaching of the nuclear-polyhedrosis virus of *Trichoplusia ni* from soil. *J. Invertebr. Pathol.* *13*, 256-263.
Jaques, R. P. (1970a). Application of viruses to soil and foliage for control of the cabbage looper and imported cabbageworm. *J. Invertebr. Pathol.* *15*, 328-340.
Jaques, R. P. (1970b). Natural occurrence of viruses of the cabbage looper in field plots. *Can. Entomol.* *102*, 36-41.
Jaques, R. P. (1971a). Control of cabbage insects by viruses. *Proc. Entomol. Soc. Ontario* *101*, 28-34.
Jaques, R. P. (1971b). Tests on protectants for foliar deposits of a polyhedrosis virus. *J. Invertebr. Pathol.* *17*, 9-16.
Jaques, R. P. (1972a). Control of the cabbage looper and the imported cabbageworm by viruses and bacteria. *J. Econ. Entomol.* *65*, 757-760.
Jaques, R. P. (1972b). The inactivation of foliar deposits of viruses of *Trichoplusia ni* (Lepidoptera: Noctuidae) and *Pieris rapae* (Lepidoptera: Pieridae) and tests on protectant additives. *Can. Entomol.* *104*, 1985-1994.
Jaques, R. P. (1973a). Tests on microbial and chemical insecticides for control of *Trichoplusia ni* and *Pieris rapae* on cabbage. *Can. Entomol.* *105*, 21-27.
Jaques, R. P. (1973b). Methods and effectiveness of distribution of microbial insecticides. *Ann. N.Y. Acad. Sci.* *217*, 109-119.
Jaques, R. P. (1974a). Occurrence and accumulation of viruses of *Trichoplusia ni* in treated field plots. *J. Invertebr. Pathol.* *23*, 140-152.
Jaques, R. P. (1974b). Occurrence and accumulation of the granulosis virus of *Pieris rapae* in treated field plots. *J. Invertebr. Pathol.* *23*, 351-359.
Jaques, R. P. (1975). Persistence, accumulation, and denaturation of nuclear polyhedrosis and granulosis viruses. *In* "Baculoviruses for Insect Pest Control: Safety Considerations" (M. Summers, R. Engler, L. A. Falcon, and P. V. Vail, eds.), pp. 90-99. Amer. Soc. Microbiol., Washington, D.C.
Jaques, R. P. (1977a). Field efficacy of viruses infectious to the cabbage looper and imported cabbageworm on late cabbage. *J. Econ. Entomol.* *70*, 111-118.
Jaques, R. P. (1977b). Stability of entomopathogenic viruses. *Misc. Publ. Entomol. Soc. Amer.* *10*(3), 99-117.
Jaques, R. P. (1978). Manipulation of the environment to increase effectiveness of microbial agents. *In* "Microbial Control of Insect Pests: Future Strategies in Pest Management Systems" (G. E. Allen, C. M. Ignoffo, and R. P. Jaques, eds.), pp. 72-85. Proc. NSF-USDA-U. of Florida Workshop, Jan., 1978. 290 pp.
Jaques, R. P., and Harcourt, D. G. (1971). Viruses of *Trichoplusia ni* (Lepidoptera: Noctuidae) and *Pieris rapae*

(Lepidoptera: Pieridae) in soil in fields of crucifers in southern Ontario. *Can. Entomol. 103*, 1285-1290.

Jaques, R. P., and Huston, F. (1969). Tests on microbial decomposition of polyhedra of the nuclear polyhedrosis virus of the cabbage looper, *Trichoplusia ni*. *J. Invertebr. Pathol. 14*, 289-290.

Jaques, R. P., and Laing, D. R. (1976). The effect of sauerkraut fermentation and processing on activity of *Bacillus thuringiensis* and viruses of *Trichoplusia ni* and *Pieris rapae*. *Environ. Entomol. 5*, 302-306.

Jaques, R. P., and Laing, D. R. (1978). Efficacy of mixtures of *Bacillus thuringiensis*, viruses, and chlordimeform against insects on cabbage. *Can. Entomol. 110*, 443-448.

Jaques, R. P., and Morris, O. N. (1981). Compatibility with other methods of pest control and with different crops. *In* "Microbial Control of Pests and Plant Diseases 1970-1980" (H. D. Burges, ed.), pp. 695-715. Academic Press, New York.

Jaques, R. P., and Stultz, H. T. (1966). The influence of a virus disease and parasites on *Spilonota ocellana* in apple orchards. *Can. Entomol. 98*, 1035-1045.

Jaques, R. P., MacLellan, C. R., Sanford, K. H., Proverbs, M. D., and Hagley, E. A. C. (1977). Preliminary orchard tests on control of codling moth larvae by a granulosis virus. *Can. Entomol. 109*, 1079-1081.

Jaques, R. P., Laing, J. E., MacLellan, C. R., Proverbs, M. D., Sanford, K. H., and Trottier, R. (1981). Apple orchard tests on the efficacy of the granulosis virus of the codling moth, *Laspeyresia pomonella* (Lepidoptera: Olethreutidae). *Entomophaga 26*, 11-118.

Johnson, D. R. (1982). Suppression of *Heliothis* spp. on cotton by using *Bacillus thuringiensis*, *Baculovirus heliothis*, and two feeding adjuvants. *J. Econ. Entomol. 75*, 207-210.

Johnson, T. B., and Lewis, L. C. (1982). Evaluation of *Rachiplusia ou* and *Autographa californica* nuclear polyhedrosis viruses in suppressing black cutworm damage to seedling corn in greenhouse and field. *J. Econ. Entomol. 75*, 401-404.

Katagiri, K., and Iwata, Z. (1976). Control of *Dendrolimus spectabilis* with a mixture of cytoplasmic polyhedrosis virus and *Bacillus thuringiensis*. *Appl. Entomol. Zool. 11*, 363-364.

Kaya, H. K., and Tanada, Y. (1972). Response of *Apanteles militaris* to a toxin produced in a granulosis-virus-infected host. *J. Invertebr. Pathol. 19*, 1-17.

Kaya, H. K., and Tanada, Y. (1973). Hemolymph factor in armyworm larvae infected with a nuclear polyhedrosis virus toxic to *Apanteles militaris*. *J. Invertebr. Pathol. 21*, 211-214.

Kaupp, W. J. (1980). A simple method of assessing the quantities of nuclear polyhedrosis virus entering the soil from diseased populations of European pine sawfly. *Can. Forest Serv. Bi-Mo. Res. Notes, Envir. Can. 36*, 31.

Kawarabata, T., Funakoshi, M., and Aratake, Y. (1980). Purification and properties of the *Bombyx mori* nuclear polyhedrosis virus liberated from polyhedra by dissolution with silkworm gut juice. *J. Invertebr. Pathol. 35*, 34-42.

Kinsinger, R. A., and McGaughey, W. H. (1976). Stability of *Bacillus thuringiensis* and a granulosis virus of *Plodia interpunctella* on stored wheat. *J. Econ. Entomol. 69*, 149-154.

Kobayashi, M., Inagaki, S., and Kawase, S. (1981). Effect of high temperature on the development of nuclear polyhedrosis virus in the silkworm, *Bombyx mori*. *J. Invertebr. Pathol. 38*, 386-394.

Komarek, J., and Breindl, V. (1924). Die Wipfelkrankheit der Nonne und der Erreger derselben. *Z. Angew. Entomol. 10*, 99-162.

Krieg, A., Groner, A., Huber, J., and Matter, M. (1980). Uber die Wirkung von mittel-und langwellen ultravioletter Strahlen (UV-B and UV-A) auf insektenpathogene Bakterien und Viren und deren Beeinflussung durch UV-Schutzstoffe. *Nachrichtenbl. Deut. Pflanzenschutzdienst (Braunschweig) 32*, 100-105.

Krieg, A., Gröner, A., Huber, J., and Zimmerman, G. (1981). Inaktivierung von verschidenen Insektenpathogenen durch ultraviolette Strahlen. *Z. Pflanzenkr. Pflanzenschutz 88*, 38-48.

Ladd, T. L., and McCabe, P. J. (1967). Persistence of spores of *Bacillus popilliae*, the causal organism of Type A milky disease of Japanese beetle larvae, in New Jersey soils. *J. Econ. Entomol. 60*, 493-495.

Lautenschlager, R. A., and Podgwaite, J. D. (1979). Passage of nucleopolyhedrosis virus by avian and mammalian predators of the gypsy moth, *Lymantria dispar* L. *Environ. Entomol. 8*, 210-214.

Lautenschlager, R. A., Rothenbacher, H., and Podgwaite, J. D. (1978). Response of small mammals to aerial applications of the nucleopolyhedrosis virus of the gypsy moth, *Lymantria dispar*. *Environ. Entomol. 7*, 676-684.

Lautenschlager, R. A., Rothenbacher, H., and Podgwaite, J. D. (1979). Response of birds to aerial applications of nucleopolyhedrosis virus of the gypsy moth. *Environ. Entomol. 8*, 760-764.

Lautenschlager, R. A., Podgwaite, J. D., and Watson, D. E. (1980). Natural occurrence of the nucleopolyhedrosis virus of the gypsy moth, *Lymantria dispar* (Lep.: Lymantriidae) in wild birds and mammals. *Entomophaga 25*, 261-267.

Levin, D. B., Laing, J. E., and Jaques, R. P. (1979). Transmission of granulosis virus by *Apanteles glomeratus* to its host *Pieris rapae*. *J. Invertebr. Pathol.* *34*, 317-318.

Levin, D. B., Laing, J. E., and Jaques, R. P. (1981). Interactions between *Apanteles glomeratus* (L.) (Hymenoptera: Braconidae) and granulosis virus in *Pieris rapae* (L.) (Lepidoptera: Pieridae). *Environ. Entomol.* *10*, 65-68.

Levin, D. B., Laing, J. E., Jaques, R. P., and Corrigan, J. E. (1983). Transmission of the granulosis virus of *Pieris rapae* (Lepidoptera: Pieridae) by the parasitoid *Apanteles glomeratus* (Hymenoptera: Braconidae). *Environ. Entomol.* *12*, 166-170.

Lewis, F. B. (1981). Control of the gypsy moth by a Baculovirus. *In* "Microbial Control of Pests and Plant Disease 1970-1980" (H. D. Burgess, ed.), pp. 363-377. Academic Press, New York.

Lingg, A. J., and Donaldson, M. D. (1981). Biotic and abiotic factors affecting stability of *Beauveria bassiana* conidia in soil. *J. Invertebr. Pathol.* *38*, 191-200.

Luttrell, R. G., Yearian, W. C., and Young, S. Y. (1979). Laboratory and field studies on the efficacy of selected chemical insecticide-Elcar (*Baculovirus heliothis*) combinations against *Heliothis* spp. *J. Econ. Entomol.* *72*, 57-60.

Luttrell, R. G., Yearian, W. C., and Young, S. Y. (1983). Effect of spray adjuvants on *Heliothis zea* (Lepidoptera: Noctuidae) nuclear polyhedrosis virus efficacy. *J. Econ. Entomol.* *76*, 162-167.

MacCollom, G. B., and Reed, E. M. (1971). A nuclear-polyhedrosis virus of the light brown apple moth. *Epiphyas postvittana*. *J. Invertebr. Pathol.* *18*, 337-343.

McGaughey, W. H. (1975). Compatibility of *Bacillus thuringiensis* and granulosis virus treatments of stored grain with four grain fumigants. *J. Invertebr. Pathol.* *26*, 247-250.

McKinley, D. J., Brown, D. A., Payne, C. C., and Harrap, K. A. (1981). Cross-infectivity and activation studies with four baculoviruses. *Entomophaga* *26*, 79-80.

McLaughlin, R. E., Andrews, G., and Bell, M. R. (1971). Field tests for control of *Heliothis* spp. with a nuclear polyhedrosis virus included in a boll weevil bait. *J. Invertebr. Pathol.* *18*, 304-305.

McLeod, P. J., Yearian, W. C., and Young, S. Y. (1977). Inactivation of *Baculovirus heliothis* by ultraviolet irradiation, dew, and temperature. *J. Invertebr. Pathol.* *30*, 237-241.

McLeod, P. J., Young, S. Y., and Yearian, W. C. (1982). Application of a baculovirus of *Pseudoplusia includens* to soybean: Efficacy and seasonal persistence. *Environ. Entomol.* *11*, 412-416.

McVay, J. R., Gudauskas, R. T., and Harper, J. D. (1977). Effects of *Bacillus thuringiensis* nuclear-polyhedrosis virus mixtures on *Trichoplusia ni* larvae. *J. Invertebr. Pathol. 29*, 367-372.

Magnoler, A. (1974). Field dissemination of a nucleopolyhedrosis virus against the gypsy moth, *Lymantria dispar* L. *Z. Pflanzenkr. Pflanzenschutz 81*, 497-511.

Martignoni, M. E., and Iwai, P. J. (1977). Thermal inactivation characteristics of two strains of nucleopolyhedrosis virus (Baculovirus subgroup A) pathogenic for *Orgyia pseudotsugata*. *J. Invertebr. Pathol. 30*, 255-262.

Martignoni, M. E., and Iwai, P. J. (1981). A catalogue of viral diseases of insects, mites and ticks. *In* "Microbial Control of Pests and Plant Diseases 1970-1980" (H. D. Burges, ed.), pp. 895-911. Academic Press, New York.

Mohamed, M. A., Coppel, H. C., and Podgwaite, J. C. (1982). Persistence in soil and on foliage of nucleopolyhedrosis virus of the European pine sawfly, *Neodiprion sertifer* (Hymenoptera: Diprionidae). *Environ. Entomol. 11*, 1116-1118.

Morris, O. N. (1971). The effect of sunlight, ultraviolet and gamma radiations and temperature on the infectivity of nuclear polyhedrosis virus. *J. Invertebr. Pathol. 18*, 292-294.

Morris, O. N. (1972). Inhibitory effects of foliage extracts of some forest trees on commercial *Bacillus thuringiensis*. *Can. Entomol. 104*, 1357-1361.

Morris, O. N. (1977). Long-term effects of aerial applications of virus-fenitrothion combinations against the spruce budworm, *Choristoneura fumiferana* (Lepidoptera: Tortricidae). *Can. Entomol. 109*, 9-14.

Morris, O. N. (1980). Entomopathogenic viruses: Strategies for use in forest insect pest management. *Can. Entomol. 112*, 573-584.

Morris, O. N., and Moore, A. (1975). Studies on the protection of insect pathogens from sunlight inactivation. II. Preliminary field trials. *Chem. Cont. Res. Inst. Envir. Can. Rept.*, CC-X-113.

Morris, O. N., Armstrong, J. A., Howse, G. M., and Cunningham, J. C. (1974). A 2-year study of virus-chemical insecticide combinations in the integrated control of the spruce budworm (Torticidae: Lepidoptera). *Can. Entomol. 106*, 813-824.

Neelgund, Y. F., and Mathad, S. B. (1978). Transmission of nuclear polyhedrosis virus in laboratory populations of the armyworm, *Mythimma (Pseudaletia) separata*. *J. Invertebr. Pathol. 31*, 143-147.

Neilson, M. M., and Elgee, D. E. (1960). The effect of storage on the virulence of a polyhedrosis virus. *J. Insect Pathol. 2*, 165-171.

Pavan, O. H., Boucias, D. G., and Pendland, J. C. (1981). The effects of serial passage of a nucleopolyhedrosis virus through an alternate host system. *Entomophaga 26*, 99-108.

Pawar, V. W., and Ramakrishnan, N. (1971). Investigations on the nuclear-polyhedrosis of *Prodenia litura* Fabricius. II. Effect of surface disinfectants, temperature and alkalies on the virus. *Indian J. Entomol. 33*, 428-432.

Payne, C. C. (1981). The susceptibility of the pea moth, *Cydia nigricana*, to infection by the granulosis virus of the codling moth, *Cydia pomonella*. *J. Invertebr. Pathol. 38*, 71-77.

Pfrimmer, T. R. (1979). *Heliothis* spp.: Control on cotton with pyrethroids, carbamates, organophosphates, and biological insecticides. *J. Econ. Entomol. 72*, 593-598.

Podgwaite, J. C., Shields, K. S., Zerillo, R. T., and Bruen, R. B. (1979). Environmental persistence of the nucleopolyhedrosis virus of the gypsy moth, *Lymantria dispar*. *Environ. Entomol. 8*, 528-536.

Prell, H. (1926). Die Polyederkrankheiten der Insekten. *3rd Intern. Kongr. Entomol. Verh. Zurich, 1925* Vol. 2, 145-168.

Raimo, B., Reardon, R. C., and Podgwaite, J. D. (1977). Vectoring gypsy moth nuclear polyhedrosis virus by *Apanteles melanoscelus*. *Entomophaga 22*, 207-251.

Reardon, R., and Podgwaite, J. D. (1976). Disease-parasitoid relationships in natural populations of *Lymantria dispar* [Lep.: Lymantriidae] in the northern United States. *Entomophaga 21*, 333-341.

Reed, D. K. (1974). Effects of temperature on virus-host relationships and on activity of the noninclusion virus of citrus red mites, *Panonychus citri*. *J. Invertebr. Pathol. 24*, 218-223.

Reed, D. K., Hendrickson, R. M., Rich, J. R., and Shaw, J. G. (1973). Laboratory evaluation of extenders for the noninclusion virus of the citrus red mite. *J. Invertebr. Pathol. 22*, 182-185.

Ritter, K. S., and Tanada, Y. (1978). Interference between two nuclear polyhedrosis viruses of the armyworm, *Pseudaletia unipuncta*. *Entomophaga 23*, 349-359.

Roberts, D. W., and Campbell, A. S. (1977). Stability of entomopathogenic fungi. *In* "Environmental Stability of Microbial Insecticides" (D. L. Hostetter and C. M. Ignoffo, eds.). *Misc. Publ. Entomol. Soc. Amer. 10*(3), 19-76.

Roome, R. E., and Daoust, R. A. (1976). Survival of the nuclear polyhedrosis virus of *Heliothis armigera* on crops and in soil in Botswana. *J. Invertebr. Pathol. 27*, 7-12.

Saleh, S. M., Harris, R. F., and Allen, O. N. (1970). Recovery of *Bacillus thuringiensis* var. *thuringiensis* from field soils. *J. Invertebr. Pathol. 15*, 55-59.

Sears, M. K., Jaques, R. P., and Laing, J. E. (1983). Utilization of action thresholds for microbial and chemical control of lepidopterous pests (Lepidoptera: Noctuidae, Pieridae) on cabbage. *J. Econ. Entomol. 76*, 368-374.

Shapiro, M., and Bell, R. A. (1982). Enhanced effectiveness of *Lymantria dispar* nucleopolyhedrosis virus formulated with boric acid. *Ann. Entomol. Soc. Amer. 75*, 346-349.

Shapiro, M., and Ignoffo, C. M. (1969). Nuclear polyhedrosis of *Heliothis*: Stability and relative infectivity of virions. *J. Invertebr. Pathol. 14*, 130-134.

Shapiro, M., Martignoni, M. E., Cunningham, J. C., and Goodwin, R. H. (1982). Potential use of the saltmarsh caterpillar as a production host for nuclear polyhedrosis viruses. *J. Econ. Entomol. 75*, 69-71.

Shaw, J. G., Chambers, D. L., and Tashiro, H. (1968). Introducing and establishing the noninclusion virus of the citrus red mite in citrus groves. *J. Econ. Entomol. 61*, 1352-1355.

Shaw, J. G., Rich, J. E., and Reed, D. K. (1972). Effect of prolonged storage on a noninclusion virus of the citrus red mite. *J. Econ. Entomol. 65*, 1512.

Sikorowski, P. P., Andrews, G. L., and Broome, J. R. (1973). Trans-ovum transmission of a cytoplasmic polyhedrosis virus of *Heliothis virescens*. *J. Invertebr. Pathol. 21*, 41-45.

Smirnoff, W. A. (1967). Effect of gamma radiation on the larvae and the nuclear-polyhedrosis virus of the eastern tent caterpillar, *Malacosma americanum*. *J. Invertebr. Pathol. 9*, 264-266.

Smirnoff, W. A. (1968). Effects of volatile substances released by foliage of various plants on the entomopathogenic *Bacillus cereus* group. *J. Invertebr. Pathol. 11*, 513-515.

Smirnoff, W. A. (1972). The effect of sunlight on the nuclear-polyhedrosis virus of *Neodiprion swainei* with measurement of solar energy received. *J. Invertebr. Pathol. 19*, 179-188.

Smirnoff, W. A., and MacLeod, C. F. (1964). Apparent lack of effects of orally introduced polyhedrosis virus on mice and of pathogenicity of rodent-passed virus for insects. *J. Insect Pathol. 6*, 537-538.

Smith, D. B., and Hostetter, D. L. (1982). Laboratory and field evaluations of pathogen-adjuvant treatments. *J. Econ. Entomol. 75*, 472-476.

Smith, D. B., Hostetter, D. L., and Ignoffo, C. M. (1978). Formulation and equipment effects on application of a viral *(Baculovirus heliothis)* insecticide. *J. Econ. Entomol. 71*, 814-817.

Smith, D. B., Hostetter, D. L., and Pinnell, R. E. (1980). Laboratory formulation comparisons for a bacterial *(Bacillus*

thuringiensis) and a viral *(Baculovirus heliothis)* insecticide. *J. Econ. Entomol. 73*, 18-21.
Smith, D. B., Hostetter, D. L., Pinnell, R. E., and Ignoffo, C. M. (1982). Laboratory studies of viral adjuvants: Formulation development. *J. Econ. Entomol. 75*, 16-20.
Stacey, A. L., Yearian, W. C., and Young, S. Y. (1977). Evaluation of *Baculovirus heliothis* with feeding stimulants for control of *Heliothis* larvae on cotton. *J. Econ. Entomol. 70*, 779-784.
Stairs, G. R. (1966). Transmission of virus in tent caterpillar populations. *Can. Entomol. 98*, 1110-1114.
Stairs, G. R. (1968). Inclusion-type insect viruses. *Current Topics Microbiol. Immunol. 42*, 1-23.
Stairs, G. R. (1980). Comparative infectivity of nonoccluded virions, polyhedra, and virions released from polyhedra for larvae of *Galleria mellonella. J. Invertebr. Pathol. 36*, 281-282.
Stairs, G. R., and Milligan, S. E. (1979). Effects of heat on nonoccluded nuclear polyhedrosis virus *(Baculovirus)* from *Galleria mellonella* larvae. *Environ. Entomol. 8*, 756-759.
Stairs, G. R., and Milligan, S. E. (1980). Quantitative effects of heat on *Baculoviruses* virions released from polyhedra (Nuclear-polyhedrosis virus-*Galleria mellonella*). *Environ. Entomol. 9*, 586-588.
Stairs, G. R., Fraser, T., and Fraser, M. (1981). Changes in growth and virulence of a nuclear polyhedrosis virus of *Choristoneura fumiferana* after passage in *Trichoplusia ni* and *Galleria mellonella. J. Invertebr. Pathol. 38*, 230-235.
Steinhaus, E. A. (1948). Polyhedrosis ("wilt disease") of the alfalfa caterpillar. *J. Econ. Entomol. 41*, 859-865.
Steinhaus, E. A. (1960). The duration of viability and infectivity of certain insect pathogens. *J. Insect Pathol. 2*, 225-229.
Stelzer, M. J., Neisses, J., and Thompson, C. G. (1975). Aerial application of a nucleopolyhedrosis virus and *Bacillus thuringiensis* against the Douglas-fir tussock moth. *J. Econ. Entomol. 68*, 269-272.
Stuermer, C. W., and Bullock, H. R. (1968). Thermal inactivation of *Heliothis* nuclear-polyhedrosis virus. *J. Invertebr. Pathol. 12*, 473-474.
Summers, M. D. (1971). Electron microscopic observations on granulosis virus entry, uncoating, and replication processes during infection of the midgut cells of *Trichoplusia ni. J. Ultrastr. Res. 35*, 606-625.
Tanada, Y. (1953). Description and characteristics of a granulosis virus of the imported cabbageworm. *Proc. Hawaiian Entomol. Soc. 15*, 235-260.

Tanada, Y. (1959). Synergism between two viruses of the armyworm, *Pseudaletia unipuncta* (Haworth) (Lepidoptera: Noctuidae). *J. Insect Pathol. 1*, 215-231.

Tanada, Y. (1971). Persistence of entomogenous viruses in the insect ecosystem. "Entomological Essays to Commemorate the Retirement of Professor K. Yasumatsu," pp. 367-379. Hokuryukan Publishing Co. Ltd., Tokyo.

Tanada, Y. (1973). Environmental factors external to the host. *Ann. N.Y. Acad. Sci. 217*, 120-130.

Tanada, Y. (1975). Persistence of pathogens in the aquatic environment. *In* "Impact of the Use of Microorganisms on the Aquatic Environment" (A. W. Bourquin, D. G. Ahearn, and S. P. Meyers, eds.), pp. 83-99. *Proc. E.P.A. Workshop, Gulf Breeze, Florida, Apr., 1974.* Publ. E.P.A. 660-3-75-001.

Tanada, Y., and Chang, G. Y. (1968). Resistance of the alfalfa caterpillar, *Colias eurytheme*, at high temperatures to a cytoplasmic-polyhedrosis virus and thermal inactivation point of the virus. *J. Invertebr. Pathol. 10*, 79-83.

Tanada, Y., and Hara, S. (1975). Enzyme synergistic for insect viruses. *Nature (London) 254*, 328-329.

Tanada, Y., and Omi, E. M. (1974). Persistence of insect viruses in field populations of alfalfa insects. *J. Invertebr. Pathol. 23*, 360-365.

Tanada, Y., Inoue, H., Hess, R. T., and Omi, E. M. (1980). Site of action of a synergistic factor of a granulosis virus of the armyworm, *Pseudaletia unipuncta*. *J. Invertebr. Pathol. 34*, 249-255.

Tashiro, H., Beavers, J. B., Grosa, M., and Moffit, C. (1970). Persistence of a nonoccluded virus of the citrus red mite on lemons and in intact dead mites. *J. Invertebr. Pathol. 16*, 63-68.

Thomas, E. D., Reichelderfer, C. F., and Heimpel, A. M. (1972). Accumulation and persistence of a nuclear-polyhedrosis virus of the cabbage looper in the field. *J. Invertebr. Pathol. 20*, 157-164.

Thomas, E. D., Reichelderfer, C. F., and Heimpel, A. M. (1973). The effect of soil pH on the persistence of cabbage looper nuclear polyhedrosis virus in soil. *J. Invertebr. Pathol. 21*, 21-25.

Thomas, E. D., Heimpel, A. M., and Adams, J. R. (1974). Determination of the active nuclear polyhedrosis virus content of untreated cabbages. *Environ. Entomol. 3*, 908-910.

Thompson, C. G. (1959). Thermal inhibition of certain polyhedrosis virus diseases. *J. Insect Pathol. 1*, 189-192.

Thompson, C. G., and Scott, D. W. (1979). Production and persistence of the nuclear polyhedrosis virus of the Douglas-fir tussock moth, *Orgyia pseudotsugata* (Lepidoptera: Lymantriidae), in the forest ecosystem. *J. Invertebr. Pathol. 33*, 57-65.

Thompson, C. G., and Steinhaus, E. A. (1950). Further tests using a polyhedrosis virus to control the alfalfa caterpillar. *Hilgardia 19*, 411-445.

Thompson, C. G., Scott, D. W., and Wickman, B. E. (1981). Long-term persistence of the nuclear polyhedrosis virus of the Douglas-fir tussock moth, *Orgyia pseudotsugata* (Lepidoptera: Lymantriidae), in forest soil. *Environ. Entomol. 10*, 254-255.

Tompkins, G. J., and Cantwell, G. E. (1975). The use of ethylene oxide to inactivate insect viruses in insectaries. *J. Invertebr. Pathol. 25*, 139-140.

Thompkins, G. J., Vaughn, J. L., Adams, J. R., and Reichelderfer, C. F. (1981). Effects of propagating *Autographa californica* nuclear polyhedrosis virus and its *Trichoplusia ni* variant in different hosts. *Environ. Entomol. 10*, 801-806.

Vail, P. V. (1981). Cabbage looper nuclear polyhedrosis virus-parasitoid interactions. *Environ. Entomol. 10*, 517-520.

Vail, P. V., Henneberry, T. J., Kishaba, A. N., and Arakawa, K. Y. (1968). Sodium hypochlorite and formalin as antiviral agents against nuclear-polyhedrosis virus in larvae of the cabbage looper. *J. Invertebr. Pathol. 10*, 84-93.

Vail, P. V., Sutter, G., Jay, D. L., and Gough, D. (1971). Reciprocal infectivity of nuclear polyhedrosis viruses of the cabbage looper and alfalfa looper. *J. Invertebr. Pathol. 17*, 383-388.

Vail, P. V., Jay, D. L., Stewart, F. D., Martinez, A. J., and Dulmage, H. T. (1978). Comparative susceptibility of *Heliothis virescens* and *H. zea* to the nuclear polyhedrosis virus isolated from *Autographa californica*. *J. Econ. Entomol. 71*, 293-296.

Vail, P. V., Romine, C. L., and Vaughn, J. L. (1979). Infectivity of nuclear polyhedrosis virus extracted with digestive juices. *J. Invertebr. Pathol. 33*, 328-330.

Vail, P. V., Seay, R. E., and DeBolt, J. (1980). Microbial and chemical control of the cabbage looper on fall lettuce. *J. Econ. Entomol. 73*, 72-75.

Vail, P. V., Knell, J. D., Summers, M. D., and Cowan, D. K. (1982). *In vivo* infectivity of Baculovirus isolates, variants, and natural recominants in alternate hosts. *Environ. Entomol. 11*, 1187-1192.

Vaughn, J. L. (1972). Long-term storage of hemolymph from insects infected with nuclear polyhedrosis virus. *J. Invertebr. Pathol. 20*, 367-368.

Versoi, P. L., and Yendol, W. G. (1982). Discrimination by the parasite, *Apanteles melanoscelus*, between healthy and virus-infected gypsy moth larvae. *Environ. Entomol. 11*, 42-45.

Volkman, L. E., and Summers, M. D. (1977). *Autographa californica* nuclear polyhedrosis virus: Comparative infectivity of the occluded, alkali-liberated, and nonoccluded forms. *J. Invertebr. Pathol. 30*, 102-103.

Watanabe, S. (1951). Studies on the grasserie virus of the silkworm, *Bombyx mori*. IV. Physical and chemical effects upon the virus. *Jap. J. Exp. Med. 21*, 299-313.

Witt, D. J., and Hink, W. F. (1979). Selection of *Autographa californica* nuclear polyhedrosis virus for resistance to inactivation by near ultraviolet, far ultraviolet, and thermal radiation. *J. Invertebr. Pathol. 33*, 222-232.

Witt, D. J., and Stairs, G. R. (1975). The effects of ultraviolet irradiation on a Baculovirus infecting *Galleria mellonella*. *J. Invertebr. Pathol. 26*, 321-327.

Wollam, J. D., Yendol, W. G., and Lewis, F. B. (1978). Evaluation of aerially-applied nuclear polyhedrosis virus for suppression of the gypsy moth, *Lymantria dispar* L. *U.S.D.A. Forest Serv. Res. Paper* NE-396, 8 pp.

Wood, H. A., Hughes, P. R., Johnston, L. B., and Langridge, W. H. R. (1981). Increased virulence of *Autographa californica* nuclear polyhedrosis virus by mutagenesis. *J. Invertebr. Pathol. 38*, 236-241.

Yamamoto, T., and Tanada, Y. (1978). Phospholipid, an enhancing component in the synergistic factor of a granulosis virus of the armyworm, *Pseudaletia unipuncta*. *J. Invertebr. Pathol. 31*, 48-56.

Yamamoto, T., and Tanada, Y. (1979). Comparative analyses of the enveloped virions of two granulosis viruses of the armyworm, *Pseudaletia unipuncta*. *Virology 94*, 71-81.

Yamamoto, T., and Tanada, Y. (1980). Acylamines enhance the infection of a Baculovirus of the armyworm, *Pseudaletia unipuncta* (Noctuidae, Lepidoptera). *J. Invertebr. Pathol. 35*, 265-268.

Yendol, W. G., and Hamlen, R. A. (1973). Ecology of entomogenous viruses and fungi. *Ann. N.Y. Acad. Sci. 217*, 18-30.

Yendol, W. G., Hedlund, R. C., and Lewis, F. B. (1977). Field investigation of a Baculovirus of the gypsy moth. *J. Econ. Entomol. 70*, 598-602.

Young, E. C., and Longworth, J. F. (1981). The epizootiology of the Baculovirus of the coconut palm rhinoceros beetle *(Oryctes rhinoceros)* in Tonga. *J. Invertebr. Pathol. 38*, 362-369.

Young, S. Y., and Yearian, W. C. (1974). Persistence of *Heliothis* NPV on foliage of cotton, soybean, and tomato. *Environ. Entomol. 3*, 253-255.

Young, S. Y., and Yearian, W. C. (1979). Soil application of *Pseudoplusia* NPV: Persistence and incidence of infection in soybean looper caged on soybean. *Environ. Entomol. 8*, 860-864.

Young, S. Y., Yearian, W. C., and Kim, K. S. (1977). Effect of dew from cotton and soybean foliage on activity of *Heliothis* nuclear polyhedrosis virus. *J. Invertebr. Pathol.* *29*, 105-111.

Zelazny, B. (1972). Studies on *Rhabdionvirus oryctes*. I. Effect on larvae of *Oryctes rhinoceros* and inactivation of the virus. *J. Invertebr. Pathol.* *20*, 235-241.

Zelazny, B. (1976). Transmission of a Baculovirus in populations of *Oryctes rhinoceros*. *J. Invertebr. Pathol.* *27*, 221-227.

Zether, O. (1980). Control of *Agrolis segetum* (Lep: Noctuidae) in root crops by granulosis virus. *Entomophaga* *25*, 27-35.

DEVELOPMENT OF VIRAL RESISTANCE IN INSECT POPULATIONS

D. T. BRIESE

CSIRO Division of Entomology
Canberra, Australia

J. D. PODGWAITE

Center for Biological Control of Northeastern
Forest Insects and Diseases
Hamden, Connecticut

I. INTRODUCTION

Studies on insect viruses, as candidate agents for biological control, have tended to progress as follows. Initial *preapplication* studies have centered on identification of the virus and pathogenicity testing against various target organisms, or against the different stages of a particular host insect (Carner et al., 1979; Evans, 1981). Should the virus prove promising, experimental work has usually been channelled into studies of more fundamental aspects of the biology, pathology, and genetics of the virus itself (Faulkner, 1981; Smith and Summers, 1978; Harrap and Payne, 1979), and/or into studies concerned specifically with the *application* of the virus in the field, such as field trials of virus efficacy (Reed and Springett, 1971), the development of effective delivery systems (Yearian, 1978; Couch and Ignoffo, 1981), safety considerations and registration (Burges, 1981), mass production of the virus (Shapiro et al., 1981), and predictive modelling of treatment effects (Brand and Pinnock, 1981; Payne, 1982; Flückiger, 1982). This stage probably represents the state of the art for most viruses under study, and only recently have concerted efforts begun to examine *postapplication* aspects, such as the short- and long-term interactions between the life systems of insect and pathogen in the field.

ISBN 0-12-470295-3

While progress has been made in studies of the population dynamics of insect-pathogen interactions (Anderson, 1982), relatively little effort has been directed toward understanding evolutionary effects, such as the development of viral resistance in insect populations. In view of the fact that one of the more important incentives for looking toward pathogens for insect control is the problem of insects developing resistance to chemical insecticides (Tinsley, 1979), it is curious that little attention has been paid to the possibility of a similar response in pathogens. In fact, of a series of recommendations for future research priorities on insect pest management with microbial agents made at a recent workshop (IPRC, 1980), none referred directly to the problem of resistance. This chapter demonstrates that studies on resistance should form an important part of any program aimed at developing and using viruses as biocontrol agents.

In general, studies reporting resistance or differential susceptibility to viruses have arisen ad hoc, which has led to scattered, unconnected reports (see Briese, 1981a). Only in a few cases has anything approaching a systematic study been undertaken. In this chapter we will examine five such case histories, involving four pest species and one domesticated species, and their associated viruses:

1. The silkworm, *Bombyx mori* L., is a domesticated insect used for the commercial production of silk. Sericulture has been an important branch of agriculture in Japan, China, and other Asian countries for several hundred years. In Japan, there are now some 700 genetically distinct strains of *B. mori* (Yokoyama, 1979) and an ongoing program of selection operates to improve traits in silk quality and quantity, insect vigor, and the prevention of disease. As a result, this insect has provided a rich source of examples of intraspecific variability in response to pathogens. These include several viruses: a nuclear polyhedrosis virus (NPV), cytoplasmic polyhedrosis virus (CPV), densonucleosis virus (DNV), and infectious flacherie virus (IFV). Viral resistance has been most intensively studied in *B. mori* and may serve, therefore, as a paradigm for similar studies involving pest insects.

2. The large white butterfly, *Pieris brassicae* (L.), is a widespread pest of cruciferous food plants in the Northern Hemisphere. This is one of the earliest reported cases of differential susceptibility to a virus in a pest species, involving a granulosis virus (GV). This GV was short-listed by FAO (1973) as a promising candidate for biological control.

3. The light-brown apple moth, *Epiphyas postvittana* (Walker), is an extremely polyphagous pest in Australia and New Zealand. It is affected by endemic nuclear polyhedrosis virus (NPV). Studies on resistance to this virus are historically linked to studies on the potato moth.

4. The potato moth, *Phthorimaea operculella* (Zeller), is a cosmopolitan pest of solanaceous crop plants. A GV, which appears to be endemic, is one of four such viruses listed by FAO (1973) as worthy of further investigation for insect control.

5. The gypsy moth, *Lymantria dispar* (L.), is an important pest of deciduous forests in the Northern Hemisphere. The NPV of this species has been registered and is currently used in control programs under the commercial name of Gypchek.

A detailed look at specific case histories will not only examine viral resistance per se, but will offer insights into the development of methodologies designed to understand the phenomenon. Each case involves different approaches whose findings may inform us not only of the implications that the development of viral resistance might have for biological control, but also of the type of studies needed to elucidate these implications.

II. CASE HISTORIES

A. *THE SILKWORM*

Even before the isolation and identification of causative agents, some strains of *Bombyx mori* were known to be more susceptible to certain diseases than others. Four viruses have now been isolated which are responsible for some of the more important diseases in commercial silkworm farms, and early studies were directed at quantifying the differences in susceptibility of the various strains to these organisms. Considerable variability in the mean dosage required either to infect or to kill larvae of different strains was observed for the cytoplasmic polyhedrosis virus (CPV), the nuclear polyhedrosis virus (NPV), and a nonoccluded RNA virus known as infectious flacherie (IFV) (Table I). Interstrain differences were found at all stages of larval development, although a comparison of the data in Table I suggests that such differences might increase as the larvae mature. A further point of interest is that differences in response to the nonoccluded IFV seem to be considerably greater than differences in response to the two occluded viruses, CPV and NPV. In fact, some of the strains exposed to IFV during the third instar were not susceptible to the highest concentrations of virus given. Such nonsusceptible strains have also been found for a recently isolated nonoccluded DNA virus, densonucleosis virus (DNV) (Watanabe and Maeda, 1978, 1981), which causes flacherie-like symptoms.

TABLE I. *Variability in Response to Several Viruses Shown by Different Strains of Bombyx mori Exposed at Different Stages of Larval Development*

Virus	No. of strains compared (including hybrids)	Instar tested	Response[a] range	Maximum difference in response[b]	Reference
CPV	13	1st	4.07 - 5.94	74	Aruga and Watanabe (1964)
	7	2nd	4.11 - 6.90	617	
	11	3rd	5.72 - 8.43	513	
	13	4th	5.98 - 8.99	1,023	
	13	5th	6.27 - 9.35	1,202	
CPV	12	3rd	5.45 - 8.60	1,410	Watanabe (1966a)
	14	4th	6.55 - 8.95	250	
CPV	16	3rd	5.70 - >8.84	>1,380	Watanabe et al. (1974)
NPV	10	1st	3.25 - 5.05	63	Aratake (1973a)
NPV	12	3rd	6.35 - 8.90	355	Watanabe (1966a)
NPV	14	4th	6.75 - 9.55	630	
IFV	18	1st	3.2 - 7.0	6,300	Aratake (1973b)
IFV	13	3rd	<3.0 - 6.1	>1,260	
IFV	15	3rd	<3.0 - 7.10	>10,000	Watanabe et al. (1974)

[a]Response is measured as log (mean infective dosage or concentration of virus) for CPV and NPV and as log (mean lethal dosage or concentration of virus) for IFV.

[b]Ratio of responses of most and least resistant strains.

TABLE II. Genetic Mechanisms Found to Control Intraspecific Differences in Susceptibility to Viruses in *Bombyx mori*

Virus	Resistant strain	Susceptible strain	Difference in susceptibility	Genetic mechanism	Reference
CPV	Daizo	Okusa	>400	Dominant major gene	Watanabe (1965)
IFV	NG	H4	Nonsusceptible	Recessive major gene	Funuda (1968)
SFV[a]	C-124	N-124	Nonsusceptible	Recessive major gene	Furuta (1978)
DNV	C-124	N-124	Nonsusceptible	Recessive major gene	Watanabe and Maeda (1978)
DNV	Daizo	N-124	Nonsusceptible	Recessive major gene	Watanabe and Maeda (1981)

[a]SFV is an earlier name for DNV, from which it is indistinguishable (Watanabe, personal communication).

While cases of completely resistant versus susceptible strains are relatively clear-cut, there are problems in interpreting cases involving strains with different levels of response. Unfortunately, the early published data on comparative resistance of strains did not include the slopes of the dosage-mortality curves used to derive mean response levels. Thus it is not possible to distinguish between those cases in which a difference in response might be due to reduced variability following the elimination of more susceptible individuals from those cases in which one strain might possess a defense mechanism which reduces the susceptibility of the population as a whole.

In the case of strains showing more extreme responses, subsequent work has resolved this problem. A comparison of the responses to CPV of hybrids and backcrosses of Daizo, the most resistant strain tested, and Okusa, a very susceptible strain, showed that a dominant major gene was controlling the increased resistance shown by Daizo to CPV (Watanabe, 1965) (Table II). Nonsusceptibility of strains to the two non-occluded viruses was also found to be controlled by a recessive major gene (Table II). Other genes might also contribute to some of the smaller differences in response found between strains, for marked F_1 heterosis in response to virus was observed in hybrid strains exposed to CPV (Aruga and Watanabe, 1964), NPV (Aratake, 1973a), and IFV (Aratake, 1973b).

Of particular relevance to the sericulture industry is whether increased resistance to a virus can be selected for in particular strains of *B. mori*. Uzigawa and Aruga (1966) exposed the susceptible N-124 strain to IFV over five generations and observed a 12- to 30-fold change in LD_{50}. However, they did not publish regression slopes, so it is impossible to say whether susceptible individuals had merely been eliminated. Funada (1968) also carried out selection trials with IFV over successive generations of a hybrid strain (NG × NT). While this work also was not adequately quantified, Funada had previously determined that NG carried a recessive major gene for resistance to IFV whereas NT was susceptible. The inbred hybrid line derived from these parents would therefore contain the resistance gene, predominantly as heterozygotes, but with a small proportion of homozygous individuals. It would, therefore, show mainly a susceptible phenotypic response to IFV. Selection of survivors from virus-fed larvae of this line led rapidly to the phenotypic expression of resistance after six generations, whereas the control line remained quite susceptible (mean mortality of 7 versus 83% at the same virus concentration). These data suggest that continued exposure to IFV has led to an increase in frequency of the resistance gene.

The most detailed selection trials were carried out by Watanabe (1967) using two susceptible strains of *B. mori* and two strains of CPV. The courses of selection over eight generations are shown in Fig. 1. In all four cases they resulted in a 10- to 16-fold increase in the mean level of resistance to peroral infection by CPV by the sixth generation. This level of resistance remained at a plateau over the last few generations of selection (Fig. 1). While there were slight increases in the slopes of the dosage-mortality curves, these did not seem to be very significant, for variations in slope between generations were similar to those between treatments. In fact, in the last generation tested, two of the selected lines had lower slopes than their respective controls. Interestingly, hybrid crosses of the two control lines showed marked heterosis in resistance to CPV, whereas none of the hybrid crosses of the selected lines were more resistant than their parents. Watanabe (1967) interpreted this as indicating that selection had acted on somewhat different polygenic backgrounds in the two strains to produce selected lines which were more homozygous for resistance, and, consequently, did not exhibit the same degree of heterosis.

A slightly different aspect of viral resistance had been investigated by Aizawa and co-workers. It had been known that if *B. mori* was stressed by cold, starvation, or certain chemicals, this could induce the development of an apparently latent viral infection in larvae that had not been fed virus. Selection of a line for resistance to such virus induction was undertaken by subjecting the larvae to cold stress for 24 hours following each molt, and breeding from the survivors who did not succumb to the induced virus. After eleven generations consistently fewer larvae were dying from viral infection in the selected line than in the corresponding control line (Aizawa et al., 1961). This difference was maintained in subsequent generations (Aizawa and Furuta, 1962, 1964). While all three viruses occurred in stressed individuals, the selection process seems to have primarily led to an increased resistance to NPV induction, for the incidence of CPV and IFV remained the same in both the stressed line and unstressed control. Aizawa et al. (1961) found no correlation between the level of resistance shown to virus induction and that shown to peroral infection with NPV, which indicates that the mechanisms controlling each mode of resistance were independent. This would seem plausible if one considers that a virus particle eaten by a larva has to pass through the midgut epithelium before it can begin to multiply, whereas a "latent" virus particle would already be present in the cell structure.

There is strong indirect evidence that resistance to peroral infection of viruses depends on the inhibition of virus particles invading the midgut cells, rather than the suppression

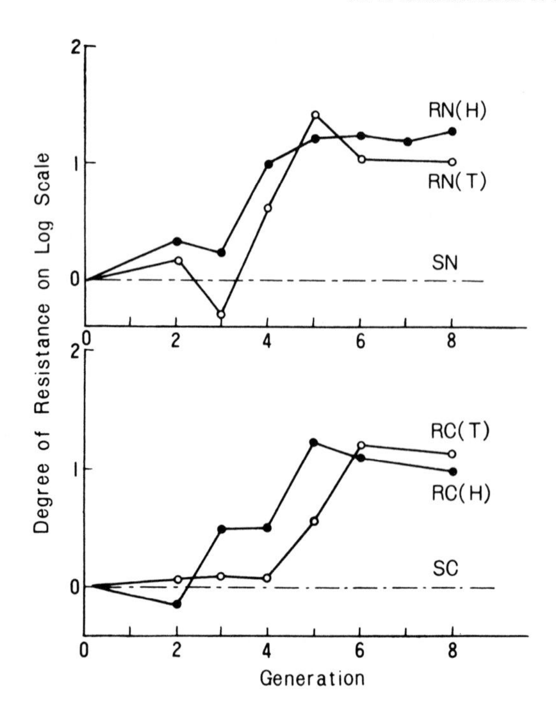

FIGURE 1. Development of resistance in the 4th instar larvae to infection with CPV in successive generations of Bombyx mori selected with the virus (Watanabe, 1967). RN(H) and RN(T) are selected Japanese strains with HC virus, which is occluded in tetragonal polyhedra, respectively, from an SN control strain. RC(H) and RC(T) are selected Chinese strains with HC virus and TC virus, respectively, from SC control strain. The degree of resistance is presented as a logarithmic ratio obtained by dividing the log ED_{50} of the selected strain by the log ED_{50} of the unselected strain.

of subsequent virus multiplication within a cell. Watanabe (1966b) found up to a 10,000-fold reduction in LD_{50} values for *B. mori* strains given a subcutaneous injection of CPV compared to CPV given perorally. Moreover, interstrain differences in susceptibility to subcutaneous infection by CPV were quite small relative to differences between strains given CPV perorally, and there was no correlation between the two methods of administering the virus (Watanabe, 1971).

More recently, direct evidence has been published which supports the argument that the main defense mechanism against infection by virus particles involves the midgut epithelial cells. Early observations on the pathogenesis of IFV showed that, although there was a relationship between the level of resistance shown by a particular strain and the overall rate of multiplication of the virus, IFV still multiplied actively in a highly resistant strain such as Daizo, without producing symptoms of flacherie disease (Tanaka, 1969). Using immuno-fluorescence techniques, Inoue (1974) found that the virus multiplied at the same rate in larvae of both strains. However, in the resistant strains the infected goblet cells were discharged into the gut lumen at each molt, leading to a large, albeit temporary, reduction in infection levels. Regenerative cells located in nidi developed into new goblet cells to repair the midgut epithelium. This did not occur in susceptible strains, where the number of infected cells continued to increase, suggesting that the ability to regenerate infected cells might be a major mechanism contributing to IFV resistance.

A similar mechanism may occur in the case of CPV resistance, for Inoue and Miyagawa (1978) found that infected columnar cells were discharged during a molt and replaced by regenerated cells. Furthermore, the degree of epithelial regeneration varied with the strain of *B. mori* and the dosage of the virus. In *B. mori* the new cells are subsequently reinfected. However, when the fall webworm, *Hyphantria cunea*, was given *B. mori* CPV, the discharged columnar cells were replaced by new cells that were not susceptible to the virus (Yamaguchi, 1979). Thus, both the extent of epithelial regeneration following infection and the susceptibility of the new cells to subsequent infection seem to be important mechanisms in the resistance of *B. mori* to CPV.

While the defense mechanisms to CPV and IFV appear superficially similar, there may be important differences, since Watanabe et al. (1974) found no correlation between the susceptibility of different strains to the two viruses. Moreover, different genetic mechanisms have been found to exercise control over resistance to these viruses (see Table II). This may be due to the fact that CPV and IFV attack primarily different sites in the midgut--the columnar and goblet cells, respectively. On the other hand, there was a strong positive correlation for interstrain susceptibility of different strains to CPV and NPV, which also enters the columnar cells prior to passing on to susceptible tissues (Watanabe, 1966a). It thus seems likely that a common resistance mechanism involving those cells might be important for both these viruses.

The mechanism causing nonsusceptibility to DNV is not yet known, although it appears to be quite different from that for the other nonoccluded virus (IFV). In susceptible strains DNV infects midgut columnar cells which are discharged into the gut lumen, without accompanying cell regeneration, leading to rapid death (Watanabe, 1981). In nonsusceptible strains,

Watanabe (personal communication) found that neither peroral nor hemocoelic virus inoculation caused any infection of the midgut cells, leading him to speculate that the recessive gene for nonsusceptibility may control one of the host cell enzymes concerned with DNV replication or with receptor synthesis.

Smaller variations in susceptibility between strains may be produced by factors such as the antiviral substances found in the digestive fluid of *B. mori* (Aizawa, 1962), which may inactivate both CPV (Aratake et al., 1974) and NPV (Aratake and Ueno, 1973). One such substance, a red fluorescent protein which is synthesized *de novo* in the midgut tissue (Hayashiya et al., 1971), has been isolated and found to agglutinate both NPV and IFV, although it did not react to *Autographa* NPV (Hayashiya et al., 1978). The antiviral activity of the digestive fluid varied between strains (Aruga and Watanabe, 1964; Aratake et al., 1974), but this did not account for the large differences between strains in susceptibility to peroral infection.

Certain developmental factors may influence the manifestation of resistance. Kobara et al. (1967) found that the response to CPV of fourth instar larvae varied markedly in both resistant and susceptible strains, depending upon the development of the instar infected. Susceptibility was greatest immediately after and before molts, but decreased considerably during the intervening period. Kobayashi et al. (1969) also found a within-instar increase in resistance to NPV in larvae fed virus. This was greater in a resistant strain than in a susceptible one. Moreover, the increase was not as apparent in larvae given subcutaneous injections of NPV, which again suggests that such resistance is due to an increase in the inhibitory mechanisms of the midgut epithelium. Such changes may be controlled by hormone levels, since both diapausing pupae (Watanabe and Aruga, 1970) and diapausing embryos (Kobayashi, 1982) have been shown to be considerably more resistant to NPV than nondiapausing individuals. In both cases, the addition of ecdysone analogs to the diapausing individuals increased the latter's susceptibility to the virus.

An examination of data given by Aruga and Watanabe (1964) shows that the rate of increase in resistance to CPV between successive instars of three resistant strains of *B. mori* was considerably more rapid than that of two susceptible strains. This might suggest that the increased resistance shown by some strains is due to an acceleration of the normal process of maturation resistance as larvae age.

The environment to which the insect is exposed also plays an important role in the manifestation of resistance. Stress factors, such as abnormal temperatures, nutritional deficiencies, or exposure to certain chemicals, may increase susceptibility to viral infection (Watanabe, 1971). Silkworm larvae reared in different seasons tended to show differences in susceptibility to NPV and IFV, independent of interstrain differences (Aratake, 1973a,b). Such differences may be due in part to the quality of the mulberry leaves upon which they feed (Watanabe, 1971), and, in artificial diets, may involve components such as sucrose or protein levels (Watanabe and Imanishi, 1980), the latter affecting directly the antiviral activity and protease activity of the digestive fluid (Watanabe, personal communication). Light may have an indirect effect on the response of *B. mori* to a virus since it is required for the synthesis of the antiviral protein (Hayashiya et al., 1976). It may also affect directly the midgut defense barrier, since larvae reared in constant light were more resistant to both NPV and CPV than those reared in continuous darkness (Watanabe and Takamiya, 1976). The former larvae also exhibited a higher proportion of columnar cells in the midgut, which may have enabled them better to survive infection by these viruses, both of which invade these particular cells. It thus seems possible that environmental, developmental, and genetic controls might involve very similar defense mechanisms in governing the phenotypic expression of viral resistance in *B. mori*.

B. *THE LARGE WHITE BUTTERFLY*

During the late 1950s, the large white butterfly, *Pieris brassicae*, and its GV became the objects of the first concerted investigations into virus resistance in a pest species. The problem was first suspected when one particular laboratory-reared strain ("Cambridge"), which had suffered an epizootic of GV in 1955 that killed about 90% of the larvae, showed reduced mortality levels in subsequent virus outbreaks, both in absolute terms and relative to other strains of *P. brassicae* that were held at that time (David and Gardiner, 1960).

Early time-mortality experiments, using crude virus extracts, confirmed that fewer larvae of the Cambridge strain died when given the same single dosage than did larvae of several other discrete strains (Rivers, 1959; Sidor, 1959; David and Gardiner, 1960). David and Gardiner (1965) subsequently compared the Cambridge strain with two other ["virus-free" (VF) and "cheiranthi"], using serial dilutions of crude virus extract. Significant differences in mortality were again confirmed (Table III), which demonstrated variability in response to GV between strains--Cambridge being more resistant than VF, which was, in turn, more resistant than cheiranthi.

TABLE III. Virus-Induced Mortality in Three Stocks of *Pieris brassicae* Larvae Fed with Serial Dilutions of Crude GV Extract Derived from the VF Stock[a]

Insect stock	Replication	Mortality (%) dilution of virus stock suspension				Significance of differences between stocks
		10^{-7}	10^{-6}	10^{-5}	10^{-4}	
VF vs. Cambridge						
VF	1	-	0	29	67	
	2	-	13	73	100	
	3	-	52	86	100	
Cambridge	1	-	11	4	22	
	2	-	0	26	73	
	3	-	13	35	96	$p < 0.001$
VF vs. Cheiranthi						
VF	1	-	-	9	19	
	2	-	-	68	86	
	3	0	8	20	-	
Cheiranthi	1	-	20	100	100	
	2	-	100	100	100	
	3	48	100	-	-	$p < 0.001$

[a]Data from David and Gardiner, 1965.

Unfortunately, because of the unsophisticated techniques available at that time, the data were not adequate to produce the dosage-mortality curves required to accurately quantify the relative susceptibilities of the different strains. The high resistance level shown by the Cambridge strain might have been due merely to the elimination of more susceptible forms (David, 1978). However, Ripa (1978) was subsequently able to produce convincing results for VF and cheiranthi, as well as two other strains (Table IV). By feeding known dosages of a purified GV extract individually to larvae on disks of cabbage leaf, Ripa found a 12-fold difference in susceptibility between VF and the three other strains. Using similar methods, Payne et al. (1981) reported no significant difference between VF and a strain recently collected from the field ("Rimpton"). A comparison of all sets of data suggests that there are probably three distinguishable levels of susceptibility in the populations of *P. brassicae* tested to date: Cambridge was the most resistant; VF and Rimpton showed moderate resistance; and the cheiranthi, French and Dutch strains exhibited the greatest susceptibility to GV.

The question of whether the high level of resistance found in the Cambridge strain was due to selection of a gene or genes during laboratory culture remains unresolved. Some evidence exists for this since Rivers (1959) observed that increasing amounts of crude virus extract were required to produce high levels of infection in generations following the original out-

TABLE IV. Differences in Susceptibility to GV between Several Stocks of Pieris brassicae

Stock	*Log LD_{50}*	*Slope*	*Resistance factor*[c]
Using 4th instar larvae[a]			
Cheiranthi	*6.70 a*	*1.44*	*1 a*
Dutch	*6.95 a*	*1.59*	*1.8 a*
French	*7.00 a*	*1.25*	*2.0 a*
VF	*7.79 b*	*1.50*	*12.2 b*
Using 3rd instar larvae[b]			
Rimpton	*5.42 a*	*1.21*	*1 a*
VF	*5.68 a*	*0.99*	*1.8 a*

[a]*Data from Ripa, 1978.*

[b]*Data from Payne et al., 1981.*

[c]*Values followed by the same letter (a) (b) do not differ significantly.*

break of virus in the Cambridge stock. By the ninth generation a virus suspension which had previously killed all exposed larvae, now killed only very few larvae. However, as mentioned earlier, in the absence of adequate dose-response data one cannot eliminate the possibility that the virus outbreaks were merely eliminating more susceptible forms, so increasing the mean resistance level. Whatever the etiology of the high level of resistance shown, it persisted for many generations within the inbreeding Cambridge stock until at least 1965 (David and Gardiner, 1965).

Unfortunately, the subsequent history of this strain is somewhat clouded. Huber (1974, personal communication) reported that progeny of David's resistant strain of *P. brassicae* were sent to the Entomological Institute, ETH, in Zurich where they showed no response to very high dosages of GV, whose virulence had been demonstrated previously using larvae from a local Swiss population. At the time quantitative dosage-mortality bioassays were not carried out. When such bioassays were undertaken two years later, an increased level of resistance could not be demonstrated (Benz, cited in Huber, 1974). This raises the interesting possibility of a reversion from low to high susceptibility within a strain that had maintained a high level of resistance to GV over many generations. The cause of such a reversion is open to considerable speculation. It may have resulted from selection against a resistance gene in the inbreeding population, because of reduced fitness in the absence of GV under particular conditions. Alternatively, it may have been due to a return to the normal variability of response to GV within a single population, with a consequent reduction in the determined LD_{50} levels.

No concerted studies have been carried out on the genetics or physiological mechanisms controlling differential susceptibility to GV in *P. brassicae*. However, David (1978) did report that differences between VF and two more susceptible strains persisted when free virions were injected into the hemolymph, instead of virus particles being fed to the larvae. This suggests that, unlike *B. mori*, factors other than those operating in the gut during the initial stages of injection might be involved in this resistance to GV.

C. *THE LIGHT-BROWN APPLE MOTH*

The case of the light-brown apple moth, *Epiphyas postvittana*, is unusual, in that the initial studies involving its NPV were not directly concerned with the implementation of biological control, but formed part of a long-term investigation into factors affecting the life system of the species in southeastern Australia (see Geier and Briese, 1981). One of the more important findings was the large variability shown by the insect over its distribution range, both in certain morphological characters (Geier and Springett, 1976) and in its demographic performance (Geier and Briese, 1980). It was also realized that NPV was an important part of the insect's life system, in view of its ability to cause heavy mortality in local populations and to impair the reproductive capability of survivors (Geier and Oswald, 1977), and because it was endemic throughout the range of *E. postvittana* (Geier and Briese, 1979).

In order to determine whether the variable nature of *E. postvittana* extended to the response of this species to NPV, Geier and Briese (1979) exposed twelve populations of F_1 progeny of field specimens collected throughout southeastern Australia to two dosage levels of NPV, applied to the surface of an artificial diet. Because the experiment had been designed for a different purpose, the data generated was insufficient to produce LD_{50} values and response slopes. However, they did indicate considerable and significant differences in NPV-induced mortality between discrete breeding populations exposed to the same dosage of virus (Table V).

Moreover, a difference of more than 100-fold in the minimum dosage required to kill all exposed larvae was found between some field populations and a population (CAN) that had been reared for many generations in the laboratory. Briese et al. (1980) subsequently verified these differences by quantifying the response of the more resistant CAN strain relative to a field strain (AD) and a susceptible laboratory strain (BAR) (Table VI). The dosage-mortality response slopes for the three populations did not differ significantly, while significant differences in LD_{50} values were found (Table VI), showing that there was a true difference in the susceptibility of these populations to NPV (*sensu* Burges, 1971).

A series of bioassays with hybrids and backcrosses of the AD and CAN strains produced response slopes intermediate to those of the parents (Fig. 2), a result which suggested that the genetic control of this increased resistance was not due to a single autosomal gene but rather to a gene complex (Briese et al., 1980). Control by such a complex could help to explain the wide variability in response to NPV observed in the field populations.

The origin of resistance in the CAN strain can only be speculated upon. However, like the Cambridge stock of *P. brassicae*, it is known to have been reared in an NPV-contaminated environment and to have suffered several outbreaks of NPV, whereas the susceptible BAR laboratory strain had never shown signs of NPV infection. This raises the possibility that

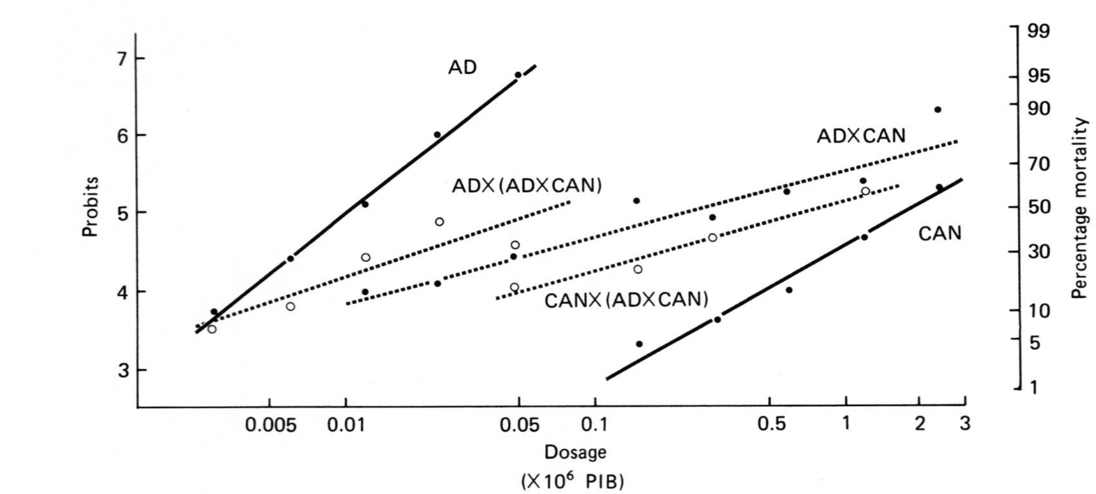

FIGURE 2. Dosage-mortality regressions for the CAN and AD strains of Epiphyas postvittana, their hybrid (AD × CAN), and backcrosses, exposed to NPV (Briese et al., 1980).

TABLE V. Variability in NPV-Induced Mortality between Samples of Epiphyas postvittana Populations Collected Throughout Southeastern Australia[a]

Origin of population sample	Mortality due to NPV (%) dosage level	
	4.5×10^3 PIB[b]/ml	144×10^3 PIB/ml
New South Wales		
Narrabri	19.3 ab[c]	57.8 cd
Bathurst	21.5 abc	64.0 d
Batlow	39.0 bc	87.7 e
Albury	43.9 c	78.7 de
Yanco	47.2 c	31.9 b
Dubbo	84.3 d	78.6 de
Victoria		
Ballarat	4.3 a	9.2 a
Kerang	14.8 ab	34.9 bc
Melbourne	19.1 b	20.8 ab
Orbost	37.1 bc	49.2 c
Kerang	44.8 c	64.1 cd
Tasmania		
Huonville	15.5 ab	47.6 c

[a]Data from Geier and Briese, 1979.
[b]PIB, particle inclusion bodies.
[c]Values followed by the same letter (a-e) do not differ significantly.

TABLE VI. Comparison of dosage-mortality lines for three strains of Epiphyas postvittana[a]

Strain	Log LD_{50}	Slope	Resistance factor[b]
AD (ex field)	4.00	2.56	1
BAR (ex laboratory)	4.51	2.02	3.2
CAN (ex laboratory)	6.26	1.77	163.6

[a]Derived from Briese et al., 1980.
[b]Based on common slope, since individual slopes did not differ significantly.

selection for increased resistance might have occurred in the laboratory. Support for this possibility comes from the observation by Geier and Oswald (1977) that progeny of survivors of an NPV infection showed greater resistance when subsequently exposed to the virus.

Some preliminary investigations were carried out on the physiology of virus resistance in *E. postvittana*. R. F. Powning, W. J. Davidson, and H. Mende (personal communication) investigated differences between the virus-resistant and virus-susceptible strains, and found that although they could be distinguished by isoenzyme patterns, there were no significant differences in larval gut pH [a property which was considered to account for instar differences in viral susceptibility in *P. brassicae* (Ripa, 1978)], oxidation-reduction potentials, or certain enzyme systems thought to be involved in the infection process.

D. *THE POTATO MOTH*

While early field trials of GV against the potato moth, *Phthorimaea operculella*, in southwest Western Australia resulted in infection levels of close to 100% (Reed and Springett, 1971), subsequent trials using the same dosage rates in similar locations could only achieve infection levels of 53 to 85% (Matthiessen et al., 1978). Several factors, such as the density and age structure of the target populations and the prevailing climatic conditions, could have contributed to these results. However, in view of the recently discovered variability in response to a virus within populations of *E. postvittana* (see previous section), it was considered that genetic factors could have contributed to the variable success of field trials.

To investigate this possibility, Briese and Mende (1981) ran a series of bioassays of a highly purified preparation of GV against sixteen field populations and one laboratory population of *P. operculella*. The techniques used for *E. postvittana* were improved, and known dosages were fed to individual 9-day-old larvae on disks of potato. This enabled LD_{50} values, regression slopes, and relative resistance levels to be determined (Table VII). These data clearly indicated significant, albeit relatively small (up to 11.6-fold), differences in susceptibility between the F_1 generations of field-derived populations. Furthermore, the laboratory strain showed the greatest level of resistance of all those tested, being some 30-fold more resistant to GV than some field populations.

TABLE VII. LD_{50} Values, Regression Line Slopes, and Resistance Factors for Sixteen Field Populations and One Laboratory Population of Phthorimaea operculella[a]

Origin of population sample	Log LD_{50}	Slope	Resistance factor[b]
Group 1			
Virginia	5.03 a	0.63	3.6 a
Koroit	4.93 ab	0.56	2.3 a
Koo-Wee-Rup	4.48 ab	0.71	1.2 a
Toolangi	4.35 b	0.88	1.0 a
Group 2			
Tolga	4.74 a	0.95	10.5 a
Northdown	4.70 a	0.60	6.2 ab
Manjimup	4.47 a	0.78	4.8 abc
Ballarat	4.32 ab	0.76	3.4 abc
Finley	3.94 b	0.84	1.4 bc
Dardanup	3.81 b	0.92	1.0 c
Group 3			
Guyra	4.92 a	0.91	11.6 a
Grantham	4.67 a	0.85	5.6 ab
Ginninderra	4.12 b	1.03	1.6 bc
Dorrigo	4.03 b	0.69	1.3 bc
Maitland	3.90 b	0.76	1.0 c
Lawrence	3.89 b	0.70	1.0 c
Group 4			
LAB	6.01 a	0.86	30.9 a
Guyra	5.28 b	0.85	5.7 b
Toolangi	5.33 b	0.70	4.5 b
Finley	4.49 c	0.97	1.0 c

[a]Comparisons are only valid within groups of tested populations, not between them. After Briese and Mende, 1981.

[b]Based on common slope, since individual slopes did not differ significantly. Values followed by the same letter (a-c) do not differ significantly.

These findings paralleled those reported for *E. postvittana* (see Section II,C).

Briese (1982) was subsequently able to establish that the increased resistance to GV in the laboratory strain of *P. operculella* was controlled by a single dominant autosomal gene; the F_1 hybrid between the resistant and a susceptible strain showed the same response to GV as the more resistant parent (Fig. 3a), while the F_2 hybrid produced a response which showed segregation of resistant to susceptible pheno-

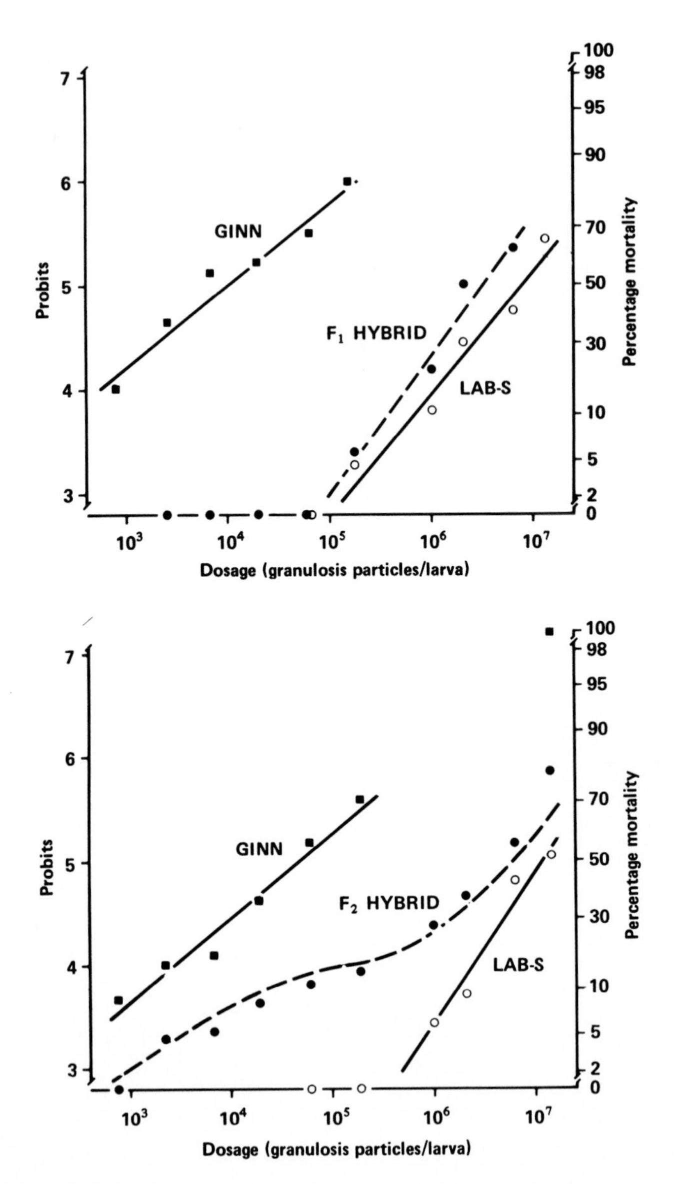

FIGURE 3. Dosage-mortality regressions for the GINN and LAB-S strains of *Phthorimaea operculella*, compared with (a) an F_1 hybrid (GINN ♂ × LAB-S ♀), and (b) the F_2 hybrid, concurrently exposed to GV. The dashed line is the predicted response curve for an F_2 hybrid assuming normal segregation of alleles from heterozygous F, parents (Briese, 1982).

types in a simple Mendelian 3:1 ratio (Fig. 3b). No sex linkage of the gene was observed.

Interestingly, the laboratory strain of *P. operculella* had a similar history to that of both the laboratory strains of *E. postvittana* and *P. brassicae*, which showed increased resistance to NPV and GV, respectively. This strain also suffered several outbreaks of GV while in culture, during which selection for the resistance gene could have occurred. This prompted Briese and Mende (1983) to attempt to emulate the course of events suspected of having produced the resistant strain, by exposing a field population to GV over several generations. This population (GINN) was collected from the area where the laboratory population had originated. Within six generations there was a 140-fold increase in LD_{50} value, without a significant increase in slope (Table VIII), suggesting a change in the frequency of a resistance gene within that population.

An attempt to select for even greater resistance in the laboratory strain produced somewhat different results. While there was a slight increase in LD_{50} after ten generations, the slope had increased significantly as well (Table VIII), indicating that the result was due to an elimination of more susceptible forms and consequent reduction in variability rather than a true shift in mean resistance to GV. No heterosis effects were detected in crosses between the laboratory population and the newly selected field population, suggesting that the resistance mechanisms in the two populations were probably the same (Briese and Mende, 1983).

Recently, Griffith (personal communication) also recorded differential susceptibility of the GINN and LAB populations, treated as 9-day-old larvae. By contrast, however, he found no difference in response to GV between the two populations when exposed to the virus as neonates. These results suggest that there is a difference in the rate of development of resistance between GINN and LAB, similar to that reported for strains of *B. mori* (see Section II,A). The data provide evidence for the proposition by Briese (1981a) that the increase in resistance of larvae with age and the intraspecific differences in levels of innate resistance shown by larvae of different populations are two facets of the same phenomenon.

GV appears to be endemic throughout populations of *P. operculella* in Australia, occasionally developing into epizootic outbreaks under favorable conditions (Briese, 1981b). A tentative model, proposed by Briese and Mende (1981, 1983), suggests that the patterns of variability in response to GV found among field populations of *P. operculella* probably reflect the past history of exposure to the virus. Periodic epizootics could select for resistance genes present in the populations,

TABLE VIII. Comparison of Dosage-Mortality Parameters for Two Populations of *Phthorimaea operculella* Selected for Increased Resistance to GV by Serial Exposure for Several Generations[a]

No. of generations of selection	Substrain	Log LD_{50}	Log LD_{99}	Slope	LD_{50} ratio
LAB population (ex laboratory; original resistance level high)					
5	Selected (LAB-S)	6.03	7.77	0.93	6.0
	Control (LAB)	5.25[b]	7.86	1.27	
10	Selected (LAB-S)	6.21	8.23	0.78	7.1
	Control (LAB)	5.36[b]	8.33	1.15[b]	
GINN population (ex Field; original resistance level low)					
6	Selected (GINN-S)	6.08	8.38	0.79	
	Control (GINN)	3.91[b]	6.85*	1.02	140

[a]Data from Briese and Mende, 1983.
[b]Indicates that difference between selected and control lines are significantly different.

while, in the absence of GV, selection against a less fit virus-resistant phenotype and/or immigration of susceptible individuals might result in a shift back to increased susceptibility. Thus, within a given population, there would be a periodic ebb and flow of resistance build-up and reversion, which over a wider area would lead to a changing pattern of differential responses between populations. The end product would be a stable, oscillating system of coexistence between *P. operculella* and its GV.

E. THE GYPSY MOTH

The gypsy moth, *Lymantria dispar* (L.), is subject to a naturally occurring NPV disease which is generally considered to be the primary natural regulator in dense North American populations of the insect (Reiff, 1911; Glaser and Chapman, 1913; Doane, 1970; Podgwaite, 1981). After many years of research and development, the NPV product, Gypchek, was registered with the United States Environmental Protection Agency for use in gypsy moth management programs (Lewis et al., 1979).

As implied in the introduction, many reports of resistance have arisen somewhat unsystematically out of studies not necessarily designed to show the development of resistance in a particular population, but rather to show the instantaneous susceptibility of a particular population to a particular virus preparation. In the case of *L. dispar*, it must be stated that the majority of what is known in this regard has been generated as "spin-off" from other studies, particularly those designed either to determine the most virulent virus or to determine dose rates for field application. Nevertheless, the existing information, mostly based on extensive, albeit variably performed, bioassays, does indicate that populations of *L. dispar* are variable in their resistance to NPV.

While searching for the most virulent virus strain for development, several geographical isolates of NPV were tested against laboratory cultures of the insect, as well as against larvae reared from egg masses of several geographically distinct populations of *L. dispar* in the eastern United States (Rollinson and Lewis, 1973). Results of these tests, and those obtained by other workers (Vasiljevic and Injac, 1973; Magnoler, 1974), indicated that there was a wide variation in the potency of NPV isolates tested against larvae of similar developmental stage and weight (Table IXA). Further, larvae from different geographical populations of the insect responded differently to the same virus source (Table IXB). The strain of virus that was eventually selected for development generally exhibited strong potency (as assessed from LC_{50} values)

TABLE IX. A. *Differences in the Activity of NPV Isolates Against the Same Population of Lymantria dispar*[a]

Source of virus	Larval source	Instar	Log LC_{50}	Relative potency[c]	Reference
Yugoslavia	Michigan	2nd	7.54[d]	1	Rollinson and Lewis (1973)
Japan			6.63	8	
Hamden K-rotor[b]			3.81	5370	
USSR	Yugoslavia	2nd	3.71[e]	1	Vasiljevic and Injac (1973)
Hamden K-rotor			3.48	1.7	
Yugoslavia	Italy	3rd	5.34[e]	1	Magnoler (1974)
Hamden K-rotor			4.24	12.7	

B. *Differences in Response of Lymantria dispar Populations to the Same Preparation of NPV*

Source of virus		Instar	Log LC_{50}	Resistance factor[b,c]	Reference
Hamden K-rotor[b]	Massachusetts	2nd	3.65[d]	1	Rollinson and Lewis (1973)
	Michigan	2nd	3.81	1.4	
	New Jersey	2nd	4.92	18.6	
	Connecticut	2nd	5.76	129	
Yugoslavia	Serbia	2nd	3.51[e]	1	Vasiljevic and Injac (1973)
	Croatia	2nd	3.57	1.2	
	Slovenia	2nd	3.66	1.4	

USSR	Bulgaria (1969)	2nd	5.00[f]	1	
	Yugoslavia (1970)	2nd	7.36	230	
Serbia	Bulgaria (1969)	2nd	5.70[f]	1	Vasiljevic and Injac (1973)
	Yugoslavia (1970)	2nd	8.46	580	
Hamden K-rotor	Bulgaria (1969)	2nd	5.48[f]	1	
	Yugoslavia (1970)	2nd	5.90	2.6	
Hamden K-rotor	Pennsylvania	2nd	2.34[e]	Not comparable	Hedlund (1974)
	New Jersey	2nd	2.67		Lewis et al. (1981)
	Yugoslavia	2nd	3.48		Vasiljevic and Injac (1973)

[a]Modified from Lewis et al., 1981.

[b]Connecticut isolate, purified by procedures of Breillatt et al. (1972) and designated the Hamden standard.

[c]Ratio of particular Log LC_{50} with lowest Log LC_{50} in each set of data.

[d]Measured as PIBs/ml of artificial diet.

[e]Measured as PIBs/insect.

[f]Measured as PIBs/ml of suspension applied to leaf surface.

against United States populations of *L. dispar*. This strain was isolated from a Connecticut population suffering a natural NPV epizootic, and has been characterized and designated as the "Hamden standard."

Vasiljevic and Injac (1973) found that larvae from Yugoslavia were less susceptible to NPVs from the USSR and from Serbia than were larvae from Bulgaria. However, the same authors tested the Connecticut NPV (Hamden standard) and found little difference in the LD_{50}'s of the Bulgarian and Yugoslavian larvae (Table IXB). Such results might be due to strain-specific differences in response to the virus. An alternative possibility is that the results occur because the strains from different regions are at different stages of their normal cycle of population growth and decline. Vasiljevic (1961) suggested that larvae that are not in the same "gradation cycle" are likely to respond differently to the same NPV isolate. However reasonable, there is little data available to either support or refute this hypothesis. Laboratory studies referred to by Vasiljevic and Injac (1973) showed that after several successive generations reared in the laboratory, the sensitivity to a particular NPV isolate changed. This trend has also been observed after testing the Hamden standard NPV against a laboratory culture of *L. dispar* (New Jersey strain) over 11 generations, F_{14}-F_{24} (Table X). Examination of the slope and LC_{50} data presented using a multivariate routine indicated no differences in slopes ($p > 0.5$), but a significant change in LC_{50} values ($p = 0.012$) over time. This increase in LC_{50} from the earliest to the latest bioassay was highly significant and indicated a trend, since successive changes in LC_{50} showed no significant increase. Further, although there was no difference in slopes or LC_{50} value between the Connecticut wild bioassays, they differed very significantly in slope ($p = 0.002$) but not significantly in LC_{50} from the New Jersey larvae, thus indicating differences in resistance to the same virus.

Unfortunately, there have been no systematic studies on the development of resistance to NPV in natural populations of *L. dispar* following either periodic natural epizootics, or after several years of introduction of the NPV for control of the pest. Such studies are necessary, not only to better understand the evolutionary aspects of such phenomena, but also to provide information relative to the prudent and efficacious use of this NPV as a microbial control agent.

TABLE X. Sensitivity[a] of the Laboratory Strain (NJ) and Connecticut (CT) Wild Populations of Lymantria dispar to Hamden standard NPV

Year	Strain	Slope	Log LC_{50}[b]
1983	NJ F_{24}	1.26	5.06
1982	NJ F_{24}	1.46	4.78
	NJ F_{24}	1.46	4.53
	NJ F_{24}	1.57	4.33
	NJ F_{24}	1.34	4.52
	NJ F_{23}	1.40	4.67
	NJ F_{23}	1.45	4.54
	NJ F_{23}	1.46	4.93
1981	NJ F_{22}	1.47	4.45
1979	NJ F_{18}	1.56	3.89
1977	NJ F_{14}	1.33	3.97
	NJ F_{14}	1.52	4.18
1980	CT wild	1.87	4.10
1981	CT wild	1.74	4.25
	CT wild	1.61	4.21
	CT wild	1.73	4.14

[a]W. D. Rollinson (unpublished data).
[b]Bioassays performed according to Lewis and Rollinson (1978).

or in large-scale field trials (Burges, 1971; Briese, 1981a). However, the existence of heritable differences in susceptibility to viruses demonstrates that in the long term, as usage becomes heavier and more widespread, the possibility of selecting for increased resistance must be considered. Moreover, Payne et al. (1981) have pointed out the importance of knowing dosage-mortality responses for effective control under field conditions. Therefore, in the short term it will be necessary to determine the extent of any differences in response that may exist between target populations in order to adjust any predetermined control strategies.

II. VIRAL RESISTANCE AND BIOLOGICAL CONTROL

A. *THE NATURE OF THE PROBLEM*

The evidence presented here, which covers the most intensively investigated cases, shows that the study of insect resistance to viruses is still at an early stage, particularly in the case of pest species. Importantly though, several prerequisites for the development of resistance have been established. These include the presence of variability in response to a particular virus between individuals of a species, evidence that certain components of this variability are under genetic control, and evidence that selection pressures can lead to a change in the frequency of individuals showing different levels of resistance.

From the case histories examined here, it can be seen that, with the exception of the small nonoccluded viruses infecting *B. mori*, the phenomenon of resistance involves a shift in the mean response level of a population to a virus rather than the appearance of nonsusceptibility. Certainly the baculoviruses, which are the most likely candidates as control agents, fall into this category. As well as the examples described in this chapter, Briese (1981a) lists an additional five species in which such intraspecific differences have been recorded. Differences in these mean response levels range from a few to several thousandfold. The majority have involved laboratory-reared populations, and only recently have examples of differential susceptibility extended to populations originating from the field (e.g., *E. postvittana* and *P. operculella*). However, the wide variability found between populations of *B. mori* demonstrates the potential that exist in insects for differences in response to viral attack.

Most viruses being considered for insecticidal use are endemic throughout their host's range and have coevolved with their hosts over long periods. Therefore it is reasonable to expect that resistance genes, which may arise *de novo* from time to time (see Whitten and McKenzie, 1982), will have become fixed in discrete populations of many host species. Under natural conditions, the frequency of such genes would be expected to fluctuate according to the intensity of selective forces exerted by the virus on its host (Anderson, 1982). Indiscriminate use of viral insecticides might change the pattern of selection on these genes, thereby destabilizing the system and promoting the development of resistance. To date, there is no published evidence to suggest that this has occurred for viruses currently being used either commercially

B. RESEARCH REQUIREMENTS

The above considerations demonstrate that studies of the phenomenon of viral resistance are necessary for the development of effective control procedures. Such studies should be pursued in several directions. First, a methodology should be developed for the systematic detection of differences in response to a virus in field populations. Second, the more fundamental aspects of the genetics and physiology of defense mechanisms must be investigated. Third, our knowledge of the interactions between the life systems of virus and insect host in the field, especially of factors that might contribute to the manifestation of resistance in the field, must be increased.

As has been repeatedly mentioned in the case histories reported here, many of the earlier studies provided data which were not adequately quantified, leading to doubts about their interpretation. Variations in methods of virus purification and administration and subsequent analyses of dosage-mortality curves have made many sets of data incomparable. Solutions to the first two problems are largely specific to each host-virus interaction, and the sophistication of techniques currently available should ensure that adequate and comparable techniques can be established for the systematic testing of populations. With regard to analysis, it is important to reiterate Burges' (1971) plea for the presentation of all dosage-mortality parameters in published data. In particular, failure to realize the importance of the slope of dosage-mortality curves has, in some instances, led to an inability to distinguish between shifts in the variability of response and the development of resistance. The importance of such a distinction for control procedures is obvious.

Significant progress in understanding the physiological defense mechanisms involved in resistance has been made only for the silkworm, *B. mori*. The evidence suggests that the major obstacle to viral infection lies in the midgut barrier, although inhibition of viral particles can occur in the digestive juices and, after initial infection has taken place, in the hemolymph. There are also indications that insects may possess common defense mechanisms against different viruses, and studies of pest species, similar to those carried out on *B. mori*, should be undertaken to elucidate the physiology of resistance. As Granados (1980) has pointed out, once these factors are identified it might be possible to manipulate the host-virus system to effect better control.

Evidence from *B. mori* indicates that some of these defense mechanisms are under genetic control. Further, while large shifts in resistance tend to be controlled by single locus genes, part of the variability seems to be of polygenic origin and is linked to the overall genetic architecture of the strains. The nature of the controlling gene system is considered to be one of the primary determinants of the development of resistance to chemical insecticides (Tabashnik and Croft, 1982), and therefore, it will be important to understand the formal genetics of resistance to viruses intended for use as control agents. This might also help us understand whether drifts in population response levels, such as those mentioned in the case of *L. dispar*, have a genetic origin or otherwise.

Developmental factors such as growth rate, hormonal levels, stage of development, and stages of population cycle have all been implicated in affecting the level of susceptibility shown by an insect, as have environmental factors, such as, food type, light, temperature, and physical stress. The particular response to a virus shown by any given field population of insect will be determined by the interaction between genetic, environmental, and developmental factors (Briese, 1984), and further studies are needed to understand these effects and how they influence the course of selection.

Other factors involving the population dynamics of the insect-virus system will also determine the extent of any shift in population response to a virus; the frequency and level of exposure to the virus, rates of immigration, and the reproductive fitness of resistant phenotypes under "normal" conditions. It will eventually be necessary to understand the role of these factors, so that control programs can be planned to minimize or forestall shifts in response to an applied virus. Several models have recently been produced which examine the development of resistance to chemical insecticides (Georghiou and Taylor, 1977a,b; Comins, 1979; Whitten and McKenzie, 1982; Tabashnik and Croft, 1982). They have helped to determine key factors and formulate recommendations for minimizing the risk of increasing resistance. At present, those models could provide useful guidelines for the usage of viral insecticides in the field. Eventually though, it might prove fruitful to modify them to better accommodate viruses. Areas of difference between viral and chemical insecticides which may have to be considered include the ability of viruses to reproduce and their potential for changes in virulence.

In conclusion, it is important to view the issue of resistance in context, as one of several equally important problems (see Burges, 1981) facing the development of viral control agents. If an adequate understanding of how resistance develops can be obtained, the risk of acquired resistance should not impede the successful use of viruses as control agents.

ACKNOWLEDGMENTS

We would like to thank the following people for their advice on certain aspects of the case histories reported here: Dr. H. Watanabe for the silkworm, Dr. P. W. Geier for the light-brown apple moth, Dr. C. C. Payne, and Dr. J. Huber for the large white butterfly.

REFERENCES

Aizawa, K. (1962). Antiviral substance in the gut-juice of the silkworm *Bombyx mori* (Linneus). *J. Insect Pathol. 4*, 72-76.

Aizawa, K., and Furuta, Y. (1962). Resistance to virus induction in F_1 hybrids between resistant and common strains in the silkworm *Bombyx mori* L. *J. Sericult. Sci. Japan 31*, 245-252.

Aizawa, K., and Furuta, Y. (1964). Resistance to polyhedroses in F_1 hybrids between resistant and original strains in the silkworm, *Bombyx mori*. *J. Sericult. Sci. Japan 33*, 403-406.

Aizawa, K., Furuta, Y., and Nakamura, K. (1961). Selection of a resistant strain to virus induction in the silkworm *Bombyx mori*. *J. Sericult. Sci. Japan 30*, 405-412. (in Japan.)

Anderson, R. M. (1982). Theoretical basis for the use of pathogens as biological control agents of pest species. *Parasitology 84*, 3-34.

Aratake, Y. (1973a). Strain differences of the silkworm *Bombyx mori* L. in the resistance to a nuclear polyhedrosis virus. *J. Sericult. Sci. Japan 42*, 230-238. (in Japan.)

Aratake, Y. (1973b). Difference of the resistance to infectious flacherie virus between the strains of the silkworm, *Bombyx mori* L. *Sansi-Kenkyu 86*, 48-57. (in Japan.)

Aratake, Y., and Ueno, H. (1973). Inactivation of a nuclear-polyhedrosis virus by the gut-juice of the silkworm, *Bombyx mori* L. *J. Sericult. Sci. Japan 42*, 279-284. (in Japan.)

Aratake, Y., Kayamura, T., and Watanabe, H. (1974). Inactivation of a cytoplasmic-polyhedrosis virus by gut-juice of the silkworm, *Bombyx mori* L. *J. Sericult. Sci. Japan 43*, 41-44. (in Japan.)

Aruga, H., and Watanabe, H. (1964). Resistance to *per os* infection with cytoplasmic-polyhedrosis virus in the silkworm, *Bombyx mori* (Linnaeus). *J. Insect Pathol. 6*, 387-394.

Brand, R. J., and Pinnock, D. E. (1981). Application of biostatistical modelling to forecasting the results of microbial control trials. *In* "Microbial Control of Pests and Plant Diseases 1970-1980" (H. D. Burges, ed.), pp. 667-693. Academic Press, London.

Breillatt, J. P., Brantley, J. N., Mazzone, H. M., Martignoni, M. E., Franklin, J. E., and Anderson, N. G. (1972). Mass purification of nucleopolyhedrosis virus inclusion bodies in the K-series centrifuge. *Appl. Microbiol. 23*, 923-930.

Briese, D. T. (1981a). Resistance of insect species to microbial pathogens. *In* "Pathogenesis of Invertebrate Microbial Diseases" (E. W. Davidson, ed.), pp. 511-545. Allenheld-Osmun, Totowa.

Briese, D. T. (1981b). The incidence of parasitism and disease in field populations of the potato moth, *Phthorimaea operculella* (Zeller) in Australia. *J. Aust. Entomol. Soc. 20*, 319-326.

Briese, D. T. (1982). Genetic basis for resistance to a granulosis virus in the potato moth, *Phthorimaea operculella*. *J. Invertebr. Pathol. 39*, 215-218.

Briese, D. T. (1984). Insect resistance to baculoviruses. *In* "Biology of Baculoviruses" (R. Granados and B. Federici, eds.), in press. CRC Press, Boca Raton.

Briese, D. T., and Mende, H. A. (1981). Differences in susceptibility to a granulosis virus between populations of the potato moth *Phthorimaea operculella* (Zeller). *Bull. Entomol. Res. 71*, 11-18.

Briese, D. T., and Mende, H. A. (1983). Selection for increased resistance to a granulosis virus in the potato moth, *Phthorimaea operculella* (Zeller). *Bull. Entomol. Res. 73*, in press.

Briese, D. T., Mende, H. A., Grace, T. D. C., and Geier, P. W. (1980). Resistance to a nuclear polyhedrosis virus in the light-brown apple moth, *Epiphyas postvittana* (Lepidoptera: Tortricidae). *J. Invertebr. Pathol. 36*, 211-215.

Burges, H. D. (1971). Possibilities of pest resistance to microbial control agents. *In* "Microbial Control of Insects and Mites" (H. D. Burges and N. W. Hussey, eds.), pp. 445-457. Academic Press, London.

Burges, H. D. (1981). Safety, safety testing and quality control of microbial pesticides. *In* "Microbial Control of Pests and Plant Diseases 1970-1980" (H. D. Burges, ed.), pp. 737-768. Academic Press, London.

Carner, G. R., Hudson, J. S., and Barnett, O. W. (1979). The infectivity of a nuclear polyhedrosis virus of the velvetbean caterpillar for eight noctuid hosts. *J. Invertebr. Pathol. 33*, 211-216.

Comins, H. N. (1979). The management of pesticide resistance: models. *In* "Genetics in Relation to Insect Management" (M. A. Hoy and J. J. McKelvey, Jr., eds.), pp. 55-68.

Couch, T. L., and Ignoffo, C. M. (1981). Formulation of insect pathogens. *In* "Microbial Control of Pests and Plant Diseases 1970-1980" (H. D. Burges, ed.), pp. 621-634. Academic Press, London.

David, W. A. L. (1978). The granulosis virus of *Pieris brassicae* (L.) and its relationship with its host. *Advan. Virus Res. 22*, 112-161.

David, W. A. L., and Gardiner, B. O. C. (1960). A *Pieris brassicae* (Linnaeus) culture resistant to a granulosis. *J. Insect Pathol. 2.*, 106-114.

David, W. A. L. and Gardiner, B. O. C. (1965). Resistance of *Pieris brassicae* (Linnaeus) to granulosis virus and the virulence of a virus from different host races. *J. Invertebr. Pathol. 7*, 285-290.

Doane, C. C. (1970). Primary pathogens and their role in the development of an epizootic in the gypsy moth. *J. Invertebr. Pathol. 15*, 21-33.

Evans, H. F. (1981). Quantitative assessment of the relationships between dosage and response of the nuclear polyhedrosis virus of *Mamestra brassicae*. *J. Invertebr. Pathol. 37*, 101-109.

FAO (1973). Use of insect viruses in integrated pest control. *Plant Protein Bull. 21*, 142-143.

Faulkner, P. (1981). Baculoviruses. *In* "Pathogenesis of Invertebrate Microbial Diseases" (E. W. Davidson, ed.), pp. 5-37. Allenheld-Osmun, Totowa.

Flückiger, C. R. (1982). Untersuchungen über drei Baculovirus-Isolate des Schalenwicklers, *Adoxophyes orana* F.V.R. (Lep., Tortricidae), dessen Phänologie und erste Feldversuche, als Grundlagen zur Mikrobiologischen Bekämpfung diese Obstschädlings. *Mitt. Schweiz. Entomol. Ges. 55*, 241-288.

Funada, T. (1968). Genetic resistance of the silkworm *Bombyx mori* L. to an infection of a flacherie virus. *J. Sericult. Sci. Japan 37*, 281-287. (in Japan.)

Furuta, Y. (1978). On the heredity of the resistance to small flacherie virus in the silkworm, *Bombyx mori*. *J. Sericult. Sci. Japan 47*, 241-242. (in Japan.)

Geier, P. W. and Briese, D. T. (1979). The light-brown apple moth, *Epiphyas postvittana* (Walker): 3. Differences in susceptibility to a nuclear polyhedrosis virus. *Aust. J. Ecol. 4*, 187-194.

Geier, P. W., and Briese, D. T. (1980). The light-brown apple moth, *Epiphyas postvittana* (Walker): 5. Variability of demographic characteristics in field populations of southeastern Australia. *Aust. J. Ecol. 5*, 135-142.

Geier, P. W., and Briese, D. T. (1981). The light-brown apple moth, *Epiphyas postvittana* (Walker): A native insect fostered by European settlement. *In* "The Ecology of Pests: Some Australian Case Histories" (R. L. Kitching and R. E. Jones, eds.), pp. 131-154. CSIRO, Melbourne.

Geier, P. W., and Oswald, L. T. (1977). The light-brown apple moth, *Epiphyas postvittana* (Walker): 1. Effects associated with contamination by a nuclear polyhedrosis virus on the demographic performance of a laboratory strain. *Aust. J. Ecol. 2*, 9-29.

Geier, P. W., and Springett, B. P. (1976). Population characteristics of Australian leaf-rollers infesting orchards (*Epiphyas* spp., Lepidoptera). *Aust. J. Ecol. 1*, 129-144.

Georghiou, G. P., and Taylor, C. E. (1977a). Genetic and biological influences in the evolution of insecticide resistance. *J. Econ. Entomol. 70*, 319-323.

Georghiou, G. P., and Taylor, C. E. (1977b). Operational influences in the evolution of insecticidal resistance. *J. Econ. Entomol. 70*, 653-658.

Glaser, R. W., and Chapman, J. W. (1913). The wild disease of gypsy moth caterpillars. *J. Econ. Entomol. 6*, 479-488.

Granados, R. R. (1980). Infectivity and mode of action of baculoviruses. *Biotech. Bioeng. 22*, 1377-1405.

Harrap, K. A., and Payne, C. C. (1979). The structural properties and identification of insect viruses. *Advan. Virus Res. 25*, 273-356.

Hayashiya, K., Nishida, J., and Kawamoto, F. (1971). On the biosynthesis of the red fluorescent protein in the digestive juice of the silkworm larvae. *Jap. J. Appl. Entomol. Zool. 15*, 109-114. (in Japan.)

Hayashiya, K., Nishida, J., and Uchida, Y. (1976). The mechanism of formation of red fluorescent protein in the digestive juice of silkworm larvae: the formation of chlorophyllide-a. *Jap. J. Appl. Entomol. Zool. 20*, 37-43. (in Japan.)

Hayashiya, K., Uchida, Y., and Himeno, M. (1978). Mechanism of antiviral action of red fluorescent protein (RFP) on nuclear-polyhedrosis virus (NPV) in silkworm larvae. *Jap. J. Appl. Entomol. Zool. 22*, 238-242. (in Japan.)

Hedlund, R. C. (1974). Field and laboratory investigations of a nuclear polyhedrosis virus of the gypsy moth, *Porthetria dispar* (L.). Ph.D. Thesis, Pennsylvania State Univ., University Park, Pennsylvania.

Horie, Y., and Watanabe, H. (1980). Recent advances in sericulture. *Amer. Rev. Entomol. 25*, 49-71.

Huber, J. (1974). Selektion einer Resistenz gegen perorale infektion mit einem Granulosisvirus bei einem Laborstamm des Apfelwicklers, *Lespeyresis pomonella* L. Diss. Nr. 5044, Eidgenössich. Techn. Hochs. Zurich, pp. 45.

Inoue, H. (1974). Multiplication of an infectious-flacherie virus in the resistant and susceptible strains of the silkworm *Bombyx mori*. *J. Sericult. Sci. Japan 43*, 318-324. (in Japan.)

Inoue, H., and Miyagawa, M. (1978). Regeneration of midgut epithelial cells in the silkworm, *Bombyx mori*, infected with viruses. *J. Invertebr. Pathol. 32*, 373-380.

IPRC (1980). Recommendations. *Proc. Workshop Insect Pest Management with Microbial Agents: Recent Achievements, Deficiencies, and Innovations*, pp. 53-66. Insect Pathology Resource Center, Ithaca, N.Y.

Kobara, R., Aruga, H., and Watanabe, H. (1967). Effect of larval growth on the susceptibility of silkworm, *Bombyx mori* L., to a cytoplasmic polyhedrosis virus. *J. Sericult. Soc. Jap. 36*, 165-168. (in Japan.)

Kobayashi, M. (1982). Factors affecting the susceptibility of the cultured silkworm embryo, *Bombyx mori* L., to nuclear polyhedrosis virus infection. *Jap. J. Appl. Entomol. Zool. 26*, 68-73. (in Japan.)

Kobayashi, M., Yamaguchi, S., and Yokoyama, Y. (1969). Influences of the larval development on the susceptibility of the silkworm *Bombyx mori* L. to the nuclear-polyhedrosis virus. *J. Sericult. Sci. Japan 38*, 481-487. (in Japan.)

Lewis, F. B., and Rollinson, W. D. (1978). Effect of storage on the virulence of gypsy moth nucleopolyhedrosis inclusion bodies. *J. Econ. Entomol. 71*, 719-722.

Lewis, F. B., McManus, M. L., and Schneeberger, M. F. (1979). Guidelines for the use of Gypchek® to control the gypsy moth. U.S. Dept. Agr. Res. Paper NE-441, pp9.

Lewis, F. B., Rollinson, W. D., and Yendol, W. G. (1981). Gypsy moth nucleopolyhedrosis virus: Laboratory evaluations. *In* "The Gypsy Moth: Research toward Integrated Pest Management" (C. C. Doane and M. L. McManus, eds.), pp. 455-459. USDA Tech. Bull. 1584. pp. 757.

Magnoler, A. (1974). Bioassay of a nucleopolyhedrosis virus of the gypsy moth, *Porthetria dispar*. *J. Invertebr. Pathol. 23*, 190-196.

Magnoler, A. (1974). Bioassay of a nucleopolyhedrosis virus of the gypsy moth, *Porthetria dispar*. *J. Invertebr. Pathol. 23*, 190-196.

Matthiessen, J. N., Christian, R. L., Grace, T. D. C., and Filshie, B. K. (1978). Large-scale field propagation and the purification of the granulosis virus of the potato moth, *Phthorimaea operculella* (Zeller) (Lepidoptera: Gelechiidae). *Bull. Entomol. Res. 61*, 207-222.

Payne, C. M. (1982). Insect viruses as control agents. *Parasitology 84*, 35-78.

Payne, C. C., Tatchell, G. M., and Williams, C. F. (1981). The comparative susceptibilities of *Pieris brassicae* and *Pieris rapae* to a granulosis virus from *P. brassicae*. *J. Invertebr. Pathol. 38*, 273-280.

Podgwaite, J. D. (1981). Natural disease within dense gypsy moth populations. *In* "The Gypsy Moth: Research toward Integrated Pest Management." (C. C. Doane and M. L. McManus, eds.), pp. 125-134. USDA Tech. Bull. 1584. pp. 757.

Reed, E. M., and Springett, B. P. (1971). Large-scale field testing of a granulosis virus for the control of the potato moth (*Phthorimaea operculella* (Zell.) (Lep., Gelechiidae)). *Bull. Entomol. Res. 61*, 223-233.

Reiff, W. (1911). "The Wilt Disease, or Flacherie of the Gypsy Moth," 60 pp. Wright and Potter Printing Co., Boston, MA.

Ripa, R. (1978). Studies of the susceptibility of *Pieris brassicae* (L.) to a granulosis virus. Ph.D. Thesis, University of London.

Rivers, C. F. (1959). Virus resistance in larvae of *Pieris brassicae* (L.). *Trans. 1st Intern. Conf. Insect Pathology Biol. Control, Praha, 1958*, pp. 205-210.

Rollinson, W. D., and Lewis, F. B. (1973). Susceptibility of gypsy moth larvae to *Lymantria* spp. nuclear and cytoplasmic polyhedrosis virus. *Zast. Bilja 24*, 163-168.

Shapiro, M., Owens, C. D., Bell, R. A., and Wood, H. A. (1981). Simplified efficient system for *in vivo* mass production of gypsy moth nuclear polyhedrosis virus. *J. Econ. Entomol. 74*, 341-343.

Sidor, C. (1959). Susceptibility of larvae of the large white butterfly (*Pieris brassicae* L.) to two virus diseases. *Ann. Appl. Biol. 47*, 109-113.

Smith, G. E., and Summers, M. D. (1978). Analysis of baculovirus genomes with restriction endonucleases. *Virology 89*, 517-527.

Tabashnik, B. E., and Croft, B. A. (1982). Managing pesticide resistance in crop-arthropod complexes: Interactions between biological and operational factors. *Environ. Entomol. 11*, 1137-1144.

Tanaka, S. (1969). Comparison of the mode of multiplication of the flacherie virus in silkworm strains having different susceptibilities. *J. Sericult. Soc. Jap. 38*, 21-27. (in Japan.)

Tinsley, T. W. (1979). The potential of insect pathogenic viruses as pesticidal agents. *Annu. Rev. Entomol. 24*, 63-87.

Uzigawa, K., and Aruga, H. (1966). On the selection of resistant strains to the infectious flacherie virus in the silkworm *Bombyx mori* L. *J. Sericult. Sci. Japan 35*, 23-27. (in Japan.)

Vasiljevic, L. (1961). Influence of various phases of gradation upon the susceptibility of gypsy moth caterpillars to polyhedrosis. *J. Sci. Agr. Res. 14*, 1-16.

Vasiljevic, L., and Injac, M. (1973). A study of gypsy moth viruses originating from different geographical regions. *Zast. Bilja 24*, 169-186.

Watanabe, H. (1965). Resistance to peroral infection by the cytoplasmic-polyhedrosis virus in the silkworm, *Bombyx mori* (Linnaeus). *J. Invertebr. Pathol. 7*, 257-258.

Watanabe, H. (1966a). Relative virulence of polyhedrosis viruses and host-resistance in the silkworm *Bombyx mori* L. (Lepidoptera: Bombycidae). *Appl. Entomol. Zool. 1*, 139-144.

Watanabe, H. (1966b). Some aspects on the mechanism of resistance to peroral infection by cytoplasmic-polyhedrosis virus in the silkworm, *Bombyx mori* L. *J. Sericult. Sci. Japan 35*, 411-417. (in Japan.

Watanabe, H. (1967). Development of resistance in the silkworm, *Bombyx mori*, to peroral infection of a cytoplasmic-polyhedrosis virus. *J. Invertebr. Pathol. 9*, 474-479.

Watanabe, H. (1971). Resistance of the silkworm to cytoplasmic-polyhedrosis virus. *In* "The Cytoplasmic-Polyhedrosis Virus of the Silkworm" (H. Aruga and Y. Tanada, eds.), pp. 169-184. University of Tokyo Press, Tokyo.

Watanabe, H. (1981). Characteristics of densonucleosis in the silkworm, *Bombyx mori*. *JARQ: Jap. Agric. Res. Quart. 15*, 133-136.

Watanabe, H., and Aruga, H. (1970). Susceptibility of dauerpupa of the silkworm, *Bombyx mori* L. (Lepidoptera: Bombycidae), to a nuclear polyhedrosis virus. *Appl. Entomol. Zool. 5*, 118-120.

Watanabe, H., and Imanishi, S. (1980). The effect of the content of certain ingredients in an artificial diet on the susceptibility to virus infection in the silkworm, *Bombyx mori*. *J. Sericult. Sci. Japan 49*, 404-409. (in Japan.)

Watanabe, H., and Maeda, S. (1978). Genetic resistance to peroral infection with a densonucleosis virus in the silkworm, *Bombyx mori*. *J. Sericult. Sci. Japan 47*, 209-214. (in Japan.)

Watanabe, H., and Maeda, S. (1981). Genetically determined non-susceptibility of the silkworm, *Bombyx mori*, to infection with a densonucleosis virus (*Densovirus*). *J. Invertebr. Pathol. 38*, 370-373.

Watanabe, H., and Takamiya, K. (1976). Susceptibility of the silkworm larvae, *Bombyx mori*, reared under different light conditions to polyhedrosis viruses. *J. Sericult. Sci. Japan 45*, 403-406. (in Japan.)

Watanabe, H., Tanaka, S., and Simizu, T. (1974). Interstrain difference in the resistance of the silkworm *Bombyx mori* to a flacherie and a cyroplasmic polyhedrosis virus. *J. Sericult. Sci. Japan 43*, 98-100. (in Japan.)

Whitten, M. J., and McKenzie, J. A. (1982). The genetic basis for pesticide resistance. *Proc. 3rd Australasian Conf. Grassl. Invert. Ecol.*, pp. 1-16. S.A. Govt. Printer, Adelaide.

Yamaguchi, K. (1979). Natural recovery of the fall webworm, *Hyphantria cunea* to infection by a cytoplasmic-polyhedrosis virus of the silkworm, *Bombyx mori*. *J. Invertebr. Pathol. 33*, 126-128.

Yearian, W. C. (1978). Application technology to increase effectiveness of entomopathogens. *In* "Microbial Control of Insect Pests: Future Strategies in Pest Management Systems" (G. E. Allen, C. M. Ignoffo, and R. P. Jaques, eds.), pp. 100-110. NSF, USDA and University of Florida.

Yokoyama, T. (1979). Silkworm selection and hybridisation. *In* "Genetics in Relation to Insect Management" (M. A. Hoy and J. J. McKelvey, Jr., eds.), pp. 71-83. Rockefeller Foundation, New York.

THE SAFETY OF INSECT VIRUSES AS BIOLOGICAL CONTROL AGENTS

GABRIELE DÖLLER

Department for Medical Virology and Epidemiology of Virus Diseases
Hygiene-Institute
University of Tübingen
Federal Republic of Germany

I. INTRODUCTION

Many problems arise in connection with the damage caused to many economically important plants in agriculture and forestry by insects. One example is the reduction of food production. The effective control of insect pests has therefore become important, especially in developing countries whose populations are increasing (Woods, 1974; Falcon, 1978; Harrap and Tinsley, 1978). Despite human intervention, insects, whether as direct or indirect vectors, develop quickly. Fifty years ago when large-scale chemical pest control came into use relatively good results were obtained. However, it is now well documented that chemical pesticides cause many side effects in nontarget organisms with the result that great physiological and genetical damage, or even death, may be induced by these compounds. Pest control using pesticides became a matter for concern for numerous reasons, including: chemical resistance of target insects, the accumulation of harmful chemical residues in the environment, and disruption of balanced insect communities. In addition was the need of a rapidly growing world population requiring more food. About 300 insect species are already more or less resistant to chemical insecticides (WHO, 1973a). These increasing difficulties have stimulated investigations by several laboratories into new ways of

ISBN 0-12-470295-3

controlling insect pests. The use of biological insecticides such as viruses, fungi, nematodes, bacteria, and microsporidia has been an important step in this direction (Franz and Krieg, 1976). Since insect viruses have long been known to cause epizootics (Smith, 1967), they rapidly became a necessity in integrated control programs using parasites, predators, or chemicals.

The expression "insect viruses" is used here to indicate the viruses infecting insects. Two representatives of the occluded viruses (Ignoffo, 1968), namely, the nuclear polyhedrosis viruses (NPVs) and granulosis viruses (GVs), designated together as Baculoviridae (Wildly, 1971) seemed to be very well suited for application in pest control. In both cases the DNA-containing rod-shaped virions are generally occluded in protein crystals called inclusion bodies.

Although baculoviruses are isolated from naturally infected, dead insects, their use as insecticides in many countries is subject to regulations and proof is required to show that they are safe, efficacious, and without detrimental impact upon the environment. Thus, the question which arises is how safe is the application of pathogenic insect viruses. In other words, is there potential harm to mammals, man, and to nontarget organisms when baculoviruses are sprayed as pesticides? Several workshops have been held in the United States and in Europe concerning this question. As far back as 1973, the World Health Organization had recommended test procedures for safety testing of pathogenic insect viruses (WHO, 1973). The United States Environmental Protection Agency (EPA) published guidelines for safety testing in 1975 (Summers et al., 1975). Two years later, an EPA conference at Myrtle Beach, North Carolina, recommended the use of a battery of test systems, including techniques in molecular genetics, biochemistry, serology, virology, and cell biology (Summers and Kawanishi, 1978). In 1978, another international workshop organized by the Federal Ministry for Research and Technology of the Federal Republic of Germany (FRG), held at Jülich (FRG), resulted in similar recommendations (Miltenburger, 1978). The working group on the safety of Microbial Control Agents of the Society for Invertebrate Pathology prepared a statement on the present knowledge of the efficacy, specificity, and safety of baculoviruses (Krieg et al., 1980). In 1981, a WHO memorandum outlined the safety tests necessary to assess the effects of biological agents in pest control on mammals, and hence to perform a risk/benefit analysis for each of the different uses envisaged for the agents. The recommendations took into account and generally accepted the principles embodied in several recent consultative documents on the testing of the safety of microbial pathogens of invertebrates (Kurstak et al., 1978; Burges et al., 1980a; Burges, 1982; Krieg et al., 1980).

It is very difficult to test the safety of baculoviruses, because they do not infect nontarget organisms. Therefore, in contrast to chemical insecticides a LD_{50} level cannot be obtained. It is, therefore, desirable to test quantities of the active, unformulated virus product in dosages related to dosages used in the field. Between 1970-1980 many safety tests were carried out with baculoviruses. In these studies, NPV proved harmless to microorganisms, noninsect invertebrate cell lines, vertebrate cell lines, vertebrates, plants, and nonarthropod invertebrates. Replication was unusual in insects other than the insect family in which the virus was first discovered. For example, GVs occur only in Lepidoptera, most are believed to be very specific, and none have replicated in cell lines from other insects or animals. Summers et al. (1975), Kurstak et al. (1978), Burges et al. (1980a), Miltenburger (1978), and Burges (1982) have summarized all of the safety tests carried out until 1980. During this period the NPV of *Heliothis zea* underwent a series of tests as thorough as those required for chemicals by the U.S. Environmental Protection Agency (EPA) and by guidelines recommended by WHO, including long-term carcinogenicity and teratogenicity tests and tests on primates and man. Since 1975, the NPV of *Heliothis zea* has been registrated in the United States as a viral insecticide (Elcar).

New methods at the molecular level have permitted a more detailed study of the viral genome and its behavior. It is now possible to investigate persistence and expression of the viral genome or its parts in nontarget systems. These methods (DNA hybridization and restriction endonuclease analysis) were also used to control for the safety of baculoviruses. Furthermore, safety studies in regard to baculovirus replication in nontarget organisms have been carried out with more sensitive antigen respectively antibody detection systems like radio or enzyme immunoassay.

In this chapter the methods and results primarily obtained by experiments published since 1978 will be summarized.

II. SAFETY TESTS WITH NUCLEAR POLYHEDROSIS VIRUS (Fam.: Baculoviridae) FROM DIFFERENT HOSTS

A. *TESTS WITH INVERTEBRATES: HOST RANGE*

The host range of baculoviruses has been studied both *in vivo* and *in vitro*. The nuclear polyhedrosis viruses (NPVs) are reported to be very host specific and replicate in only one, or in a few, related host species (Vaughn et al., 1977).

Typical infection *in vivo* and *in vitro* results in the production of proteinaceous crystals called polyhedra inclusion bodies (PIBs) into which a considerable proportion of virus particles became occluded. PIBs are, therefore, considered one marker of virus replication. However, Sherman and McIntosh (1979) reported that the baculovirus from the lepidopteran host *Autographa californica* was shown to replicate in dipteran cell lines without the production of characteristic PIBs. The low level of replication could not be detected by 50% tissue culture infective dose titrations, but was apparent by [^{3}H]thymidine labeling of the viral genome. Immunoprecipitation of the radioactive product confirmed baculovirus production. The results of these experiments indicate that the *Autographa californica* NPV is being produced in low amounts by a mosquito cell line. This was the first demonstration that baculoviruses may not be as host specific *in vitro* as was previously thought. In these experiments, *Autographa californica* NPV not only crossed species lines but also replicated in another order of insects. The authors believed that the given results point out the need for a reevaluation of the host ranges and virus specificities of the baculoviruses. In the last few years many attempts have been made to clarify host specificity, recombination, and mutation of baculoviruses in invertebrates.

Tjia et al. (1979) and other investigators (Lee and Miller, 1978; Smith and Summers, 1978) have found that stock virus preparations generally considered to be *Autographa californica* NPV do not necessarily constitute a genetically homogeneous population. Moreover, *Autographa californica* NPV isolates used in different laboratories exhibit distinct differences in the restriction analysis of their DNA's. It will, therefore, be important to characterize virus preparations used in the laboratory or even field studies. Since differences in the cleavage pattern of the *Autographa californica* NPV DNA, derived from different plaque isolates of a given population, are not very extensive, it remains to be established that the apparent genetic heterogeneity may not arise by a high frequency of recombination between the virus and host genomes. Tjia et al. (1979) concluded tentatively from their own findings that integration of *Autographa californica* NPV DNA into the genome of the homologous insect host is not a frequent event. In reality, until now no definite data exist which would exclude the recombination of NPV in homologous or heterologous insect hosts.

The resistance of the light brown apple moth *Epiphyas postvittana* (Lepidoptera: Tortricidae) to NPV was described by Briese et al. (1980). Measurements were made of the relative susceptibility of three populations of *Epiphyas postvittana* to NPV: a resistant laboratory strain, a susceptible laboratory strain, and a field population. The resistant laboratory strain was found to be 50 times more resistant than the susceptible laboratory strain and 160 times more resistant than the field line. These measurements in *Epiphyas postvittana* revealed mechanisms of the resistance which differed from the classical case of determination by a dominant major gene, such as was found in the silkworm, *Bombyx mori*, by Watanabe (1965). This indicates that genetic resistance to NPV can arise in several ways. The question which remained was whether the resistant laboratory strain was naturally resistant or whether this resistance was as a result of laboratory rearing. In addition, is the question as to whether resistance might have developed to viruses used as agents of biological insect control. Briese et al. (1980) reported that in contrast to the susceptible strain the resistant laboratory strain had been reared in a NPV-contaminated environment and had suffered several outbreaks of NPV, which suggests that resistance might have developed in the laboratory. This is supported by evidence, from rearing experiments, that the progeny of survivors of a NPV infection were more resistant to subsequent infection. If selection for resistance could be effected in the laboratory, the same could occur in the field upon continued exposure to the virus. With *Epiphyas postvittana* the natural selection of a less susceptible form may be possibly enhanced by the variability in response to NPV (Geier and Briese, 1979). The situation could be further complicated by variability of the NPV. For example, Vlak et al. (1981) discussed the genetic variability of six naturally occurring NPVs of the cabbage moth, *Mamestra brassicae*. Six wild isolates of *Mamestra brassicae* NPV were obtained from the Netherlands (NL), Federal Republic of Germany (D), Yugoslavia (Yu), Bulgaria (B), and two regions of the Soviet Union, the Ukraine (Uk) and Siberia (S). The polyhedrins (matrix protein), virion polypeptides, and DNA's of the six isolates were analyzed and compared. On SDS-polyacrylamide gels the electrophoretic mobilities of virion polypeptides were similar, but were distinct for each of the six isolates. All polyhedrins had molecular weights of about 34,000. Partial digestion with V8 protease indicated that the polyhedrins of *Mamestra brassicae* NPV-Uk and -B had similar digestion products, which differed from the polyhedrins of the other *Mamestra brassicae* NPVs. Restriction endonuclease analysis and southern-blot hybridization showed that the viral DNA's of 5 out of 6 isolates had different but related nucleotide sequences. These results indicate that there is DNA heterogeneity and suggest that these isolates contain variant viruses. It is not yet known whether the main genotype in one geographical isolate is a minor variant or whether all major and minor variants in a given isolate possess unique genotypes (Vlak, 1982). Summers et al. (1980) showed that baculovirus recombinants can be obtained from doubly infected cells without the use of biological selection for recombinants. Based on these results, it is not surprising that Vlak et al. (1981) found a new NPV from *Spodoptera exigua*.

To test the sensitivity to *Spodoptera exigua* NPV, first instar larvae were fed polyhedra. Virus could cause a polyhedrosis in *Spodoptera exigua* but not in the congenetic species *Spodoptera littoralis*. Cross-transmission of *Spodoptera exigua* NPV to other noctuids, *Mamestra brassicae*, *Mamestra oleraceae*, *Agrotis segetum*, or members of other families, *Ostrinia nubilalis*, *Galleria mellonella*, *Plutella maculipennis*, *Adoxophyes orana*, and *Laspeyresia pomonella* did not occur. *Spodoptera exigua* NPV was further characterized by analysis of the viral DNA with restriction endonucleases. DNA heterogeneity in the virus preparation, as seen with *Autographa californica* NPV (Smith and Summers, 1978), was not apparent but cannot be excluded until *Spodoptera exigua* NPV has been plaque-purified. In addition, the *Eco*RI digestion pattern of this *Spodoptera exigua* DNA did not resemble that of other baculovirus DNA's. The authors concluded that *Spodoptera exigua* NPV described, is a new, unique baculovirus.

Allaway (1981) controlled the host range of *Mamestra brassicae* NPV and *Agrotis segetum* NPV for first instar larvae of *Pieris rapae*, *Agrotis segetum*, *Agrotis exclamationis*, *Noctua pronuba*, *Mamestra brassicae*, *Lacanobia oleracea*, and *Plusia gamma*. *Mamestra brassicae* NPV infected all the species tested and was highly virulent for all but *Agrotis* spp., whereas *Agrotis segetum* NPV infected only the three cutworm species and were most virulent for *Agrotis segetum*.

In vitro studies with NPV of *Autographa californica* showed that *Spodoptera littoralis* cells were as susceptible to infection as the *Spodoptera frugiperda* cells (Roberts and Naser, 1981). Pawar (1982) investigated bioassays with the two closely related species, *Spodoptera litura* and *Spodoptera littoralis* by testing the progeny virus obtained by homologous and reciprocal infections; *Spodoptera litura* NPV was cross-infective to *Spodoptera littoralis*. However, *Spodoptera littoralis* NPV caused expression of latent virus in the *Spodoptera litura* and thus was not truly cross-infective. Crawford and Sheehan (1982) showed that serial passage of *Spodoptera littoralis* NPV in *Spodoptera frugiperda* cells for 21 passages reduced the amount of infectious virus approximately 50-fold. This drop in ability to produce infectious virus was accompanied by an ability of the cell culture fluids to interfere with replication of wild-type virus. The authors suggested that this interference was the result of the production of defective interfering (DI) particles. Burand et al. (1983) also reported about defective particles from a persistent baculovirus infection in *Trichoplusia ni* tissue culture cells. An *Heliothis zea* baculovirus clone (HZ-1) isolated from persistently infected *Trichoplusia ni* (TN-368) tissue culture cells was homogenous as determined by particle morphology and restriction enzyme analysis of virus DNA. Reestablishment of a persistent infection of TN-368 cells with this clone (B1) was unsuccessful. There was a 99% decrease in infectious virus titer following 10 passages in tissue culture at a high multiplicity of infection, and the 10th passage virus population was capable of establishing a persistently infected cell line (TN-B1). Based on particle morphology, restriction enzyme analysis, and focus-forming plaque assays, the virus population recovered from TN-B1 cells included a high percentage of defective virus particles which did not interfere with the infection of second-passage B1 virus particles. The authors considered the possibility that the lack of interference may have been due to evolution of infectious HZ-1 virus particles, such as in the B1 clone resistant to the interfering properties of the defective particles. Another possibility is that HZ-1 persistence is connected to a temperature-sensitive mutation. Burand et al. (1983) intend to investigate this possibility. Furthermore, the authors discussed that as with certain diseases in animals, several insect diseases have been associated with persistent or latent viruses. It is thought that under certain conditions these latent insect viruses become activated and result in a productive virus infection. However, neither the state of these viruses nor the mechanism of activation are known.

Kislev and Edelman (1982) compared the DNA restriction-pattern from two NPVs, isolated from *Spodoptera littoralis* larvae collected from several areas in Israel. It was found that the viruses had no detectable sequence homology and it was suggested that these viruses are not closely related.

Thus far, recombination and specificity of closely related baculoviruses have been studied using methods derived from molecular biology. Martignoni et al. (1980), however, differentiated baculovirus isolates with a virus neutralization test, which is one of the most specific and sensitive serologic tests and generally considered to be the decisive criterion for identifying viral isolates. The authors tested the serum neutralization of NPVs pathogenic for *Orgyia pseudotsugata*. They found that the viruses were antigenically distinguishable. Neutralization tests could also be useful in quality control of commercial preparations. Langridge et al. (1981) described another method for controlling the relatedness of baculoviruses. The structural proteins of *Autographa californica* and *Heliothis zea* NPVs were detected by indirect enzyme-linked immunosorbent assay (ELISA). The immunoassay detected less than 1 ng of *Autographa californica* NPV protein. The extent of immunological relatedness between *Autographa californica* NPV occluded virus and *Autographa californica* NPV polyhedrin, *Autographa californica* NPV nonoccluded virus, *Estigmene acrea* granulosis virus, *Amsacta moorei* entomopoxvirus, *Heliothis zea* NPV, and *Lymantria dispar* NPV was determined. The authors detected no immunological relatedness between *Heliothis zea* NPV, *Autographa californica* NPV, and a persistent rod-shaped virus isolated from the *Heliothis zea* cell line (JMC-HZ-1). The polyhedrins of *Heliothis zea* NPV and *Autographa californica* NPV were found to be immunologically identical. Roberts and Naser (1982) characterized monoclonal antibodies to the *Autographa californica* NPV and noted that the use of monoclonal antibodies against various *Autographa californica* NPV proteins and other baculoviruses should prove useful in virus identification and for studying virus replication. Yamada and Maramorosch (1981) described a plaque assay with NPV of *Heliothis zea* using a mixed solid overlay, another accurate technique for the determination of virus concentration.

Gettig and McCarthy (1982) reviewed genotypic variation among wild isolates of *Heliothis* spp NPVs from different geographical regions. A number of NPVs have been isolated worldwide from species in the genus *Heliothis* (Lep.: Noctuidae), including such agricultural pests as *Heliothis zea*, *Heliothis virescens*, *Heliothis armigera*, and *Heliothis punctigera* (Ignoffo and Couch, 1982). Ten NPVs from *Heliothis* were characterized using restriction endonuclease analysis. Digestion of the DNA genomes with *Bam*HI, *Eco*RI, and *Hin*dIII resulted in fragmentation profiles which separated the wild-type isolates into two major genotypes, corresponding to the morphology of the virions. Submolar fragments present in a number of the digests suggest that these wild-type isolates might themselves consist of two or more genotypic variants. However, when Ignoffo and Couch (1982) bioassayed the ten *Heliothis* NPVs, significant differences in virulence were observed. This suggests a relationship between genotype and biological activity and indicates that genotypic variants of a particular NPV may be the result of geographical and/or host influences on the genome. Kelly et al. (1983) reported that the major nucleocapsid protein of *Heliothis zea* single nucleocapsid NPV is a low molecular weight, basic, DNA-binding protein present in the core of the capsid. The protein enhances the infectivity of baculovirus DNA and may be important biologically. The protein is rich in arginine and helix-destabilizing residues, although it possesses neither lysine nor hydrophobic residues. In many respects it is similar to the basic protein described by Tweenten et al. (1980b) in a granulosis virus, although amino acid analysis shows that it is quite distinct from that protein. Kelly et al. (1983) discussed that since the protein is presumably intimately involved in collapsing DNA into a form which can be packaged within the capsid, one might expect that its gene would be highly conserved among various baculoviruses. Amino acid analysis of the *Heliothis zea* NPV protein and the basic protein of four other baculoviruses suggests considerable divergence in composition, although the function of the protein is presumably conserved by the abundance of arginine residues. Nevertheless, it is probable that the protein interacts in a specific fashion with DNA phosphate of the baculovirus genome assuming a stable and structured form. Therefore, NPV of *Heliothis armiger* was tested against *Heliothis zea* and *Spodoptera frugiperda*, which often occur together on crops such as corn and sorghum. More information on the host range of *Heliothis armiger* NPV was obtained by comparing the susceptibility of *Heliothis virescens* and *Spodoptera exigua* (Hamm, 1982). The results of bioassays of *Heliothis armiger* on *Heliothis zea*, *Heliothis virescens*, *Spodoptera exigua*, and *Spodoptera frugiperda* were as following: *Heliothis zea* was most susceptible, *Heliothis virescens* and *Spodoptera exigua* were less susceptible, and *Spodoptera frugiperda* was the least susceptible. Although its pathogenicity to both *Heliothis* and *Spodoptera* species made *Heliothis armiger* a potential candidate for use in microbial control of these parts, the requirement of using about 420 times as many polyhedra to produce 50% mortality in *Spodoptera frugiperda* as in *Heliothis zea* makes it unlikely that this virus will be helpful in controlling *Spodoptera frugiperda*. Wood et al. (1981) noted an increase in the virulence of *Autographa californica* NPV by mutagenesis. Based on the results, it was clear that phenotypic variants with different levels of virulence existed in the wild virus populations as well as did genotypic variants (Lee and Miller, 1978).

Many attempts have been made to isolate virus and to eventually relate DNA sequences with specific biological parameters such as virulence and specificity (Brown et al., 1980; Burand et al., 1980; Carstens et al., 1980; Jewell and Miller, 1980; Vlak and Gröner, 1980; Kelly, 1981; Miller, 1981; Miller et al., 1981; Burand and Summers (1982); Carstens, 1982; Cochran et al., 1982; Fraser and Hink, 1982; Jun-Chuan and Weaver, 1982; Maruniak and Summers, 1981; Miller and Miller, 1982; Rohrmann et al., 1982; Vlak and Smith, 1982; Vlak and van der Krol, 1982; Wood et al., 1982). Restriction endonuclease analysis is indispensable in gaining fundamental knowledge of baculovirus DNA structure and genetic information. Genetic changes in the virus when exposed in the natural environment can eventually be monitored with restriction endonuclease analysis of viral DNA. The availability of these enzymes to facilitate DNA sequencing and molecular cloning has brought the study of baculovirus molecular biology into higher gear. Target populations in areas where viral insecticides are applied should be monitored routinely to ensure early detection of any trend

TABLE I. Safety Tests for Controlling the Replication of NPV of *Mamestra brassicae*

Application method	Concentration gm/body wt.	Antibody response		Species	Number of animals
		polyhedrin	virions		
per oval	1×10^8 NPV	Neg. (within 60 day postinfection)	Neg. (within 60 day postinfection)	Mice	120
	Biol. active virions from				
	1×10^9 NPV	Neg. (within 60 d.p.i.)	Neg. (within 60 d.p.i.)	Mice	90
	2×10^9 NPV			Mice	90
	UV-inactivated virions from 1×10^9 NPV	Neg. (within 60 d.p.i.)	Neg. (within 60 d.p.i.)	Mice	90
	1×10^9 NPV to mice before fertilization	Neg. (sera of the progeny were tested at the age of 20 days)	Neg. (sera of the progeny were tested at the age of 20 days)	Mice	61
	to pregnant mice				44
	5×10^7 NPV	Neg. (within 42 d.p.i.)	n.t.[b]	Swine	5
	1×10^7 NPV	Neg. (within 60 d.p.i.)	n.t.	Chickens	15

Aerosol	1×10^7 NPV	Neg. (within 60 d.p.i.)	Neg. (within 60 d.p.i.)	Mice	20
Intra-venously	1×10^{10} NPV_{alk}[a]	Neg. (14 d.p.i.)	4 Mice pos. (14 d.p.i.)	Mice	5
	1×10^9 NPV_{alk}	Neg. (14 d.p.i.)	3 Mice pos. (14 d.p.i.)	Mice	5
	1×10^8 NPV_{alk}	Neg. (14 d.p.i.)	4 Mice pos. (14 d.p.i.)	Mice	5

[a] NPV_{alk} = NPV dissociated in 0.05 M Na_2CO_3-0.05 M NaCl buffer, pH 11.
[b] n.t. = not tested.

toward developing resistance or loss of specificity. With these current methods we are now better able to control the behavior of the applicated NPVs. Furthermore, we are able to control the hygienic risk of NPVs to invertebrates at the molecular level.

B. TESTS WITH VERTEBRATES

1. Control of Virus Replication in Nontarget Organisms

a. Animal Experiments and Tissue Culture Experiments. i. Mammals. Many *in vivo* tests conducted with nuclear polyhedrosis viruses (NPVs) have been done (Burges et al., 1980) and no adverse effects due to the virus have been reported. NPVs from the following types of insects were tested: *Heliothis zea, H. armigera, Trichoplusia ni, Orgyia pseudotsugata, Spodoptera exigua, S. frugiperda, S. exempta, S. littoralis, Choristoneura fumiferana, Neodiprion swainei, N. sertifer N. lecontei, Autographa californica, Malacosoma disstria, Lymantria dispar, Galleria mellonella, Erranis filliara, Oryctes rhinocerus.* However, the NPV from *Mamestra brassicae*, the cabbage moth, was not tested extensively (Gröner et al., 1978). We investigated as to whether NPV from *Mamestra brassicae* (Lepidoptera: Noctuidae) replicated in mammals. Tests were generally conducted at doses 10 to 1000 times the per-acre field rate equated to a 70-kg man. The production of antibodies against virions and/or polyhedrin (matrix protein) was considered to be an indication of virus replication. Virus-specific antibodies were detected by radioimmunoassay (RIA) (Döller, 1978). In the RIA, viral proteins were labeled with ^{125}I (Hunter and Greenwood, 1962).

All tests, in which vertebrate species were exposed to NPV of *Mamestra brassicae* are summarized in Table I.

In these experiments it was shown that the laboratory mice (NMRI-strain/Naval Medical Research Institute-strain) produced virus-specific antibodies after intravenous injection of virions. The antibody response was weak and it was concluded that the immune response was produced by a nonreplicating foreign protein and not by virus replication (Döller and Gröner, 1981, 1982).

In addition to the determination of virus-induced antibodies, NPV infectivity in organs of inoculated animals was investigated by feeding the organs to larvae (bioassay). Heart, liver, spleen, brain, and kidney from NPV-fed animals were tested in the bioassay using highly sensible first instar larvae of *Mamestra brassicae*. The larvae were checked for mortality three times a week and dead larvae were examined by light microscopy for the presence of polyhedra. In the experiments no virus activity was detected in organs of mice and swine.

The liberation of virions from polyhedra into the alimentary tract of vertebrates is a prerequisite of a hypothetical infection of these nontarget organisms, and the safety tests should include experiments showing whether this occurs. For this, the passage of NPV through the alimentary tract of mice and swine was studied. After oral NPV application to the test animals, the feces were collected and combined daily for 8 days postfeeding. In order to determine the form of the virus in the feces (i.e., free virions or polyhedra), all samples were treated with chloroform, which inactivates free virions of *Mamestra brassicae*; the treatment has less effect on virions occluded within the polyhedra. The results showed that virus activity in the chloroform-treated feces had dropped drastically, so the virus activity depended on the presence of liberated virions in the feces (inactivated by chloroform) (Döller et al., 1983).

Swine were included in safety tests because they are more closely related to humans physiologically than rodents and because of their agricultural importance. Pathogenicity tests (influence on body temperature, number of leukocytes, swelling of lymph nodes) were also performed. Four of the NPV inoculated animals showed a slight temperature rise at the second day postfeeding to 40.1°C. This phenomenon may be due to the manipulation and a possible slight disturbance in the digestion tract caused by the inoculum. After the fourth day postfeeding, no differences were observed in comparison to the controls. The number of leukocytes decreased in all animals, which is considered normal and due to adaptation and growth. In summary, there was no evidence of an inapparent virus replication in swine and mice of *Mamestra brassicae* NPV.

Since *Autographa californica* NPV is also being considered for registration, it seemed to be the appropriate NPV to test for potential replication in vertebrate cells. McIntosh and Shamy (1980) controlled the replication of NPV from *Autographa californica* in a mammalian cell line. The authors reported that in two of five experiments, replication was observed following inoculation of chinese hamster cells (CHO) with *Autographa californica* NPV. CHO is routinely grown at 35°C, whereas most insect cell lines are grown at lower temperatures (28°-30°C). *Autographa californica* NPV-inoculated CHO cultures were incubated at the lower temperature of 28°C; no cell replication took place at this temperature and the authors were unable to adapt the CHO line to grow at 28°C. For replication of *Autographa californica* NPV in invertebrate cell line, a permissive *Trichoplusia ni* cell line was taken as "positive control." Radiolabeling experiments with CHO cells using [^{3}H]thymidine consistently resulted in a labeled entity which banded in sucrose gradients at a density of 1.24 to 1.25 g/ml, which is similar to that observed for *Autographa californica* NPV in the permissive *Trichoplusia ni* cell line. The labeled entity is believed to be *Autographa californica* NPV, based on infectivity and serological data. The infectious center assay revealed that at a multiplicity of infection (MOI) of 3, only 0.38% of chinese hamster cells were infected compared with 28% for cabbage looper cells. However, 4.2% of chinese hamster cells were infected at a MOI of 41. Based on these results the authors required more sensitive techniques to be employed for the *in vitro* safety evaluation of NPVs.

In 1982, Tjia et al. reported about studies on the persistence of *Autographa californica* NPV and its genome in mammalian cells. The authors set out to study whether *Autographa californica* NPV can persist or replicate in a number of established mammalian cell lines. Human HeLa cells or primary human embryonic kidney cells, simian CV1 cells, hamster BHK 21 (B3) cells or Muntiacus muntjak cells growing in monolayer cultures were used in those studies. Cells were inoculated with *Autographa californica* NPV at multiplicities ranging from 0.1 to 100 PFU/cell. Subsequently, the inoculated cells were investigated for virus production and for the replication and the persistence of the viral DNA. Extracts of inoculated cells were also screened for the occurrence of *Autographa californica* NPV-specific RNA.

First, the authors proved the stability of *Autographa californica* NPV at 27° and 37°C. *Autographa californica* NPV was kept at 27° or 37°C in Dulbecco-modified Eagle's medium, supplemented with 10% fetal calf serum, since mammalian cells are usually propagated in this medium. *Autographa californica* NPV did not appreciably loose infectivity within 24 hours at either temperature under these conditions. It was concluded that *Autographa californica* NPV was stable at 37°C for at least 24 hours in medium used for the propagation and maintenance of mammalian cells. However, *Autographa californica* NPV does not multiply in any of the cell lines studied. Viral DNA replication or transcription could not be detected by blotting and nucleic acid hybridization experiments using nick-translated, cloned viral probes. Furthermore, there was no evidence for the persistence of viral DNA or of fragments of viral DNA in mass cultures of mammalian cells. Since mammalian cells can take up and integrate any foreign DNA at very low frequency, it cannot be ruled out by the approach chosen that a very small number of cells might have incorporated and fixed viral DNA in their genomes. As concluded by Tjia et al. (1983), as this caveat is always pertinent for any population of cells exposed to foreign DNA, this reservation does not appear to be of particular significance in safety considerations when working with baculoviruses, or any virus for that matter. Unfortunately, only the two reports mentioned above concerning baculovirus replication in mammalian cell lines exist. Since the results of these reports differ further investigations are necessary.

ii. Birds. As in mammals, the production of antibodies against virions and/or polyhedrin was considered to be an indication of a virus replication in birds. Virus-specific antibodies were detected with RIA as mentioned in Section II,B,1,a. For controlling a *Mamestra brassicae* NPV replication in birds, chickens were used as test animals. Fifteen chickens received a single feeding dose of 1×10^8 NPV/gm body weight, applied on a piece of bread. The same number of control animals received buffer-treated bread. The feces were collected for 6 days. For each sample of lyophilized feces, an aliquot equal to 16.7% of the sample was incorporated into semisynthetic diet for *Mamestra brassicae* larvae at 40°C. These contaminated diets were tested in a standardized bioassay with 4 replications of 50 second instar larvae per treatment.

In order to determine the form of the virus in the feces (i.e., free or occluded), aliquots of the feces were treated with 10% chloroform (Gröner, 1978a). Treatment with chloroform inactivates free virions of *Mamestra brassicae* NPV but has less effect on virions occluded within the polyhedra. After incubation for 5 hours at 4°C, the chloroform was removed and these samples were incorporated into the diet and treated in bioassay. The results showed that virus activity could be detected in the feces until 5 and 6 days postfeeding. The treatment of the chicken feces with chloroform had no deleterious effect on the virus activity in the feces. Therefore, the polyhedra were presumably not solubilized in the alimentary tract of the chickens (Döller and Gröner, 1982). The reason for the behavior of polyhedra in the alimentary tract of the birds may be due to the low acidity in the stomach of birds. Within 60 days postfeeding, no virus-induced antibodies could be detected using RIA in the sera from test animals. From these experiments it was concluded that no virus replication had taken place in chickens.

iii. Fishes. Until now, few experiments have been done to exclude an inapparent baculovirus infection of aquatic animals. As in experiments with mammals and birds, it was assumed that virus replication in fishes is accompanied by antibody production. The ability of rainbow trout and carp to produce specific baculovirus antibodies after immunization was investigated.

Rainbow trout and carp were immunized with polyhedrin by an intraperitoneally injection of 8 mg viral protein with complete Freund's adjuvant. Fishes were challenged 14 days after the first injection with the same amount of viral protein. Blood was withdrawn from the caudal vein 6 weeks after the first injection. Specific antisera from rabbits served as controls. Virus-specific antibodies were detected by Ouchterlony gel diffusion tests. Sera from immunized rainbow trout, carp, and rabbits (as controls) showed identical precipitation lines (Döller and Enzmann, 1982). These experiments showed that trout and carp can produce detectable baculovirus-specific antibodies, providing the basis for safety studies to investigate virus replication *in vivo* by determination of virus-specific antibodies. Until now 7.5×10^9 NPV/animal were applicated orally to ten rainbow trouts. Feces were collected daily until 6 days postfeeding. Feces were purified by sucrose density centrifugation and NPV was detected by electron microscopy and by immunodiffusion test until the fourth day postfeeding. In the alimentary tract of rainbow trout, NPV was not dissociated into free virions.

2. *Control with Respect to Virus-Induced Chromosome Aberration Rates and Sister Chromatid Exchanges*

It is important to know whether there is a potential physiological and/or genetic hazard for nontarget organisms living in an area where baculoviruses are sprayed for pest control. Viruses like herpes simplex virus, hepatitis virus, adenovirus, and others can induce heavy structural and numerical chromosome aberrations (Sharma and Polasa, 1978). Basic studies have been carried out since 1978 on the serious genetic alterations (Miltenburger, 1978; Miltenburger and Reimann, 1979) caused by baculoviruses. It is surprising that quantitative genetic investigations with baculoviruses have been more or less neglected.

At present it seems justified to assume that increased rates of sister chromatid exchanges (SCEs) and chromosomal aberrations point out a high probability for mutagenic and/or carcinogenic capabilities of an agent. Both criteria represent a good measure for estimating possible risks for mammalian cells at the cytogenetic level.

Reimann and Miltenburger (1982a,b; 1983) have performed experiments using the following NPVs: *Autographa californica* because of its relatively wide host range, including 28 insect species in several families and its multiplication in several insect cell lines (Van der Beek, 1980) and *Mamestra brassicae*, which has a smaller host range (Gröner, 1978b). Chinese hamsters and NMRI-mice were used as test animals. The virus was applied either intraperitoneally or orally. Bone marrow cells from both femora were prepared 24 hours after the last virus treatment. For the preparation of SCEs, 5-bromouridinedeoxyribose was used.

Reimann and Miltenburger (1982b) carried out the following experiments with *Mamestra brassicae* NPVs and chinese hamster as test animals. The NPVs were inoculated per os as follows: single doses during 24 hours, 90 single doses during 90 days, and control: distilled water. The chromosome preparations were done 12 or 24 hours after the last treatment. No significant change of SCE rates could be detected in any of the treated groups.

The following experiments were carried out with *Autographa californica* NPVs and chinese hamster as test animals: 3 applications within 24 hours (orally), 14 applications within 14 days (orally), 2 injections within 24 hours (intraperitoneally), 3 injections within 36 hours (intraperitoneally). With regard to cytogenetic data there was no difference between the treated animals and the controls.

The following *in vitro* experiments were done with *Autographa californica* NPV: control of inducing SCEs with cell lines from the Indian Muntjak and from the mouse, control of cytogenetic data from cells of chinese hamster and human lymphocytes after treatment with infectious DNA from *Autographa californica* NPV, and control of chromosome aberrations in chinese hamster cell line and in human leukocytes. In all experiments there was no difference between the treated groups and the control. In addition, according to all available data, the authors exclude the possibility that in their experiments the NPVs had been replicated in the mammalian cells. They observed no signs of early or complete virus replication. Further, they concluded that NPV neither induced structural chromosome aberrations nor SCE events.

3. *Control on the Retrovirus Inducing Activity Potential of NPV in Mammalian Cultures*

The basis of the studies is the fact that several mammalian species carry stable inherited viral genomes in their DNA (Aaronson and Stephenson, 1976). The viral genomes are transmitted through the germ line and code for the production of infectious C-type RNA viruses. The endogenous viruses have similar properties to known oncogenic RNA tumor viruses. The role of the endogenous viruses is not known but it is suggested that they are involved in embryogenesis and in differentiation. The isolation of oncogenic viruses from mouse fibroblasts treated with halogenated pyrimidine analogons supports this idea (Stephenson et al., 1974). There are several agents which are known to induce endogenous viruses. The most important ones are mutagenic and carcinogenic chemical compounds like, e.g., benzpyrene, methylcholanthrene or mitomycin C (Teich et al., 1973); the endogenous viruses can also be induced by transforming DNA viruses such as herpes simplex virus type 2 (Isom et al., 1978).

McIntosh et al. (1979) inoculated a vertebrate viper cell line, known to be able to replicate an insect Chilo irridescent virus with NPV of *Autographa californica*. Cultures were incubated at 28°C for 5 days and examined by electron microscopy. The authors found NPV particles intracellularly among the C-type particles which were carried by this line. Furthermore, NPV particles with envelope partially removed could be seen extracellularly with many C-type virions. No increase in virus titer could be observed over the 5-day incubation period. However, it was observed that many more C-type particles were present in inoculated cultures than in unchallenged cultures, suggesting an induction of C-type particles in the presence of baculovirus proteins. Schmidt and Erfle (1982) also investigated the activation of endogenous C-type retroviruses in cell cultures of four mammalian species: mouse, rat, monkey, and human. Cells were treated with NPVs from *Autographa californica*, *Mamestra brassicae*, *Lymantria dispar*, and with virions of *Autographa californica* (nonoccluded virus), baculovirus DNA, C-type retrovirus-activating chemicals (halogenated pyrimidine analogons), and chemical insecticides alone and in combination. The activation of retroviral genomes was tested by the determination of reverse transcriptase activity in concentrated cell culture supernatants and by the demonstration of the intracellular localization of retrovirus structural protein p30 using the indirect immunoperoxidase technique. In NPV-treated cell cultures no C-type retrovirus activation was detectable. C-type retroviruses were activated in mouse cells only by the halogenated pyrimidine analogon iododeoxyuridine. In simultaneous treatments of the cells with NPVs and chemicals, no potentiating effects by NPVs could be detected. Virions of NPV in cell cultures showed unaltered morphology and upon reisolation remained infectious in homologous insect cell cultures for a long time. No influence on growth or morphology of the treated mammalian cells could be observed. The authors concluded that application of NPVs for pest control is uncritical with respect to their retrovirus-inducing activity potential in mammalian cells.

4. *Nonimmunological Interaction of NPV with Mammalian Sera*

To evaluate possible medical risks of nuclear polyhedrosis virus (NPV) in biological pest control, it is important to know whether NPV-specific antibodies are present in human sera since this would indicate a virus replication in humans. For these studies NPV of larvae from *Mamestra brassicae* was used.

Human sera seemed to show positive immunoreactions to *Mamestra brassicae* NPV proteins in the radioimmunoassay (RIA) (Döller and Matthaeus, 1980). However, differences could not be detected between persons who had and had not been in contact with *Mamestra brassicae* NPV. Based on these observations it was decided to determine whether human sera contained *Mamestra brassicae* NPV-specific antibodies. The detection of specific antibodies in human sera would suggest that baculoviruses are infectious for humans and are therefore not suitable for insect pest control.

Two possibilities for the reaction of human sera with baculovirus proteins were considered: (1) Specific reaction--in this case the baculovirus component might be considered as being ubiquitous antigen. Consequently, an immune response against this antigen in the human sera would be expected. (2) Nonspecific reaction.

Based on these possibilities, human sera of different age groups were tested using RIA. The results showed that 11% of the sera of newborns showed a positive reaction with *Mamestra brassicae* NPV proteins. The positive reaction slowly increased according to age group, from 4 to 8 years up to 86% by 13 to 17 years. However, no geographical differences between sera from Europe, Brazil, and Greenland was found. After tenfold concentration, all negative sera yielded a positive reaction.

The serum proteins that interact with viral proteins were determined. The proteins in human sera reacting with baculovirus proteins were separated in 19 and 7 S fractions by gel filtration. The 19 S region showed only a weak reaction in the RIA. By immunoelectrophoretic analysis low level precipitation with anti-human-IgM and anti-human-IgA and a strong precipitation with anti-human-IgG was found. Further purification by ion-exchange chromatography and affinity chromatography with protein A sepharose CL-4B revealed that IgG was the main reacting immunoglobulin (Döller et al., 1983b). This raised the possibility that there might indeed be a repeated exposure to the viral antigen or a cross-reacting antigen which is widely distributed in the environment. Longworth et al. (1973) reported reactions between an insect picornavirus and naturally occurring IgM antibodies in mammalian species, which was interpreted as an infection by agents sharing common antigens with insect viruses. A similar reaction between NPV from *Autographa californica* and human sera was detected by Röder and Gröner (1980) using immunoelectrophoretic methods, but the authors had no explanation for the phenomenon. Further studies by Döller (unpublished results), using RIA, human immunoglobulins of the IgD and IgE class showed weak positive reactions with baculovirus proteins. All human sera tested using RIA reacted positively with *Mamestra brassicae* NPV proteins. However, the positive reaction was dependent only on the relative concentration of IgG in the human sera. This prompted the authors to suggest a nonimmunological interaction of immunoglobulins with the baculovirus proteins.

MacCallum et al. (1979) reported that precipitating antibodies to an insect pathogenic RNA virus of *Darna trima* from East Malaysia have been found in a small percentage of human sera from several different groups of persons in West Malaysia and the United Kingdom. No associated illness was identified. The authors used a micro double-diffusion test as a detection method. However this method is not as sensitive as the RIA, so it may be possible that with a more sensitive method the authors would have found more positive reactions. Although MacCallum et al. (1979) did not suggest that the positive results from either country would indicate antigenic stimulation or infection by *Darna trima*, they did note that an antigenically related virus or viruses could have been present in the environment that may have been associated with symptomless or inapparent infections in man.

To demonstrate the specificity in the positive reaction of binding baculovirus proteins to human immunoglobulins, the binding site of *Mamestra brassicae* NPV proteins to human IgG was determined by incubation of baculovirus proteins with Fab fragments of purified human immunoglobulins. Human Fab fragments were obtained by papain digestion, and it was found that the viral proteins bound to the Fc fragment of human IgG (Döller et al., 1981). To determine if the viral proteins interacted with the antigen binding site of human IgG, a competition assay for the detection of antibodies against hepatitis A virus, which was used as model, was performed in the presence and absence of viral proteins. Based on the results, the antigen binding site directed against hepatitis A virus of human anti-hepatitis A virus IgG was not blocked by the baculovirus proteins. However, RIA binding of immunoglobulins to the viral proteins could be measured, however no influence on the function of immunoglobulins was observed. Experiments were also performed to investigate if human immunoglobulins, which reacted positive with *Mamestra brassicae* NPV proteins in the RIA, also possessed neutralizing activity, which would indicate a specific immunoreaction. The neutralization activity studies were done with *Spodoptera littoralis* cells. The experiments showed that human immunoglobulins, which gave a positive RIA reaction did not possess neutralization activity against baculovirus proteins (Döller et al., 1983c). These results enhance the suggestion that the positive reaction found in the RIA is not a specific immunoresponse. These experiments confirm and extend the observation that the binding of baculovirus proteins to human immunoglobulins does not involve the antigen binding site of the IgG molecule but rather the Fc fragment.

Field sera from horses, cattle, pigs, and sheep also showed a positive reaction to baculovirus proteins. To quantify the binding reaction, immunoglobulins of the IgG class were tested for their binding capacities with ^{125}I-labeled *Mamestra brassicae* NPV proteins. IgG's of rabbit and goat reacted strongly with the viral proteins, while IgG's of horse, bovine, and rat reacted weakly. The results showed that all IgG's reacted positively with baculovirus proteins, although the animals were not infected with *Mamestra brassicae* NPV (Döller, 1983). Scotti and Longworth (1980) reported on naturally occurring IgM antibodies to a small insect virus in some mammalian sera in New Zealand. Antibodies to cricket paralysis virus were demonstrated in sera from a pig, a horse, and numerous cattle in New Zealand. The reactions in immunodiffusion tests were variable, but significant and consistent reactions were obtained with these sera in virus neutralization assays. The sedimentation coefficient of the antibodies was 19 S, and their activity was destroyed by treatment with 2-mercaptoethanol, indicating that they were of the IgM class. The phenomenon of binding viral proteins to mammalian sera is not a new observation. Although poliovirus is unable to infect cattle, neutralizing antibodies as well as precipitating antibodies have been detected in cattle sera (Pagano et al., 1965; Sabin and Fieldsteel, 1953; Urusawa et al., 1968). Svehag and Mandel (1962), in studies on the formation of polio virus neutralizing antibodies in rabbits, proposed that macroglobulin antibodies in normal sera were a response to repeated exposure of minute amounts of cross-reacting or identical antigens. Scotti and Longworth (1980) discussed also that it is likely that the neutralizing antibodies detected in the above mentioned study were stimulated by transient exposures to low levels of cricket paralysis virus. Actually the authors did not know why these antibodies were present. In conclusion, not only did baculovirus proteins show positive reactions with mammalian sera, but also another insect pathogenic virus, cricket paralysis virus.

The possible interaction of baculovirus proteins with the Fc fragment was investigated. This was done by competition of viral proteins with the binding site of *Staphylococcus aureus* to human IgG. After preincubation of baculovirus proteins to ^{125}I-labeled human IgG, the precipitation by *Staphylococcus aureus* was found to be reduced by 11%. Thus, although baculovirus proteins bind to the Fc fragment of human IgG, it appears that the viral proteins bind only to a subclass of the human IgG. This inhibition, however, is sufficient to recognize positive binding reactions in all human sera using a sensitive antibody detection system like RIA (Döller et al., 1983b). The study was continued by controlling whether the binding of *Mamestra brassicae* NPV proteins to the Fc fragments prevented complement function on the Fc fragment. For this, a complement-dependent cytotoxicity test was used (Döller et al., 1982). In this test system, human erythrocytes (A) were lysed in the presence of specific anti-erythrocyte (A) antibodies (from mice) and complement. The results showed an inhibition of the complement-dependent cytotoxicity test by *Mamestra brassicae* NPV proteins (in the range of 2.5 to 0.15 mg/ml). It was concluded that functions mediated by the Fc fragment were inhibited However, with the complement-dependent cytotoxicity test it was not possible to test if *Mamestra brassicae* NPV proteins were binding to the complement components or to the complement receptors on the Fc fragment. For further characterization of the binding position the test system was modified. First, human erythrocytes (A) were incubated with specific antiserum. These reaction mixtures were then washed and the viral proteins were added, incubated, and the reaction mixtures rewashed. In this step, the proteins were allowed to bind to the Fc fragment. If the proteins bound to the complement receptors located on the Fc portion, the added complement could not bind and no lysis could be measured. The results showed that the viral proteins indeed bound to the complement receptors, since protein concentrations in the range of 250 μg/ml-0.9 μg/ml were found. When the binding capacity of protein A from *Staphylococcus aureus* was compared to the binding capacity of *Mamestra brassicae* NPV proteins it was found that in the test system protein A showed the same inhibition reaction as *Mamestra brassicae* NPV proteins (Döller and Koszinowski, unpublished results). These experiments showed that *Mamestra brassicae* NPV proteins bound to the complement receptors on the Fc portion of IgG. This reaction confirmed the findings of the positive binding reactions of *Mamestra brassicae* NPV proteins with mammalian sera using RIA. However, this should not be interpreted to mean that all interactions of immunoglobulins with *Mamestra brassicae* NPV proteins are of nonimmunological nature. Specific immunoresponse may be induced in rabbits and mice, although they are difficult to differentiate from the binding phenomenon described. This makes it almost impossible to exclude that in human sera, in addition to the nonimmunological interaction, there might be also cases of some specific antibody production toward *Mamestra brassicae* NPV.

III. SAFETY TESTS WITH GRANULOSIS VIRUS (GV) (fam.: *Baculoviridae*)

A. TESTS WITH INVERTEBRATES: HOST RANGE

Predictive control methods are necessary to understand both the dosage-mortality responses of an insect to virus infection and virus epizootiology. Therefore, Payne et al. (1981) compared the dosage-mortality responses of larvae of *Pieris brassicae* and *Pieris rapae* to infection by *Pieris brassicae* GV. Bioassays with first, second, third, and fourth instar larvae of both species revealed a marked difference in susceptibility between instars and species. The authors found that medium lethal dosages (LD_{50}'s) for *Pieris rapae* larvae ranged from 5 capsules for the first instar to 662 capsules for the fourth instar. With *Pieris brassicae* this range extended from 66 capsules to 2.3×10^7 capsules. The time-mortality responses of the two species were similar when fed virus dosages equivalent to an LD_{90}. Median lethal times (LT_{50}'s) ranged from 5 days for first instar larvae to 7-8 days for fourth instar larvae. A comparison between a long-established laboratory stock of *Pieris brassicae* and a stock acquired from the field showed no significant differences in their susceptibility to GV. The differences in susceptibility between *Pieris rapae* and *Pieris brassicae* to GV infection are of considerable practical significance for the field control of these species by virus. Much higher virus dosages would be required for field control of *Pieris brassicae* than for *Pieris rapae*.

Briese and Mende (1981) reported about the differences in susceptibility to a GV between 16 field populations of the potato moth *Phthorimaea operculella* (Zeller) (Lepidoptera: Gelechiidae). A laboratory bioassay technique was used for the comparison. A difference by a factor of 11.6 times was found between the most and the least susceptible population; a laboratory strain was over 30 times as resistant as some field populations. The authors suggested that this variability might reflect the past history of exposure of different populations to the virus, which appears to be endemic. The existence of widespread variability between populations gives evidence for either the selection of increased resistance or the spread of resistant genes already present. Briese (1982) discussed that in *Pieris operculella* other genes may play a role either by modifying the action directly or affecting the resistant phenotype. There are several ways in which insects can defend themselves against viruses that could be subject to genetic variability. Hence, the overall response of a species to a virus is likely to be polygenic. The possibility that insect pests might also acquire resistance to viral insecticides has received little attention particularly regarding the application of GV as a biological control agent (Briese, 1981).

Crook (1981a) studied the structure and biological properties of five GV isolates from *Pieris* spp. (one from *Pieris brassicae*, one from *Pieris napi*, and three from *Pieris rapae*). Dosage-mortality studies showed that all five isolates were highly virulent for *Pieris rapae* with LD_{50} values for third instar larvae ranging from $10^{2.0}$ to $10^{2.6}$ capsules. With third instar *Pieris brassicae* larvae, LD_{50} values were not only much higher for all the isolates but also varied over a much wider range. One isolate failed to infect any *Pieris brassicae* larvae at a dose of $10^{10.1}$ capsules, whereas the most virulent isolate had an LD_{50} of $10^{5.8}$ capsules. These results demonstrate the importance of correct identification of GVs. In addition, Crook (1981b) used serological methods (immunodiffusion and enzyme-linked immunosorbent assay) for the comparison of the GVs. *Pieris brassicae* GV and one of the *Pieris rapae* GV isolates showed that a very strong serological cross-reaction occurred both between capsules and virus particles. SDS-polyacrylamide gel electrophoresis of virus particles confirmed that all five isolates were closely related, but that small differences could be detected. Using a more sensitive method, the agarose gel electrophoresis of DNA fragments produced by digestion with *Eco*RI, *Bam*HI, and *Hin*dIII endonucleases showed that four of the five isolates were very closely related but small differences could be detected. For correct identification of GVs sensitive methods are thus required.

Hotchkin (1981) compared the electrophoretic mobilities of granulin from GV of *Pseudoletia unipunctata* and of *Spodoptera exigua*. *Pseudoletia unipunctata* infects *Spodoptera exigua* and other noctuid hosts. Hotchkin reported that the granulin is unchanged when GV of *Pseudoletia unipunctata* is cross-transmitted to different host species.

Huber (1982) reported on the GV of the codling moth, *Cydia pomonella*. The *Cydia pomonella* GV is highly infective for codling moth and it is rather specific. Of the two dozen species tested, it infected only the congeneric species *Cydia nigricana* and a few other closely related tortricides such as *Rhyacionia buoliana*, *Grapholitha molesta*, *Grapholitha funebrana* and *Lathronympha strigana*. Many field studies were carried out with GVs. The *Cydia pomonella* GV used in the world-wide testing seemed to originate from the same source, the Berkeley isolate of the virus, and was propagated *in vivo*.

The biochemical and biophysical characteristics of the closely related *Diacrisia virginica* GV and *Hyphantria cunea* GV isolates were examined by Boncias and Nordin (1980). Sucrose gradient sedimentation patterns of alkali-solubilized

Diacrisia virginica GV and *Hyphantria cunea* GV capsules were identical. Electrophoretic analysis of alkaline-solubilized granulin extracts demonstrated that both viruses contained alkaline proteolytic activity. Electrophoretic separation of the virus proteins demonstrated some quantitative differences between the two GVs. The enveloped nucleocapsids and the nucleocapsids of the two viruses were morphologically indistinguishable. Perhaps Huber (1982) had explained this phenomenon by the observation that all known GVs seemed to originate from the Berkeley isolate. However, Tweenten et al. (1980a) carried out a comparison of the genomes of *Plodia interpunctella* GV and *Pieris rapae* GV using restriction endonuclease fragment pattern. Obviously, the DNA's from the viruses were clearly distinguishable. Two years later, Cattano and Langridge (1982) characterized the DNA from GV of *Estigme acrea* (Lepidoptera). In addition, the authors compared the DNA pattern of *Estigme acrea* GV to that of *Plodia interpunctella* GV and *Pieris brassicae* GV. The patterns and numbers of DNA fragments of the three DNA's were different. The authors concluded that *Estigme acrea* GV DNA is not closely related to the DNA of the other GVs.

The use of GV in the field demonstrated the existence of other problems, namely, their relative persistence on leaf surfaces. It is generally considered that the ultraviolet (uv) component of sunlight is the key factor in virus inactivation. Richards and Payne (1982) tested the persistence of GVs on leaf surfaces when purified suspensions of GV capsules were applied to crops by conventional high-volume spray application methods. The amount of infective virus on the leaf surfaces decayed rapidly. The decay of *Pieris* GV infectivity on cabbage varies at different times of the year. It is most rapid in June, at height of summer, and slowest in October. Nonetheless, more than 90% of the virus was inactivated in 7 days. The significance of such results is twofold: (1) The long-term persistence of a small proportion of virus on leaves could provide extended pest control for significant periods. (2) It may be possible to select virus strains which are more resistant to uv-inactivation.

Improved spray application methods and timing could reduce the need for highly persistent formulations by directing a greater proportion of the virus to the sites where larvae are feeding. The authors concluded that for field application further work is required on the most appropriate droplet size, virus concentration, and formulation needed for successful control.

B. *TESTS WITH VERTEBRATES*

1. *Control of Virus Replication in Nontarget Organisms*

a. *Animal Experiments.*

i. Mammals. In contrast to NPV, few tests have been done to prove the safety of GV in regard to virus replication. Bailey et al. (1982) carried out field studies with *Cydia pomonella*. Two methods of application were used, mist and hand spray. Longworth traps were set in the test areas before, during, and after spraying and were examined daily for the presence of small mammals. Test animals were identified and released after collecting blood and fecal samples. Sixty one individuals of the wood mouse (*Apodemus silvaticus*) and 13 individuals of the bank vole (*Clethrionomys glareolus*) were caught. Antibody to *Cydia pomonella* GV was detected by indirect enzyme-linked immunosorbent assay (ELISA) in all sera of animals caught in the mist-sprayed orchard. In this area virus at a concentration of 2×10^{10} capsules per liter was sprayed at the end of June and middle of July at a rate of 1000 liters per hectare. The authors found that antibody titers increased with time after spraying. Bailey and Hunter (1982) continued their studies with laboratory experiments. They used the wood mouse in the experiments designed to simulate, as closely as possible, the type of exposure to virus that would be experienced in the wild. The authors administered GV of *Cydia pomonella* both orally or intranasally. Both infective virus and virus inactivated by ultraviolet irradiation were used. Doses containing 100 μg, 1 μg, or 10 ng protein were administered either once, or at daily or weekly intervals for a minimum period of 12 weeks. Each animal was bled weekly and sera were tested for the presence of antibodies against *Cydia pomonella* GV using the indirect ELISA. All sera collected from animals inoculated intranasally were positive. IgG_1 predominated and the degree of response depended on both concentration and frequency of dose. The best response was recorded for animals receiving daily doses of 100 μg each. Sera from mice given virus orally were only positive when multiple doses of 100 μg were given.

In addition, Bailey and Hunter (1982) collected fecal samples daily for 3 weeks after exposure of mice to virus. These were tested for infectivity by feeding to larvae of *Cydia pomonella*, and the presence of virus antigens was investigated using a range of serological tests. These results are in agreement with the results of Döller and Huber (1983), where 5×10^{11} capsules were applicated to NMRI-mice. Within 6 days postfeeding, feces were collected and tested in bioassay by feeding the feces to larvae. Biological active virus was detected. Thus, on the second day postfeeding, 4.7×10^9 GVs were detected, or 1% of the concentration applied. The production of antibodies against virions and/or granulin (matrix protein) was considered to be an indication of virus replication. Virus-specific antibodies should be detected by radioimmunoassay (RIA). Within 80 days postfeeding, no virus-specific antibodies could be detected using RIA. To examine the possibility of vertical virus transmission, GV virions (a total of 0.2 mg) were administered orally to mice before fertilization and to pregnant mice. Sera from young mice were obtained at the 28th day postfeeding. No virus-specific antibodies were detected. It was concluded that virus transmission had not taken place in the investigated animals after exposure to single doses of GV. These results agree with those of Bailey and Hunter (1982), who found virus-specific antibodies only after exposure to multiple doses of virus. In NMRI-mice virus-specific antibodies were detected after two intraperitoneally or intramuscularly injections of granulin. Based on the reported results, further investigations are necessary to decide if GV of *Laspeyresia pomonella* replicate in wood mouse, a naturally occurring animal species, which would be in contrast to NMRI mice, which is a laboratory strain.

Thus far, it has not been possible to prove the safety of GV in *in vitro* experiments. GV replication has been investigated by Blumer-Wolf (1982). For this purpose organ cultures were produced from fat body removed from larvae that had been infected (*in vivo*) 6 hours earlier. Electron microscopic investigations showed that, within 4 days, all the different virosis stages developed *in vitro* in a manner similar to that which occurs *in vivo*. The number of normal occlusion bodies (capsules) per cell *in vitro* was similar to that *in vivo*. However, the number of large abnormal occlusion bodies was higher *in vitro* than *in vivo*, especially in older cultures. Enveloped nucleocapsids and nucleocapsids were also found within the intercellular spaces of the fat body. Blumer-Wolf (1982) carried out autoradiographic studies on DNA metabolism. They found that uninfected fat bodies did not incorporate [^{3}H]thymidine *in vitro*, whereas this did occur in infected fat body cultures. The rate of incorporation depended on the time after infection, the age of culture, and culture conditions, such as, variation in the composition of the media, including the addition of juvenile hormone. Saldanha and Hunter (1982) reported on attempts to propagate GVs in cell cultures. The authors inoculated *Spodoptera frugiperda* cell cultures with virus particles prepared from *Plodia interpunctella* GV capsules or with hemolymph from GV-infected *Plodia interpunctella* larvae. Cell cultures were also treated with 5'-iododeoxyuridine, heat, ultraviolet irradiation, or hypertonic solutions before inoculation with hemolymph in order to increase the susceptibility of the cells to infection. Transfection of cell cultures with DNA from *Plodia interpunctella* GV or *Spodoptera littoralis* GV was also attempted. All these techniques failed to produce a productive infection *in vitro*. Cell fusion between *Spodoptera littoralis* cell cultures and hemocytes from GV-infected *Spodoptera littoralis* larvae was attempted by Saldanha and Hunter (1982) using uv-inactivated Sendai virus. Following incubation at 26°C, cell fusion was not detected, nor were hybrid cells isolated. Similar attempts on cell fusion using *Spodoptera littoralis* or *Spodoptera frugiperda* cell cultures and hemocytes from GV-infected *Spodoptera interpunctella*, *Spodoptera frugiperda* or *Carpocapsa pomonella* larvae failed to produce either cell fusion or replication of a GV *in vitro*.

ii. Fishes. To exclude an inapparent GV replication in rainbow trout and carp, safety studies were conducted with GV from *Laspeyresia pomonella*. In order to investigate if the test animals were able to produce detectable virus-specific antibodies, immunization experiments were carried out. The animals were immunized with granulin by intramuscular injection with complete Freund's adjuvant. Fishes were challenged 14 days after the first injection. Virus-specific antibodies could be detected by immunodiffusion test (Döller and Enzmann, 1982), showing that trout and carp are able to produce detectable granulin-specific antibodies. For controlling GV replication in rainbow trout, 1×10^{12} capsules/animal were force-fed to the animals. Feces were collected daily for 6 days postfeeding and purified with sucrose gradient centrifugation. Until 3 days postfeeding GV was detectable in the feces by electron microscopy and with immunodiffusion tests. Bioassay showed that viruses had already infected the larvae of *Laspeyresia pomonella* (Döller and Huber, unpublished results). Within 80 days postfeeding, no virus-specific antibodies could be detected in the sera of test animals using the immunodiffusion test. From these experiments it was concluded, that no virus replication had taken place in the test animals.

2. *Control with Respect to Virus-Induced Chromosome Aberration Rates and Sister Chromatid Exchanges*

As already mentioned in Section I,B,3 chromosome aberrations and sister chromatid exchanges should represent a good measure for estimating possible risks for mammalian at the cytogenetic level. Reimann and Miltenburger (1982b) controlled whether the GV of *Laspeyresia pomonella* had an effect on the chromosomal structure in nontarget organisms. Feeding experiments were done with Chinese hamsters; 1×10^{12} of *Laspeyresia pomonella* GV capsules were given per animal. The capsules were placed directly on the feeding diet in one acute dose or in 90 split doses of 1.7×10^{10} over a 90-day period. The authors found neither an increase of the sister chromatid exchanges rates nor of the chromosomal aberration rates. There was no difference between the treated animals and the controls.

3. *Nonimmunological Interaction of GV with Mammalian Sera*

It is important to know whether GV-specific antibodies are present in human sera since this would indicate virus replication in humans. For these studies GV from larvae of *Laspeyresia pomonella* were investigated.

First, sera of persons exposed to GV by handling the virus in the laboratory were tested for the detection of GV-specific antibodies in the radioimmunoassay (RIA). Most of the tested sera showed positive reactions with *Laspeyresia pomonella* GV proteins. By testing sera from humans who had not been in contact with GV in the laboratory, positive reactions were also obtained. In order to test if it is a general phenomenon that humans acquire immunity to *Laspeyresia pomonella* GV, sera from humans (uninfected normal population) of different ages were assayed. Ninety-five percent of sera of newborns showed a positive reaction with *Laspeyresia pomonella* GV proteins. One hundred percent of the age group of 3- to 12-month-old showed a positive reaction; 89% of the age group of 1 to 3 years, 87% of the age group of 4 to 8 years, 88% of the age group of 9 to 12 years, and 100% of the age group of 13 to 17 years showed a positive reaction. No differences between sera from Europe and Brazil were found. Based on these results it was concluded that humans had not acquired immunity against *Laspeyresia pomonella* GV (Döller, 1981).

In order to define the protein responsible for the binding capacity of the human sera, the following analysis was carried out: treatment with chloroform, ammonium sulfate precipitation, and ion-exchange chromatography on DEAE-cellulose. Immunoelectrophoretic analysis with class-specific anti-human sera showed that human IgG reacted strongly with *Laspeyresia pomonella* GV proteins. In comparison to the experiments carried out with *Mamestra brassicae* NPV (Section II,B,4), immunoglobulins from negative sera were concentrated and found positive. Therefore, a positive reaction was dependent mainly on the relative IgG concentration of the human sera. In contrast to *Mamestra brassicae* NPV proteins, immunoglobulins of the IgA and IgD class did not react with *Laspeyresia pomonella* GV proteins; IgE and IgM reacted weakly (Döller and Flehmig, 1983).

The characterization of the binding site of *Laspeyresia pomonella* GV proteins to human IgG showed that *Laspeyresia pomonella* GV proteins reacted with human Fab fragments, but not with the antigen binding site directed against hepatitis A virus, which was taken as a model for a human system. It was concluded that no function of the Fab fragment was inhibited by binding the *Laspeyresia pomonella* GV proteins to the fragment. Also, as with *Mamestra brassicae* NPV, a nonimmunological phenomenon was suggested.

Furthermore, the possible interaction of *Laspeyresia pomonella* GV proteins with the Fc fragment of immunoglobulins was investigated. It was controlled if *Laspeyresia pomonella* GV proteins prevented the function of the complement on the Fc fragment. For this purpose, a complement-dependent cytotoxicity test was used (Döller et al., 1982). The results showed an inhibition of the complement-dependent cytotoxicity test by *Laspeyresia pomonella* GV proteins in the range of 2.5 mg/ml-0.15 mg/ml. *Laspeyresia pomonella* GV proteins showed the same reactions as *Mamestra brassicae* NPV proteins. The experiments were continued as described with *Mamestra brassicae* NPV proteins (Section II,B,4). It was found that *Laspeyresia pomonella* GV proteins do bind to the complement receptor on the Fc fragment with protein concentrations in the range of 250 μg/ml-0.9 μg/ml. It was also found that in the test system used, protein A from *Staphylococcus aureus* inhibited the reaction in the same manner as viral proteins. This reaction confirmed the findings of the positive binding reactions of *Laspeyresia pomonella* GV with human sera using RIA.

Not only were positive reactions found with *Laspeyresia pomonella* GV proteins in human sera but also in sera from uninfected animals as horses, cattle, sheep, and swine.

It is very difficult to differentiate between a specific and unspecific immunoreaction. This makes it almost impossible to exclude that underlying the nonimmunological interaction in human sera there might be also cases of some specific antibody production toward *Laspeyresia pomonella* GV.

Interestingly, when *Mamestra brassicae* NPV and *Laspeyresia pomonella* GV proteins were compared in regard to the binding capacities to mammalian sera, few differences were found, e.g., *Laspeyresia pomonella* GV proteins did not bind to human immunoglobulins of the IgA and IgD class. However, by controlling the influence on the functions of the Fc fragment, both viral proteins reacted identically.

IV. CONCLUSION

Although baculoviruses (NPV and GV) are isolated from naturally infected, deceased insects, their use as insecticides require proof that they are safe. NPVs are reported to be very host specific and replicate in only one, or in a few, related host species. There now exist results indicating that the *Autographa californica* NPV is being produced in small quantities by a mosquito cell line. This would indicate that NPV not only crossed species lines but also replicated in another order of insects. Laboratory experiments showed selection for resistance, but the same could occur in the field upon continued exposure to the virus, enhanced by the genetic variability of naturally occurring NPVs. Genetic changes in virus can eventually be monitored with restriction endonuclease analysis of viral DNA. We are now better able to control the behavior of the applicated NPV or GV using these methods.

No measurable virus replication could be detected in non-target organisms like mammals, birds, and fishes. In the feces of the test animals, however, biologically active virus was found which indicates that, for example, birds are able to spread the virus over the fields. Experiments on virus-induced chromosome aberration rates and sister chromatic exchanges revealed negative results. Experiments on retrovirus-inducing activity potential in mammalian cultures also showed negative results.

Detection of baculovirus-specific antibodies in human sera would suggest that baculoviruses are infectious for humans and, therefore, may not be suitable for insect pest control. Surprisingly, human sera showed positive immunoreactions to baculovirus proteins. No differences were found between human sera from Europe, Brazil and Greenland. Further experiments showed that the positive reaction was dependent only on the relative concentration of immunoglobulins. Therefore, a nonimmunological interaction between human immunoglobulins and baculovirus proteins was suggested. Baculovirus proteins bind nonspecifically to the Fab fragment of human IgG, but not to its antigen binding site. In addition, the viral proteins also bind to the Fc fragment, which is comparable to the binding of protein A from *Staphylococcus aureus*. This makes it almost impossible to exclude a specific reaction from this nonspecific reaction.

Based on the results summarized in this chapter, it is very difficult to decide whether there is a potential risk when spraying baculoviruses as biological control agents in the field.

In summary, I would like to refer to Harrap (1982), who gave a comprehensive account of the evolution of registration guidelines and safety testing procedures for the use of viruses for pest control. The guidelines are still evolving. The final version is likely to have a significant impact in gaining international acceptance of registration criteria.

REFERENCES

Aaronson, S. A., and Stephenson, J. R. (1976). Endogenous type C-RNA viruses of mammalian cells. *Biochem. Biophys. Acta 458*, 323-354.

Allaway, G. P. (1981). Baculovirus host range and inclusion body dissolution. *Proc. 5th Intern. Congr. Virol.*, p. 291.

Bailey, M. J., and Hunter, F. R. (1982). Environmental impact of spraying apple orchards with the granulosis virus of the codling moth (*Cydia pomonella*) (2) Laboratory studies. *Proc. 3rd Intern. Colloq. Invertebr. Pathol., University of Sussex*, p. 183.

Bailey, M. J., Field, A. M., and Hunter, F. R. (1982). Environmental impact of spraying apple orchards with the granulosis virus of the codling moth (*Cydia pomonella*) (1) Field studies. *Proc. 3rd Intern. Colloq. Invertebr. Pathol., University of Sussex*, p. 182.

Beek van der, C. P. (1980). On the origin of the polyhedral protein of the nuclear polyhedrosis virus of *Autographa californica*. *Meded. Landbouwhogesch. Wageningen*, pp. 2-4.

Blumer-Wolf, A. (1982). Observations on early stages of granulosis virus development *in vivo* and *in vitro*. *Proc. 3rd Intern. Colloq. Invertebr. Pathol., University of Sussex*, p. 171.

Boucias, D. G., and Nordin, G. L. (1980). Comparative analysis of the alkali-liberated components of the *Hyphantria cunea* and the *Diacrisia virginica* granulosis viruses. *J. Invertebr. Pathol. 36*, 264-272.

Briese, D. T. (1981). Resistance of insect species to microbial pathogens. *In* "Pathogenesis of Invertebrate Microbial Diseases" (E. W. Davidson, ed.), pp. 511-545. Allanheld-Osmun, Montclair, New Jersey.

Briese, D. T. (1982). Genetic basis for resistance to a granulosis virus in the potato moth *Phthorimaea operculella*. *J. Invertebr. Pathol. 39*, 215-218.

Briese, D. T., and Mende, H. A. (1981). Differences in susceptibility to a granulosis virus between field populations of the potato moth, *Phthorimaea operculella* (Zeller) (Lepidoptera: Gelechiidae). *Bull. Entomol. Res. 71*, 1-18.

Briese, D. T., Mende, H. A., Grace, T. C., and Geier, P. W. (1980). Resistance to a nuclear polyhedrosis virus in the light-brown apple moth *Epiphyas postvittana* (Lepidoptera: Tortricidae). *J. Invertebr. Pathol. 36*, 211-215.

Brown, D. A., Allen, C. J., and Bignell, G. N. (1982). The use of a protein A conjugate in an indirect enzyme-linked immunosorbent assay (ELISA) of four closely related baculoviruses from *Spodoptera* species. *J. Gen. Virol. 62*, 375-378.

Brown, M., Faulkner, P., Cochran, M. A., and Chung, K. L. (1980). Characterization of two morphology mutants of *Autographa californica* nuclear polyhedrosis virus with large cuboidal inclusion bodies. *J. Gen. Virol. 50*, 309-316.

Burand, J. P., and Summers, M. D. (1982). Alteration of *Autographa californica* nuclear polyhedrosis virus DNA upon serial passage in cell culture. *Virology 119*, 223-229.

Burand, J. P., Summers, M. D., and Smith, G. E. (1980). Transfection with baculovirus DNA. *Virology 101*, 286-290.

Burand, J. P., Wood, H. A., and Summers, M. D. (1983). Defective particles from a persistent baculovirus infection in *Trichoplusia ni* tissue culture cells. *J. Gen. Virol. 64*, 391-398.

Burges, H. D. (1982). Safety, safety testing and quality control of microbial pesticide. *In* "Microbial Control of Pests and Plant Diseases, 1970-1980" (H. D. Burges, ed.), pp. 737-767. Academic Press, New York.

Burges, H. D., Croizier, G., and Huber, J. (1980a). A Review of Safety Tests on Baculoviruses. *Entomophage 25*(4), 329-340.

Carstens, E. B. (1982). Mapping the mutation site of an *Autographa californica* nuclear polyhedrosis virus polyhedron morphology mutant. *J. Virol. 43*(3), 809-818.

Carstens, E. B., Tjia, S. T., and Doerfler, W. (1980). Infectious DNA from *Autographa californica* nuclear polyhedrosis virus. *Virology 101*, 311-314.

Cattano, S. P., and Langridge, W. H. R. (1982). Characterization of DNA from a granulosis virus of *Estigmene acrea* (Lepidoptera). *Virology 119*, 199-203.

Cochran, M. A., Carstens, E. B., Eaton, B. T., and Faulkner, P. (1982). Molecular cloning and physical mapping of restriction endonuclease fragments of *Autographa californica* nuclear polyhedrosis virus DNA. *J. Virol. 41*(3), 940-946.

Crawford, A. M., and Sheehan, C. M. (1982). Serial passage of *Spodoptera littoralis* NPV in *Spodoptera frugiperda* cells: The production of interfering virus particles. *Proc. 3rd Intern. Colloq. Invertebr. Pathol., University of Sussex*, p. 170.

Crook, N. E. (1981a). Genetic variability and virulence characteristics of granulosis viruses isolated from *Pieris* spp. *Proc. 5th Intern. Congr. Virol., Strasbourg*, p. 291.

Crook, N. E. (1981b). A comparison of the granulosis virus from *Pieris brassicae* and *Pieris rapae*. *Virology 115*, 173-181.

Döller, G. (1978). Solid Phase Radioimmunoassay for the Detection of Polyhedrin Antibodies. *In* "Safety Aspects of B aculoviruses as Biological Insecticides" (H. G. Miltenburger, ed.), pp. 203-208. Symposium Proceedings, Federal Ministry for Research and Technology, Bonn, FRG.

Döller, G. (1981). Unspecific interaction between granulosis virus and mammalian immunoglobulins. *Naturwissenschaften 68*, 573-574.

Döller, G. (1983). Kernpolyedervirus aus *Mamestra brassicae:* Versuche zur Prüfing einer Vermehrung in Vertebraten. *Mitt. Deut. Ges., allg. angew. Ent. 4*, 65-69.

Döller, G., and Enzmann, H.-J. (1982). Induction of baculovirus specific antibodies in rainbow trout and carp. *Bull. Europ. Assoc. Fish Pathol. 2*, 53-55.

Döller, G., and Flehmig, B. (1983). Granulosevirus aus *Laspeyresia pomonella:* Charakterisierung der Bindungsreaktion zwischen Granulin und Humanseren. *Mitt. Deut. allg. angew. Ent. 4*, 70-75.

Döller, G., and Gröner, A. (1981). Sicherheitsstudie zur Prüfung einer Virusvermehrung des Kernpolyedervirus aus *Mamestra brassicae* in Vertebraten. *Z. Angew. Entomol. 92*, 1, 99-105.

Döller, G., and Gröner, A. (1982). Safety-studies for the control of baculovirus replication in vertebrates. *Proc. 3rd Intern. Colloq. Invertebr. Pathol., University of Sussex*, p. 198.

Döller, G., and Huber, J. (1983). Sicherheitsstudie zur Prüfung einer Vermehrung des Granulosevirus aus *Laspeyresia pomonella* in Säugern. *Z. Angew. Entomol. 95*, 64-69.

Döller, G., and Matthaeus, W. (1980). Insectpathogenic baculovirus antibodies: Detection in human sera. *Zentrablat. Bakteriol. Parasitenk. Infektionskr. Hyg., Abt. I. Orig. A248*, 25.

Döller, G., Flehmig, B., and Dietzschold, B. (1981). Characterization of the binding site(s) in human IgG of baculovirus components. *Proc. 5th Intern. Congr. Virol., Strasbourg*, p. 219.

Döller, G., Glatthaar, B., and Koszinowski, U. (1982). Inhibition of the complement dependent cytotoxicity by baculovirus proteins. *Zentrablat Bakteriol. Parasitenk. Infektionskr. Hyg., Abt. I. Orig. A253*, 11.

Döller, G., Gröner, A., and Straub, O. C. (1983a). Safety-evaluation of nuclear polyhedrosis virus replication in pigs. *Appl. Environ. Microbiol. 45*, 1229-1233.

Döller, G., Matthaeus, W., Flehmig, B., and Lorenz, R. J. (1983b). Non-immunological interactions between baculovirus-antigen and human immunoglobulins. *Z. Angew. Entomol. 95*, 379-389.

Döller, G., Reimann, R., and Gröner, A. (1983c). Baculovirus proteins binding to human immunoglobulins without neutralizing activity. *Naturwissenschaften 70*, 371-372.

Erfle, V., and Schmidt, J. (1978). Studies on the activation on endogenous mammalian C-type retroviruses by baculoviruses: principles and test systems. *In* "Safety Aspects of Baculoviruses as Biological Insecticides" (H. G. Miltenburger, ed.), pp. 231-238. Symposium Proceedings, Federal Ministry for Research and Technology, Bonn, FRG.

Falcon, L. A. (1978). Viruses as alternatives to chemical pesticides. *In* "Viral Pesticides: Present Knowledge and Potential Effects on Public and Environmental Health" (M. D. Summers and C. Y. Kawanishi, eds.), pp. 11-23. Health Effects Research Laboratory, Office of Health and Ecological Effects, U.S. Environmental Protection Agency, Research Triangle Park, North Carolina.

Franz, J. M., and Krieg, A. (1976). "Biologische Schädlingsbekämpfung," p. 222. Verlag Paul Parey, Berlin-Hamburg (FRG).

Fraser, M. J., and Hink, W. F. (1982). The isolation and characterization of the MP and FP plaque variants of *Galleria mellonella* nuclear polyhedrosis virus. *Virology 117*, 366-378.

Geier, P. W., and Briese, D. T. (1979). The light-brown apple moth *Epiphyas postvittana* (Walker). 3. Differences in susceptibility to a nuclear polyhedrosis virus. *Aust. J. Ecol. 4*, 187-194.

Gettig, R. R., and McCarthy, W. J. (1982). Genotypic variation among wild isolates of *Heliothis* spp. nuclear polyhedrosis viruses from different geographical regions. *Virology 117*, 245-252.

Gröner, A. (1978a). Qualitätskontrolle von *in vivo* produziertem Kernpolyedervirus der Kohleule *Mamestra brassicae* (L.). *Mitt. Deut. Entomol. Ges. 1*, 127-131.

Gröner, A. (1978b). Studies on the specificity of the nuclear polyhedrosis virus of *Mamestra brassicae* (L.) (Lep.: Noctuidae). *In* "Safety Aspects of Baculoviruses as Biological Insecticides" (H. Miltenburger, ed.), pp. 265-269. Symposium Proceedings, Federal Ministry for Research and Technology, Bonn, FRG.

Gröner, A., Huber, J., and Krieg, A. (1978). Untersuchungen mit Baculoviren an Säugetieren. *Z. Angew. Zool. 65*, 69-80.

Hamm, J. J. (1982). Relative susceptibility of several noctuid species to a nuclear polyhedrosis virus from *Heliothis armiger*. *J. Invertebr. Pathol. 39*, 255-256.

Harrap, K. A. (1982). Assessment of the human and ecological hazards of microbial insecticides. *Parasitology 84*, 269-296.

Harrap, K. A., and Tinsley, T. W. (1978). The international virus research in controlling pests. *In* "Viral Pesticides: Present Knowledge and Potential Effects on Public and Environmental Health" (M. D. Summers and C. Y. Kawanishi, eds.), pp. 27-40. Health Effects Research Laboratory, Office of Health and Ecological Effects, U.S. Environmental Protection Agency, Research Triangle Park, North Carolina.

Harrap, K. A., Payne, C. C., and J. S. Robertson (1977). The properties of three baculoviruses from closely related hosts. *Virology 79*, 14-31.

Hotchkin, P. G. (1981). Comparison of virion proteins and granulin from granulosis virus produced in two host species. *J. Invertebr. Pathol. 38*, 303-304.

Huber, J. (1982). The baculoviruses of *Cydia pomonella* and other Tortricides. *Proc. 3rd Intern. Colloq. Invertebr. Pathol., University of Sussex*, pp. 119-124.

Hunter, W. M., and Greenwood, F. C. (1962). Preparation of iodine-131 labelled growth hormone of high specific activity. *Nature (London) 194*, 495.

Ignoffo, C. M. (1968). Specificity of insect viruses. *Bull. Entomol. Soc. Amer. 14*, 265-276.

Ignoffo, C. M., and Couch, T. L. (1982). The nuclear polyhedrosis virus of *Heliothis* species as a microbial insecticide. *In* "Microbial Control of Pests and Plant Diseases, 1970-1980: (H. D. Burges, ed.), pp. 329-362. Academic Press, New York.

Isom, H., Colberg, A., Reed, C., and Rapp, F. (1978). Conditions required for induction of murine p30 by Herpes simplex virus. *Intern. J. Cancer 22*, 22-27.

Jewell, J. E., and Miller, L. K. (1980). DNA sequence homology relationships among six lepidopteran nuclear polyhedrosis viruses. *J. Gen. Virol. 48*, 161-175.

Jun-Chuan, Q., and Weaver, R. F. (1982). Capping of viral RNA in cultured *Spodoptera frugiperda* cells infected with *Autographa californica* nuclear polyhedrosis virus. *J. Virol. 43*(1), 234-240.

Kelly, D. C. (1981). Baculovirus replication: Stimulation of thymidine kinase and DNA polymerase activities in *Spodoptera frugiperda* cells infected with *Trichoplusia ni* polyhedrosis virus. *J. Gen. Virol. 52*, 313-339.

Kelly, D. C., Brown, D. A., Ayres, M. D., Allen, C. J., and Walker, J. O. (1983). Properties of the major nucleocapsid protein of *Heliothis zea* singly enveloped nuclear polyhedrosis virus. *J. Gen. Virol. 64*, 399-408.

Kislev, N., and Edelman, M. (1982). DNA Restriction-pattern differences from geographic isolates of *Spodoptera littoralis* nuclear polyhedrosis virus. *Virology 119*, 219-222.

Krieg, A., Franz, J. M., Gröner, A., Huber, J., and Miltenburger H. G. (1980). Safety of entomopathogenic viruses for control of insect pests. *Environ. Conserv. 7*, 158-160.

Kurstak, E., Tijssen, P., and Maramorosch, K. (1978). Safety considerations and development problems make an ecological approach of biocontrol by viral insecticides imperative. *In* "Viruses and Environment" (E. Kurstak and K. Maramorosch, eds.), pp. 571-592. Academic Press, New York.

Langridge, W. H. R., Granados, R. R., and Greenberg, (1981). Detection of *Autographa californica* and *Heliothis zea* baculovirus proteins by enzyme-linked immunosorbent assay (ELISA). *J. Invertebr. Pathol. 38*, 242-250.

Lee, H. H., and Miller, L. K. (1978). Isolation of the genotypic variants of *Autographa californica* nuclear polyhedrosis virus. *J. Virol. 27*, 754-767.

Longworth, J. F., Robertson, J. S., Tinsley, T. W., Rowlands, D. J., and Brown, F. (1973). Reactions between an insect picornavirus and naturally occurring IgM antibodies in several mammalian species. *Nature (London) 242*, 314-316.

MacCallum, F., Grown, G., and Tinsley, T. (1979). Antibodies in human sera reacting with an insect pathogenic virus. *Intervirology 11*, 234-237.

McIntosh, A. H., and Shamy, R. (1980). Biological studies of a baculovirus in mammalian cell line. *Intervirology 13*, 331-341.

McIntosh, A. H., Maramorosch, K., and Riscoe, R. (1979). *Autographa californica* nuclear polyhedrosis virus (NPV) in a vertebrate cell line: Localization by electron microscopy. *NY Entomol. Soc. 87(1)*, 55-58.

Martignoni, M. E., Iwai, P. J., and Rohrmann, G. F. (1980). Serum neutralization of nucleopolyhedrosis viruses (Baculovirus Subgroup A) pathogenic for *Orgyia pseudosugata*. *J. Invertebr. Pathol. 36*, 12-20.

Maruniak, J. E., and Summers, M. D. (1981). *Autographa californica* nuclear polyhedrosis virus phosphoproteins and synthesis of intracellular proteins after virus infection. *Virology 109*, 25-34.

Miller, D. W., and Miller, L. K. (1982). A virus mutant with an insertion of a copia-like transposable element. *Nature (London) 299*, 562-564.

Miller, L. K. (1981). Construction of a genetic map of the baculovirus *Autographa californica* nuclear polyhedrosis virus by marker rescue of temperature-sensitive mutants. *J. Virol. 39*(3), 973-976.

Miller, L. K., Jewell, J. E., and Brown, D. (1981). Baculovirus induction of a DNA polymerase. *J. Virol. 40*(1), 305-308.

Miltenburger, H. G. (1978). "Safety Aspects of Baculoviruses as Biological Insecticides" (H. G. Miltenburger, ed.). Symposium Proceedings, Federal Ministry for Research on Technology, Bonn, FRG.

Miltenburger, H. G. (1978). No effect of NPV on mammalian cells *in vivo* and *in vitro* (cell proliferation, chromosome structure). *In* "Safety Aspects of Baculoviruses as Biological Insecticides" (H. G. Miltenburger, ed.), pp. 185-200. Symposium Proceedings, Federal Ministry for Research and Technology, Bonn, FRG.

Miltenburger, H. G., and Reimann, R. (1979). Viral pesticides: biohazard evaluation on the cytogenetic level. *Develop. Biol. Stand. 46*, 217-222.

Pawar, V. M. (1982). Comparative studies on homologous and reciprocal baculovirus infection in two closely related lepidoptera. *Proc. 3rd Intern. Colloq. Invertebr. Pathol., University of Sussex*, p. 82.

Payne, C. C., Tatchell, M. G., and Williams, C. F. (1981). The comparative susceptibilities of *Pieris brassicae* and *P. rapae* to a granulosis virus from *P. brassicae*. *J. Invertebr. Pathol. 38*, 273-280.

Pagano, J. S., Gilden, R. V., and Sedwick, W. D. (1965). The specificity and interaction with poliovirus of an inhibitory bovine serum. *J. Immun. 95*, 909-917.

Reimann, R., and Miltenburger, H. G. (1982a). Cytogenetic investigations in mammalian cells *in vivo* and *in vitro* after treatment with insect pathogenic virus (Baculoviridae). *Proc. 3rd Intern. Colloq. Invertebr. Pathol., University of Sussex*, p. 236.

Reimann, R., and Miltenburger, H. G. (1982b). Cytogenetic studies in mammalian cells after treatment with insect pathogenic viruses (Baculoviridae). I. *In vivo* studies with rodents. *Entomophaga 27*, 25-37.

Reimann, R., and Miltenburger, H. G. (1983). Cytogenetic studies in mammalian cells after treatment with insect pathogenic viruses (Baculoviridae). II. *In vitro* studies with mammalian cell lines. *Entomophaga 28*, 33-44.

Richards, M. G., and Payne, C. C. (1982). Persistence of baculoviruses on leaf surfaces. *Proc. 3rd Intern. Colloq. Invertebr. Pathol., University of Sussex*, pp. 296-301.

Roberts, P. L., and Naser, W. (1981). Replication of *Autographa californica* NPV in a *Spodoptera littoralis* cell line. *Proc. 5th Intern. Congr. Virol., Strasbourg*, p. 292.

Roberts, P. L., and Naser, W. (1982). Characterization of monoclonal antibodies to the *Autographa californica* nuclear polyhedrosis virus. *Virology 122*, 424-430.

Röder, A., and Gröner, A. (1980). Immunreaktion von menschlichen und tierischen Seren mit Baculoviren. *Naturwissenschaften 67*, 49-50.

Rohrmann, G. F., Martignoni, M. E., and Beaudreau, G. D. (1982). DNA sequence homology between *Autographa californica* and *Orgya pseudosugata* nuclear polyhedrosis virus. *J. Gen. Virol. 62*, 137-143.

Sabin, A., and Fieldsteel, A. H. (1953). Nature of spontaneously occurring neutralizing substances for 3 types of poliomyelitis virus in bovine sera. *Proc. 6th Intern. Congr. Microbiol. 2*, 373-375.

Saldanha, J. A., and Hunter, F. R. (1982). Attempted propagation of granulosis viruses in cell cultures resulting in activation of an occult nuclear polyhedrosis virus. *Proc. 3rd Intern. Colloq. Invertebr. Pathol., University of Sussex*, p. 243.

Schmidt, J., and Erfle, V. (1982a). Insect pathogenic baculoviruses: Studies of the activation of endogenous C-type retroviruses in mammalian cell cultures. *Zentrabl. Bakteriol. Hyg. Parasitenk. Infektionskr., Abt. I. Orig. A251*, 425.

Schmidt, J., and Erfle, V. (1982b). Untersuchungen über das Retrovirus aktivierende Potential von Kernpolyederviren in Säugetier-Zellkulturen. *Zentrabl. Bakteriol. Hyg. Parasitenk. Infektionskr., Abt. I. Orig. A252*, 438-455.

Scotti, P. D., and Longworth, J. F. (1980). Naturally occurring IgM antibodies to a small RNA insect virus in some mammalian sera in New Zealand. *Intervirology 13*, 186-191.

Sharma, G., and Polasa, H. (1978). Cytogenetic effects of influenza virus infection on male germ cells of mice. *Human Genet. 45*, 179-187.

Sherman, K. E., and McIntosh, A. H. (1979). Baculovirus replication in a mosquito (dipteran) cell line. *Inf. Immunity 26*(1), 232-234.

Smith, K. M. (1967). "Insect Virology." Academic Press, New York.

Smith, G. E., and Summers, M. D. (1978). Analysis of baculovirus genomes with restriction endonucleases. *Virology 89*, 517-527.

Stephenson, J. R., Greenberger, J. S., and Aaronson, S. A. (1974). Oncogenicity of an endogenous C-type virus chemically activated from mouse cells in cultures. *J. Virol. 13*, 237-240.

Summers, M. D., and Kawanishi, C. Y. (1978). "Viral Pesticides: Present Knowledge on Potential Effects on Public and Environmental Health." Research Triangle Park, North Carolina: Health Effects Research Laboratory, Office of Health and Ecological Effects, U.S. Environmental Protection Agency.

Summers, M. D., Engler, R., Falcon, L. A., and Vail, P. V. (1975). "Baculoviruses for Insect Pest Control: Safety Considerations." American Society for Microbiology, Washington, D.C.

Summers, M. D., Smith, G. E., Knell, J. D., and Burand, J. P. (1980). Physical maps of *Autographa californica* and

Rachiplusia ou nuclear polyhedrosis virus recombinants. *J. Virol. 34*(3), 693-703.

Svehag, S.-V., and Mandel, B. (1962). The production and properties of poliovirus neutralizing antibody of rabbit origin. *Virology 18*, 508-510.

Reich, N., Lowy, D. R., Hartley, J. W., and Row, W. P. (1973). Studies on the mechanism of induction of infectious murine leukemia virus from AKR mouse embryo cell lines by 5-iododeoxyuridine and 5-bromodeoxyuridine. *Virology 51*, 163-173.

Tjia, S. T., Carstens, E. B., and Doerfler, W. (1979). Infection of *Spodoptera frugiperda* cells with *Autographa californica* nuclear polyhedrosis virus. *Virology 99*, 399-409.

Tjia, S. T., Lübbert, H., Kruczek, J., Meyer z. Altenschildesche, G., and Doerfler, W. (1982). Studies on the persistence of *Autographa californica* nuclear polyhedrosis virus and its genome in mammalian cells. *Proc. 5th Intern. Congr. Virol., Strasbourg*, p. 289.

Tjia, S. T., Meyer z. Altenschildesche, G., and Doerfler, W. (1983). *Autographa californica* nuclear polyhedrosis virus (AcNPV) DNA does not persist in mass cultures of mammalian cells. *Virology 125*(1), 107-117.

Tweenten, K. A., Bulla, L. A., Jr., and Consigli, R. A. (1980a). Restriction enzyme analysis of the genomes of *Plodia interpunctella* and *Pieris rapae* granulosis viruses. *Virology 104*, 514-519.

Tweenten, K. A., Bulla, L. A., and Consigli, R. A. (1980b). Characteristics of an extremely basic protein derived from granulosis virus nucleocapsids. *J. Virol. 33*, 866-876.

Urusawa, S., Urusawa, T., Chiba, S., and Kanamitsu, M. (1968). Studies on poliovirus inhibitors in sera of domestic animals. III. A comparison of physiochemical properties of poliovirus inhibitors and specific antibodies. *Japan J. Med. Sci. Biol. 21*, 173-183.

Vaughn, J. L., Goodwin, R. H., Tomkins, G. J., and McCawley, P. M. (1977). The establishment of two cell lines from the insect *Spodoptera frugiperda* (Lepidoptera: Noctuidae). *In Vitro 13*, 213-217.

Vlak, J. M. (1982). Restriction endonucleases as tools in baculovirus identification. *Proc. 3rd Intern. Colloq. Invertebr. Pathol., University of Sussex*, pp. 218-225.

Vlak, J. M., and Smith, G. E. (1982). Orientation of the genome of *Autographa californica* nuclear polyhedrosis virus: a proposal. *J. Virol. 41*(3), 1118-1121.

Vlak, J. M., and van der Krol, S. (1982). Transcription of the *Autographa californica* nuclear polyhedrosis genome: Location of late cytoplasmic mRNA. *Virology 123*, 222-228.

Vlak, J. M., and Gröner, A. (1980). Identification of two nuclear polyhedrosis viruses from the cabbage moth, *Mamestra brassicae* (Lepidoptera: Noctuidae). *J. Invertebr. Pathol. 35*, 269-278.

Vlak, J. M., Gröner, A., Smith, G. E., and Summers, M. D. (1981a). Genetic variability of six naturally occuring NPVs of the cabbage moth, *Mamestra brassicae. Proc. 5th Intern. Congr. Virol., Strasbourg*, p. 288.

Vlak, J. M., van Frankenhuyzen, K., Peters, D., and Gröner, A. (1981b). Identification of a new nuclear polyhedrosis virus from *Spodoptera exigua. J. Invertebr. Pathol. 38*, 297-298.

Watanabe, H. (1965). Resistance to peroral infection of the cytoplasmic polyhedrosis virus in the silkworm *Bombyx mori* (Linnaeus). *J. Invertebr. Pathol. 7*, 257-270.

Wildly, P. (1971). Classification and nomenclature of viruses. *Monogr. Virol. 5*, 32.

Wood, H. A., Hughes, P. R., Johnston, L. B., and Langridge, W. H. R. (1981). Increased virulence of *Autographa californica* nuclear polyhedrosis virus by mutagenesis. *J. Invertebr. Pathol. 38*, 236-241.

Wood, H. A., Johnston, L. B., and Burand, J. P. (1982). Inhibition of *Autographa californica* nuclear polyhedrosis virus replication in high-density *Trichoplusia ni* cell cultures. *Virology 119*, 245-254.

Woods, A. (1974). "Pest Control: A Survey," p. 407. McGraw-Hill, London.

World Health Organization (1973a). The use of viruses for the control of insect pests and disease vectors. *Tech. Rep. Ser. No. 531*, 1-48.

World Health Organization (1973b). "Conference on the Safety of Biological Agents for Arthropod Control." Center for Disease Control, Atlanta, Georgia.

World Health Organization (1981). Mammalian safety of microbial agents for vector control: a WHO memorandum. *Bull. WHO Organ. 59*(6), 857-863.

Yamada, K., and Maramorosch, K. (1981). Plaque assay of *Heliothis zea* baculovirus employing a mixed agarose overlay. *Arch. Virol. 67*, 187-189.

THE ROLE OF VIRUSES IN THE ECOSYSTEM

W. J. KAUPP AND S. S. SOHI

Forest Pest Management Institute
P.O. Box 490
Sault Ste. Marie
Ontario, Canada

I. INTRODUCTION

It has been stated that the role of insect viruses in the ecosystem is unknown (Kurstak and Tijssen, 1982). However, if one accepts the principle that existence is the prime objective of all organisms, the argument becomes one of not why insect viruses exist but rather how they exist in the ecosystem. These viruses assert pressure on insect populations as microparasites and it is this factor which has prompted much research into their potential as pathogens, and, ultimately, their use as bioinsecticides.

Viruses which have been shown to naturally infect insects can be organized into seven distinct groups based on morphological and physical properties. Of these, viruses of the Baculoviridae, Reoviridae, and Spheruloviridae have been employed as viral insecticides and, hence, have generated the most information concerning the epizootiological relationships of viral pathogens. It is to these viruses, commonly known as the "occluded insect viruses," that the information presented in this chapter is restricted. Other viruses, notably members of the Iridoviridae, Parvoviridae, Rhabdoviridae, and Picornaviridae have been observed in insect hosts but little information is available concerning their prevalence in host populations (see Wigley and Scotti, 1983) and safety considerations

ISBN 0-12-470295-3

make their use as bioinsecticides limited, if not totally undesirable.

Because of the nature of the infection cycle, viruses can be described as parasites at the genetic level (Luria et al., 1978). However, the inert nature of insect viruses in the host-pathogen relationship must be emphasized in order to understand the epizootiology of these entomopathogens. Studies on the persistence, infectivity, production, and dispersal of viral pathogens are necessary to fully comprehend the role of insect viruses. It is the purpose of this chapter to elucidate the nature of insect viruses in the ecosystem by briefly highlighting their potential as entomopathogens, describing the characteristics of virus epizootics, investigating the quantities of virus produced in epizootics, and reviewing some hypotheses explaining observed host-virus relationships.

II. INSECT VIRUSES AS PATHOGENS

With the potential to control insect populations through many different facets of host physiology and survival, viruses were seen as potential candidates for environmentally safe, effective, and selective insecticides. Experiments, primarily conducted to assess the efficacy of these biocontrol agents, progressed rapidly and culminated in the recommendation of their use against several selected insect species. The numerous efficacy trials conducted over the past several years partially reflected the trends in insect control which assumed that viruses could be applied using methods similar to those used for chemical insecticides and the responsibility of the scientific community for the development of a safe and effective insecticide. As a result of these trials many of the candidate viruses were found unsuitable as bioinsecticides based solely on current efficacy data while others were widely used despite a lack of knowledge on the host-virus dynamics. A study of the case histories of insect control trials with viruses would be quite an exercise in learing about the many areas of insect control, such as methods of viral dissemination, insect population assessment, efficacy and virus epizootiology. When one considers that Martignoni and Iwai (1981) have recorded 1271 cases of viruses infecting insects, but thus far only six viruses have been registered as biocontrol agents, it must be conceded that somewhere in the science of insect pathology a problem exists in assessing the potential of viruses as insecticides. The following is a brief review of these biological agents.

A. *CYTOPLASMIC POLYHEDROSIS VIRUSES*

Only a few cytoplasmic polyhedrosis viruses (CPVs) have been successfully employed to control insect pests. Although less virulent than nuclear polyhedrosis viruses (NPVs) in natural situations, CPVs represent important mortality factors throughout the life cycle of numerous insect species. Larval mortality is recognized as a poor criterion for assessing the efficacy of CPVs (Katagiri, 1981). The potential of CPV infections to regulate insect populations has been observed in several insects of economic importance including *Lymantria fumida* (Koyama and Katigiri, 1967), *Colias eurytheme* (Tanada and Chang, 1964), *Operophtera brumata* (Cunningham et al., 1981), *Choristoneura fumiferana* (Bird and Whalen, 1954), and *Malacosoma disstria* (Bird, 1969), but CPVs have been used successfully to control only the pine processionary caterpillar, *Thaumetopoea pityocampa* (Dusaussoy and Geri, 1969), and the pine caterpillar, *Dendrolimus spectabilis* (Katagiri, 1969). The level of infection produced both by CPV spray and indirectly by secondary infection processes was found directly proportional to host population density and the pathogen dosage applied (Katagiri, 1969). This led to the conclusion that pest population cycles should be considered in spray applications and that *D. spectabilis* CPV was most effective when applied before peak population density.

In 1974, *D. spectabilis* CPV was registered as a microbial insecticide in Japan for the control of *D. spectabilis*. Although interest in developing CPVs as insecticides seems to be dwindling, research into their potential should continue. In agricultural systems where rapid insect death is required CPVs, due to their high rate of infection, could be used in conjunction with chemicals (Katagiri, 1981). In the forest ecosystem where long-term control is desirable and CPVs have been observed more frequently, this virus may have a use once the host-virus relationship has been investigated more intensively.

B. *ENTOMOPOX VIRUSES*

Successful instances of the use of entomopox viruses (EPVs) as microbial agents are very limited and potential for development has been restricted to the EPV infections of *C. fumiferana* and the cockchafer, *Melolantha melolantha* (Jaques, 1983). However, EPV infections have been studied in several insect species including the saltmarsh caterpillar, *Estigmene acrea* (Granados and Roberts, 1970), the winter moth, *O. brumata* (Weiser and Vago, 1966; Weiser, 1970), *Choristoneura biennis*

(Bird et al., 1971), *Choristoneura diversana* (Katagiri, 1973), *Choristoneura conflictana* (Cunningham and McPhee, 1973), and *Wiseana* spp. (Moore et al., 1973). The use of EPVs as biocontrol agents has been restricted because of their similarity to vertebrate pox viruses but, since they are present naturally in some insect populations, their role in insect regulation should be investigated further.

In New South Wales, Australia, a naturally occurring EPV infection was found responsible for a dramatic decrease in populations of *Oncopera alboguttata* (Milner, 1977). Over a 3-year outbreak period, the incidence of infection was observed to peak in the second year with an average of 54, 85, and 10% infection of samples collected from three sites. In the third year, disease incidence was at a very low level in substantially reduced populations of this insect. These populations were also infected with the microsporidian parasite *Pleistophora oncopere*. Although EPV infection caused heavy mortality during the prepupal stage, it was not evident in the adults, and the population collapse was not attributed solely to virus disease. The author suggested that the combination of viral and microsporidian infections acting in a density-dependent manner were responsible for the collapse of the insect populations. Further, Milner (1977) stressed that if a chemical insecticide was used which had a tendency to reduce host population density, it could interfere with the dynamics of EPV and microsporidian infections, and, hence, prolong the insect outbreak beyond its normal 2-3 year duration.

C. *BACULOVIRUSES*

1. *Nuclear Polyhedrosis Viruses*

The use of nuclear polyhedrosis viruses (NPVs) to control pests of both forestry and agriculture has met with limited success despite their potential. A complete résumé of these trials is published elsewhere (Yearian and Young, 1982; Cunningham, 1982). NPVs to control *Orgyia pseudotsugata*, *Neodiprion sertifer*, *N. lecontei*, *L. dispar*, and *Heliothis zea* have all been granted registration in varying degrees for use as bioinsecticides. However, despite this breakthrough, the use of NPVs for pest control has been declining in recent years. Notable successes utilizing NPVs to control insects are seen in the regulation of populations of *Gilpinia hercyniae* (Bird and Burke, 1961; Entwistle et al., 1977a,b), *N. sertifer* (Kaupp, 1981; Cunningham et al., 1975), and *N. lecontei* (de Groot and Cunningham, 1983) with only the studies on *G. hercyniae* and *N. sertifer* contributing to the knowledge of host-virus dynamics. Other studies involving the NPVs affecting *L. dispar*, *Mamestra brassicae*, *Trichoplusia ni*, *Spodoptera unipuncta*, *S. frugiperda*, *C. fumiferana*, and *Heliothis* spp. have illustrated that viruses cannot be used successfully if applied in the same manner as chemical insecticides. Knowledge of a great many ecological factors must be first understood before successful insect control can be achieved with viruses. A classic example of one study undertaken to understand the complexities of host-pathogen relationship is illustrated in the NPV control of the pasture pests, *Wiseana* spp. (Crawford and Kalmakoff, 1977; Kalmakoff and Crawford, 1982). After exhaustive studies, these authors concluded that enzootic control of this pest could be achieved naturally if farm management practices were adopted that would manipulate the environment in favor of maximizing the survival of NPV on the host habitat. Pasture cultivation and the application of chemical insecticides, once thought the only way of controlling this pest, were found antagonistic to the host-pathogen dynamics. Similarly, with more intense investigation into the ecological dynamics of virus diseases, perhaps more NPVs would be found suitable as potential bioinsecticides.

2. *Granulosis Viruses*

The use of granulosis viruses (GVs) in pest control has been adequately reviewed by Yearian and Young (1982), and Entwistle (1983). Although GV infections have been documented from 97 species, only 16 GVs have been employed to control forest and agricultural pests (Martignoni and Iwai, 1981). The majority of these trials have utilized GVs more as a bio-insecticide, varying application rate and technology as well as timing to achieve a high degree of insect mortality.

In experiments designed to evaluate the control potential of the GV of the summerfruit tortrix *Adoxophyes orana*, Shiga et al. (1973) demonstrated that the GV persisted on trees after virus application and members of the subsequent generation became infected. The summerfruit tortrix is multivoltine. The successive infection of the resident population maintained the GV in the environment. During periods of host scarcity, the amount of persisting virus is continually reduced and may reach such a low level that the incidence of disease becomes inapparent in the host population. The presence of an acceptable host density to provide impetus to the infection cycle by adult immigration/emmigration is also revealed in the *A. orana* dynamics, and is noted as one possible method of virus dispersal.

Similarly, the potential for GV to regulate populations of *Pseudaletia unipuncta* (Tanada, 1961, 1967), *Pierris brassicae* and *P. rapae* (Jaques, 1970b; Kelsey, 1958), and *Laspeyresia pomonella* (Tanada, 1964; Falcon et al., 1968) has been demonstrated. All these studies suggest that a better understanding of the host-virus relationship will lead to a more efficient use of the virus as an insecticide.

3. *Nonoccluded Baculoviruses*

The nonoccluded baculovirus infecting the rhinoceros palm beetle *Oryctes rhinoceros* has been demonstrated to be an effective bioinsecticide to control outbreaks of this pest (Marschall, 1970; Marschall and Doane, 1982; Young, 1974). Care taken in assessing the potential of this pathogen for pest control and in investigating the host-virus relationship resulted in effective virus dissemination and pest control. Such success stories are rare but should serve as an example to all investigators.

III. THE NATURE OF EPIZOOTICS

Since the work of Cornelia and Maestri in the early 10th century on the so-called jaundice of the silkworm *Bombyx mori*, it has been apparent that virus diseases can occur in insect populations resulting in high mortality. Most observations of insect mortality have been restricted to recording the disease when prevalent in high-density populations, a phenomenon now commonly known as an epizootic.

The occurrence of epizootics has been known to be influenced by many factors, such as the pathogenicity of the virus, habits of the insect, host density, the presence of parasites and predators, and other biotic and abiotic agents which act to sustain the pathogen in the host's habitat. In addition, of paramount importance, is the distribution of the host insect, since in the absence of the host, the real extent of the distribution of the pathogen in the environment and its potential as a mortality factor are unknown. Because of the intimate host-pathogen relationship, the occurrence of epizootics in insect populations has been characterized by three observations:

1. Epizootics become noticeable in high-density host populations and tend to act in a density-dependent or delayed density-dependent manner (Tanada, 1963; Kaupp, 1981; Milner, 1977).
2. Epizootics result in severe population mortality.

3. There is considerable release of the virus into the host's habitat as a result of the disintegration of virus-killed larval cadavers (Thompson and Scott, 1979; Podgwaite et al., 1979; Kaupp, 1983b; Entwistle et al., 1983).

To fully understand the nature of epizootics, it is necessary to appreciate that the initiation and progress of disease through host populations consists of two components: (a) a temporal component which is concerned with the localized changes in the incidence of infection with time, and (b) a spatial component which is concerned with the movement of disease through the host population. The spatial component can be seen as the pattern of infection within the entire host population resulting from the combination of many localized foci of infection and the changes in them with time. Hence, there can be little temporal change which is not reflected to some degree in a change in the spatial patterns of infection (Watt, 1968). Excellent reviews on the development and progress of epizootics in insect populations are found in Entwistle et al. (1983), Cunningham and Entwistle (1981), and Kaupp (1981), and will only be briefly discussed here.

A. *TEMPORAL GEOMETRY*

The typical pattern of growth of infection in a vertebrate host population which differs significantly from an insect population in possessing an immune system, can be described graphically by a curve designated as the epizootic wave (Fig. 1). This curve is a function of the numbers of susceptible hosts available and the number of sources of infection present (Burnet, 1953; Tanada, 1963). The temporal pattern of virus disease observed in insect populations is of the same form as found in vertebrates but in this case the epizootic wave is expressed when the buildup of infection is expressed over several insect generations.

The onset of infection in an insect population usually begins early in the larval stage, when larvae freshly emerged from eggs encounter virus in the environment. This virus, which has either persisted between insect generations or been dispersed onto their habitat, is referred to as primary inoculum. Some of these young, infected larvae die producing more virus (secondary inoculum) to infect other members of the population which increases the incidence of infection. The number of early instar larvae becoming infected and the quantity of primary inoculum present have been recognized as important factors in the development of epizootics since they both tend to increase the quantity of secondary inoculum available for infection (Doane, 1975; Kaupp, 1981, 1983b).

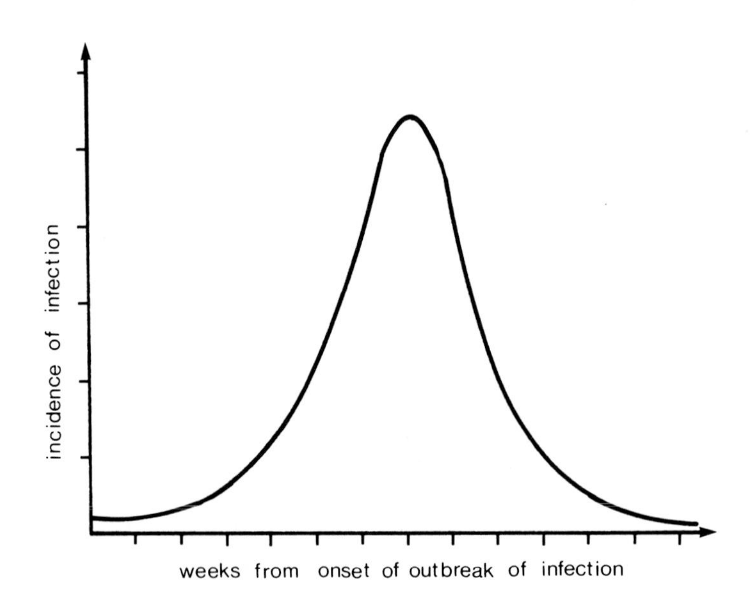

FIGURE 1. Illustration of a typical epizootic wave.

Depending upon the degree of change in larval susceptibility, two distinct patterns of increasing infection can be documented to occur annually in host populations. Where the increase in larval resistance with age is comparatively small in relation to the quantity of secondary inoculum produced, a sigmoid pattern in the incidence of infection is known to occur. Such patterns of infection have been observed in populations of *Neodiprion swainei* (Smirnoff, 1961), *G. hercyniae* (Evans and Entwistle, 1976), *N. sertifer* (Kaupp, 1981), *N. lecontei* (de Groot and Cunningham, 1983), and in populations of the armyworm, *Pseudaletia unipuncta* (Haworth) (Tanada, 1961). However, in some populations of lepidopterous insects, a wavelike pattern of the seasonal increase of infection has been observed (Stairs, 1965; Wigley, 1976). In this case, after an initial period of increase, a seasonal fall in infection occurs, probably reflecting the fact that accumulating inoculum is unable to cause continuously increasing level of infection due to a very large decrease in host susceptibility with age. In subsequent insect generations the incidence of infection increases so that wide-scale mortality is observed, and then tends to decrease due to substantial reduction in host density, persisting inoculum, and changes in population susceptibility (Martignoni and Schmid, 1961). Regardless of the pattern of incidence of infection that follows, a distinct

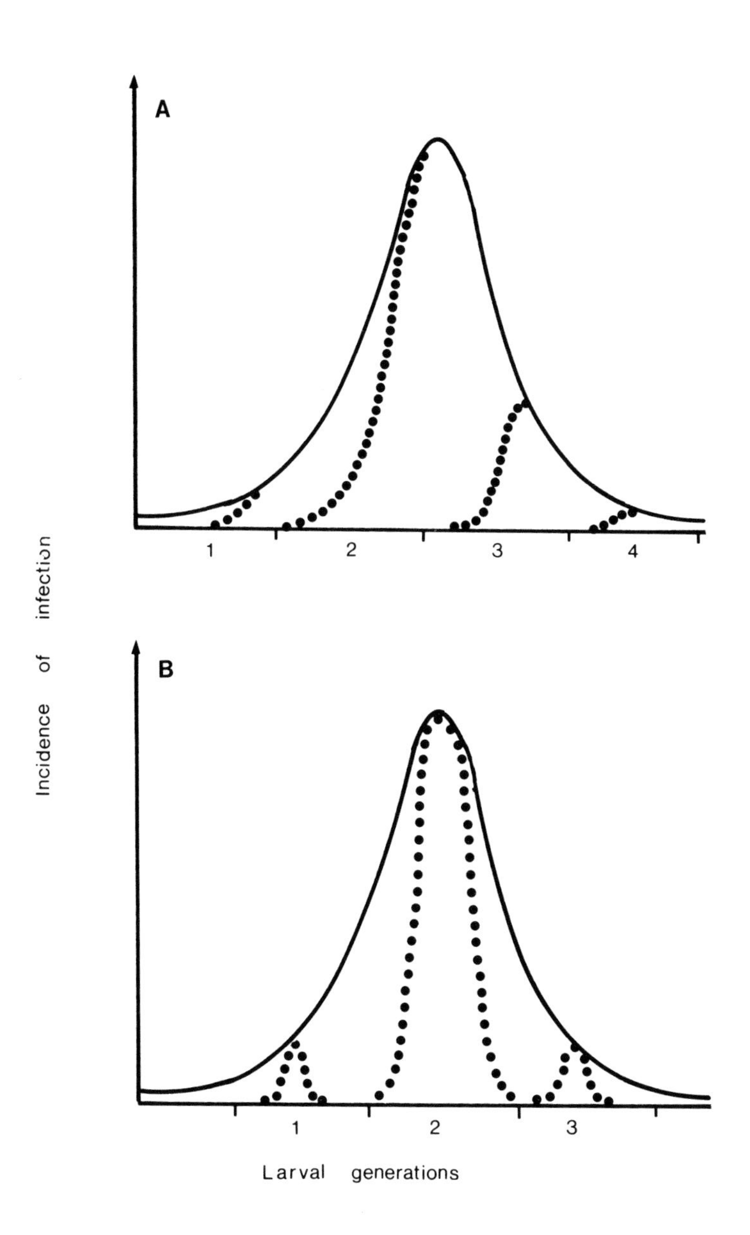

FIGURE 2. Development of the epizootic wave over several larval generations in hosts having no appreciable reduction in susceptibility (A), and exhibiting a decrease in susceptibility to infection in older larvae (B).

overall wave form of the temporal component of the epizootic is developed over several successive host generations (Fig. 2).

To compare the seasonal increase of infection in insect populations, the sigmoid curve or the ascending phase of the "wavelike" infection growth pattern observed over a single insect generation, can be normalized by the application of the transformation

$$\log_{10} x/(1-x) \qquad (1)$$

or

$$\log_{10} 1/(1-x) \qquad (2)$$

where x is the proportion of the population infected at a specific time (Van der Plank, 1963; Entwistle et al., 1983; Evans and Entwistle, 1976; Kaupp, 1981). Equation (2) is employed in situations where insects infected early do not contribute to later infections, or when the sources of inoculum available to the insect population are relatively constant (Kaupp, 1981). The daily infection rate, r, can be derived by calculating the slope of the normalized line or from the equation:

$$r = 1/t \log_{10} [x_2(1-x_1)]/[x_1(1-x_2)] \qquad (3)$$

where x_1 and x_2 are sequential levels of infection and t is the time between them. This rate is expressed as "per unit per day" accounting for the influence of many related, inseparable factors contributed by the host, virus, and environment on the rate of infection (Van der Plank, 1963). This single comprehensive figure estimating the rate of infection has been used to make comparisons between increases in infection observed in different insect species. Rates for infection development in *G. hercyniae* in Wales varied between 0.07 to 0.13 as compared to 0.12 observed in populations of this sawfly in Canada (Evans and Entwistle, 1976; Entwistle et al., 1983). Kaupp (1981) found values that ranged from 0.01 to 0.08 for the development of naturally occurring virus disease epizootics in *N. sertifer* in United Kingdom, attributing differences in development rates to changes in density of the host population.

B. SPATIAL GEOMETRY

Observations on the spread of infection have been restricted to studies involving virus epizootics in populations of *G. hercyniae* (Bird and Burke, 1961), *M. disstria* (Stairs,

1965), and *O. rhinoceros* (Young, 1974), and are adequately described by Entwistle et al. (1983). Data describing dispersal distances, x, and disease incidence, y, can be converted to straight lines by logarithmic transformation and tend to conform to a line expressed by Eq. (4):

$$\log y = a - b \log x \quad (4)$$

where the regression coefficient, b, expresses the gradient of dispersal and can be used for comparative purposes.

Generally, experiments designed to characterize spread of infection are difficult to execute because of the patchy distribution of the host, interference from naturally occurring pathogens, and problems in identifying the disease. However, despite these difficulties, the spread of infection throughout the host population is an important facet of viral epizootiology, and understanding of this spatial geometry is paramount to the effective use of viruses as insecticides. Unfortunately, most programs that have utilized viral pathogens have assumed that the disease will spread throughout the entire host population. Little thought is given to the inert nature of the virus, and that their dispersal is primarily the result of biotic and abiotic agents which must be characterized to be completely exploited.

C. INOCULUM RELEASE AND ITS EFFECT ON EPIZOOTIC DEVELOPMENT

The successive recurrence of epizootics in insect populations was first attributed to transovum/transovarial transmission of the viral pathogen or to the activation of an occult virus present in the host population (Bird, 1961; Smirnoff, 1961; Tanada, 1961). Little attention was paid to the possibility of inoculum persisting in the host environment contributing to the onset of epizootics, except in agricultural situations where virus persisting on the soil was found capable of initiating epizootics (Jaques, 1964, 1967a,b). Such information about viruses in the forest ecosystem has been lacking because viruses were thought incapable of persisting in the foliar environment since polyhedra were readily inactivated by uv radiation (Smirnoff, 1961; Bird, 1961). Recently, reviews of literature on transovum/transovarial transmission indicate that for NPVs and GVs there is no unequivocal proof supporting this mechanism (David, 1978) and reanalysis of many experiments of the reactivation of occult virus have indicated that the possibility of occult viruses has yet to be proved (Allaway, 1982). Consequently, the role of inoculum persisting in the host habitat in initiating epizootics is being reinvestigated. Studies by Clark (1958), Thompson and Scott (1979), Jaques (1964), Podgewaite et al. (1979), Mohamed et al. (1982), and Kaupp (1983a) have all indicated that inclusion bodies of NPVs, GVs, and CPVs are capable of persisting in the environment to initiate subsequent epizootics. As a result of these findings the quantity of inoculum released from the large numbers of larvae dying in epizootics and its relationship to and influence on the nature of successive epizootics is now considered an essential part of the research required in the study of epizootiology.

There is copious information available regarding the quantity of virus produced by single larva; most have been provided by studies of *in vivo* large-scale virus production facilities. The quantity of virus contained in each insect was found dependent upon a number of factors including nature of virus infection and host species. Further, it has been observed that older larvae produce more virus than the younger ones (Kaupp, 1981).

Little information is available on quantifying the amount of inoculum produced in epizootics in insect populations. One of the earliest reports concerns the quantity produced in populations of *N. sertifer* artificially infected with an NPV (Franz and Niklas, 1954). They observed that larger quantities of polyhedra were released into the forst ecosystem in treated areas as compared to the untreated areas. The plots treated before egg eclosion produced less NPV than the plots treated when sawfly larvae were in the first and second instars and died of infection in the final instar. However, no mention of the significance of these results to epizootiology of the disease was mentioned.

In investigations into the production and persistence of the NPV of Douglas fir tussock moth, *O. pseudotsugata*, in the forst ecosystem, it was found that as high as 1.6×10^{15} polyhedral inclusion bodies (PIBs) were produced from populations affected by naturally occurring epizootics (Thompson and Scott, 1979). The majority of the NPV production was again attributed to late instar larvae. Two areas experimentally treated with the virus produced only 6.87×10^{14} and 7.36×10^{14} PIBs/acre, respectively, mainly a result of the death of early instar larvae. The quantity of viable NPV found to persist in the soil of these plots after 1 year was related to the quantity of NPV produced previously, with 0.16% remaining active on the untreated area, which experienced a natural epizootic, while 0.007 and 0.14% remained viable in each of the treated plots. It was concluded that high dosage applications of virus increased the rate of epizootic development resulting in less total inoculum produced in the ecosystem. Since this NPV was found to persist in the soil for up to 40 years (Thompson et al. (1981) and was capable of being reintroduced into the host habitat to cause infection (Thompson and Scott, 1979), any reduction in the quantity of NPV produced in epizootics was seen to be reflected in the reduced potential for the occurrence of successive epizootics.

In epizootiological studies of a nuclear polyhedrosis virus disease of the winter moth, *O. brumata* L., Wigley (1976) found that virus on the egg surface was the source of infection for each generation. The origin of this inoculum was found to be virus from the previous larval generation persisting on the moss covering the tree trunks which was transferred by abiotic agents to the egg surface exclusive of adult-mediated transmission. In these studies large quantities of NPV (2.16×10^{12} to 3.50×10^{10} PIBs) were calculated to have been released from the diseased population observed on each of eight trees, and these quantities decreased over 3 years, undoubtedly as a response to the declining larval population. The projected cumulative polyhedral population produced was expressed in terms of the LD_{50} doses required to infect each population at specific sample periods and was called the "population infective unit." This transformation indicated that much of the inoculum produced was not effective in terms of late instar infection due to its loss from the host habitat and changes in the relative susceptibility of the larval population. However, further examination of the calculated population infective units indicated that their value increased substantially in the early stages of epizootic development in the final study year. This was probably caused by large quantities of NPV overwintering in the ecosystem which resulted in an increase in the initial incidence of infection in the population. Briefly, Wigley (1976) identified (1) total virus production, (2) changes in the relative susceptibility of the larval population with time, and (3) changes in the potential LD_{50} requirement of the host population, as the major components of the winter moth-virus interaction. He concluded that trends observed in the dynamics of the epizootics studied reflected fluctuations in the pattern of host-virus interactions, because of changes in larval susceptibility and virus population.

Recent studies into the production of NPV from diseased populations of *N. sertifer* have also provided much needed data elucidating the virus-host relationship in this species (Kaupp, 1981, 1983b). The quantity of virus measured as the number of PIBs produced at death and liberated from two diseased populations over a 3-year period was determined (Table I). In one population, as high as 2.3×10^{15} PIBs/ha were produced; this quantity decreased in subsequent years as a direct result of the reduction in sawfly population. This NPV has been shown to persist over winter in the foliar environment (Kaupp, 1983a) and it was found to alter the nature of

TABLE I. Estimated production of nuclear polyhedral inclusion bodies (PIBs) per hectare from insect death in two plots caused by naturally occurring epizootics[a]

Year	Plot 1	Plot 2
1978	3.3×10^{14}	2.3×10^{15}
1979	4.4×10^{12}	3.9×10^{14}
1980	3.1×10^{11}	9.8×10^{13}

[a]*From Kaupp (1981).*

subsequent epizootics by causing the occurrence of virus infection at an earlier stage in larval development than previously observed. This resulted in a decrease in the total amount of inoculum produced by the death of the insect population. This shift in the temporal component of the epizootic resulted in an increase in the rate of infection due in part to the increased quantity of secondary inoculum produced from the death of early instar larvae. More subtle changes in *N. sertifer* population levels during this study point out that changes in the virus-host population dynamics can alter the character of disease epizootics resulting in an increase in the pest population.

Although not directly quantifying the amount of virus liberated in epizootics, other researchers have established the importance of virus production resulting in environmental contamination in the recurrence of disease in host populations. The development of disease in populations of *T. ni* has been attributed to persistence of its NPV in the soil environment (Jaques, 1967a,b, 1970a, 1974; Jaques and Harcourt, 1971). Tanada and Omi (1974a,b) suggested that epizootics occur in low-density populations of the beet armyworm, *Spodoptera exigua* (Hübner), and the alfalfa looper, *Autographa californica*, because viruses are present in the soil and contaminate the environment of both species. Quantities of virus produced and liberated during epizootics have been implicated in the recurrence of disease in *Malacosoma fragile* (Clark, 1958), *L. dispar* (Podgwaite et al., 1979; Doane, 1976), *N. sertifer* (Mohamed et al., 1982), and *Spodoptera mauritia acrynyotoides* (Chon, 1982).

D. POSTULATES ON EPIZOOTIC DEVELOPMENT

Several postulates on the occurrence and expression of virus diseases in host insect populations have been published, all attempting to explain how, when, where, and why populations suddenly collapse due to virus disease (Franz, 1961; Krieg, 1961).

Inherent in any discussion of the postulates of epizootics are the mechanisms for the maintenance and survival of a pathogen population when the host insect is either at very low levels, or absent. Conditions describing the former situation where disease is at a low level or perhaps unobservable, have been termed "enzootics" and have been claimed to be responsible for the maintenance of disease in many pest populations (Tanada, 1961). Concomitant with this has been the suggestion of the presence of an occult virus or latent infection which becomes apparent when triggered by various types of population or environmental stresses causing increases in the incidence of disease, ultimately leading to an epizootic (Wellington, 1962). On the other hand, long-term persistence of environmentally stable virus becomes one of the main methods of maintaining a viable pathogen population in the absence of the host population. As the host population increases, so does the probability of infection from these foci of persisting virus, resulting in an epizootic (Kaupp, 1981, 1983a). It is the suggestion of the authors that all the above factors are important in maintaining host-virus relationships. However, extensive reliable experimentation is needed before many of these mechanisms of persistence can be confidently incorporated into present-day epizootiological models.

One explanation of the recurrence of natural epizootics in fluctuating insect populations is based on the interaction between the tolerance of insects to disease, population density, and the presence of a viable virus population (Martignoni and Schmid, 1961). According to these authors virus epizootics are thought to cause an increase in the proportion of individuals resistant to disease and hence a decrease in the heterogeneity of response of the population. This greatly reduces the epizootic potential of the available virus population. During the postepizootic period the disease ceases to act as selection pressure and the heterogeneity in the response of the host insect population gradually returns. This reduces the resistance of the population to disease and once again predisposes the host population to an epizootic of the disease. This explanation has been widely accepted in view of those reports of increased resistance to infection observed in insect populations repeatedly challenged by virus disease (Briesse et al., 1980).

In epizootiological studies on the NPVs and CPVs of *L. fumida*, Katagiri (1977) elaborates on how changes in the host population characteristics cause cyclic fluctuations in the incidence of disease. Initial expression of infection in *L. fumida* populations was considered to be a result of the induction of an occult virus. In subsequent generations, the disease existed in an epizootic state in high host populations causing high mortality and the liberation of large quantities of polyhedra into the environment. The insect population ultimately collapsed. Year-to-year transmission of the disease was considered the result of virus persisting in the host habitat. In the postepizootic period of the cycle, disease disappeared and was assumed to have become occult. Analogous to the conclusions of Martignoni and Schmid (1961), increase in the population density affected population quality and was seen related to increases in the incidence of latent infections predisposing the population to increases in the expressions of apparent infection. All of these factors led to the formation of foci of infection capable of initiating an epizootic.

A modification of the density-dependent regulation of insect populations has been noted by Doane (1976) working with *L. dispar* and by Kaupp (1981) working with *N. sertifer*. In both these cases, the occurrence and the nature of epizootics were altered by the environmental persistence of the virus. In *L. dispar* populations, Doane (1976) noted that NPV infection, newly introduced or present at a very low level, increased in a density-dependent manner until the host population collapsed due to the epizootic. The environment of the gypsy moth then became heavily contaminated with NPV and the dynamics of the host-virus interaction was observed not to act in a density-dependent manner. The reason for this was that the large-scale environmental contamination resulted in widespread infection of all larvae present which was not mediated by larva to larva spread. With the passage of time most of the NPV on the contaminated surfaces weathered and disappeared and some of it was covered by vegetation allowing population buildup, creating a situation suitable for the occurrence of another epizootic cycle. Although initial expression of disease was density-dependent, massive environmental contamination allowed for an accelerated collapse of the host population.

Based on a 3-year study, Kaupp (1981) postulated that the development of epizootics in populations of *N. sertifer* was driven mainly by host population density, and long-term persistence of the pathogen (Fig. 3). The epizootic cycle begins with a disease-free population infesting an area with a past history of epizootics (Fig. 3A). The foliage is contaminated with virus from a previous epizootic. The polyhedra having been redistributed all over the canopy persist as foci of infection. As the host population density increases, the chance of larval colonies encountering these foci of infection increases. Occasional larval infection and death result in a gradual buildup of inoculum, some of which remains as further foliar contamination. A large proportion of this inoculum enters the forest soil. Tree biomass increases during this period, but has a negligible effect in diluting the epizootic

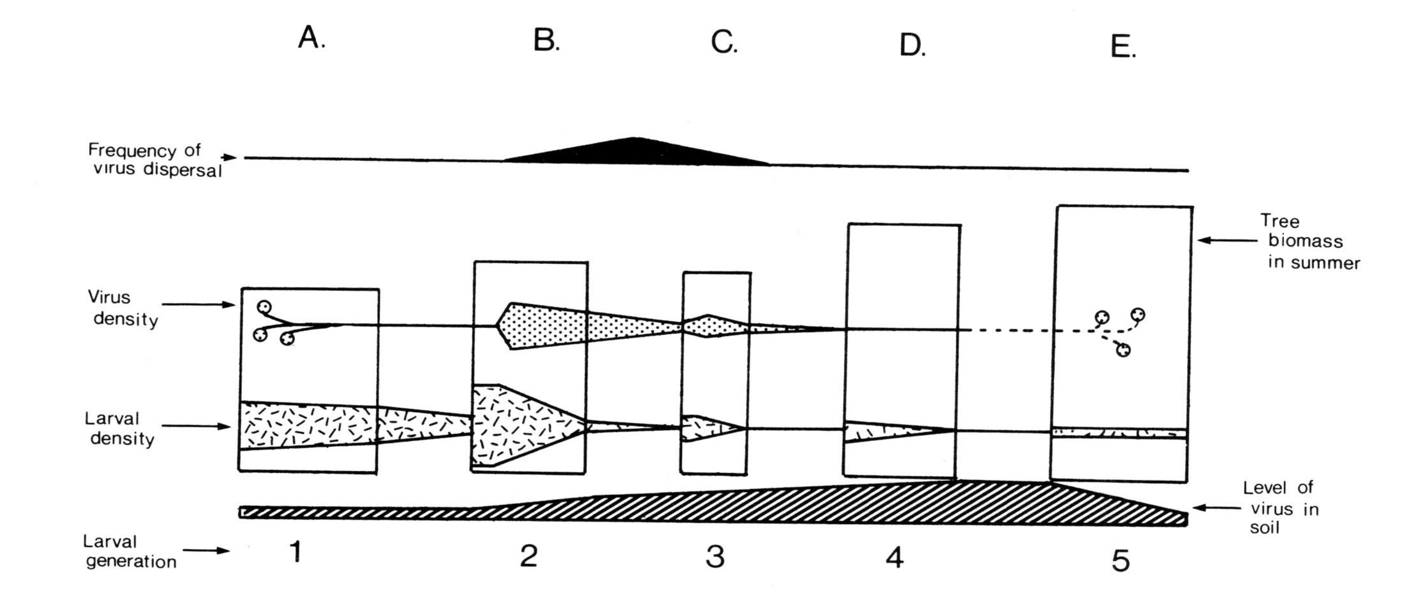

FIGURE 3. Schematic representation of the development of nuclear polyhedrosis virus epizootics in N. sertifer populations (Kaupp, 1981). Successive summer periods are represented from A through to E.

potential at this stage, because host density and the quantity of inoculum persisting in the canopy are both still sufficient to initiate an epizootic. Other population mortality factors, such as egg and cocoon predation, although reducing the number of nonlarval hosts, do not influence epizootic development.

In successive generations, severe epizootics occur in high-density sawfly populations (Fig. 3B and C), and heavy defoliation of the trees occurs. Also, since most of the mortality results from secondary infection, most larvae die just prior to pupation, having consumed almost the same amount of foliage as healthy larvae. This results in short-term loss of forest biomass, large-scale foliar contamination with polyhedra, and an increase in the quantities of polyhedra found in the soil. Increases in quantities of polyhedra produced by secondary infection are represented in Fig. 3B and C by increases in the quantity of polyhedra in the canopy during the larval period. The frequency of dispersal of polyhedra by biotic and abiotic agents is probably highest during the major epizootic period, although theoretically it can occur in any host generation.

The epizootic soon enters a decline phase, where both virus and host population are decreasing (Fig. 3D). In contrast to the early stages of this cycle, the effectiveness of primary inoculum in initiating infection now decreases following increased foliage production by the host trees which dilutes both the persisting inoculum and already sparse host population.

As the epizootic continues to decline, the quantity of inoculum in the environment is reduced further and a situation similar to the start of this cycle develops (Fig. 3E). The quantity of polyhedra in the soil gradually decreases as they move downward or are inactivated. Subsequent increases in tree biomass, in addition to the loss of polyhedra from the canopy, reduces the epizootic potential of this virus population and results in the establishment of foci of polyhedra capable of initiating epizootics when suitable conditions are recreated.

IV. CONCLUSIONS

It is evident that the development of complex epizootiological models has paralleled our understanding of the nature of epizootics and of host-virus relationships. The viruses manage to survive in the ecosystem by a complex of pathways and interrelationships with the environment and host population. It has been observed that virus diseases can cause high mortality in insect populations, change the quality of host populations (Wellington, 1962), and that they can persist in the physical environment for long periods. It would appear, therefore, that the role of insect viruses in the ecosystem cannot be viewed as just mortality factors, but rather must be recognized as an integral relationship exhibited by an entity existing within a complex ecosystem.

Insect viruses can indeed cause heavy insect mortality as discussed in this chapter and in the reviews by Cunningham and Entwistle (1981), Cunningham (1982), Yearian and Young (1982), David (1978), and Katagiri (1981). Viruses also have an extraordinary capacity to persist both within the physical environment and in their respective host populations.

As a result of these two factors, as well as a combination of the various responses of insect populations to disease, epizootics have been observed which have temporal and spatial characteristics that appeal to the proponents of microbial control of insects. These characteristics are disease dispersal and the production of large quantities of inoculum through virus-mediated insect mortality to perpetuate the epizootic. Unfortunately, adherence to the implications of these factors for pest control have not always provided successful results by way of developing epizootics. Many programs have shown poor results with no carry-over effect on subsequent populations when economically large quantities of virus have been applied to pest populations.

It is apparent from the results of Kaupp (1981), Wigley (1976), and Thompson and Scott (1979) that a large proportion of the virus produced in epizootics is ineffective in increasing virus infection but rather seems to be more efficiently utilized to completely contaminate the environment ensuring virus survival to the next suitable host generation. As it has already been discussed, the quantities of inoculum produced under natural conditions are very high, and the mechanisms of persistence of the virus in the host-virus relationship are not well understood. It is simply not enough to introduce a quantity of virus, deemed economically large in production terms, into a population and hope to create a situation suitable for an epizootic. More intensive work must be done on the host-virus relationship as is evidenced from this chapter, to achieve the full potential of viruses as insect control agents, as well as to determine which of these viruses are unsuitable as bioinsecticides. It has been premature to pursue insect viruses as utopian insect control agents without fully investigating their role within the rest of the ecosystem, including the host population.

REFERENCES

Allaway, G. P. (1982). "Infectivity of some occluded insect viruses." Ph.D. Thesis. 274 pp. University of London, U.K.

Bird, F. T. (1961). Transmission of some insect viruses with particular reference to ovarial transmission and its importance in the development of epizootics. *J. Insect Pathol. 3*, 352-380.

Bird, F. T. (1969). Infection and mortality of spruce budworm, *Choristoneura fumiferana*, and forest tent caterpillar, *Malacosoma disstria*, caused by nuclear and cytoplasmic polyhedrosis viruses. *Can. Entomol. 101*, 1269-1285.

Bird, F. T., and Burke, J. M. (1961). Artificially disseminated virus as a factor controlling the European spruce sawfly, *Diprion hercyniae* (Htg.) in the absence of introduced parasites. *Can. Entomol. 93*, 228-238.

Bird, F. T., and Whalen, M. M. (1954). A nuclear and a cytoplasmic polyhedral virus disease of the spruce budworm. *Can. J. Zool. 32*, 82-86.

Bird, F. T., Sanders, C. J., and Burke, J. M. (1971). A newly discovered virus disease of the spruce budworm, *Choristoneura biennis* (Lepidoptera: Tortricidae). *J. Invertebr. Pathol. 18*, 159-161.

Briese, D. T., Mende, H. A., Grace, T. D. C., and Geier, P. W. (1980). Resistance to a nuclear polyhedrosis virus in the light-brown apple moth, *Epiphyas postvittana* (Lepidoptera: Tortricidae). *J. Invertebr. Pathol. 36*, 211-215.

Burnet, F. M. (1953). "Natural History of Infectious Disease," 2nd ed., 356 pp. Cambridge University Press, Cambridge.

Chon, T. S. (1982). "A systems approach utilizing simulation modeling for the management of the lawn armyworm with its nuclear polyhedrosis virus." Ph.D. Thesis. 245 pp. University of Hawaii, Hawaii.

Clark, E. C. (1958). Ecology of the polyhedrosis of tent caterpillars. *Ecology 39*, 132-139.

Cunningham, J. C. (1982). Field trials with baculoviruses: Control of forest insect pests. *In* "Microbial and Viral Insecticides" (E. Kurstak, ed.), pp. 335-386. Dekker, New York.

Cunningham, J. C., and Entwistle, P. F. (1981). Control of sawflies by baculoviruses. *In* "Microbial Control of Pests and Plant Diseases 1970-1980" (H. D. Burges, ed.), pp. 379-408. Academic Press, New York.

Cunningham, J. C., and McPhee, J. (1973). Aerial application of entomopoxvirus and nuclear polyhedrosis virus against spruce budworm at Chapleau, Ont. 1972. *Can. Forest. Serv. Inform. Rept. IP-X-3*, 27 pp.

Cunningham, J. C., Kaupp, W. J., McPhee, J., Sippell, W. L., and Barnes, C. A. (1975). Aerial application of nuclear polyhedrosis virus to control European pine sawfly, *Neodiprion sertifer* (Geoff.) at Sandbanks Provincial Park, Quinty Island, Ontario in 1975. *Can. Forest. Serv. Inform. Rept. IP-X-7*, 18 pp.

Cunningham, J. C., Tonks, N. V., and Kaupp, W. J. (1981). Viruses to control winter moth, *Operophtera brumata* (Lepidoptera: Geometridae). *J. Entomol. Soc. Brit. Columbia 78*, 17-24.

Crawford, A. M., and Kalmakoff, J. (1977). A host-virus interaction in a pasture habitat: *Wiseana* spp. (Lepidoptera: Hepialidae) and its baculovirus. *J. Invertebr. Pathol. 29*, 81-87.

David, W. (1978). The granulosis virus of *Pieris brassicae* (L.) and its relationship with its host. *Advan. Virus Res. 22*, 111-161.

de Groot, P., and Cunningham, J. C. (1983). Aerial spray trials with baculovirus to control redheaded pine sawfly in Ontario in 1979 and 1980. *Can. Forest. Serv. Inform. Rept. FPM-X-63*, 12 pp.

Doane, C. C. (1975). Infectious sources of nuclear polyhedrosis persisting in natural habitats of the gypsy moth. *Environ. Entomol. 4*, 392-394.

Doane, C. C. (1976). Ecology of pathogens of the gypsy moth. *In* "Perspectives in Forest Entomology" (J. F. Anderson and H. K. Kaya, eds.), pp. 265-284. Academic Press, New York.

Dusaussoy, G., and Geri, C. (1969). Étude des fluctuations du niveau de population de la processionnaire du pin dans la Vallée du Niolo en Corse. *Ann. Sci. Forest. 26*, 103-125.

Entwistle, P. F. (1983). Control of insects by viral diseases. *C.I.B.C. Biocontrol News Inform. 4*, 203-224.

Entwistle, P. F., Adams, P. H. W., and Evans, H. F. (1977a). Epizootiology of a nuclear polyhedrosis virus in European spruce sawfly (*Gilpinia hercyniae*): The status of birds as dispersal agents of the virus during larval season. *J. Invertebr. Pathol. 29*, 354-360.

Entwistle, P. F., Adams, P. H. W., and Evans, H. F. (1977b). Epizootiology of a nuclear polyhedrosis virus in European spruce sawfly, *Gilpinia hercyniae:* Birds as dispersal agents of the virus during winter. *J. Invertebr. Pathol. 30*, 15-19.

Entwistle, P. F., Adams, P. H. W., Evans, H. F., and Rivers, C. F. (1983). Epizootiology of a nuclear polyhedrosis virus (Baculoviridae) in European spruce sawfly (*Gilpinia*

hercyniae): Spread of disease from small epicenters in comparison with spread of baculovirus disease in other hosts. *J. Appl. Ecol. 20*, 473-487.
Evans, H. F., and Entwistle, P. F. (1976). The development of infection during a virus epizootic in spruce sawfly populations in mid-Whales. *Proc. 1st Intern. Colloq. Invertebr. Pathol., Kingston, Ont., Can., 1976*, pp. 350-351.
Falcon, L. A., Kane, W. R., and Bethell, R. S. (1968). Preliminary evaluation of a granulosis virus for control of the codling moth. *J. Econ. Entomol. 61*, 1208-1213.
Franz, J. M. (1961). Biologishe Schädlingsbekampfung. *In* "Handbuch de Pflanzenkrankheiten" (H. Richter, ed.), 2nd ed., Vol. 6, pp. 1-302. Paul Parey, Berlin.
Franz, J., and Niklas, O. F. (1954). Feldversuche zur Bekämpfung der roten Kiefernbusch horn blattwespe (*Neodiprion sertifer* [Geoff.]) durch Künstliche Verbreitung einer Virusseuche. *Nachrichtenbl. Deut. Pflanzenschutz. 6*, 131-134.
Granados, R. R., and Roberts, D. W. (1970). Electron microscopy of a pox-like virus infecting an invertebrate host. *Virology 40*, 230-243.
Jaques, R. P. (1964). The persistence of a nuclear polyhedrosis virus in soil. *J. Invertebr. Pathol. 6*, 251-254.
Jaques, R. P. (1967a). The persistence of a nuclear polyhedrosis virus in the habitat of the host insect, *Trichoplusia ni*. I. Polyhedra deposited on foliage. *Can. Entomol. 99*, 785-794.
Jaques, R. P. (1967b). The persistence of a nuclear polyhedrosis virus in the habitat of the host insect, *Trichoplusia ni*. II. Polyhedra in soil. *Can. Entomol. 99*, 820-829.
Jaques, R. P. (1970a). Natural occurrence of virus of the cabbage looper in field plots. *Can. Entomol. 102*, 36-41.
Jaques, R. P. (1970b). Application of viruses to soil and foliage for control of the cabbage looper and imported cabbageworm. *J. Invertebr. Pathol. 15*, 328-340.
Jaques, R. P. (1974). Occurrence and accumulation of viruses of *Trichoplusia ni* in treated soil plots. *J. Invertebr. Pathol. 23*, 140-152.
Jaques, R. P. (1983). The potential of pathogens for pest control. *Agric. Ecosyst. Environ. 10*, 101-126.
(T. L. V. Ulbricht, ed.). Agricultural Research Council, London, U.K., in press.
Jaques, R. P., and Harcourt, D. G. (1971). Viruses of *Trichoplusia ni* (Lepidoptera: Noctuidae) and *Pieris rapae* (Lepidoptera: Pieridae) in soils in fields of crucifers in southern Ontario. *Can. Entomol. 103*, 1285-1290.
Kalmakoff, J., and Crawford, A. M. (1982). Enzootic virus control of *Wiseana* spp. in the pasture environment. *In*

"Microbial and Viral Pesticides" (E. Kurstak, ed.), pp. 435-448. Dekker, New York.

Katagiri, K. (1969). Review of microbial control of insect pests in forests in Japan. *Entomophaga 14*, 203-214.

Katagiri, K. (1973). A newly discovered entomopox virus of *Choristoneura diversana* (Lepidoptera: Tortricidae). *J. Invertebr. Pathol. 22*, 300-302.

Katagiri, K. (1977). Epizootiological studies on the nuclear and cytoplasmic polyhedrosis of the red belly tussock moth, *Lymantria fumida* (Lepidoptera: Lymantriidae). *Bull. Govt. For. Exp. Stat., Tokyo, Japan No. 294*, pp. 85-135.

Katagiri, K. (1981). Pest control by cytoplasmic polyhedrosis viruses. *In* "Microbial Control of Pest and Plant Diseases" (H. D. Burgess, ed.), pp. 433-440. Academic Press, New York.

Kaupp, W. J. (1981). Studies on the ecology of the nuclear polyhedrosis virus of the European pine sawfly, *Neodiprion sertifer* (Geoff.). Ph.D. Thesis, 363 pp. University of Oxford, Oxford.

Kaupp, W. J. (1983a). Persistence of *Neodiprion sertifer* (Hymenoptera: Diprionidae) nuclear polyhedrosis virus on *Pinus contorta* foliage. *Can. Entomol. 115*, 869-873.

Kaupp, W. J. (1983b). Estimation of nuclear polyhedrosis virus produced in field populations of the European pine sawfly, *Neodiprion sertifer* (Geoff.) (Hymenoptera: Diprionidae). *Can. J. Zool. 61*, 1857-1861.

Kelsey, R. P. (1958). Control of *Pieris rapae* by granulosis virus. *N. Zeal. J. Agr. Res. 1*, 778-782.

Koyama, R., and Katagiri, K. (1967). On application of nuclear and cytoplasmic polyhedrosis viruses against *Lymantria fumida* Butler (Lepid. Lymantridae). *Bull. Govt. Forest. Exp. Stat. Tokyo, Japan No. 207*, pp. 1-10.

Kurstak, E., and Tijssen, P. (1982). Microbial and viral pesticides: modes of action, safety and future prospects. *In* "Microbial and Viral Pesticides" (E. Kurstak, ed.), pp. 3-32. Dekker, New York.

Krieg, A. (1961). "Grundlagen der Insektenpathologie, Viren-Rickettsien-und Bakterieninfekionen," 304 pp. Steinkopff, Darmstadt.

Luria, S. E., Darnell, J. E., Baltimore, D., and Campbell, A. (1978). Introduction: The science of virology. *In* "General Virology," 3rd ed., pp. 1-20. Wiley, New York.

Marschall, K. J. (1970). Introduction of a new virus disėase of the coconut rhinocerous beetle in Western Samoa. *Nature (London) 225*, 288-289.

Marschall, K. J., and Doane, I. (1982). The effect of re-release of *Oryctes rhinoceros* baculovirus in the biological control of rhinocerous beetles in Western Samoa. *J. Invertebr. Pathol. 39*, 267-276.

Martignoni, M. E., and Iwai, P. J. (1981). A catalogue of viral diseases of insects, mite and ticks. *In* "Microbial Control of Pests and Plant Diseases 1970-1980" (H. D. Burges, ed.), Appendix 2, pp. 897-911. Academic Press, New York.

Martignoni, M. E., and Schmid, P. (1961). Studies on the resistance to virus infections in natural populations of Lepidoptera. *J. Insect Pathol. 3*, 62-74.

Milner, R. J. (1977). The role of disease during an outbreak of *Oncopera alboguttata* Tindale and *O. rufobrunnea* Tindale (Lepidoptera: Hepialidae) in the Ebor/Dorrigo Region of N.S.W. *J. Aust. Entomol. Soc. 16*, 21-26.

Mohamed, M. A., Coppel, H. C., and Podgwaite, J. D. (1982). Persistence in soil and on foliage of nuclear polyhedrosis virus of the European pine sawfly, *Neodiprion sertifer* (Hymenoptera: Diprionidae). *Environ. Entomol. 11*, 1116-1118.

Moore, S. G., Kalmakoff, J., and Miles, J. A. R. (1973). Virus diseases of Porina (*Wiseana* spp. Lepidoptera: Hepialidae). *N. Zeal. J. Sci. 16*, 139-153.

Podgwaite, J. D., Stoneshields, K., Zerillo, R. T., and Bruen, R. B. (1979). Environmental persistence of the nuclear polyhedrosis virus of the gypsy moth, *Lymantria dispar*. *Environ. Entomol. 8*, 528-536.

Shiga, M., Yamada, H., Oho, N., Nakazawa, H., and Ito, Y. (1973). A granulosis virus, possible biological agent for control of *Adoxophyes orana* (Lepidoptera: Tortricidae) in apple orchards. *J. Invertebr. Pathol. 21*, 149-157.

Smirnoff, V. A. (1961). A virus disease of *Neodiprion swainei* Middleton. *J. Invertebr. Pathol. 3*, 29-46.

Stairs, G. R. (1965). Artificial initiation of virus epizootics in forest tent caterpillar populations. *Can. Entomol. 97*, 1059-1962.

Tanada, Y. (1961). The epizootiology of virus diseases in field populations of the armyworm, *Pseudaletia unipuncta* (Haworth). *J. Invertebr. Pathol. 3*, 310-323.

Tanada, Y. (1963). Epizootiology of infectious diseases. *In* "Insect Pathology: An Advanced Treatise" (E. A. Steinhaus, ed.), Vol. 2, pp. 423-475. Academic Press, New York.

Tanada, Y. (1964). A granulosis virus of the codling moth, *Carpocapsa pomonella* (Linnaeus) (Olethreutidae, Lepidoptera). *J. Insect. Pathol. 6*, 378-380.

Tanada, Y. (1967). The role of virus in the regulation of the population of the armyworm, *Pseudaletia unipuncta* (Haworth). *Proc. Joint U.S.-Japan Sem. Microbial Control of Insect Pests, Fukuoka, April 21-23*, pp. 25-31.

Tanada, Y., and Chang, G. Y. (1964). Interaction of two cytoplasmic-polyhedrosis viruses in three insect species. *J. Invertebr. Pathol. 6*, 500-516.

Tanada, Y., and Omi, E. M. (1974a). Epizootiology of virus diseases in three lepidopterous insect species in alfalfa. *Res. Population Ecol. 16*, 59-68.
Tanada, Y., and Omi, E. M. (1974b). Persistence of insect viruses in field populations of alfalfa insects. *J. Invertebr. Pathol. 23*, 360-365.
Thompson, C. G., and Scott, D. W. (1979). Production and resistance of the nuclear polyhedrosis virus of the douglas fir tussock moth *Orgyia pseudotsugata* (Lepidoptera: Lymantriidae) in the forest ecosystem. *J. Invertebr. Pathol. 33*, 57-65.
Thompson, C. G., Scott, D. W., and Wickman, B. E. (1981). Long-term persistence of the nuclear polyhedrosis virus of the douglas fir tussock moth, *Orgyia pseudotsugata* (Lepidoptera: Lymantriidae) in forest soil. *Environ. Entomol. 10*, 254-255.
Van der Plank, J. E. (1963). "Plant Diseases: Epidemics and Control," 349 pp. Academic Press, New York.
Watt, K. E. F. (1968). "Ecology and Resource Management: A Quantitative Approach," 450 pp. McGraw-Hill, Toronto.
Weiser, J. (1970). Ultrastructure of the spindle-shaped virus *Vagoiavirus operophterae*, in the winter moth, *Operophtera brumata* L. *Acta Virol. 14*, 314-317.
Weiser, J., and Vago, C. (1966). A newly described virus of the winter moth *Operophtera brumata* Hübner (Lepidoptera, Geometridae). *J. Invertebr. Pathol. 8*, 314-319.
Wellington, W. G. (1962). Population quality and maintenance of a nuclear polyhedrosis between outbreaks of *Malacosoma pluviale* (Dyan.). *J. Invertebr. Pathol. 4*, 285-305.
Wigley, P. J. (1976). "The epizootiology of a nuclear polyhedrosis virus disease of the winter moth, *Operophtera brumata* L. at Wistman's Wood, Dartmoor." Ph.D. Thesis, 185 pp. University of Oxford, Oxford.
Wigley, P. J., and Scotti, P. D. (1983). The seasonal incidence of cricket paralysis virus in a population of the New Zealand small field cricket, *Pteronemobius nigrovus* (Orthoptera: Gryllidae). *J. Invertebr. Pathol. 41*, 378-380.
Yearian, W. C., and Young, S. Y. (1982). Control of insect pests of agricultural importance by viral insecticides. *In* "Microbial and Viral Pesticides" (E. Kurstak, ed.), pp. 387-424. Dekker, New York.
Young, E. C. (1974). The epizootiology of two pathogens of the coconut palm rhinocerous beetle. *J. Invertebr. Pathol. 24*, 82-92.

IV.

Physical, Biological and Chemical Characteristics

THE STRUCTURE AND PHYSICAL CHARACTERISTICS OF BACULOVIRUSES

D. C. KELLY

Natural Environment Research Council
Institute of Virology
Mansfield Road, Oxford

Baculoviruses are large structurally complex viruses. Their very complexity has long intrigued electron microscopists and most evidence concerning the structure of the viruses derives from electron microscope studies. Other biophysical techniques have been applied to study baculovirus architecture and, in some instances, these approaches have proved significant insights into the structural organization of the viruses.

Baculovirus particles are, in most cases, packaged into large protein crystals variously termed as polyhedra, capsules, granules, whole inclusion bodies or polyhedral inclusion bodies. The "occlusion" or packaging of virus particles appears to be a genetically defined trait with some virus types containing just one virus particle per crystal whereas other types package many virus particles within a protein crystal.

Baculoviruses have been classified according to their structure into nuclear polyhedrosis viruses, granulosis viruses, and nonoccluded baculoviruses (Matthews, 1982). Nonoccluded baculoviruses, typified by a baculovirus from *Oryctes rhinoceros* (Payne, 1974) fail to synthesize polyhedra (Huger, 1966; Kelly, 1975) and, consequently, never become packaged in protein crystals. Granulosis viruses (GVs) typically comprise small crystals (capsules or granules--hence the term granulosis) containing a solitary virus particle. Nuclear polyhedrosis viruses (NPVs), on the other hand, contain numerous virus particles within a given "polyhedron." They take

VIRAL INSECTICIDES
FOR BIOLOGICAL CONTROL

ISBN 0-12-470295-3

their name from the presence of polyhedra in an infected cell nucleus. Generally speaking polyhedra or granules act as vectors of virus particles from one insect larva to another and, in certain instances, where one generation of larvae is temporally spaced from the succeeding generation, facilitates the survival of the virus.

The actual virus particles are rod shaped and comprise a nucleocapsid surrounded by an envelope. The nucleocapsid comprises a DNA-protein complex (the deoxyribonucleoprotein or DNP) contained within a protein shell (the capsid). In the case of nuclear polyhedrosis viruses more than one nucleocapsid may occur within an envelope and the NPVs may be described as single nucleocapsid NPVs (*S*NPV) or multiple nucleocapsid NPVs (*M*NPV).

This chapter is organized so that the "core" of the virus is considered first and the subsequent elaboration of functional layers is then related to preceding structures. The genome of the virus is infectious and so all the complex superstructure associated with the genome is merely involved in protection and transport of the viral genes from one insect to another or from one cell or organ to another. In many respects the structures associated with a virus are a direct reflection of their function in the replication of the virus. Consequently, many aspects of baculovirus replication are referred to in this chapter. Detailed and comprehensive reviews of baculovirus replication are found in the chapter by and in reviews by Granados (1980), Faulkner (1981), Tweeten et al. (1981), and Kelly (1982).

I. STRUCTURE OF BACULOVIRUS DNA

All baculoviruses contain double-stranded DNA. Usually the DNA extracted from virus particles is circular and some molecules are covalently closed (i.e., the two chains of DNA run antiparallel and each chain is joined only to itself: Crick, 1976). Figure 1 shows a "relaxed" circular DNA molecule from the MNPV of *Trichoplusia ni*. A relaxed molecule has "nicks" in one or both DNA strands and is referred to as a nicked circular or ncDNA in contrast to the covalently closed or ccDNA. When nicks occur coincidentally in opposite strands a linear molecule results. Most fresh preparations of baculovirus DNA contain less than 1% linear molecules and rarely more than 30% ccDNA molecules. Baculovirus DNA is about 40 μm long, equivalent to a molecular weight of about 80 million (corresponding to 80 to 120 genes). The size of

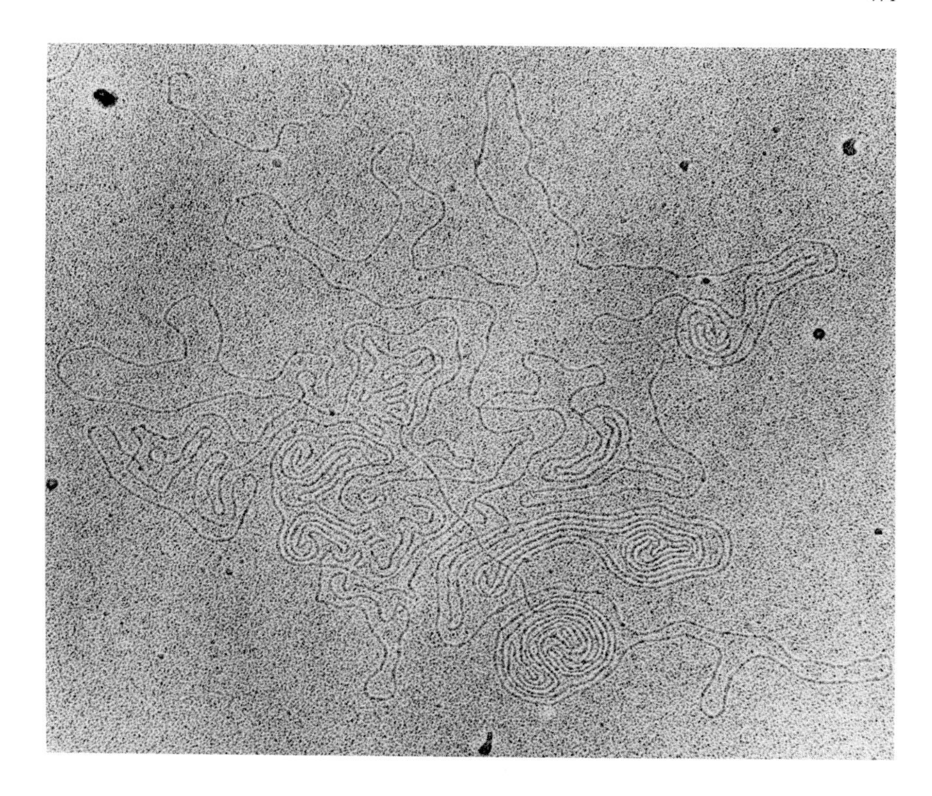

FIGURE 1. A relaxed circular molecule of double-stranded DNA extracted from Trichoplusia ni M nuclear polyhedrosis virus particles. Magnification 60,000×.

the DNA is fixed for a given isolate, although different isolates vary in size, and apparently reliable estimates of 60 to 110 million have been determined (Bud and Kelly, 1977; Burgess, 1977; Tweeten et al., 1981; Kelly et al., 1980; Loh et al., 1982; Kelly et al., 1980). These sizes, obtained by electron microscopy, have been confirmed by other biophysical methods such as agarose gel electrophoresis of fragmented DNA generated with restriction enzymes or DNA reassociation kinetics (Kelly, 1977; Rohrmann et al., 1978; Tweeten et al., 1980; Smith and Summers, 1978).

Each nucleocapsid contains a single molecule of DNA. Double- and triple-length nucleocapsids have been observed (Burley et al., 1982; Skuratovskaya et al., 1977) and Skuratovskaya et al. (1982) have claimed that oligomeric circular DNA (i.e., the circles of DNA are linked as if in a chain) is contained by multiple-length nucleocapsids. Their evidence rests on electron microscopy and the presence of DNA of intermediate density detected on ethidium bromide/cesium chloride gradients. cc- and ncDNA molecules differ in their ability to bind the intercalating dye ethidium bromide, and the DNA-dye complex differs in density which permits their separation on cesium chloride density gradients (Bauer and Vinograd, 1968). Baculovirus DNA of a density intermediate between cc- and ncDNA was first observed by Summers and Anderson (1972) and it has been interpreted as cc and nc molecules linked to create an entity of intermediate density. Two cc and one nc molecule linked or two nc and one cc molecule linked would also be resolved by this method and has been observed (Skuratovskaya et al., 1977; Kunjeku, 1982; E. Kunjeku, M. D. Ayres, H. F. Evans, and D. C. Kelly, unpublished). Skuratovskaya et al. (1977) correlated the presence of multiple-length nucleocapsids with the presence of oligomeric forms. Kunjeku (1982), however, also found multiple bands in pure preparations of single-length nucleocapsids and it is possible that some bands of intermediate density represents nc- and ccDNA draped and entangled, rather than chained, together. It is nevertheless an attractive concept that oligomeric chains of DNA are contained in multiple-length nucleocapsids.

Revet and Guelpa (1979) showed that the superhelical density of baculovirus ccDNA is very low and approximates to zero. It is, therefore, equivalent to relaxed DNA in certain properties. The superhelical density is an index of the twisting of a DNA molecule. In their study of *Tipula paludosa* NPV, they used ethidium bromide binding to titrate the superhelical density. This phenomenon has now been observed with other baculovirus DNA's (Kelly and Wang, 1981). The low superhelical density (i.e., the number of supertwists per unit length of DNA) will undoubtedly affect the configuration that baculovirus DNA can assume since it influences both the torsion and writhing of a ccNDA molecule (Crick, 1976). It is also probably better to consider baculovirus ccNDA not to be supercoiled. Within the nucleocapsid circular dichroism studies have shown that the DNA is in the classical B form of DNA (Kelly et al., 1983).

The DNA is infectious (Skuratovskaya et al., 1977; Bud and Kelly, 1980 ; Burand et al., 1980; Carstens et al., 1980; Kelly and Wang, 1981). To be infectious the DNA must be circular, although there is little difference in infectivity between cc- and ncDNA (Kelly and Wang, 1981). The infectivity is enhanced by protection with basic DNA binding proteins, including the major nucleocapsid protein (Kelly and Wang, 1981; Kelly et al., 1983)

Recently, Miller and Miller (1982) showed that baculovirus DNA may possess host as well as viral genes and in their study the presence of a *copia*-like transposable element was shown to be inserted in a baculovirus genome.

II. STRUCTURE OF THE DEOXYRIBONUCLEOPROTEIN

The DNA in baculovirus particles is associated with basic proteins, constituting a deoxyribonucleoprotein (Monsarrat et al., 1975; Tweeten et al., 1980 ; Bud and Kelly, 1980 ; Burley et al., 1982; Kelly et al., 1983). No chromatin-like structures have been observed in the DNP extruded from nucleocapsids (Monsarrat et al., 1975; Bud and Kelly, 1980 ; Tweeten et al., 1980). No histone proteins are associated with baculovirus DNA (Kelly et al., 1983) and this probably explains the lack of nucleosomal structure. In *Heliothis zea* SNPV a single basic DNA binding protein was found (Kelly et al., 1983). with properties similar to the acid-soluble arginine-rich protein in *Plodia interpunctella* GV (Tweeten et al., 1980). Both proteins have protamine-like properties although they are far larger than protamines. Few of the polyamines present in virus particles are associated with the nucleocapsids (Elliott and Kelly, 1977, 1979) and so they play no role in the structure of the DNP.

The structure of the DNP exuding from nucleocapsids of *H. zea* SNPV is shown in Fig. 2. The DNP has not yet been satisfactorily purified to elucidate the proteins associated with the DNA. With *H. zea* SNPV the DNA binding protein represents the major component of the DNP, whereas with *T. ni* MNPV a number of minor DNA binding proteins are detected besides the major protein (M. D. Ayres, D. A. Brown, and D. C. Kelly, unpublished). The major basic protein appears to undergo a substantial conformational change in a highly charged environment (Kelly et al., 1983). This change involves tyrosine side chains and these may play a role in DNA binding by intercalating between the base pairs as occurs in the interaction between DNA and the gene 5 protein of bacteriophage fd5 (McPherson et al., 1979).

The DNP within the capsid comprises DNA and protein associated heterogenously and occupies a central core 32 nm in diameter (Burley et al., 1982). The DNA is approximately 120 times longer than the nucleocapsid and so the DNP within the capsid contains DNA compacted in an orderly fashion. The actual order is not known. It has been suggested that the DNA is supercoiled and resupercoiled to the final dimensions of the nucleocapsid core (Koslov and Alexenko, 1967; Schvedchikova et al., 1969; Revet and Guelpa, 1979). Krieg (1961) suggested that the DNA is intertwined with capsid protein in a fashion similar to that of tobacco mosaic virus but this model does not match the structure of nucleocapsids

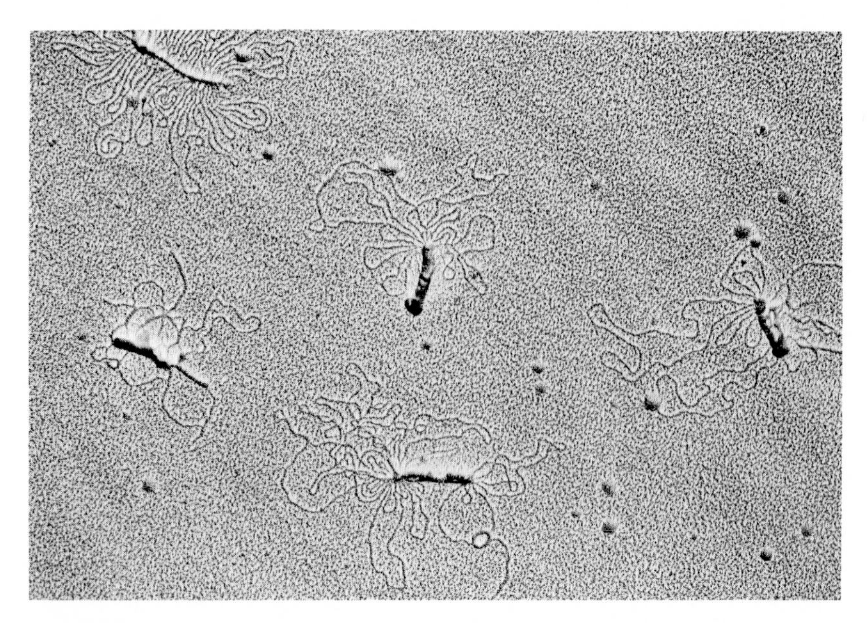

FIGURE 2. The deoxyribonucleoprotein exuding from nucleocapsids of Heliothis zea SNPV nucleocapsids. Note that loops of DNP emerge from the capsids and that there is no beaded structure typical of nucleosomes associated with chromatin. The nucleocapsids are spread using the Kleinschmidt technique. Magnification 32,000×.

as they are known (Burley et al., 1982). Bud and Kelly (1980) proposed a folded DNA model for NPV DNA where DNA is folded parallel to the long axis of the nucleocapsid rod without multiorder supercoiling. Compaction experiments using the basic polyamino acid polylysine collapses baculovirus DNA into rods of similar size to the DNP core (Bud and Kelly, 1980) and since DNA is present in the nucleocapsid in B form (Kelly et al., 1983) this model may in part be correct. The outer DNP strands may run parallel to the helical twist of the capsid subunits thus creating a toroid.

III. STRUCTURE OF THE NUCLEOCAPSID

Nucleocapsids are rod-shaped entities about 40 nm in diameter and up to 350 nm long. A few isolates have been observed with curved nucleocapsids (Entwistle and Robertson, 1968). The outer shell or sheath is the capsid (i.e., the

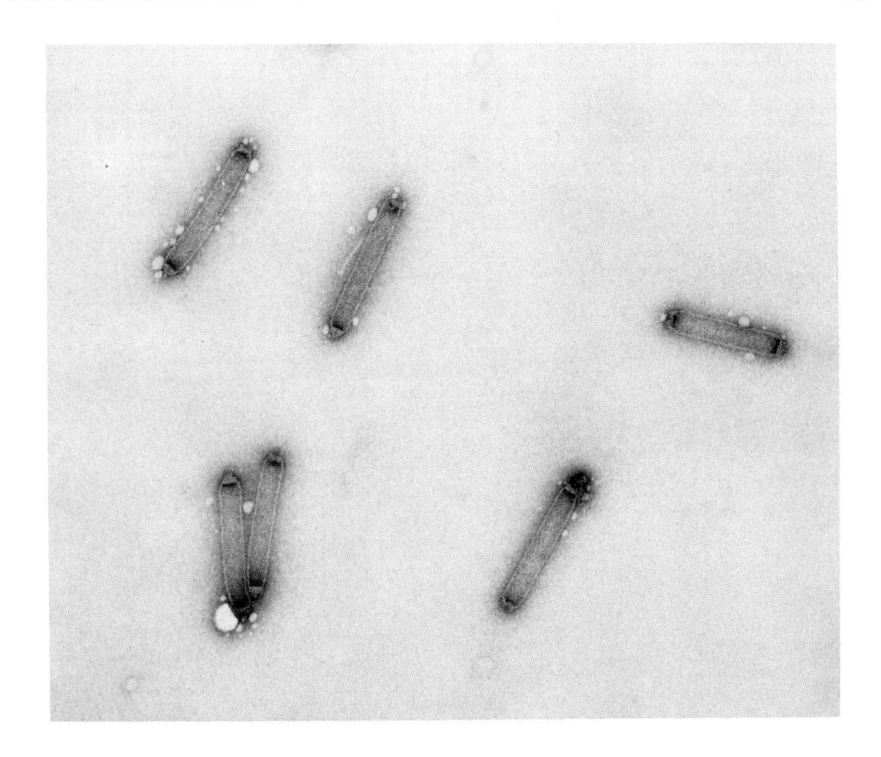

FIGURE 3. Capsids from Heliothis zea SNPV negatively stained. The empty nucleocapsids are rod shaped and show defined structural features at the ends. Magnification 66,000×.

nucleocapsid minus the DNP). This shell has been described as an envelope or as an inner or intimate membrane. Capsids artificially produced retain the cylindrical rod-shaped features of the nucleocapsids and so the structural organization of the nucleocapsids is strongly influenced by capsid structure. The structure is not a rigid one, however, since doughnut-shaped nucleocapsids have been observed within envelopes (Yamamoto and Tanada, 1979). Electron micrographs of capsids and nucleocapsids are shown in Figs. 3 and 4.

The capsids possess a number of structural features. Beaton and Filshie (1976) using optical diffraction showed that 4.5 nm subunits were arranged in a stacked disk formation and that the arrangement was similar for GVs and NPVs. Harrap (1972) showed that distinct substructure was distinguishable on nucleocapsids. Burley et al. (1982) confirmed these observations and showed that the essential features of the capsid is that of a 12-start helix system of monomers giving rise to an open stacked ring structure repeating in three rings at a

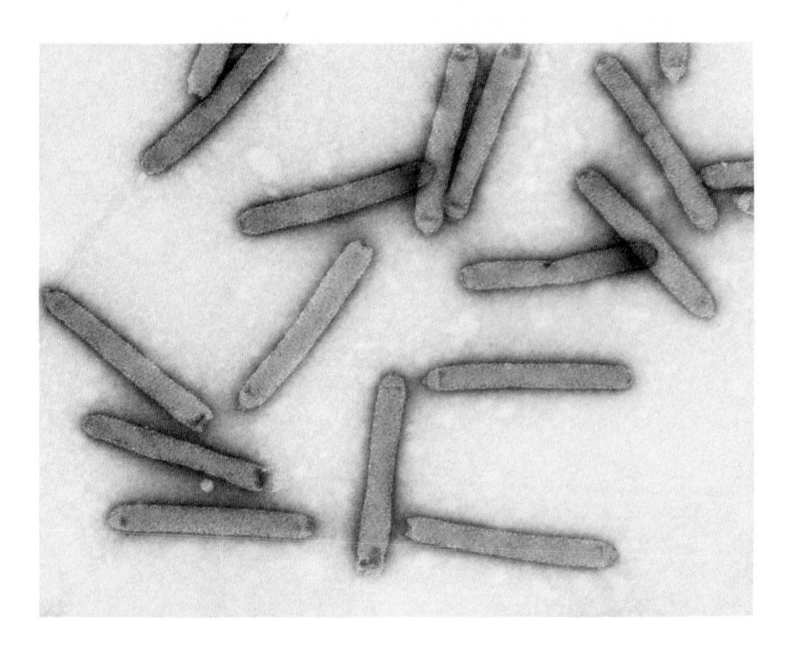

FIGURE 4. Nucleocapsids from Heliothis armigera SNPV showing defined capsid structure. Magnification 100,000×.

distance of 13.2 nm. Using contrast variation electron microscopy, Burley et al. (1982) showed that the DNA is contained in more or less uniformly packed cylinder about 32 nm in diameter. A protein shell 4 nm thick surrounds this heterogenous mixture of DNA and protein.

Double- and triple-length nucleocapsids are sometimes observed (Skuratovskaya et al., 1982). The integrity of the capsid is retained along the length of the nucleocapsid and they do not represent two nucleocapsids end on (Burley et al., 1982). Various structures are observed at the ends of the nucleocapsids revealed by differential staining. With the NOBV from *Oryctes rhinoceros*, a tail about 10 nm wide and up to 300 nm long is observed (Monsarrat et al., 1975; Payne et al., 1977) and this has been considered to be the DNP partially extruded from the nucleocapsid. The tails have been observed folded or coiled between the nucleocapsids and virus particle envelope, although it was concluded by Payne et al. (1977) that the tails were not normally external to the nucleocapsid. Similar structures are observed with the unusual nuclear polyhedrosis virus from *Tipula paludosa* (Bergoin and Guelpa, 1977).

Other structures are observed at the ends of nucleocapsids of NPVs and GVs (Koslov and Alexeenko, 1967; Summers and

Paschke, 1970). These structures comprise two distinct components: "nipples" and "claws" (Teakle, 1969). The claw was frequently at both ends whereas the nipple was found at just one end. Their function is not known, nor is it known if one structure is derived from another, i.e., the nipple is a derivative of the claw. Polarity in the nucleocapsid is an attractive concept. In assembly, one end is possibly complete prior to insertion of the DNP. Also sealing of the nucleocapsid at the conclusion of insertion may also require a structurally different arrangement. This polarity may also aid uncoating of the viral genome. It is not clear when and where uncoating occurs. Summers (1969) showed that some GV nucleocapsids invading cells align with nuclear pores and shows electron micrographs of both capsids and nucleocapsids at this site suggesting that the DNP may be released at this point. Granados and Lawler (1981) have, however, observed intact nucleocapsids in the nuclei of NPV-infected cells and shows uncoating occurs there. Since *in vitro* DNP is released primarily from the ends (Bud and Kelly, 1980 ; Monsarrat et al., 1975; Tweeten et al., 1980 ; Revet and Guelpa, 1979), the "caps" of the nucleocapsids are undoubtedly important. The caps arenot merely extensions of the sheath. They appear to represent structurally distinct proteins (Teakle, 1969; Burley et al., 1982) although the proteins associated with the caps are not known as yet. Most nucleocapsids comprise two main proteins which represent the core DNP protein and the sheath capsid protein, and from three to eight minor polypeptides some of which may represent "cap" proteins (Summers and Smith, 1978).

IV. STRUCTURE OF VIRUS PARTICLES

The virus particles comprise enveloped nucleocapsids. They are also rod-shaped structures and most isolates contain one nucleocapsid per envelope. Some nuclear polyhedrosis viruses isolated from lepidoptera contain more than one nucleocapsid and over 30 nucleocapsids per envelope may be found (Kawamoto and Asayama, 1975). The difference between SNPV and MNPV has not been determined. Presumably, in SNPV, GVs, and NOBVs the nucleocapsids are unable to interact with each other to partly crystallize prior to acquisition of an envelope. Interestingly, with MNPV the nucleocapsids align lengthwise exactly.

The envelope surrounding the nucleocapsids may be acquired either *de novo* within the nucleus (Stoltz et al., 1973) or by budding at the plasma membrane (Hess and Falcon, 1978). No difference in nucleocapsids acquiring envelopes at the two sites has yet been described. Most virus particles acquiring envelopes at the plasma membrane are either single or double, though rarely more (Knudson and Harrap, 1976; Kelly, 1982; Adams et al., 1977). With MNPVs the multiple nucleocapsid virus particles are formed in the nucleus. There are pronounced biochemical and serological differences between virus enveloped within the nucleus and at the plasma membrane (Summers and Volkman, 1976; Roberts, 1983), reflecting both differences in origin and future function. Virus occluded within polyhedra are destined to infect another insect and play no further role in the pathogenesis within the host insect. Virus acquiring plasma membrane envelopes infect additional cells in the original host. Virus occluded within polyhedra have design features which permit occlusion and the ability to fuse with plasma membranes of the gut at alkaline pH. Plasma membrane-derived virus infects cells within the host at neutral or fractionally acid pH.

Little is known about envelope proteins. Some of the proteins are glycosylated (Goldstein and McIntosh, 1980; Kelly and Lescott, 1983) and glycosylation of the proteins may be an essential prerequisite for assembly of the envelope (Kelly and Lescott, 1983). The envelope proteins of plasma membrane-derived virus differ from those found in occluded virus (Summers and Volkman, 1976; Kelly and Lescott, 1983).

Morphologically, the envelope of virus particles are not very interesting. No convincing regular substantial repeating subunit has been described. Plasma membrane virus derived with peplomers or spikes at one end of the virus particle has been observed in thin sections (Adams et al., 1977). Summers and Volkman (1976) also claim to have observed peplomers on negatively stained plasma membrane-derived virus particles but not on occluded virus particles. Harrap (1972) also showed structure on envelopes. Consequently, it is not known if transmembrane proteins exist in virus particles of either origin and whether there is a regular order to these proteins which match the regular repeat on the nucleocapsid surface. There is no evidence for a protein linking the nucleocapsid to the envelope protein, i.e., a matrix protein akin to that found in rhabdoviruses.

A space is observed between the nucleocapsid and the envelope although the contents of the space is not known and may be artifactual. The envelope contains lipid as evidenced from the ether and detergent solubility of the virus particles and the use of detergents to routinely prepare nucleocapsids. No detailed analysis of baculovirus lipids has been presented. Assuming that the polypeptides present in virus particles and not those present in nucleocapsids, there are eight to twelve

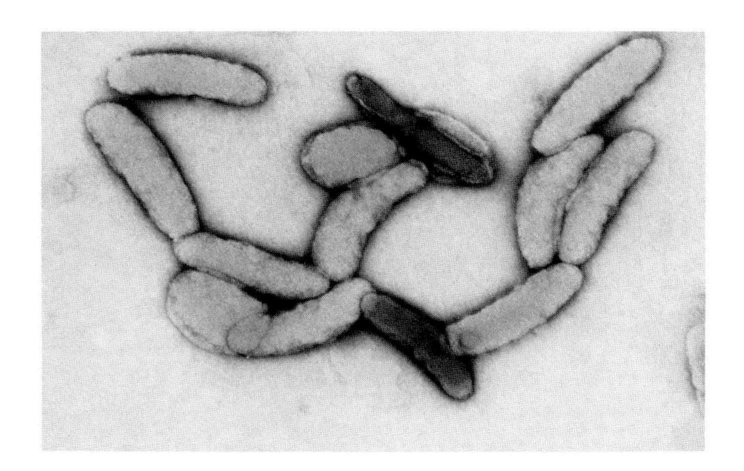

FIGURE 5. Virus particles from Pieris brassicae GV showing morphology typical of a single nucleocapsid virus particle. Magnification 80,000×.

polypeptides present in the virus envelope. Protease activity has been found in virus particles (Payne and Kalmakoff, 1978; Wood, 1980) isolated from occluded virus.

The morphology of single nucleocapsid and multiple nucleocapsid virus particles is shown in Figs. 5 and 6.

V. STRUCTURE OF POLYHEDRA AND GRANULES

The proteinaceous crystals which occlude baculoviruses are the most conspicuous elements of the baculovirus structure. Certainly these entities were those first observed in viral infections when polyhedra were observed in nuclei of cells in gypsy moth caterpillars suffering from wilt (Glaser, 1915) and granules were observed in nuclei of diseased cabbage white butterfly caterpillars (Paillot, 1926). Baculoviruses are the only viruses which induce intranuclear protein crystals although both cytoplasmic polyhedrosis and pox viruses are assembled within cytoplasmic protein crystals.

Paillot's observations that the granules differed from polyhedra associated with wilt disease is remarkable, since the subdivision of baculoviruses into granulosis and nuclear polyhedrosis viruses is still recognized. Polyhedra are large crystals commonly ranging in size from 1 to 4 μm in size. They are cubic or quasispherical in shape, although in the case of *T. paludosa* NPV they are crescent shaped (Bergoin and Guelpa,

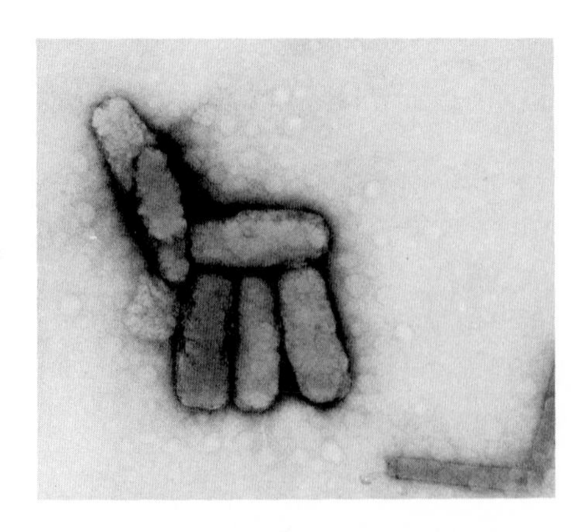

FIGURE 6. Virus particles from Trichoplusia ni MNPV showing morphology typical of a multiple nucleocapsid virus particle. Magnification 52,000×.

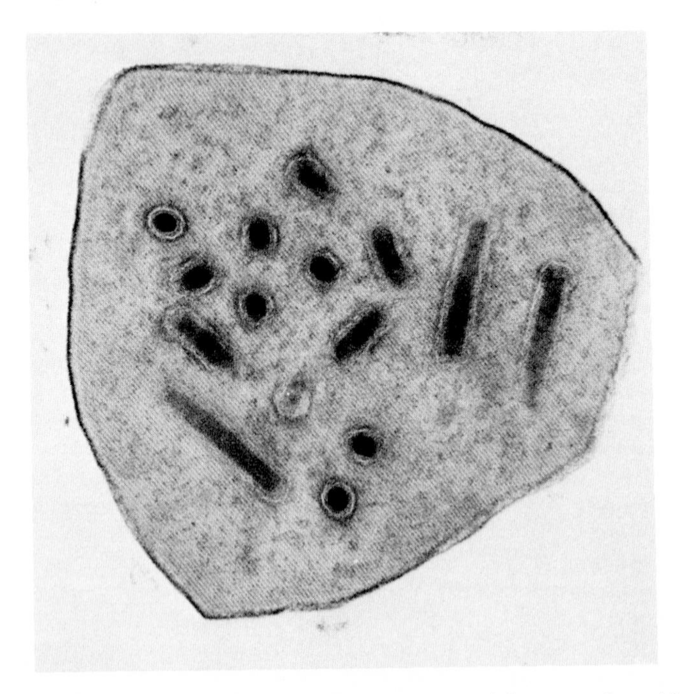

FIGURE 7. Electron micrograph of a thin section through a polyhedron containing single nucleocapsid virus particles. Note the crystalline lattice is uninterrupted by the virus particles oriented at random. Polyhedra are from Trichoplusia orichalcea. Magnification 100,000×.

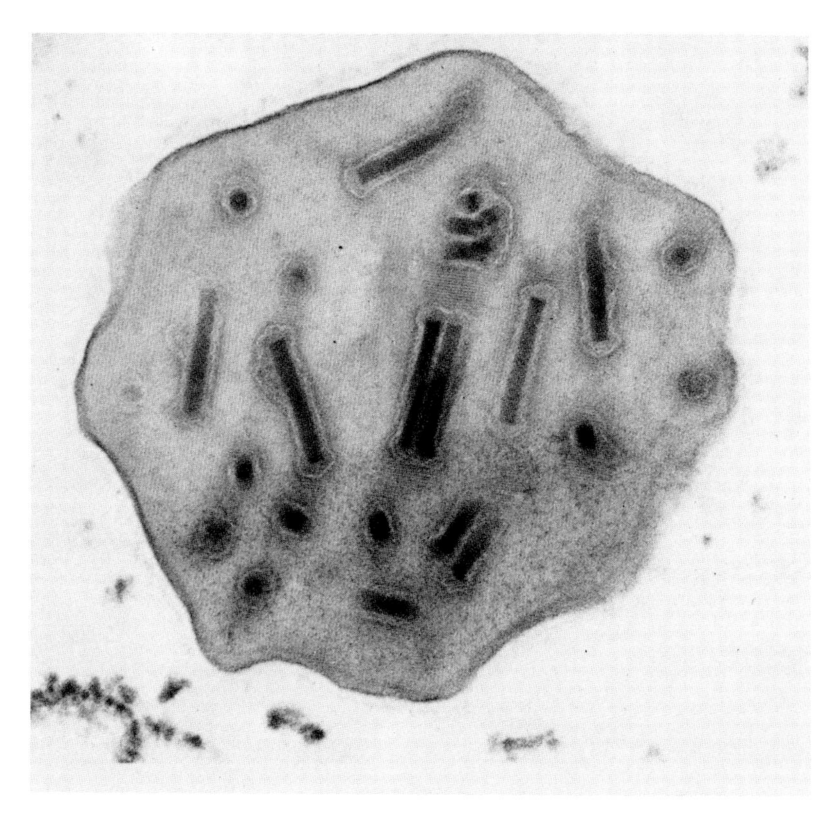

FIGURE 8. Electron micrograph of a thin section through a polyhedron containing multiple nucleocapsid virus particles. The polyhedron is from Spodoptera exempta. The polyhedron envelope is conspicuous at the polyhedron surface. Magnification 80,000×.

1977). The polyhedra contain numerous virus particles and as many as a hundred virus particles may be occluded within a single polyhedron (Figs. 7 and 8). By contrast, granules are ovocylindrical in shape, some 0.3 to 0.5 μm in length and 0.1 to 0.4 μm in width, and contain a single virus particle (Fig. 9). Stairs (1964) has described cuboidal granules in *Galleria mellonella*. Recently, the granules of some GVs have been shown to occasionally contain two to five virus particles (Crook et al., 1982). Although polyhedra and granules are distinct, their structure is in essence remarkably similar.

The most striking feature of a polyhedron is the crystalline lattice (Bergold, 1963; Harrap, 1972). It is a face-centered 4 nm cubic lattice. The polyhedron protein has a molecular weight of 27,000 to 34,000 varying with the isolate. It has a marked tendency to aggregate and it is likely that

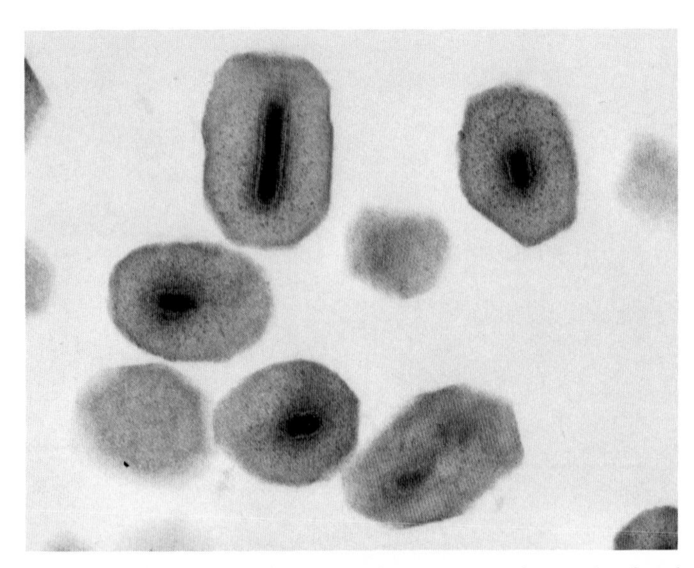

FIGURE 9. Electron micrograph of granules of Pieris brassicae granulosis virus. The granules contain a single virus particle of the single nucleocapsid type. Magnification 60,000×.

the polyhedron contains octomers of the polypeptides as subunits (Rohrmann, 1977). The lattice is not distorted by the virus particles occluded within it. The virus particles apparently lie at random within the lattice. The polyhedron protein apparently interacts directly with the virus particle envelope. It is not known if there is a direct interlocking of the polyhedron protein with envelope proteins.

The main physical characteristic of polyhedra is that they are stable at moderate pH and are alkali labile. Polyhedra dissolve at pH 9.5 or above. *In vitro* dissolution solubilizes the polyhedra leaving virus particles intact. Protease activity present in the polyhedron is activated and facilitates breakdown (Eppstein and Thoma, 1975). In the insect gut where the juices are alkaline and contain proteases a similar breakdown obviously occurs (Faust and Adams, 1966). The polyhedron protein has an isoelectric point of about 5.8 and will recrystallize at this and lower pH's.

"Mature" polyhedra are surrounded by an outer morphologically distinct layer sometimes referred to as the polyhedron envelope (Minion et al., 1979). Its function and composition have not been convincingly established. Functionally it may offer further protection against desiccation and uv irradiation thus complementing part of the physical function of the polyhedron. It may also play a role in the interaction of the polyhedron with surfaces. Jurkicova (1979) suggested that the polyhedron envelope is proteinaceous, whereas Minion et al. (1979), who purified the envelope, showed that it is composed mainly of carbohydrates. Treatment of polyhedra with alkali disrupts the polyhedron but leaves the polyhedron envelope virtually intact (Harrap, 1972). The envelope is resistant to proteases (Gipson and Scott, 1975). The envelope is 6 to 10 nm thick and has no obvious structure.

VI. CONCLUDING REMARKS

This chapter illustrates the complexity of baculoviruses. Many facets of the assembly of the virus both within the nucleus and at the plasma membrane are incompletely understood. An understanding of the structure of baculoviruses and the assembly and disassembly of the components will lead to improved methods of diagnosis and detection of baculoviruses and possibly aid the genetic engineering of baculoviruses defined in host range and virulence.

REFERENCES

Adams, J. R., Goodwin, R. H., and Wilcox, T. A. (1977). Electron microscopic investigations on invasion and replication of insect baculoviruses *in vivo* and *in vitro*. *Biol. Cell.* *28*, 261-268.

Bauer, W., and Vinograd, J. (1968). The interaction of closed circular DNA with intercalative dyes. I. The superhelix density of SV40 DNA in the presence and absence of dye.

Beaton, C. D., and Filshie, B. K. (1976). Comparative ultrastructural studies of insect granulosis and nuclear polyhedrosis viruses. *J. Gen. Virol.* *31*, 151-161.

Bergold, G. H. (1963). The molecular structure of some insect virus inclusion bodies. *J. Ultrastr. Res.* *8*, 360-366.

Bergoin, M., and Guelpa, B. (1977). Dissolution des inclusions du virus de la polyédrose nucléaire du diptere *Tipula paludosa* Meig. Etude ultrastructurale du virion. *Arch. Virol.* *53*, 243-254.

Bud, H. M., and Kelly, D. C. (1977). The DNA contained by nuclear polyhedrosis viruses isolated from four *Spodoptera* spp. (Lepidoptera, Noctuidae): genome size and configuration assessed by electron microscopy. *J. Gen. Virol.* *37*, 135-143.

Bud, H. M., and Kelly, D. C. (1980a). Nuclear polyhedrosis virus DNA is infectious. *Microbiologica 3*, 103-108.

Bud, H. M., and Kelly, D. C. (1980b). An electron microscope study of partially lysed baculovirus nucleocapsids: the intranucleocapsid packaging of viral DNA. *J. Ultrastr. Res. 73*, 361-368.

Burand, J. P., Summers, M. D., and Smith, G. E. (1980). Transfection with baculovirus DNA. *Virology 101*, 286-290.

Burgess, S. (1977). Molecular weights of lepidopteran baculovirus DNAs: derivation by electron microscopy. *J. Gen. Virol. 37*, 501-510.

Burley, K., Miller, A., Harrap, K. A., and Kelly, D. C. (1982). Structure of the baculovirus nucleocapsid. *Virology 120*, 433-440.

Carstens, E. B., Tjia, S. T., and Doefler, W. (1980). Infectious DNA from *Autographa californica* nuclear polyhedrosis virus. *Virology 101*, 311-314.

Crick, F. H. C. (1976). Linking numbers and nucleosomes. *Proc. Natl. Acad. Sci. US 273*, 2639-2643.

Crook, N. E. (1981). A comparison of the granulosis virus from *Pieris brassicae* and *P. rapae*. *Virology 118*, 173-181.

Crook, N. E., Brown, J. D., and Foster, G. N. (1982). Isolation and characterisation of a granulosis virus from the Tomato moth *Lacanobia deracea* and its potential as a control agent. *J. Invertebr. Pathol. 40*, 221-227.

Elliott, R. M., and Kelly, D. C. (1977). The polyamine content of a nuclear polyhedrosis virus from *Spodoptera littoralis*. *Virology 76*, 472-474.

Elliott, R. M., and Kelly, D. C. (1979). Compartmentalization of the polyamines contained by a nuclear polyhedrosis virus from *Heliothis zea*. *Microbiologica 2*, 409-413.

Engstrom, A., and Kilkson, R. (1968). Molecular organisation in the polyhedra of *Porthetria dispar* nuclear polyhedrosis virus. *Exp. Cell Res. 53*, 305-310.

Entwistle, P. F., and Robertson, J. S. (1968). An unusual nuclear polyhedrosis virus from larvae of a Lepialid moth. *J. Invertebr. Pathol. 11*, 487-495.

Eppstein, D. A., and Thoma, J. A. (1975). Alkaline protease associated with the matrix protein of a virus infecting the cabbage looper. *Biochem. Biophys. Res. Commun. 62*, 478-484.

Faulkner, P. F. (1981). Baculovirus. *In* "Pathogenesis of Invertebrate Microbial Diseases" (E. W. Davidson, ed.), pp. 5-38. Allanheld, Osmun, New Jersey.

Faust, R. M., and Adams, J. R. (1966). The silicon content of nuclear and cytoplasmic viral inclusion bodies causing polyhedrosis in lepidoptera. *J. Invertebr. Pathol. 8*, 526-530.

Gibson, I., and Scott, H. A. (1975). An electron microscope study of the effect of various fixatives and thin section enzyme treatments on a nuclear polyhedrosis virus.

Glaser, R. W. (1915). Wilt of gypsy moth caterpillar. *J. Agr. Res. 4*, 101-112.

Goldstein, N. I., and McIntosh, A. H. (1980). Glycoproteins of nuclear polyhedrosis viruses. *Arch. Virol. 64*, 119-126.

Granados, R. R. (1978). Early events in baculovirus infection of *Heliothis zea*. *Virology 108*, 297-308.

Granados, R. R. (1980). Infectivity and mode of action of baculoviruses. *Biotechnol. and Bioeng. 22*, 1377-1405.

Granados, R. R., and Lawler, K. A. (1981). *In vivo* pathway of *Autographa californica* baculovirus invasion and infection. *Virology 108*, 297-308.

Harrap, K. A. (1972a). The structure of nuclear polyhedrosis viruses. I. The inclusion body. *Virology 50*, 114-123.

Harrap, K. A. (1972b). The structure of nuclear polyhedrosis viruses. II. The virus particle. *Virology 50*, 124-132.

Hess, R. T., and Falcon, L. A. (1978). Electron microscope observations of the membrane surrounding polyhedral inclusion bodies of insects. *Arch. Virol. 56*, 169-176.

Huger, A. M. (1966). A virus disease of the Indian rhinoceros beetle, *Oryctes rhinoceros* (Linnaeus), caused by a new type of insect virus. *J. Invertebr. Pathol. 8*, 38-51.

Jurkicova, A. M. (1979). Characterisation of nuclear polyhedrosis viruses obtained from *Adoxophyes orana* and from *Barathra brassicae*. Ph.D. Thesis. Wageningen, Netherlands.

Kawamoto, F., and Awayama, T. (1975). Studies on the arrangement patterns of nucleocapsids with envelopes of nuclear polyhedrosis virus in fat body cells of the brown tail moth *Euproctis similis*. *J. Invertebr. Pathol. 26*, 47-55.

Kelly, D. C. (1975). "*Oryctes*" virus replication: electron microscopic observations of infected moth and mosquito cells. *Virology 69*, 596-606.

Kelly, D. C. (1977). The DNA contained by nuclear polyhedrosis viruses isolated from four *Spodoptera* spp. (Lepidoptera, Noctuidae). *Virology 76*, 468-471.

Kelly, D. C. (1982). Baculovirus replication. *J. Gen. Virol. 63*, 1-13.

Kelly, D. C., and Lescott, T. (1983). Baculovirus replication: glycosylation of polypeptides synthesised in *Tricoplusia ni* nuclear polyhedrosis virus infected cells and the effect of Tunicamycin. *J. Gen. Virol.*, in press.

Kelly, D. C., and Wang, X. (1981). The infectivity of nuclear polyhedrosis virus DNA. *Ann. Virol. 132E*, 247-259.

Kelly, D. C., Brown, D. A., Robertson, J. S., and Harrap, K. A. (1980). Biochemical, biophysical and serological properties of two singly enveloped nuclear polyhedrosis viruses from

Heliothis armigera and *Heliothis zea*. *Microbiologica 3*, 319-331.

Kelly, D. C., Brown, D. A., Ayres, M. D., Allen, C. J., and Walker, I. O. (1983). Properties of the major nucleocapsid protein of *Heliothis zea* singly enveloped nuclear polyhedrosis virus. *J. Gen. Virol. 104*, 399-408.

Koslov, E. A., and Alexeenko, I. P. (1967). Electron microscope investigation of the structure of the nuclear polyhedrosis virus of the silkworm *Bombyx mori*. *J. Insect. Pathol. 9*, 413-420.

Krieg, A. (1961). Uber den Aufbau und die Vermehnungsmoglichkeiton von stabchen formigen Insekten Viren. II. *Z. Naturforsch. 16b*, 115-117.

Kunjeku, E. (1982). M. Sc. Thesis. University of London.

Loh, L. C., Hamm, J. J., Kawanishi, C., and Huang, E. S. (1982). Analysis of the *Spodoptera frugiperda* nuclear polyhedrosis virus genome by restriction endonucleases and electron microscopy. *J. Virol. 44*, 747-751.

McPherson, A., Jurnak, F., Wang, A., Molineux, I., and Rich, A. (1979). Preliminary molecular replacement results for a crystalline gene 5 protein deoxynucleotide complex. *J. Supramolec. Struct. 10*, 457-465.

Matthews, R. E. F. (1982). "Classification and Nomenclature of Viruses." Fourth report of the International Committee on Taxonomy of Viruses.

Miller, L. K., and Dawes, K. P. (1978). Restriction enzyme analysis to distinguish two closely related nuclear polyhedrosis viruses: *Autographa californica* MNPV and *Trichoplusia ni* MNPV. *Appl. Environ. Microbiol. 35*, 1206-1210.

Miller, D. W., and Miller, L. K. (1982). A virus mutant with an insertion of a copia-like transposable element. *Nature (London) 299*, 562-564.

Minion, F. C., Coons, L. B., and Broome, J. R. (1979). Characterisation of the polyhedral envelope of the nuclear polyhedrosis virus of *Heliothis virescens*. *J. Invertebr. Pathol. 34*, 303-307.

Monsarrat, P., Revet, B., and Gourevitch, I. (1975). Mise en evidence stabilization et purification d'une structure nucleoproteique intracapsidaire chez le baculovirus d'*Oryctes rhinoceros* L. *C.R. Acad. Sci. 281*, 1439-1442.

Payne, C. C. (1974). The isolation and characterisation of a virus from *Oryctes rhinoceros*.

Paillot, A. (1926). *Sci. Paris 181*, 180-185.

Payne, C. C., and Kalmakoff, J. (1978). Alkaline protease associated with virus particles of a nuclear polyhedrosis virus: assay, purification and properties. *J. Virol. 26*, 84-92.

Payne, C. C., Compson, D., and De Looze, S. M. (1977). Properties of the nucleocapsids of a virus isolated from *Oryctes rhinoceros*. *Virology 77*, 269-280.

Revet, B. M. J., and Guelpa, B. (1979). The genome of a baculovirus infecting *Tipula paludosa* (Meig.) (Diptera): a high molecular weight closed circular DNA of zero superhelix density. *Virology 96*, 633-639.

Roberts, P. L. (1983). Neutralisation studies on *Autographa californica* nuclear polyhedrosis virus. *Arch. Virol. 75*, 147-150.

Rohrmann, G. F. (1977). Characterisation of N-polyhedrin of two baculovirus strains pathogenic for *Orgyia pseudotsugata*. *Biochemistry 16*, 1631-1634.

Rohrmann, G. F., McParland, R. H., Martignoni, M. E., and Beaudreau, G. S. (1978). Genetic relatedness of two nucleopolyhedrosis viruses pathogenic for *Orgyia pseudotsugata*. *Virology 84*, 213-217.

Schvedchikova, N. G., Vlanov, B. P., and Rarasevich, L. M. (1969). On the structure of granulosis virus of the Siberian silkworm *Dendrolimus sibiricus*. *Mol. Biol. 3*, 361-364.

Skuratovskaya, I. N., Strokovskaya, L. I., Zherebtsova, E. N., and Gudz-Gorban, A. P. (1977). Supercoiled DNA of nuclear polyhedrosis virus of *Galleria mellonella* L. *Arch. Virol. 53*, 79-86.

Skuratovskaya, I. N., Fodor, I., and Strokovskaya, L. J. (1982). Properties of nuclear polyhedrosis virus of the Great War Moth: oligomeric circular DNA and the characteristics of the genome. *Virology 120*, 465-471.

Smith, K. M., and Xeros, N. (1954). An unusual virus disease of a dipterous larva. *Nature (London) 173*, 866-867.

Smith, G. E., and Summers, M. D. (1978). Analysis of baculovirus genomes with restriction endonucleases. *Virology 89*, 517-527.

Stairs, G. R. (1964). Selection of a strain of insect granulosis virus producing only cubic inclusion bodies. *Virology 24*, 520-521.

Stoltz, D. B., Pavan, C., and Da Cunha, A. B. (1973). Nuclear polyhedrosis virus. A possible example of '*de novo*' intranuclear membrane morphogenesis. *J. Gen. Virol. 19*, 145-150.

Summers, M. D. (1969). Apparent *in vivo* pathway of granulosis virus invasion and infection. *J. Virol. 4*, 188-190.

Summers, M. D., and Anderson, D. L. (1972). Granulosis virus deoxyribonucleic acid: a closed circular double stranded molecule. *J. Virol. 9*, 710-713.

Summers, M. D., and Paschke, J. D. (1970). Alkali liberated granulosis virus of *Trichoplusia ni*. *J. Invertebr. Pathol. 16*, 227-240.

Summers, M. D., and Smith, G. E. (1978). Baculovirus structural proteins. *Virology 84*, 390-402.

Summers, M. D., and Volkman, L. E. (1976). Comparison of biophysical and morphological properties of occluded and non-occluded baculoviruses. *J. Virol. 17*, 962-972.

Teakle, R. E. (1969). A nuclear polyhedrosis virus of *Anthela varia* (Lepidoptera: Anthelidae). *J. Invertebr. Pathol. 14*, 18-27.

Tweeten, K. A., Bulla, L. A., and Consigli, R. A. (1980a). Restriction enzyme analysis of the genomes of *Plodia interpunctella* and *Pieris brassicae* granulosis viruses. *Virology 104*, 514-519.

Tweeten, K. A., Bulla, L. A., and Consigli, R. A. (1980b). Characterisation of an extremely basic protein derived from granulosis virus nucleocapsids. *J. Virol. 33*, 866-876.

Tweeten, K. A., Bulla, L. A., and Consigli, R. A. (1981). Applied and molecular aspects of insect granulosis viruses. *Microbiol. Rev. 45*, 379-408.

Wood, H. A. (1980). Protease degradation of *Autographa californica* nuclear polyhedrosis virus proteins. *Virology 103*, 392-399.

Yamamoto, T., and Tanada, Y. (1979). Structural damage to the granulosis virus of the armyworm *Pseudaletia unipuncta* caused by sucrose density gradient centrifugation. *J. Invertebr. Pathol. 34*, 312-314.

THE NATURE OF POLYHEDRIN

JUST M. VLAK

Department of Virology
Agricultural University
Wageningen, The Netherlands

GEORGE F. ROHRMANN

Department of Agricultural Chemistry
Oregon State University
Corvallis, Oregon

I. INTRODUCTION

Baculoviruses cause fatal diseases in insects. Such infections usually result in the production of massive amounts of large protein crystals (occlusion bodies). These occlusion bodies are termed polyhedra for nuclear polyhedrosis viruses (NPV) and granula for granulosis viruses (GV). They have attracted scientific attention since their presence was associated with disease (Cornalia, 1856; Maestri, 1856) and later when Bolle (1894) demonstrated that the occlusion bodies were the causative agents. Von Prowazek (1907) and Glaser and Chapman (1916) disclosed the viral nature of the disease by showing that the infectious agent from extracts of infected tissue could pass through an ultrafilter which excluded the occlusion bodies. Occlusion bodies appeared in insects upon infection with these extracts.

The biological and chemical nature of the causative agent of the disease remained a matter of dispute for many years (Bergold, 1953). Von Prowazek (1907) erroneously postulated that occlusion bodies were the result of an apparently unsuccessful defense reaction of the insect host to an ultrafilterable disease agent. Bergold and Schramm (1942) contended that the protein in the occlusion bodies was responsible for causing

VIRAL INSECTICIDES
FOR BIOLOGICAL CONTROL

ISBN 0-12-470295-3

the disease; the particles now known as virions were thought to be aggregation products of the polyhedral protein (Bergold, 1943). It was not until 1947 that Bergold demonstrated that his previous conclusions were false: the rod-shaped virus particles encapsulated in the occlusion body were the infectious agents (Bergold, 1947). Subsequently, Bergold (1953) showed that DNA was an important element of the virus particle. It is now well established that the genetic information of baculoviruses is contained in DNA.

The major component of the occlusion body is a single polypeptide with a molecular weight of about 29,000 daltons. It is present in a multimeric conformation and forms a paracrystalline lattice around the virus particles. The occlusion body protein of all baculoviruses is soluble in alkaline solutions (Bergold, 1947). In insects, ingested occlusion bodies are dissolved by the alkaline fluid of the midgut thereby releasing virus particles and allowing the infection to proceed. The apparent function of the occlusion bodies is to provide stability for the virus outside the host insect, thereby enhancing the dissemination and survival of the virus in the natural environment. Infectious baculoviruses have been found in soil over 40 years after an epizootic of the virus (Thompson et al., 1981).

Baculoviruses appear to control the population size of certain susceptible insect species. Because of their stability, limited host range, and pathogenicity to a variety of major insect pests (notably Lepidoptera and Hymenoptera), these viruses were recognized relatively early as potential insect control agents (see Steinhaus, 1956, for review). However, it was not until after the environmental drawbacks of chemical insecticides became apparent that baculoviruses were seriously considered as biological insecticides.

The occlusion of viruses in a crystalline protein lattice occurs only in three groups of viruses, all of which infect invertebrates. The cytoplasmic polyhedrosis viruses (CPV) (Payne and Mertens, 1983) are members of the family Reoviridae which also includes pathogens of plants and vertebrates. The entomopox viruses (EPV) (Granados, 1981) are in the family Poxviridae, which are pathogenic for invertebrates and a wide range of vertebrates. In contrast, baculoviruses are limited to a number of insect orders and two species of shrimp (Table I). Baculoviruses have not been found in any vertebrates, including man.

Although the occlusion body proteins of baculoviruses and CPVs are similar in size (25,000-30,000 daltons), chemical properties (alkali solubility), and apparent biological function (protection of virus particles), they are produced by virus with different types of genetic material (Reoviridae: double-stranded RNA; Baculoviridae: double-stranded DNA).

TABLE I. Baculovirus Distribution[a]

Arthropoda	*Virus*[b]	*Morphotype*	*Number of affected species*
Crustacea			
Decapoda	*NPV*	*S*	*2*
Insecta			
Trichoptera	*NPV*	*S*	*1*
Hymenoptera	*NPV*	*S*	*26*
Diptera	*NPV*	*S*	*22*
Lepidoptera	*NPV*	*S,M*	*355*
	GV	*S*	*113*
	NOBV	*S*	*2*
Coleoptera	*NOBV*	*S*	*16*

[a]Occurrence of baculoviruses in various arthropod orders and the number of affected species (updated by Martignoni, 1983). NPV = nuclear polyhedrosis virus; GV = granulosis virus; NOBV = nonoccluded baculovirus; S = single nucleocapsid per envelope; M = many nucleocapsids per envelope.

Serological, tryptic peptide, and amino acid sequencing investigations have revealed no relatedness between CPV and baculovirus occlusion body protein (Rohrmann et al., 1980). The occlusion body protein of EPV is considerably larger (102,000 daltons) (Bilimoria and Arif, 1979) than those from CPV and baculoviruses. Therefore, occlusion body proteins may have evolved independently at least three times, suggesting that occlusion confers a powerful selective advantage on those viruses which possess it.

In this chapter we shall consider the nature of baculovirus occlusion body protein. This protein has been termed occlusion body protein, polyhedral protein, and, more recently, polyhedrin for NPVs and granulin for GVs. It is now apparent that polyhedrins and granulins all belong to one group of related proteins (Rohrmann et al., 1981; Smith and Summers, 1981). Since some polyhedrins are more closely related to granulins than to other polyhedrins (Rohrmann et al., 1981), we will not distinguish between these proteins and will call them all polyhedrin.

In addition to the nature of polyhedrin itself, we shall discuss the origin, location, structure, and expression of the polyhedrin gene. Our understanding of these aspects progressed greatly in recent years due to the advances in molecular biology and molecular genetics. We shall also consider the diversity of polyhedrins in terms of baculovirus evolution and the relationship of baculovirus evolution to the phylogeny of their insect hosts.

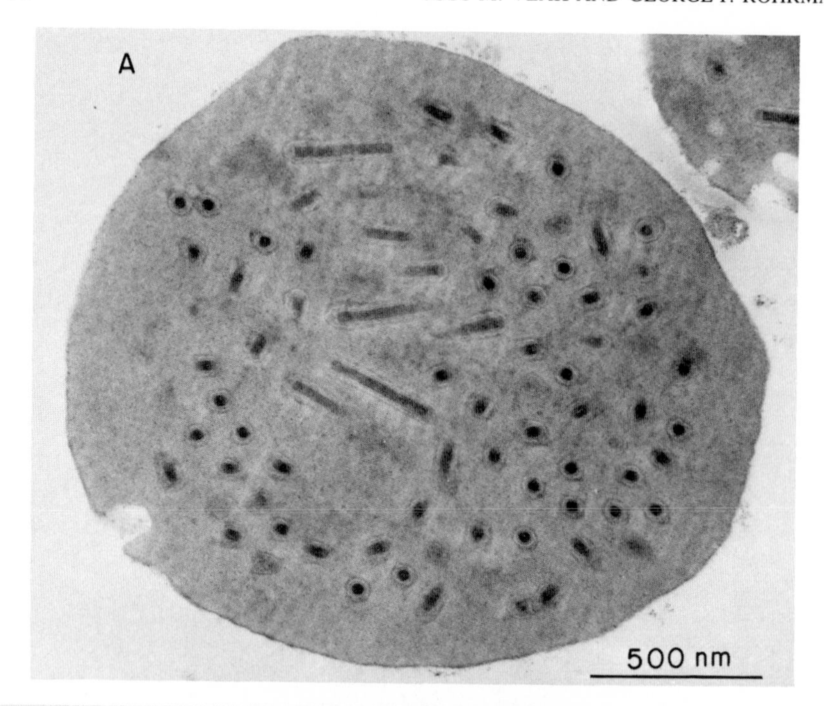
A
500 nm

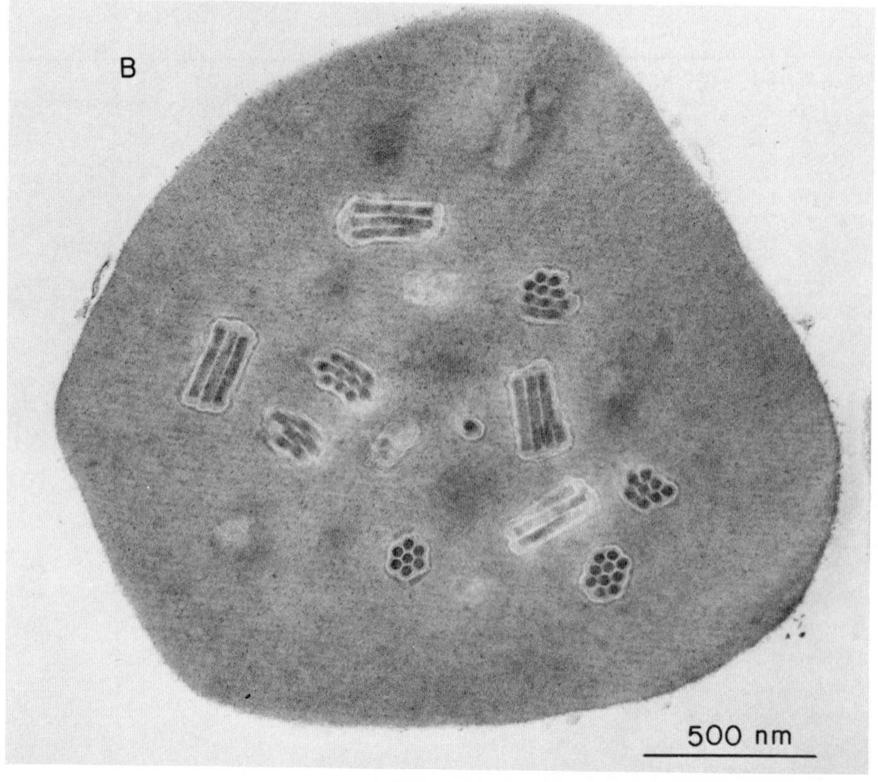
B
500 nm

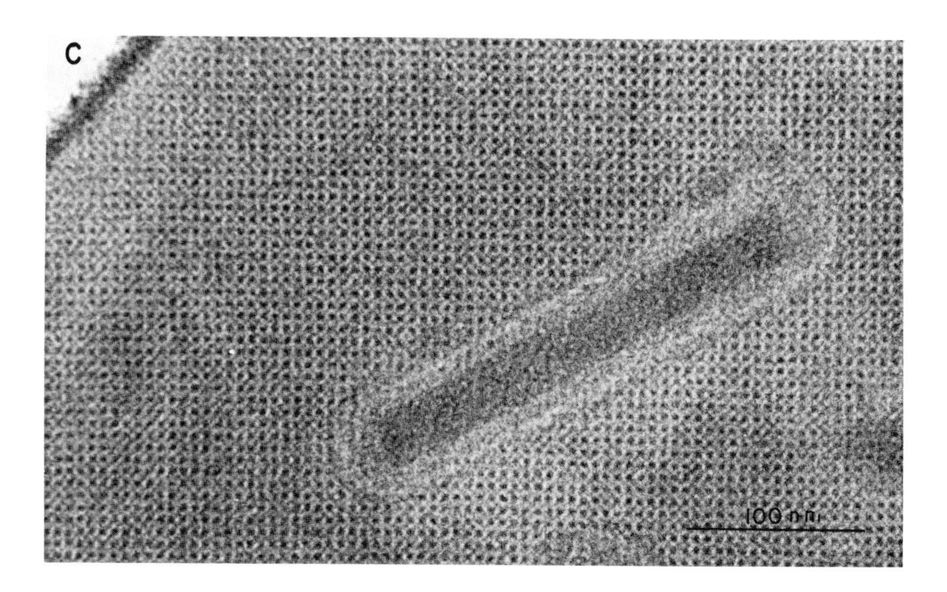

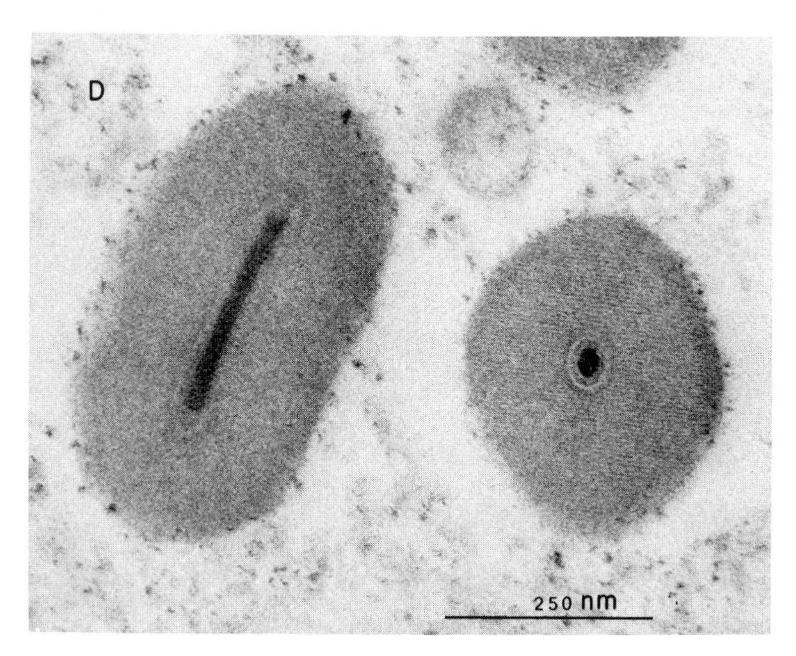

FIGURE 1. Sections of polyhedra of (A) Orgyia pseudotsugata SNPV, (B) O. pseudotsugata MNPV, (C) Pseudohazis eglanterina SNPV, and (D) granula from Choristoneura vindis. Electron micrographs by K. M. Hughes, Forestry Sciences Laboratory, USDA, Forest Service, Corvallis, Oregon.

II. BACULOVIRUSES

A. *OCCURRENCE AND CLASSIFICATION*

Baculoviruses are members of the family Baculoviridae (Matthews, 1982) and are pathogenic for arthropods, mainly insects (Table I). More than 500 baculoviruses have been reported in the literature (Martignoni and Iwai, 1977). Their occurrence is limited to a number of holometabolous insect orders (Trichoptera, Diptera, Hymenoptera, Lepidoptera, Coleoptera) and two crustacean species (*Peneus duararum*, Couch, 1974; *P. monodon*, Lightner and Redman, 1981). The virions consist of rod-shaped (baculum = rod) nucleocapsids, surrounded by a lipoprotein membrane (Tinsley and Harrap, 1978). In occluded baculoviruses these virions are embedded in a proteinaceous crystal, 0.1-15 μm in size. These protein bodies are refractile in the light microscope and are present in massive amounts in moribund or decreased insects. They are called occlusion bodies, inclusion bodies, capsules, polyhedra or granula. We will refer to them as occlusion bodies in this chapter.

The family Baculoviridae consists of a single genus, *Baculovirus*. The members of this genus are divided into three subgroups, A, B, and C. In the nuclear polyhedrosis viruses (NPV), which form subgroup A, nucleocapsids are either enveloped by a lipoprotein membrane singly (SNPV) or in multiples (MNPV) of up to 39 nucleocapsids (Kawamoto and Asayama, 1975) per common envelope (Fig. 1A-C). Up to 100 virion packages are embedded in an occlusion body, polyhedral to cuboidal in shape, and 1-15 μm in diameter. Subgroup B comprises the granulosis viruses (GV), where one singly-enveloped nucleocapsid is embedded per occlusion body, round to ellipsoidal in shape, and 0.1-1 μm in size (Fig. 1D). Baculoviruses of subgroup C are not occluded (NOBV = nonoccluded baculoviruses) in an occlusion body. The occurrence of MNPVs and GVs is limited to Lepidoptera (Table I).

At present, baculoviruses are named after the host in which they were originally found. The Latin binomial of the host is added to the virus group name (e.g., *Autographa californica* MNPV or *Orgyia pseudotsugata* SNPV, abbreviated herein as AcMNPV and OpSNPV, respectively). However, many baculoviruses can infect more than one host. Therefore, in the future, the nomenclature should be revised when the identity of the virus has been unequivocally established. This is now possible by the application of new techniques, such as enzyme-linked immunosorbent assays (ELISA) in combination with monoclonal antibodies, and restriction endonuclease analyses (REN) of viral DNA (Payne and Kelly, 1981; Vlak, 1982).

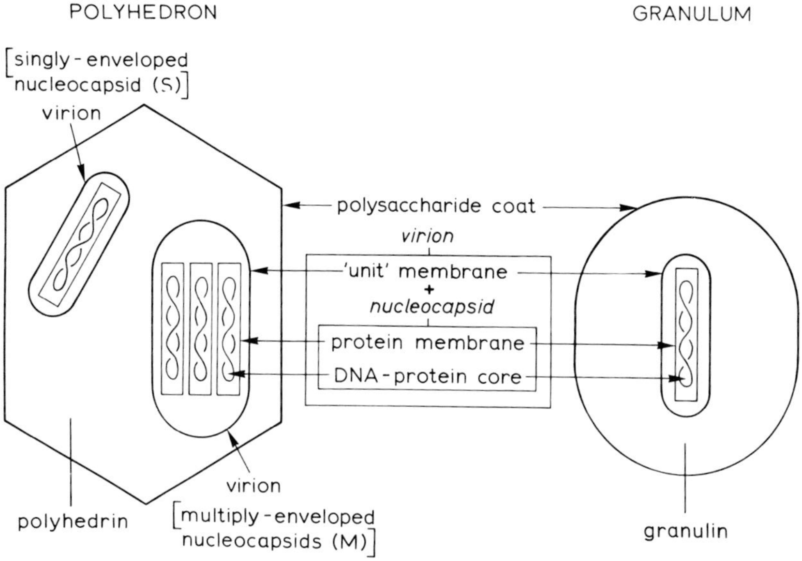

FIGURE 2. Schematic representation of the structure of polyhedra and granula.

B. STRUCTURE AND COMPOSITION

Baculovirus particles consist of a DNA-containing nucleoprotein core, surrounded by a protein capsid which together compromise the nucleocapsid (Fig. 2). The nucleocapsids, in turn, are enveloped either singly (SNPV and GV) or in multiples (MNPV) in a "unit" membrane. The virions are occluded in a paracrystalline protein matrix which forms a refractile body, polyhedral, cuboidal or spherical in shape. The multimeric polyhedrin protein is composed of 29,000 dalton monomers and comprises 95% of the occlusion bodies. The occlusion body is surrounded by a polysaccharide membrane (Fig. 2). A detailed review on the structure of baculoviruses is provided by Tinsley and Harrap (1978) and Harrap and Payne (1979).

Baculoviruses contain a double-stranded, circular, supercoiled DNA molecule (Bergold, 1953; Kok et al., 1968; Summers and Anderson, 1972) which comprises approximately 13% of the virion by weight. The molecular weights of the various baculovirus DNA's range from 60 to 110 × 10^6 daltons, as determined by a variety of biophysical and biochemical techniques, such as, electron microscopy, sedimentation analyses, reassociation kinetics, and restriction endonuclease analyses (see Harrap and Payne, 1979; Faulkner, 1981; Vlak, 1982; and references therein). Repetitive sequences amounting to 1-2% of the DNA

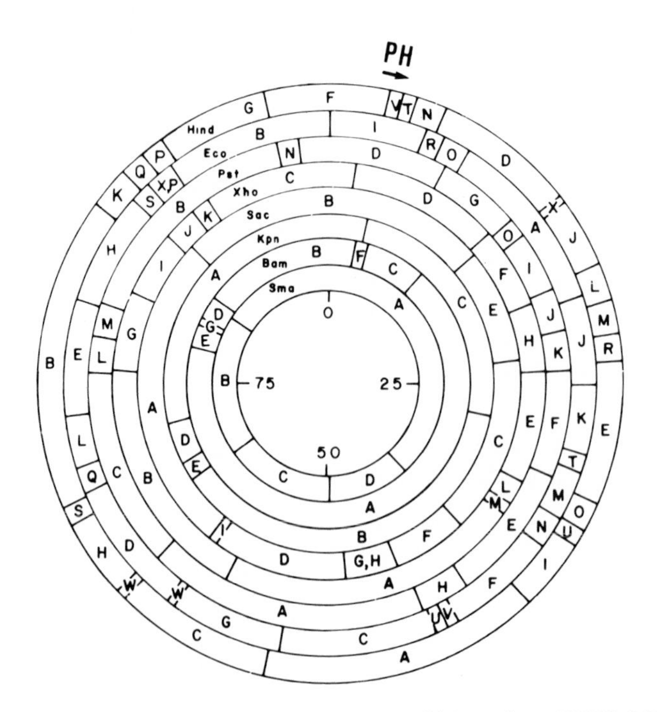

FIGURE 3. Physical map of A. californica MNPV DNA (strain E2; Summers and Smith, 1978) recognition sites of restriction endonucleases HindIII, EcoRI, PstI, XhoI, SacI, KpnI, BamHI and SmaI according to the adopted orientation (Vlak and Smith, 1982). Fragments for each nuclease are named with letters of the alphabet in order of decreasing size. The arrow indicates the location and direction of transcription of the polyhedrin gene.

have been demonstrated in NPV DNA's of *Spodoptera* spp. (Kelly, 1977) and in AcMNPV DNA (Cochran and Faulkner, 1983).

A more detailed structure of baculovirus DNA is obtained from a so called physical map, i.e., the physical arrangement within the circular genome of DNA fragments generated by various restriction endonucleases. At present, physical maps of AcMNPV and its variants, *Galleria mellonella* (Gm-MNPV), *Trichoplusia ni* (TnMNPV), and *Spodoptera frugiperda* (SfMNPV), have been constructed (Miller and Dawes, 1979; Smith and Summers, 1979; Vlak, 1980; Lübbert et al., 1981; Loh et al., 1981; Cochran et al., 1982). As an example, the physical map of AcMNPV (strain E2) DNA is presented for various restriction endonucleases (Fig. 3) (Vlak and Smith, 1982). Physical maps of other baculovirus DNA will undoubtedly be reported in the near future as they form a starting point for detailed comparative studies on baculovirus transcription, replication, genetic variation, and pathogenicity.

Baculovirus relatedness has been investigated by DNA-DNA hybridization (Kelly, 1977; Rohrmann et al., 1978b; Jurkovícová et al., 1979; Jewell and Miller, 1980; Rohrmann et al., 1982a; Smith and Summers, 1982). The disadvantage of this method is that it measures the formation of stable DNA duplexes and these values do not correlate directly with DNA sequence homology. Even under the least stringent hybridization conditions, DNA's must have greater than 67% homology in order to produce stable hybrids (Howley et al., 1979). Most of the experiments on baculoviruses have been performed under such stringent conditions that DNA sequence homology of less than 87% will not lead to any detectable duplex formation. Therefore, although many reports indicate very low relatedness, such values do not rule out actual base sequence homology of up to 87% or more (Rohrmann et al., 1982a). Values such as those for four *Spodoptera* MNPVs of 15 to 75% (Kelly, 1977) are probably suggestive of very close relatedness. Blot hybridization (Southern, 1975) has provided useful qualitative information on the relatedness of certain regions of the genome and the conservation of certain sequences (e.g., the polyhedrin gene) relative to other DNA segments (Jewell and Miller, 1980; Rohrmann et al., 1982a; Smith and Summers, 1982).

Based on the size of their genomes, baculoviruses have the capacity to code for up to 100 proteins. As reviewed by Harrap and Payne (1979), 25 polypeptides, ranging in size from 9 to 160×10^6 daltons, have been identified for a number of viruses on SDS polyacrylamide gels using staining techniques. Autoradiographic techniques have allowed the identification of 35 polypeptides for cell culture-derived AcMNPV (Vlak, 1979). Virion protein patterns on SDS-polyacrylamide gels appear to be characteristic for a given isolate. Further resolution of virion polypeptides can be achieved by two-dimensional gel electrophoresis (O'Farrell, 1975). Using this technique, Singh et al. (1983) distinguished more than 80 polypeptides in occluded virions of AcMNPV and *Lymantria dispar* (Ld)MNPV. It is not known how many of those are glycosylated or phosphorylated polypeptide isomers. This technique in combination with the highly sensitive silver staining method (Oakly et al., 1980) or radioimmunological techniques may allow resolution of additional virion polypeptides.

In a few instances virion envelopes have been removed from nucleocapsids by treatment with detergents (Harrap et al., 1977; Summers and Smith, 1978). One or more of the major virion polypeptides remain with the nucleocapsid, while most of the high molecular weight polypeptides are associated with the envelope. Recent work of Smith and Summers (1981) and Knell et al. (1983) suggest that the major virion polypeptides residing in the nucleocapsids of several different baculoviruses share antigenic determinants, while the high molecular weight, presumably glycosylated polypeptides of the envelope appear to contain virus-specific antigens.

Although it was reported earlier (Singh et al., 1979; Pritchett et al., 1979; McCarthy and Lambiase, 1979) on the basis of immunodiffusion tests that subgroup A and B viruses were antigenically distinct, the versatile technique of protein blot radioimmunoassay has now demonstrated common antigenic determinants between the nucleocapsids of viruses of subgroup A and B (Smith and Summers, 1981; Knell et al., 1983). The availability of monoclonal antibodies against virion polypeptides should help to resolve more detailed questions on the antigenic relationships among these baculoviruses (see chapter by L. E. Volkman, this volume).

Polyhedrin, the major constituent of the occlusion body, is the subject of a separate section (Section III).

C. MULTIPLICATION AND PATHOGENESIS

Two morphogenically and functionally distinct baculovirus forms are produced by infected insect cells. The nonoccluded virus (NOV) form serves to spread the infection from cell to cell within the insect. The occluded virus (OV) form is stable within the occlusion body for many years outside the insect and disseminates the disease in the insect population. The synthesis of these two forms is under stringent temporal control of the virus. The NOV is synthesized at early times after infection; the nucleocapsids are assembled in the nucleus and acquire an envelope as they bud through the plasma membrane (Faulkner and Henderson, 1972). Later in infection, nucleocapsids are synthesized and acquire *de novo* a synthesized envelope in the nucleus where the virions are then occluded. These two virus forms display different polypeptide patterns on SDS-polyacrylamide gels (Summers and Smith, 1978), and show only limited antigenic relatedness (Smith and Summers, 1981; Volkman, 1983). In addition, the NOV has a higher infectivity *in vitro* (Volkman and Summers, 1977). The marked differences between OVs and NOVs are a reflection of their unique morphogenesis.

Infection of the larvae is initiated when the occlusion bodies are ingested, usually as a contaminant of the food. Due to the alkalinity of the midgut juice, occlusion bodies dissolve and the liberated virus particles are then able to infect the midgut epithelial cells. An alkaline protease present in the insect gut is thought to enhance the liberation of virions. In most infected insect orders, the infected midgut epithelial cells produce both OVs and NOVs. However, in Lepidoptera, the infected midgut cells produce only NOVs. The NOVs are subsequently shed into the circulating hemolymph and infect, in turn, a variety of tissues including hemocytes and fat body. The nuclei of fat body cells produce massive amounts of occlusion bodies that are released in the environment following insect decay. Viral occlusion bodies may account for up to one-half of the insect body weight (10^9 occlusion bodies per larva).

The pathogenesis of a baculovirus-infected insect is marked initially by a reduced appetite, slow locomotion, and an increased incidence of secondary infections (bacteria, fungi). Later on, the larvae become sluggish, stop eating, and eventually die. Baculovirus multiplication and pathogenesis in insect hosts and tissue culture cells are extensively discussed in some recent reviews (Tinsley and Harrap, 1978; Granados, 1980; Faulkner, 1981; Tweeten et al., 1981; Kelly, 1982).

III. THE NATURE OF POLYHEDRIN

A. *POLYHEDRAL ENVELOPE*

The polyhedra of NPVs are surrounded by an electron-dense envelope (Harrap, 1972). The envelope is only slightly disrupted by alkali dissolution of the polyhedron. Many of the virions are trapped within these envelopes and can be observed by dark field microscopy (Martignoni, 1972). Minion et al. (1979) demonstrated that the polyhedral envelope of *Heliothis virescens* NPV is composed of carbohydrate, predominantly hexoses (60.9%), pentoses (29.1%), uronic acid (8.4%), and hexosamine (1.6%). No protein was detected.

B. *ALKALINE PROTEASE PRESENT IN POLYHEDRA*

The presence of a protease associated with polyhedra was initially reported by Yamafuji et al. (1958). The effects of this enzyme on solubilized polyhedrin prevented efforts to characterize the polyhedral protein for a number of years. It was found that the protease can be inactivated by heat (70°C for 20 min) or by dissolving the polyhedra in either acetic acid or urea, or in the presence of mercuric chloride or diisopropyl fluorophosphate (DFP) (Kozlov et al., 1975; Summers and Smith, 1975a). Subsequently, protease activity was shown to be present in all insect-derived baculovirus occlusion bodies examined but not in polyhedra derived from tissue culture cells (Zummer and Faulkner, 1979; McCarthy and DiCapua, 1979; Wood, 1980a). It was recently demonstrated that baculovirus-associated proteases are similar in properties to those found in the insect gut (Rubinstein and Polson, 1983). This indicates that these proteases are either gut-produced host enzymes or are produced by bacteria in the insect gut and become associated with polyhedra during disintegration of the insect.

Eppstein and Thoma (1975) determined that the protease of *T. ni* NPV polyhedra has a pH optimum of 9.5 and is a serine protease (i.e., it has a serine in its active site). Tweeten et al. (1978) found the protease for a GV of *Plodia interpunctella* (Pi) has a pH optimum of 10.5, a temperature optimum of 40°C, and a molecular weight of 14,000. The protease was found to be associated with polyhedrin. In contrast, while Payne and Kalmakoff (1978) demonstrated the protease of *S. littoralis* NPV had similar pH (9.6) and temperature optima (30°-40°C), they found it to have a molecular weight of 71,000 and to be associated with virions. The protease of the NPV of *T. ni* cleaves the polyhedrin into two major fragments of 16,000 and 9430 molecular weight (Eppstein and Thoma, 1977). Kozlov et al. (1981), examining polyhedrins from *Bombyx mori*, *Lymantria dispar*, and *G. mellonella*, found they had common cleavage sites in five places. Four of these sites had tyrosine and one had tryptophan at the N-terminal side of the cleavage site. In addition, there was one unique site occurring in *B. mori* polyhedrin (amino acids 124-125; Ala-Asn) and one in *G. mellonella* (amino acids 140-141; Glu-His). The protease was found to be required for complete removal of polyhedrin from viral envelopes during alkali dissolution in *A. californica* NPV (Wood, 1980a). Proteases have also been reported to be associated with occlusion bodies of cytoplasmic polyhedrosis viruses (Rohrmann et al., 1980) and entomopox viruses (Bilimoria and Arif, 1979).

C. *SIZE, ISOELECTRIC POINT, AND AMINO ACID COMPOSITION*

Once methods were developed for inactivating the proteases of polyhedra, accurate molecular weights for polyhedrins could be measured. Kozlov et al. (1974) initially reported a molecular weight for the *B. mori* SNPV polyhedrin monomer of about 28,000 daltons. Following their report, numerous investigators reported similar molecular weights for a large number of other polyhedrins. In the most comprehensive study to date, Croizier and Croizier (1977) have measured the molecular weights of 23 polyhedrins including representative baculoviruses pathogenic for the following orders: one NPV from Diptera, two NPVs from Hymenoptera, and fourteen NPVs and six GVs from Lepidoptera.

TABLE II. Amino Acid Composition of Polyhedrins[a]

	AcMNPV	GmMNPV	OpMNPV	LdMNPV	BmSNPV	OpSNPV	OpCPV
Ala	12	10	9	11	10	12	17
Asn	14	31[b]	15	14	19	16	38[b]
Asp	15		14	10	11	11	
Arg	13	12	13	14	11	15	15
Cys	3	3	3	3	3	3	2
Gln	4	26[b]	3	10	6	5	24[b]
Glu	21		22	20	21	23	
Gly	12	12	11	11	12	11	27
His	5	4	7	7	5	6	6
Ile	14	14	15	13	16	11	8
Leu	17	19	17	18	18	21	12
Lys	19	21	19	18	21	17	8
Met	5	4	6	4	6	6	7
Phe	14	13	13	12	12	14	12
Pro	17	15	16	14	15	15	11
Ser	9	8	11	10	10	10	16
Thr	11	11	10	13	10	12	9
Trp	4	4	4	2	4	3	0
Tyr	15	13	15	16	16	16	15
Val	20	21	21	18	19	19	17
	244	241	244	238	245	246	244
Mol. wt.	28,764	28,415	28,956	28,169	29,031	29,202	27,936[b]

[a]*Amino acid composition of polyhedrins from A. californica MNPV, G. mellonella MNPV, L. dispar MNPV, O. pseudotsugata MNPV, SNPV and CPV, and B. mori SNPV. The amino acid composition of GmMNPV, LdMNPV, and BmSNPV polyhedrin is from the sequence data of Kozlov et al. (1981) except three instead of two cysteines were tabulated (see discussion in legend for Fig. 5). Amino acid compositions of AcMNPV, OpMNPV, and OpSNPV polyhedrins were derived from the DNA sequences of their polyhedrin genes (see Figs. 5 and 9). The CPV polyhedrin amino acid composition is taken from Rohrmann et al. (1980). The listed molecular weights were calculated from the amino acid molecular weights.*

[b]*When Gln/Glu and Asp/Asn were not distinguished, the average molecular weight for these amino acids was used.*

They determined molecular weights of 25,200-31,260. In another comprehensive study (Summers and Smith, 1978) polyhedrins from lepidopteran baculoviruses, including four MNPVs, two SNPVs, and two GVs, were investigated, and molecular weights of 25,500-31,000 were found. For further references on polyhedrin molecular weights, see the review by Harrap and Payne (1979).

Based on the recently reported protein sequencing work by Kozlov et al. (1981), and DNA sequences shown below, the molecular weights of the six sequenced polyhedrins ranged from 28,000-29,000 daltons (Table II). Although molecular weights up to 33,000 have been commonly reported for AcMNPV (Wood, 1980a; Vlak et al., 1981), the value of 29,000 determined from the sequence is likely to be a more accurate value as no evidence of substantial glycosylation or phosphorylation of this protein has been detected (Maruniak and Summers, 1981).

Isoelectric points of 5.3-6.5 have been determined for a variety of polyhedrins (Bergold and Schramm, 1942; Eppstein and Thoma, 1977; Brown et al., 1979). Scharnhorst and Weaver (1980) examined the pI for *H. zea* NPV polyhedrin under denaturing conditions and found four polyhedrin species with pIs of 6.08, 6.15, 6.22, and 6.3, indicating they were each separated by one charge. The origin of this charge difference was not determined. Under native conditions they determined a pI of 5.9. They suggested that the protein aggregated extensively under native conditions and that a difference of a few charges between the high molecular weight aggregate would not be resolvable.

Exact amino acid compositions derived from six polyhedrins which have been sequenced are shown in Table II and compared to an unrelated CPV polyhedrin. The polyhedrins are equally rich in both strongly basic amino acids (Lys, Arg), and strongly acidic amino acids (Asp, Glu). However, the protein has a net acidic character as indicated by the pI of 5.3-6.50.

D. PHOSPHORYLATION AND GLYCOSYLATION

Polyhedrins have been demonstrated to be slightly phosphorylated in both GVs and NPVs. The polyhedrin of the NPV of *T. ni* contains about one phosphate residue per polyhedrin molecule (Eppstein and Thoma, 1977). Rohrmann et al. (1979) reported that the polyhedrin from the MNPV of *Orgyia pseudotsugata* (Op) was phosphorylated but contained less than 0.5 phosphate residue per molecule. In addition, Maruniak and Summers (1981) and Summers and Smith (1975a) demonstrated that polyhedrins from the MNPVs of *A. californica* and *Rachoplusia ou* (Ro) are phosphorylated with serine as the principle phosphoryl acceptor. Dobos and Cochran (1980) indicated that AcMNPV polyhedrin is weakly phosphorylated. It has also been reported that polyhedrins from the GVs of *T. ni* and *P. interpuntella* are phosphorylated (Summers and Smith, 1975b; Tweeten et al., 1980). No one has yet attributed a function for phosphorylation of these molecules; it is unknown whether only certain specific amino acid residues are phosphorylated or that many residues are randomly phosphorylated at a low level. Eppstein and Thoma (1977) reported in their study that phosphorylation was confined to a fragment which comprised two-thirds of the polyhedrin molecule.

Summers and Smith (1975a) could detect no carbohydrate associated with polyhedrins they examined. Eppstein and Thoma (1977) demonstrated the presence of 3% carbohydrate in a TnNPV. It included the sugars mannose, glucose, and galactose, present in a ratio of 2:1:1, respectively. Galactosamine was also present but sialic acid was not detected. However, it was not determined whether the carbohydrate was covalently linked to polyhedrin.

E. AGGREGATION, CRYSTALLIZATION, AND SOLUBILITY

It is generally agreed that polyhedrin molecules have a sedimentation velocity of 11-13 S after alkali dissolution of polyhedra. A variety of molecular weights from 200,000 to 374,000 daltons has been calculated for these aggregates (Bergold, 1947, 1948; Harrap, 1972; Eppstein and Thoma, 1977; Rohrmann, 1977). Scharnhorst and Weaver (1980) used the cross-linking agent dimethyl suberimidate to demonstrate with the NPV of *H. zea* that different degrees of cross-linking between polyhedrins result in aggregates of up to 12 subunits, with a total molecular weight of about 336,000.

By means of x-ray diffraction, Engstrom (1974) determined that polyhedrin is crystallized in a body-centered cubic lattice. Harrap (1972) has suggested that the arrangement of subunits observed in electron microscopic images of polyhedrin crystals is consistent with a six-armed nodal unit. If each arm was composed of two subunits, this arrangement would be in agreement with the dodecameric pattern found in cross-linking studies.

There are a number of features of polyhedrin which could account for the alkali solubility of polyhedrin crystals. If carboxyl groups are combined with divalent cations, then an increase in OH^- ions (pH) could cause a precipitation of these cations and the solubilization of the crystals. Under these conditions a chelating agent would be able to solubilize the crystals. However, this has not been observed. One hypothesis that has been advanced is that the tyrosine which appears to be clustered near the N-terminus of the protein influenced their solubility (Rohrmann et al., 1979). Normal tyrosine residues have pK_a values between 9.5 and 10.5, which are within the range at which polyhedra normally dissolve. It is possible that this concentration of tyrosine is responsible for the

```
Lepidoptera

AcMNPV            Pro Asp Tyr Ser Tyr Arg Pro Thr Ile Gly Arg Thr Tyr Val Tyr Asp Asn Lys Tyr Tyr Lys Asn Leu Gly Ala Val Ile Lys Asn Ala Lys Arg Lys Lys

GmMNPV             .  Asn  .   .   .  (.   .   .   .   .)  .   .   .   .   .   .   .   .   .   .   .   .   .   .   .   .   .   .   .   .   .   .   .   .

OpMNPV             .   .   .   .   .   .   .   .   .   .   .   .   .   .   .   .   .   .   .   .   .   .   .   .  Ser  .   .   .   .   .   .   .   .   .

LdMNPV    Met Lys Asn  .   .  ––– –––  .  Ala Leu  .  Lys  .   .   .   .   .   .   .   .   .   .   .   .  Thr  .   .   .  Gln  .   .   .  Gln  .

BmSNPV             .  Asn  .   .   .  Asn  .   .   .   .   .   .   .   .   .   .   .   .   .   .   .   .   .   .  Gly Leu  .   .   .   .   .   .   .   .

OpSNPV    Met Tyr Thr Arg  .   .   .  Asn  .  Ser Leu  .   .   .   .   .   .   .   .   .   .   .   .   .   .   .   .   .   .   .   .   .   .   .   .

PbGV                          Gly  .  Asn Arg Ala Leu  .  Pro  .  X2   .  Ile  .   .  Gln His  .   .  X1  ND   .   .   .  Leu  .  Asp Val  .  His  .   .
                                          └─Ala Tyr X1 Lys His Glu─┘

Hymenoptera

NsSNPV             .  Asn Leu Gly  .  Gln ND  Ser ––– Ala Lys Ser  .  Ile  .   .   .   .   .   .   .  Gly  .   .  Asp Ile  .  (.) Ser  .  (.)
                          └Ala Gln┘

Diptera

TpSNPV           Gln  .   .  Gly  .  Glu  .  Asn Val Asp Tyr Pro Asn Leu  .   .  Arg  .  Pro  .  Val Asp(His)Asp ND Tyr(Pro)
             └Tyr Glu Val Asn┘                └(Ala Ser His Ala Gly)┘ └(Ser)┘└(Gln)┘
```

FIGURE 4. Amino acid sequences of the N-terminus of nine polyhedrins from singly-enveloped (S) and multiply-enveloped (M) nuclear polyhedrosis viruses (NPV), and from a granulosis virus (GV). Data from G. mellonella MNPV, L. dispar MNPV, and B. mori SNPV are taken from Kozlov et al. (1981). N-terminal sequences from O. pseudotsugata SNPV and MNPV, P. brassicae GV, N. sertifer SNPV, and T. paludosa SNPV are from Rohrmann et al. (1979, 1981). The amino acid sequences for A. californica MNPV, OpSNPV and OpMNPV are derived from the nucleotide sequences of their polyhedrin genes (Hooft van Iddekinge et al., 1983; G. F. Rohrmann, personal communication; see also Fig. 9). The proteins are aligned to give maximal amino acid homology. The dots refer to the amino acids indicated for AcMNPV polyhedrin. Probable insertions are displayed below the line and indicated by (   ). Tentatively identified amino acids are indicated by brackets, while those not determined are indicated by (ND). Unidentified amino acid derivatives are noted as X_1, X_2, etc.

										10										20										30
AcMNPV	[Met]	---	Pro	Asp	Tyr	Ser	Tyr	Arg	Pro	Thr	Ile	Gly	Arg	Thr	Tyr	Val	Tyr	Asp	Asn	Lys	Tyr	Tyr	Lys	Asn	Leu	Gly	Ala	Val	Ile	Lys
GmMNPV	[Met]	---	.	Asn	.	.	.	(.	.	.	.	.)	.	.	.	.	.	.	.	.	.	.	.	.	.	.	.	.	.	.
OpMNPV	[Met]	---	.	.	.	.	.	.	.	.	.	.	.	.	.	.	.	.	.	.	.	.	.	.	.	.	Ser	.	.	.
LdMNPV	Met	---	Lys	Asn	.	.	---	---	.	Ala	Leu	.	Lys	.	.	.	.	.	.	.	.	.	.	.	.	.	Thr	.	.	.
BmSNPV	[Met]	---	.	Asn	.	.	.	Asn	.	.	.	.	.	.	.	.	.	.	.	.	.	.	.	.	.	.	Gly	Leu	.	.
OpSNPV	Met	Tyr	Thr	Arg	.	.	.	Asn	.	Ser	Leu	.	.	.	.	.	.	.	.	.	.	.	.	.	.	.	.	.	.	.

										40										50										60
AcMNPV	Asn	Ala	Lys	Arg	Lys	Lys	His	Phe	Ala	Glu	His	Glu	Ile	Glu	Glu	Ala	Thr	Leu	Asp	Pro	Leu	Asp	Asn	Tyr	Leu	Val	Ala	Glu	Asp	Pro
GmMNPV	.	.	.	.	.	.	(.	Leu	Glx	Glx	Glx	His	.	Glx	Glx	Lys	Asx	.	Asx	Val	.	Asx	Asx	Gly)	.	.	.	.	.	.
OpMNPV	.	.	.	.	.	.	.	Leu	Leu	.	.	.	Glu	Asp	.	Lys	His	.	.	.	.	.	His	.	Met	.	.	.	.	.
LdMNPV	Gln	.	.	.	Gln	.	.	Leu	Gln	.	Glu	His	.	.	.	Arg	Ser	.	.	His	.	.	Arg	.	.	.	.	.	.	.
BmSNPV	.	.	.	.	.	.	.	Leu	Ile	.	Glu	His	Lys	.	.	Lys	Gln	Trp	.	Leu	.	.	.	.	Met	.	.	.	.	.
OpSNPV	.	.	.	.	.	.	.	Gln	Ile	.	.	.	Ala	.	.	His	.	.	.	.	.	.	Lys	.	.	.	.	.	.	.

										70										80										90
AcMNPV	Phe	Leu	Gly	Pro	Gly	Lys	Asn	Gln	Lys	Leu	Thr	Leu	Phe	Lys	Glu	Ile	Arg	Asn	Val	Lys	Pro	Asp	Thr	Met	Lys	Leu	Val,	Val	Gly	Trp
GmMNPV	.	.	(.	.	.)	.	.	.	.	.	.	.	.	.	.	.	.	.	.	.	.	.	.	.	.	.	.	.	Asn	.
OpMNPV	.	.	.	.	.	.	.	.	.	.	.	.	.	.	.	.	.	.	.	.	.	.	.	.	.	.	Ile	.	Asn	.
LdMNPV	.	Tyr	.	.	.	.	.	.	.	.	.	.	.	.	.	.	.	.	.	.	.	.	.	.	.	.	.	.	Asn	.
BmSNPV	.	.	.	.	.	.	.	.	.	.	.	.	.	.	.	Val	.	.	.	.	.	.	.	.	.	.	Ile	.	Asn	.
OpSNPV	.	Phe	.	.	.	.	.	.	.	.	.	.	.	.	.	.	.	.	.	.	.	.	.	.	.	.	.	.	Asn	.

										100										110										120
AcMNPV	Lys	Gly	Lys	Glu	Phe	Tyr	Arg	Glu	Thr	Trp	Thr	Arg	Phe	Met	Glu	Asp	Ser	Phe	Pro	Ile	Val	Asn	Asp	Gln	Glu	---	Val	Met	Asp	Val
GmMNPV	.	Ser	.	.	.	.	.	.	.	.	.	.	.	.	.	Asn	.	.	.	.	.	Asx	Asx	Glx	.	Val	.	.	.	.
OpMNPV	Ser	.	.	.	.	Leu	.	.	.	.	.	.	.	Val	.	.	.	.	.	.	.	.	.	.	.	---	.	.	.	.
LdMNPV	Ser	.	.	.	.	Leu	.	.	.	.	.	.	.	.	.	Asn	.	.	.	.	.	.	.	.	.	Glu	.	.	.	Ile
BmSNPV	Ser	.	.	.	.	Leu	.	.	.	.	.	.	.	Val	.	.	.	.	.	.	.	.	.	.	.	Val	.	.	.	.
OpSNPV	Ser	.	.	.	.	Leu	.	Gln	.	.	.	.	.	.	.	.	.	.	Leu	.	.	.	.	.	.	---	Ile	.	.	.

```
                                                130                                               140                                               150
AcMNPV   Phe Leu Val Val Asn Met Arg Pro Thr Arg Pro Asn Arg Cys Tyr Lys Phe Leu Ala Gln His Ala Leu Arg Cys Asp Pro Asp Tyr Val
GmMNPV   .   .   .   Arg( .  Leu)Lys .   .   .   .   .   .   .   .   .   .   .   .   .   .   .   .   .   *   .   Asp .   .   .
OpMNPV   .   .   .   .   .   .   .   .   .   .   .   .   .   .   .   .   .   .   .   .   .   .   .   .   Trp .   Cys .   .   .
LdMNPV   Tyr .   Thr Ile .   Val .   .   .   .   .   .   .   .   .   .   .   Val .   .   .   .   .   .   *   Gln Asp Gly .   .
BmSNPV   Tyr .   .   Ala .   Leu Lys .   .   .   .   .   .   .   .   .   .   .   .   .   .   .   .   .   *   Gln Asn .   .   .
OpSNPV   .   .   .   Ile .   .   .   .   .   .   .   .   .   .   Phe Arg .   .   .   .   .   .   .   .   .   .   .   Glu .   .

                                                160                                               170                                               180
AcMNPV   Pro His Asp Val Ile Arg Ile Val Glu Pro Ser Trp Val Gly Ser Asn Asn Glu Tyr Arg Ile Ser Leu Ala Lys Lys Gly Gly Gly Cys
GmMNPV   .   .   Glu .   .   .   .   .   .   .   .   .   .   .   .   Asx Asx Glx .   .   .   .   .   .   .   .   .   .   .   .
OpMNPV   .   .   Glu .   .   .   .   .   .   .   .   Tyr .   .   Met .   .   .   .   .   .   .   .   .   .   .   .   .   .   .
LdMNPV   .   .   Glu .   .   .   .   .   .   Thr .   Tyr .   Asn Gln Pro .   .   .   .   .   .   .   .   .   Arg .   .   .   .
BmSNPV   .   .   Glu .   .   .   .   Met .   .   .   Tyr .   .   Met .   .   .   .   .   .   .   .   .   .   .   .   .   .   .
OpSNPV   .   .   Glu .   .   .   .   .   .   .   .   Tyr .   .   .   .   .   .   .   .   Leu Val .   .   .   Arg .   .   .   .

                                                190                                               200                                               210
AcMNPV   Pro Ile Met Asn Leu His Ser Glu Tyr Thr Asn Ser Phe Glu Gln Phe Ile Asp Arg Val Ile Trp Glu Asn Phe Tyr Lys Pro Ile Val
GmMNPV   .   .   .   .   .   --- --- --- --- .   Asx .   .   Glx Glx .   .   .   .   .   .   .   Glx Asx .   .   .   .   .   .
OpMNPV   .   .   .   .   .   Ile .   Ala .   .   .   .   .   .   Ser .   Val Asn .   .   .   .   .   .   .   .   .   .   .   .
LdMNPV   .   .   Arg .   .   .   .   Ala .   .   His .   .   .   Thr .   --- --- --- --- --- --- Asp .   .   .   .   .   .   .
BmSNPV   .   .   .   .   .   Ile .   .   .   .   .   .   .   .   Ser .   Val Asn .   .   .   .   .   .   .   .   .   .   .   .
OpSNPV   .   Val .   .   .   Thr Ala .   .   .   .   .   .   .   Glu .   .   Asn .   .   His .   .   .   .   .   .   .   Thr .
```

	211									220										230										240
AcMNPV	Tyr	Ile	Gly	Thr	Asp	Ser	Ala	Glu	Glu	Glu	Glu	Ile	Leu	Leu	Glu	Val	Ser	Leu	Val	Phe	Lys	Val	Lys	Glu	Phe	Ala	Pro	Asp	Ala	Pro
GmMNPV	.	.	.	.	.	Thr	Ser(	.	.	.	Glx)	Leu	Ile	.	.	.	.	.	.	.	.	.	.	.	.	.	.	.	.	.
OpMNPV	.	.	.	.	.	.	Ser	.	.	.	.	.	.	Ile	.	.	.	.	.	.	.	.	.	.	.	.	.	.	.	.
LdMNPV	.	Val	.	.	Thr	Ala	Ser(	.	.	.	Gln)	.	.	.	.	.	.	.	.	.	.	Ile	.	.	.	.	.	.	.	.
BmSNPV	.	.	.	.	.	Ala	Ser	.	.	.	Gln	.	.	Ile	.	.	.	.	.	.	.	Ile	.	.	.	.	.	.	.	.
OpSNPV	.	Val	.	.	.	.	.	.	.	.	.	.	.	.	.	.	.	.	Leu	.	.	Ile	.	.	.	.	.	.	.	.

							247
AcMNPV	Leu	Phe	Thr	Gly	Pro	Ala	Tyr
GmMNPV	.	.	.	.	.	.	.
OpMNPV	.	.	.	.	.	.	.
LdMNPV	.	.	Gln	.	.	.	.
BmSNPV	.	.	.	.	.	.	.
OpSNPV	.	Tyr	Ser	.	.	.	.

FIGURE 5. Amino acid sequences of polyhedrins from A. californica MNPV, G. mellonella MNPV, L. dispar MNPV, O. pseudotsugata MNPV and SNPV, and B. mori SNPV. Data for GmMNPV, LdMNPV, and BmSNPV are taken from Kozlov et al. (1981). The amino acid sequences for AcMNPV, OpMNPV, and OpSNPV are derived from the nucleotide sequence of their respective polyhedrin genes (Hooft van Iddekinge et al., 1983; G. F. Rohrmann, personal communication; E. H. Carstens, personal communication; see also Fig. 9). The proteins are aligned in order to give maximum amino acid homology. Deletions in the sequence are indicated (---). Conventional abbreviations for amino acids are used. Asx and Glx are used where Glu/Gln and Asp/Asn were not discriminated (Kozlov et al., 1981). The dots refer to the amino acids indicated for AcMNPV.

Three areas in the proteins derived from DNA sequences differ from the protein sequences reported by Kozlov et al. (1981). In all the genes sequenced, amino acids 41-42 are His·Glu whereas in the proteins sequenced these amino acids are reported in the opposite order (Glu·His). At amino acid 116 all the genes sequenced lacked an amino acid present in the proteins sequenced. In addition, at amino acid 145 all the genes sequenced demonstrated cysteine or tryptophan, whereas the proteins sequenced show a deletion at this position denoted by (*). This may be due to the use by Kozlov et al. (1981) of a mixture of chymotrypsin and trypsin. Trypsin would cleave at arginine (144) and chymotrypsin may cleave at cysteine (Dognin and Wittmann-Liebold, 1977). This single amino acid would be lost in the peptide purification procedure.

solubility of polyhedrin (Rohrmann et al., 1979). It has also been suggested that crystallization results from salt bridges formed by amino acid groups lysine and arginine with carboxyl groups from glutamine and asparagine (I. P. Griffith, personal communication). Such ionic bonds would be disrupted by a strongly alkaline or acidic pH.

F. TRYPTIC PEPTIDE ANALYSIS

The cleavage of polyhedrins with trypsin and the two-dimensional separation of the cleavage fragments using mobility based on size in one direction and charge in another, results in tryptic peptide fingerprints which can be used to compare the primary structure of these proteins. Kozlov et al. (1975), using these procedures, demonstrated that the polyhedrins from NPVs of *B. mori* and *G. mellonella* could be distinguished. Summers and Smith (1975a) and Maruniak and Summers (1978) compared polyhedrins from four MNPVs, three SNPVs, and two GVs and although they had many fragments in common, they all gave distinct tryptic peptide fingerprints. Similar findings were reported for five NPVs examined by Cibulsky et al. (1977), although their preparations may not have been free of protease activity. Rohrmann et al. (1979) performed cation exchange resin chromatography on radio-labeled tryptic peptides of OpSNPV and OpMNPV polyhedrins. They could find six differences out of 19 major peaks between their elution profiles. Although these procedures yield distinctive patterns, they are cumbersome and therefore are not commonly used for viral identification.

G. AMINO ACID SEQUENCE

Although the N-terminal amino acid sequence of polyhedrins from dipteran and hymenopteran NPVs, and a lepidopteran GV demonstrate substantial variation (Fig. 4) (Rohrmann et al., 1981), the six lepidopteran NPV polyhedrins which have now been completely sequenced (Fig. 5) form a closely related group of similar size (238-246 amino acids) and molecular weight (28,000-29,000 daltons) in which 80-90% of the amino acids are conserved (see Table III). The nucleotide sequencing of polyhedrin genes of several lepidopteran NPVs confirms the extent of sequence conservation reported earlier on polyhedrin amino acid sequences (Kozlov et al., 1981).

Within these lepidopteran polyhedrins there are several regions which are highly conserved, e.g., amino acids 15-26 and 58-86. It is likely that these regions are involved in those properties of polyhedrin such as occlusion, alkali solubility, and the formation of the paracrystalline lattice which are characteristic of these molecules. The region between amino acid 38-55 which is rich in hydrophilic amino acids (Glu and Asp) is intensely variable. Since hydrophilic amino acids are usually on the external surface of molecules and are highly antigenic (Hopp and Woods, 1981), the variation observed in this region may account for their distinct antigenicity. Other sites of amino acid variation are the N-terminal region, and regions 120-127, 145-148, 165, 195, and 216. The production of antibodies against chemically synthesized peptides representative of these variable regions (Walter et al., 1980) may lead to a highly specific method to differentiate between baculovirus polyhedrins.

The studies of Jones and Kafatos (1982) indicate that the duplication of short amino acid stretches in the most likely mechanism for production of segmental mutations. Several regions of polyhedrins appear to have resulted from such duplications. The sequence beginning at amino acid 127 (Arg·Pro·Thr·Arg·Pro·Asn) could have arisen by a nine-nucleotide duplication followed by only a single additional base change to convert the Thr to Asn. Likewise, a duplication at position 236 (Ala·Pro·Asp·Ala·Pro·Leu) could have occurred; the Asp and Leu change would have required an additional two nucleotide alteration. This process may explain the increase in polyhedrin size from the ancestral dipteran type of polyhedrin of 25,000 to the lepidopteran polyhedrin of 29,000 daltons (see Section V).

H. ANTIGENICITY AND ANTIGENIC RELATEDNESS

A molecule as large as polyhedrin has the potential for containing a variety of antigenic determinants arising not only from its own amino acid sequence but also from the interaction of neighbouring polyhedrin molecules. Recently it has been demonstrated by Hopp and Woods (1981) that antigenic sites on a molecule can be predicted from its primary structure. This is accomplished by assigning a hydrophilicity value to each amino acid. Amino acids such as Arg, Asp, Glu, and Lys, which have hydrophilic side chains, are given the highest values. Such amino acids are frequently found in regions of molecules which have high degrees of exposure to solvents. In addition, such charged hydrophilic side chains are common features of antigenic determinants. Therefore, one can calculate a six-amino acid average of the hydrophilicity values. The region which has the highest hydrophilicity value is invariably a major antigen. Additional antigenic sites can be

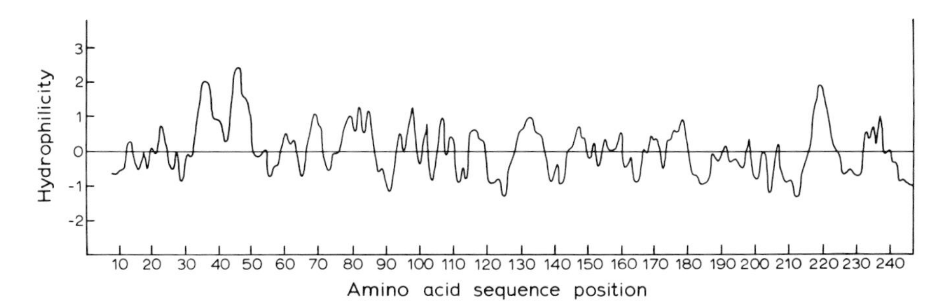

FIGURE 6. Hydrophilicity values for amino acid positions in B. mori SNPV polyhedrin. The values were obtained using the amino acid hydrophilicity values and methods of Hopp and Woods (1981) and the sequence data of Kozlov et al. (1981). The values are determined for the amino acids in groups of 6. The peak values suggest the location of antigenic determinants.

predicted to be present in areas with high hydrophilicity values. When the hydrophilicity values for *B. mori* NPV polyhedrin are plotted there are up to 20 strongly hydrophilic regions which are potential antigenic sites (Fig. 6). Additional antigenic sites may be formed by interactions of more than one polyhedrin molecule creating higher ordered aggregates. Using immunodiffusion, one or two antigens have generally been detected in polyhedrins. Using antisera against alkali-solubilized polyhedrin, Rohrmann (1977) detected two major polyhedrin antigens. Only one of these appeared to be present when polyhedrin was reacted with antisera produced against the monomeric polyhedrin. This suggests that the second antigen was formed by an aggregation of polyhedrin subunits. Eppstein and Thoma (1977) were able to chromatographically separate two distinct polyhedrin components, one of 100,000 and the other larger than 200,000 daltons, which contained noncross-reacting antigens. They suggested that the antigens were derived from higher ordered aggregates of polyhedrin.

Although it has long been evident that all lepidopteran NPV polyhedrins are antigenically related, the relatedness of polyhedrins from lepidopteran, hymenopteran, and dipteran NPVs and GVs was not clear. It is now known that all baculovirus polyhedrins have regions of common amino acid sequence and are, therefore, evolutionarily related (Rohrmann et al., 1981). The polyhedrin from the NPV of the dipteran *T. paludosa* (Tp) is the most distantly related and although the N-terminal amino acid sequence shows distant relatedness, relatedness could not be detected by radioimmunoassay (Rohrmann et al., 1981) or by immunodiffusion (Guelpa et al., 1977). In a comprehensive study, Krywienczyk and Bergold (1960), using complement fixation, demonstrated three groups of polyhedrins including those containing: (1) hymenopteran NPV polyhedrins, (2) two GV polyhedrins from *Choristoneura murinana* (Cm) and *C. fumiferana* (Cf), and (3) two other GV polyhedrins and eleven lepidopteran NPV polyhedrins. Rohrmann et al. (1981), using radioimmunoassay, demonstrated two lepidopteran NPV polyhedrins were closely related and polyhedrins from a GV, a hymenopteran NPV, and a dipteran NPV were more distantly related to the lepidopteran NPV polyhedrin in that order. For further discussion on antigenic relatedness, the reader is referred to the chapter by L. E. Volkman in this volume.

I. MONOCLONAL ANTIBODIES

The recently developed methodology for production of monoclonal antibodies in hybridoma cell lines is being applied to study polyhedrins. Roberts and Naser (1982) described two monoclonal antibodies that cross-reacted with polyhedrin from six lepidopteran baculoviruses, including four MNPVs, one SNPV, and one GV. Hohmann and Faulkner (1983) have also produced two monoclonal antibodies against polyhedrin. One of these appears to react only with the native polyhedrin while the other reacts with previously denatured and renatured polyhedrin and several protease-generated digestion products. Volkman and Falcon (1982) used monoclonal antibodies against polyhedrin in an enzyme-linked immunosorbent assay to detect SNPV in *T. ni* larvae.

The use of monoclonal antibodies against polyhedrins should improve our understanding of the antigenic structure of polyhedrin and will also be of great value in the study of other baculovirus proteins (see chapter by L. E. Volkman in this volume).

IV. THE POLYHEDRIN GENE

A. VIRAL ORIGIN OF POLYHEDRIN

At the time Bergold (1947) initiated the discussion on the origin of polyhedra, it was generally believed that they were produced by the host in response to baculovirus infection. On one hand, the strong antigenic relationships demonstrable between polyhedrins and, on the other, the broad morphological variety of polyhedra in different insects, led to confusion as to the origin of this protein. Now, however, several lines of evidence clearly indicate that polyhedrin is a virus-coded protein.

Spontaneously occurring viruses with altered polyhedron morphology have been known for many years (Gershenson, 1958; Stairs, 1964; Cunningham, 1970; Aratake and Watanabe, 1973). Recently, both Lee and Miller (1979) and Brown et al. (1980) have produced temperature-sensitive (*ts*) mutants of AcMNPV that display altered polyhedron morphology. The occurrence of such mutants provides evidence that viral gene functions are involved in the determination of polyhedron morphology.

Tryptic digests, amino acid analysis, and serological comparison of polyhedrins from baculoviruses whose host specificity allowed passage through different insect species indicate that the polyhedrins are not altered upon transmission through alternate hosts (Cibulski et al., 1977; Maruniak et al., 1979; McCarthy and DiCapua, 1979). This finding provided additional evidence that polyhedrins were virus-coded.

Direct evidence for the viral origin of polyhedrin was first obtained in a study of genetic recombination between AcMNPV and RoMNPV (Summers et al., 1980), as the genes for the distinctive polyhedrins segregated out into virus recombinants. Another line of evidence came from *in vitro* translation studies of virus-specific RNA (Van der Beek et al., 1980; Vlak et al., 1981). Cytoplasmic RNA from AcMNPV-infected cells was hybridized to viral DNA that had been covalently linked to a solid phase (diazotized *m*-aminobenzylmethyl-cellulose or nitrocellulose). The hybridized RNA which was then eluted from the DNA (hybridization selection) could produce detectable polyhedrin by translation in an *in vitro* RNA translation system. Using a slightly different method (section IV, C), Rohrmann et al. (1982b) showed that the polyhedrin of OpMNPV is, as well, a virus-coded gene product.

B. *TRANSCRIPTION AND TRANSLATION OF POLYHEDRIN mRNA*

After infection of susceptible cells with AcMNPV, polyhedrin is detectable as early as 12 hr post infection but it becomes the major gene product only at late times post infection and polyhedra formation is not apparent until 24 hr post infection (Carstens et al., 1979; Dobos and Cochran, 1980; Wood, 1980; Maruniak and Summers, 1981).

There is, at present, only limited information about the viral and/or host factors involved in RNA synthesis in baculovirus-infected cells. Host RNA polymerase II, which is normally responsible for the synthesis both of cellular mRNA in eukaryotes and of mRNA of DNA viruses which replicate within animal cell nuclei, does not seem to be involved in late AcMNPV transcription (Grula et al., 1981). AcMNPV-specific RNA synthesis late after infection appears to be unaffected by the presence of α-amanitin, which effectively blocks RNA polymerase II of the insect host cell. However, the presence of a virus-induced or -coded RNA polymerase has not yet been directly demonstrated (Grula et al., 1981). The majority of the late virus-specific RNA's are capped at their 5'-terminus (Qin and Weaver, 1982) and likely includes polyhedrin mRNA since this is the major late transcript in AcMNPV-infected cells (Smith et al., 1983b).

Vlak and Van der Krol (1982) detected high levels of presumable polyhedrin mRNA in the cytoplasm of AcMNPV-infected cells at 24 hr post infection by hybridizing ^{32}P pulse-labeled cytoplasmic RNA to viral DNA restriction fragments. Hybridization was found, in particular, to fragments *Eco*RI-I and *Hin*dIII-V, now known to contain the polyhedrin gene (Fig. 3). Using a similar approach, Erlandson and Carstens (1983) showed that RNA hybridizable to the polyhedrin gene region of the genome is present at low frequency as early as 7.5 hr postinfection. Polyhedrin mRNA is present very late in infection (Smith et al., 1982). In addition to polyhedrin, a 7,000-10,000 dalton protein is also a major translation product in deteriorating cells (Smith et al., 1983b). The present data indicate that in AcMNPV-infected cells there is a strong correlation between the emergence in the cytoplasm of mRNA from the polyhedrin gene area and the onset of polyhedrin protein synthesis.

C. LOCALIZATION OF THE POLYHEDRIN GENE

Using the restriction maps of AcMNPV and RoMNPV (Smith and Summers, 1979, 1980) and observing the pattern of segregation of polyhedrins among the AcMNPV-RoMNPV recombinants, Summers et al. (1980) showed that the approximate location of the polyhedrin coding sequence mapped between position 90.0 and 8.7 of the circular AcMNPV restriction map (Fig. 3). They suggested that the AcMNPV DNA fragment *Eco*RI-I might harbor the polyhedrin gene. This hypothesis proved correct. Vlak et al. (1981) and later Esche et al. (1982) demonstrated that cytoplasmic RNA selected by hybridization to AcMNPV DNA fragment *Eco*RI-I could be translated into polyhedrin.

By taking smaller DNA fragments contained within *Eco*RI-I (Fig. 3) and following the same approach, Smith et al. (1982) and Rohel et al. (1983) showed that fragments *Hin*dIII-V and *Bam*HI-F could hybrid-select polyhedrin mRNA and, therefore, must contain segments of the polyhedrin gene. Via a slightly different approach, Adang and Miller (1982) mapped cDNA clones that selected for polyhedrin mRNA. They demonstrated that the polyhedrin gene was of viral origin and located in fragment *Eco*RI-I and *Hind*III-V of AcMNPV DNA. Similarly, Rohrmann et al. (1982b) isolated mRNA from *O. pseudotsugata* larvae infected with OpMNPV, fractionated it on a sucrose gradient, and produced cDNA from that RNA fraction, which was shown to be enriched for polyhedrin mRNA by *in vitro* translation. The cDNA hybridized to a specific OpMNPV DNA fragment (*Xho*I-J). This fragment was also able to hybrid-select polyhedrin mRNA.

Taking the available information together, there is conclusive evidence that the polyhedrin genes are located as single copies in very specific areas of baculovirus genomes. Nucleotide sequencing (Section IV, E) has disclosed the exact location of the polyhedrin gene.

D. SIZE AND STRUCTURE OF POLYHEDRIN mRNA

In order to determine the size of AcMNPV polyhedrin mRNA, cytoplasmic polyadenylated RNA was separated on agarose-formaldehyde (Smith et al., 1983b) or agarose-methyl mercury (Rohel et al., 1983) gels and transferred to nitrocellulose or diazobenzyloxymethyl paper. The polyhedrin mRNA was detected by hybridization with ^{32}P-labeled AcMNPV DNA fragments containing sequences of the polyhedrin gene (*Bam*HI-F or *Hind*III-V). The size of the polyhedrin mRNA was estimated to be 1200 (Smith et al., 1983b) to 1400 (Rohel et al., 1983) bases in length. The polyhedrin message of OpMNPV appeared to be about 1000 bases in length as derived from R-loop experiments (Rohrmann et al., 1982b) (Fig. 7). Therefore, the size of polyhedrin mRNA is well over 750 bases, which is the minimum genetic information that would be required to code for about 250 amino acids which is the size of polyhedrin (Table II).

Like many other viral and cellular messengers in eukaryotes (Brawerman, 1981), the polyhedrin mRNA contains a tail of poly(A) residues at its 3'-terminus (Fig. 7). Polyadenylated RNA obtained from infected cells or larvae by chromatography of cytoplasmic RNA on oligo(dT)-cellulose, can be translated into polyhedrin. The product has been identified by immunoprecipitation (Vlak et al., 1981; Rohrmann et al., 1982b) and by V8 digestion (Rohel et al., 1983). In fact, most if not all baculovirus-specific RNA's present in the cytoplasm are polyadenylated (Vlak et al., 1981). The length of the poly(A) extension of the 3'-terminus of polyhedrin mRNA is not known. A poly(A) tail was not observed on the OpMNPV polyhedrin mRNA in electron micrographs used for R-loop mapping. Although this mRNA was isolated by retention to oligo(dT)-cellulose,

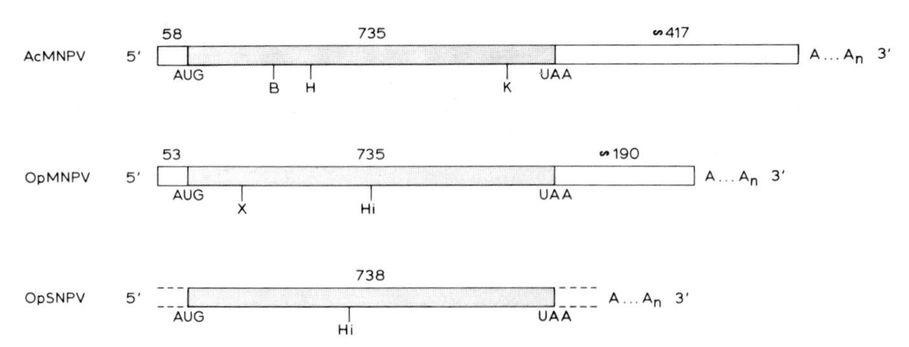

FIGURE 7. Structure of the polyhedrin mRNA from A. californica MNPV and O. pseudotsugata SNPV and MNPV as determined by physical mapping of the mRNA (Rohrmann et al., 1982b; Smith et al., 1983b) and sequencing data of the polyhedrin gene (Hooft van Iddekinge et al., 1983; G. F. Rohrmann, personal communication). The dotted areas correspond to the coding sequences. The initiation (AUG) and termination (UAA) sites of the coding sequence, as well as the 5' and 3' end of the message, are indicated. The numbers (in bases) refer to the length of nucleotide sequence contained in the coding area and the 3' end and 5' flanking regions. The BamHI (B), HindIII (H), KpnI (K), HincII (Hi), and XbaI (X) restriction sites of the DNA complement are indicated.

its poly(A) tail may be very small (Rohrmann et al., 1982b). It is, however, likely to be at least 15 nucleotides long as this is the minimum size required for retention by oligo(dT)-cellulose columns (Nudel et al., 1976). The AcMNPV polyhedrin mRNA has been shown by S_1 mapping to be at least 1210 nucleotides long (Smith et al., 1983b) with a size of 1400 bases on Northern gels (Rohel et al., 1983). This suggests that the poly(A) extension could be as large as 150-200 bases in polyhedrin mRNA.

Most eukaryotic mRNAs also contain a so-called cap structure at their 5'-terminus (Furuichi and Miura, 1975; Shatkin, 1976). In the analysis of late RNA from AcMNPV-infected *S. frugiperda* cells, Qin and Weaver (1982) determined that virus-specific RNA contained structures, presumably a cap 1 structure ($m^7GpppX_p^mYp$). It is likely that polyhedrin mRNA also has a cap structure, as it is the most abundant viral message in the cytoplasm at late times after infection (Adang and Miller, 1982; Smith et al., 1983b). The location of the cap sites of AcMNPV (Smith et al., 1983b) and OpMNPV (G. F. Rohrmann, personal communication) polyhedrin mRNA have been shown to be at position -58 and -53, respectively (Fig. 8), by

```
           -170      -160      -150      -140      -130      -120      -110      -100
AcMNPV     AATAACAGCCATTGTAATGAGACGCACAAACTAATAATCACAAACTGGAAAATGTCTATCAATATATAGTTGTCGATATC

OpMNPV     CTTGGGAACCGGCACTCCGCCAAAGGTTTAATGTTTGAGCATATGTTTTGCTGCTTACGAATTTATGTACAACAAAAAAT

OpSNPV     TCTCGAAATGTCAATAATACTAATGCGCGTTATCTCAAACTACGTAATATATTTGTGTCGATAGCGCGGGCTGAGTAATT

-90       -80       -70       -60       -50       -40       -30       -20       -10
ATGGAGATAATTAAAATGATAACCATCTCGCAAATAAATAAGTATTTTACTGTTTTCGTAACAGTTTTGTAATAAAAAAACCTATAAATA ATG

AAAACACTAGTTACTATTGGCGTTTCGTTTTTTATTAATAAGTAATTTCCTGTTATTGTAACAATTTTGTAAAAAAATTTCCCATAACC ATG

CGATTTTTTGCGTTGAGAATTTCAACGACACACCTCAATAAGTATTTTTGTTCCTTCGTAAACATTGTGAAATTTCAATACACCATA ATG
                                                            T
```

FIGURE 8. 5' Noncoding region of the polyhedrin genes from A. californica MNPV and O. pseudotsugata MNPV and SNPV. The sequence from the AcMNPV polyhedrin gene was obtained from Hooft van Iddekinge et al. (1983). Those of OpMNPV and OpSMNPV polyhedrin genes were provided by G. F. Rohrmann (personal communication). The sequences are aligned to demonstrate homologous regions. The boxes in solid lines indicate the strongly homologous regions; the deviating bases are underlined. The dashed boxes point to the potential TATA sequences. Possible CAAT boxes are underlined (——). The direct repeats are indicated by parentheses (   ). The arrows indicate the cap site. The ATG initiation codons are underlined (——).

S_1 nuclease protection experiments. There appear to be minor cap sites one nucleotide on either side of these positions. The cap site for OpSNPV polyhedrin mRNA is not yet known.

The polyhedrin mRNA (Fig. 7) does not appear to be spliced (Rohrmann et al., 1982b; Smith et al., 1983b). Many viral and cellular genes contain sequences that are not represented in the final translated mRNA. These so-called introns or intervening sequences are transcribed, but later removed from the message in an excision process (splicing). When the mRNA is hybridized to single-stranded genomic DNA, intron regions remain single-stranded, as they cannot find complementary sequences in the mRNA. These intron sequences then remain single-stranded and become sensitive to S_1 nuclease (Berk and Sharp, 1978) or form loops which are visible in the electron microscope. With the S_1 nuclease technique, Smith et al. (1983b) found that polyhedrin mRNA protected a contiguous stretch of about 1210 nucleotides of AcMNPV DNA fragment *Eco*RI-I and, therefore, is an unspliced message. They also demonstrated polyhedrin mRNA is transcribed clockwise on the circular AcMNPV genome (Fig. 3). None of the R-loops examined suggested that the OpMNPV polyhedrin gene contained intervening sequences (Rohrmann et al., 1982b).

The splicing process is thought to enhance transportation and stability of the mRNA during transport across nuclear membranes (Gruss et al., 1981). One may speculate that shut-off of host syntheses due to viral infection, as seen in insect cells (Carstens et al., 1979; Vlak et al., 1981), could eliminate the system for splicing. Since polyhedrin mRNA is transcribed late in infection in large quantities, it follows that polyhedrin mRNA synthesis would be severely impaired if dependent on a splicing mechanism. More interesting, however, is the observation that all baculovirus mRNAs, including early ones, may be unspliced (Lübbert and Doerfler, 1984; Mulder et al., 1984). Therefore, the absence of splicing may be a common feature of these baculovirus mRNAs.

The noncoding 5' leader sequences of polyhedrin mRNAs are 50-60 nucleotides in length (Figs. 7 and 8). From sequence data, it is clear that the size of the coding region of AcMNPV, OpMNPV, and OpSNPV polyhedrin mRNA transcripts is nearly constant in size and encompasses about 735 bases. These figures combined with the measurements of mRNA size, suggest that the AcMNPV and OpMNPV mRNAs contain 3'-flanking regions of about 470 and 170 nucleotides, respectively. These long noncoding regions may play a role in the stability of these mRNAs as a compensation for the absence of splicing.

E. NUCLEOTIDE SEQUENCE OF POLYHEDRIN GENES

For several baculoviruses, the nucleotide sequences of genes that code for polyhedrin and some of its flanking regions have recently been determined (Hooft van Iddekinge et al., 1983; E. H. Carstens, personal communication; G. F. Rohrmann, personal communication). The polyhedrin coding sequences for AcMNPV, OpMNPV, and OpSNPV are very similar in size and sequence (Fig. 9), encompassing about 735 bases. About 75% of the nucleotides are conserved, while most of the variable nucleotides are in wobble positions (i.e., changes in these positions do not affect the amino acid sequence). This high conservation of nucleotides correlates well with the amino acid sequence data (Fig. 5) of these and three other polyhedrins, which show over 80% of the amino acid sequence homology. The consequences of these findings with regard to polyhedrin evolution are discussed in Section V.

Comparison of amino acid and nucleotide sequences clearly indicate that there are no intervening sequences (introns) in the coding region of polyhedrin genes and confirms the unspliced nature of polyhedrin mRNA. The sequence data also show that the G+C contents of the OpMNPV and OpSNPV polyhedrin genes, 53 and 45%, respectively, are close to those of the respective entire virus genomes, 54 and 44%, as determined by T_m measurements and CsCl density gradient centrifugation (Rohrmann et al., 1978a). However, the G+C content of AcMNPV polyhedrin gene (50%; Fig. 9) is dissimilar to that of the complete genome (42%; Vlak and Odink, 1979). Thus, G+C content is not a conserved character of polyhedrin genes.

The DNA sequence data also show that about 50 out of the 61 possible amino acid codons are present in the mRNA (Table III). Nonrandom codon use is frequently observed with genes that are strongly expressed and efficiently translated (Bennetzen and Hall, 1982). Thus, although polyhedrin is one of the most strongly expressed eukaryotic genes, it does not display an extreme codon preference.

The 5'-flanking region (Fig. 8) between the cap site at about position -55 and the AUG initiation codon, is extremely rich in A+T (over 80%). Similar regions rich in A+T have been observed in this region of many genes of prokaryotes and lower eukaryotes, and are considered to be ribosome binding sites. Although initiation of translation in eukaryotes occurs by scanning the 5' leader rather than binding to it (Kozak, 1983), the two "boxed" areas are possible polymerase or ribosome recognition sites in the leader which may enhance the transcriptional and translational efficiency of polyhedrin mRNA.

The sequences of the first 55 bases of the 5' leader region are highly conserved in the polyhedrin genes examined,

```
                                                  10                                                          20                                                          30
AcMNPV  ATG --- CCG GAT TAT TCA TAC CGT CCC ACC ATC GGG CGT ACC TAC GTG TAC GAC AAC AAG TAC TAC AAA AAT TTA GGT GCC GTT ATC AAG   90
OpMNPV  ATG --- ..A ... ..C ..G ... ..G ..G ... ..T ..T ..C ... ... ... ... ... ... ..A ... ... ... ..C ..G ..C T.. ..C ... ..A
OpSNPV  ATG TAT A.T CGA ..C AGC ... AAC ..G T.A C.G ..T ..C ..T ..T ... ... ... ... ... ..T ... ... ... ... ... ... ..C ..T ...

                                                  40                                                          50                                                          60
AcMNPV  AAC GCT AAG CGC AAG AAG CAC TTC GCC GAA CAT GAG ATC GAA GAG GCT ACC CTC GAC CCC CTA GAC AAC TAC CTA GTG GCT GAG GAT CCT  180
OpMNPV  ... ..C ... ... ... ... ..T C.T CTA ... ..C ..A GAG ..T ..A AAG CA. ..G ... ..G ... ... C.. ... A.G ..C ..C ... ..C ..C
OpSNPV  ..T ... ... ... ... ..A ... CAG ATT ..G ..C ..A GCT ... ... CAC ..G ... ... ..A ..G ... ..A ... ... ..C ..C ..A ... ...

                                                  70                                                          80                                                          90
AcMNPV  TTC CTG GGA CCC GGC AAG AAC CAA AAA CTC ACT CTC TTC AAG GAA ATC CGT AAT GTT AAA CCC GAC ACG ATG AAG CTT GTC GTT GGA TGG  270
OpMNPV  ... T.. ..C ... ... ..A ... ... ... ..G ..C ..G ..T ..A ... ..T ..C ..C ..C ..G ... ... ... ... ... ..C A.. ..C AAC ...
OpSNPV  ..T T.T ... ..T ... ..A ... ... ... ... ... T.G ..T ..A ... ..T ..C ... ..A ... ... ... ..T ... ..A T.A ..A ..A AAC ...

                                                  100                                                         110                                                         120
AcMNPV  AAA GGA AAA GAG TTC TAC AGG GAA ACT TGG ACC CGC TTC ATG GAA GAC AGC TTC CCC ATT GTT AAC GAC CAA GAA --- GTG ATG GAT GTT  360
OpMNPV  .GC ..C ..G ... ..T .TA C.C ... ... ... ..G ... ... G.. ... ... ... ... ... ..C ..C ... ... ... ... --- ..A ... ..C ..G
OpSNPV  .GC ..C ... ... ..T CTG ... C.. ... ... ..G ... ... ... ..G ... ..T ..T .T. ... ... ... ..T ... ... --- A.C ... ..C ..G

                                                  130                                                         140                                                         150
AcMNPV  TTC CTT GTT GTC AAC ATG CGT CCC ACT AGA CCC AAC CGT TGT TAC AAA TTC CTG GCC CAA CAC GCT CTG CGT TGC GAC CCC GAC TAT GTA  450
OpMNPV  ..T ..C ..C ... ... ... ... ... ..G C.C ... ... ..C ..C ... ... ... T.. ..G ... ... ..A ..C A.G ..G ... TG. ... ..C ..G
OpSNPV  ..T T.A ... A.T ... ... ... ... ..C ..G ... ... ..C ... .T. .G. ..T T.. ..T ... ... ... T.. ... ... ..T ... ..G ..C ..T

                                                  160                                                         170                                                         180
AcMNPV  CCT CAT GAC GTG ATT AGG ATC GTC GAG CCT TCA TGG GTG GGC AGC AAC AAC GAG TAC CGC ATC AGC CTG GCT AAG AAG GGC GGC GGC TGC  540
OpMNPV  ..C ..C ..G ... ... ... ..T ... ... ..G ..G .AC ... ... .TG ... ... ... ... ... ... ... ... ..C ..A ..A ... ... ... ...
OpSNPV  ..C ..C ..A ..A ... C.C ..T ..A ... ..C AGC .AC ..A ... ... ... ... ... ... A.A T.A GTT ... ... ..A CGC ... ... ..T ..T
```

```
                                         190                                    200                                    210
AcMNPV   CCA ATA ATG AAC CTT CAC TCT GAG TAC ACC AAC TCG TTC GAA CAG TTC ATC GAT CGT GTC ATC TGG GAG AAC TTC TAC AAG CCC ATC GTT  630
OpMNPV   ..T ..C ... ... A.. ... G.C G.A ... ... ... ... ..T ... TC. ... G.A A.C ..C ... ... ... ... ... ..T ..T ... ... ... ..G
OpSNPV   ..C G.G ... ... T.A ACA G.C ... ... ... ... ... ..T ..T ..G G.. ..T ..T A.C ..C ..T CAT ... ... ... ... ... ..A ... .CT ..G

                                         220                                    230                                    240
AcMNPV   TAC ATC GGT ACC GAC TCT GCT GAA GAG GAG GAA ATT CTC CTT GAA GTT TCC CTG GTG TTC AAA GTA AAG GAG TTT GCA CCA GAC GCA CCT  720
OpMNPV   ... ..T ..C ..G ..T ..G AGC ..G ... ... ... ... ... A.C ..G ..G ..G ..C ... ..T ..G ... ..A ... ... ..G ..C ... ..G ..G
OpSNPV   ... G.. ..C ..A ..T ... ..C ..G ..A ..A ... ..A T.A ... ... ... ..G ..T C.. ... ... A.T ..A ..A ... ... ... ... ... ..C

                             247
AcMNPV   CTG TTC ACT GGT CCG GCG TAT  741
OpMNPV   T.. ... ..C ..C ... ... ...
OpSNPV   T.. .A. TCA ..C ... ..T ...
```

FIGURE 9. Nucleotide sequence of A. californica MNPV and O. pseudotsugate MNPV and SNPV polyhedrin genes. Sequence data were provided by Hooft van Iddekinge et al. (1983) and by G. F. Rohrmann (personal communication). The dots refer to the nucleotides indicated for AcMNPV. Deletions in the nucleotide or amino acid sequences (Fig. 5) are indicated (---).

TABLE III. Codon Usage in Polyhedrin Genes[a]

	AcMNPV	OpMNPV	OpSNPV		AcMNPV	OpMNPV	OpSNPV
UUU Phe	1	6	10	UCU Ser	2	0	2
UUC Phe	13	7	4	UCC Ser	1	1	2
UUA Leu	1	1	6	UCA Ser	2	0	2
UUG Leu	0	3	4	UCG Ser	1	6	1
CUU Leu	4	2	2	CCU Pro	5	2	2
CUC Leu	4	5	3	CCC Pro	9	7	9
CUA Leu	2	2	1	CCA Pro	2	2	2
CUG Leu	6	4	5	CCG Pro	2	5	2
AUU Ile	3	7	9	ACU Thr	4	1	6
AUC Ile	10	8	1	ACC Thr	6	5	2
AUA Ile	1	0	1	ACA Thr	0	0	2
AUG Met	6	7	6	ACG Thr	1	4	2
GUU Val	8	1	6	GCU Ala	5	0	6
GUC Val	4	9	3	GCC Ala	1	4	0
GUA Val	2	3	6	GCA Ala	2	2	2
GUG Val	6	6	4	GCG Ala	3	3	4

[a]*The codon usage for polyhedrin genes of A. californica MNPV, O. pseudotsugata MNPV and SNPV is listed as the number of times each triplet appears in the plus strand of the DNA sequence (Fig. 9). The DNA triplets are converted into RNA codons.*

with some regions having almost identical sequences. The size of this 5' leader sequence is similar (40-80 nucleotides) to that for most cellular messages reported so far (Kozak, 1981). The cap sites on the DNA terminate the highly homologous, A-T-rich 5' flanking region. Upstream from the cap sites, the sequences in different viruses diverge considerably. The nucleotides flanking the AUG initiation codon only partially follow the $^{G}_{A}$NNATGG consensus sequence that is thought to be the optimal context for efficient initiation in eukaryotes (Kozak, 1983). The -3 position is a purine (A), but at +4 only pyrimidines are found.

Although Carstens et al. (1979) reported processing of polyhedrin protein, an ATG codon was not found in any of the 5' flanking regions. Therefore, processing of polyhedrin appears to be unlikely.

In AcMNPV, sequences similar to TATA boxes that are possibly involved in recognition by RNA polymerase, are found in this upstream region at -20 and -50 from the cap site. TATA-like sequences are also found in OpMNPV and OpSNPV at about -35 and -65 nucleotides upstream from the cap site, respectively. The role of these TATA-like sequences in polyhedrin mRNA transcription remains to be determined. In addition, in AcMNPV

	AcMNPV	*OpMNPV*	*OpSNPV*		*AcMNPV*	*OpMNPV*	*OpSNPV*
UAU Tyr	*3*	*2*	*4*	*UGU Cys*	*1*	*0*	*2*
UAC Tyr	*12*	*13*	*12*	*UGC Cys*	*2*	*3*	*1*
UAA -	-	-	-	*UGA -*	-	-	-
UAG -	-	-	-	*UGG Trp*	*4*	*4*	*3*
CAU His	*2*	*1*	*1*	*CGU Arg*	*7*	*1*	*2*
CAC His	*2*	*6*	*5*	*CGC Arg*	*3*	*9*	*0*
CAA Gln	*3*	*3*	*4*	*CGA Arg*	*0*	*0*	*1*
CAG Gln	*2*	*0*	*1*	*CGG Arg*	*0*	*1*	*0*
AAU Asn	*2*	*0*	*3*	*AGU Ser*	*0*	*0*	*1*
AAC Asn	*12*	*15*	*13*	*AGC Ser*	*3*	*4*	*4*
AAA Lys	*7*	*10*	*13*	*AGA Arg*	*1*	*0*	*2*
AAG Lys	*12*	*9*	*4*	*AGG Arg*	*2*	*2*	*2*
GAU Asp	*4*	*3*	*4*	*GGU Gly*	*3*	*1*	*3*
GAC Asp	*11*	*11*	*7*	*GGC Gly*	*5*	*10*	*7*
GAA Glu	*10*	*10*	*11*	*GGA Gly*	*3*	*0*	*1*
GAG Glu	*11*	*12*	*12*	*GGG Gly*	*1*	*0*	*0*

two direct repeats are found in the 5' flanking region about 100 bases upstream from the cap site. These direct repeats may be involved in efficient and accurate transcription (Dierks et al., 1983). Tentative CAAT boxes, whose role in transcription is still obscure, are found distal to the tentative TATA boxes. In addition, there are frequently observed repeats of DNA segments varying from 2 to 7 nucleotides in the noncoding regions. The incorporation of such small repeats into the genome may have contributed to the expansion of the baculovirus genome. Such duplications are frequently seen in other genes, e.g., insect chorion protein genes (Jones and Kafatos, 1982).

The 3' regions flanking the polyhedrin gene at AcMNPV, OpMNPV, and OpSNPV are of diverse sequence and are of different lengths. The OpSNPV polyhedrin mRNA may have a noncoding region as short as 20 nucleotides before the poly(A) tail begins, compared to approximately 450 for AcMNPV and 170 for OpMNPV polyhedrin (Fig. 7). As yet the meaning of this variation is unclear. Several AATAA-like termination signals have been found in the OpSNPV flanking region (not shown), but the establishment of the actual termination site awaits mapping of the mRNA on the viral chromosome. The 3' flanking regions of other polyhedrin genes have not been sequenced to completion.

It should be emphasized here that the 5' flanking region containing the promoter for polyhedrin mRNA must have some peculiar features that make it one of the strongest eukaryotic promoters. Alteration of the promoter region by introducing mutations or deletions at specific sites might result in measurable effects on polyhedrin mRNA synthesis, and thereby may shed some light on the working mechanism of this active promoter.

F. POLYHEDRIN MUTANTS

Mutants of AcMNPV that affect polyhedrin formation or morphology have been produced by treatment with mutagens (Brown et al., 1979; Lee and Miller, 1979; Duncan and Faulkner, 1982) or by genetic engineering (Smith et al., 1983a). Several types of mutants have been described: (1) temperature-sensitive (*ts*) mutants, (2) morphology (*m*) mutants, and (3) deletion (*d*) mutants.

The *ts* mutants of AcMNPV produce polyhedra in a normal way at the permissive temperature (27°C), while the production of polyhedra at the nonpermissive temperature (33°C) is altered, reduced, or impaired. Most of these mutants are sorted into several complementation groups (Grown et al., 1979; Lee and Miller, 1979). Brown and Faulkner (1980) produced a partial genetic map of their mutants involving a series of two-factor crosses, while Miller (1981) constructed a genetic map of her *ts* mutants by marker-rescue with DNA restriction fragments with known positions on the AcMNPV genome. From these data it is apparent that gene products of more than one complementation groups are involved in polyhedron formation. One of the *ts* mutants (*ts* B821) described by Miller (1981) was thought to be a putative polyhedrin mutant, but it now appears to be defective in an early function (Miller et al., 1983). It is not known whether any of the described *ts* mutants code for a temperature-sensitive polyhedrin.

To date, four mutants have been described (Brown et al., 1979; Duncan and Faulkner, 1982), that have an altered polyhedron morphology at the normal temperature (27°C). Two of these mutants, designated *m-5* and *m-6*, produce one large cuboidal inclusion per cell (Brown et al., 1979), while the other two show irregular condensation of polyhedrin into irregularly shaped or amorphous structures in the nucleus (Duncan and Faulkner, 1982).

The polyhedron morphology mutant *m*-5 appeared to code for an altered polyhedrin of slightly lower molecular weight than wild-type (wt) polyhedrin and with a different peptide map following digestion with proteolytic enzymes (Brown et al., 1980).

M-5 is not infectious for insect larvae and the cuboidal occlusion body contains no virions. Analysis of the viral DNA of this mutant indicated that, among other peculiarities, the *Bam*HI restriction site at map position 3.0 (between fragments *Bam*HI-F and -B) (Fig. 3) has disappeared (Carstens, 1982). It has recently been demonstrated that this mutation is due to a single nucleotide change in the *Bam*HI site converting a codon for proline (CCT) to leucine (CTT) at position 60 (Fig. 9) (E. H. Carstens, personal communication). Proline is often associated with breaking α helices or β sheets in proteins (Chou and Fasman, 1978) and it is likely that the loss of this proline substantially alters the secondary structure. This does not explain the lower molecular weight of the *m-5* polyhedrin as reported by Brown et al. (1980).

A spontaneous revertant of this *m-5* mutant, designated *m-5R*, reverted to the wild-type polyhedron morphology and also possessed a polyhedrin similar in size to wild type. The mutant, however, apparently retained some other characteristic of the *m-5* mutant, i.e., a 400 base pair insertion elsewhere on the viral genome (Carstens, 1982). The other polyhedron morphology mutant, *m-6*, formed a similar large cuboidal inclusion body as *m-5*. However, differences in polyhedrin amino acid sequence compared to the wild-type polyhedrin could not be detected by the conventional fingerprinting techniques (Brown et al., 1980). Amino acid or DNA sequencing of *m-6*, and the revertant of *m-5*, *m-5R*, should indicate which other amino acids are important in forming the correct paracrystalline lattice of the wild-type polyhedron.

Recently, a category of genetically engineered deletion (*d*) mutants have been produced that will provide a powerful tool in the study of polyhedrin gene structure and function (Smith et al., 1983a). These *d*-mutants were generated by deleting with endonuclease *Bal*31 parts of the polyhedrin gene at the *Kpn*I site (Fig. 3). These *d*-mutants were isolated by selecting for plaques that were, upon transfection with wild-type DNA, devoid of polyhedra. It appears that truncated polyhedrins are synthesized in cells infected with these *d*-mutants. In addition, polyhedrin genes from SfMNPV and TnSNPV have transferred to the AcMNPV *d*-mutant genome by cotransfection of insect cells followed by *in vivo* recombination. The Sf/AcMNPV recombinant showed reduced efficiency of occlusion of AcMNPV virions as compared to Ac/Ac virions, whereas, the recombinant TnSNPV polyhedrin gene produced crystals devoid of virions (G. E. Smith, personal communication) This suggests that occlusion involves a delicate relationship between the virion (envelope) and its homologous polyhedrin.

V. POLYHEDRIN EVOLUTION

Comparison of the amino acid sequences of proteins which have the same function (orthologous proteins) in different organisms reveals that phylogenies derived from the degree of sequence homology are similar to those derived by other means, such as analysis of the fossil record. Consequently, molecular phylogenies reflect the phylogenies of the whole organisms (Wilson et al., 1977). Moreover, it has been found that orthologous proteins evolve at constant rates although each family of orthologous protein has its own distinct rate constant. For example, histone H4 is very stable and changes 1% in amino acid sequence every 400×10^6 years since it diverged from a common ancestor, whereas cytochrome c evolves at 1% every 15×10^6 years (Wilson et al., 1977). Since the polyhedrin gene is located in the baculovirus genome and codes for a protein with the same function in all baculoviruses, comparison of polyhedrin sequences offers a means for establishing a baculovirus phylogeny. Such an approach has also been used to examine papovavirus phylogeny and it was demonstrated that these viruses diverged and evolved with their hosts (Soeda et al., 1980).

Baculovirus phylogeny, based on molecular-evolutionary analysis of polyhedrin sequences, is consistent with baculovirus distribution (Table I) and appears to be related to the evolution of insects. For example, the Lepidoptera are the most recently evolved order of insects (Reik, 1970). They first appeared $40\text{-}60 \times 10^6$ years ago and subsequently underwent extensive speciation. This proliferation of lepidopteran viruses may indicate that the divergence of the Lepidoptera provided NPVs with a variety of new hosts which facilitated their evolution. The multiply-enveloped NPVs are present only in the Lepidoptera (Table I) and are, therefore, likely to have appeared since the Lepidoptera diverged. In addition, sequence data and antigenic relatedness (Rohrmann et al., 1981) of lepidopteran polyhedrins indicate a closer phylogenetic relatedness to one another than to polyhedrins from the other two insect orders. Analysis of amino acid homologies (Table IV) indicates the AcMNPV, GmMNPV, and OpMNPV polyhedrins are closely related (87-92%). AcMNPV and GmMNPV are variants of the same virus as indicated by their similar restriction enzyme fragment patterns (Smith and Summers, 1979), and ability to form recombinants *in vivo* (Croizier et al., 1980). In addition, when Rohrmann et al. (1982a) compared DNA homologies between AcMNPV, OpMNPV, and OpSNPV, they found the AcMNPV and OpMNPV to be more closely related to each other than to the OpSNPV. These homology studies are consistent with the phylogeny determined from polyhedrin sequences (Fig. 10). Both the

TABLE IV. Amino Acid Homology Percentage[a]

	GmMNPV	OpMNPV	LdMNPV	BmSNPV	OpSNPV
AcMNPV	92	90	79	84	84
GmMNPV		87	82	87	80
OpMNPV			78	85	85
LdMNPV				80	77
BmSNPV					80

[a]The amino acid homology between the polyhedrins of A. californica MNPV, G. mellonella MNPV, L. dispar MNPV, O. pseudotsugata MNPV and SNPV, and B. mori SNPV is expressed as the percentage of amino acids conserved in the sequence (see Fig. 5). Deletions or insertions occurring in the polyhedrins were not taken into account in the comparisons.

Op- and BmSNPVs appear to have diverged earlier than the Op-, Gm-, and AcMNPVs. The LdMNPV appears least related to all these viruses. This would suggest either that MNPVs evolved more than once from SNPVs, or a LdMNPV virus obtained an SNPV polyhedrin through recombination with a SNPV.

Granulosis virus evolution is integrally involved with NPV phylogeny. The GVs appear to have diverged from a SNPV after the divergence of the Hymenoptera viruses but much earlier than the appearance of MNPVs (Fig. 10). The presence of the large number of GVs within the Lepidoptera suggests that they may have undergone a speciation similar to that postulated for the lepidopteran NPVs.

In contrast to the recent speciation of the Lepidoptera, the Diptera and Hymenoptera are ancient insect orders which have been in existence 190-230 × 10^6 years (Reik, 1970). The presence of only SNPVs in the Diptera and Hymenoptera (Table I) and the large phylogenetic distance between them and the lepidopteran NPVs (Fig. 10) suggests that NPVs have either evolved along with, or have had an ancient relationship with their insect hosts and that the infectivity of baculoviruses has been confined to their respective order since the divergence occurred. The data from the molecular evolution of polyhedrin are not consistent with a recent horizontal invasion by baculovirus of the insect orders or a random exchange of the polyhedrin gene between viruses of different orders, because this would result in similar phylogenetic distances between the proteins from the three orders (Rohrmann et al., 1981).

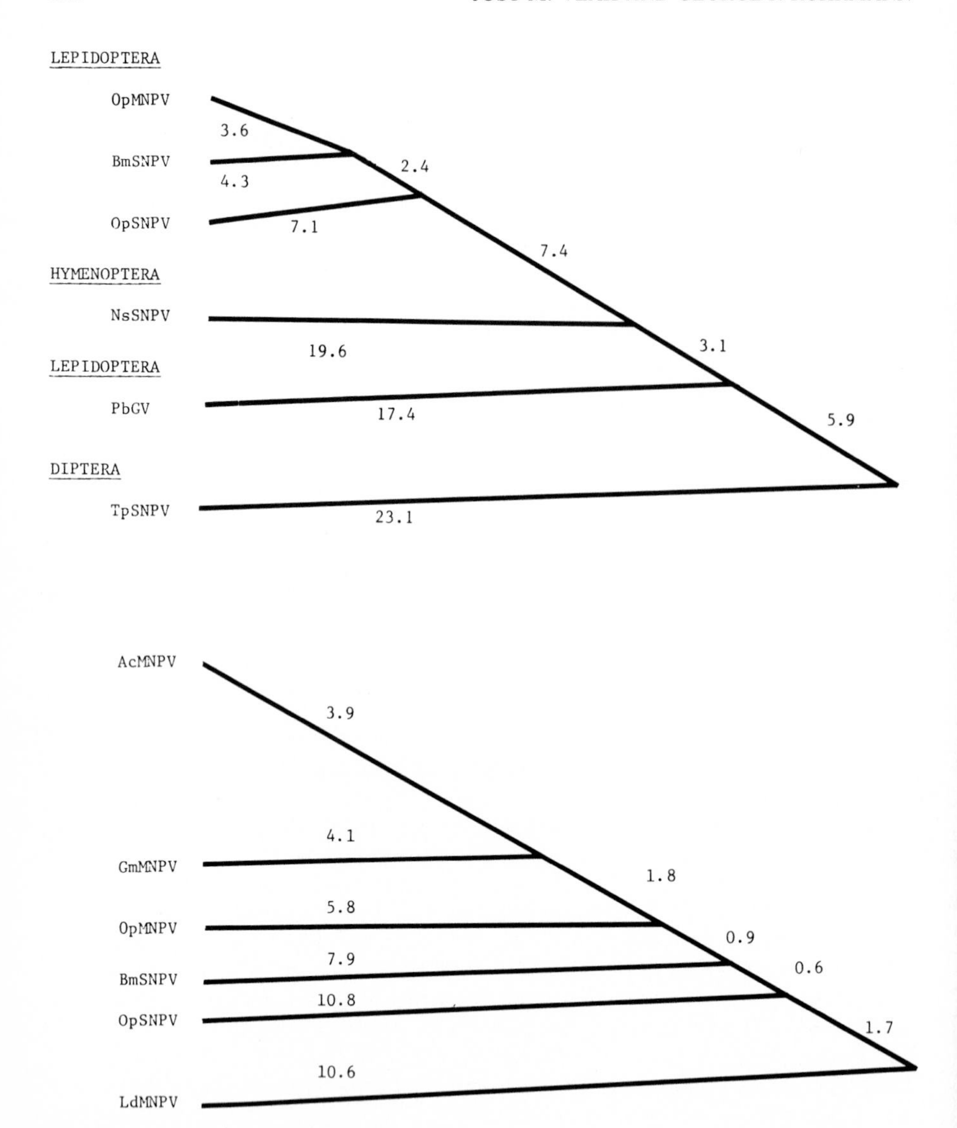

FIGURE 10. Molecular phylogeny of baculovirus polyhedrins. (Top) Phylogenies determined from N-terminal amino acid sequences. The numbers represent the average phyletic distance in nucleotide substitutions between points of branching (see Rohrmann et al., 1981). (Bottom) Polyhedrin phylogeny determined from six complete polyhedrin sequences. The numbers represent the average phyletic distance in percentage amino acid differences between points of branching. Phylogenies were determined using the method of Fitch (1977). For abbreviations see Fig. 4.

VI. CONCLUDING REMARKS

Polyhedrin, which constitutes the matrix of baculovirus occlusion bodies, plays a fascinating role in the replicative cycle of baculoviruses. It has evolved two highly specialized structural properties that allow it to function in the virus life cycle. First, it has the ability to form a protective crystal around the virions in the insect cell nucleus; second, it resists solubilization except under strongly alkaline conditions similar to that found in the insect gut. Both of these properties allow the virions to remain viable for many years outside the insect host. In addition to these conserved structural properties, it also has conserved a regulatory process which results in the synthesis of massive amounts of polyhedrin in a highly coordinated fashion. This process may be governed by an α-amanitin-resistant RNA polymerase which may appear during viral infection. It is possible to derive molecular phylogenies from polyhedrin sequences. These phylogenies may prove to be the most sensitive methods for understanding the relatedness of this diverse virus group, and may elucidate the evolution of these viruses in relation to their insect hosts. These unique features of polyhedrin were sufficient incentive to devote an entire review to this protein.

At present, our knowledge of the physical structure of polyhedrin and its multimers is limited and the actual three-dimensional conformation of polyhedrin in the paracrystalline matrix is unknown. The primary amino acid sequences available (Fig. 5) now make it possible to investigate the secondary and tertiary structure of polyhedrin using computer modeling and could also assist in definitive x-ray crystallographic studies. In particular, the role of the three cysteines is of interest. Two of these can form intramolecular S—S bridges, while the third one is available for intermolecular S—S bridge formation. Such studies may shed light on the inter- and intramolecular linkages in polyhedrin complexes and eventually on its structure in the paracrystalline occlusion body lattice.

The formation of the polyhedrin lattice and the occlusion of virus particles in polyhedra is an accurate and highly ordered process. Certain regions of the protein are highly conserved whereas others vary considerably. A single amino acid change, altering proline to leucine, in one of the conserved regions of the *m-5* mutant of brown et al. (1980) influences not only the crystallization process and ultimately the morphology of the polyhedrin, but also prevents the virus particles from proper occlusion in a polyhedron (E. H. Carstens, personal communication). Using recombinant DNA technology, it should be possible to construct polyhedrin mutants with point mutations or deletions in specific regions of the polyhedrin gene in order to determine which amino acid sequences are essential for aggregation, crystal formation, and for viral occlusion.

An interesting aspect in the synthesis of polyhedrin is the role of virus-coded factors in the efficient transcription of this gene. Polyhedrin is under stringent temporal control of the virus and its synthesis follows the production of infectious virus late in the replication cycle. At that time polyhedrin is abundantly expressed, suggesting that its gene contains a strong promoter. The gene is not transcribed in the early phase of infection (Erlandson and Carstens, 1983) when the majority of the viral genome is transcribed, presumably with host enzymes. As late transcription takes place, a new α-amanitrin-insensitive RNA polymerase activity has been identified (Grula et al., 1981). It is possible that the polyhedrin promoter is specifically recognized by this new RNA polymerase. In order to understand the high level of expression of polyhedrin it will be necessary to identify and understand the function of recognition sequences in the promoter region of this gene.

A recent and important advance in baculovirology is the engineering of recombinant baculovirus vectors using the polyhedrin gene as the acceptor for foreign genes. As tabulated by Miller et al. (1983), a number of unique features make baculoviruses attractive as recombinant DNA vector systems. (1) Baculoviruses are replicative entities with a double-stranded DNA genome; (2) the genome and rod-shaped particle seem expandable; and (3) baculoviruses contain at least one gene (polyhedrin) that is highly expressed, but nonessential for the production of infectious virus, and thus deletable (Smith et al., 1983a). Recently, the genes for human β-interferon (Smith et al., 1983c) and bacterial β-glactosidase (Pennock et al., 1984) have been transferred using recombinant DNA techniques into the polyhedrin gene locus of AcMNPV DNA and brought under the control of the potent polyhedrin promoter. Both β-interferon and β-glactosidase are apparently produced at levels comparable to polyhedrin. The potential of these recombinant baculoviruses as expression vectors for foreign genes as well as their commercial application will become clear in the near future. The potential of baculoviruses as insect biocontrol agents may also be expanded. For example, one can envisage recombinant baculoviruses produced which contain genes for proteinaceous toxins (e.g., *Bacillus thuringiensis* δ-endotoxin). Upon expression, these toxins would enhance or accelerate the insecticidal effect of the virus.

ACKNOWLEDGMENTS

We wish to thank Marja van Steenbergen and Kitty Dougherty for excellent typing support during the preparation of this manuscript. We are also indebted to Drs. Dick Peters and Michael Nesson for critically reading the manuscript. This paper was supported by a grant from the Stichting "Fonds Landbouw Export Bureau 1916/1918" to JMV and NIH Grant ES02129 to GFR. We thank Gale E. Smith and Eric Carstens for sharing the nucleotide sequence of AcMNPV polyhedrin gene prior to publication.

REFERENCES

Adang, M. J., and Miller, L. K. (1982). Molecular cloning of DNA complementary to mRNA of the baculovirus *Autographa californica* nuclear polyhedrosis virus: Location and gene products of RNA transcripts found late in infection. *J. Virol. 44*, 782-793.

Aratake, Y., and Watanabe, H. (1973). A newly discovered polyhedrosis of the small blackish cochlid, *Scopelodes contracta* Walker. *Japan J. Appl. Entomol. Zool. 17*, 132-136.

Bennetzen, J. L., and Hall, B. D. (1982). Codon selection in yeast. *J. Biol. Chem. 257*, 3026-3031.

Bergold, G. H. (1943). Ueber Polyederkrankheiten bei Insekten. *Biol. Zentralbl. 63*, 1-55.

Bergold, G. H. (1947). Die Isolierung des Polyedervirus und die Natur der Polyeder. *Z. Naturforsch 2b*, 122-143.

Bergold, G. H. (1948). Ueber die Kapselvirus-krankheit. *Z. Naturforsch. 3b*, 338-342.

Bergold, G. H. (1953). Insect viruses. *Advan. Virus Res. 1*, 91-139.

Bergold, G. H., and Schramm, G. (1942). Biochemische Charakterisierung von Insektenviren. *Biol. Zentralbl. 62*, 105-118.

Berk, A. J., and Sharp, P. A. (1978). Spliced early mRNA of simian virus 40. *Proc. Natl. Acad. Sci. U.S. 76*, 1274-1278.

Bilimoria, S. L., and Arif, B. M. (1979). Subunit protein and alkaline protease of entomopox spheroids. *Virology 96*, 596-603.

Bolle, J. (1894). Il giallume o il mal del grasso del baco da seta. Notizia preliminare. *Atti Mem. I.R. Soc. Agr. Gorizia 33*, 133-136.

Braweman, G. (1981). The role of the poly(A) sequence in mammalian messenger RNA. *CRC Critical Rev. Biochem. 10*, 1-38.

Brown, M., and Faulkner, P. (1980). A partial genetic map of the baculovirus, *Autographa californica* nuclear polyhedrosis virus, based on recombination studies with *ts* mutants. *J. Gen. Virol. 48*, 247-251.

Brown, M., Crawford, A. M., and Faulkner, P. (1979). Genetic analysis of a baculovirus, *Autographa californica* nuclear polyhedrosis virus. I. Isolation of temperature-sensitive mutants and assortment into complementation groups. *J. Virol. 31*, 190-198.

Brown, M., Faulkner, P., Cochran, M. A., and Chung, K. L. (1980). Characterization of two morphology mutants of *Autographa californica* nuclear polyhedrosis virus with large cuboidal inclusion bodies. *J. Gen. Virol. 50*, 309-316.

Carstens, E. H. (1982). Mapping the mutation site of an *Autographa californica* nuclear polyhedrosis virus polyhedron morphology mutant. *J. Virol. 43*, 809-818.

Carstens, E. B., Tjia, S. T., and Doerfler, W. (1979). Infection of *Spodoptera frugiperda* cells with *Autographa californica* nuclear polyhedrosis virus. I. Synthesis of intracellular proteins after virus infection. *Virology 99*, 386-398.

Chou, P. Y., and Fasman, G. D. (1978). Prediction of the secondary structure of proteins from their amino acid sequence. *Advan. Enzymol. 47*, 45-148.

Cibulski, R. J., Harper, J. D., and Gudauskas, R. T. (1977). Biochemical comparison of polyhedral protein from five nuclear polyhedrosis viruses infecting *Plusiine* larvae (Lepidoptera: Noctuidae). *J. Invertebr. Pathol. 29*, 182-191.

Cochran, M., and Faulkner, P. (1983). Location of homologous DNA sequences interspersed at five regions in the *Baculovirus* AcMNPV genome. *Virology 45*, 961-970.

Cochran, M. A., Carstens, E. B., Eaton, B. T., and Faulkner, P. (1982). Molecular cloning and physical mapping of restriction endonuclease fragments of *Autographa californica* nuclear polyhedrosis virus DNA. *J. Virol. 41*, 940-946.

Cornalia, E. (1856). Monografia del bombice del gelso. *Mem. Rend. Inst. Lombardo Sci. Lett. Arte 6*, 1-387. (Parte 4: Patol. del baco, pp. 322-366.)

Couch, J. A. (1974). An enzootic nuclear polyhedrosis virus of pink shrimp: ultrastructure, prevalence, and enhancement. *J. Invertebr. Pathol. 24*, 311-331.

Croizier, G., and Croizier, L. (1977). Évaluation du poids moléculaire de la protéine des corps d'inclusion de divers baculovirus d'insectes. *Arch. Virol.* *55*, 247-250.

Croizier, G., Godse, D., and Vlak, J. (1980). Sélection de types viraux dans les infections doubles a Baculovirus chez les larves de Lépidoptère. *C.R. Acad. Sci., serie D, 290*, 579-582.

Cunningham, J. C. (1970). Strains of nuclear polyhedrosis viruses displaying different inclusion body shapes. *J. Invertebr. Pathol.* *16*, 299-300.

Dierks, P., Van Ooyen, A., Cochran, M. D., Dobkin, C., Reiser, J., and Weissman, C. (1983). Three regions upstream from the cap site are required for efficient and accurate transcription of the rabbit β globin gene in mouse 3T6 cells. *Cell* *32*, 695-706.

Dobos, P., and Cochran, M. A. (1980). Protein synthesis in cells infected by *Autographa californica* nuclear polyhedrosis virus (Ac-NPV): The effect of cytosine arabinoside. *Virology* *103*, 446-464.

Dognin, M. J., and Wittmann-Liebold, B. (1977). The primary structure of L 11, the most heavily methylated protein from *Escherichia coli* ribosomes. *FEBS Lett.* *84*, 342-346.

Duncan, R., and Faulkner, P. (1982). Bromodeoxyuridine-induced mutants of *Autographa californica* nuclear polyhedrosis virus defective in occlusion body formation. *J. Gen. Virol.* *62*, 369-373.

Engstrom, A. (1974). The arrangement of the protein molecules in nuclear-polyhedrosis inclusions. *Biochem. Exp. Biol.* *11*, 7-13.

Eppstein, D. A., and Thoma, J. A. (1975). Alkaline protease associated with the matrix protein of a virus infecting the cabbage looper. *Biochem. Biophys. Res. Commun.* *62*, 478-484.

Eppstein, D. A., and Thoma, J. A. (1977). Characterization and serology of the matrix protein from a nuclear-polyhedrosis virus before and after degradation by an endogenous protease. *Biochem. J.* *167*, 321-332.

Eppstein, D. A., Thoma, J. A., Scott, H. A., and Young III, S. Y. (1975). Degradation of matrix protein from a nuclear polyhedrosis virus of *Trichoplusia ni* by an endogenous protease. *Virology* *67*, 591-594.

Erlandson, M. A., and Carstens, E. B. (1983). Mapping of early transcription products of *Autographa californica* nuclear polyhedrosis virus. *Virology* *126*, 398-402.

Esche, H., Lübbert, H., Siegmann, B., and Doerfler, W. (1982). The translational map of the *Autographa californica* nuclear polyhedrosis virus (AcNPV) genome. *EMBO J.* *1*, 1629-1633.

Faulkner, P. (1981). Baculoviruses. *In* "Pathogenesis of Invertebrate Microbial Diseases" (E. W. Davidson, ed.), pp. 3-37. Allanheld, New Jersey.

Faulkner, P., and Henderson, J. F. (1972). Serial passage of a nuclear polyhedrosis disease virus of the cabbage looper (*Trichoplusia ni*) in a continuous tissue culture cell line. *Virology 50*, 920-924.

Fitch, W. M. (1977). The phyletic interpretation of macromolecular sequence information: Simple methods. *In* "Major Patterns in Vertebrate Evolution" (M. K. Hecht, P. C. Goody, and B. M. Hecht, eds.), pp. 169-204. Plenum Press, New York.

Furuichi, Y., and Mirua, K. (1975). A blocked structure at the 5' terminus of mRNA from cytoplasmic polyhedrosis virus. *Nature (London) 253*, 374-375.

Gershenson, S. (1958). The variability of polyhedral viruses. *Trans. 1st Intern. Conf. Insect Pathol. Biol. Control, Prague*, pp. 198-200.

Glaser, R. W., and Chapman, J. W. (1916). The nature of the polyhedral bodies found in insects. *Biol. Bull. 30*, 367-391.

Granados, R. R. (1980). Infectivity and mode of action of baculoviruses. *Biotechnol. Bioeng. 22*, 1377-1405.

Granados, R. R. (1981). Entomopoxvirus infections in insects. *In* "Pathogenesis in Invertebrate Microbial Diseases" (E. W. Davidson, ed.), pp. 101-126. Allanheld, New Jersey.

Grula, M. A., Buller, P. L., and Weaver, R. F. (1981). α-Amanitin-resistant viral RNA synthesis in nuclei isolated from nuclear polyhedrosis virus-infected *Heliothis zea* larvae and *Spodoptera frugiperda* cells. *J. Virol. 38*, 916-921.

Gruss, P., Efstratiadis, A., Karatha Nasis, S., König, M., and Khoury, G. (1981). Synthesis of stable unspliced mRNA from an intronless simian virus 40-rat preproinsulin gene recombinant. *Proc. Natl. Acad. Sci. U.S. 78*, 6091-6095.

Guelpa, B., Bergoin, M., and Croizier, G. (1977). La protéine d'inclusion et les protéines du virion du baculovirus du diptère *Tipula paludosa* (Meigen). *C.R. Acad. Sci., serie D, 284*, 779-782.

Harrap, K. A. (1972). The structure of nuclear polyhedrosis viruses. II. The virus particle. *Virology 50*, 124-132.

Harrap, K. A., and Payne, C. C. (1979). The structural properties and identification of insect viruses. *Advan. Virus Res. 25*, 273-355.

Harrap, K. A., Payne, C. C., and Robertson, J. S. (1977). The properties of three baculoviruses from closely related hosts. *Virology 79*, 14-31.

Hohmann, A. W., and Faulkner, P. (1983). Monoclonal antibodies to baculovirus structural proteins: Determination of specificities by Western blot analysis. *Virology 125*, 432-444.

Hooft van Iddekinge, B. J. L., Smith, G. E., and Summers, M. D. (1983) Nucleotide sequence of the polyhedrin gene of *Autographa californica* nuclear polyhedrosis virus. *Virology 131*, 561-565.

Hopp, T. P., and Woods, K. R. (1981). Prediction of protein antigenic determinants from amino acid sequences. *Proc. Natl. Acad. Sci. U.S. 79*, 3824-3828.

Howley, P. M., Israel, M. A., Law, M., and Martin, M. A. (1979). A rapid method for detecting and mapping homology between heterologous DNAs. *J. Biol. Chem. 254*, 4876-4883.

Jewell, J. E., and Miller, L. K. (1980). DNA sequence homology relationships among six lepidopteran nuclear polyhedrosis viruses. *J. Gen. Virol. 48*, 161-175.

Jones, C. W., and Kafatos, F. C. (1982). Accepted mutations in a gene family: Evolutionary diversification of duplicated DNA. *J. Mol. Evol. 19*, 87-103.

Jurkovícová, M., Van Touw, J. H., Sussenbach, J. S., and Ter Schegget, J. (1979). Characterization of the nuclear polyhedrosis virus DNA of *Adoxophyes orana and of Barathra brassicae*. *Virology 93*, 8-19.

Kawamoto, F., and Asayama, T. (1975). Studies on the arrangement patterns of nucleocapsids within the envelope of nuclear polyhedrosis virus in the fat body cells of the brown tail moth *Euproctis similis*. *J. Invertebr. Pathol. 26*, 47-55.

Kelly, D. C. (1977). The DNA contained by nuclear polyhedrosis viruses isolated from four *Spodoptera* sp. (Lepidoptera: Noctuidae): Genome size and homology assessed by DNA renaturation kinetics. *Virology 76*, 468-471.

Kelly, D. C. (1982). Baculovirus replication. *J. Gen. Virol. 63*, 1-13.

Knell, J. D., Summers, M. D., and Smith, G. E. (1983). Serological analysis of 17 baculoviruses from Subgroups A and B using protein blot radioimmunoassay. *Virology 125*, 381-392.

Kok, I. P., Chistyakova, A. V., Gudz-Gorban, S. P., and Solomko, A. P. (1968). Infectivity and structure of DNA of virus of nuclear polyhedrosis of the silkworm. *Proc. 13th Intern. Congr. Entomol., Moscow*, p. 127.

Kozak, M. (1981). Mechanism of mRNA recognition by eukaryotic ribosomes during initiation of protein synthesis. *Curr. Topics Microbiol. Immunol. 93*, 81-123.

Kozak, M. (1983). Comparison of initiation of protein synthesis in procaryotes, eukaryotes and organelles. *Microbiol. Rev. 47*, 1-45.

Kozlov, E. A., Sidorova, N. M., and Serebryani, S. B. (1974). The isolation and physicochemical properties of polyhedral protein of nuclear polyhedrosis virus of large wax moth (*G. mellonella*). *Biokhimiya 39*, 130-134.

Kozlov, E. A., Sidorova, N. M., and Serebryani, S. B. (1975). Proteolytic cleavage of polyhedral protein during dissolution of inclusion bodies of the nuclear polyhedrosis viruses of *Bombyx mori* and *Galleria mellonella* under alkaline conditions. *J. Invertebr. Pathol. 25*, 97-101.

Kozlov, E. A., Levitina, T. L., Gusak, N. M., and Serebryani, S. B. (1981). Comparison of the amino acid sequence of inclusion body proteins of nuclear polyhedrosis viruses *Bombyx mori*, *Porthetria dispar* and *Galleria mellonella*. *Bioorg. Khim. 7*, 1008-1015.

Krywienczyk, J., and Bergold, G. H. (1960). Serological relationship between inclusion body protein of some lepidoptera and lymenoptera. *J. Immunol. 84*, 404-408.

Lee, H. H., and Miller, L. K. (1979). Isolation, complementation, and initial characterization of temperature-sensitive mutants of the baculovirus *Autographa californica* nuclear polyhedrosis virus. *J. Virol. 31*, 240-252.

Lightner, D. V., and Redman, R. M. (1981). A baculovirus-caused disease of the Penaeid Shrimp *Penaeus monodon*. *J. Invertebr. Pathol. 38*, 299-302.

Loh, L. C., Hamm, J. J., and Huang, E.-S. (1981). *Spodoptera frugiperda* nuclear polyhedrosis virus genome: Physical maps for restriction endonucleases *Bam*HI and *Hin*dIII. *J. Virol. 38*, 922-931.

Lübbert, H., Kruczek, I., Tjia, S. T., and Doerfler, W. (1981). The cloned *Eco*RI fragments of *Autographa californica* nuclear polyhedrosis virus. *Gene 16*, 343-345.

McCarthy, W. J., and DiCapua, R. A. (1979). Characterization of solubilized proteins from tissue culture and host-derived nuclear polyhedra of *Lymantria dispar* and *Autographa californica*. *Intervirology 11*, 174-181.

McCarthy, W. J., and Lambiase, J. T. (1979). Serological relationships among *Plusiinae* baculoviruses. *J. Invertebr. Pathol. 34*, 170-177.

Maestri, A. (1856). Frammenti anatomici, fisiologici, e pathologici sul baco da seta. *Fusi, Pavia*, pp. 172. (Parte 5: Malattie, pp. 111-134.)

Martignoni, M. E. (1972). A rapid method for the identification of nucleopolyhedron types. *J. Invertebr. Pathol. 19*, 281-283.

Martignoni, M. E. (1983). Baculovirus: An attractive biological alternative. *In* "Implication of Chemical and Biological Control Agents in Forests" (J. Harvey and W. Y. Garner, eds.), Amer. Chem. Soc., Washington, D.C., pp. 55-67.

Martignoni, M. E., and Iwai, P. J. (1977). "A Catalog of Viral Diseases of Insects and Mites," 2nd ed. U.S.D.A.

Forest Ser. Gen. Tech. Report PNW-40. U.S. Dept. Agriculture, Washington, D.C.

Maruniak, J. E., and Summers, M. D. (1978). Comparative peptide mapping of baculovirus polyhedrins. *J. Invertebr. Pathol. 32*, 196-201.

Maruniak, J. E., and Summers, M. D. (1981). *Autographa californica* nuclear polyhedrosis virus phosphoproteins and synthesis of intracellular proteins after virus infection. *Virology 109*, 25-34.

Maruniak, J. E., Summers, M. D., Falcon, L., and Smith, G. E. (1979). *Autographa californica* nuclear polyhedrosis virus structural proteins compared from *in vivo* and *in vitro* sources. *Intervirology 11*, 82-88.

Matthews, R. E. F. (1982). Classification and nomenclature of viruses. *Intervirology 17*, 1-199.

Miller, L. K. (1981). Construction of a genetic map of the baculovirus *Autographa californica* nuclear polyhedrosis virus by marker rescue of temperature-sensitive mutants. *J. Virol. 39*, 973-976.

Miller, L. K., and Dawes, K. P. (1979). Physical map of *Autographa californica* nuclear polyhedrosis virus. *J. Virol. 29*, 240-252.

Miller, L. K., Trimarchi, R. E., Browne, D., and Pennock, G. D. (1983). A temperature-sensitive mutant of the baculovirus *Autographa californica* nuclear polyhedrosis virus defective in an early function required for further gene expression. *Virology 126*, 376-380.

Minion, F. C., Coons, L. B., and Broome, J. R. (1979). Characterization of the polyhedral envelope of the nuclear polyhedrosis virus of *Heliothis virescens*. *J. Invertebr. Pathol. 34*, 303-307.

Mulder, G. H., Vos, P. J., and Vlak, J. M. (1984). Transcription of *Autographa californica* nuclear polyhedrosis virus genome: Mapping of the major transcript from DNA fragment *Eco*RI-J. In preparation.

Nudel, U., Soreq, H., Littauer, U. Z., Marbaix, G., Huez, G., Leclercq, M., Hubert, E., and Chantrenne, H. (1976). Globin mRNA species containing poly(A) segments of different length. Their functional stability in Xenopus oocytes. *Eur. J. Biochem. 64*, 115.

Oakley, B. R., Kirsch, D. R., and Morris, N. R. (1980). A simplified ultrasensitive silver stain for detecting proteins in polyacrylamide gels. *Anal. Biochem. 105*, 361-363.

O'Farrell, P. H. (1975). High resolution two-dimensional electrophoresis of proteins. *J. Biol. Chem. 250*, 4007-4021.

Payne, C. C., and Kalmakoff, Y. (1978). Alkaline protease activity associated with virus particles of a nuclear polyhedrosis virus: Assay, purification and properties. *J. Virol. 26*, 84-92.

Payne, C. C., and Kelly, D. C. (1981). Identification of insect and mite viruses. *In* "Microbial Control of Pests and Plant Diseases" (H. D. Burges, ed.), pp. 61-91. Academic Press, New York.

Payne, C. C., and Mertens, P. P. C. (1983). Cytoplasmic polyhedrosis viruses. *In* "The Reoviridae" (W. K. Joklik, ed.), pp. 425-504. Plenum Press, New York.

Pritchett, D. W., Scott, H. A., and Young, S. Y. (1979). Serological relationships of five nuclear polyhedrosis viruses from lepidopterous species. *J. Invertebr. Pathol. 33*, 183-188.

Qin, Jun-Chuan and Weaver, R. F. (1982). Capping of viral RNA in cultured *Spodoptera frugiperda* cells infected with *Autographa californica* nuclear polyhedrosis virus. *J. Virol. 43*, 234-240.

Reik, E. F. (1970). Fossil history. "The Insects of Australia," pp. 168-186. Melbourne University Press, Melbourne.

Roberts, P. L., and Naser, W. (1982). Characterization of monoclonal antibodies to the *Autographa californica* nuclear polyhedrosis virus. *Virology 122*, 424-430.

Rohel, D. Z., Carstens, E. H., and Faulkner, P. (1983). Characterization of two abundant mRNAs of *Autographa californica* nuclear polyhedrosis virus present late in infection. *Virology 124*, 357-365.

Rohrmann, G. F. (1977). Characterization of N-polyhedrin of two baculovirus strains pathogenic for *Orgyia pseudotsugata*. *Biochemistry 16*, 1631-1634.

Rohrmann, G. F., Martignoni, M. E., and Beaudreau, G. S. (1978a). Quantification of two viruses in technical preparations of *Orgyia pseudotsugata Baculovirus* by means of buoyant density centrifugation of viral deoxyribonucleic acid. *Appl. Environ. Microbiol. 35*, 690-693.

Rohrmann, G. F., McParland, R. H., Martignoni, M. E., and Beaudreau, G. S. (1978b). Genetic relatedness of two nucleopolyhedrosis viruses pathogenic for *Orgyia pseudotsugata*. *Virology 84*, 213-217.

Rohrmann, G. F., Bailey, T. J., Brimhall, B., Becker, R. R., and Beaudreau, G. S. (1979). Tryptic peptide analysis and NH_2-terminal amino acid sequences of polyhedrins of two baculoviruses from *Orgyia pseudotsugata*. *Proc. Natl. Acad. Sci. U.S. 76*, 4976-4980.

Rohrmann, G. F., Bailey, T. J., Becker, R. R., and Beaudreau, G. S. (1980). Comparison of the structure of C- and N-polyhedrins from two occluded viruses pathogenic for *Orgyia pseudotsugata*. *J. Virol. 34*, 360-365.

Rohrmann, G. F., Pearson, M. N., Bailey, T. J., Becker, R. R., and Beaudreau, G. S. (1981). N-terminal polyhedrin sequences and occluded *Baculovirus* evolution. *J. Mol. Evol. 17*, 329-333.

Rohrmann, G. F., Martignoni, M. E., and Beaudreau, G. S. (1982a). DNA sequence homology between *Autographa californica* and *Orgyia pseudotsugata* nuclear polyhedrosis viruses. *J. Gen. Virol. 62*, 137-143.

Rohrmann, G. F., Leisy, D. J., Chow, K.-C., Pearson, G. D., and Beaudreau, G. S. (1982b). Identification, cloning and R-loop mapping of the polyhedrin gene from the multicapsid nuclear polyhedrosis virus of *Orgyia pseudotsugata*. *Virology 121*, 51-60.

Rubinstein, R., and Polson, A. (1983). Midgut and viral associated proteases of *Heliothis armigera*. *Intervirology 19*, 16-25.

Scharnhorst, D. W., and Weaver, R. F. (1980). Structural analysis of the matrix protein from the nuclear polyhedrosis virus of *Heliothis zea*. *Virology 102*, 468-472.

Shatkin, A. J. (1976). Capping of eukaryotic mRNAs. *Cell 9*, 645-653.

Singh, S. P., Gudauskas, R. T., and Harper, J. D. (1979). Serological comparison of polyhedron protein and virions from four nuclear polyhedrosis viruses of plusiine larvae (Lepidoptera: Noctuidae). *J. Invertebr. Pathol. 33*, 19-30.

Singh, S. P., Gudauskas, R. T., and Harper, J. D. (1983). High resolution two-dimensional gel electrophoresis of structural proteins of baculoviruses of *Autographa californica* and *Porthetria (Lymantria) dispar*. *Virology 125*, 370-380.

Smith, G. E., and Summers, M. D. (1979). Restriction maps of five *Autographa californica* MNPV variants, *Trichoplusia ni* MNPV, and *Galleria mellonella* MNPV DNAs with endonucleases *SmaI, KpnI, BamHI, SacI, XhoI, and Eco*RI. *J. Virol. 30*, 828-838.

Smith, G. E., and Summers, M. D. (1980). Restriction map of *Rachiplusia ou* and *Autographa californica* baculovirus recombinants. *J. Virol. 33*, 311-319.

Smith, G. E., and Summers, M. D. (1981). Application of a novel radioimmunoassay to identify *Baculovirus* structural antigens that share interspecies antigenic determinants. *J. Virol. 39*, 125-137.

Smith, G. E., and Summers, M. D. (1982). DNA homology among subgroup A, B and C baculoviruses. *Virology 123*, 393-406.

Smith, G. E., Vlak, J. M., and Summers, M. D. (1982). *In vitro* translation of *Autographa californica* nuclear polyhedrosis virus early and late mRNAs. *J. Virol. 44*, 199-208.

Smith, G. E., Fraser, M. J., and Summers, M. D. (1983a). Molecular engineering of the *Autographa californica* nuclear polyhedrosis virus genome: Deletion mutations within the polyhedrin gene. *J. Virol. 46*, 584-593.

Smith, G. E., Vlak, J. M., and Summers, M. D. (1983b). Physical analysis of *Autographa californica* nuclear polyhedrosis virus transcripts for polyhedrin and 10,000-molecular-weight protein. *J. Virol. 45*, 215-225.

Smith, G. E., Summers, M. D., and Fraser, M. J. (1983c). Production of human beta interferon in insect cells infected with a baculovirus expression vector. *Mol. Cell. Biol. 3*, 2156-2165.

Soeda, E., Maruyama, T., Arrand, J. R., and Griffin, B. E. (1980). Host-dependent evolution of three papoviruses. *Nature (London) 285*, 165-167.

Southern, E. M. (1975). Detection of specific sequences among DNA fragments separated by gel electrophoresis. *J. Mol. Biol. 98*, 503-517.

Stairs, G. R. (1964). Selection of a strain of insect granulosis virus producing only cubic inclusion bodies. *Virology 24*, 520-521.

Steinhaus, E. A. (1956). Microbial control--the emergence of an idea. *Hilgardia 26*, 107-159.

Summers, M. D., and Anderson, D. L. (1972). Granulosis virus deoxyribonucleic acid: a closed, double-stranded molecule. *J. Virol. 9*, 710-713.

Summers, M. D., and Smith, G. E. (1975a). Comparative studies of baculovirus granulins and polyhedrins. *Intervirology 6*, 168-180.

Summers, M. D., and Smith, G. E. (1975b). *Trichoplusia ni* granulosis virus granulin: a phenol-soluble, phosphorylated protein. *J. Virol. 16*, 1108-1116.

Summers, M. D., and Smith, G. E. (1978). Baculovirus structural polypeptides. *Virology 84*, 390-402.

Summers, M. D., Smith, G. E., Knell, J. D., and Burand, J. P. (1980). Physical maps of *Autographa californica* and *Rachiplusia ou* nuclear polyhedrosis virus recombinants. *J. Virol. 34*, 693-703.

Thompson, C. G., Scott, D. W., and Wickman, B. E. (1981). Long-term persistence of the nuclear polyhedrosis virus of the Douglas-fir tussock moth, *Orgyia pseudotsugata* (Lepidoptera: Lymantriidae), in forest soil. *Environ. Entomol. 10*, 254-255.

Tinsley, T. W., and Harrap, K. A. (1978). Viruses of invertebrates. *In* "Comprehensive Virology" (H. Fraenkel-Conrat and R. R. Wagner, eds.), Vol. 12, pp. 1-101. Plenum Press, New York.

Tweeten, K. A., Bulla, L. A., Jr., and Consigli, R. A. (1978). Characterization of an alkaline protease associated with a granulosis virus of *Plodia interpunctella*. *J. Virol. 26*, 702-711.

Tweeten, K. A., Bulla, L. A., Jr., and Consigli, R. A. (1980). Structural polypeptides of the granulosis virus of *Plodia interpunctella*. *J. Virol. 33*, 877-886.

Tweeten, K. A., Bulla, L. A., Jr., and Consigli, R. A. (1981). Applied and molecular aspects of insect granulosis viruses. *Microbiol. Rev. 45*, 379-408.

Van der Beek, C. P., Saayer-Riep, J. D., and Vlak, J. M. (1980). On the origin of the polyhedrin protein of *Autographa californica* nuclear polyhedrosis virus. *Virology 100*, 326-333.

Vlak, J. M. (1979). The proteins of nonoccluded *Autographa californica* nuclear polyhedrosis virus produced is an established cell line of *Spodoptera frugiperda*. *J. Invertebr. Pathol. 34*, 110-118.

Vlak, J. M. (1980). Mapping of *Bam*HI and *Sma*I DNA restriction sites on the genome of the nuclear polyhedrosis virus of the alfalfa looper, *Autographa californica*. *J. Invertebr. Pathol. 36*, 409-414.

Vlak, J. M. (1982). Restriction endonucleases as tools in baculovirus identification. *Proc. 3rd Intern. Colloq. Invertebr. Pathol., Brighton*, pp. 218-225.

Vlak, J. M., and Odink, K. G. (1979). Characterization of *Autographa californica* nuclear polyhedrosis virus deoxyribonucleic acid. *J. Gen. Virol. 44*, 333-347.

Vlak, J. M., and Smith, G. E. (1982). Orientation of the genome of AcMNPV: A proposal. *J. Virol. 41*, 1118-1121.

Vlak, J. M., and Van der Krol, S. (1982). Transcription of the *Autographa californica* nuclear polyhedrosis virus genome: location of late cytoplasmic RNA. *Virology 123*, 222-228.

Vlak, J. M., Smith, G. E., and Summers, M. D. (1981). Hybridization selection and *in vitro* translation of *Autographa californica* nuclear polyhedrosis virus mRNA. *J. Virol. 40*, 762-771.

Volkman, L. E. (1983). Occluded and budded *Autographa californica* nuclear polyhedrosis virus: Immunological relatedness of structural proteins. *J. Virol. 46*, 221-229.

Volkman, L. E., and Falcon, L. A. (1982). Use of monoclonal antibody in an enzyme-linked immunosorbent assay to detect the presence of *Trichoplusia ni* (Lepidoptera: Noctuidae)S nuclear polyhedrosis virus in *T. ni* larvae. *J. Econ. Entomol. 75*, 868-871.

Volkman, L. E., and Summers, M. D. (1977). *Autographa californica* nuclear polyhedrosis virus: comparative infectivity of the occluded, alkali-liberated, and nonoccluded forms. *J. Invertebr. Pathol. 30*, 102-103.

Von Prowazek, S. (1907). Chlamydozoa. II. Gelbsucht der Seidenraupen. *Arch. Protistenk. 10*, 358-364.

Walter, G., Scheidtmann, K., Carbone, A., Laudano, A. P., and Doolittle, R. F. (1980). Antibodies for the carboxy- and amino-terminal regions of simian virus 40 large tumor antigen. *Proc. Natl. Acad. Sci. U.S. 77*, 5197-5200.

Wilson, A. C., Carlson, S. S., and White, T. J. (1977). Biochemical evolution. *Annu. Rev. Biochem. 46*, 573-639.

Wood, H. A. (1980a). Protease degradation of *Autographa californica* nuclear polyhedrosis virus proteins. *Virology 103*, 392-399.

Wood, H. A. (1980b). *Autographa californica* nuclear polyhedrosis virus induced proteins in tissue culture. *Virology 102*, 21-27.

Yamafuji, K., Yoshihara, F., and Hirayama, K. (1958). Protease and deoxyribonuclease in viral polyhedral crystals. *Enzymologia 19*, 53.

Zummer, M., and Faulkner, P. (1979). Absence of protease in baculovirus polyhedral bodies propagated *in vitro*. *J. Invertebr. Pathol. 33*, 383-384.

V.

Replication

THE REPLICATION OF IRIDOVIRUSES IN HOST CELLS

Peter E. Lee

Department of Biology
Carleton University
Ottawa, Ontario

INTRODUCTION

Insects are in constant competition with man. Thus, it is necessary to maintain the natural balance of beneficial insect populations while at the same time control insect pests which devastate food supplies. Pesticides, such as dichlorodiphenyltrichloroethane (DDT), which was, at one time, very effective in the control of crop pests and insect vectors have not recently been used to any extent because of insect resistance to commonly used insecticides and because some have been shown, even in minute quantities, to be very toxic to man and his food sources by accumulating in body tissues.

At present biological control is receiving wide attention as a method for pest control. A major limitation of biological control is that to be effective, it must be very specific or have a narrow host range. One area of biological control for which there is a growing interest is the use of viruses to control insect pests. To be an effective pest control method, several factors must be known, including: (1) the host range of the virus, (2) the developmental stage of the pest that the virus can invade and replicate in, which finally results in its destruction, and (3) the resistance of the virus to environmental factors.

ISBN 0-12-470295-3

The iridoviruses are a group of cytoplasmic double-stranded DNA viruses (Stoltz, 1971), a large number of which infect invertebrates, particularly the class Insecta (see Table 3, Tinsley and Harrap, 1978; Goorha and Granoff, 1979). Some viruses, like *Tipula* iridescent virus (TIV), are stable when purified and virus suspensions retain their infectivity for some time when stored at 4°C. This stability is an important consideration when choosing them as a means of biological control.

This chapter is concerned with the replication of iridoviruses in insect host cells. Recently, two excellent reviews have been published. The first review on viruses of invertebrates (Tinsley and Harrap, 1978) has a section on iridescent viruses, while the second on icosahedral cytoplasmic deoxyriboviruses (Goorha and Granoff, 1979) presents the salient features of frog virus 3 (FV3) replication. Although FV3 is not an insect pathogen, it is in the iridovirus group and because of many similarities with the iridescent viruses infecting insects, information on FV3 replication is included in this chapter.

ENTRY OF THE VIRAL INOCULUM IN HOST CELLS

Some of the earlier studies on virus development of *Sericesthis* iridescent virus (SIV) and *Tipula* iridescent viruses were initiated by injecting insect larvae with purified virus suspensions (Leutenegger, 1967; Younghusband and Lee, 1969; Yule and Lee, 1973). This method of inoculation has a major drawback, namely the asychrony of infection of susceptible cells due to the use of intact insects. Nonetheless, some information has been obtained using this technique of mechanical inoculation. Leutenegger (1967) demonstrated that SIV adsorbed to the plasma membrane of hemocytes within 2 hours postinoculation (p.i.). Virus inoculum was ingested by phagocytosis and was observed in lysosome-like vesicles. Younghusband and Lee (1969) found that with TIV viral inoculum could accumulate in membrane-bound vesicles of hemocytes within 1 hour p.i., and there was cytochemical evidence of localization of TIV inoculum particles in lysosomes determined by the Gomori assay of Barba and Anderson (1963) at the electron microscopic level. Acid phosphatase activity was observed in these single membrane-bound vesicles containing virions when hemocyte samples from experimentally inoculated larvae were treated with the complete Gomori medium. Acid phosphatase activity was indicated by an electron-dense precipitate in the lysosome-like structures, but was absent in similar hemocyte samples suspended in Gomori medium lacking glycerophosphate

(Younghusband and Lee, 1970). In the period proceeding viral synthesis, virions within these single membrane cytoplasmic bodies could be observed in various stages of disintegration for both SIV and TIV (Leutenegger, 1967; Younghusband and Lee, 1969, 1970). In studying virus infection of two iridoviruses, SIV and *Chilo* iridescent virus in continuous cell cultures, Kelly and Tinsley (1974a) reported the uptake of virus by pinocytosis; this process was observed up to 3 hours p.i. These viruses could then be found in the cytoplasm either enclosed in single membranes or free in the cytoplasm. This differs from the findings of Leutenegger (1967) who studied SIV in experimentally inoculated larvae and for TIV studied in experimentally inoculated larvae and in continuous cell cultures (Younghusband and Lee, 1969, 1970; Mathieson and Lee, 1981).

By inoculating hemocytes of the saltmarsh caterpillar, *Estigmene acrea*, grown in continuous suspension cultures, thereby surmounting the problem of asynchrony in inoculation, Mathieson and Lee (1981) studied the infection pathway of TIV. Inoculum virus particles were immediately adsorbed to the plasma membrane. This was followed by viropexis in which cytoplasmic protusions engulfed virions resulting in uptake within pinocytotic vesicles. These vesicles coalesced and no free inoculating virions could be found in the cytoplasm. The initial stages of infection of cells in suspension cultures is in agreement with observations made of cells obtained from experimentally infected larvae using SIV and TIV (Leutenegger, 1967; Younghusband and Lee, 1970). The internalization of TIV into *E. acrea* culture cells is presented in Fig. 1. Following adsorption at the cell membrane (Fig. 1a), viropexis occurred. Cytoplasmic protrusions engulfed virions and were localized in the vesicles (Fig. 1b,c). After viropexis, cytoplasmic vesicles containing loose virions coalesced into lysome-like structures packed with virions (Fig. 1d).

SYNTHESIS OF MACROMOLECULES IN VIRUS-INFECTED CELLS

Kelly and Tinsley (1974a) studied the replication of *Sericesthis* iridescent virus (SIV) and *Chilo* iridescent virus (CIV) (Iridoviruses types 2 and 6) by microscopy in continuous cell cultures of *Aedes aegypti* and *Antheraea eucalypti*. They were able to obtain synchronous inoculation of cells. These results differ from the earlier work with SIV and TIV (Leutennegger, 1967; Younghusband and Lee, 1969, 1970). Both cell species were reduced to one-half of their normal size 72 hours p.i. and the mosquito cells which were normally attached to the flasks were in suspension 96 hours p.i. with

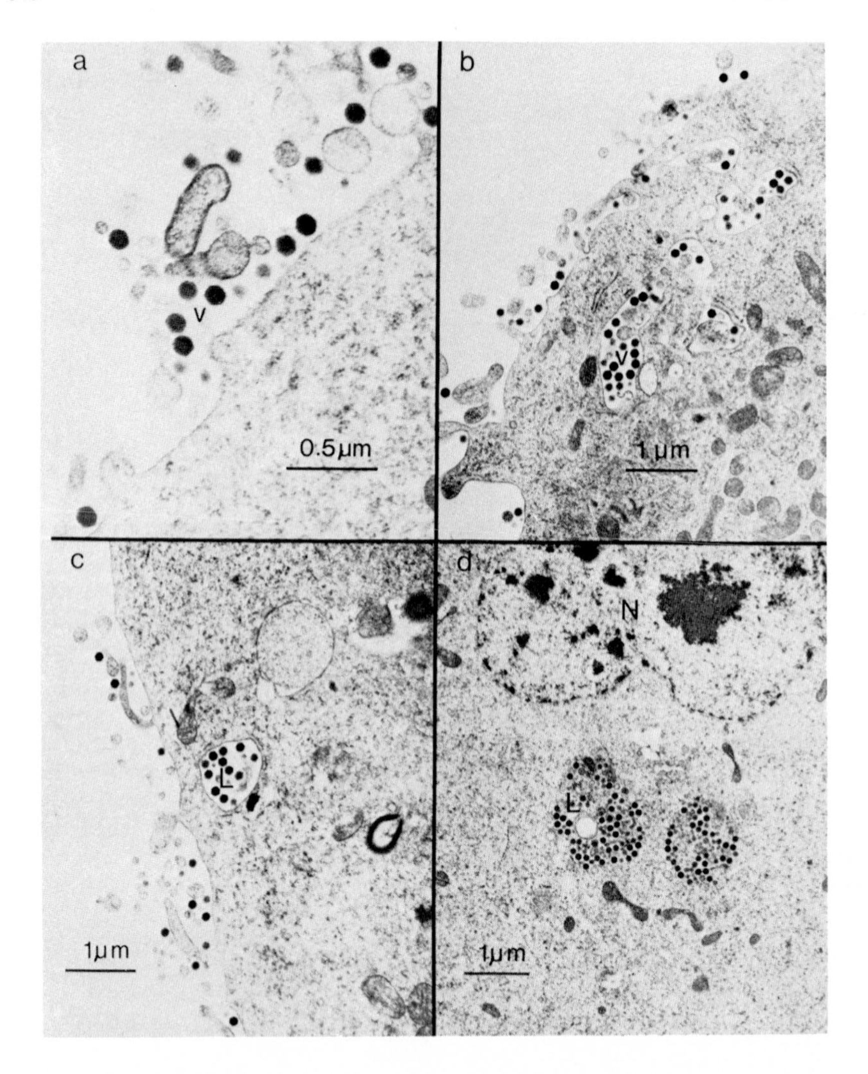

FIGURE 1. Initial steps of TIV infection in E. acrea cells. (a) Adsorption of virus to plasma membrane. (b) Phagocytic engulfment of inoculum (viropexis). (c and d) Accumulation of virions into small vesicles (c) and coalescing into larger packed vesicles in the cytoplasm (d). v, Virus inoculum; N, nucleus; L, lysosome. Fig. 1a and b from Mathieson and Lee (1981).

both type 2 and type 6 viruses. The authors state that detachment is not due to virus multiplicity input and it is likely that this is a cytotoxic effect rather than cytopathic one. Using the indirect immunofluorescence method, Kelly and Tinsley

(1974a) found that the cytoplasm was the site of antigen localization but that specific stain was not observed before 96 hours p.i. with 30 and 40% of cells staining 144 hours p.i.

On rare occasions SIV-infected mosquito cells contained two closely similar paracrystalline structures in the cytoplasm, with repeating units of 65 ± 5 nm and 64 ± 2 nm diameters. The crystal of the latter dimension subunit was found to be closely associated with SIV accumulation. Although the nature of these crystalline structures is not known, Kelly and Tinsley (1974a) suggest they may be involved with host-coded protein synthesis. However, this does not seem likely since no host protein synthesis occurs late in infection. Nucleic acid synthesis of these two iridoviruses was also studied by Kelly and Tinsley (1974b). Using [^{3}H]thymidine, and [^{3}H]uridine, respectively, and assaying the trichloroacetic acid insoluble products of both infected and control cells, it was found that both DNA and RNA synthesis was depressed up to 48 hours. This depression was followed by an increase in the synthesis of both nucleic acids. To further elucidate nucleic acid synthesis in the virus/cell systems, nucleic acid hybridizations were conducted. From these experiments, Kelly and Tinsley (1974b) found early V-RNA synthesis which, they suggested, is probably directed by the viral genome, since V-DNA is not detected prior to 96 hours p.i. However, the location of V-DNA synthesis was not determined.

Using light and electron microscope autoradiography, Mathieson and Lee (1981) studied DNA synthesis in *E. acrea* cells infected with TIV. Although this technique does not irrevocably indicate the sites of synthesis, as detected by the presence of silver grains in the autoradiogram, it does indicate where the newly synthesized nucleic acid accumulates. From this study the first viroplasmic center was detected by light microscope autoradiography at 6 hours p.i., with 14% of the inoculated cells containing newly synthesized DNA in cytoplasmic loci as detected by the presence of silver grains by 8.5 hours p.i. (Fig. 2). Under these experimental conditions, 74% of inoculated cells contained viroplasms by 24 hours p.i. (Mathieson, 1978). At the electron microscope level, silver grains were located over the dense matrix of the viroplasms, while the newly assembled virions were in the electron-lucent matrix of these cytoplasmic centers (Fig. 3 and 4). Thus, in the TIV/*E. acrea* system, loci of DNA accumulation are distinct from loci of virion assembly in the viroplasm. TIV has an inhibitory effect on host DNA synthesis which was detectable at 4 hours p.i. and by 29 hours thymidine incorporation was reduced to 20% in control cells (Fig. 5).

Changes in the rates of cellular RNA and protein synthesis were also monitored using light microscope autoradiography of incorporated [^{3}H]uridine and [^{3}H]leucine, respectively (Mathieson and Lee, 1981). There was a rapid decline in

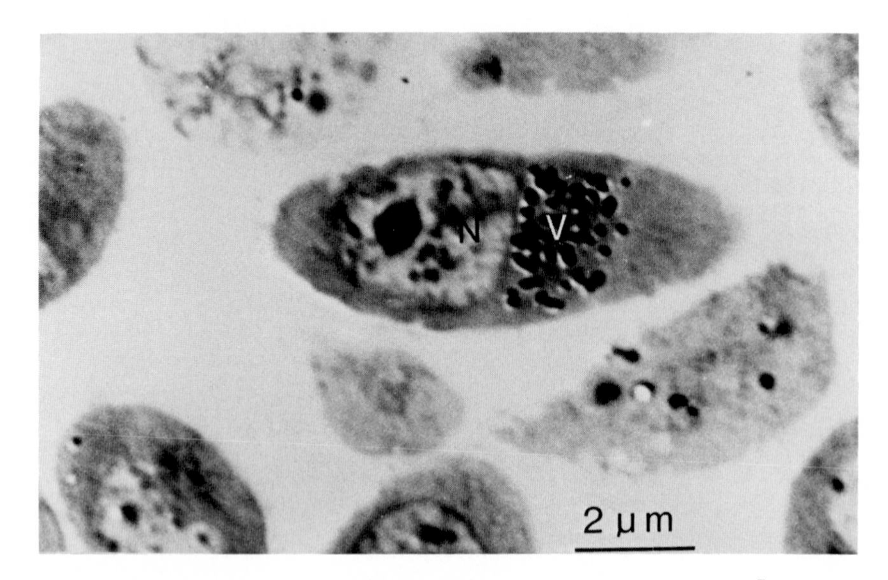

FIGURE 2. Light microscope autoradiography of (^{3}H) thymidine incorporation into TIV-infected cells at 8.5 hours p.i. Micrograph taken with partial interference optics. N, nucleus; V, viroplasm. From Mathieson and Lee (1981).

[^{3}H]uridine incorporation during the initial 5 hours of infection, which is probably due to inhibition of cellular transcription. By 7 hours p.i. there was an increase of uridine incorporation as well as a larger number of viroplasmic centers with silver grains. This increase was probably due to V-RNA synthesis, which was indistinguishable from cellular RNA synthesis. At the same time [^{3}H]leucine incorporation in protein synthesis declined, reaching 50% inhibition at 6 hours p.i. The initial decline up to 5 hours p.i. probably was due to a decrease in host protein synthesis and the subsequent leveling off between 5 to 25 hours p.i. was a combination of the decline and synthesis of host and viral synthesis, respectively (Fig. 6).

Control cells never showed newly synthesized DNA in the cytoplasm. The *E. acrea* cells used in the TIV experiments were asynchronized and under the experimental conditions used (with [^{3}H]thymidine), 30% of nuclei in these controls synthesized DNA (Mathieson and Lee, 1981; Lee and Brownrigg, 1982). Like frog virus 3 (FV3), which aborts host cell DNA synthesis and is probably induced by a structural component of the virus (Goorha and Granoff, 1974), TIV inhibits nuclear DNA synthesis of *E. acrea* cells.

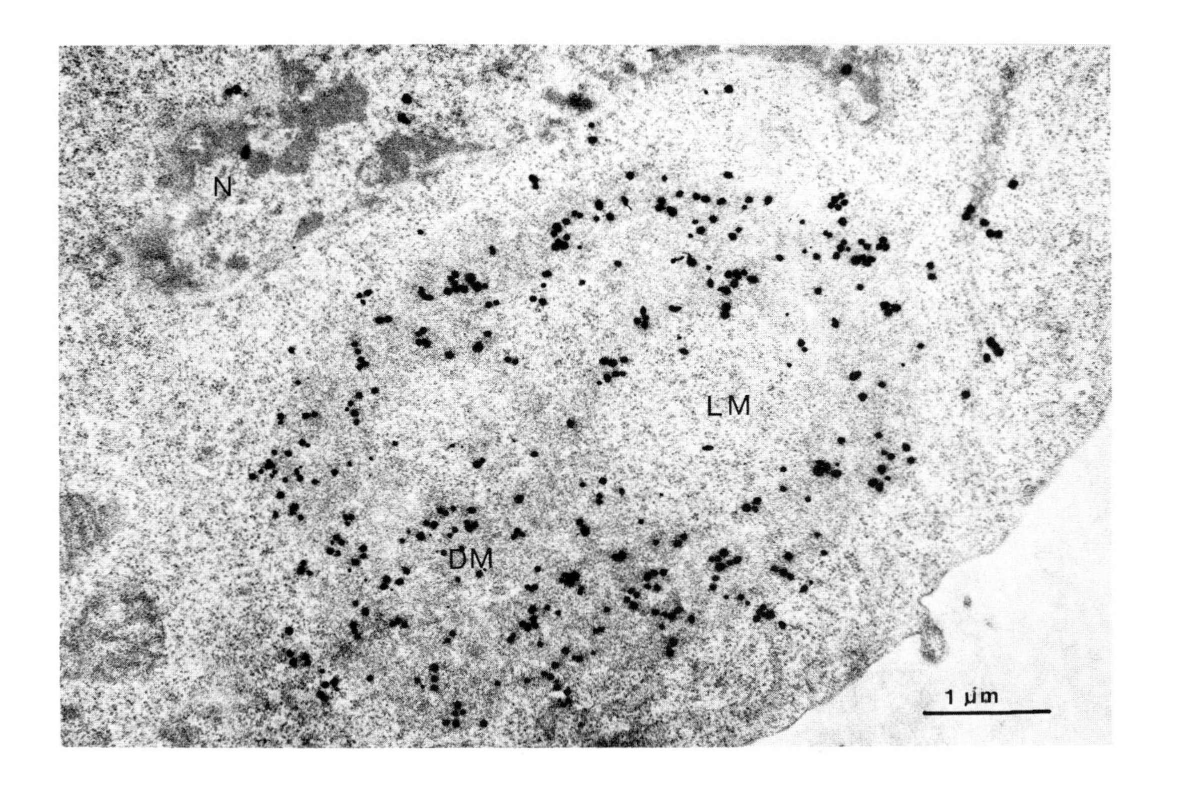

FIGURE 3. Electron microscope autoradiography of [^{3}H]thymidine incorporation into TIV-infected cells. Silver grains in the cytoplasmic viroplasm indicate DNA label localized over dense matrix areas (DM), but loose matrices (LM) were devoid of silver grains. From Mathieson and Lee (1981).

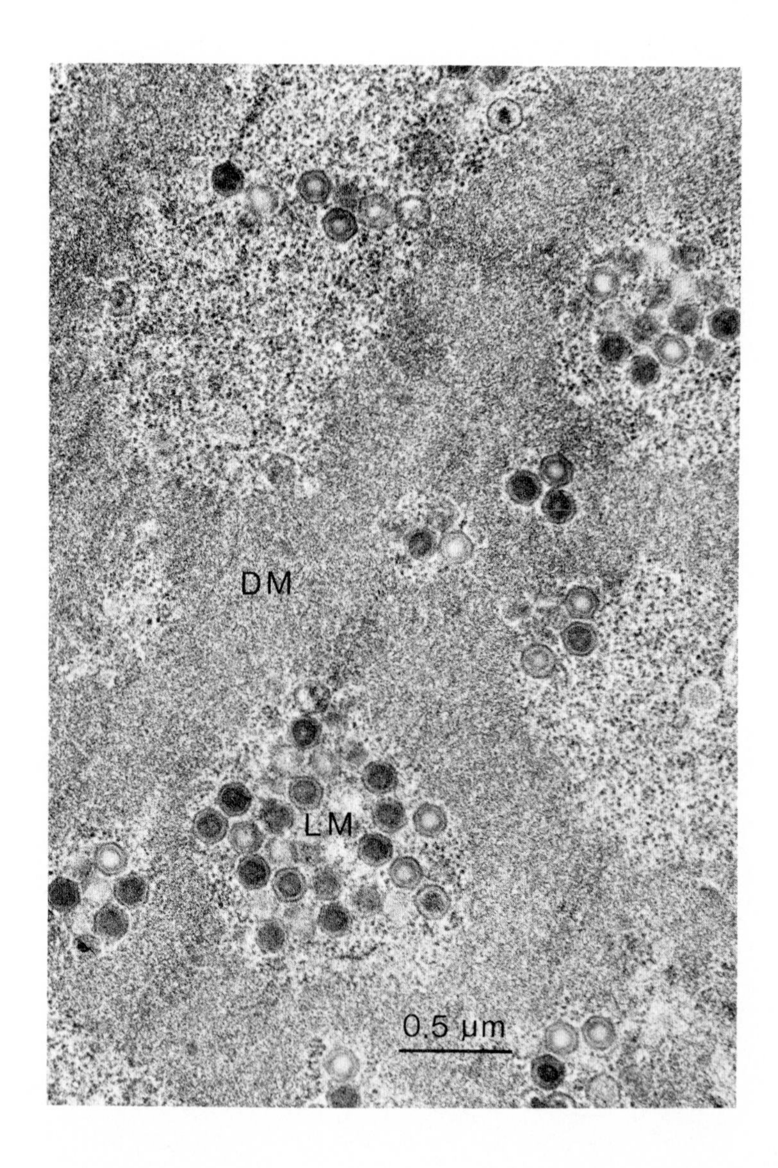

FIGURE 4. Electron microscopy of E. acrea hemocyte cytoplasm infected with TIV for 16 hours p.i. Loose matrix (LM) contains ribosomes and is the site for viral assembly. Dense matrix (DM) appears devoid of ribosomes and consisted of a homogeneous nucleoprotein material. From Mathieson and Lee (1981).

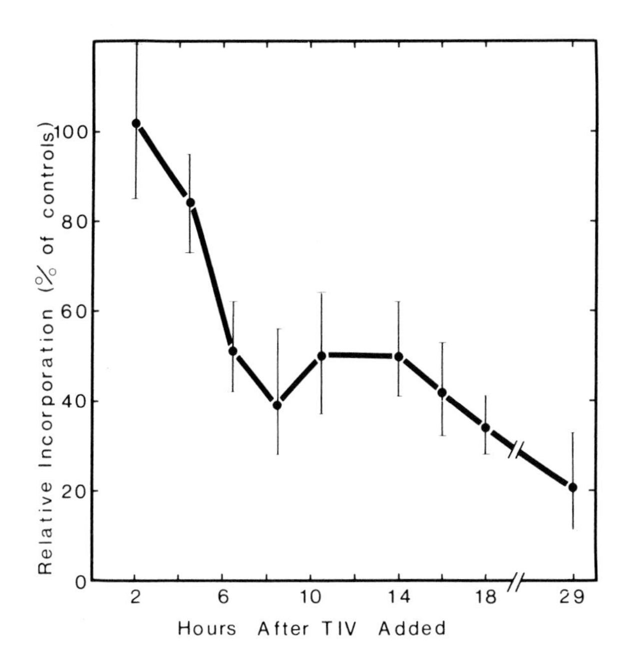

FIGURE 5. Autoradiographic analysis of [^3H]thymidine incorporation into nuclei of TIV-infected hemocytes. Cultures were incubated for 2 hours with [^3H]thymidine (15 μCi/ml, 20 Ci/mmole) at various times p.i. The average number of silver grains over nuclei of (infected/control) × 100 was used to calculate the relative incorporation values. Error bars correspond to the standard error of the mean. From Mathieson and Lee (1981).

Some innovative experiments have been conducted on nucleic acid synthesis of FV3 and its relationship to the host cell. Compelling evidence for nuclear involvement in FV3 DNA replication has been presented (Goorha et al., 1977, 1978). For example, when cells are enucleated or treated by uv irradiation, replication of FV3 does not occur.

NUCLEAR INVOLVEMENT IN V-DNA SYNTHESIS AND VIRUS EFFECTS ON THE NUCLEUS

In further pursuing the function of the nucleus in inoculated cells, Goorha et al. (1978) found by electron microscope autoradiography that 30% of V-DNA was initially synthesized in

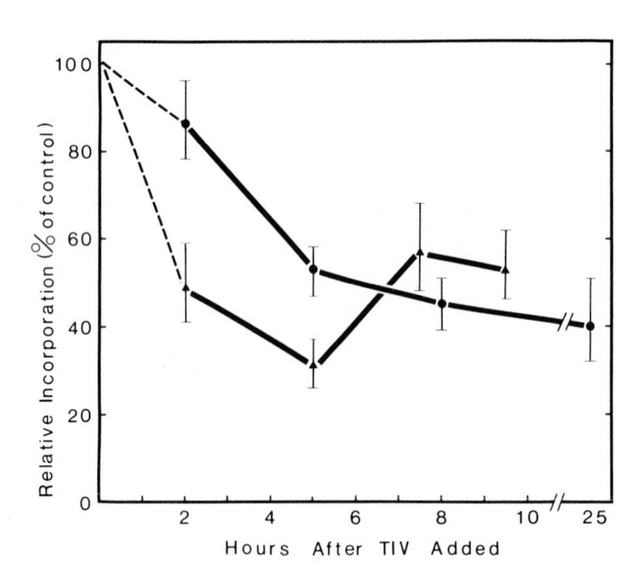

FIGURE 6. Autoradiographic analysis of [^{3}H]thymidine incorporation into nuclei of TIV-infected hemocytes. Cultures were incubated for 2 hours with [^{3}H]thymidine (15 μCi/ml, 20 Ci/mmole) at various times p.i. The average number of silver grains over nuclei of (infected/control) × 100 was used to calculate the relative incorporation values. Error bars correspond to the standard error of the mean. From Mathieson and Lee (1981).

the nucleus and, as the infection progressed, V-DNA was transported to the cytoplasm where silver grains were seen in association with the viroplasmic centers. These pulse-chase autoradiographic experiments provided information on whether the DNA was transient or stable. RNA synthesis in infected FV3 cells followed a similar pattern. By alkaline sucrose gradient comparisons of nuclear V-DNA and cytoplasmic V-DNA with DNA from purified FV3 particles, it was determined that nuclear V-DNA was one-sixth the size of DNA from purified virus, and that cytoplasmic V-DNA was the same size or larger than DNA from virus particles. From these data and pulse-chase results, it was proposed that V-DNA is synthesized in subgenomic fragments in the nucleus and transported to the cytoplasm. The intact genome size of FV3 is then produced by ligation or elongation.

Further evidence was presented for this concept by adding the protein synthesis inhibitor cycloheximide to cells synthesizing FV3 DNA; cycloheximide inhibits the initiation of V-DNA synthesis by blocking viral protein synthesis (Kang et al., 1971; Yu et al., 1975). This caused a significant inhibition

TABLE I. Incorporation of [^{3}H]Thymidine in Nuclei of E. acrea Cells Following Inoculation with Various Sources of TIV[a]

	Nuclei		Percentage nuclei labeled
	Labeled	Unlabeled	
TIV	0	45	0
Partially filled TIV	0	64	0
Empty capsids	0	48	0
UV TIV	0	48	0
BPL TIV	17	42	28.9
UV BPL TIV	17	38	30.9
Control	53	108	32.7

[a]Cells were labeled at 16 hours p.i. for 2 hours with ^{3}H thymidine (20 Ci/mmole, 20 μCi/ml), then washed three times with 1 mM thymidine, and at 20 hours p.i. prepared for light, microscope autoradiography. From Lee and Brownrigg (1982).

of nuclear V-DNA, but less inhibition of cytoplasmic V-DNA. These results were interpreted as the completion of nucleotide synthesis in the nucleus, which had been already initiated when cycloheximide was added, but that nucleic acid synthesis did not occur in the nucleus after addition of the inhibitor.

Another aspect of iridovirus infection of susceptible cells is the virus effect on cells. Here the nucleus is the organelle which is significantly affected. When virus is heat-inactivated, in addition to a lack of viral replication in FV3, host macromolecular synthesis is markedly reduced by up to 90% (Goorha et al., 1978). Although the site of TIV DNA synthesis is not known, Lee and Brownrigg (1982) studied the relationship of both infectious and noninfectious virus on the host cell with particular emphasis on the nucleus. The range of inocula studied were: infectious TIV, noninfectious partially filled virions, empty capsids, and infectious TIV treated with uv irradiation and/or β-propiolactone (β-PL). Results of [^{3}H]thymidine incorporation in nuclei of cells following inoculation with these sources of TIV is presented in Table I and Fig. 7. No nuclei incorporated [^{3}H]thymidine when cells were exposed to TIV, partially filled TIV, empty capsids, and uv-treated TIV, but when the protein moiety of the virus was affected, nuclear incorporation was similar to control cells. Thus, like FV3, a protein(s) of the virion is responsible for inhibiting host DNA synthesis.

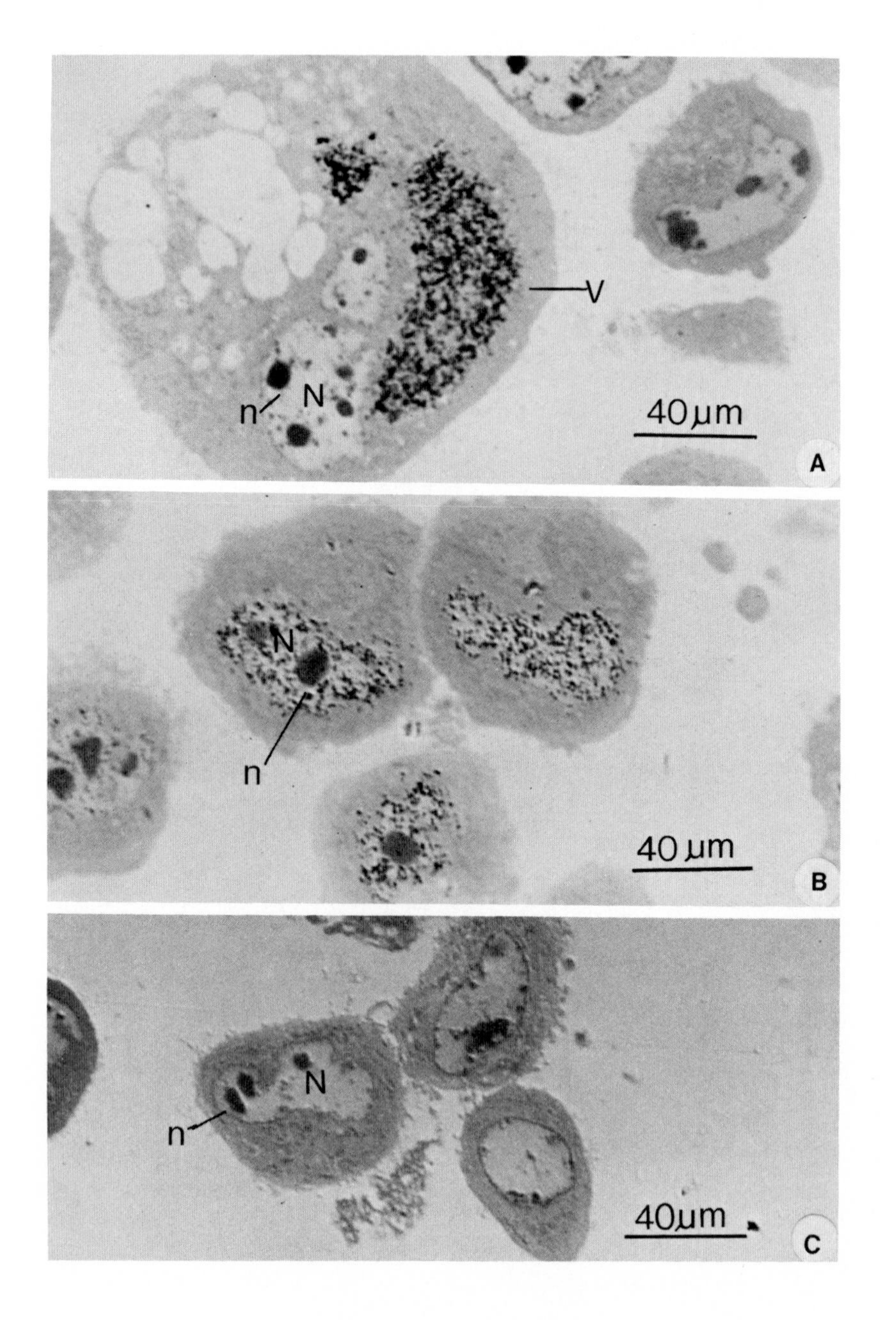

FIGURE 7. Light microscope autoradiography of E. acrea cells. In (A) cells were infected with TIV resulting in no

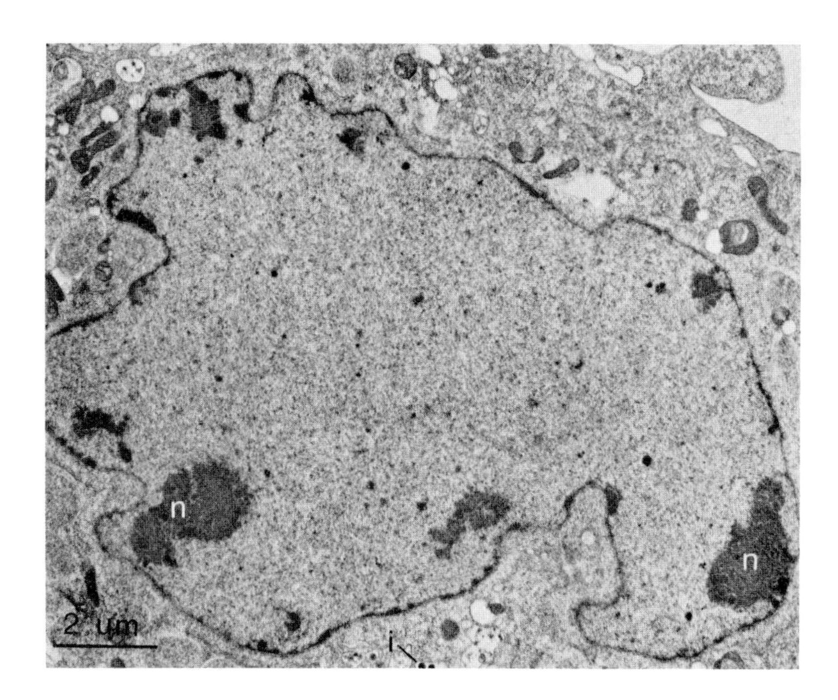

FIGURE 8. Nucleus from E. acrea cell inoculated with uv-inactivated TIV showing very little condensed chromatin and well-defined nucleoli (n) at the nuclear membrane. Particles (i) from inoculum in cytoplasm. From Lee and Brownrigg (1982).

In addition to studying (^{3}H) thymidine incorporation by light microscope autoradiography, the fine structure of these nuclei were studied using the electron microscope. Nuclei that were inhibited from synthesizing DNA by specific sources of virus inoculum (see Table I), contained sparse condensed chromatin when compared to control cells (Figs. 8 and 9). However, cells inoculated with BPL-treated TIV have nuclei with heterochromatin similar to control cells. It has been reported that β-PL inactivates viruses by modifying proteins

(Figure 7 continued) [^{3}H]thymidine incorporation in nuclei (N) but cytoplasmic incorporation in the viroplastic center (V). n, Nucleolus. (B) Nuclear incorporation of [^{3}H]thymidine in normal cells. When cells were inoculated with noninfectious partially filled TIV particles (C), there was no [^{3}H]thymidine incorporation in the nucleus or cytoplasm. Photomicrography by particl interference optics. From Lee and Brownrigg (1982).

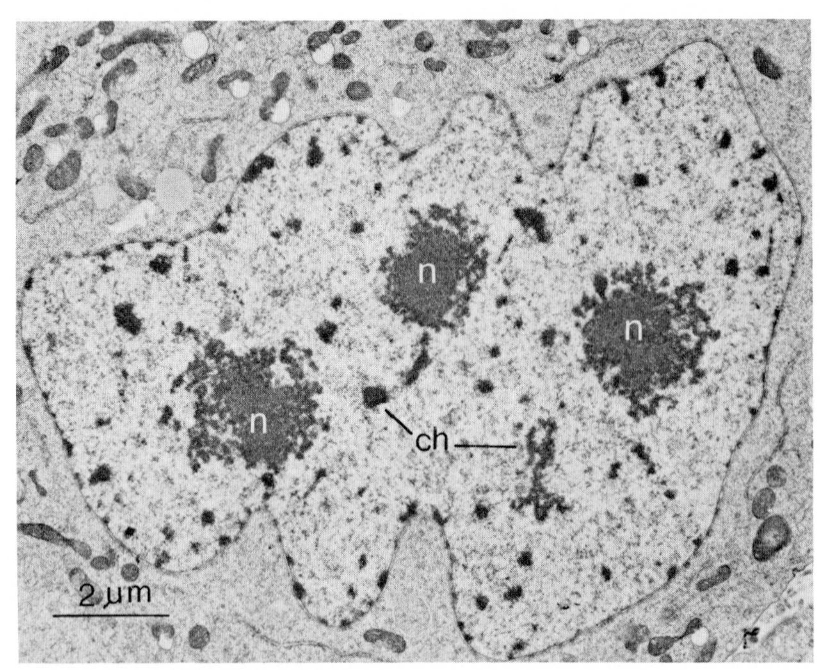

FIGURE 9. Nucleus of normal cell showing three nucleoli (n) and aggregated chromatin distributed throughout the nucleus. Ch, chromatin. From Lee and Brownrigg (1982).

because of a lactone which reacts with both alkyl and acyl bonds (Lo Grippo, 1960).

These findings are similar to those reported for poxvirus inhibition of host DNA synthesis (Pogo and Dales, 1973), since infectious or uv-treated vaccinia virus inhibited nuclear DNA synthesis. The failure of nuclei to incorporate [^{3}H]thymidine and to be generally devoid of condensed chromatin following the inoculation of cells with infectious TIV, partially filled capsids, empty capsids or uv-treated virus may be explained by the hypothesis proposed by Setterfield et al. (1978) using temperature-sensitive mutant lines of mouse cells to study DNA synthesis. Nuclear DNA synthesis occurs in a cycle in which chromatin is decondensed prior to DNA synthesis; after replication, chromatin is recondensed. In mutant lines of mouse cells, a shift from 34° to 38.5°C was the limiting factor. At this temperature, condensed chromatin is in the central nuclear region. This is not due entirely to decondensation of chromatin fibrils but also to dispersal into small chromatin aggregates. When the cells were returned to 34°C, DNA synthesis resumed with aggregation of chromatin. With TIV nuclear inhibition, viral protein(s) may have a similar effect to that of temperatures of 38.5°C on mouse host cells. However, it does not follow that inhibition of DNA synthesis in *E. acrea* cells by TIV protein(s) is reversible.

THE FINE STRUCTURE OF VIROPLASMIC CENTER OR ASSEMBLY SITE

The possible interplay of viral DNA, RNA, and proteins with the viroplasmic center has already been mentioned. The structure of these centers, which are not membrane bound and are localized in the cytoplasm of infected cells will now be described. Stoltz et al. (1968) reported on a field-collected infection in larvae of *Chironomus plumosus*. This larval disease was shown to be virus associated, with fat body cells in the cytoplasm containing large numbers of virus particles. Although the virus was morphologically similar to the iridoviruses and was found in association with DNA-containing inclusions, as determined by light microscope cytology, it did not cause iridescence. Fine filamentous structures (fibrils) were seen associated with the capsid using electron microscopy. These fibrils were also retained by extracellular virus particles. Stoltz et al. (1968) suggest that these filamentous structures prevent virus particles from forming microcrystals, which are responsible for iridescence because of the periodic spacing in the crystal (Fig. 10).

The viroplasmic centers which develop in TIV-infected *E. acrea* cells have been studied in detail (Mathieson and Lee, 1981). Electron microscopy of these loci show two well-defined areas; electron dense ones, which electron microscope autoradiography show are the centers of V-DNA accumulation, and electron-lucent ones, which are interspersed in those areas in which new virions assemble (Figs. 3 and 4). These latter areas contain ribosomes and, on occasion, mitochondria (see Fig. 4, Kloc and Lee, 1983) and most probably are areas of cell cytoplasm which result from the infolding of the v-DNA-containing zones which thereby entraps cytoplasm. Based on this interpretation the bona fide viroplasm are the electron-dense areas. Virions with intact capsid shells are seen in various stages of core density in the electron-lucent zones (Fig. 4). This is taken as further evidence that the capsid shell is first assembled prior to entry of the DNA nucleoid through an opening in the shell. This proposal was suggested earlier for TIV assembly (Yule and Lee, 1973) from a ferritin antibody study in which ferritin was seen closely aligned to the inner faces of various stages of empty capsid assembly. However, this is not the only possibility for the assembly of mature virions, since Bird (1961), 1962) proposed that viral DNA forms first and

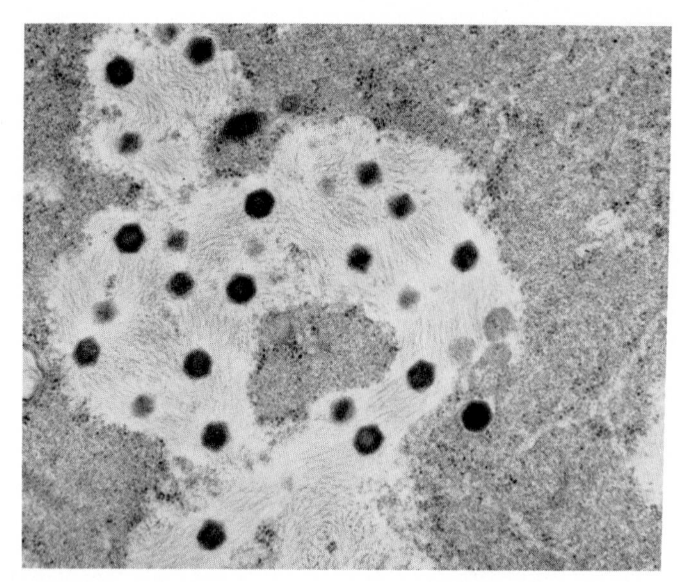

FIGURE 10. Chironomus plumosus iridovirus in fat body cell cytoplasm. Note fine fibrils associated with virions. From D. B. Stoltz, personal communication.

then is encapsulated by protein. This proposal has been favorably accepted by some investigators (Kelly and Tinsley, 1974a). Although Fig. 4 is a two-dimensional view, the evidence is compelling for capsid formation before DNA nucleoid entry.

The fine structure of viroplasmic centers in fathead minnow (FHM) cells infected with FV3 is totally different from that of TIV-infected *E. acrea* cells. Although these loci in FHM cells contain FV3 virions in various stages of development, the virus assembly sites were lower in electron density than the surrounding cytoplasm and cytoplasmic organelles, including ribosomes, were absent. From electron microscopy (Figs. 2A and B, Murti and Goorha, 1983), these viroplasmic centers or assembly sites, although not contiguous with the surrounding cytoplasm, were more diffuse with the cytoplasm than those of TIV centers in *E. acrea* cells. Murti and Goorha (1983) also studied these virus assembly sites by electron microscope immunoperoxidase labeling, in addition to fluorescence microscope immunolabeling of both FV3 virus and the cytoskeleton of infected cells. The latter will be discussed later in this chapter.

In an earlier study, Tripier-Darcy et al. (1980) found microtubules in the nucleus of chick embryo fibroblasts infected with FV3. Although it was not fully explained why microtubules would be located in the nuclei of FV3-infected chick embryo fibroblasts and never in control cell nuclei, the authors proposed the following explanation. Infection with FV3 causes the nuclear membrane to become permeable to tubulin which can then polymerize in the nucleoplasm as microtubules. This theory, the authors suggest, is indirectly supported by earlier experiments of Goorha et al. (1977, 1978) who showed that the nucleus is involved in FV3 nucleic acid synthesis. These data (Goorha et al., 1977, 1978; Tripier-Darcy et al., 1980), indicate that there is a two-way movement of macromolecules between the cytoplasm and nucleus. Although Tripier-Darcy et al. (1980) did not report on the relationship of microtubules to viroplasmic centers in chick embryo fibroblasts, their observations are included here since they perpend to the assembly of new virions in the cytoplasm.

POSSIBLE CYTOSKELETON INVOLVEMENT IN THE VIROPLASMIC CENTER

Viroplasmic centers are usually discrete cytoplasmic loci. Thus, for viruses like TIV there is a well-defined difference in ultrastructure with electron-dense DNA accumulating regions, as detected by electron microscope autoradiography, interspersed with electron-lucent or cytoplasmic regions where virions are found in different stages of assembly (Mathieson and Lee, 1981). Of interest are the following: (1) what factor(s) is responsible for maintaining the integrity of these centers since they lack a membrane; and, in conjunction with this query, (2) what role does the cytoskeleton perform in cells infected with iridoviruses?

There is ample evidence that elements of the cytoskeleton may be involved in virus assembly, in addition to the virus causing cytoskeletal alterations (Ball and Singer, 1981; Hynes, 1980; Lenk and Penman, 1979). In vaccinia virus-infected cells, it has been shown that the cytoskeleton is closely associated with the assembly sites (Hiller et al., 1981; Stokes, 1976). Murti and Goorha (1983) studied the possible relationship between cytoskeletal elements and FV3 in BHK and FHM cells. There was a dimunition and reorganization of microtubules in FV3-infected cells when compared to control BHK cells which was determined by the use of tubulin antibodies (Fig. 11). The decrease in microtubule number was coincidental with the appearance of viroplasmic centers. Figure 11B and C show infected cells 6 and 10 hours p.i., respectively. At these times it is questionable whether the reduction in number and reorganization of microtubules are due to the deleterious effects of FV3 infection.

Intermediate filaments are normally found distributed throughout the cytoplasm of fibroblastic cells in association

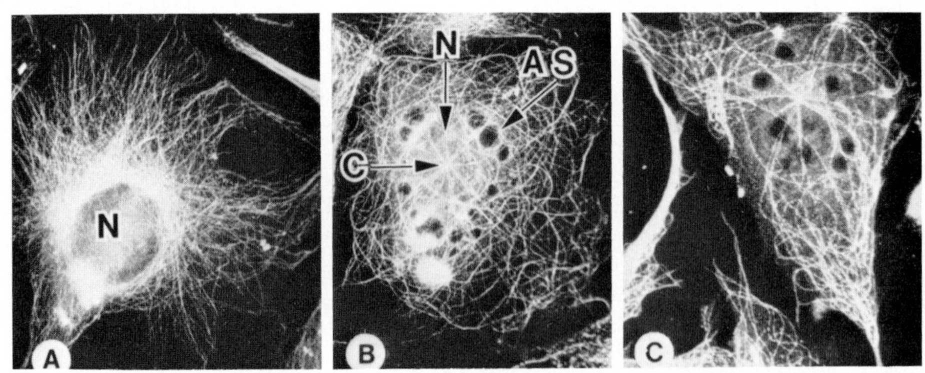

FIGURE 11. Indirect immunofluorescence with tubulin antibodies on BHK cells. (A) Uninfected cells; (B) FV3 infected cell at 6 hours, and (C) 10 hours. N, nucleus; AS, assembly site; label C, focal point. From Murti and Goorha (1983).

with microtubules and microfilaments. They are 10 nm in diameter and are divided into five classes, depending on subunit structure. An excellent review on intermediate filaments has been presented by Lazarides (1980) and to which the reader is referred for further information on this component of the cytoskeleton. Murti and Goorha (1983) found that when cells were infected with FV3 virus at 6 hours p.i., intermediate filaments were no longer found in the periphery of cells and appeared to coerce around the assembly sites. This observation was more pronounced at 8 hours p.i. (Fig. 12). These observations were obtained from indirect immunofluorescence with antibodies to vimentin, and were verified with transmission electron microscopy of FV3-infected FHM cells in which intermediate filaments were seen surrounding assembly sites (Fig. 7; Murti and Goorha, 1983). The authors suggest two possibilities for the changes observed in the intermediate filament pattern: (1) change in intermediate filaments is secondary and is brought about by the disruption of microtubules; and (2) the intermediate filaments play an active role in the formation and maintenance of the membrane-free assembly sites (Murti and Goorha 1983).

Microfilaments are normally seen in the cytoplasm as stress fibers or thin filaments, the former being thick bundles of filaments. This component of the cytoskeleton, according to the observations of Murti and Goorha (1983), also undergoes drastic changes in cells following FV3 infection, which has been determined by immunofluorescence and rhodamine-labeled phalloidin of infected cells. Stress fibers were seen traversing uninfected cells, but at 6 hours p.i. when viroplasmic centers began to appear in the cytoplasm, they were absent; at 8 hours p.i. the infected cell surface contained many fluores-

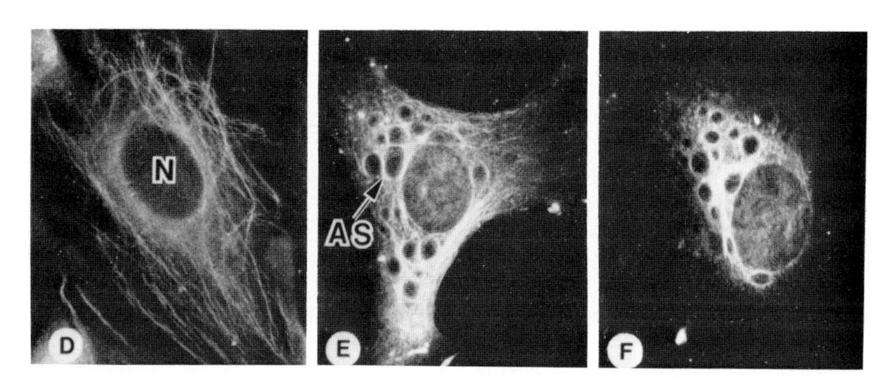

FIGURE 12. Indirect immunofluorescence with 58,000-dalton (vimentin) antibodies on BHK cells. (A) Uninfected cells; (B) cells infected with FV3 for 6 hours; and (C) for 8 hours. N, nucleus; AS, assembly site. From Murti and Goorha (1983).

cent projections which were absent in earlier times of infection or in normal cells (Fig. 13). FV3-infected cells undergo changes in both cell shape and adhesion and the modifications seen in microfilaments may be due to these changes (Kelly and Atkinson, 1975; Maes et al., 1967; Murti and Goorha, 1983). However, Murti and Goorha (1983) suggest also that in the latter stages of infection, actin may be responsible for the release of newly synthesized virions from the cell. In summary, Murti and Goorha (1983) propose that the cytoskeleton of infected cells is actively involved in viral functions from the formation of the assembly sites to the maturation and release of new virions.

The observations on the possible role of the cytoskeleton in FV3-infected cells (Murti and Goorha, 1983) acted as a catalyst for a study of TIV-infected *E. acrea* cells to determine whether cytoskeleton components were implicated in TIV synthesis (Seagull and Lee, in preparation). The reader should bear in mind that *E. acrea* cells are grown in suspension culture and not as fibroblasts like BHK and FHM cells. This difference is mentioned because unlike fibroblasts it is impossible to obtain a well-focused suspension cell due to the thickness of the cell. Healthy E. acrea cells are teardrop in shape and microtubules are found in these cells both by immunofluorescense, using tubulin antibodies, and by transmission microscopy. Microtubules are not associated with the cell membrane, although there is some clustering in small groups as shown by electron microscopy. The use of fluorescence microscopy has shown the tubules to run parallel to the long axis of the cell. The teardrop shape is easily disrupted and usually occurs in the seeding of flasks to establish new cultures. The cells

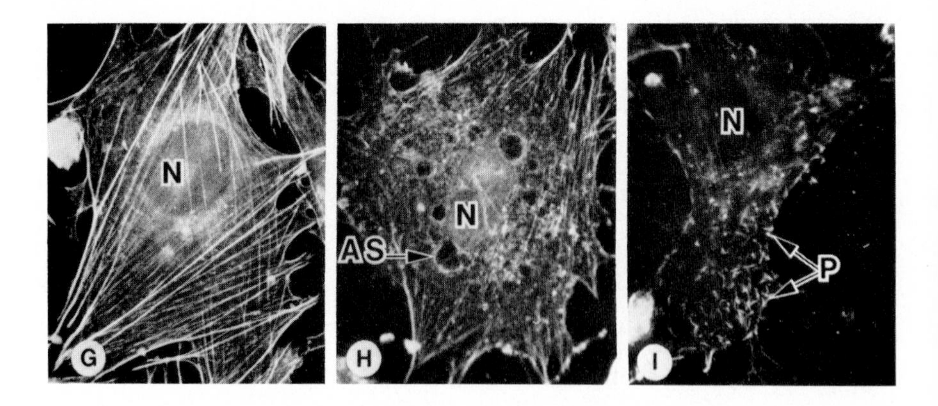

FIGURE 13. Immunofluorescence with rhodamine-conjugated-phalloidin on BHK cells. (A) Uninfected cells; (B) FV3-infected cells at 6 hours; and (C) 10 hours. N, nucleus; AS, assembly site; P, surface projections. From Murti and Goorha (1983).

then are rounded and contain very few microtubules. As determined by fluorescence microscopy, actin has been found to form microspikes at the cell membrane and, unlike microtubules, the organization of actin does not change with cell shape. When cells are infected with TIV, there is no change in the number or pattern of microtubules or is there a change in actin as detected by immunofluorescence. These observations of infected cells were made 12 hours p.i. when viroplasmic centers are visible by phase microscopy or fluorescent staining with Hoechst 33258. At 24 hours p.i., infected *E. acrea* cells show no change in microtubule pattern and, at this time, actin can either be found localized around viroplasmic centers or not. These observations of TIV-infected *E. acrea* cells suggest that neither microtubules nor microfilaments play a role in the formation and maintenance of the assembly sites in infected *E. acrea* cells.

NUCLEARLIKE PROTEINS IN VIROPLASMIC CENTERS

Are there any alternate suggestions which may be responsible for the formation and maintenance of viroplasmic centers in infected cells? Bladon et al. (1983) have raised antibodies against non-histone nuclear proteins from mouse lymphocytes. It was determined from the immune responses when assayed against mouse lymphocytes by indirect immunofluorescence, that there were at least four monoclonal antibodies which were immunologically positive with nuclear skeletal proteins of mouse lymphocytes. These proteins are highly conserved since two of the antibodies reacted with nuclei of *E. acrea* cells in both control and virus-infected cells. In addition, the same antibodies also reacted with viroplasmic centers of TIV-infected *E. acrea* cells (Bladon et al., in preparation). Nuclear skeletal proteins may be responsible for the assembly and maintenance of the chromatin content and configuration in the nucleus (Long et al., 1979; Agutter and Richardson, 1980; Hancock, 1982). Since antibodies to two of these mouse lymphocyte nuclear skeletal proteins reacted positively with viroplasmic centers of TIV infected *E. acrea* cells, it is possible that proteins similar to the matrix type found in nuclei may be responsible for the formation and maintenance of these virus assembly sites and may play a role in the packaging of the viral nucleoid in the empty capsid.

ACKNOWLEDGMENTS

I wish to thank Drs. D. B. Stoltz, K. G. Murti, and R. Goorha for providing some of the photographic plates used for this review. I also thank Dr. R. W. Seagull for helpful discussions during the preparation of this manuscript. The investigations of the author were supported by NSERC Grants A-2911 and E-2134.

REFERENCES

Agutter, P. S., and Richardson, J. C. (1980). Nuclear non-chromatin proteinaceous structures: their role in the organization and function of the interphase nucleus. *J. Cell Sci. 44*, 395-435.

Ball, E. H., and Singer, S. J. (1981). Association of microtubules and intermediate filaments in normal fibroblasts and its disruption upon transformation by a temperature-sensitive mutant of Rous sarcoma virus. *Proc. Natl. Acad. Sci. U.S. 78*, 6986-6990.

Barba, T., and Anderson, P. J. (1963). *"Histochemistry."* Harper and Row, New York.

Bird, F. T. (1961). The development of *Tipula* iridescent virus in the crane fly *Tipula paludosa* Meig. and the wax moth *Galleria mellonella* L. *Can. J. Microbiol. 7*, 827-830.

Bird, F. T. (1962). On the development of the *Tipula* iridescent virus particle. *Can. J. Microbiol. 8*, 533-534.

Blaton, T., Chaly, N., Setterfield, G., Brash, K., and Brown, D. L. (1983). Monoclonal antibodies against lymphocyte nuclear matrix. *Proc. 23rd Meeting Soc. Cell Biol.*

Bladon, T., Sabour, M., Lee, P. E., and Setterfield, G. Viroplasmic centers in *Tipula* iridescent virus-infected cells respond to antibodies raised against nuclear matrix. In preparation.

Goorha, R., and Granoff, A. (1974). Macromolecular synthesis in cells infected by frog virus 3. I. Virus-specific protein synthesis and its regulation. *Virology 60*, 237-250.

Goorha, R., and Granoff, A. (1979). Icosahedral cytoplasmic deoxyriboviruses. *Comprehensive Virol. 14*, 347-399.

Goorha, R., Willis, D. B., and Granoff, A. (1977). Macromolecular synthesis in cells infected by frog virus 3. VI. Frog virus 3 replication is dependent on the cell nucleus. *J. Virol. 21*, 802-805.

Goorha, R., Murti, G., Granoff, A., and Tirey, R. (1978). Macromolecular synthesis in cells infected by frog virus 3. VIII. The nucleus is a site of frog virus 3 DNA and RNA synthesis. *Virology 84*, 32-50.

Hancock, R. (1982). Topological organization of interphase DNA: the nuclear matrix and other skeletal structures. *Biol. Cell. 46*, 105-122.

Hiller, G., Jungwirth, C., and Weber, K. (1981). Fluorescence microscopical analysis of the life cycle of vaccinia virus in chick embryo fibroblasts. *Exp. Cell Res. 132*, 81-87.

Hynes, R. (1980). Cellular location of viral transforming proteins. *Cell 21*, 601-602.

Kang, H. S., Eshbach, T. B., White, D. A., and Levine, A. J. (1971). DNA replication in SV40-infected cells. IV. Two different requirements for protein synthesis during SV40 DNA replication. *J. Virol. 7*, 112-120.

Kelly, D. C., and Atkinson, M. A. (1975). Frog virus 3 replication: electron microscope observations on the terminal stages of infection in chronically infected cell cultures. *J. Cell. Sci. 28*, 391-407.

Kelly, D. C., and Tinsley, T. W. (1974a). Iridescent virus replication: A microscope study of *Aedes aegypti* and *Antherea eucalypti* cells in culture infected with iridescent virus types 2 and 6. *Microbios 9*, 75-93.

Kelly, D. C., and Tinsley, T. W. (1974b). Iridescent virus replication: Patterns of nucleic acid synthesis in insect cells infected with iridescent virus types 2 and 6. *J. Invertebr. Pathol. 24*, 169-178.

Kloc, M., and Lee, P. E. (1983). The effect of hydroxyurea on DNA synthesis of *Tipula* iridescent virus and *Estigmene acrea* cells. *Can. J. Microbiol. 29*, 254-260.

Lazarides, E. (1980). Intermediate filaments as mechanical integrations of cellular space. *Nature (London) 283*, 249-256.

Lee, P. E., and Brownrigg, S. P. (1982). Effect of virus inactivation on *Tipula* iridescent virus-cell relationships. *J. Ultrastruct. Res. 79*, 189-197.

Lenk, R., and Penman, S. (1979). The cytoskeletal framework and poliovirus metabolism. *Cell 16*, 289-301.

Leutenegger, R. (1967). Early events of *Sericesthis* iridescent virus infection in hemocytes of *Galleria mellonella* (L.). *Virology 32*, 109-116.

Lo Grippo, G. A. (1960). Investigations of the use of beta-propiolactone in virus inactivation. *Ann. N.Y. Acad. Sci. 83*, 574-594.

Long, B. H., Huang, C-Y., and Pogo, A. O. (1979). Isolation and characterization of the nuclear matrix in Friend erythroleukemia cells: chromatin and hn RNA interactions with the nuclear matrix. *Cell 18*, 1079-1090.

Maes, R., Granoff, A., and Smith, W. R. (1967). Viruses and renal carcinoma of *Rana pipiens*. III. The relationship between input multiplicity of infection and inclusion body formation in frog virus 3 infected cells. *Virology 33*, 137-144.

Mathieson, B. (1978). A cytological and biochemical study of *Estigmene acrea* continuous cell cultures infected with *Tipula* iridescent virus. M.Sc. Thesis, Carleton University Archives, Ottawa, Canada.

Mathieson, B., and Lee, P. E. (1981). Cytology and autoradiography of *Tipula* iridescent virus infection of insect suspension cell cultures. *J. Ultrastruct. Res. 74*, 59-68.

Murti, K. G., and Goorha, R. (1983). Interaction of frog virus-3 with the cytoskeleton. I. Altered organization of microtubules, intermediate filaments and microfilaments. *J. Cell Biol. 96*, 1248-1257.

Pogo, B., and Dales, S. (1973). Biogenesis of poxvirus: Inactivation of host DNA polymerase by a component of the invading inoculum particle. *Proc. Nat. Acad. Sci. U.S. 70*, 1726-1729.

Seagull, R. W., and Lee, P. E. The cytoskeleton and its role in *Tipula* iridescent virus-infected *Estigmene acrea* cells. In preparation.

Setterfield, G., Sheinin, R., Dardick, I., Kiss, G., and Dubsky, M. (1978). Structure of interphase nuclei in relation to the cell cycle. Chromosome organization in mouse L-cells temperature sensitive for DNA replication. *J. Cell Biol. 77*, 246-263.

Stokes, G. V. (1976). High-voltage electron microscope study of the release of vaccine virus from whole cells. *J. Virol. 18*, 636-643.

Stoltz, D. B. (1971). The structure of icosahedral cytoplasmic deoxyriboviruses. *J. Ultrastruct. Res. 37*, 219-239.

Stoltz, D. B., Hilsenhoff, W. L., and Stich, H. F. (1968). A virus disease in *Chironomus plumosus*. *J. Invertebr. Pathol. 12*, 118-128.

Tinsley, T. W., and Harrap, K. A. (1978). Viruses of invertebrates. *Comprehensive Virol. 12*, 1-101.

Tripier-Darcy, F., Braunwald, J., and Kirn, A. (1980). Induction of intranuclear microtubules in chick embryo fibroblasts by frog virus 3. *Cell Tissue Res. 209*, 271-277.

Younghusband, H. B., and Lee, P. E. (1969). Virus-cell studies of *Tipula* iridescent virus in *Galleria mellonella* (L.). *Virology 38*, 247-254.

Younghusband, H. B., and Lee, P. E. (1970). The cytochemistry and autoradiography of *Tipula* iridescent virus in *Galleria mellonella*. *Virology 40*, 757-760.

Yu, K., Kowalski, J., and Cheevers, W. (1975). DNA synthesis in polyoma virus infection. III. Mechanisms of inhibition of viral DNA replication of cycloheximide. *J. Virol. 15*, 1409-1417.

Yule, B. G., and Lee, P. E. (1973). A cytological and immunological study of *Tipula* iridescent virus-infected *Galleria mellonella* larval hemocytes. *Virology 51*, 409-423.

THE REPLICATION OF BACULOVIRUSES

JAMES L. VAUGHN
EDWARD M. DOUGHERTY

Insect Pathology Laboratory
Plant Protection Institute
BARC, ARS, USDA
Beltsville, MD

I. INTRODUCTION

The diverse morphology, the complex ultrastructure, and the varied host patterns of the baculoviruses have made the study of their replication both scientifically intriguing and technically challenging. The virulence of these viruses for their insect hosts and the devastating effects of the viruses on some insect populations have made them prime candidates for use in the control of insect pests and have thus provided an additional, practical stimulus for the investigation of the infection and replication processes. In order to establish a theoretical basis for the safety of these viruses for nontarget animals and plants it is necessary to understand both the normal replication in permissive hosts and the incomplete or abnormal replication in nonpermissive hosts. The available information on the replication processes has been developed through a series of investigations using infectivity tests and ultrastructural studies in infected insects and ultrastructural and biochemical studies in infected cells grown under closely controlled conditions using the most sensitive and discriminating technology possible. The early research in infected insects has been reviewed by Aizawa (1963), Bergold (1963), Huger (1963), and Smith (1977); interested readers are referred to these reviews for detailed information.

ISBN 0-12-470295-3

These early studies established the general concept of the limited host range of the baculovirus, e.g., granulosis viruses (GV) were found to infect and replicate only in closely related species within a single genus and to have quite highly defined tissue tropisms, e.g., some granulosis were shown to replicate only in the tissues of the midgut. Nuclear polyhedrosis viruses (NPV), however, had a highly variable host range pattern. Some replicated only in closely related species within a single genus, e.g., the *Heliothis* NPVs, while others infected and replicated in species from several families within an order of insects. The NPV isolated from the alfalfa looper, *Autographa californica*, infects species in several families of Lepidoptera and nearly all tissues within the insect are susceptible to infection and support some replication.

In the decade from 1965 to 1975 great progress was made in the development of the methodology required to culture insect cells *in vitro*. New insect cell lines were developed that provided the means for the *in vitro* growth and replication of the nuclear polyhedrosis viruses. The development of plaque assays (Hink and Vail, 1973; Wood, 1977; Knudson, 1979) provided a means of quantitating the infectivity of a virus suspension in cell cultures. The results of these early studies that established the reliability of the *in vitro* methods for the study of NPV replication have been reviewed by Granados (1976) and Knudson and Buckley (1977).

Following the establishment of acceptable *in vitro* methods for the replication of the NPV, studies on these viruses expanded rapidly. This chapter will concentrate on the developments in this area since 1976 using the replication of the NPV from *A. californica* as the model system. In order to provide as complete as possible a picture of all the stages of infection and replication, results from *in vivo* studies will be integrated with the results of *in vitro* studies wherever practical. A short review of the recent developments in cell culture methods and the cell culture factors that could influence the virus replication will be discussed first. Then the viral replication process will be discussed chronologically: (1) viral attachment, penetration, and uncoating of the nucleic acid, (2) early replication events, (3) synthesis of virus structural components, and (4) the assembly and release of progeny viruses. Studies on the replication of the Group B and C baculoviruses will be reviewed to compare their replication with similar processes in the nuclear polyhedrosis viruses. The fourth and more recently discovered baculoviruses are the viruses isolated from the lumen of the calyx region of ovaries in the parasitoid wasps. The replication of these new viruses appears to be similar to that of other baculoviruses but is complicated by the fact that the virus multiplies both in the wasp and in the body of the parasitized host. The limited information on this life cycle and the relationship of the virus to the interactions between the parasite and its host have been thoroughly reviewed by Stoltz and Vinson (1979) and will not be covered here.

II. CELL CULTURE SYSTEMS FOR THE STUDY OF BACULOVIRUS REPLICATION

A. *CHOICE OF CELL LINES*

The study of such events as viral control of cell functions, the time sequence of the synthesis of new viral components, and mechanisms by which these events are regulated must be done in experimental systems where the infectious process can be accurately timed and factors influencing these events can be closely controlled. This can only be achieved in cell culture systems. The availability of suitable cell systems has made possible a large number of such studies with the NPV from Lepidoptera. The absence of suitable cell systems in which to grow the GV has greatly limited the study of the replication of these viruses even though there is a considerable interest in them because of their potential for controlling insect pests. Since both of these baculoviruses occur either predominantly or exclusively in Lepidoptera it seems unusual that a successful cell system for the GV has not been developed.

The most recent compilation indicates that there are 15 new cell lines from Lepidoptera (Hink, 1980), in addition to the 24 previously listed (Hink, 1976). In their review, Knudson and Buckley (1977) list 16 different NPV that have been replicated in one or more cell lines. Although this is a very small percentage of the reported isolates of NPV, it seems reasonable to assume that any NPV can be grown in a cell culture. The methodology for producing a cell line from a species of Lepidoptera is well enough established that cell lines could be produced from a suitable host species if none of the existing lines permitted replication of the desired virus.

Development of several cell lines from a single species may at times be necessary because not all cell lines will permit replication of a NPV even though it is from the same species as the cell line. Ignoffo et al. (1971) found that the *Heliothis zea* NPV would not replicate completely in a cell line derived from ovarian tissue of that insect. They were not able to detect infectious virus or polyhedra although the virus was passed through the cell line seven consecutive times using infected cells from the previous pass as the inoculum. However, later cultures from *H. zea* established by Goodwin et al. (1973) and McIntosh and Ignoffo (1981a) did permit complete replication of this NPV. McIntosh and Ignoffo also reported that even though the *Heliothis* NPV replicated well in three different *H. zea* cell lines they had developed, the virus replicated very poorly in a cell line derived from similar tissues in pupal ovaries from *Heliothis virescens*. Only 3% of the cells in the virescens line showed polyhedral formation compared with 25 to 50% of the cells in the three *H. zea* cell lines. Similar differences in susceptibility were reported among several cell lines developed from pupal ovarian tissue from *Lymantria dispar* (Goodwin et al., 1978). They reported that two of the lines, IPLB-LO-65 and IPLB-LD-67, supported complete replication of the *L. dispar* NPV. However, in some of the sublines of the IPLB-LD-65, only a few cells produced polyhedra after infection. Two other cell lines, IPLB-LD-64 and IPLB-LD-66, did not produce polyhedra at all when infected with the NPV. Whether or not an incomplete replication of the virus occurred was not reported.

In some instances the ability of a cell line to replicate one or more viruses appears to be rather narrowly defined. Differences in the ability of cell lines to support replication of four related *Spodoptera* NPV from *S. exempta, S. exigua, S. frugiperda,* and *S. littoralis* have been reported (Knudson and Buckley, 1977). The IPLB-SF-21AE cell line from *S. frugiperda* would support replication of all four NPVs, but a second *Spodoptera* cell line, from *S. littoralis*, supported replication of only the *S. frugiperda* and *S. littoralis* NPVs. According to the results of nucleotide sequence homology, the *S. exempta* and *S. exigua* NPVs are more closely related to each other than to either the *S. frugiperda* or *S. littoralis* NPV. The *S. frugiperda* and *S. littoralis* NPVs are distantly related to each other and also share 30-40% common sequences with *S. exempta* and *S. exigua* NPVs (Kelly, 1977).

These qualitative studies indicate that there is a degree of variability in the permissiveness of cell lines for replication of the NPV (Group A baculoviruses). Some, but not all, cell lines will support the NPV pathogenic for the insect species from which a line was derived. Other cell lines will support several NPV. The GV (Group B baculoviruses) and the *Oryctes* virus (Group C baculoviruses) appear to represent the extremes. The GV apparently are highly tissue specific and none of the available cell lines will support replication of these viruses. At the other extreme, the *Oryctes* virus has been shown to replicate in cell lines from three different orders, Coleoptera (Quiot et al., 1973), Lepidoptera, and Diptera (Kelly, 1976).

Quantitative evaluation of the replication of the *A. californica* NPV in five different cell lines has been done by Lynn and Hink (1980). The five cell lines tested were BTI-EAA

(*Estigmene acrea*), IPLB-LD-64BA (*L. dispar*), IZD-MB-0503 (*Mamestra brassicae*), IPLB-SF-1254 (*S. frugiperda*), and TN-368 (*Trichoplusia ni*). The IZD-MB-0503 cell line produced statistically significant higher numbers of polyhedra per cell than any of the other four cell lines. The second highest number of polyhedra per cell was produced by the TN-368 cell line. When the quality of the polyhedra produced was determined by feeding to first instar *T. ni* larvae, the polyhedra produced in the TN-368 cells were significantly more virulent than the polyhedra produced in any of the other four cell lines. The IPLB-SF-1254 cell line produced noticeably less nonoccluded virus as measured by plaque assay than did any of the others.

The sensitivity of each cell line to infection was measured by using each one in a plaque assay system. The highest number of plaque forming units/milliliter (PFU/ml) in the standard virus suspension was obtained with the IZD-MB-0503 and almost no plaques were formed in the BTI-EAA cell line. The maximum number of PFU was reached in cell cultures 3 days postinfection. To determine if the variation in plaque number was due to the relative number of susceptible cells in each culture, cultures of each cell line were inoculated with equal amounts of virus. At 4 days postinfection only 59% of the ILPB-LD-64BA cells contained polyhedra, whereas 96% of the IZD-MB-0503 cells had formed polyhedra. The percentages of polyhedra-containing cells in the other three cell lines were above 80%.

Volkman and Goldsmith (1982) used an immunoperoxidase staining technique rather than polyhedra formation to detect infected cells after challenge with *A. californica* NPV. They also found considerable variation in the response of several cell lines to virus challenge. When the IPLB-SF-21AE cell line was challenged, 21% of the stained foci consisted of single cells 64 hr postinfection. In a *Bombyx mori* cell line, more than 99% of the detectable foci were single cells and in a *Manduca sexta* cell line, single infected cells accounted for 80% of the detectable foci. When the *A. californica* NPV was titered in *L. dispar* cells, more than 99% of the foci consisted of single cells as late as 6 days postinfection. Since the virus used was the cloned E2 variant of AcNPV, the observed differences were most likely to result of differences in the host cells.

Such differences in the percentage of cells infected, the number of polyhedra per cell, or the titer of nonoccluded virus (NOV) produced could indicate a high level of variability within the population of a cell culture. Several investigators have attempted to reduce this apparent cellular variation by using cloned cell strains. Brown and Faulkner (1975) cloned the TN-368 cell line and tested the clones for the yield of polyhedra per cell produced. In the uncloned cell line the yield varied from 5 to 200 polyhedra per cell; similar variation was observed in all of the clones. Furthermore, the mean numbers of polyhedra per cell were also similar for all of the clones.

Volkman and Summers (1975) cloned the same cell line in an attempt to improve the sensitivity of their plaque assay system. Most of the cloned strains developed were less sensitive as indicators than was the uncloned parent line. However, one clone was obtained that was as good an indicator as the parent cultures. The authors did observe considerable variation in the time that elapsed from inoculation until the appearance of polyhedra in the cultures. The elapsed time was the shortest for the clone that was the best indicator. This observation did not correlate with the population doubling time of the clones as these were all quite uniform.

Thus, to date, cloning of insect cell lines has not produced any evidence to explain the relatively large variation observed in insect cell cultures. Furthermore, it should be noted that the cloning studies reported here deal only with a single cell line that had been in culture for some time before the cloning studies were done. Therefore, perhaps the cell population already had undergone vigorous selection during the many passages *in vitro* and cell variation was limited. There are, however, some significant differences in the ability of cell lines to support replication of the nuclear polyhedrosis viruses and selection of the most appropriate cell line should be the first task of any new study of these viruses *in vitro*. It should also be kept clearly in mind that most of the existing cell lines were derived from ovarian tissue from one or another stage of Lepidoptera. A few cell lines originated from embryos and one or two originated from hemocytes. There are no cell lines available from midgut epithelium or fat body, both of which are important in the replication cycle of these viruses. Recently, Lynn et al. (1982) reported a new cell line from imaginal wing discs from *T. ni*, while Lynn and Oberlander (1983) reported a new cell line from these tissues in *S. frugiperda* and *Plodia interpunctella*. Lynn et al. (1983) demonstrated that the new cell lines from *T. ni* and *S. frugiperda* permitted complete replication of the NPV from *A. californica, T. ni,* and *Anticarsia gemmatalis*. These are, hopefully, the first of the new cell lines from a wider variety of tissues within the insect. Such new cell lines will provide additional cell types in which baculovirus replication can be studied and perhaps will provide a more complete understanding of the *in vitro* replication process.

B. EFFECT OF MEDIA AND CELL NUTRITION ON BACULOVIRUS REPLICATION

Most of the media used for the growth of insect cells for virus studies are modifications of the medium used by Grace (1962) to establish the first cell lines. The medium is normally substituted with fetal bovine serum and other sources of peptides or vitamins. Goodwin et al. (1970) demonstrated that the addition of glycerol, soybean lecithin (Type II-S), triinolein, trilinolenin, α-tocopherol acetate, and cholesterol to a modified Grace's medium significantly improved the replication of the *S. frugiperda* NPV in a *S. frugiperda* cell line. However, the medium also was supplemented with 3 ml *B. mori* hemolymph, 5 ml fetal bovine serum, and 5 ml egg ultrafiltrate, all of which added potentially critical nutrients. This work did, however, indicate that the cell nutrition as well as the cell itself might have some influence on the replication of the baculoviruses in cell culture. A study on the use of conditioned medium by Stockdale and Gardiner (1977) revealed that medium taken from cultures in the later stages of growth suppressed polyhedra production when used to culture infected early log-phase cells. They concluded from this that their medium containing lactalbumin hydrolysate, tryptose broth, fetal bovine serum, and TC yeastolate (Difco Laboratories, Detroit, MI), might be depleted of essential nutrients during viral replication.

The presence of serum and other complex nutrient sources in these media made the analysis of the effect of specific nutrients difficult, if not impossible, to evaluate. Successful replication of baculoviruses was soon reported on media that did not contain serum (Goodwin, 1976; Vail et al., 1976; Hink et al., 1977a). All of these media contained one or more peptones or hydrolysates and yeast extracts which presumably replaced the fetal bovine serum as a source of unidentified nutrients. Goodwin and Adams (1980) determined some nutrient effects on virus replication in a series of studies with the *L. dispar* NPV in cell lines from *L. dispar* on one such medium. Their initial medium formulation supported cell growth but not the replication of the NPV. The addition of methyl oleate, Tween-80, and cholesterol to the medium resulted in the partial replication of the NPV. Virogenic stroma developed, followed by the formation of enveloped nucleocapsids. Although some polyhedra formed, they only rarely contained virions. Further refinement of the medium and supplementation with α-tocopherol acetate, cholesterol, methyl oleate, and Tween-80 gave what appeared to be normal replication of the gypsy moth NPV.

However, this medium of Goodwin and Adams, although serum-free, contained peptic peptone, liver digest, and TC yeastolate

TABLE I. *Chemically Defined Medium for Insect Cells*[a,b]

Ingredient	mg/L	Ingredient	mg/L
KCl	2870	Putrescine	1.0
$CaCl_2 \cdot 2\ H_2O$	1320	Spermidine	1.0
$MgCl_2 \cdot 6\ H_2O$	2280	Spermine·4 HCl	1.0
$MgSO_4 \cdot 7\ H_2O$	2780	Carnitine	1.0
$NaH_2PO_4 \cdot 2\ H_2O$	1140	α-Amino-n-butyric acid	1.0
$NaHCO_3$	350	D-Phosphorylethanolamine	2.0
		Taurine	1.0
L-Arginine	550	Riboflavin	0.2
L-Aspartate (K^+ salt)	450	p-Aminobenzoic acid	2.0
L-Asparagine	350	Folic acid	1.0
L-Alanine	225	D-Biotin	0.05
L-Glutamate (K^+ salt) $\cdot H_2$)	829	Ca^{2+} D-pantothenate	1.2
L-Glutamine	600	i-Inositol	2.0
Glycine	650	Ascorbic acid	0.2
L-Histidine·HCl·H_2O	3380	Cyanocobalamin	1.0
L-Isoleucine	50	Nicotinamide	1.2
L-Leucine	75	Thiamine·HCl	2.0
L-Lysine·HCl	625	Pyridoxine·HCl	1.0
L-Methionine	50	Choline chloride	20.0
L-Proline	350	Hypoxanthine	10.0
L-Phenylalanine	150		
L-Threonine	175	$FeSO_4(NH_4)_2SO_4 \cdot 6\ H_2O$	5.0
L-Valine	100	$ZnSO_4 \cdot 7\ H_2O$	0.44
L-Serine	550	$CuSO_4 \cdot 5\ H_2O$	0.39
L-Cystine	75	$MnCl_2 \cdot 4\ H_2O$	0.35
L-Tyrosine	70		
L-Tryptophan	100		
α-D-Glucose	4000		
Methylcellulose(15 cps)	2000		
Stearic acid[c]	0.100	Trilinolein	0.1
Myristic acid	0.100	Trilinolenin	0.1
Oleic acid	0.100	Phosphatidylcholine	0.2
Linoleic acid	0.100	Cholesterol	1.0
Palmitic acid	0.100	β-Sitosterol	1.0
Palmitoleic acid	0.100	Stigmasterol	1.0
Arachidonic acid	0.020		
		Tween-80	20.0
		Ethanol	(2 ml)

[a]From Wilkie et al. (1980).

[b]Adjust pH to 6.3 and osmotic pressure to 330 mOsm with KCl.

[c]Values for fatty acids corrected from original table based on information provided by authors.

as possible sources of undefined, critical nutrients. Recently, Wilkie and his colleagues published the first chemically defined medium reported to support replication of a baculovirus in cell culture (Wilkie et al., 1980). This formulation supported four passages of the *S. frugiperda* NPV in the IPLB-SF-21AE cell and large numbers of polyhedra were produced in most cells. The formulation, shown in Table I, is quite complex. However, it makes possible an entirely new area of study--the effect of nutrition on the replication of the baculoviruses in cell culture. The results of these studies will certainly be of importance in the production of baculoviruses for experimental or control use.

In addition to the necessary nutrients the medium provides a number of factors that effect the efficiency of virus infection *in vitro* (Dougherty et al., 1981). Divalent ions (Mg^{2+} and Ca^{2+}) at concentrations of 10^{-2} and 10^{-3} M promoted optimal virus attachment as determined by plaque assays but monovalent ions resulted in poor infection of the TN-368 cells. The medium used in this study, Hink's TNM-FH medium, contained these ions in approximately optimal amounts. The marked response to divalent cations prompted the authors to investigate the effects of various polyanions and polycations. Poly-l-ornithine, and diethylaminoethane more than doubled the rate of NOV infection. In this study, heparin, which normally inhibits the attachment of viruses, enhanced the infection. Neither hypotonic nor hypertonic conditions, within a range tolerated by the cells, appeared to alter the infection process. The NOV were unstable below pH 6.0 but were stable from pH 6 to 10. However, infectivity was markedly reduced at a pH of 8.0 and above.

C. *OTHER CELL-VIRUS FACTORS THAT EFFECT THE EFFICIENCY OF NPV INFECTION*

The phase of cell growth at the time of virus inoculation appears to influence the susceptibility of the cell to infection by baculoviruses. Several studies have shown that plaque assays of the NPV were the most sensitive when cells in the log phase of growth were used (Knudson and Tinsley, 1974; Volkman and Summers, 1975; Wood, 1977). Summers and Volkman found that two or three times more plaques were obtained in monolayers of TN-368 cells with log phase than with lag phase cells to approximately two times more than the number with stationary phase cells. Similar results were obtained with a clone of the parent cell line indicating that it was not due to cell variation in the uncloned culture. Wood (1977) also found that TN-368 cells in log phase were more sensitive than cells in stationary phase in his plaque method but the difference was only 25% more.

However, his method involved centrifugation of the monolayer cells and the virus suspension before the overlay was added. This increased the number of plaques detected and may have reduced the differences in susceptibility between the two stages.

Attempts to increase the production of polyhedra in cell cultures have also shown that the polyhedra yield was influenced by the growth stage of cells when virus was added. Vaughn (1976) found that the confluent monolayers of *S. frugiperda* cells in roller bottles produced no polyhedra when inoculated with infectious virus. However, if the cells were infected 3 days earlier, when presumably they were in log phase, 6.38×10^7 polyhedra per ml were produced. To separate the effects of cell age from medium depletion, Stockdale and Gardiner (1977) inoculated different cell densities and then fed the attached cell cultures with media recovered from cultures that had reached different phases of growth. In their studies, as the cell concentration increased the number of polyhedra produced per cell declined, regardless of the age of the medium used. After mid-log phase virtually no polyhedra production was observed in cultures fed either with fresh medium or medium taken from a stationary phase culture.

Hink et al. (1977b) obtained different results with the TN-368 line in their studies of cells maintained in suspension culture. At 24-hr intervals from 0- to 168-hr cells were removed from the suspension, transferred to flasks, and after attachment and infection, were fed with fresh medium. The percentage of cells infected was not significantly different for any of the samples and the numbers of polyhedra per cell declined significantly only in the 168-hr sample. Since the cells were growing slowly in the fermentor and did not reach maximum cell density until the 168-hr sample, they may not have reached the stationary phase until then.

None of these studies were done on synchronized cells and the growth phases referred to represent the phase of the population rather than the condition of the individual cells. Lynn and Hink (1978) produced synchronized cultures using a modified double thymidine block technique to determine more accurately the effect of the cell growth stage on virus susceptibility. At 6 hr postblock they reported that 30% of the cells were in S phase and 70% were in G_2. The cultures were inoculated with a multiplicity of infection (MOI) of 5 plaque forming units (PFU) per cell and the results were evaluated 25 hr postinfection. The highest percentage of infection was 87.7% at 4.5 hr postblock and the lowest was 74.6% at 12 hr postblock. Thus, the time at which the synchronized cells were most susceptible to virus infection corresponded to the middle and late S phase of the cell cycle.

The yields of both NOV and polyhedra were not significantly different among cultures infected at all phases of the cell cycle. These results indicated that while the phase of the cell cycle at which insect cells were challenged with baculoviruses had some effect on the level of infection, the cell cycle did not influence virus replication if the cell density at inoculation was not equal to confluency.

The effects of cell density on viral replication were recently studied in TN-368 cells by Wood et al. (1982) with some interesting results. Replication was monitored in terms of viral peptide synthesis under conditions of low (1.4×10^5 cells/cm^2) and high cell density (5.7×10^5 cells/cm^2). Under conditions of low cell density, an MOI of 10 resulted in over 90% infection as determined by the presence of polyhedra 36 hr postinfection. However, under high cell density less than 0.1% of the cells contained polyhedra. In addition, the production of NOV per cell was reduced more than 98% at high cell density. Also, at high cell density the infected cells did not synthesize a 30K early protein that was produced in low density conditions. When high density conditions were changed to low density by dilution, the synthesis of the 30K protein began within 2 hr of the dilution. Physical contact with glass beads did not show the same inhibiting effect as did cell to cell contact. Wood and his colleagues determined that cell to cell contact also inhibited cell DNA synthesis but that following reduction in cell density DNA synthesis resumed. Conversion of low to high density conditions by the addition of healthy cells as late as 9 hr postinfection resulted in a significant reduction in virus yield. The replacement of fresh medium with conditioned medium in a low density culture did not inhibit virus replication. Thus, the contributing factor did not appear to be extracellular. Since similar results were obtained in *S. frugiperda* cells as well, Wood et al. postulated that the shut off mechanism might be a common feature of NPV replication and must be considered when establishing conditions for the study or the production of NPV in cell cultures.

The multiplicity of infection (MOI) appears to have significant effects on several aspects of NPV replication *in vitro*. Several estimates have been made on the MOI needed to achieve synchronized infection (Volkman and Summers, 1975; Brown and Faulkner, 1975; Hink et al., 1977b; McCarthy et al., 1980; Dougherty et al., 1981). Most authors have reported that an MOI of five provides greater than 90% infection. McCarthy et al. (1980), however, obtained less than 50% of the cells initially infected with an MOI of 1 to 5. Since all of the authors were using the TN-368 cell line and the *A. californica* NPV or the closely related *T. ni* NPV the reason for the differences found by McCarthy et al. was not apparent. It is important to establish this parameter for each cell-virus system before studies on the replication process are begun in order to avoid the confusion of nonsynchronous virus replication and additional cell division.

Brown and Faulkner (1975) reported that the yields of both NOV and polyhedra were affected by the MOI. The maximum NOV production occurred at MOI's of 0.01 to 4 and decreased as the MOI increased from 10 to 300. Polyhedra yield, on the other hand, was lowest at MOI's below 4. At a MOI of 10 or more the polyhedra yield doubled. Extremely high MOI's, greater than 75, resulted in a decrease of polyhedra yield. McCarthy et al. (1980) found that the maximum titer of extracellular NOV increased tenfold or more as the MOI was increased from 5 to 25 or 50. Other markers of infection such as the appearance of nucleocapsid antigen occurred at 6 hr postinfection and antigens for polyhedron protein appeared at 12 hr postinfection regardless of the MOI used. However, the rates at which the percentage of cells showing these indications of infection varied considerably depending upon the MOI used. These findings further strengthen the idea that any studies on the replication of baculoviruses must be done with synchronized infections or interpretation of the data becomes very difficult.

III. THE PROCESS OF INFECTION

A. *THE INFECTIOUS AGENT*

Early studies on the infectivity of the baculovirus revealed that the infectious agents could be transmitted in three ways: (1) by the occlusion body which, when carefully purified, was infectious by feeding but not by injecting, (2) by either feeding or injecting the virions (OV) released by dissolving the occlusion body, and (3) by the injection of hemolymph from diseased insects free of occlusion bodies (Bergold, 1953). Early attempts to infect primary cultures of cells and tissues demonstrated that only the hemolymph or cell-free extracts of infected tissues were reliable sources of the virus for inoculating these cultures (Vago and Bergoin, 1963; Vaughn and Faulkner, 1963). These reports were followed by several studies to identify the infectious agent in the hemolymph and to determine the differences that could account for the inability of the released virions to infect cell cultures. Granados (1976) has reviewed these studies quite thoroughly. He came to the following conclusions based on his review of the available data: (1) several laboratories had reported the existence of subgenomic infectious units and the nature of these infectious units needed to be clarified. (2) The infectious units (OV) obtained by dissolving polyhedra were enveloped and nonenveloped virus rods. The small spherical particles found in such preparations were noninfectious and probably resulted from the degradation of structural components of the virus. (3) Nonenveloped virus particles appeared to be the infective unit in infectious hemolymph from diseased insects. (4) Both enveloped and nonenveloped virions were possibly the infectious units in the supernatants of infected cell cultures. Interested readers are referred to Granados' review for the background for these conclusions.

In late 1975, Dougherty and his co-workers published the results of a study comparing some of the characteristics of the NOV from both tissue culture supernatants and the hemolymph of diseased insects (Dougherty et al., 1975). Virions from both sources were infectious by intrahemocoelic but not by oral injection, stable over a pH range of 11 to 6, labile when treated for 30 min at 50°C or when exposed to acid treatment at pH 3.0 for 30 min, and were inactivated by a variety of organic solvents including ether. The latter is characteristic of enveloped viruses. Dougherty and his colleagues were able to demonstrate the presence of enveloped virions in an electron micrograph of virus-infected cell cultures. This confirmed the earlier studies of Henderson et al. (1974) who reported that the infectivity of the virions from the supernatant of infected cell cultures was destroyed by treatment with deoxycholate or Tween-ether. After such treatment the buoyant density of the virion in CsCl increased from 1.26 to 1.35 g/ml to 1.42. These results were also indicative of an enveloped virion. However, micrographs of completely enveloped nucleocapsids were not obtained which suggested that the envelope must be quite fragile.

Summers and Volkman (1976) compared the biophysical and morphological properties of the NOV with those of the OV and reported that both were enveloped. The NOV from either hemolymph of infected insects or from infected tissue culture supernatants exhibited similar densities of approximately 1.17 to 1.18 g/ml in sucrose gradients. Electron micrographs of these virions revealed single nucleocapsids with an envelope that varied in amount and in the degree of fit. The envelope in a large number of negatively stained preparations showed varying degrees of disruption unless great care was taken in the handling and preparation of the grids. Occluded virions, on the other hand, exhibited the typical multiple banding characteristic of envelopes containing varying numbers of nucleocapsids. However, the density of the single enveloped nucleocapsid of the OV was greater than that of the single enveloped nucleocapsid of the NOV.

Thus, it now seems clearly established that the infectious unit of the baculovirus is an enveloped nucleocapsid. Electron micrographs by several workers have established its presence and also its fragile nature on the NOV. The loss of infectivity as a result of treatment of the NOV with either deoxycholate or ether confirms that an envelope is essential for infective virions.

Several investigators (Potter and Miller, 1980a; Burand et al., 1980; Carstens et al., 1980; Bud and Kelly, 1980; Kelly and Wang, 1981) have established that cells in culture can be infected by transfection with baculovirus DNA. However, there has been no recent evidence that infectious DNA occurs either in the hemolymph of infected insects or in the supernatants of infected cell cultures. Neither has there been any recent evidence for infectious subviral particles from either infected insects or cell cultures.

Despite the fact that the infectious form of both the OV and the NOV has been shown to be enveloped nucleocapsids, the NOV is the preferred infectious form for cell culture studies because of its higher infectivity. Knudson and Tinsley (1974) estimated that the nonoccluded virions from cell culture had a mean particle to $TCID_{50}$ ratio of 186 ± 124. Later, Volkman et al. (1976) estimated this ratio to be 128 particles per PFU. They estimated that the ratio for the single nucleocapsid per envelope OV was 2.4×10^5 particles per PFU and for the multiple nucleocapsid per envelope OV was 2.2×10^5 particles per PFU. When the NOV was subjected to alkali treatment by a procedure similar to the dissolution procedure used to free the OV, the particle to PFU ratio increased to 256:1 indicating relatively minor damage to the virion in the alkali. Volkman and Summers (1977) studied the question further by quantitatively comparing the particle to infectious dose ratios of the two types of virions under a variety of conditions. The number of particles was estimated as genome equivalents calculated from determinations of the DNA and protein of the virions. These results indicated that the NOV was more infectious by hemocoelic injection than was the OV (genome equivalents for OV per $LD_{50} = 7.05 \times 10^4$ and for NOV $= 2.8 \times 10^2$). The reverse was true when infection was per os (OV genome equivalents per $LD_{50} = 2.3 \times 10^4$ and for NOV $= 8.8 \times 10^6$). When the insects were infected by feeding polyhedra, the genome equivalent per LD_{50} was 4.3×10^3. The difference between the ratio of OV fed to polyhedra fed was just over fivefold indicating that little damage was done to the virion by the dissolution procedure. There was very poor infectivity with the NOV when assayed per os (8.8×10^6 genomic equivalents per LD_{50}) compared to NOV assayed either by intrahemocoelic injection (274 genomic equivalents per LD_{50}) or in cell culture (86 genomic equivalents per $TCID_{50}$). These results are in agreement with the earlier work which resulted in the use of the NOV for cell culture studies and with the work of Dougherty et al. (1975) which had shown that NOV from neither hemolymph nor cell culture supernatant was infectious per os. Stairs (1980) obtained no deaths when third instar larvae of *Galleria mellonella* were fed small pieces of food to which dilutions of the NOV from *G. mellonella* NPV had been added. However, the suspension had a LD_{50} titer of 10^{-5} when injected into fifth instar larvae. When the whole polyhedra were fed to third instar larvae, the LD_{50} was 8×10^2 polyhedra. Unlike the results of Volkman and Summers above, when the OV were released from the polyhedra and assayed by feeding to similar larvae, the LD_{50} increased 10,000-fold to 8×10^6 polyhedra equivalents. However, the methods for dissolving the polyhedra, for administering the test material, and the age and species of the insect used differed. Thus, the two tests could not be equated because any of these factors could have altered the results. Occluded virions freed by dissolving in gut juice either recovered from the fed insect (Granados and Lawler, 1981) or *in vitro* (Vail et al., 1979), also demonstrated infectivity in cell culture but the data from Vail et al. was indicative of a high particle to infectious unit ratio (2.75×10^4 polyhedra per PFU).

These differences in infectivity indicate some significant differences between the NOV and OV. The electron micrographs discussed earlier indicated a difference in the fit of the envelope surrounding the nucleocapsid and perhaps some difference in the stability as well. The envelope of the NOV also contains specialized structures, peplomers, which have not been seen on the envelopes of the OV (Summers and Volkman, 1976; Adams et al., 1977; Granados and Lawler, 1981). However, no specific function has been identified for them. The envelopes of the two virions also appear to have somewhat different antigenic properties (Volkman et al., 1976). Antiserum raised against the NOV was capable of neutralizing the NOV and OV but antiserum against the OV only neutralized that virion and had little or no cross reaction with the NOV. When the anti-NOV serum was adsorbed with OV the ability to neutralize NOV was unchanged. Thus, it was concluded that the two types of virions had one or more antigen in common and each also had at least one antigen that was unique.

Kawarabata and Aratake (1978) compared the ability of the anti-NOV and anti-OV sera to neutralize the two forms of the *B. mori* NPV using as detection systems *B.* mori larvae or pupae. They found that anti-NOV serum was highly effective for neutralization of both NOV and OV when tested by intrahemocoelic injection. Anti-OV serum did not significantly neutralize either the OV or the NOV in this test system. When the neutralizing ability of the antisera was tested by per os injection, the occluded virus was not neutralized by its homologus antiserum, even though the antiserum had a high complement fixation titer. The anti-NOV serum had no neutralizing effect on the occluded virus when tested per os despite its high neutralizing activity against the OV when tested by intrahemocoelic injection and the infectivity of the NOV was so low that the neutralizing activity could not be determined.

The authors concluded that the neutralization of the intrahemocoelic infectivity was because the infectious particle responsible was the nucleocapsid only, and that the nucleocapsids of either the NOV or OV were antigenically homogenous. These findings that antiserum against NOV neutralized both NOV and UV agree with the earlier findings of Volkman et al. (1976) that the NOV and the OV each have at least one antigen that is unique. However, some differences in the constituent peptides of the two forms were recently demonstrated (Section IV,2,b). The functional significance of these differences has not been defined and their role in cell attachment, if any, is unknown.

B. CELL ATTACHMENT AND PENETRATION

Perhaps one of the more difficult stages of the viral replication cycle to study is the penetration of the virus into a host cell. This event occurs shortly after exposure of the cell to the virus and appears to be of short duration for most animal viruses. Most of what is known about this phase of insect virus replication has been learned from the study of electron micrographs, which, as pointed out by Granados (1976), can contain artifacts and must be evaluated with caution. The process in insect cells is further confused by the difficulty in preparing morphologically uniform suspensions of purified NOV. This, as discussed above, has resulted in disagreement as to the nature of the infectious particle. It is easier to prepare morphologically uniform suspensions of the OV form; the natural route of infection for the OV is through the midgut epithelium. Since no cultures of midgut epithelial cells exist and the OV infectivity for existing cell lines is very poor, the attachment and penetration of the OV has, of necessity, been studied *in vivo*.

Harrap (1970) studied the infection of larvae *Aglais urticae* feeding on plants that had been coated with a suspension of polyhedra. He found that enveloped virus particles in the lumen of the midgut were occasionally seen closely associated with the microvilli at the apex of the columnar epithelial cells. He observed that the virus had an envelope that loosely surrounded such particles as opposed to the tightly fitting envelope seen in polyhedra. No evidence for fusion of the viral envelope and the cell plasma membrane was obtained; however, virus particles seen inside the microvilli did not contain the envelope. Fusion of the viral envelope and the cell plasma membrane was observed by Kawanishi et al. (1972) in larvae of *Rachiplusia ou*. No particular orientation could be determined as a prerequisite for fusion as the virions were observed aligned with the cell both tip-to-tip and side-by-side. As had been reported by Harrap, they observed that the virions lost their envelope upon entering the microvilli. However, the cluster configuration of the multicapsid virions was retained.

The first attempts to determine the time frame in which these events occurred was reported by Granados (1978) and Granados and Lawler (1981). In his first study, Granados used *H. zea* larvae micro-fed with large numbers of polyhedra (2.2×10^7/larvae); midintestines were periodically removed for examination. He found many free enveloped virus particles among the microvilli at 1, 2, and 4 hr postinfection. Non-enveloped or empty capsids were rarely seen, indicating that the virion was stable in the midgut environment. Fusion was observed within 2 hr and the nucleocapsid entered the microvilli without the virion envelope, as previously observed by others. In the second study, Granados and Lawler did not observe fusion of the virion of *A. californica* NOV with the microvilli of *T. ni*. However, nonenveloped particles were seen only rarely in the midgut lumen and only nucleocapsids were seen within the microvilli. These nucleocapsids were observed in the microvilli within 30 minutes postinfection. Multiple nucleocapsids entered the microvilli together and had a tendency to remain clustered until exiting the microvilli. Granados and Lawler reported that they did not observe virions entering cells by viropexis or did the virions appear to pass into the hemocoel through intercellular spaces. In none of these studies with NPV were virus infections observed in the goblet cells of the midgut. Thus it appears that the initial infection occurs only in the columnar epithelial cells.

The attachment and penetration of GV *in vivo* is, in some ways, similar to that described for the NPV--the columnar epithelial cells are the site of infection in the midgut and goblet cells are not involved, enveloped virus particles quickly appear adjacent to the microvilli, and only nucleocapsids are seen inside the microvilli. Summers (1969, 1971) presents considerable evidence that the steps in attachment and penetration of the GV in *T. ni* involve fusion of the viral envelope and the plasma membrane of the microvilli and that the virus particle moves into the microvilli with concomitant loss of the viral envelope. He thought that the virion entered at the tip of the microvilli, unlike the NPV, and few particles were seen side-by-side with the microvilli. Summers also found no evidence that phagocytosis might be a means by which the virions entered a susceptible cell.

Tanada and Leutenegger (1970) found virus particles apparently enclosed in phagocytic vesicles in *T. ni* larvae, which is evidence for phagocytosis. They also found numerous virus particles aligned in the intercellular spaces between midgut cells. The virus particles tended to accumulate near the basement membrane 24 hr after injestion of the polyhedra. However, no direct evidence was obtained as to whether or not these particles were able to enter the hemocoel through the basement membrane. Tanada and Leutenegger reported that no evidence of viral infection was seen at 6 hr postinfection. This would appear to indicate that GV infection proceeds much slower than NPV, where nucleocapsids were observed in the microvilli within 30 min postinfection (Granados and Lawler, 1981). *In vitro* studies provide the most accurate means of determining the time sequence of such events. Unfortunately, these excellent *in vivo* studies cannot be duplicated *in vitro* because midgut epithelial cell cultures do not yet exist and for GV none of the existing cell lines permit replication.

There have been limited studies of the invasion by NPV of insect cells in cultures. In one of the first studies of the infection process *in vitro*, Vaughn et al. (1972) reported that when *S. frugiperda* cells were inoculated with OV released from the polyhedra, particles were observed apparently undergoing phagocytosis at the plasma membrane and that within 12 hr postinoculation numerous virus particles both in vacuoles and free were seen in the cytoplasm. However, no further evidence of infection in the cells was observed. When similar cultures of these cells were inoculated with NOV, complete virus replication occurred but virus entry was not observed.

Virus entry was observed by Raghow and Grace (1974) in cultures of *B. mori* cells. Enveloped virus particles were observed to attach by one end to the plasma membrane. This step was followed by an active invagination of the plasma membrane and phagocytosis. It was noted that all entering particles had their membranes intact and at no stage were nucleocapsids seen penetrating the plasma membranes. These events took place between 0 and 8 hr postinoculation and viral particles seen in the cytoplasm 4-8 hr postinoculation showed only part of the viral membrane.

These findings have been confirmed by several others (Hirumi et al., 1975; Knudson and Harrap, 1976; Adams et al., 1977). All of the latter authors reported that the virions in the cytoplasm were contained within vesicles that were lost shortly after the virion had entered the cell. Attachment to the plasma membrane was observed as early as 10 min postinoculation (Hirumi et al., 1975) and was nearly always seen to occur at one end of the virion. All of the attached virus particles were enveloped and there was some evidence that fusion occurred. These observations could be evidence that the peplomers at one end of the viral membrane have a specific role in the attachment. However, the electron microscope evidence is inconclusive.

Peplomers are normally associated with glycosylated proteins. The protein portion is virus specific while the carbohydrate moiety is directed by cellular transferases. Glycoproteins have been described in the Baculoviridae (see Section V,2), however, no specific localization of glycoprotein has been made with peplomer structure. Although the nonoccluded form of the virus is the only form having peplomers, only the occluded virus hemagglutinates (HA). Previous studies have shown that both virions (Shapiro and Ignoffo, 1970) and polyhedrin protein (Reichelderfer, 1974) are capable of causing HA. However, the strong agglutinating activity of the polyhedrin protein could be responsible for much of the virion HA activity due to incomplete dissolution of polyhedrin from the virions. Hapten inhibition studies (Peters and DiCapua, 1978) have shown that *N*-acetylgalactosamine and *N*-acetylglucosamine greatly inhibit polyhedrin agglutination; galactosamine, glucosamine, and fucose inhibit to a lesser extent. HA has not been effectively demonstrated with the nonoccluded form of the virus. If the peplomers are responsible for HA activity, their polar position may not allow for bridging between erythrocytes, a condition necessary for HA. Radiolabeling nonoccluded virus and reacting it with several erythrocytes from several animal species and buffer systems (pH, ionic strength, and ion species variation) has not shown binding of virus to blood cells (EMD, unpublished data). Future experiments are needed for determining optimum conditions for performing HA and other biological and biochemical tests for the presence of virions and virion components (peplomers). A recent preliminary report (Stiles et al., 1983) has shown that the glycosylation inhibitor tunicamycin reduces the infectivity of the nonoccluded virus but does not alter the infectivity of the occluded virus form. The basis of this inhibition may be the beginning of understanding the biochemical mechanisms of initiation of infection.

C. NUCLEOCAPSID UNCOATING

In *in vivo* studies, Kawanishi et al. (1972) observed that NPV nucleocapsids in *R. ou* epithelial cells were present in the nucleus within 4 hr after exposure of the larvae to the virus. Granados (1978) also found that nucleocapsids appeared in the nuclei of infected *H. zea* midgut cells within 2 to 4 hr postinoculation. Empty or partially empty capsids were observed in the nucleus as well, indicating that uncoating of the NPV nucleocapsids occurred within the nucleus. However, Tanada and Hess (1976) observed apparently empty capsids adjacent to the nuclear pores in the infected midgut cells of the armyworm, *Pseudaletia unipuncta*, implying uncoating outside the nucleus.

In studies *in vitro* partially electron-dense nucleocapsids were seen by Raghow and Grace (1974) attached at the nuclear pore in cultures of infected *B. mori* cells prior to the appearance of any recognizable change in the nucleus. Hirumi et al. (1975) and Knudson and Harrap (1976), observed entire nucleocapsids in the nuclei of infected cultured cells. Neither were able to detect empty capsids at the nuclear pore. The nucleocapsids were present in the nuclei within 1 to 3 hr postinoculation and nucleocapsids aligned "end on" with the nuclear pores were rarely observed by these workers. Similarly, the evidence obtained by Adams et al. (1977) indicates that uncoating occurred within the nucleus of infected cells in culture. They suggested that nucleocapsids might bud into the nucleus by an unfolding of the nuclear membrane in a manner similar to the process by which the virion enters the cytoplasm. Thus the preponderance of evidence, both *in vivo* and *in vitro*, is that the NPV virion is uncoated as early as 1 to 4 hr postinfection within the nucleus of an infected cell.

The first evidence for the mechanism of nucleocapsid uncoating for the GV was provided by Summers (1969, 1971). He observed that in midgut cells infected with GV the nucleocapsids migrated through the cytoplasm to the nucleus where they appeared to attach at the nuclear pores. Partially empty and empty capsids were also observed at the nuclear pores, indicating that uncoating had taken place at the nuclear membrane and the viral DNA had been released through the pore into the nucleus.

Tweeten et al. (1980) achieved extrusion of nucleoprotein from nucleocapsids by combining purified nucleocapsids with 0.01 *M* EDTA and 0.005 *M* dithiothreitol, a chelator, and a reducing agent. Subterminal openings in nucleocapsids developed, through which rod-shaping structures possessing the compactness and staining properties characteristic of a DNA-protein complex were extruded. Recent morphological studies on nucleocapsids by Burley et al. (1982) utilizing contrast variation methods in electron microscopy and low angle X-ray solution scattering provide evidence for the existence of caps on the *S. litura* granulosis viruses. Structural proteins composing the open stacked ring structure running parallel to the axis of the cylinder appear distinct from structural proteins at the ends of the cylinder, which have decreasing diameters as they approach the tips. Structural and function analysis in cells remain to be determined to see if the test tube experiments of Tweeten and the morphological studies of Burley do indeed describe the uncoating mechanism for the GV.

Quiot et al. (1973) studied the replication of the *Oryctes* virus in cultures of cells from coleopteran host insects and observed partially empty and empty nucleocapsids aligned on the outside of the nucleus. It appears, therefore, that this baculovirus uncoated outside of the nucleus in a manner very similar to that of the GV. In a study of the replication of this virus in the dipteran cell line from *Aedes albopictus*, Kelly (1976) observed that virus replication occurred entirely within the cytoplasm and the nuclei of infected cells appeared normal. Therefore, in this cell system uncoating must take place in the cytoplasm although no evidence for this was reported.

IV. EARLY EVENTS

A. *SHUTDOWN OF HOST SYNTHETIC PROCESSES*

1. *Nucleic Acids*

The early events of a virus infection are critical for the eventual success of the infection and often play a definitive role in establishing the severity of the virulence. In several animal virus systems the infecting virus is capable of inhibiting the host cell capabilities for macromolecular synthesis. The cessation of nucleic acid (Adenovirus, Herpesvirus, Poxvirus, Reovirus) and/or protein synthetic events (Adenovirus, Herpesvirus, Poxvirus, Picornavirus) in the host cell not only renders host cell functions useless but allows the synthetic events of the incoming genome (virus) to occur with little or no competition.

In vitro studies of the group A Baculoviridae, the nuclear polyhedrosis viruses (NPV), have demonstrated inhibition of both host cell DNA and protein synthesis. Although there has been no in-depth study made of the fate and future synthesis of the host chromosome after NPV infection, there is enough data to show that synthesis of chromosomal material is impeded. Knudson and Tinsley (1974) have reported that a NPV of the fall armyworm, *S. frugiperda*, replicating in its homologous cell line was capable of diminishing host cell synthesis so that by 8-12 hr postinfection only 12% of newly synthesized DNA hybridized with host cell DNA. Further, the authors state that "sucrose gradient profiles may suggest that the virus may carry or stimulate DNA nuclease activity because radioactivity was observed at the top of the gradient during early intervals." A more definitive study has to be made to determine if a nuclease is indeed present and whether it is a virus coded or an induced cell nuclease. The answer to this question would aid in determining whether the reduction in synthesis of host DNA is caused by an inhibiting molecule (other than nuclease) synthesized by the invading genome or whether the host genome is destroyed by a nuclease and thus is unable to be copied.

Knudson and Tinsley's report of the inhibition of cellular DNA synthesis in NPV-infected cells was corroborated by Brown et al. (1979) using temperature-sensitive mutants. Synthesis of total DNA in *S. frugiperda* cells infected with a temperature-sensitive mutant of *A. californica* NPV, *ts*8, deficient for viral DNA synthesis, was low compared to cells infected with wild-type virus where total DNA synthesis was maintained at a high rate. Hybridization, specific for *A. californica* NPV, showed negligible virus DNA synthesis in the *ts*8 mutant-infected cells grown at 33°C compared to wild-type infected cells. The *ts*8-infected cells, however, were unable to divide and synthesized little DNA. As with Knudson and Tinsley's study, the quality of the cell DNA produced was not characterized.

2. *Host Proteins*

There is more known about the inhibition of host protein synthesis studies by NPV's (Carstens et al., 1979; Dobos and Cochran, 1980; Kelly and Lescott, 1981; Wood, 1980a). Unlike the polio virus where host cell protein synthesis is abruptly and radically inhibited, all current reports concerning protein synthesis with NPV-infected cells state in one fashion or another that "suppression of host polypeptide synthesis is slow." Kelly and Lescott (1981), studying the replication of *T. ni* NPV in *S. frugiperda* cells, found that low levels of cytosine arabinoside were helpful in not only delineating the β and γ phases of viral protein synthesis but also showed "the bulk of the inhibition of cell polypeptide synthesis is probably due to γ phase polypeptides (virus)." The γ phase polypeptides are initiated from 10-12 hr postinfection in this cell virus system. Inspection of autoradiograms of SDS-polyacrylamide gels of NPV-infected cells (of several cell-virus systems) throughout the course of an infection shows that eventually the host cell protein synthesis was almost completely inhibited and the major labeled proteins present were virus associated.

B. APPEARANCE OF EARLY PROTEINS

Early proteins, those synthesized 4 to 6 hr postinfection, before progeny genome replication, have been reported from the NPV (Carstens et al., 1979; Dobos and Cochran, 1980; Kelly and Lescott, 1981; Wood, 1980a). There are several unique proteins which appear in NPV-infected cells prior to that time period, however, few specific functions have been attributed to them. Kelly and Lescott (1981) reported five polypeptides that were detected initially from 2 to 3 hr postinfection when *S. frugiperda* cells were infected by *T. ni* NPV. These proteins included structural components of 42 and 37K mol. wt. and nonstructural components of 121, 54, and 29K mol. wt. Wood (1980a) describing the *A. californica* NPV replication in TN-368 and IPLB-21 cell lines found four viral-induced proteins during the "early phase" of replication. These proteins were described as having molecular weights of 45, 35, 34, and 31K daltons. Specific structural, nonstructural, and enzymatic characterization was not ascribed for any of the proteins. Carstens et al. (1979), describing *A. californica* NPV replication in *S. frugiperda* cells, found three virus-induced polypeptides of 46, 30, and 29K mol. wt. Although no specific function was attributed to any of the three proteins, the authors did speculate that there was a definite temporal sequence to the synthesis of these peptides before progeny genome replication occurred. Thus, these proteins "may have some functions in priming the infected cells for later stages of DNA and protein synthesis." Finally, the work of Dobos and Cochran (1980) arbitrarily divided NPV protein replication in cultured cells into three segments with an early phase containing peptides synthesized from 0 to 12 hours. This was a longer "early time" than other authors had found and a period in which progeny DNA synthesis had already begun. During the early phase they detected seven peptides of which only a 48K mol. wt. molecule was detected at 4 to 6 hr. The other early proteins (45, 36, 32, 27K mol. wt. were not detected until 10-12 hr postinfection and thus would be grouped with later temporal areas of synthesis when compared to the findings of other authors. By increasing the efficiency of the system by adding 1% DMSO, the authors were able to step up synthetic events by 6-8 hr for most peptides. Although the authors do not give definite quantitation for new times of detection of these peptides, which originally appeared at 10 to 12 hr, visual inspection of the SDS-PAGE gels in the manuscript showed the 32K mol. wt. peptide appearing at 6 hr and at least two other peptides which may be 36 and 45K mol appearing in similarly early times more in agreement with other studies reported.

It should be noted, however, that molecular replication with NPV in *in vitro* infections show similar patterns of protein and nucleic acid synthesis. The use of different viruses, cell lines, multiplicity of infection, temperature, etc. can and does cause slight skewing of the detection times of various macromolecules. However, it seems that all components in such a system are proportionately delayed or hurriedly made.

All of the peptides described as "early times," α synthesis," "first detected," or "early phase" have no stated function except the 42 and 37K mol. wt. peptides, which were described as structural in Kelly's system. Although the Group A Baculoviridae have coding capacity for approximately three times the number of proteins ascribed to a virus infection, little progress has been made in finding virus-coded enzymes and other types of virus-coded molecules in infected cells. However, mention should be made of two recently discovered enzymes that are potentially virus encoded before moving to a review of structural synthesis. Miller et al. (1981) has reported the induction of a new aphidicolin-sensitive α-like DNA polymerase upon infection of *T. ni* larvae. The activity of the induced enzyme was compared to endogenous DNA polymerase activity at 4 days postinfection and a substantial increase in activity in infected larvae over control larvae was found. There was considerable evidence that the increased activity was the result of a new enzyme. The induced enzyme activity was much more sensitive to heat inactivation at 45°C (10 vs 70% activity: virus induced to cell), but the activity of the induced enzyme was increased and the cell enzyme decreased when either 0.25 *M* KCl (240%: 25%) or 0.1 *M* phosphate (240%: 89%) was added to the reaction solution. However, several properties were shared, including equivalent sensitivities to aphidicolin. The data shows that the behavior of the virus-induced enzyme differs from the endogenous host polymerase. However, the authors do not disregard the possibility of substantial viral modification of the preexisting host cell polymerase. The authors go on to state that "direct conformation that a new α-like polymerase is encoded fully or in part by the 8.7×10^7 dalton *A. californica* NPV DNA genome must await correlation with viral genetic mutants."

Concurrently, a study by Kelly (1981) has shown that *T. ni* NPV infection of *S. frugiperda* cells stimulates thymidine kinase and DNA polymerase activity in infected cells. Using a more defined *in vitro* system, Kelly was able to show that thymidine kinase activity was detected within 6 hr after infection, reached a plateau at 8 to 9 hr, and remained elevated for up to 24 hr. The enzyme was also stimulated in cells treated with cytosine arabinoside which inhibited formation of polypeptides synthesized after DNA replication (γ and δ peptides). The activity of thymidine kinase activity in virus-infected cells differed from endogenous activity in several ways: (1) virus induced a k_m of 5.57 μM compared to a control cell where the KM was 12.5 μM, and (2) a virus induced optimal pH activity and Mg^{2+} concentration of pH 7.5 and 0.3 μM, respectively, and an optimum pH and Mg^{2+} concentration of pH 8.5 and 1 μM, respectively. Virus-induced DNA polymerase was also reported by Kelly. The first detection of DNA polymerase stimulation was at 5 hr postinfection and activity reached a plateau at 10 to 12 hr, remaining high until 16 hr. Optimal conditions for virus-induced DNA polymerase activity differed from those for endogenous activity as follows: pH 7.5, 2.8 mM Mg^{2+}, 26.5 M KCl versus pH 7.5, 1.4 mM Mg^{2+}, and 14 mM KCl, respectively. As with the induced thymidine kinase activity, the viral-induced DNA polymerase activity was stimulated in cells treated with cytosine arabinoside. Thus, both enzymes were probably induced during α or β synthetic temporal periods. Because both induction activities were initially stimulated at 5 to 6 hr they were probably β synthetic events occurring before the onset of virus DNA synthesis (6-7 hr postinfection) and thus they could be considered as true "early proteins."

Of the two enzymes detected, the thymidine kinase activity of the virus-induced system differed greatly from endogenous activity and was considered a "novel enzyme." The induced DNA polymerase did not show great differences from endogenous activity as far as inhibition and optimal conditions, etc., and thus there was speculation about its origin. Until the activity can be ascribed to a viral gene, either entirely, or as a modification of a host DNA polymerase, the origin of the increased activity will remain unknown.

V. VIRUS STRUCTURAL SYNTHESIS

A. *NUCLEIC ACID SYNTHESIS*

The temporal qualitation and quantitation of nucleic acid synthesis in NPV-infected cells has been studied in two cell virus systems with comparable results (Knudson and Tinsley, 1978; Tjia et al., 1979). There was an eclipse period of approximately 5 hr as described by Tjia et al. (1979) in the *A. californica* NPV-infected *S. frugiperda* cell culture when only input virus could be detected. Using a high multiplicity of infection (100 PFU/cell), only small amounts of parental virus were found at 2 to 4 hr with Southern blot hybridization. The first noticeable amount of increased virus synthesis was detected at 5 hr with Southern blot and at 6 to 7 hr with DNA-

DNA filter hybridization. After the detection of initial DNA synthesis there was a progressive and rapid increase of progeny virus synthesis until 16-18 hr when a plateau was achieved. Thereafter, there was a gradual decrease in synthesis until 22 hr, which was the last time sampled. Similar results were obtained in the *S. frugiperda* NPV-infected *S. frugiperda* cells (Knudson and Tinsley, 1974). Newly synthesized progeny viral DNA was first detected in appreciable amounts at 6 to 8 hr post-infection and reached maximum intensity at 10 to 12 hr, thereafter decreasing until 36-38 hr postinfection when very little virus-specific material was detected. The DNA which was synthesized was not methylated as determined by digestion with the isoschizomeric restriction enzymes *Hpa*II and *Msp*I (Tjia et al., 1979). Their experiments, which were designed to show previous integration of progeny virus DNA into the host cell DNA, failed to show little, if any, virus sequences present in non-infected cells.

The purified genomes of the baculoviruses are a mixture of covalently closed circle, open circle, and linear molecules (Knudson and Tinsley, 1978; Smith and Summers, 1978; Tjia et al., 1979). The discrepancies in the relative configuration and various amounts of each have not been accounted for, although extraction and purification procedures probably account for nicking and relaxation of the genome. Little is known about the mechanisms involved in the processing and packaging of the progeny genomes. Tweeten et al. (1980) have reported a very basic protein in a GV of *P. interpunctella* that may be involved in packaging viral DNA into compact cores suitable for encapsidation. The very basic protein, 12K mol. wt. protein, was analyzed on PAGE gels containing 6.25 μM urea and adjusted to pH 3.2 (after 0.25 M sulfuric acid extraction from purified nucleocapsids) and found to migrate further (therefore, more basic) than calf thymus histones, but not as far (therefore, less basic) as protamine sulfate. Over one-third of the amino acid composition of the peptide was basic (27% arginine and 12% histidine with trace amounts of lysine present). The isoelectric point was found to be between pH 9.8 and 10. The protein was shown to be associated with the core structures of DNA-protein complex and was not a component of the capsid. Treatment with 0.01 M EDTA and 0.005 M dithiothreitol caused "blowing out" of the ends of the nucleocapsids and visualization of an extruded nucleoprotein core after uranyl acetate staining.

Elliott and Kelly (1977) reported the presence of polyamines in the virions contained in the polyhedral inclusion bodies of *S. frugiperda* NPV. Putrescine, spermine, and spermidine are present, although their role in nucleocapsid organization as well as their presence and relative amounts in the plasma membrane-budded nonoccluded virus is unknown.

The characterization of various classes of RNA during the replication cycle of the Group A Baculoviridae has not been extensively studied. Recently, several investigations have been reported on the various RNA's and associated enzymes necessary for a baculovirus infection. The study by Grula et al. (1981) has been the only one which identified specific enzymes used in RNA processing in baculovirus-infected cells. The authors felt that α-amanitin did not block NPV specific RNA synthesis. Therefore, the host cell RNA polymerase II, the enzyme normally responsible for transcribing cellular mRNA and the mRNA for all nuclear replicating DNA viruses, did not play a significant role in transcribing NPV mRNA. The authors demonstrated that the endogenous host cell RNA polymerase II present in extracts of infected cells late in RNA synthesis (48, 72, 96, and 120 hr postinfection) and of noninfected *S. frugiperda* cells was α-amanitin sensitive, but the transcription of viral RNA was not inhibited. The RNA polymerase II is the enzyme normally responsible for transcribing cellular RNA and mRNA for all nuclear replicating DNA viruses. However, no new RNA polymerase activity was detected on DEAE-sephadex columns although the viral genome contains ample coding capacity which is unaccounted for. It is possible that the other host cell RNA polymerases, I and III, both of which are α-amanitin resistant, play a significant role in transcribing NPV genomes. However, a lack of specific inhibitors causes this hypothesis to remain untested. Although there are reports of a nuclear DNA virus genome transcribed by a host RNA polymerase other than II, the amount of transcription was low compared to NPV infections. Thus, the enzymatic process for transcribing NPV genomes is unique among those DNA viruses replicating in cell nuclei.

The product of the viral transcription, mRNA, has been investigated by Qin and Weaver (1982) who have observed the capping process involved with NPV infections. An approximately equal amount of caps appeared in viral polyadenylated and nonpolyadenylated RNA's. The ratios of ^{32}P incorporated into the caps compared to ^{32}P incorporated into mononucleotides suggested average lengths of 1800 nucleotides for polyadenylated and 1200 nucleotides for nonpolyadenylated RNA's. The caps were found to contain both adenosine and guanosine and *in vivo* labeling with [^3H]methionine and ^{32}P followed by DEAE-cellulose chromography showed that caps with a -5 charge marker probably have the structure of cap 1 (M^7 Gppp × ^{m}p Y) while activity at -4 charger was putatively cap 0 (M^7 Gppp × p), with this structure probably being m^7 GpppA p from adenosine incorporation studies. All of these studies characterized late mRNA for *in vivo* labeling and were performed 16-20 hr postinfection.

Isolation and characterization of virus mRNA's was begun by van der Beek et al. (1980), who studied the origin of polyhedrin protein of *A. californica* NPV replicating in *S. frugi-*

perda cells. Using RNA extracts taken from cells 24 hr post-infection, the authors partially purified the virus mRNA by hybridizing viral DNA to diazotized *m*-aminobenzyloxymethyl-cellulose. Characterization of the eluted mRNA by PAGE in 98% formamide showed one dominant and two minor RNA species having molecular weights of 240, 700, and 750K, respectively. Cell-free translation of the mRNA's using a wheat germ tranlational system optimized for plant viral mRNA's gave fourfold [^{35}S]-methionine incorporation whereas uninfected control cell extracts did not stimulate the wheat germ system. Analysis of translation products using PAGE showed several discrete products, one being a protein of 30K mol. wt. Immunoprecipitation of the translation mixture with α polyhedrin and subsequent analysis of the precipitate with PAGE showed only one labeled product, the 30K mol. wt. moiety.

Further work involving characterization of late mRNA of virus-infected cells was described by Vlak et al. (1981) using *A. californica* NPV in *S. frugiperda* cells. Poly(A)$^+$ RNA (purified by oligodeoxythymidylic acid cellulose chromatography) and RNA hybridized to *A. californica* NPV from infected cells 21 hr postinfection were found to have almost identical *in vitro* translational products. Peptides having molecular weights of 41, 39, 35, 33, 32, 31, 29, and 24K were synthesized in large quantities together with a variety of low molecular weight peptides. The 33K dalton peptide was identified as polyhedrin by a variety of criteria including immunoprecipitation and comigration with ^{3}H-labeled polyhedrin in PAGE. Two other peptides possessed molecular weights (41 and 64K) similar to structural peptides found in the virus. If both poly (A)$^+$ RNA and vRNA from 21 hr postinfection were used to hybridize against Southern blots of the parental genome digested by *Bam*HI, *Eco*RI, and the double digest *Bam*HI, *Eco*RI, most fragments, if not all, found RNA, showing that 21 hr postinfection transcripts had sequences complementary to large portions of the genome, although some fragments found more vRNA than others as judged by visual inspection. Finally, virus-specific mRNA which annealed to restriction fragments of *Eco*RI (fragments I and J) were translated *in vitro* and it was found that the product of fragment I mRNA was polyhedrin (as determined by molecular weight and immunoprecipitation). Hybridization of mRNA's to various restriction fragments could be due in part to promoter sequences which raised the question of posttranscriptional modification or generation of virus mRNA's by splicing mechanisms as reported in other virus systems. Further studies are needed to determine if this is the case.

In a more recent paper the previous authors, in a shuffled sequence, (Smith et al., 1982) have determined a preliminary transcription map of the *A. californica* NPV genome. They used specific restriction fragments immobilized on nitrocellulose filters to isolate viral mRNA's at a late time, 21 hr postinfection and an early time, 4 hr postinfection. Overall, the authors found that changes occurring in protein synthesis in *A. californica* NPV infected cells early and late in the infection cycle were in response to the presence of specific mRNA. It was determined that none of the translation products of the early (4 hr) mRNA were major components of the virus. A 39K mol. wt. polypeptide was the major early peptide and was made throughout the infection cycle. It was speculated that this peptide might play an important role in the infection process. There was no detectable increase in any host protein synthesis.

Late mRNA transcripts accounted for peptides having the molecular weight and relative abundance of all early and late polypeptides reported for *A. californica* NPV-infected cells. Furthermore, the transcripts for 21 hr infected cells represent all of the major protein components for the virus and nonstructural products as well. The transcripts were shown to hybridize scattered around the genome with no apparent clustering, leading to the speculation that many of the late *A. californica* NPV genes are under control of different promoters. By 48 hr all of the protein synthesis in infected cells was due to either the 10 or the 33K peptides. Although not a major viral component, the function of the 10K peptide plays an unknown role in the infection process. The transcriptional rate of these two peptides is also of interest because it indicates that they are being regulated in a different mode at 21 than at 48 hr.

In a similar study, Adang and Miller (1982) have used a successful synthesis and cloning of DNA complimentary to late *A. californica* NPV mRNA in order to study the late transcription and translation processes. Twenty-seven hr postinfection cDNA was obtained from *S. frugiperda* mRNA via reverse transcription, annealed to the plasmid pBR 322 via homopolymer tails, and used to transform *E. coli* RR1. Approximately 20% of the resultant colonies contained *A. californica* NPV sequences, as shown by colony hybridization with the virus. Positive plasmid DNA was bound to nitrocellulose filters and used to selectively hybridize mRNA from 27 hr postinfected cultures. *In vitro* translation of these transcripts produced eleven proteins of viral origin ranging in weight from 7.2 to 60K. Again, problems of premature termination and the assignment of functions to viral proteins not found to be structural remain to be elucidated.

Corroboration of late mRNA's coding for several virus proteins including polyhedrin and a protein of approximately 8K was made by Rohel et al. (1983). They found an intense synthesis of mRNA and the concomitant 8K protein starting at 15 hr postinfection and continuing throughout the remainder of the infection. No specific role for this amply produced peptide was found. However, its abundance led to speculation that it was involved in virus morphogenesis in the nuclei of infected cells.

B. PROTEIN SYNTHESIS

The effects of virus production on host protein synthesis and the events surrounding early protein synthesis have been previously described (Section IV, B). Therefore, consideration will be given here only to those synthetic events following early or α protein synthesis. All reported studies of protein synthesis in baculovirus-infected insect cells have a similar, but not identical, pattern of replication with respect to temporal bursts and product production. According to Carstens et al. (1979), the next burst of synthetic activity, following the synthesis of early peptides, occurs at 6 to 8 hr postinfection when peptides having apparent molecular weights of 67, 65, 54, 39, 24, 22-23, and 22K appear. Variable intensity and duration of the syntheses were noted for different peptides. For example, the 54K peptide, although always present, was very faint from its initiation at 8 until 32 hr. The 22-23 K peptide bands were also weak at first. However, the 22K band varied in intensity throughout the course of the infection becoming very intense as the maximum rate of synthesis was attained at 10 to 18 hr postinfection. Relative intensities of different peptides could reflect stoichiometric relationships between protein components or the relative amount of methionine content of a protein. At 8 and 10 hr after infection, polypeptides of 125, 112, 108, 98, 93, 88, 52, 50, and 28K became visible. Starting at 10 hr postinfection, additional virus-induced bands were observed at apparent molecular weights of 36, 33, 21, 16, 15-13, and 12K. At twelve hours postinfection, peptides of 20 and 19K appear. At later times in the infection cycle there were three additional peptides being synthesized, one at 16 hr (62K) and two (17 and 17.5K) appearing very late after infection at 22 to 24 hr. Peptides were detected by [^{35}S]methionine incorporation and autoradiography of PAGE gels; therefore, proteins lacking methione are not represented in this study. A visual representation of this study and others is graphically depicted in Fig. 1. As with earlier time studies, peptides discovered at 10, 12, and 16 hr also varied in intensity and duration of synthesis; however, no significance has yet been attributed to any variations noted. The 28K peptide that appeared at 10 hr and whose synthesis continued until 65 hr postinfection was presumed to be polyhedrin. Examination of purified extracellular virions grown in the presence of [^{35}S] methionine and examined by PAGE

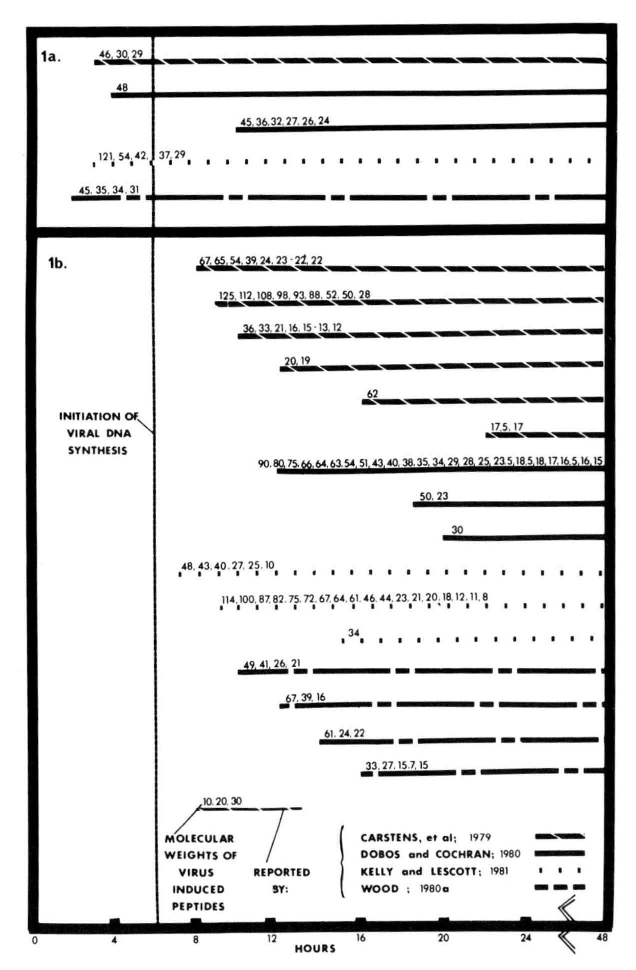

FIGURE 1. Graphic representation of nuclear polyhedrosis virus (NPV)-directed protein synthesis in infected insect cell cultures. (a) Synthesis of "early proteins" and initiation of DNA synthesis for cell-virus systems used by the authors cited. (b) The remaining virus protein synthesis occurring throughout the course of a synchronous NPV infection.

revealed "about 15 bands of radioactivity," most of which co-migrated with bands seen in infected cells during the time course study. These bands included peptides with molecular weights of 98, 93, 88, 67, 65, 52, 46, 39, 36, 30, 29.5, 28.5, 26, 23, 21, 17, 16, 15, and less than 10K. Thus, most of the structural components of the virus can be identified in a time course study using the methods and cell virus system described.

In addition to these studies, the authors examined various genotypic variants whose genomic content is known to differ, as determined by digestion fragments generated by various restriction endonucleases. No differences could be detected in the number or intensity of any presumably virus-directed peptides by labeling at 17 and 27 hr after infection.

Wood's (1980a) description of *A. californica* NPV replication in *S. frugiperda* cells is also depicted in Fig. 1 and can be summarized as follows: there are four temporal areas of protein synthesis during the infection cycle. The earliest times (0-2 hr) have been described (Section IV, B). The second group of proteins was initiated at 10 hr and had molecular weights of 49, 41, 26, and 21K. At 12 hr, proteins of 67, 39, and 16K were initiated followed by 61, 24, and 22K proteins at 14 hr and 23, 27, 15.7, and 15K at 16 hr postinfection. Once synthesis of a particular peptide was initiated there was continued synthesis of the species. Polyhedrin (33K) was first detected at 16 hr postinfection. The 18 viral-induced proteins detected at 16 hr postinfection were identical to those detected at 36 hr postinfection. Molecular weights of the reported proteins were consistent with the virus structural protein (except the 34K peptide) and were considered virus directed because identical patterns were obtained in either *S. frugiperda* or *T. ni* cells. As with the previous study, protein detection was a function of incorporation of [^{35}S]-methionine.

In a study of *A. californica* NPV replication in *S. frugiperda* cells (Dobos and Cochran (1980) the synthesis of proteins was divided in three time periods. The early protein synthetic events (up until 10 hr) thus included some protein synthesis which were in the second phase in other studies and discussed here in Section IV.B. From 12 to 18 hr, termed "middle" by Dobos and Cochran, there were 23 detectable peptides which were potentially virus directed. They possessed the following molecular weights: 90, 80, 75, 66, 64, 63, 54, 51, 43, 40, 38, 35, 34, 29, 28, 25, 23.5, 18.5, 18, 17, 16.5, 16, and 15K. Those peptides designated by the authors as "late" had apparent molecular weights of 50, 30, and 23 K and were detected from 18 to 24 hr postinfection. The authors also used [^{35}S]methionine as their label. They observed that once a protein synthesis was "switched on" it continued to be synthesized throughout the infection, with the exception of intracellular protein (IC) of 48 and IC 32K mol. wt., with IC 32 being produced only between 10 and 20 hr. As with other studies, a depiction of peptides showing initiation and duration of synthesis is presented in Fig. 1. The structural polypeptides of the virus were considered to be mostly from middle and late times based on a comparison of their electrophoretic mobilities with peptides from disassociated virions.

The relationship between viral DNA synthesis and protein synthesis was investigated by the use of cytosine arabinoside (ARA-C). In several DNA virus groups (Papaviruses, Adenoviruses, and Poxviruses), if virus DNA synthesis is arrested no late structural proteins are produced. This has been interpreted to mean that the infecting virus genome was transcribed for early proteins, but progeny genomes were transcribed for structural proteins. Herpesvirus, however, could make both early and late proteins in the presence of DNA synthesis inhibition and was able to form empty virions. In ARA-C-inhibited *A. californica* NPV-infected cells, although [^{3}H]-thymidine incorporation was inhibited by more than 99%, there was production of polyhedral inclusion bodies (70% of control) which contained some virions (Dobos and Cochran, 1980). Peptide synthesis was not affected during early times, however, middle and late times were delayed. By 42-44 hr, all peptides were made which was shown by comparing peptide patterns of occluded virus from ARA-C and non-ARA-C-treated cultures. Electron microscopic thin sections of ARA-C-treated polyhedral inclusion bodies showed both empty and full virions. The nucleic acid content of complete virions was derived from input infecting virions as determined with previously labeled DNA.

Finally, Kelly and Lescott (1981) have studied protein synthesis of *T. ni* NPV replication in *S. frugiperda* cells and have found four major temporal areas of synthesis, termed α, β, γ, and δ (Fig. 1). Using [^{35}S] methionine as a label, the authors found that α protein synthesis occurred 2-3 hr postinfection (Section IV, B). β protein synthesis involved 48, 43, 40, 27, 25, and 10K peptides and was initiated at 6 to 7 hr postinfection. The γ phase of synthesis began at 10 hr postinfection and included peptides with apparent molecular weights of 114, 100, 87, 82, 75, 72, 67, 64, 46, 44, 23, 21, 20, 18, 12, 11, and 8K. Finally, the last phase of synthesis, δ, began at 15 hr postinfection and involved synthesis of a 34K peptide, which was identified as polyhedrin.

Although protein synthesis patterns were similar, synthetic events were slower in *T. ni* cells than in *S. frugiperda* cells. Analysis of infected cells labeled with ^{14}C protein hydrolysate instead of [^{35}S]methionine showed that no methionine-deficient, virus-infected, cell-specific peptide was synthesized. Whether *T. ni* NPV has different properties from *A. californica* NPV or whether proper conditions were not met should be investigated, since Rohel et al. (1983) found a methionine-deficient 10K peptide synthesized at late times. Kelly and Lescott (1981), also used ARA-C to study the effects of DNA synthesis inhibition on protein production. They found that 50 μg/ml caused inhibition of γ and δ phases but that lower levels of ARA-C merely delayed protein synthesis γ and δ phases although α and β phases were not affected. Blocks were only allowed to progress for 18 hr. This study, therefore, is not directly comparable with the one by Dobos and Cochran (1980), who allowed the inhibition to progress for 42-44 hr. The use of another DNA synthesis inhibitor, hydroxyurea (800 μg/ml) stopped protein synthesis at the α phase. Inhibition reversal at 18 hr allowed β synthesis to progress.

Further experiments revealed that protein synthesis in baculovirus-infected cells resembled cascade induction of peptide synthesis found in herpesviruses. Not only does the previous phase of protein synthesis have to be in progress, but functional peptides must be produced in order to prime the system for the next round of synthesis. Addition of cyclohexamide, which prevents RNA translation, blocked the build up of mRNA and the synthesis of protein upon reversal of inhibition. Inhibition of one phase of protein synthesis resulted in subsequent delays in initiation of other protein synthesis phases. Addition of amino acid analogs such as fluorophenylalanine (FPA) which causes synthesis of nonfunctional proteins also resulted in delayed and/or reduced amounts of later phase protein synthesis.

Posttranslational modification of baculovirus peptides has been investigated by several groups. Modifications can include the addition of prosthetic groups by such processes as phosphorylation, glycosylation, and methylation, rapid posttranslational cleavage (e.g., picornavirus) or slow posttranslational cleavage (e.g., Herpesvirus). Dobos and Cochran's (1980) study of protein synthesis in virus-infected cells encompassed investigating the synthesis of glycoproteins using a variety of labeled sugar precursors. Purification of occluded virus from infected cells and subsequent protein analysis by PAGE showed that a viral protein doublet VP35-34 was labeled if [^{3}H]mannose was used as a precursor. Polypeptides of VP90, VP64, VP40-37, and VP25 were labeled if acetyl-[^{14}C]glucosamine was used as a precursor. No virion protein was labeled if [^{3}H]glucosamine was used as the precursor. All glycosylated peptides would be associated with what was termed in previous discussions as the "middle" temporal area of protein synthesis.

Goldstein and McIntosh (1980) have studied the glycoproteins of *A. californica* NPV replicating in both *T. ni* and *S. frugiperda* cells. Using [^{3}H]glucosamine and [^{14}C]glucosamine to label both cells and virus-infected cells, the authors reported radioactivity associated with a 70K dalton glycopeptide (Gp 70) using SDS-PAGE analysis. Gp 70 was not found in either uninfected cell line and was found irrespective of the cell line used when *A. californica* NPV and *T. ni* NPV were used to infect cell cultures. Gp 70 was associated with only one class of peptides which migrated in the G or 2500 molecular weight region of a Sephadex G-50 column, a region associated with short, incomplete carbohydrate side chains containing only a core region of glycopeptides. Whether this was a natural state of GP 70 due to the absence of necessary glycosylating enzymes or if a mechanism existed for cleaving terminal carbohydrate moieties before viral release was not determined.

Phosphorylation of viral peptides has been studied for both the occluded as well as nonoccluded form of the virus. Dobos and Cochran (1980) showed that there were 9-10 proteins present in occluded virus which were considered phosphoproteins. Peptides with estimated molecular weights of 75, 65, 35 and/or 34, 30, 27, 25, 23, 19, and 18K all incorporated ^{32}P. The 35 and 34K peptides could not be properly resolved to tell if both or only one possessed ^{32}P.

Maruniak and Summers (1981) performed a more in depth study of phosphorylation of viral peptides and the segregation of the modified proteins into cytoplasmic or nuclear fractions. In *T. ni* cells infected with *A. californica* NPV, the authors found that progeny-occluded virus (OV) and extracellular virus (EV) both contained numerous, but not identical, phosphoproteins and that at least 20 phosphorylated proteins were detected in infected cell extracts. Most phosphoproteins found in cell extracts were present in infected and control cells. Of the virus-induced phosphoproteins, the 75, 64, 27, 25, 23, and 22K moieties were detected principally in the cytoplasm and the 32 and 18K phosphoproteins were detected principally in the nucleus. Analysis of purified virus revealed extracellular virus containing phosphoproteins of 75, 64, 54, 37, 32, 23, 18.5, 18, and 16K mol. wt while occluded virus contained fourteen phosphoproteins of 115, 85, 75, 68, 64, 54, 32, 23, 22, 19, 18.5, 18, 17, and 16K mol. wt. Although some phosphoproteins were common to both virus forms (e.g., 64K), they were not present in either equal amounts or in equal amounts of phosphorylation in the two viral forms.

High-voltage electrophoresis and acid hydrolysis of ^{32}P-labeled polyhedrin showed that phosphoserine was the major phosphoamino acid while phosphothreonine was a minor one. No other proteins were analyzed for phosphoamine content.

Recently, a protein kinase activity has been associated with the extracellular and occluded forms of *A. californica* NPV (Miller et al., 1983). Occluded virions disrupted with 0.1% NP-40 (no reaction with 1.0% NP-40) and incubated with [^{32}P]ATP produced major phosphorylation of peptides of 15, 17, 30, and 34 kd, the latter two usually associated with polyhedrin, and minor phosphorylation of 42 and 54 kd proteins. Examination of purified nonoccluded virus by similar assay conditions provided a greater (10X) incorporation of ^{32}P into protein and produced predominant phosphorylation products of 98, 70, 60, 54, 37, 28, and 15 kd. No experimentation was performed to determine the origin (either viral or host) of the enzyme.

However, the authors speculated that the variable activities of the enzyme in the two viral forms could be indicative of the origin of the membranes of the virions. Also different substrates present in the two membranes could result in differential phosphorylation of the two virion forms. Analysis of target amino acids of the protein kinase showed that serine and threonine were predominantly phosphorylated. Phosphorylation of tyrosine was not detected.

There are a variety of reports concerning posttranslational cleavage of baculovirus proteins. Carstens et al. (1979) reported that there were lower molecular weight proteins which appeared at the expense of higher molecular weight proteins during the virus replication cycle. More specifically, infected cells labeled at 8 hr and chased for 15.5 hr showed an absence of 29 and 22K bands in the chased samples compared to the pulse samples. After the chase, a prominent 28K band appeared along with minor bands of 22.5 and 18.5K. At 18 hr, similar pulse-chase (22-hr chase) experiments were performed. A weak 29 and 22K band present after a pulse of 18 hr was greatly reduced after the chase. Bands of 67, 20, and 19K also seemed to disappear while new bands of 22.5 and 18.5 K appeared after the chase. Increased radioactivity was found in the 28 and 15K polypeptides compared to activity after a pulse. There were two peptides of 22.5 and 18.5K which could only be detected after chase and not during or after pulse.

Wood's (1980a) study of baculovirus replication reported no findings of posttranslational cleavage as did the study by Kelly and Lescott (1981). Addition of the protease inhibitors TPCK (chymotrypsin inhibitor) and TLCK (trypsin inhibitor), known inhibitors of poliovirus "polyprotein" cleavage, failed to produce additional peptides in *T. ni* virus-infected cells thereby indicating no rapid cleavage of precursor protein. Additional treatment with translation and cleavage interfering agents such as high salt, zinc acetate, iodoacetamide, and aprotinin also failed to show protein processing. Pulse-chase experiments performed by pulsing at 5, 9, 12, 15, and 22 hr (1 hr pulse) and chasing for 3 hr failed to show any processing. Similarly designed experiments pulsed at 12, 13, 14, 15, and 16 hr (polyhedrin synthesis onset) failed to show any protein being chased to polyhedrin. However, evidence of 21, 22, and 25K processing was found. At 12, 13, and 14 hr the 22 and 24K species disappeared and increased amounts of the 25K protein appeared; no product-precursor relationship was established. The authors felt that the lack of evidence for rapid posttranslational cleavage and little, if any, subsequent cleavage was tentative evidence that the proteins detected were primary gene products derived from monocistronic mRNA.

Finally, Dobos and Cochran (1980) also reported no rapid posttranslation cleavage, although they did find evidence for at least four posttranslational modified peptides. Both pulse label experiments and the addition of TPCK and TLCK failed to show any rapid posttranslational cleavage. To detect slow posttranslational cleavage and/or modification, infected cells were labeled at 4, 10, 16, and 18 hr postinfection followed by a 12-hr chase. Significant differences were noted between pulsed and pulse-chase samples. Modifications of a 43 to a 45K and a 24 to a 25K protein were readily evident. A potential cleavage could have occurred when diffuse radioactivity bands at 23.5 and 23K became sharper at 23K with time. Also, bands of 18.5, 18, 17, 16.5, and 16K showed changes in radioactivity, although this again was probably due to modification and not cleavage because the 16, 16.5 and 17K bands diminished while the 18 and 18.5 bands became more intense.

Thus, the replication of the Group A' Baculoviridae has been investigated at the molecular level under the very artificial condition of synchronous infection. One final aspect of possible importance to future baculovirus studies involves the presence of an alkaline protease discovered in *in vivo*-produced polyhedral inclusion bodies. The presence of alkaline protease (Yamafugi et al., 1959; Eppstein et al., 1975; Eppstein and Thoma, 1975) associated with baculoviruses is important when molecular weight determinations of structural proteins are made. Wood (1980b) has shown that the protease is capable of degrading inclusion body protein and can alter the viral membrane of virions so that it can readily be removed by NP-40. Several investigators (McCarthy and DiCapua, 1979; Wood, 1980b; Zummer and Faulkner, 1981) have reported that there was no active or apparent alkaline protease associated with *in vitro*-produced polyhedra. Tissue culture-derived polyhedra were slow to dissolve and virion membranes were not easily removed by NP-40 treatment. In addition, Wood demonstrated that the alkaline protease was capable of degrading the enveloped nucleocapsid polypeptides of *A. californica* NPV tissue culture-derived polyhedral inclusion bodies, thus giving them identical SDS-PAGE profiles as insect-derived polyhedral inclusion bodies. Several reports were cited which presented major differences in virus protein analysis which could be accounted for by alkaline protease activity.

Although the biochemical characterization of the enzyme has been performed (Eppstein and Thoma, 1977) no one has proved whether the enzyme is a bacterial, fungal or host-contaminating enzyme of *in vivo*-produced polyhedral inclusion bodies. Development of lepidopteran cell cultures from tissue sources other than ovarian tissue may replicate NPV's and GV's and produce progeny virus with alkaline protease.

VI. ASSEMBLY AND RELEASE OF PROGENY VIRUS

A. *NONOCCLUDED VIRIONS*

1. *Nucleocapsid Formation*

The first observable change in virus-infected cells is an enlargement of the nuclei and the formation of aggregates of dense granular chromatin-like material called the virogenic stroma (Smith, 1977). Initially, this granular material is dispersed throughout the nucleus, but shortly after its appearance is condensed and forms a dense ring structure in the central region of the nucleus. It is in the virogenic stroma that the progeny nucleocapsids are formed. The developing nucleocapsids are interspersed within the stromal network and develop either as single rods or in clusters of rods of varying numbers. At this time, all of the newly formed rods are closely associated with the stroma and no envelope material is present. Occasionally, empty capsid structures are seen in the nucleus prior to nucleocapsid formation [Hess and Falcon (1981) in *in vivo* studies and Hirumi et al. (1975) and Knudson and Harrap (1976) in *in vitro* studies].

The timing of the replication events appears to vary slightly depending upon the virus-cell system. Hirumi et al. (1975) found that with the rapidly replicating *A. californica* NPV and the *T. ni* cell line, TN-368, the first changes were seen in a few cells within 3 hr postinfection and in most cells, 6-12 hr postinfection. The nucleocapsids appear between 24 and 48 hr postinfection and during this time are unenveloped and closely associated with virogenic stroma. Knudson and Harrap (1976) found that the signs of nuclear change in *S. frugiperda* infected with that NPV were seen at 8 but not at 6 hr postinfection. At 9 hr postinfection, the morphogenesis of virus particles was observed. With the somewhat slower replicating *B. mori* NPV the virogenic stroma was not recognizable until 16 hr postinfection (Raghow and Grace, 1974). Eight hours later many nucleocapsids were seen associated with the stroma. The nucleocapsids were of uniform diameter but of varying lengths. The number of nucleocapsids continued to increase up to 32 hr postinfection with no evidence for envelope formation.

These *in vitro* studies reveal a morphogenesis of nucleocapsids that is very similar to that observed in the infected cells of several tissues *in vivo* (Harrap, 1972b; Tanada and Hess, 1976; Hess and Falcon, 1981). Granados and Lawler (1981), studying the infection of midgut cells in *T. ni* larvae infected with *A. californica* NPV, observed the appearance of progeny nucleocapsids in the nuclei of infected midgut cells

8 hr postinfection. This was considerably earlier than the appearance of nucleocapsids in infected cultured cells from *T. ni* as reported by Hirumi et al. (1975).

Recently, a study of morphogenesis of a NPV in the mosquito *Aedes triseriatus* was reported by Federici (1980). This study, one of the few on a morphogenesis of a NPV in the order Diptera, indicates a somewhat different type of replication pathway. Federici observed that capsids were the first viral structure assembled. The capsids occurred singly or in parallel arrays either partially or completely assembled. The mean capsid length increased as the disease progressed from less than 80 to about 300 nm. Nucleocapsids were formed by the entry of nucleoprotein into the capsids. During the early stages of infection, nucleocapsids of uniform length were formed singly. Later the nucleocapsids were formed in arrays, apparently as progenitor nucleocapsids, up to 1 μm in length. These were subsequently cleaved to give capsids of a mean length of 184 nm which had 7 nm caps at both ends when morphogenesis was completed. This sequence of steps in assembling the progeny nucleocapsids differs considerably from those observed in the Lepidoptera and needs to be further investigated in those other Diptera for which NPV's have been found.

Little is known about the formation of nucleocapsids of the granulosis viruses, but it appears that the early stages of nucleocapsid are similar to those in the NPV. Within 24 hr there is a clearing of some areas of the infected nucleus and the formation of dense virogenic stroma (Tanada and Leutenegger, 1970; Summers, 1971).

Tweeten et al. (1980) described a very basic peptide, 12K mol. wt. that was associated with the nucleoprotein component of the GV nucleocapsid. The peptide was rich in arginine and was a major component of the nucleocapsid, thus resembling protamines often found associated with viral DNA.

Burley et al. (1982) suggests that a very basic protein, as was described by Tweeten et al. (1980), participated in nucleic acid-protein interactions perhaps leading to condensation of the nucleoprotein. The presence of such a nucleoprotein condensate was suggested as necessary for the encapsidation of the nucleoprotein which occurred, first, only in the infected nucleus and, unlike the NPV replication, was in areas not occupied by the chromatin. Later, the nuclear membrane disintegrated and nucleocapsids from the nuclear region moved directly into the cytoplasm. Others were, in some unexplained manner, enclosed in vesicles, "transport envelopes," with a double wall appearance and thus moved toward the basal region of the midgut cell (Summers, 1971).

As was occasionally seen in the replication of the NPV (Fig. 2), Summers (1971) observed the formation of long unbranched tubes which he interpreted as self-assembly of capsid

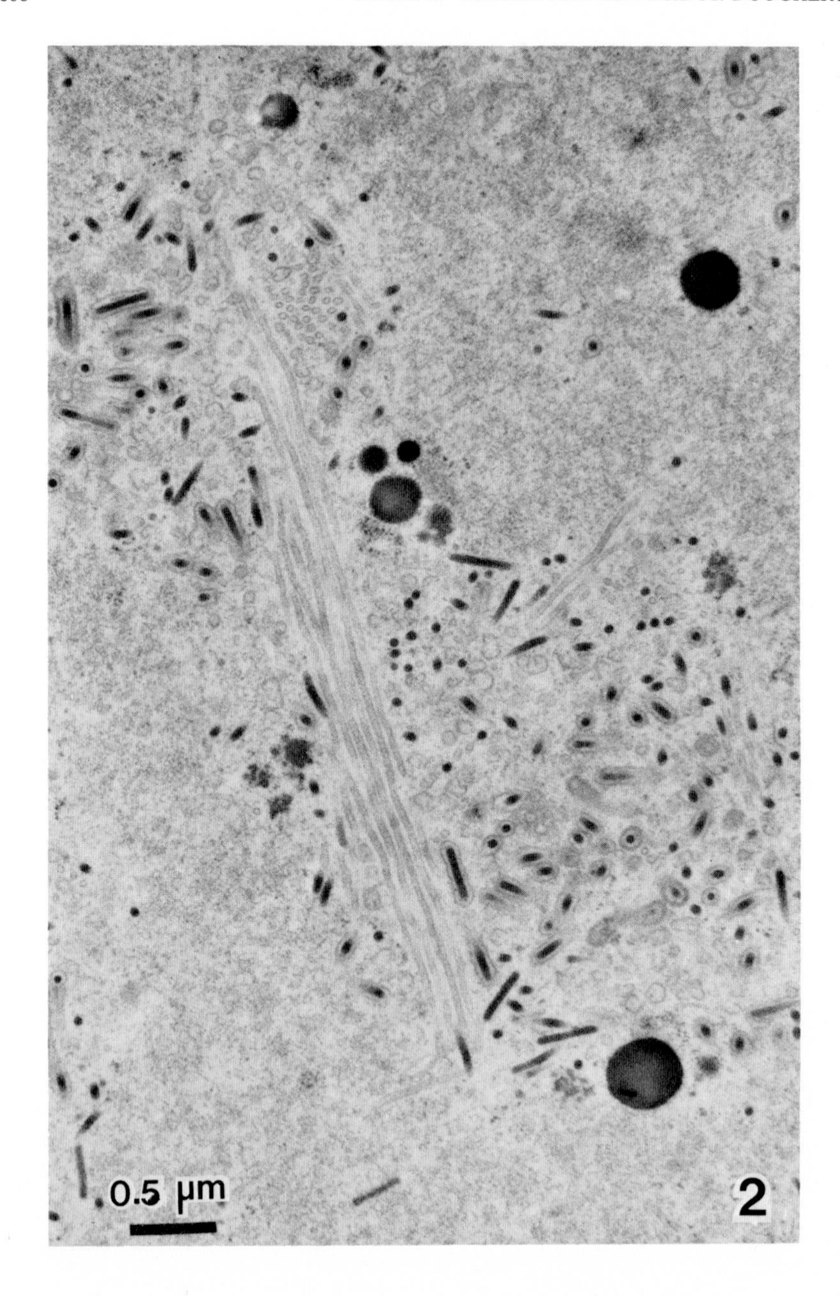

FIGURE 2. *Trichoplusia ni* midgut epithelial cell nucleus infected with *T. ni* NPV, 4 days postinfection. Note long tube-like formations of excess capsid material. Electron micrograph courtesy of Dr. Jean R. Adams.

protein in the absence of viral nucleic acid. These occurred both in infected midgut epithelial cells and in infected fat body cells. In some instances these tubes appeared partially filled with nucleoprotein. Similar, but branched, tubules had been described earlier as occurring in the cytoplasm of GV-infected fat body cells (Smith and Brown, 1965). These authors were unable to attribute any specific function to these bizarre forms but suggested they could be either an initial stage of nucleocapsid formation or a defect in the replication process. Burley et al. (1982) suggested that the long nucleocapsids in virus preparations contained two genome copies. The mechanisms of normal condensation and dimers, etc. remain to be elucidated.

In cultured cells from the host insect the nucleocapsids of the *Oryctes* virus are formed in the central region of an infected nucleus associated with an electron-dense area similar to the virogenic stroma described for the NPV (Quiot et al., 1973). A similar process of nucleocapsid formation was observed in cultured *S. frugiperda* cells by Kelly (1976). However, in the mosquito cell the nucleocapsids were formed in the cytoplasm of infected cells (Kelly, 1976). The cytoplasm appeared highly vaculated but no evidence of the typical stroma was found.

2. *Acquisition of the Envelope*

Cell culture studies of the mechanism by which the nonoccluded virion form of the NPV acquires the envelope have not produced a clear definition of the process. It appears that there are several ways that the nucleocapsids move out of the nucleus to the plasma membrane. Adams et al. (1977) suggested four: (1) protrusion of nuclear membrane extending toward the plasma membrane, (2) acquiring an envelope of inner nuclear membrane as it buds into a vesicle formed from outer nuclear membrane (a transport membrane) (3) budding through the nuclear membrane, and (4) exiting through a rupture in the nuclear membrane.

Hirumi et al. (1975) found that nucleocapsids exited the nucleus of infected *T. ni* culture cells via route 2 of Adams et al. In their cultures this vesicle membrane was lost before the nucleocapsid reached the plasma membrane, releasing the nucleocapsid into the cytoplasm. The nucleocapsid in infected *S. frugiperda* cells also appeared to bud through the inner nuclear membrane, as in the other two studies, but they did not always acquire a transport membrane. Knudson and Harrap (1976) observed that by 12 to 15 hr postinfection nucleocapsids could be seen budding through the inner nuclear membrane into an enlarged perinuclear space. Free nucleocapsids could also be seen in the cytoplasm and budding through the plasma membrane.

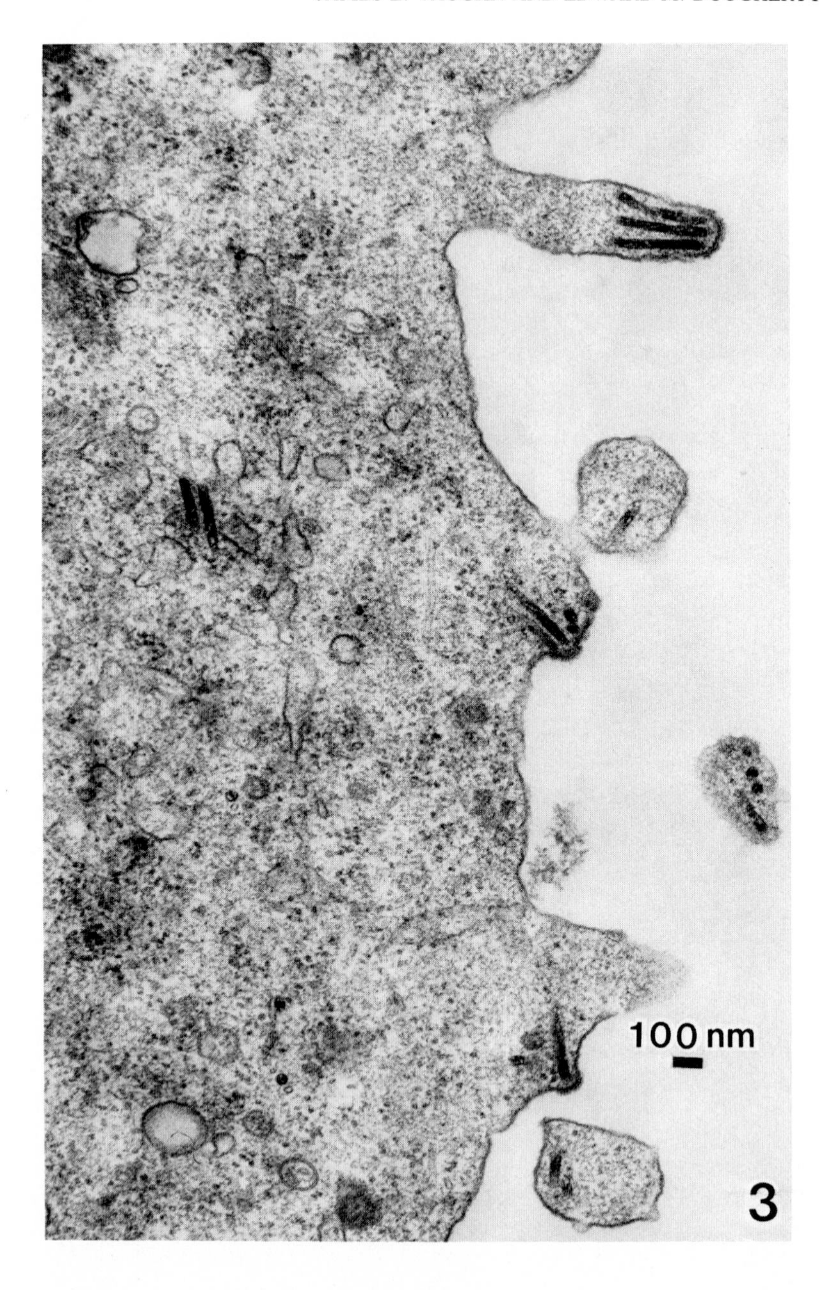

FIGURE 3. *Heliothis zea* IPLB-HZ-1075 cells inoculated with *H. zea* NPV, 8 days postinfection. Note evagination of plasma membrane containing one or more nucleocapsids. Modification of the plasma membrane to form peplomers can be seen at the points of evagination. From Adams et al., 1977.

These *in vitro* studies have generally confirmed similar observations on various tissues from infected insects. Budding of nucleocapsids through the inner nuclear membrane was observed in fat body cells of *Hyphantria cunea* (Injac et al., 1971), the midgut cells of *Pseudaletia unipuncta* (Tanada and Hess, 1976), and the hemocytes of *L. dispar* (Nappi and Hammill, 1975). Granados and Lawler (1981) reported that in *T. ni* midgut cells the nucleocapsids of *A. californica* NPV budded through the nuclear membrane and in the process acquired a double membrane consisting of the inner and outer nuclear membrane. At the same time, they also observed numerous free (naked) nucleocapsids in the cytoplasm that already had either lost the double membrane or had exited via another route, such as through breaks in the nuclear membrane. Hess and Falcon (1977) found that the nucleocapsids did, in fact, enter the cytoplasm of *A. californica* NPV-infected cells through the nuclear pores or small holes in the nuclear membrane.

The final step in the replication of the NOV is the enclosure of the nucleocapsid in an envelope and the release of the enveloped virion into the medium surrounding the infected tissues or cells--either the hemolymph of infected insects or the cell culture media. The only mechanism by which this has been observed to occur is one in which the nucleocapsid buds through the plasma membrane (Fig. 3). This process was described by Adams et al. (1977) and Hess and Falcon (1977). In both studies, once the nucleocapsids reached the cytoplasmic membrane they aligned more or less in a perpendicular position. Following the alignment of the nucleocapsids, the plasma membrane thickened by a process of infolding and morphologically distinct structures (peplomers) were formed (Adams et al., 1977). The remainder of the viral envelope was not visibly different from the cytoplasmic membrane. The altered plasma membrane evaginated until the enclosed nucleocapsid was released into the surrounding medium. The two ends of the membrane then joined forming the loose fitting, fragile envelope characteristic of the NOV.

In studies of the very early events in the infection of midgut cells, Granados and Lawler (1981) observed that numerous nucleocapsids lined the basal plasma membrane of infected cells within 0.5 hr postinfection and that as early as 2 hr postinfection they appeared to bud through the membrane acquiring an envelope in the process. This envelope had the peplomer structure characteristic of the envelope enclosing the NOV. Since ultrastructural signs of virus replication in the midgut cell indicated that replication did not begin until 8 hr postinfection, it was concluded that the nucleocapsids seen at the basal plasma membrane between 0.5 to 6 hr postinfection were from the original inoculum. If the viral membrane contains only viral-coded peptides and the peplomers are viral-specific structures, the question is raised as to whether the necessary peptides could be synthesized and inserted into the cytoplasmic membrane that quickly? Reports cited earlier by Hirumi et al. (1975) that in *in vitro* tests the first signs of virus replication could be seen in some cells as early as 3 hr postinfection and reports by Wood (1980a) that viral-specific proteins could be detected as early as 2-4 hr postinfection indicate that it may be possible for the necessary peptides to be synthesized within 6 hr. Recently, Volkman (1983) published serological evidence that viral-specific antigens appeared in the membranes of infected cells earlier than 8 hr postinfection. These antigens were demonstrated only with antiserum prepared against the NOV indicating that such peptides are associated specifically with that viral form. Clearly, this is an interesting new development in the replication of the NPV and one where further *in vitro* studies will be most useful in providing additional information.

The nucleocapsids of the GV destined to be released from the midgut into the hemocoel of an infected insect apparently also acquired their envelopes as a result of the process of budding through the plasma membrane (Tanada and Leutenegger, 1970; Summers, 1971; and Hunter et al., 1975). Following nucleocapsid formation in the nuclei, the nuclear membrane disintegrated and the nucleocapsids were released into the cytoplasm. Here some acquired a loose, circular-type vesicle, the origin of which could not be determined because of the intermixing of nuclear and cytoplasmic components. Both the free nucleocapsids and those in vesicles moved to the cytoplasmic membrane where they budded into the hemocoel. Summers (1971) observed that those nucleocapsids in vesicles budded through the plasma membrane with the vesicle intact. In infected midgut cells, free nucleocapsids budded through the cytoplasmic membrane at sites which had become evaginated around one end of the nucleocapsid, which is similar to NPV nucleocapsids. This method of acquiring an envelope on the NOV was also observed in fat body cells by Hunter et al. (1975). Thus, it appears that the processes for forming enveloped NOV are the same in both the primary site of replication, the midgut epithelium, and at secondary sites, such as, fat body cells. However, the source and the function of the intracellular vesicles needs further clarification.

The *Oryctes* virus appears to be another of the baculoviruses for which the envelopes are synthesized *de novo*. In both the coleopteran and the lepidopteran cell lines, large amounts of envelope material have been seen in the nuclei of infected cells (Quiot et al., 1973; Kelly, 1976). Kelly found that in the infected mosquito cells discrete areas of envelope synthesis occurred in the cytoplasmic. The cytoplasmic membranes were indistinct, probably as a result of cell fusion that appeared to be an early sign of infection in these cells. Kelly also found that infectious virions were released into the culture supernatant from both the moth and the mosquito lines, but no mechanisms for release of virions was reported.

B. OCCLUDED VIRIONS

1. Acquisition of the Envelope

Not all of the nucleocapsids produced in the nucleus of an infected insect cell are eventually released from the cell as NOV. In TN-368 cells infected with *A. californica* NPV, Volkman et al. (1976) have shown that the complete NPV replication cycle has two separate phases. Nucleocapsids produced in the first phase acquired an envelope at the cytoplasmic membrane and left the infected cell as NOV. In the second phase, the nucleocapsids acquired the envelope within the nucleus and became occluded in the polyhedra (Figs. 4 and 5). The first phase of extracellular virus production reached a peak at 36 hr postinfection and then decreased as nucleocapsids became enveloped within the nucleus. The slow down in NOV production began with the onset of polyhedron formation.

The development of this second intranuclear envelope material appeared to be the result of the *de novo* morphogenesis. Stoltz et al. (1973) studied serial sections of infected nuclei in which they had found intranuclear development of short, open-ended membrane profiles. They could find no points of contact between these membrane profiles and the nuclear membranes and the morphology of the membranes of infected nuclei appeared no different from those of noninfected nuclei, nor did the nuclear membrane appear to function as a template for the formation of the membrane profiles. This study was done on NPV infections of a Diptera, *Rhynchosciara angelae*, in which infection was limited to the gut epithelial cells. There is apparently no NOV produced and thus the involvement of the nuclear membrane in the events resulting in the release of NOV had not yet occurred. MacKinnon et al. (1974) studied the envelopment of nucleocapsids within the infected nuclei of the established lepidopteran cell line TN-368 and came to the same conclusions based upon similar observations. They observed that the nucleocapsids became enveloped within the interior of the nucleus within 24 to 28 hr postinfection, often in relatively large groups. However, the number of nucleocapsids per envelope was also as low as one or two in this virus-cell system.

The only detailed electron microscopic examination of the envelope indicated that the envelope of *L. dispar* NPV contained, on their surface, a regular arrangement of circular subunits, 20 nm in diameter. These subunits appeared to be exposed by

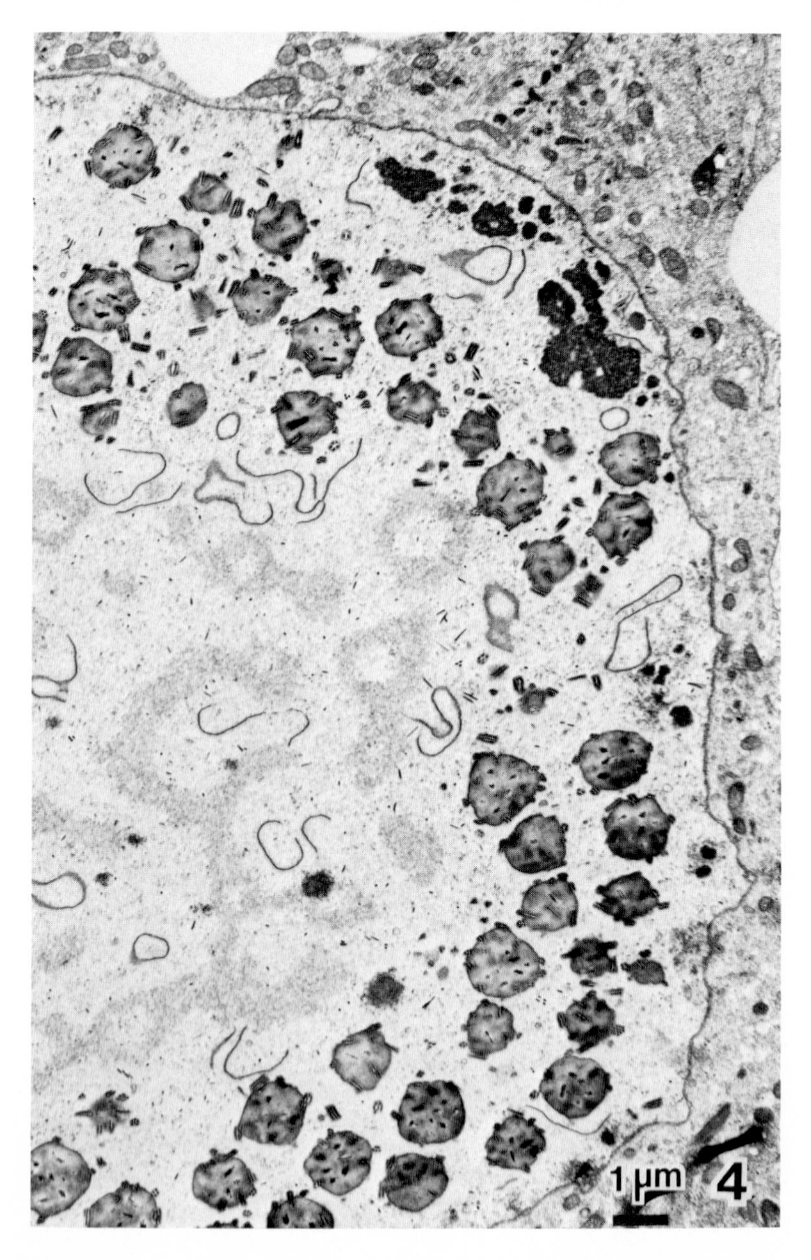

FIGURE 4. Fat body cell of beet armyworm infected with *A. californica* NPV, 6 days postinfection. Shows early stages of virion incorporation into polyhedra. Note free nucleocapsids, enveloped nucleocapsids, and envelopelike material. Electron micrograph courtesy of Dr. Jean R. Adams.

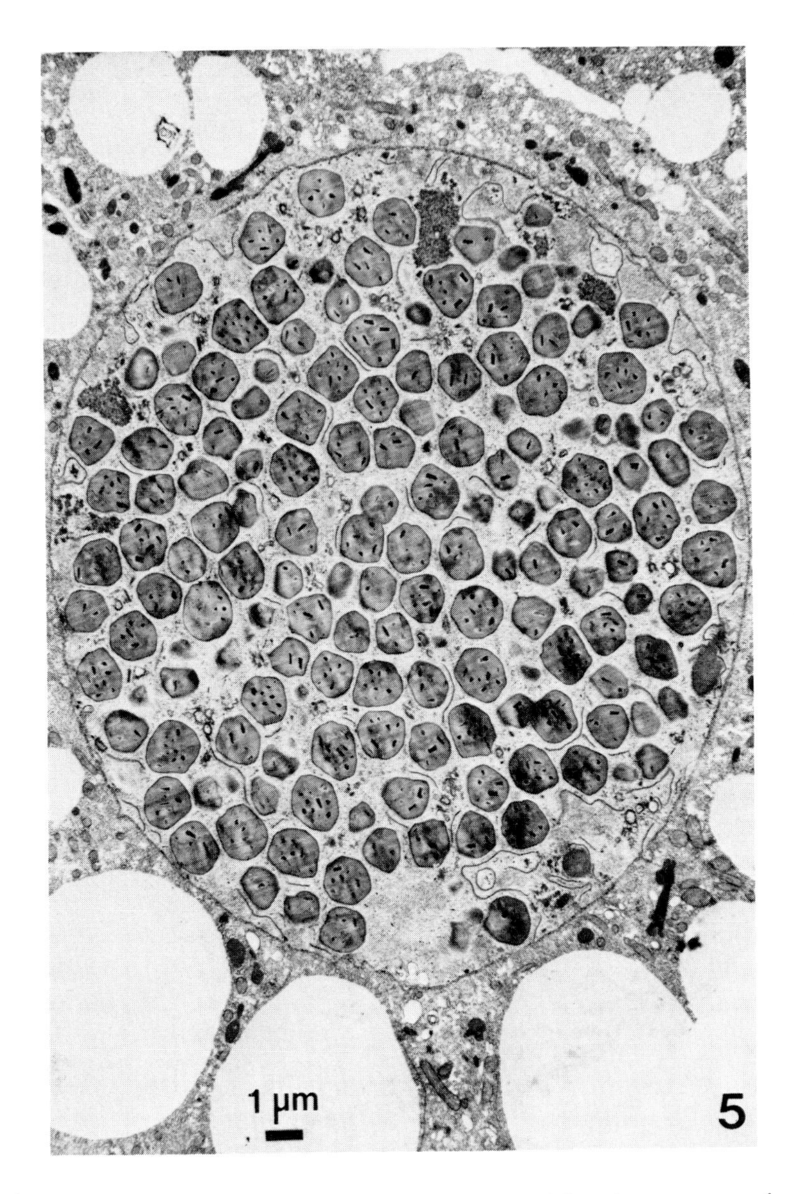

FIGURE 5. Fat body cell nucleus of beet armyworm infected with A. californica NPV, 8 days postinfection. Replication process is complete. Nucleus is filled with polyhedra-containing numerous virions. Electron micrograph courtesy of Dr. Jean R. Adams.

the removal of part of an outer layer of the envelope. The completed envelope appeared spherical and surrounded the virus particles loosely but as the occlusion body developed the envelope was pressed close to the enclosed nucleocapsid (Harrap, 1972a).

2. *Occlusion Body Formation*

It is clear that only those nucleocapsids that acquire this envelope become incorporated into the occlusion bodies (Summers and Arnott, 1969; Robertson et al., 1974; Knudson and Harrap, 1976). In cell culture studies this occlusion was seen to begin as early as 18 hr postinfection (Hirumi et al., 1975; Knudson and Harrap, 1976). As discussed in Section V, the structural peptide, polyhedrin, was synthesized beginning 10 to 16 hr postinfection and polyhedron formation occurred as the result of polymerization of this peptide. This sequence of observed events has made it attractive to speculate that the virion envelope has a specific role in the formation of the polyhedra (Summers and Arnott, 1969; Harrap, 1972b; Knudson and Harrap, 1976). These investigators have suggested that the envelope contained some type of attraction site which served as the initial point of polymerization and established the orientation of the polyhedron development. However, no evidence for such a mechanism has been reported. The first visible evidence of the polymerization may be the appearance in the nuclei of "fibrous material." This accumulation of fibrous masses was described from *in vivo* infections in *T. ni* larvae by Summers and Arnott (1969) and from *in vitro* infections of *S. frugiperda* cells by Chung et al. (1980). However, neither report showed the fibrous material associated with virions. The polyhedra increased in size and occluded additional virions as the result of the continual accumulation of polyhedrin. The final step in the morphogenesis of the polyhedra is the formation of the polyhedron membrane. Chung et al. (1980) found that this structure was formed by the condensation of the fibrous material at the interior boundary of the peripheral layer of the developing polyhedra. The completed polyhedron membrane was bounded externally by three lamellae. Although this structure is generally referred to as a polyhedron membrane, no phospholipid layer has been observed and, therefore, it is not considered a true biological membrane. These findings on the morphogenesis of the polyhedron *in vitro* are in agreement with those reported for the morphogenesis *in vivo* (Summers and Arnott, 1969).

The use of cell culture to isolate and purify temperature-sensitive virus mutants has provided clear evidence that the genetic control of the polyhedra formation is in the viral genome. Viral mutants have been isolated that do not cause the production of polyhedra in infected cells (Brown et al., 1979; Lee and Miller, 1979). Both groups were able to obtain polyhedra again when the mutant virus was combined with a second, complementary, polyhedra-negative mutant. Evidence is also accumulating that the shape of the inclusion body is controlled to some extent by the viral genome. Stairs (1964) was able to isolate a strain of the *Choristoneura fumiferana* GV that produced only a cubic inclusion body instead of the normal oval ones. Brown et al. (1980) recently described two mutants of the *A. californica* NPV that produced one single, large occlusion body per infected cell. The occlusions had a paracrystalline lattice structure similar to the wild-type virus. Peptide maps of the enzyme-digested polyhedrin from one mutant were identical to the maps of polyhedrin from the wild type but different from the polyhedrin of the second mutant. Thus, the unusual polyhedra formation was not due to a change in the polyhedrin, but the specific cause was not identified.

Wood et al. (1981) were able to isolate a mutant of this same virus that produced an unusually high number of occlusion bodies per cell and was designated "HOB." This mutant strain also had a significantly shorter LT_{50} for *T. ni* larvae and more occlusion bodies were produced in the diseased larvae. They observed no alterations in the *in vitro* replication of nonoccluded virions, occluded virus structural proteins, or DNA restriction endonuclease patterns in the HOB mutant.

Host modifications of the viral replication process also have been described for the *A. californica* NPV. Tompkins et al. (1981) found that virulence of the virus varied depending upon whether *T. ni* or *S. exigua* larvae were used to replicate the virus stock. The *T. ni*-produced occlusion bodies were ten times more virulent for neonate *T. ni* larvae than were *S. exigua*-produced occlusion bodies. When the *T. ni* NPV, a closely related variant of the *A. californica* NPV, was tested using the same regimen the reverse was true. *T. ni* NPV produced in *S. exigua* was six times more virulent for *T. ni* larvae than the *S. exigua*-produced occlusion bodies. Differences were also noted in the numbers of virions per occlusion body, numbers of nucleocapsids per virion, and the dissolution of the occlusion bodies under defined conditions. No differences in gel electrophoresis patterns of the virion polypeptides or in the serological relationships of the viruses produced under different conditions were detected. Thus, as with the virus-controlled changes, no specific mechanisms to explain the observed change have been identified.

VII. DEVIATIONS FROM THE NORM

In the preceding sections, the normal replication process for baculoviruses was described. This replication results in the production of morphologically typical, infectious viral progeny. However, as is well known in vertebrate virus-cell systems, several factors can cause incomplete or defective replication of the infecting virus. The incomplete replication caused by a shortage of lipids in the cell culture media, as described in Section II, is one example. Other causes could be lack of a suitable tissue or cell type since most of the insect cell cultures are from one or two tissue sources. The virus inoculum may contain defective particles lacking the complete genetic information required for replication.

One such incomplete replication was described by Ignoffo et al. (1971) with a *Heliothis* NPV. Cells of a continuous cell line of *H. zea* were inoculated with DNA extracted from the polyhedra or with the filtered supernatant of triturated, virus-killed larvae. After 12 days incubation the cells were examined by electron microscopy and no progeny virus or polyhedra were found in either test. However, inoculation of neonate larvae with culture supernatants produced typical infections. The infectious supernatants were successfully passed serially through seven consecutive subcultures without the appearance of typical virus progeny. The synthesis of infectious material could be blocked by 5-fluorodeoxyuridine and was inactivated by DNase. Thus, it would appear that the replication proceeded as far as the production of progeny DNA but not to the assembly of nucleocapsids. The block was apparently related to the cell type as the progeny DNA produced complete viral replication in susceptible insects.

This cell line was later shown to be persistently infected with a nonoccluded baculovirus with approximately 1% of the cells containing virions (Granados et al., 1978). This virion was then passed in another cell line (TN-368) and a persistent infection established in that cell line. The persistent virus was isolated from the TN-368 cell line, plaque purified, and examined by electron microscopy. It was shown to consist of a large number of short, defective-type particles. Burand et al. (1983) passed a clone of the persistent virus at a high multiplicity of infection in the TN-368 line. After 10 passages the infectious titer had decreased 99% and the virions were significantly shorter than normal. DNA fragment patterns, following restriction enzyme treatment, revealed minor molar fragments not present in the original clone. Thus, repeated passage at high multiplicity had produced a defective particle. However, interference was not demonstrated.

McIntosh and Ignoffo (1981b) established a persistent infection with another baculovirus-like particle in cells of *S. frugiperda*. Unlike the *H. zea* virus, this persistent virus produced some complete replication. About 3% of the cells contained occlusion bodies, 1% of the cells had unidentified infectious material, as shown by an infectious centers assay, and 70% contained viral antigen when stained by fluorescent antibody methods. The occlusion bodies formed were about 3 times less infectious than similar bodies from normal virus. The persistent virus was reported to interfere with superinfection by both homologous and heterologous viruses.

A gradual increase in abnormal patterns of replication has been observed following continuous passage of baculoviruses in cell culture. MacKinnon et al. (1974) reported that continuous passage of the *T. ni* NPV in *T. ni* cell lines resulted in a decline both in the number and in the infectivity of occlusion bodies that were produced by the 15th serial passage in cultured cells. Electron microscopic examination of the infected cells revealed some extreme abnormalities by the 20th serial passage. Some nuclei contained only virogenic stroma and membranes, while others contained unusually large amounts of capsid material. By the 40th serial passage few occlusion bodies contained virions and the virions frequently were small and otherwise abnormal. These occlusion bodies were noninfectious. Some normal NOV were produced but the quantity was greatly reduced.

Hirumi et al. (1975) passed the *A. californica* NPV in cell culture with very similar results. In cells infected with high-passage virus (p47 to p99) few occlusion bodies were seen and many holes or "empty" spaces occurred in those occlusion bodies that were formed. The nuclei contained many nucleocapsids but no viral membranes were seen. Many of the nucleocapsids were arranged around the edges of the forming occlusion bodies, which was perhaps an effect of the shortage of envelopes.

Another form of abnormal replication was first discovered as plaque variants (Hink and Vail, 1973; Potter et al., 1976; Fraser and Hink, 1982). One plaque type was composed of cells containing many polyhedron occlusion bodies (the MP variant). Direct examination of cells picked from these plaques showed that these occlusion bodies contained the typical multiple embedded virions, consisting of 4 to 10 nucleocapsids per virion. Single nucleocapsids were never occluded (Ramoska and Hink, 1974). No virion-free occlusion bodies were found in the cells. The cells contained 15 to 35 occlusion bodies ranging from 0.8 to 1.8 μm in diameter. The infected cells contained large numbers of free virions in the nucleus. The second plaque type consisted of cells containing only a few polyhedron occlusion bodies (the FP variant). All infected cells contained less than seven occlusion bodies ranging in diameter from 0.7 to

1.2 μm. Many cells contained no occlusion bodies although the nuclei were hypertrophied and contained large numbers of nucleocapsids and virions. The occlusion bodies contained fewer virions than the MP occlusion bodies and many were entirely devoid of virions. Also, unlike the MP variant, many of the occluded virions contained only a single nucleocapsid.

Attempts to stabilize the MP variant by plaque purification were not successful even though dilutions sufficient to produce only a single plaque per dish were used. Passage in cell culture after plaque purification always resulted in a gradual reappearance of the FP variants. Homogeneity could be maintained for only one to two *in vitro* passages. However, passage in insects by feeding the polyhedra from this variant resulted in an increase in the percentage of MP variants from 21 to 67% in a single passage (Hink and Strauss, 1976).

This would appear to indicate that there is some sort of selection for the FP variant during *in vitro* passage. However, Potter et al. (1978) demonstrated that the repeated passage of NOV *in vivo* resulted in the increase of FP variants with a corresponding loss in virulence of the occlusion bodies produced. The latter results indicate that the changes in the formation and occlusion of virions described by Ramoska and Hink for *in vitro* passage had occurred here as well. Homogeneity of the MP variant was retained for 15 passages in insects when occlusion bodies were used to infect the insects. Thus, the shift from the MP plaque variant to the FP variant appeared to be related to the passage of the NOV form of the virus either *in vivo* or *in vitro* and not to some factors specifically associated with *in vitro* replication.

Initial examination of the DNA with restriction enzymes demonstrated no differences in the resulting fragment patterns of DNA from the NOV produced *in vitro* and the DNA from OV purified from *in vivo*-produced polyhedra from the wild-type virus, which was presumably of the MP variant (Smith and Summers, 1978). Restriction enzyme digests of plaque-purified isolates for each infectious type revealed the presence of genetic variants, but these differences in restriction patterns could not be correlated with any of the different infectious forms. Restriction enzyme analysis of five doubly plaque-purified FP isolates indicated that two of the five isolates had restriction fragment patterns which varied from the patterns of the parent strain (Potter and Miller, 1980b). Detailed analysis of one of these two revealed that the genome contained an additional fragment of DNA located on the physical map between 74.1 and 76.2% of the genome. Since more than two restriction nuclease fragments of a unique size were produced by some enzymes, the authors concluded that there may have been incorporation of host DNA into the viral genome.

The added fragment from one of the clones has been described in more detail by Miller and Miller (1982). It was found to be a 7.3 kb sequence with direct terminal repeats of 0.27 kb that was transferred from the host genome to the viral genome. The sequence was a transposable element similar to the *copia*-like class found previously only in *Drosophila*. No specific role for this insertion in the generation of the FP mutant has been determined and since not all of the FP variants contained an insertion it would be only one of the means by which these variants would be produced.

Wood (1980c) was able to isolate an FP variant within nine passages of a cloned strain of the *A. californica* NPV *in vitro* with transfers at 0.1 to 10 PFU per cell to avoid favoring the development of defective virions. No reversion occurred following 12 transfers and independently derived FP clones did not complement one another. This indicated that the FP variant may have been a deletion mutant. The FP variants uniformly produce higher titers of NOV than do the MP variants, which would tend to give them a selective advantage either *in vitro* or *in vivo* when passage is done solely with NOV. Although no mutation rate has been estimated, the rate would have to be quite high since the FP variant develops quickly in almost all tests. Only Burand and Summers (1982) have been able to maintain the MP variant for any length of time when NOV was used as the inoculum. They maintained several strains for as many as 30 serial passages *in vitro* without the development of FP variants. Seven of the 20 plaque-purified isolates made from the E2 clone after the 30th passage contained additional restriction enzyme fragments. However, these insertions were found to be repeated sequences of the viral genome and not host genome. All of the seven isolates were phenotypically the MP type and the polyhedra produced were infectious. Thus, in this study, the genome of the virus was sometimes altered by serial passage *in vitro* but the phenotype was not. This is the exception, not the rule and defining the cause of the common, rapid change in phenotype of the NPV and developing methods to stabilize the MP phenotype in cell culture will be important advances in the use of cell cultures for the replication of the baculovirus.

VIII. CONCLUSIONS

The second decade of insect cell culture (1975-1985) has been one in which the technology developed in the 10 previous years became widely used to study complex replication processes of at least the Group A Baculoviridae. A solid beginning has been made in establishing some useful experimental approaches, defining some limiting factors, and answering some of the long-standing questions concerning the replication of these complex viruses. It has been our intent to try and provide an overall view of this replicative process and to emphasize both those areas where *in vitro* studies have been used successfully to gain information about the process and also those parts of the process where for one reason or another, little or no progress has been made. It has been shown that many factors influence the replication process: the cells used, the nutrition of these cells, the form and the amount of the virus used, and the genetics of the virus.

Perhaps, the most valuable contributions of the *in vitro* studies thus far have been those on the genetics of the viruses. Cell culture has provided the means whereby virus strains could be isolated and purified. It is now possible to study cloned viruses that are genetically homogenous and mutants can be produced, recovered, and purified. Such studies have already answered the old question of whether the polyhedron was a structural part of the virus or a response of the infected cell. Physical maps of the genome of the *A. californica* NPV have been prepared and the locations of a few genes have been determined. Hopefully, this is the basis from which studies in genetic engineering will develop to produce viruses with higher virulence, broader host range, or better environmental stability for use in pest management systems.

Continued studies on the mechanisms of virus attachment and replication will provide important information on which to develop an understanding of the host range and thus the safety of these viruses for nontarget animals and plants. In addition, such new information will aid in the development of economical, high yield *in vitro* production systems to replace the *in vivo* systems now used to produce baculoviruses for pest management. The advantages of such systems have long been recognized but are, as yet, unattainable.

These systems are not used for a number of possible reasons which serve to illustrate some areas where additional advances need to be made. One such area is the limited number of tissues from which cells can be cultured. The midgut epithelium and the fat body are two tissues that are important sites of viral replication in the insect. Continuous lines of midgut cells would permit the more detailed, closely controlled study of the initial invasive process. Cells from fat body tissue may provide a much better yield of virus than do the present cell lines that are derived mostly from ovarian tissue. Even of greater importance, cell lines from one or both of these tissues may be capable of replicating the granulosis viruses and the replication *in vitro* of this virus is perhaps one of the most important problems that should be solved in the near future.

Finally, the study of the baculoviruses needs to be expanded from the present base of knowledge developed with the *A. californica* NPV. Studies on this virus will continue to provide information with which to develop needed hypotheses to advance our understanding of the baculoviruses. It is now important to test these hypotheses with other baculoviruses and begin to expand the number of well-tested and characterized viruses available for introduction into pest management systems.

REFERENCES

Adams, J. R., Goodwin, R. H., and Wilcox, T. A. (1977). Electron microscopic investigations on invasion and replication of insect baculoviruses *in vivo* and *in vitro*. *Biol. Cell. 28*, 261-268.

Adang, M. J., and Miller, L. K. (1982). Molecular cloning of DNA complementary to mRNA of the baculovirus *Autographa californica* nuclear polyhedrosis virus: Location and gene products of RNA transcripts found late in infection. *J. Virol. 44*, 782-793.

Aizawa, K. (1963). The nature of infections caused by nuclear-polyhedrosis viruses. *In* "Insect Pathology An Advanced Treatise" (E. A. Steinhaus, ed.), Vol. I, pp. 382-412. Academic Press, New York.

Bergold, G. H. (1953). Insect viruses. *Advan. Virus Res. 1*, 91-138.

Bergold, G. H. (1963). The nature of nuclear-polyhedrosis viruses. *In* "Insect Pathology An Advanced Treatise" (E. A. Steinhaus, ed.), Vol. I, pp. 413-456. Academic Press, New York.

Brown, M., and Faulkner, P. (1975). Factors affecting the yield of virus in a cloned cell line of *Trichoplusia ni* infected with a nuclear polyhedrosis virus. *J. Invertebr. Pathol. 26*, 251-257.

Brown, M., Crawford, A. M., and Faulkner, P. (1979). Genetic analysis of a baculovirus *Autographa californica* nuclear polyhedrosis virus I. Isolation of temperature-sensitive mutants and assortment into complementation groups. *J. Virol. 31*, 190-198.

Brown, M., Faulkner, P., Cochran, M. A., and Chung, K. L. (1980). Characterization of two morphology mutants of *Autographa californica* nuclear polyhedrosis virus with large cuboidal inclusion bodies. *J. Gen. Virol. 50*, 309-316.

Bud, H. M., and Kelly, D. C. (1980). Nuclear polyhedrosis virus DNA is infectious. *Microbiologica 3*, 103-108.

Burand, J. P., and Summers, M. D. (1982). Alteration of *Autographa californica* nuclear polyhedrosis virus DNA upon serial passage in cell culture. *Virology 119*, 223-229.

Burand, J. P., Summers, M. D., and Smith, G. E. (1980). Transfection with baculovirus DNA. *Virology 101*, 286-290.

Burand, J. P., Wood, H. A., and Summers, M. D. (1983). Defective particles from a persistent baculovirus infection in *Trichoplusia ni* tissue culture cells. *J. Gen. Virol. 64*, 391-398.

Burley, S. K., Miller, A., Harrap, K. A., and Kelly, D. C. (1982). Structure of the baculovirus nucleocapsid. *Virology 120*, 433-440.

Carstens, E. B., Tjia, S. T., and Doerfler, W. (1979). Infection of *Spodoptera frugiperda* cells with *Autographa californica* nuclear polyhedrosis virus I. Synthesis of intracellular proteins after virus infection. *Virology 99*, 386-398.

Carstens, E. B., Tjia, S. T., and Doerfler, W. (1980). Infectious DNA from *Autographa californica* nuclear polyhedrosis virus. *Virology 101*, 311-314.

Chung, K. L., Brown, M., and Faulkner, P. (1980). Studies on the morphogenesis of polyhedral inclusion bodies of a baculovirus *Autographa californica* NPV. *J. Gen. Virol. 46*, 335-347.

Dobos, P., and Cochran, M. A. (1980). Protein synthesis in cells infected by *Autographa californica* nuclear polyhedrosis virus (Ac-NPV): The effect of cytosine arabinoside. *Virology 103*, 446-464.

Dougherty, E. M., Vaughn, J. L., and Reichelderfer, C. F. (1975). Characteristics of the non-occluded form of a nuclear polyhedrosis virus. *Intervirology 5*, 109-121.

Dougherty, E. M., Weiner, R. M., Vaughn, J. L., and Reichelderfer, C. F. (1981). Physical factors that affect *in vitro Autographa californica* nuclear polyhedrosis virus infection. *Appl. Environ. Microbiol. 41*, 1166-1172.

Elliott, R. M., and Kelly, D. C. (1977). The polyamine content of a nuclear polyhedrosis virus from *Spodoptera littoralis*. *Virology 76*, 472-474.

Eppstein, D., and Thoma, J. A. (1975). Alkaline protease associated with the matrix protein of a virus infecting the cabbage looper. *Biochem. Biophys. Res. Commun. 62*, 478-484.

Eppstein, D., and Thoma, J. A. (1977). Characterization and serology of the matrix protein from a nuclear polyhedrosis virus of *Trichoplusia ni* before and after degradation by an endogenous protease. *J. Biochem. 167*, 321-332.

Eppstein, D., Thoma, J. A., Scott, H. A., and Young, S. Y., III (1975). Degradation of matrix protein from a nuclear-polyhedrosis virus of *Trichoplusia ni* by an endogenous protease. *Virology 67*, 591-594.
Federici, B. A. (1980). Mosquito baculovirus: Sequence of morphogenesis and ultrastructure of the virion. *Virology 100*, 1-9.
Fraser, M. J., and Hink, W. F. (1982). The isolation and characterization of the MP and FP plaque variants of *Galleria mellonella* nuclear polyhedrosis virus. *Virology 117*, 366-378.
Goldstein, N. I., and McIntosh, A. H. (1980). Glycoproteins of nuclear polyhedrosis viruses. *Arch. Virol. 64*, 119-126.
Goodwin, R. H. (1976). Insect cell growth on serum free media. *In Vitro 12*, 303-304. (Abstr.)
Goodwin, R. H., and Adams, J. R. (1980). Nutrient factors influencing viral replication in serum-free insect cell line culture. *In* "Invertebrate Systems *In Vitro*" (E. Kurstak, K. Maramorosch, and A. Dübendorfer, eds.), pp. 493-509. Elsevier/North-Holland Biomed. Press, New York.
Goodwin, R. H., Vaughn, J. L., Adams, J. R., and Louloudes, S. J. (1970). Replication of a nuclear polyhedrosis virus in an established insect cell line. *J. Invertebr. Pathol. 16*, 284-288.
Goodwin, R. H., Vaughn, J. L., Adams, J. R., and Louloudes, S. J. (1973). The influence of insect cell lines and tissue-culture media on baculovirus polyhedra production. *Misc. Publ. Entomol. Soc. Amer. 9*, 66-72.
Goodwin, R. H., Tompkins, G. J., and McCawley, P. (1978). Gypsy moth cell lines divergent in viral susceptibility I. Culture and identification. *In Vitro 14*, 485-494.
Grace, T. D. C. (1962). Establishment of four strains of cells from insect tissues grown *in vitro*. *Nature (London) 195*, 788-789.
Granados, R. R. (1976). Infection and replication of insect pathogenic viruses in tissue culture. *Advan. Virus Res. 20*, 189-236.
Granados, R. R. (1978). Early events in the infection of *Heliothis zea* midgut cells by a baculovirus. *Virology 90*, 170-174.
Granados, R. R., and Lawler, K. A. (1981). *In vivo* pathway of *Autographa californica* baculovirus invasion and infection. *Virology 108*, 297-308.
Granados, R. R., Nguyen, T., and Cato, B. (1978). An insect cell line persistently infected with a baculovirus-like particle. *Intervirology 10*, 309-317.
Grula, M. A., Buller, P. L., and Weaver, R. F. (1981). α-Amanitin-resistant viral RNA synthesis in nuclei isolated

from nuclear polyhedrosis virus infected *Heliothis zea* larvae and *Spodoptera frugiperda* cells. *J. Virol. 38*, 916-921.
Harrap, K. A. (1970). Cell infection by a nuclear polyhedrosis virus. *Virology 42*, 311-318.
Harrap, K. A. (1972a). The structure of nuclear polyhedrosis viruses II. The virus particle. *Virology 50*, 124-132.
Harrap, K. A. (1972b). The structure of nuclear polyhedrosis viruses III. Virus assembly. *Virology 50*, 133-139.
Henderson, J. F., Faulkner, P., and MacKinnon, E. A. (1974). Some biophysical properties of virus present in tissue cultures infected with the nuclear polyhedrosis virus of *Trichoplusia ni*. *J. Gen. Virol. 22*, 143-146.
Hess, R. T., and Falcon, L. A. (1977). Observations on the interaction of baculoviruses with the plasma membrane. *J. Gen. Virol. 36*, 525-530.
Hess, R. T., and Falcon, L. A. (1981). Electron microscope observations of *Autographa californica* (Noctuidae) nuclear polyhedrosis virus replication in the midgut of the saltmarsh caterpillar, *Estigmene acrea* (Arctiidae). *J. Invertebr. Pathol. 37*, 86-90.
Hink, W. F. (1976). A compilation of invertebrate cell lines and culture media. *In* "Invertebrate Tissue Culture Research Applications" (K. Maramorosch, ed.), pp. 319-369. Academic Press, New York.
Hink, W. F. (1980). The 1979 compilation of invertebrate cell lines and media. *In* "Invertebrate Systems *In Vitro*" (E. Kurstak, K. Maramorosch, and A. Dübendorfer, eds.), pp. 553-578. Elsevier/North-Holland Biomed. Press, New York.
Hink, W. F., and Strauss, E. (1976). Replication and passage of alfalfa looper nuclear polyhedrosis virus plaque variants in cloned cell cultures and larval stages of four host species. *J. Invertebr. Pathol. 27*, 49-55.
Hink, W. F., and Vail, P. V. (1973). A plaque assay for titration of alfalfa looper nuclear polyhedrosis virus in a cabbage looper (TN-368) cell line. *J. Invertebr. Pathol. 22*, 168-174.
Hink, W. F., Strauss, E. M., and Lynn, D. E. (1977a). Growth of TN-368 insect cells in serum-free media. *In Vitro 13*, 177. (Abstr.)
Hink, W. F., Strauss, E. M., and Ramoska, W. A. (1977b). Propagation of *Autographa californica* nuclear polyhedrosis virus in cell culture: Methods for infecting cells. *J. Invertebr. Pathol. 30*, 185-191.
Hirumi, H., Hirumi, K., and McIntosh, A. H. (1975). Morphogenesis of a nuclear polyhedrosis virus of the alfalfa looper in a continuous cabbage looper cell line. *Ann. N.Y. Acad. Sci. 266*, 302-326.

Huger, A. (1963). Granuloses of insects. *In* "Insect Pathology An Advanced Treatise" (E. A. Steinhaus, ed.), Vol. I, pp. 531-575. Academic Press, New York.

Hunter, D. K., Hoffmann, D. F., and Collier, S. J. (1975). Observations on a granulosis virus of the potato tuberworm, *Phthorimaea operculella*. *J. Invertebr. Pathol. 26*, 397-400.

Ignoffo, C. M., Shapiro, M., and Hink, W. F. (1971). Replication and serial passage of infectious *Heliothis* nucleopolyhedrosis virus in an established line of *Heliothis zea* cells. *J. Invertebr. Pathol. 18*, 131-134.

Injac, M., Vago, C., Duthoit, J.-L., and Veyrunes, J.-C. (1971). Libération (release) des virions dans les polyédroses nucléaires. *C.R. Acad. Sci. D273*, 439-441.

Kawanishi, C., Summers, M. D., Stoltz, D. B., and Arnott, H. J. (1972). Entry of an insect virus *in vitro* by fusion of viral envelope and microvillus membrane. *J. Invertebr. Pathol. 20*, 104-108.

Kawarabata, T., and Aratake, Y. (1978). Functional differences between occluded and nonoccluded viruses of a nuclear polyhedrosis of the silkworm, *Bombyx mori*. *J. Invertebr. Pathol. 31*, 329-336.

Kelly, D. C. (1976). "Oryctes" virus replication: electron microscopic observations on infected moth and mosquito cells. *Virology 69*, 596-606.

Kelly, D. C. (1977). The DNA contained by nuclear polyhedrosis viruses isolated from four *Spodoptera* sp. (Lepidoptera, Noctuidae). Genome size and homology assessed by DNA reassociation kinetics. *Virology 76*, 468-471.

Kelly, D. C. (1981). Baculovirus replication: Stimulation of thymidine kinase and DNA polymerase activities in *Spodoptera frugiperda* cells infected with *Trichoplusia ni* nuclear polyhedrosis virus. *J. Gen. Virol. 52*, 313-319.

Kelly, D. C., and Lescott, T. (1981). Baculovirus replication: Protein synthesis in *Spodoptera frugiperda* cells infected with *Trichoplusia ni* nuclear polyhedrosis virus. *Microbiologica 4*, 35-57.

Kelly, D. C., and Wang, X. (1981). The infectivity of nuclear polyhedrosis virus DNA. *Ann. Virol. Inst. Pasteur 132E*, 247-259.

Knudson, D. L. (1979). Plaque assay of baculoviruses employing an agarose-nutrient overlay. *Intervirology 11*, 40-46.

Knudson, D. L., and Buckley, S. M. (1977). Invertebrate cell culture methods for the study of invertebrate-associated animal viruses. *In* "Methods in Virology" (K. Maramorosch and H. Koprowski, eds.), Vol. 6, pp. 323-391. Academic Press, New York.

Knudson, D. L., and Harrap, K. A. (1976). Replication of a nuclear polyhedrosis virus in a continuous cell culture of

Spodoptera frugiperda: Microscopy study of the sequence of events of the virus infection. *J. Virol. 17*, 274-268.

Knudson, D. L., and Tinsley, T. W. (1974). Replication of a nuclear polyhedrosis virus in a continuous cell culture of *Spodoptera frugiperda*: Purification, assay of infectivity, and growth characteristics of the virus. *J. Virol. 14*, 934-944.

Knudson, D. L., and Tinsley, T. W. (1978). Replication of a nuclear polyhedrosis virus in a continuous cell line of *Spodoptera frugiperda*: Partial characterization of the viral DNA, comparative DNA-DNA hybridization, and patterns of DNA synthesis. *Virology 87*, 42-57.

Lee, H. H., and Miller, L. K. (1979). Isolation, complementation, and initial characterization of temperature-sensitive mutants of the baculovirus *Autographa californica* nuclear polyhedrosis virus. *J. Virol. 31*, 240-252.

Lynn, D. E., and Hink, W. F. (1978). Infection of synchronized TN-368 cell cultures with alfalfa looper nuclear polyhedrosis virus. *J. Invertebr. Pathol. 32*, 1-5.

Lynn, D. E., and Hink, W. F. (1980). Comparison of nuclear polyhedrosis virus replication in five lepidopteran cell lines. *J. Invertebr. Pathol. 35*, 234-240.

Lynn, D. E., and Oberlander, H. (1983). The establishment of cell lines from imaginal wing discs of *Spodoptera frugiperda* and *Plodia interpunctella*. *J. Insect Physiol. 29*, 591-596.

Lynn, D. E., Miller, S. G., and Oberlander, H. (1982). Establishment of a cell line from lepidopteran wing imaginal discs: Induction of newly synthesized proteins by 20-hydroxyecdysone. *Proc. Natl. Acad. Sci. U.S. 79*, 2589-2593.

Lynn, D. E., Boucias, D. G., and Pendland, J. C. (1983). Nuclear polyhedrosis virus replication in epithelial cell cultures of Lepidoptera. *J. Invertebr. Pathol. 42*, 424-426.

McCarthy, W., and DiCapua, R. A. (1979). Characterization of solubilized proteins from tissue culture and host derived nuclear polyhedra of *Lymantria dispar* and *Autographa californica*. *Intervirology 11*, 174-181.

McCarthy, W. J., Lambiase, J. T., and Henchal, L. S. (1980). The effect of inoculum concentration on the development of the nuclear polyhedrosis virus of *Autographa californica* in TN-368 cells. *J. Invertebr. Pathol. 36*, 48-51.

McIntosh, A. H., and Ignoffo, C. M. (1981a). Replication and infectivity of the single-embedded nuclear polyhedrosis virus, *Baculovirus heliothis*, in homologous cell lines. *J. Invertebr. Pathol. 37*, 258-264.

McIntosh, A. H., and Ignoffo, C. M. (1981b). Establishment of a persistent baculovirus infection in a lepidopteran cell line. *J. Invertebr. Pathol. 38*, 395-403.
MacKinnon, E. A., Henderson, J. F., Stoltz, D. B., and Faulkner, P. (1974). Morphogenesis of nuclear polyhedrosis virus under conditions of prolonged passage *in vitro*. *J. Ultrastr. Res. 49*, 419-435.
Maruniak, J. E., and Summers, M. D. (1981). *Autographa californica* nuclear polyhedrosis virus phosphoproteins and synthesis of intracellular proteins after virus infection. *Virology 109*, 25-34.
Miller, D. W., and Miller, L. K. (1982). A virus mutant with an insertion of a copia-like transposable element. *Nature (London) 299*, 562-564.
Miller, L. K., Jewell, J. E., and Browne, D. (1981). Baculovirus induction of a DNA Polymerase. *J. Virol. 40*, 305-308.
Miller, L. K., Adang, M. J., and Browne, D. (1983). Protein kinase activity associated with the extracellular and occluded forms of the baculovirus *Autographa californica* nuclear polyhedrosis virus. *J. Virol. 46*, 275-278.
Nappi, A. J., and Hammill, T. M. (1975). Viral release and membrane acquisition by budding through the nuclear envelope of hemocytes of the gypsy moth, *Porthetria dispar*. *J. Invertebr. Pathol. 26*, 387-392.
Peters, J. E., and DiCapua, R. A. (1978). Immunochemical characterization of *Lymantria dispar* NPV hemagglutinin protein-carbohydrate interaction. *Intervirology 9*, 231-242.
Potter, K. N., and Miller, L. K. (1980a). Transfection of two invertebrate cell lines with DNA of *Autographa californica* nuclear polyhedrosis virus. *J. Invertebr. Pathol. 36*, 431-432.
Potter, K. N., and Miller, L. K. (1980b). Correlating genetic mutations of a baculovirus with the physical map of the DNA genome. *In* "Animal Virus Genetics" (B. N. Fields, R. Jaenisch, and C. F. Fox, eds.), pp. 71-80. Academic Press, New York.
Potter, K. N., Faulkner, P., and MacKinnon, E. A. (1976). Strain selection during serial passage of *Trichoplusia ni* nuclear polyhedrosis virus. *J. Virol. 18*, 1040-1050.
Potter, K. N., Jaques, R. P., and Faulkner, P. (1978). Modification of *Trichoplusia ni* nuclear polyhedrosis virus passaged *in vivo*. *Intervirology 9*, 76-85.
Qin, J.-C., and Weaver, R. F. (1982). Capping of viral RNA in cultured *Spodoptera frugiperda* cells infected with *Autographa californica* nuclear polyhedrosis virus. *J. Virol. 43*, 234-240.

Quiot, J. M., Monsarrat, P., Gérard, M., Croizier, G., and Vago, C. (1973). Infection des cultures cellulares de Coliópteres par le <<virus Oryctes>>. *C.R. Acad. Sci. D276*, 3229-3231.

Raghow, R., and Grace, T. D. C. (1974). Studies on a nuclear polyhedrosis virus in *Bombyx mori* cells *in vitro* I. Multiplication kinetics and ultrastructural studies. *J. Ultrastr. Res. 47*, 384-399.

Ramoska, W. A., and Hink, W. F. (1974). Electron microscope examination of two plaque variants from a nuclear polyhedrosis virus of the alfalfa looper, *Autographa californica*. *J. Invertebr. Pathol. 23*, 197-201.

Reichelderfer, C. F. (1974). A hemagglutinating antigen from a nuclear polyhedrosis virus of *Spodoptera frugiperda*. *J. Invertebr. Pathol. 23*, 46-50.

Robertson, J. S., Harrap, K. A., and Longworth, J. F. (1974). Baculovirus morphogenesis: The acquisition of the virus envelope. *J. Invertebr. Pathol. 23*, 248-251.

Rohel, D. Z., Cochran, M. A., and Faulkner, P. (1983). Characterization of two abundant mRNAs of *Autographa californica* nuclear polyhedrosis virus present late in infection. *Virology 124*, 357-365.

Shapiro, M., and Ignoffo, C. M. (1970). Hemagglutination by a nucleopolyhedrosis virus of the cotton bollworm, *Heliothis zea*. *Virology 41*, 577-579.

Smith, G. E., and Summers, M. D. (1978). Analysis of baculovirus genomes with restriction endonucleases. *Virology 89*, 517-527.

Smith, G. E., Vlak, J. M., and Summers, M. D. (1982). *In vitro* translation of *Autographa californica* nuclear polyhedrosis virus early and late mRNAs. *J. Virol. 44*, 199-208.

Smith, K. M. (1977). "Virus-Insect Relationships." 291 pp. Longman Green, New York.

Smith, K. M., and Brown, R. M., Jr. (1965). A study of the long virus rods associated with insect granuloses. *Virology 27*, 512-519.

Stairs, G. R. (1964). Selection of a strain of insect granulosis virus producing only cubic inclusion bodies. *Virology 24*, 520-521.

Stairs, G. R. (1980). Comparative infectivity of nonoccluded virions, polyhedra, and virions released from polyhedra for larvae of *Galleria mellonella*. *J. Invertebr. Pathol. 36*, 281-282.

Stiles, B., Wood, H. A., and Hughes, P. R. (1983). Effect of tunicamycin on the infectivity of *Autographa californica* nuclear polyhedrosis virus. *J. Invertebr. Pathol. 41*, 405-408.

Stockdale, H., and Gardiner, G. R. (1977). The influence of the condition of cells and medium on production of polyhedra of *Autographa californica* nuclear polyhedrosis virus *in vitro*. *J. Invertebr. Pathol. 30*, 330-336.
Stoltz, D. B., and Vinson, S. B. (1979). Viruses and parasitism in insects. *Advan. Virus Res. 24*, 125-171.
Stoltz, D. B., Pavan, C., and DaCunha, A. B. (1973). Nuclear polyhedrosis virus: a possible example of *de novo* intranuclear membrane morphogenesis. *J. Gen. Virol. 19*, 145-150.
Summers, M. D. (1969). Apparent *in vivo* pathway of granulosis virus invasion and infection. *J. Virol. 4*, 188-190.
Summers, M. D. (1971). Electron microscopic observations on granulosis virus entry, uncoating and replication processes during infection of the midgut cells of *Trichoplusia ni*. *J. Ultrastr. Res. 35*, 606-625.
Summers, M. D., and Arnott, H. J. (1969). Ultrastructural studies on inclusion formation and virus occlusion in nuclear polyhedrosis and granulosis virus-infected cells of *Trichoplusia ni* (Hübner). *J. Ultrastr. Res. 28*, 462-480.
Summers, M. D., and Volkman, L. E. (1976). Comparison of biophysical and morphological properties of occluded and extracellular nonoccluded baculovirus from *in vivo* and *in vitro* host systems. *J. Virol. 17*, 962-972.
Tanada, Y., and Hess, R. T. (1976). Development of a nuclear polyhedrosis virus in midgut cells and penetration of the virus into the hemocoel of the army-worm, *Pseudaletia unipuncta*. *J. Invertebr. Pathol. 28*, 67-76.
Tanada, Y., and Leutenegger, R. (1970). Multiplication of a granulosis virus in larval midgut cells of *Trichoplusia ni* and possible pathways of invasion into the hemocoel. *J. Ultrastr. Res. 30*, 589-600.
Tjia, S. T., Carstens, E. B., and Doerfler, W. (1979). Infection of *Spodoptera frugiperda* cells with *Autographa californica* nuclear polyhedrosis virus II. The viral DNA and the kinetics of replication. *Virology 99*, 399-409.
Tompkins, G. J., Vaughn, J. L., Adams, J. R., and Reichelderfer, C. F. (1981). Effects of propagating *Autographa californica* nuclear polyhedrosis virus and its *Trichoplusia ni* variant in different hosts. *Environ. Entomol. 10*, 801-806.
Tweeten, K. A., Bulla, L. A., Jr., and Consigli, R. A. (1980). Characterization of an extremely basic protein derived from granulosis virus nucleocapsids. *J. Virol. 33*, 866-876.
Vago, C., and Bergoin, M. (1963). Développement des virus a corps d'inclusion du Lépidoptère *Lymantria dispar* en cultures cellulaires. *Entomophaga 8*, 253-261.
Vail, P. V., Jay, D. L., and Romine, C. L. (1976). Replication of the *Autographa californica* nuclear polyhedrosis virus in insect cell lines grown in modified media. *J. Invertebr. Pathol. 28*, 263-267.

Vail, P. V., Romine, C. L., and Vaughn, J. L. (1979). Infectivity of nuclear polyhedrosis virus extracted with digestive juices. *J. Invertebr. Pathol. 33*, 328-330.

van der Beek, C. P., Saijer-Reip, J. D., and Vlak, J. M. (1980). On the origin of the polyhedral protein of *Autographa californica* nuclear polyhedrosis virus. Isolation, characterization and translation of viral messenger RNA. *Virology 100*, 326-333.

Vaughn, J. L. (1976). The production of nuclear polyhedrosis viruses in large-volume cell cultures. *J. Invertebr. Pathol. 28*, 233-237.

Vaughn, J. L., and Faulkner, P. (1963). Susceptibility of an insect tissue culture to infection by virus preparations of the nuclear polyhedrosis of the silkworm (*Bombyx mori* L.). *Virology 20*, 484-489.

Vaughn, J. L., Adams, J. R., and Wilcox, T. (1972). Infection and replication of insect viruses in tissue culture. *Monograph. Virol. 6*, 27-35.

Vlak, J. M., Smith, G. E., and Summers, M. D. (1981). Hybridization selection and *in vitro* translation of *Autographa californica* nuclear polyhedrosis virus mRNA. *J. Virol. 40*, 762-771.

Volkman, L. E. (1983). Occluded and budded *Autographa californica* nuclear polyhedrosis virus: Immunological relatedness of structural proteins. *J. Virol. 46*, 221-229.

Volkman, L. E., and Goldsmith, P. A. (1982). Generalized immunoassay for *Autographa californica* nuclear polyhedrosis virus infectivity *in vitro*. *Appl. Environ. Microbiol. 44*, 227-233.

Volkman, L. E., and Summers, M. D. (1975). Nuclear polyhedrosis virus detection: Relative capabilities of clones developed from *Trichoplusia ni* ovarian cell lines TN-368 to serve as indicator cells in a plaque assay. *J. Virol. 16*, 1630-1637.

Volkman, L. E., and Summers, M. D. (1977). *Autographa californica* nuclear polyhedrosis virus: Comparative infectivity of the occluded, alkali-liberated and nonoccluded forms. *J. Invertebr. Pathol. 30*, 102-103.

Volkman, L. E., Summers, M. D., and Hsieh, C.-H. (1976). Occluded and nonoccluded nuclear polyhedrosis virus grown in *Trichoplusia ni*: Comparative neutralization, comparative infectivity, and *in vitro* growth studies. *J. Virol. 19*, 820-832.

Wilkie, G. E. I., Stockdale, H., and Pirt, S. V. (1980). Chemically-defined media for production of insect cells and viruses *in vitro*. *Develop. Biol. Standard. 46*, 29-37.

Wood, H. A. (1977). An agar overlay plaque assay method for *Autographa californica* nuclear polyhedrosis virus. *J. Invertebr. Pathol. 29*, 304-307.

Wood, H. A. (1980a). *Autographa californica* nuclear polyhedrosis virus-induced proteins in tissue culture. *Virology 102*, 21-27.

Wood, H. A. (1980b). Protease degradation of *Autographa californica* nuclear polyhedrosis virus proteins. *Virology 103*, 392-399.

Wood, H. A. (1980c). Isolation and replication of an occlusion body-deficient mutant of the *Autographa californica* nuclear polyhedrosis virus. *Virology 105*, 338-344.

Wood, H. A., Hughes, P. R., Johnston, L. B., and Langridge, W. H. R. (1981). Increased virulence of *Autographa californica* nuclear polyhedrosis virus by mutagenesis. *J. Invertebr. Pathol. 38*, 236-241.

Wood, H. A., Johnston, L. B., and Burand, J. P. (1982). Inhibition of *Autographa californica* nuclear polyhedrosis virus replication in high-density *Trichoplusia ni* cell cultures. *Virology 119*, 245-254.

Yamafugi, K., Yoshiara, F., and Hirayama, K. (1959). Deoxyribonuclease and protease in polyhedral viral particles. *Enzymologia 22*, 1-12.

Zummer, M., and Faulkner, P. (1981). Absence of protease in baculovirus polyhedral bodies propagated *in vitro*. *J. Invertebr. Pathol. 33*, 383-384.

THE REPLICATION SCHEMES OF INSECT VIRUSES IN HOST CELLS

NORMAN F. MOORE

Institute of Virology
Oxford, United Kingdom

I. INTRODUCTION

This chapter describes the replicative events of insect viruses in cell culture systems, where these are available. It also describes these events in insect host cells as determined by electron microscopy and occasionally by immunological methods or light microscopy, in the cases of some occluded viruses. Currently the most studied replicative cycles are those of the baculoviruses because of the use of these viruses as field control agents. The replication of baculoviruses and iridoviruses is discussed in other chapters in this volume. The amount of information on replication is related to the availability of cell culture systems. It is, therefore, not surprising that relatively little work has been performed on other groups of viruses, such as the cytoplasmic polyhedrosis viruses, even though they may prove to be efficient in controlling insects.

Many viruses infecting insects are described as "arboviruses" (anthropod-borne viruses) although these are generally considered to be "vertebrate" rather than insect viruses. In many cases the disease is expressed in vertebrates, although the virus replicates in the arthropod vector. One of the major groups of these viruses is a genus within the

ISBN 0-12-470295-3

TABLE I. Virus Descriptions

Family name	Genus	Nucleic acid	Shape
Reoviridae	Cytoplasmic polyhedrosis virus group	10 pieces of linear dsRNA	Spherical 50-70 nm in diameter with 12 spikes at vertices of isosahedral. Many virions are occluded in polyhedron
	Reovirus subgroup	10 pieces of linear dsRNA	Icosahedral particles ≃ 70 nm diameter with two coats
Birnaviruses (proposed name)	-	2 pieces of linear dsRNA $\simeq 2.4 \times 10^6$	Icosahedral, 60 nm in diameter; cores seen by electron microscopy
Picornaviridae	"Insect" viruses have no genera	One molecule of infectious ssRNA; mol. wt. $\simeq 2.5 \times 10^6$	Roughly spherical, 27 nm in diameter
Nodaviridae	-	Two ssRNA molecules which are infectious together; mol. wts. $\simeq 1$ and 0.5×10^6	Spherical to icosahedral particles, 30 nm in diameter
Nudaurelia β virus group	-	One molecule of ssRNA of mol. wt. 1.8×10^6	Approximately spherical to hexagonal with a diameter of 38 nm

Caliciviridae	-	*One molecule of infectious +ve ssRNA; mol. wt.* $\simeq 2.7 \times 10^6$	*Roughly spherical, 38 nm in diameter with 32 "dark" cups on surface arranged in icosahedral symmetry*
Retroviridae	-	*Non-infective inverted dimer of linear +ve ssRNA*	*Spherical, enveloped 80-100 nm in diameter with short surface spikes; outer and inner membrane and disc core are visible*
Rhabdoviridae	*Rabies virus group*	*One molecule of noninfective ssRNA; mol. wt.* $\simeq 3 \times 10^6$	*Bullet shaped* $\simeq 200$ *nm* $\times$ *60 nm with surface projection and characteristic cross-striations*
Parvoviridae	*Densoviruses*	*Single strands of ssDNA of mol. wt.* $\simeq 2 \times 10^6$*. Strands in virions are either +ve or -ve and are complimentary, i.e., come together on extraction to form a double strand*	*Isometric particles with isosahedral symmetry* $\simeq$ *20 nm in diameter with dense cores*
Poxviridae (subfamily Entomopoxvirinae)	*A* *B* *C*	*One molecule dsDNA 130-250* $\times 10^6$	*Brick-shaped or ovoid; may be occluded in crystalline polyhedra structures*

Reoviridae, the orbiviruses (type species bluetongue virus). The viruses within this genus are transmitted by ticks, mosquitoes, and biting flies and have attracted much interest because several members of the genus cause disease in mammals. Orbiviruses have a genome which consists of 10 segments of double-stranded RNA. There is thus a great potential for change by interchange of genome segments upon multiple infection. In this chapter two groups within the *Reoviridae* will be described: (1) the cytoplasmic polyhedrosis virus group, because these are "genuine" insect viruses in that similar viruses have not been reported in mammalian species and (2) the reoviruses, which replicate in insect cells, although until recently no model system had been available for study. A number of plant reoviruses are transmitted in insects, such as leafhoppers (Grylls, 1975; Milne and Lovisolo, 1977; Hatta and Francki, 1977; Boccardo et al., 1980). In some cases it has been demonstrated that replication of these viruses occurs in the insect. However, whether the viruses are insect or plant viruses is open to discussion. If the disease is expressed to a greater extent in the plant it is potentially a less ideal type of infection for the virus than when it "inapparently" exists in the insect. Hatta and Francki (1982) provided a structural model to demonstrate a close relationship not only between the "plant" and "insect" reoviruses but the cytoplasmic polyhedrosis viruses.

Viruses in two other major groups, the members of the *Bunyaviridae* (Bishop and Shope, 1979; Bishop et al., 1980) and the *Togaviridae* (Pfefferkorn and Shapiro, 1974; Schlesinger, 1980) are transmitted by mosquitoes, ticks, horseflies, and other arthropod species. Some members of the *Togaviridae* are not arthropod transmitted, and several, which are known to be carried by arthropods, are transmitted by other mechanisms. With members of the *Bunyaviridae* and *Togaviridae*, as with the orbiviruses, there is often evidence of the presence and, in some cases proved replication of the virus in the arthropod host. However, in this chapter, for the sake of brevity, viruses which replicate in insects or insect cell culture and have no known intermediate hosts will be considered. The small RNA-containing virus, Nodamura virus, will also be considered as an insect virus although it replicates in mammalian cells and has hosts other than insects. However, in the case of the *Nodaviridae*, the type member, Nodamura virus, appears to be atypical.

Table I shows the viruses whose replication will be described in this chapter. In some cases, for example, with the birnaviruses and caliciviruses, there is only one known member which infects insects. With other families, such as the retroviruses and the genus reovirus, further research needs to be done, especially in the area of host specificity and serological interaction with the mammalian members of the genus or family.

II. DOUBLE-STRANDED RNA VIRUSES

A. *CYTOPLASMIC POLYHEDROSIS VIRUSES (CPVs)*

Studies on the replication of CPVs have been hindered by the absence of suitable cell culture systems. Although complete polyhedra have been observed in several infected established cell lines, the numbers of infected cells are normally low, and there appear to be relatively few free virions in the infected cell supernatant, making passage of the virus difficult. Although CPVs were demonstrated to replicate in culture systems over 10 years ago (Granados et al., 1974), only recently have suitable cell systems been established to assay virus infectivity (Belloncik and Chagnon, 1980; Longworth, 1981), which is a prerequisite for a thorough study of replicative events. Belloncik and Bellemare (1980) described a phenomenon, also reported by other workers, that larval CPV polyhedra were heterogeneous but cell culture-produced polyhedra were regular. When they examined the polyhedra of the CPV from *Euxoa scandens* grown in *Lymantria dispar* cells they were exclusively cubical. However, when the cell culture-derived polyhedra were repassaged through the larvae the heterogeneous population reappeared.

Continuous cell cultures are normally infected with the nonoccluded viruses derived from the insect and, with one exception (Quoit and Belloncik, 1977), low infection rates were obtained even with high multiplicities of infection (Longworth, 1981). Belloncik and Arella (1981) found that when *L.dispar* cells infected with *Euxoa scandens* CPV were exposed to γ radiation the number of polyhedra per cell increased rapidly. As yet there have been no significant reports of the production of polypeptides or RNA in cell culture.

In the absence of suitable cell culture systems, the gross replicative events of CPVs have been studied by light and electron microscopy. With few exceptions CPV replication was associated with the tissues of the alimentary tract. The normal mechanism of CPV infection is by ingestion of polyhedra, and in the alkaline conditions of the gut the polyhedra are solubilized to release the virus particles. Virions have diameters of approximately 50 to 70 nm and there are twelve, possibly hollow, spikes at the isosahedral vertices (see Table I, and Payne and Harrap, 1977; Harrap and Payne, 1979 for reviews on structure of cytoplasmic polyhedrosis viruses). The mechanism of attachment and entry is unknown and the only infected cells are those of the columnar midgut epithelium. The first signs of infection by cytoplasmic polyhedrosis viruses are opaque areas called the virogenic stroma (Xeros, 1956, 1966; Kobayashi, 1971). Polyhedra appear within 20 hr of infection. Very little is known about the mechanism of polyhedra formation, or why a proportion of the virus particles are not assembled into these structures. An early serological study on the multiplication of CPV in the silkworm *Bombyx mori* demonstrated the presence of virus antigens in the cytoplasm near the striated border of the epithelium within 6 hr of infection. While no fluorescence was detectable in the nucleus, some cells were fluorescent throughout the cytoplasm; in the majority of the cells the viral antigens appeared to be located near the striated border (Kawase and Miyajima, 1969). It appears from several studies that virus particles develop in the cytoplasm in areas in which virogenic stroma are found (Arnott et al., 1968; Stoltz and Hilsenhoff, 1969; Quoit and Belloncik, 1977; Lipa, 1977), although Andreadis (1981) found, with a CPV from the salt-marsh mosquito, that virus particles developed in areas of cytoplasm where the virogenic stroma was sparse or absent. However, the polyhedra formation occurred within the virogenic stroma as with other CPV infections.

Little is known about the synthesis of viral nucleic acid within infected insects. Viral RNA synthesis can be detected in the midguts of NPV-infected *Malacosoma disstria* and silkworms (Hayashi and Kawarabata, 1970; Kawase and Furusawa, 1971; Furusawa and Kawase, 1971). When CPV-infected midguts of *Orgyia leucostigma* and *Malacosoma disstria* were treated with actinomycin D, the virus-induced ribonuclease-sensitive RNA's had buoyant densities of 22 and 15 S, while ribonuclease-resistant RNA sedimented at 15 S (Hayashi, 1970; Hayashi and Donaghue, 1971). Similar information was obtained by Furusawa and Kawase (1973) when they examined virus-specific RNA synthesis in the midgut of infected silkworm. It was also demonstrated that the ribonuclease-sensitive single-stranded RNA annealed to heat-denatured viral RNA (Furusawa and Kawase, 1973; Payne and Kalmakoff, 1973). There was some evidence to suggest that the cell nucleus was involved in viral RNA synthesis (Watanabe, 1967; Hayashi and Retnakaran, 1970).

Much of the information on the replication of CPVs has been derived from *in vitro* studies on the synthesis of messenger RNA from the virus genome and the *in vitro* synthesis of protein from this messenger. CPV contains a dsRNA-dependent polymerase which is active without any alteration in the virus capsid structure (Furuichi, 1974). The synthesis of ssRNA by virions has been studied in detail and the polymerase required the presence of four trinucleotides and magnesium ions

(Shimotohno and Miura, 1973; Mertens and Payne, 1978; Mertens, 1979). In addition to the polymerase, enzymes are present which are involved in capping the 5'-termini of the mRNA: methyltransferases (Furuichi, 1974, 1978; Mertens, 1979), nucleotide phosphohydrolase (Storer et al., 1973/74) and guanylyl transferase (Shimotohno and Miura, 1976; Furuichi, 1978). These enzymes catalyzed the synthesis of ten messenger RNAs with a common 5'-terminal of m^7GpppA^mpGpU (Furuichi and Miura, 1975). The CPV polymerase remained stable for relatively long synthesis times and the ssRNA's hybridized to the ten viral genome segments, but not to themselves (Furuichi, 1974; Mertens, 1979; Smith and Furuichi, 1980). Under optimum conditions the synthesized mRNAs were full sized and the mRNA's could be separated by agarose gel electrophoresis (Smith and Furuichi, 1980). The time required for transcription of each segment was related to the length of each viral RNA; they were all transcribed at the same actual rate. It appears, therefore, that the segments are independently transcribed with elongation being the limiting step for the formation of mRNA's, at least *in vivo* (Smith and Furuichi, 1982). This implies that the mechanism that determines synthesis of different amounts of viral structural proteins (and presumably nonstructural proteins) operates at a translational level, unless host functions have a significant effect on the virus enzymes involved in the formation of single-stranded RNA. The latter seems unlikely because of the relative inaccessibility of the nucleotides to the virus-functions--it is unnecessary to degrade the virus to activate the transcriptase activity.

Limited evidence is available on the translation of the mRNA's derived from transcription of the dsRNA. As yet the individual mRNA's have not been separated and translated independently. However, McCrae (1982) recently compared the *in vitro* translation products from unfractionated viral *Bombyx mori* mRNA and the denatured viral genome RNA using the DMSO denaturation technique developed with reovirus (McCrae and Joklik, 1978). The products were found to be very similar. Thus, by comparing the products of the translation with purified virions and polyhedral proteins, tentative assignments were possible for the virus structural proteins, polyhedral proteins, and the nonstructural proteins. There was considerable evidence for translational control in the production of proteins since there were considerable differences in the total amounts synthesized (McCrae, 1982). By separating the individual viral dsRNA segments and translating them individually in a rabbit reticulocyte lysate, McCrae (1982) was able to make a number of RNA/protein assignments. The putative polyhedral protein was synthesized in relatively small amounts from the lowest molecular weight segment. Hence, the question of why this protein is produced in such large amounts in the insect has not been answered.

B. *REOVIRUSES*

There have been comparatively few reports of the presence of reoviruses in insects. Probable reoviruses have been isolated from the mosquitoes, *Aedes vigilax* and *Culex fatigans*, and an infant mouse which was attacked by mosquitoes (Parker et al., 1965). Moussa (1978) described reoviruses isolated from a laboratory colony of houseflies (*Musca domestica*). This virus was lethal when injected into flies, with mortalities occurring within 24 hr after infection; all infected flies died within 10 days. Virions were found in extracts of midguts and hemocytes of infected flies. The isolated particles had a spherical capsid with two shells and the dimensions of the particles were very similar to those reported for other reoviruses. Moussa (1978) found that the midgut cells of infected flies become enlarged but was unable to find reovirus particles associated with these cells. Instead, he found threads or filaments of variable length and diameter in the cytoplasm of midgut cells. The threads were associated with ribosomes and, in some cases, rough endoplasmic reticulum and lysosomes. Viruses were detected only in the cytoplasm of hemocytes and a number of formative stages of the virions appeared to be associated with the infected blood cells.

There is considerable evidence for persistent infections of *Drosophila* cell lines by reovirus-like particles (Teninges et al., 1979b; Brun and Plus, 1980). Gateff et al. (1980) reported the occurrence of reovirus in the cytoplasm of tumorous blood cell lines. The 60 nm virions were present singly, in groups, and in crystalline arrays. Virions were spherical, had a buoyant density of 1.38 g/ml in CsCl, and the capsids were composed of two protein layers. Determination of the protein and nucleic acid content showed the presence of eight proteins and ten segments of double-stranded RNA. Haars et al. (1980) reported that the reovirus from blood III cell line of *Drosophila melanogaster* contained double-stranded RNA in ten major segments with a molecular weight range of 0.55×10^6 - 2.75×10^6. Serological evidence showed that this reovirus was probably not of human or avian origin. Alatortsev et al. (1981) have reported similar information on the structural components of another reovirus isolated from an established *Drosophila* cell line. More recently, Plus et al. (1981) compared a reovirus found in the dipteran *Ceratitis capitata* with five isolates of reovirus from *Drosophila*, including the viruses from the blood cell line (Gateff et al., 1980). They found immunological evidence demonstrating that *Drosophila* reoviruses should belong to one serological group of which the *Ceratitis* virus should not be a member. One genome segment also appeared to be missing in the *Ceratitis* virus.

C. BIRNAVIRUSES

Drosophila X virus (DXV) has been identified in *Drosophila melanogaster* flies and as a contaminant of established cell lines derived from them (Plus, 1978, 1980; Teninges, 1979; Teninges et al., 1979a). Plus (1978) presented evidence which indicated that DXV was present in some fetal bovine sera used for cell culture, and this led to the hypothesis that the virus in the culture systems arose as a contaminant from the sera. Teninges et al. (1979a) demonstrated that DXV was morphologically very similar to infectious pancreatic necrosis (IPN) virus of the trout, infectious bursal disease virus of the chicken and the virus found in the bivalve mollusk *Tellina tenius* (Cohen et al., 1973; Hill, 1976; Nick et al., 1976; Dobos et al., 1977). Teninges (1979) showed that the purified virions had six proteins and while the virions contain some 5 S RNA the major species was 14 S. The latter RNA was resolved into two equimolar fractions by polyacrylamide gel electrophoresis with an average mol. wt. of 2.2×10^6. Dobos et al. (1979) found that the two segments had mol. wt. of 2.5 and 2.6×10^6. This is the same genome arrangement as is apparently found with IPN, infectious bursal disease, Thirlmere, and *Tellina tenius* viruses, all of which have two segments of double-stranded RNA with similar molecular weights (Dobos, 1976; Nick et al., 1976; Dobos et al., 1977; MacDonald and Yamamoto, 1977; Underwood et al., 1977; Kelly et al., 1982). The bisegmented form of the dsRNA has resulted in the proposed name of *Birnaviridae* (Dobos et al., 1979). Recently, Revet and Delain (1982) presented evidence to show that DXV RNA had a different arrangement than had been previously described. Using 4 *M* urea, they released an RNA protein complex which consisted of only one segment of double-stranded RNA which is maintained within the virus by a protein of molecular weight 67×10^3. These authors found that the virus contained six major proteins with molecular weight of between 27×10^3 and 100×10^3 giving a total protein output in excess of double the coding capacity of the single RNA species. They postulated overlapping genes, protein processing, or cellular-encoded proteins in the purified virions.

It was conclusively demonstrated that IPNV has two segments of double-stranded RNA. Using hybrid virus derived from two parental serotypes, it was shown that the larger segment coded for three primary genome products and the smaller segment for a single high molecular weight protein (MacDonald and

Dobos, 1981). Only two species of single-stranded RNA were produced from the two double-stranded genome RNA's (Somogyi and Dobos, 1980; Mertens et al., 1982). Mertens and Dobos (1982) separated the two genome strands, translated them independently *in vitro*, and showed that the larger segment probably gave rise to mRNA which is polycistronic, simultaneous initiation occurring at three independent sites. It should therefore be possible to show whether DXV contains one or two pieces of unique dsRNA using hybrids of two isolates or translation of the segments isolated from gels.

The birnaviruses appear to be unusual dsRNA viruses in that there are multiple products from one genome segment. The picture of intracellular polypeptides is further complicated in that the primary gene products undergo processing to form smaller proteins (Dobos, 1977; Dobos and Rowe, 1977; Müller and Becht, 1982). With IPNV the three protein derivatives from the largest genome segment undergo further cleavages (Dobos, 1977; Dobos and Rowe, 1977; MacDonald and Dobos, 1981). Because of the lack of evidence on the intracellular synthesis of DXV proteins, it is impossible to say which of the proteins found in the mature virion (Teninges, 1979; Revet and Delain, 1982) are functional virus-coded products. There is the possibility of incorporation of host protein, primary gene products from the dsRNA, and the functional cleaved products.

The ribonucleic acid polymerase of birnaviruses has been examined by several workers (Cohen, 1975; Bernard and Petitjean, 1978; Bernard, 1980; Mertens et al., 1982). Bernard (1980) demonstrated an RNA polymerase associated with DXV and found the products to be indistinguishable from the genome with respect to size, sedimentation coefficient, electrophoretic mobility, and resistance to ribonuclease. This was strong evidence that the mechanism of DXV RNA synthesis was semiconservative, i.e., both strands of the genome RNA were being copied. Mertens et al. (1982) obtained similar information with IPNV, although they also reported the synthesis of a variable amount of RNase-sensitive RNA in the size range 5-7 S. This was probably attributable to ribonuclease activity particularly since no subgenomic mRNAs were found in the infected cell (Somogyi and Dobos, 1980).

III. SINGLE-STRANDED RNA VIRUSES

A. *PICORNAVIRUSES*

Several small RNA viruses of insects have been isolated which are potential members of the *Picornaviridae*. Three of these viruses, Cricket paralysis virus (CrPV), *Drosophila* C virus (DCV), and *Gonometa* virus have been classified as members of the family *Picornaviridae*, although they have not yet assigned to genera (Matthews, 1982). A classical mammalian picornavirus contains one molecule of infective +ve single-stranded RNA of molecular weight $\simeq 2.5 \times 10^6$, with a poly-adenylic region at the 3' terminus end and a small protein (VpG) covalently linked to the 5' terminus. There are four structural proteins in the virion, three of molecular weight $\simeq 30 \times 10^3$ and the fourth of molecular weight $\simeq 10 \times 10^3$. The virus replicates via dsRNA replicative intermediates, and viral structural proteins (and other proteins of unknown function) are produced by a series of proteolytic cleavages from a polyprotein of molecular weight $\simeq 200 \times 10^3$, which is produced from the positive strand RNA. The coat proteins are produced from the 5'-end of the message, while VpG, protease, and polymerase appear to be formed at the 3'-end (Matthews, 1982).

The two well-studied insect picornaviruses, DCV and CrPV, appear to have many of the biophysical properties of, and have similar replicative strategies to, the mammalian picornaviruses. Cricket paralysis virus was first identified in the Australian field crickets, *Teleogryllus oceanicus* and *T. commodus*. Rear leg paralysis was followed by death and the virus was transmissible after extraction (Reinganum et al., 1970). A number of *Drosophila* C viruses have been isolated from laboratory and natural populations of *Drosophila*, the first being identified by Jousset et al. (1972) in dead laboratory flies (Plus et al., 1978; Moore et al., 1982b). Both CrPV and DCV have a broad host range (Jousset, 1976; Plus et al., 1978; Scotti et al., 1981), but can be distinguished both by cell culture and insect specificity (Plus et al., 1978). The virions appear to replicate in the cytoplasm of infected cells in the intact insect and in tissue culture. Plus (1978) reported that DCV can exist in a persistent infection in *D. melanogaster* cell culture system, but CrPV is extremely cytolytic (Scotti, 1976, 1977; Moore et al., 1980, 1981c). The sites of infection in the insect and the possible modes of transmission have been reviewed (Scotti et al., 1981).

With the introduction of CrPV and DCV into cell culture systems (Scotti, 1976, 1977; Plus et al., 1978), it has been possible to study the replication events of these viruses in some detail and compare the production of RNA and proteins with those of mammalian picornaviruses (Reavy and Moore, 1981). As with mammalian picornaviruses, CrPV and DCV have a diameter of 27 nm, buoyant densities of 1.34 g/ml, are acid resistant, contain four structural proteins, and each contain a single piece of RNA with a molecular weight of 2.5×10^6-3.0×10^6 (Reinganum, 1973; Jousset et al., 1977; Moore et al., 1981c). Furthermore, the RNA has a 3'-terminal poly(A) tract (Eaton and Steacie, 1980; Pullin et al., 1982), although it is not known if it contains a genome-linked protein (VpG). The RNA is translatable in a cell-free system (Reavy and Moore, 1981b) and codes for high molecular weight proteins which are cleaved to give viral proteins (Reavy and Moore, 1981a,b). In the infected cell, RNA species are found which correspond in molecular weight to the genome RNA as well as high molecular weight RNA's which are partially resistant to ribonuclease (Reavy et al., 1983; Reavy, 1982). All of these results are comparable with those obtained with mammalian picornaviruses. CrPV can be plaque purified (Moore and Pullin, 1982), thus facilitating detailed examination of intracellular events in the absence of other contaminating viruses (although this should be treated with considerable caution with the number of reoviruses, DCVs, black beetle virus-related particles, and possible retroviruses that are being found in tissue culture systems). Although the structural, precursor and low molecular weight proteins produced by both viruses in the infected cell differ (Moore et al., 1980; Moore et al., 1981c,e,f,g) the basic pattern of production is the same. High molecular weight proteins are produced which can be chased into viral proteins. However, viral proteins are produced extremely rapidly as evidenced by using short pulses of radioactivity. Two or three methods of production of viral proteins were considered possible (including fast cleavage of precursors) but the evidence from inhibitors, pulse chase, and pactamycin mapping, etc., seems to indicate that the production of proteins and general replicative events are very similar to those found with mammalian picornaviruses (Reavy, 1982; Reavy and Moore, 1982; Moore et al., 1980; 1981a,c,d,e,f,g; Moore and Pullin, 1983).

Oligonucleotide maps of CrPV and the different isolates of DCV (Pullin et al., 1982; Clewley et al., 1983) have demonstrated that it is possible to distinguish the different isolates. DCV and CrPV are the most distantly related, which is supported by serological information. It should soon be possible to look for genotypic mixing of the different isolates and it will then be important to have detailed knowledge of the comparative host spectrum to consider the effects of the mix.

Several minor and a few important questions remain to be considered with regard to the insect picornaviruses: (1) it is not known whether they have a VpG, (2) the full map of the intracellular proteins on the genome is unknown, and (3) several gaps exist in our knowledge of the replication strategy. These questions are not yet answered for the mammalian viruses (e.g., mechanisms of assembly, mechanism of shut-off of host cell function, etc.).

Several other viruses have been identified which are potential members of the *Picornaviridae*, for example, a virus of the aphid *Rhopalosiphum padi* (D'Arcy et al., 1981), Kawino virus (Pudney et al., 1978), Flacherie virus from the silkworm, *Bombyx mori* (Hashimoto and Kawase, 1983), and several viruses of the honey bee (Bailey, 1976; Bailey and Woods, 1977). However, insufficient information is available on these viruses and also on *Gonometa* virus (Longworth et al., 1973a) because suitable cell culture systems are not available. It is unsufficient to classify a virus as being a member of the *Picornaviridae* on the basis of biophysical properties and nucleic acid and protein content.

B. *THE NUDAURELIA β GROUP OF VIRUSES*

This small group of ssRNA-containing viruses has recently been conferred the status of family (Matthews, 1982). The first member of the group was isolated in Australia from the emperor gum moth, *Antherea eucalypti* (Grace and Mercer, 1965). However, considerably more effort has been expended on the type member of the group (*Nudaurelia* β virus) isolated from larvae of the pine tree emperor moth, *Nudaurelia cytherea capensis* (Lepidoptera: Saturniidae) in South Africa (Juckes, 1970; Polson et al., 1970; Struthers, 1973; Struthers and Hendry, 1974; Finch et al., 1974; Juckes, 1979). Other viruses of this group have been isolated from two limacodids, *Darna trima* and *Thosea asigna*, and the saturniid, *Philosamia cynthia x ricini* (Reinganum et al., 1978). Nine new isolates have recently been recorded on the basis of their interaction with antiserum against *Nudaurelia* β virus (Greenwood and Moore, 1982b).

The viruses of the *Nudaurelia* β group have a single strand of nucleic acid with a molecular weight of $1.8\text{-}2.1 \times 10^6$ and also contain a single protein with a molecular weight of $60\text{-}70 \times 10^3$ (Moore and Tinsley, 1982). The virions have a buoyant density in CsCl of $\simeq 1.3$ g/cm^3, a sedimentation coefficient of $\simeq 200$, and are 35-38 nm in diameter. Indirect evidence that the viruses do not contain lipid is demonstrated by the fact that infective, and apparently intact, virus is produced after treatment with organic solvents. Study of the replication of this group of viruses has been limited by the absence of a suitable permissive cell system in which to observe multiplication. Tripconey (1970) infected primary cultures of the ovaries of larvae of the silk worm, *Bombyx mori*, and the pine emperor moth, *N.capensis*, with *Nudaurelia* β virus. Ten percent of the *N.capensis* cells and 20% of the *B.mori* cells were infected. All the fluorescent antibody-stained preparations showed fluorescence in the cytoplasm. Recent investigations with two members of this group, *Dasychira pudibunda* virus (Greenwood and Moore, 1981a,b; 1982b) and *Trichoplusia ni* virus (Hess et al., 1978, Morris et al., 1979; Moore et al., 1981a) have demonstrated the sites of virus replication in the insect. Using electron microscopy, Hess et al. (1977, 1978) examined midguts of *Autographa calfornica* and *Trichoplusia ni* infected with a naturally occurring mixed inoculum containing *Autographa* nuclear polyhedrosis virus and a number of "nonoccluded" viruses including a member of the *Nudaurelia* β group. The latter was found in the cytoplasm of the midgut and appeared to be concentrated in vacuoles. Surprisingly, there was evidence of the occurrence of these virions in the nuclei, and even incorporation of the virus particles into polyhedra of the baculovirus (Hess et al., 1977, 1978). The reported occlusion of small RNA viruses into polyhedra will obviously be of interest to the commercial producers of insect control agents based on nuclear polyhedrosis viruses. However, the study was based essentially on electron microscopy and further evidence is necessary to show the incorporation of infective small RNA viruses into polyhedra. Greenwood and Moore (unpublished information) dissected *T.ni* and *D.pudibunda* larvae which had been infected with their homologous and heterologous viruses. Examination of head capsule, foregut, midgut, hindgut, elementary gonads, malpighian tubules, and other organs by electron microscopy and enzyme-linked immunosorbent assay (ELISA) showed that the virus particles were associated with the gut tissues. Virus were also found by ELISA to be associated with the epidermis but this was attributed to the relatively large quantities of virus detectable in the feces. Virus was found associated with the mid-, fore-, and hindgut, although the latter can also be attributed to the passage of contaminated feces. Electron microscopic examination of the infected tissues showed the virus to be present in large vacuoles in the cytoplasm of infected cells. There was no evidence of nuclear accumulation and a relatively small number of single particles were visible in the cytoplasm. Late in infection, when large arrays of virus were apparent in the vacuoles, it was, of course, not obvious whether the single particles were being assembled in the cytoplasm and then accumulated in the vacuoles, or whether the single particles were "leaking" from the vacuoles.

While little information can be gained about the replication of these viruses in the absence of a suitable cell system, one approach in understanding the multiplication of these virions is to employ cell-free translation of the virion RNA.

Preliminary evidence for translation of the *T.ni* virus RNA in rabbit reticulocyte lysate demonstrated a protein with a similar molecular weight to the capsid protein and also the presence of a higher molecular weight protein (Reavy, 1982). By drawing parallels with the mammalian picornaviruses, the virions need to code for capsid protein as well as an RNA polymerase. In the case of the split genome nodaviruses, a postulated polymerase is synthesized with molecular weight ≈ 100,000, which could be similar to the larger protein manufactured in the translation system in response to *T. ni* RNA. As yet it is difficult to positively identify even the capsid protein in the translation systems without having a suitably labeled virus structural protein. However, immunoprecipitation of the translation products should distinguish capsid proteins (and possible precursors and early terminators) from other products of the RNA. The translation system will at least be able to distinguish between the production of a polyprotein which is subsequently cleaved, or the production of proteins of the correct molecular weight. The third possibility is, of course, the production of proteins which undergo a small morphogenic cleavage as apparently occurs with the capsid protein of the nodaviruses .

C. NODAMURA VIRUS GROUP (NODAVIRIDAE)

The type species of this group of viruses, Nodamura virus, is an unusual "insect" virus in that it can replicate in mammalian cells as well as insect cells, causing paralysis and death when injected into suckling mice. Nodamura virus was isolated from *Culex tritaeniorhynchus* mosquitoes in Japan (Scherer and Hurlbut, 1967; Scherer et al., 1968). The virus could be serially propagated in soft ticks (*Ornithodoros savignyi*), Indian meal moth larva (*Plodia interpunctella*), and mosquitoes (*Aedes aegypti*). The virus was transmissible to suckling mice by exposure to infected mosquitoes and mortality occurred after initial hind limb paralysis. In addition to its pathogenicity for suckling mice, Nodamura virus was found to be fatal for adult honey bees (*Apis mellifera*) and the wax moth (*Galleria mellonella*) (Bailey and Scott, 1973). Using this capacity to kill adult bees and wax moth larvae, Bailey et al. (1975) were able to demonstrate, using LD_{50} assays, that the virus multiplied in baby hamster kidney (BHK) cell cultures as well as mosquito cell cultures (*Aedes albopictus* and *Aedes aegypti*). These cells yielded high titer virus without apparent cytopathic effect. More recently it was demonstrated that the mortality of mosquitoes was dependent on the route of infection. When Nodamura virus was injected into the thoraces of adult *Aedes albopictus* and *Toxorhynchites amboinesis* mosquitoes, paralysis and death was the result; "normal" ingestion of the virus resulted in no mortality (Tesh, 1980). For some unknown reason, the amount of Nodamura virus in the brain of *A.albopictus* was the same with mosquitoes injected or fed, although the latter did not result in mortality. It is, of course, a greater advantage for the virus not to kill its host as a potentially greater distribution of the virus will occur. Hence it follows that natural routes of infection should result in a reduced incidence of mortality.

Nodamura has many characteristics in common with mammalian picornaviruses, such as size, buoyant density, and resistance to acidic conditions (Murphy et al., 1970). However, when Newman and Brown (1973) added [^{32}P]phosphate to infected mice they were able to extract virus from muscle tissue, which contained two species of RNA with molecular weights of 1.0 and 0.5×10^6. Later evidence by the same workers strongly suggested that unlike many plant viruses the two RNA's were contained within the same particle (Newman and Brown, 1977). To further distinguish the virions from mammalian picornaviruses, Newman and Brown (1976) also demonstrated the absence of polyadenylate (poly A) at the 3'-end of the RNA.

In the absence of a suitable cell system to examine the intracellular proteins expressed by Nodamura virus, Newman et al. (1978) resorted to *in vitro* translation of the RNA species to enable them to elucidate the replicative mechanisms of this virus. They translated the separated RNA species in wheat embryo extract systems and found that the larger RNA generated a protein of molecular weight 105×10^3 and the small RNA gave rise to a protein of molecular weight 43×10^3. The 43×10^3 protein was larger than the capsid protein of the mature virus (40×10^3) and the authors suggested that it was a precursor of the capsid protein.

The second member of the *Nodaviridae* was found during a survey for pathogens of black beetle [*Heteronychus arator* Fabr. (Coleoptera: Scarabaeidae)] (Longworth and Archibald, 1975; Longworth and Carey, 1976). The occurrence of the virus was possibly related to high mortality in a population of the beetles in New Zealand. The virus was found to develop in the cytoplasm of the gut and fat body cells. Surprisingly, the virus was not infective for *H.arator* adults when injected into the body cavity, but it was infective for larvae of *Aphodius tasmaniae* (Coleoptera: Aphodiinae), *Pericoptus truncatus* (Coleoptera: Scarabaeidae), *Pseudaletia separata* (Lepidoptera: Noctuidae), and *Galleria mellonella* (Lepidoptera: Galleridae). Hence, as with Nodamura virus, black beetle virus (BBV) was fairly nonspecific in its host range (Longworth and Archibald, 1975). Further examination of black beetle virus demonstrated similar physicochemical properties, as well as RNA and protein content, to Nodamura virus. However, the virus was not infective to mice.

The demonstration that black beetle virus could multiply readily in Schneider's line 1 of *Drosophila* cells has allowed a detailed study of the replication of this virus. The studies were initially complicated by the presence of BBV-related particles present in one subline of *Drosophila* cells, and the origin of these particles, which caused the cells to be resistant to infection by BBV, was not elucidated (Friesen et al., 1980). It was suggested that the particles were a "maturation-defective form of BBV."

Examination of the structural proteins of BBV showed, in addition to the capsid protein of molecular weight approximately 40×10^3, two other proteins, small amounts of a protein which was slightly larger than the capsid protein, and a low molecular weight protein $\approx 5 \times 10^3$ (Friesen and Rueckert, 1981). When the intracellular proteins expressed by BBV in *Drosophila* cells were examined, five proteins were found to be expressed (Crump and Moore, 1981a,b,c; Friesen and Reuckert, 1981). These were a high molecular weight protein (104×10^3-110×10^3 molecular weight), a capsid protein precursor, capsid protein, and two low molecular weight proteins. One low molecular weight protein corresponded to a species (5×10^3) found in the virion, but the second, with a molecular weight of 10×10^3, was nonstructural (Friesen and Rueckert, 1981). Translation of the separated RNA's of the purified virions in rabbit reticulocyte lysates demonstrated that the highest molecular weight RNA (1×10^6) gave rise to the high molecular weight protein of infected cells (104×10^3-110×10^3) and the low molecular weight RNA (0.5×10^6) gave rise to the capsid protein precursor (Friesen and Rueckert, 1982). Similar evidence was obtained when the separated RNA species of BBV were translated in cell-free lysates derived from *Drosophila melanogaster* cells, although the apparent molecular weight of the protein coded for by the highest molecular weight RNA appeared to be somewhat higher than previously reported (120×10^3 instead of 104-110×10^3) (Guarino et al., 1981). Pulse chase experiments suggested that the lowest molecular weight protein found in infected cells was derived from a morphogenic cleavage during the formation of the capsid protein from its immediate precursor. The absence of relationship between the high molecular weight proteins to the capsid proteins was demonstrated by limited proteolysis as was the relationship of the capsid protein precursor to the capsid protein (Crump and Moore, 1981a).

The origin of all the proteins expressed in *Drosophila* cells by BBV has been possibly accounted for by relation to RNA species, with the exception of the 10×10^3 molecular weight protein. Friesen and Rueckert (1982) presented evidence to suggest that this protein was formed from a third RNA found in infected cells which was selectively present in a polysomal fraction. They also presented evidence of a number of other subgenomic RNA's in infected cells. While they were unable to provide evidence for the intracellular RNAs which were slightly less than the 1×10^6 and 0.5×10^6 species, they suggested that the low molecular weight subgenomic RNA was derived from either the 0.5×10^6 or 1×10^6 molecular weight RNA's. Clewley et al. (1982) found submolar amounts of oligonucleotides, after ribonuclease T_1 digestion, that were attributed to a third virus-specific RNA of higher molecular weight, but the origin of this species was unknown. Crump et al. (1983) were unable to find subgenomic forms of the viral RNA and demonstrated the presence of two double-stranded replicative forms of the virion RNA. The function of the 10×10^3 molecular weight protein is unknown and the suggested function of the high molecular weight protein found in infected cells is that of an RNA-dependent RNA polymerase. Guarino and Kaesberg (1981) presented preliminary information that this protein copurifies with a replicative complex. A few other members of the *Nodaviridae* have been identified, including a virus from the grass grub, *Costelytra zealandica* (Coleoptera: Scarabaeidae) (Dearing et al., 1980; Scotti et al., 1983) and a virus from *Lymantria ninayi* (Crump and Moore, 1982; Greenwood and Moore, 1982a; Reavy et al., 1982). Serological tests showed that both viruses were related but not identical to BBV.

Compared with other small RNA viruses of insects a considerable amount of information is available on the replication of the nodaviruses. However, many questions remain to be answered. The origin of the 10×10^3 molecular weight protein of infected cells has not been demonstrated conclusively, the morphogenic cleavage of the capsid protein precursor needs to be associated with an assembly event, and there are considerable gaps in our understanding of the intracellular appearance of several RNA species. With the availability of the virus from grass grubs, which replicate in *Drosophila* cells (Dearing et al., 1980), it should be possible to examine genotypic mixing with BBV. These viruses have the advantage of having only two genome segments, which should simplify our understanding of how viruses can exchange genome segments.

D. CALICIVIRUSES

The term calicivirus was derived from the Latin *calyx* meaning cup or chalice. When negative-stained caliciviruses are observed by electron microscopy the ≃ 40 nm diameter "spherical" particles are seen to have dark areas approximately 10 nm in diameter. Until recently it appeared that caliciviruses were restricted to mammals including feline, primate (man), porcine, and pinniped (see Studdert, 1978; Schaffer, 1979; Schaffer et al., 1980; Matthews, 1982 for reviews and summaries of the group). However, Kellen and Hoffman (1981) found a virus associated with the granular hemocytes of diseased navel orangeworms, *Amyelois transitella* (Walker) collected from almonds in California that had many of the characteristics of a caliciviruses (Hillman et al., 1982). The virus was designated chronic stunt virus. Infected larvae were found to produce two types of particles, a 185 S, 38 nm cupped calici-like particle having a single major capsid protein of 70×10^3 molecular weight and a 165 S, 28 nm smooth particle with a capsid protein of 29×10^3 molecular weight. The larval frass gave rise to a mixture of both particles. These authors also provided evidence that the 38 nm particles could be converted into the 28 nm particles by proteolytic cleavage.

The larger "undegraded" viruses described by Hillman et al. (1982) are very similar to caliciviruses in that they have the correct morphology, structural protein content, and also contain a single piece of single-stranded RNA of $\simeq 2.5 \times 10^6$. The authors did not investigate if the RNA had a VpG as has been found with other caliciviruses (Burroughs and Brown, 1978; Schaffer et al., 1980).

The mode of replication of caliciviruses is still under investigation. Presumptive replicative forms and intermediates are found as well as genome-sized ssRNA and subgenomic ssRNAs (Black et al., 1978; Ehresmann and Schaffer, 1979). The virion RNA is infective, polyadenylated, and is not capped. Relatively little work has been performed on the intracellular proteins (Black and Brown, 1978; Fretz and Schaffer, 1978) and the subgenomic RNA is possibly a messenger for a precursor to the single major capsid protein.

E. RETROVIRUSES

There have been two reports of a possible retrovirus associated with *Drosophila melanogaster* cell cultures. Heine et al. (1980) demonstrated the presence of torroidal forms within *Drosophila* cells which resembled the structures found with intracytoplasmic A-type retrovirus particles. They also purified a subcellular fraction on sucrose density gradients with a density of 1.22 g/ml that had several of the characteristics of reverse transcriptase and that contained the torroidal forms. However, the *Drosophila* particles were smaller (40 nm diameter) than normally found with intracytoplasmic A-type particles. Shiba and Saigo (1983) also described retrolike particles in *Drosophila* cells. They found high molecular weight RNA's, 15 to 35 S, associated with viruslike particles purified from the nuclei of cultured *Drosophila* cells and a distinct band of this RNA was of approximately 5 kb. In addition, the virus-like particles appeared to contain reverse transcriptase activity. While the authors were able to purify the viruslike particles, they were unable to show that they were infectious for *Drosophila* cells up to an input multiplicity of 100 virus-like particles per cell. However, the authors concluded that their viruslike particles and the transposable *copia* elements were direct derivatives of viral particles and proviral forms of copia retrovirus, a putative *Drosophila* retrovirus.

F. RHABDOVIRUSES

Exposure of a laboratory colony of the fruit fly *Drosophila melanogaster* to carbon dioxide anesthetizes them. When insects are returned to normal conditions full recovery results. However, similar treatment of flies which are infected with the rhabdovirus, Sigma, results in a fatal paralysis (L'Héritier and Teissier, 1937; Berkaloff et al., 1965). While this phenomenon provides a useful tool to demonstrate virus-infected flies it has no obvious function for the flies or, indeed, the virus.

The most studied aspect of the infection of *Drosophila* by Sigma is the hereditary passage of the virus as evidenced by CO_2 sensitivity. With few exceptions, the female *Drosophila* fly transmits the virus, and the male used in mating rarely has any influence on the transmission by the offspring. Male flies transmit to only a few of the offspring. The genetics, hereditary transmission, and other aspects of Sigma virus, including other species in which it has been identified, have been comprehensively reviewed (Printz, 1973; Teninges et al., 1980). Considering Sigma virus from a purely replicative viewpoint is very interesting in that the virus can exist in a truly inapparent state with little (or no) apparent virus production and no effect on the host, i.e., the persistent infection of female germ cells has no visible effect. The constant production in the insects is reflected in Sigma infection of *Drosophila* tissue culture cells where virus is produced through many generations with no apparent cytopathic effect (Ohanessian and Echalier, 1967; Ohanessian, 1971; Printz, 1973).

Sigma virus has a length of ≈ 200 nm and a diameter of 75 nm inclusive of surface spikes, i.e., it has a standard bullet shape with surface projections of ≈ 8 nm typical of rhabdoviruses. A coiled structure comprising 32 turns is visible within the virus using electron microscopy (Teninges et al., 1980). Virus appears to bud through the plasma membrane of cells. By drawing parallels with rabies virus it is probable that Sigma has at least five structural proteins, an outer glycoprotein (G) and an inner core consisting of a major nucleocapsid protein (N) with which is associated two proteins in much smaller amounts, the large protein (L), and the protein equivalent to NS of vesicular stomatitis virus (VSV). Between the lipid envelope, in which the glycoprotein is inserted, and the nucleocapsid protein is probably the matrix protein (M). Other members of the *Rhabdoviridae*, including VSV, contain one molecule of noninfectious single-stranded RNA with a molecular weight $\simeq 3 \times 10^6$. In the infected cell the negative sense viral RNA is transcribed, using the viral polymerase, into individual messengers for the structural proteins. The L and NS proteins are involved in transcription. The glycoprotein is then inserted into the plasma membrane and the ribonucleoprotein core containing the transcriptase buds through the plasma membrane. The steps by which this occurs, including identification and incorporation of the matrix protein, are not fully understood (Morrison, 1980). It is clear that the glycoprotein is involved in attachment of rhabdoviruses to cell membranes, although the host receptor is as yet not known. Fusion between the membrane of VS virus and the cell plasma membrane has been demonstrated as well as phagocytic uptake of whole virions. Again, even with the much studied mammalian rhabdoviruses, the infection procedure is incompletely understood (Repik, 1980).

IV. DNA VIRUSES

A. *DENSONUCLEOSIS VIRUSES*

These viruses form a genus within the *Parvoviridae*. They are approximately 20 nm in diameter and contain linear single-stranded DNA's, which are complimentary (in separate particles), of molecular weight $\simeq 1.85 \times 10^6$ (Barwise and Walker, 1970; Kurstak et al., 1971, 1973; Kurstak, 1972; Kelly et al., 1977; Kelly and Bud, 1978; Vernoux and Kurstak, 1972). Kelly and Elliott (1977) demonstrated polyamines associated with the nucleic acid and a number of workers have described the protein content and structure of the viruses (Tijssen et al., 1976, 1977; Kelly et al., 1980b). There are four structural proteins which appear to be related, as demonstrated by limited proteolysis (Tijssen and Kurstak, 1979; Moore and Kelly, 1980). Several densonucleosis viruses (DNVs) have been isolated including the viruses from *Galleria mellonella* (Meynadier et al., 1964), *Junonia coenia* (Rivers and Longworth, 1972), and *Agraulis vanillae* (Kelly et al., 1980a). The members and candidate members of the group have been listed by Tijssen et al. (1982).

Comparatively little is known about the replication of these viruses as there is no continuous cell culture system in which they multiply (see Longworth, 1978). Hence, many of the studies have been performed by a range of techniques using either the whole insect or primary cell cultures (Kurstak, 1972). There is considerable difference between the abilities of the viruses to attack the tissues of the insect. The DNV of *Bombyx* is restricted to the midgut while the *Galleria mellonella* virus occurs in most tissues except the midgut (Garzon and Kurstak, 1976). These virions apparently have no need for helper virus. There is no evidence to suggest a requirement for cells to be in late S or early G_2 phase for successful virus replication, as is found with the genus *Parvovirus*.

Obviously the mode of replication cannot be elucidated until a suitable cell culture becomes available. Garzon and Kurstak (1976) have demonstrated the formation of DNA inclusions in the nuclei of infected cells, an increase in free ribosomes, and changes in the nucleolus. Using immunoperoxidase, autoradiography, and fluorescence, the appearance of DNA and antigens has been demonstrated in cells infected with the *Galleria* virus (Kurstak, 1972). The introduction of a suitable cell culture system will explain how the apparently interrelated proteins are formed and why the DNA coding capacity does not correspond to the large amount of protein produced.

B. ENTOMOPOXVIRUSES

The entomopoxviruses are a heterogenous group of viruses which are found in four insect orders, the Diptera, Coleoptera, Lepidoptera and Orthoptera (see Granados, 1973a, 1978, 1981 and Kurstak and Garzon, 1977 for reviews on this group of viruses). These viruses are classified as the subfamily Entomopoxvirinae within the *Poxviridae* and divided into three genera based on virus morphology, host range, and the molecular weight of the single strand of dsRNA (Matthews, 1982). The entomopoxviruses are normally occluded into ovoid or spherical polyhedronlike structures like the baculoviruses. Spindle-shaped "polyhedra" are also produced during the course of infection and these do not contain virus particles. Excellent reviews covering host specificity, tissue specificity, virus and inclusion body morphology, and replicative cycles as well as a description of the three genera have already appeared in the literature (Kurstak and Garzon, 1977; Granados, 1981). Therefore, this description of their replication will be very brief.

Virus infection occurs either by fusion of the virus membrane with the membrane of gut epithelial cells and release of the core into the cytoplasm or by a process of invagination (Devauchelle et al., 1971; Granados, 1973b; Kurstak and Garzon, 1977) followed by formation of a core structure in the vacuole. Formation of virus particles occurs within electron-dense amorphous areas or areas of granular material (Stoltz and Summers, 1972; Granados and Roberts, 1970; Kurstak and Garzon, 1977; Granados, 1981). When the virions mature in discrete regions of the cytoplasm they are either moved to other areas to be occluded or are released from the cell by budding through the plasma membrane. During the infection procedure of many entomopoxviruses, virion-sized inclusion bodies are found (Spindles) which contain no virus particles (Bird, 1974; Bergoin et al., 1976; Kurstak and Garzon, 1977). The origin or function of the spindles is unknown.

V. DISCUSSION

Our knowledge of the mechanism of replication of many insect viruses has remained at a rudimentary level because of the absence of suitable cell culture systems. There has been little urgency to introduce insect viruses into culture as they initially appeared to have little economic importance. Most effort is being concentrated on the NPVs because they have frequently been demonstrated to provide safe and effective control of insect pests. Work on CPVs has lagged behind because of the absence of susceptible culture systems as well as the inability, in many instances, of CPVs to cause the catastrophic effect on insect populations seen with NPVs. The major interest in replication of CPVs has been academic with considerable knowledge being accumulated on the structure of the nucleic acids and the mechanism of formation of single-stranded messengers from the ds genome RNA. Another major area of interest in replication control concerns how the massive amounts of polyhedral protein are manufactured in the infected cell from a dsRNA which is present in approximately equimolar amounts to the other viral RNA's.

Other viruses such as the small RNA viruses have been studied primarily because of the availability of suitable cell culture systems. In addition, the members of the *Nodaviridae* may provide suitable models for the interchange of genome segments as they are much simpler than multisegmented animal viruses such as influenza. Another reason for the study of replication of the *Nodaviridae* is that the type member of the group, Nodamura virus, appears to replicate in mammalian tissues. Other small RNA viruses including CrPV and members of the *Nudaurelia* β react with sera from animals, including man, and it is not known whether this is attributable to replication or passive exposure to the viruses (Longworth et al., 1973b; MacCallum et al., 1979; Scotti and Longworth, 1980; Moore et al., 1981b). Attempts to introduce these viruses into mammalian cell culture systems have been unsuccessful. It is possible that a nonspecific reaction occurs between the insect viruses and the sera of vertebrates.

In the case of mammalian viruses, circulating antibody is considered to be a major factor in causing both antigenic shift and drift of the viruses, i.e., the interchange of whole segments of genomes during multiple infection as well as the occurrence of point mutations within individual segments. However, there is an enormous variety of insect viruses. Is the variety of genome segments found with, for example, CPVs entirely attributable to the huge number of insect species which have developed separately in different environments with different stress conditions? With single genome segment viruses such as the different DCV isolates and CrPV there are considerable differences in their RNA's as demonstrated by oligonucleotide mapping. It is probable that these viruses can undergo genomic mixing as has been demonstrated with the mammalian picornaviruses (King et al., 1982). Whether these viruses have other hosts, with insects being the transmitting agents or even the reservoirs, remains an open question.

Many insect viruses have not been considered in this chapter because so little is known about their replicative mechanisms. Numerous viruses have been isolated from *Drosophila* and honey bees as a result of the efforts of N. Plus and colleagues and L. Bailey and collaborators (Plus and Duthoit, 1969; Plus et al., 1975, 1976; Bailey, 1976; Bailey and Woods, 1977; Brun and Plus, 1980). Chronic bee paralysis virus, for example, is a complex mixture of five single-stranded RNA components, four of which appear to be related (Overton et al., 1982). The virus is further complicated by the presence of a "satellite" virus, chronic bee paralysis virus-associated, which contains three RNA's which appear to be very similar to the same molecular weight RNA's in the chronic bee paralysis virions (Overton et al., 1982). The associate particle is unable to multiply unless present with the other particles. It thus appears that they exist as satellite and helper viruses (Bailey et al., 1980). Numerous other small particles have been found associated with known insect viruses and in the majority of cases little effort has been made to establish the relationship between the particles. Occasionally it has been suggested that the very small particles can replicate independently (Jutila et al., 1970). An unusual virus has been isolated from a *Drosophila melano-*

gaster cell line. It was 36 nm in diameter and contained a single segment of dsRNA and two proteins with molecular weight 120×10^3 and 200×10^3 (Scott et al., 1980). The former protein was the major component of the virus capsid. It is unusual in having both an extremely large capsid protein and a single piece of dsRNA, and is possibly the first member of yet another new group of insect viruses. Undoubtedly other new groups of viruses will be identified in insects or in continuous cell lines derived from them.

ACKNOWLEDGMENTS

My thanks to Gail Davies and Steven Eley for assistance in the preparation of this chapter.

REFERENCES

Alatortsev, V. E., Ananiev, E. V., Gushchina, E. A., Grigoriev, V. B., and Gushchin, B. V. (1981). A virus of the *Reoviridae* in established cell lines of *Drosophila melanogaster*. *J. Gen. Virol. 54*, 23-31.

Andreadis, T. G. (1981). A new cytoplasmic polyhedrosis virus from the salt marsh mosquito, *Aedes cantator* (Diptera: Culicidae). *J. Invertebr. Pathol. 37*, 160-167.

Arnott, H. J., Smith, K. M., and Fullilove, S. L. (1968). Ultrastructure of a cytoplasmic polyhedrosis virus affecting the monarch butterfly, *Danaus plexippus*. *J. Ultrastr. Res. 26*, 31-34.

Bailey, L. (1976). Viruses attacking the honey bee. *Advan. Virus Res. 20*, 271-304.

Bailey, L. and Scott, H. A. (1973). The pathogenicity of Nodamura virus for insects. *Nature (London) 241*, 545.

Bailey, L. and Woods, R. D. (1977). Bee viruses. *In* "The Atlas of Insect and Plant Viruses" (K. Maramorosch, ed.), pp. 141-156. Academic Press, New York.

Bailey, L., Newman, J. F. E., and Porterfield, J. S. (1975). The multiplication of Nodamura virus in insect and mammalian cell cultures. *J. Gen. Virol. 26*, 15-20.

Bailey, L., Ball, B. V., Carpenter, J. M., and Woods, R. D. (1980). Small virus-like particles in honey bees associated with chronic paralysis virus and with a previously undescribed disease. *J. Gen. Virol. 46*, 149-155.

Barwise, A. H. and Walker, I. O. (1970). Studies on the DNA of a virus from *Galleria mellonella*. *FEBS Lett.* *6*, 13-16.
Belloncik, S. and Arella, M. (1981). Production of cytoplasmic polyhedrosis virus (CPV) polyhedra in a gamma irradiated *Lymantria dispar* cell line. *Arch. Virol.* *68*, 303-308.
Belloncik, S. and Bellemare, N. (1980). Polyèdres du CPV d'*Euxoa scandens* (lépidoptère: Noctuidae) produits *in vivo* et sur cellules cultivées *in vitro*. Etudes comparatives. *Entomophaga* *25*, 199-207.
Belloncik, S. and Chagnon, A. (1980). Titration of a cytoplasmic polyhedrosis virus by a microculture assay. Some applications. *Intervirology* *13*, 28-32.
Bergoin, M., Devauchelle, G., and Vago, C. (1976). Les inclusions fusiformes associeés à l'entomopoxvirus du coléoptère *Melolontha melolontha*. *J. Ultrastr. Res.* *55*, 17-30.
Berkaloff, A., Bregliano, J. C., and Ohanessian, A. (1965). Mise en évidence de virions dans des *Drosophiles* infectées par le virus héréditaire sigma. *C. R. Acad. Sci.* *260*, 5956-5959.
Bernard, J. (1980). *Drosophila* X virus RNA polymerase: tentative model for *in vitro* replication of the double-stranded virion RNA. *J. Virol.* *33*, 717-723.
Bernard, J. and Petitjean, A. M. (1978). *In vitro* synthesis of double-stranded RNA by *Drosophila* X virus purified virions. *Biochem. Biophys. Res. Commun.* *83*, 763-770.
Bird, F. T. (1974). The development of spindle inclusions of *Choristoneura fumiferana* (Lepidoptera: Tortricidae) Infected with entomopoxvirus. *J. Invertebr. Pathol.* *23*, 325-332.
Bishop, D. H. L. and Shope, R. E. (1979). Bunyaviridae. *In* "Comprehensive Virology" (H. Fraenkel-Conrat and R. R. Wagner, eds.), Vol. 14, pp. 1-156. Plenum Press, New York.
Bishop, D. H. L., Calisher, C. H., Casals, J., Chumakov, M. P., Gaidamovich, S. Ya., Hannoun, C., Lvov, D. K., Marshall, I. D., Oker-Blom, N., Pettersson, R. F., Porterfield, J. S., Russell, P. K., Shope, R. E., and Westaway, E. G. (1980). Bunyaviridae. *Intervirology* *14*, 125-143.
Black, D. N. and Brown, F. (1978). Proteins induced by infection with caliciviruses. *J. Gen. Virol.* *38*, 75-82.
Black, D. N., Burroughs, J. N., Harris, T. J. R., and Brown, F. (1978). The structure and replication of calicivirus RNA. *Nature (London)* *274*, 614-615.
Boccardo, G., Hatta, T., Francki, R. I. B., and Grivell, C. J. (1980). Purification and some properties of reovirus-like particles from leafhoppers and their possible involvement in Wallaby ear disease of maize. *Virology* *100*, 300-313.
Brun, G. and Plus, N. (1980a). The viruses of *Drosophila*. *In* "The Genetics and Biology of *Drosophila*" (M. Ashburner and

T. R. F. Wright, eds.), pp. 625-702. Academic Press, New York.

Burroughs, J. N. and Brown, F. (1978). Presence of a covalently linked protein on calicivirus RNA. *J. Gen. Virol. 41*, 443-446.

Clewley, J. P., Crump, W. A. L., Avery, R. J., and Moore, N. F. (1982). Two unique RNA species of the Nodavirus, black beetle virus. *J. Virol. 44*, 767-771.

Clewley, J. P., Pullin, J. S. K., Avery, R. J., and Moore, N. F. (1983). Oligonucleotide fingerprinting of the RNA species obtained from six *Drosophila* C virus isolates. *J. Gen. Virol. 64*, 503-506.

Cohen, J. (1975). Ribonucleic acid polymerase activity in purified infectious pancreatic necrosis virus of trout. *Biochem. Biophys. Res. Commun. 62*, 689-695.

Cohen, J., Poinsard, A., and Scherrer, R. (1973). Physico-chemical and morphological features of infectious pancreatic necrosis virus. *J. Gen. Virol. 21*, 485-498.

Crump, W. A. L. and Moore, N. F. (1981a). *In vivo* and *in vitro* synthesis of the proteins expressed by the RNA of black beetle virus. *Arch. Virol. 69*, 131-139.

Crump, W. A. L. and Moore, N. F. (1981b). The polypeptides induced in *Drosophila* cells by a virus of *Heteronychus arator*. *J. Gen. Virol. 52*, 173-176.

Crump, W. A. L. and Moore, N. F. (1981c). The proteins induced in black beetle virus-infected *Drosophila* cells at elevated temperature. *Ann. Virol. Inst. Pasteur 132E*, 495-501.

Crump, W. A. L. and Moore, N. F. (1982). Split genome ssRNA-viruses of insects. *Proc. 3rd intern. Colloq. Invertebr. Pathol.*, pp. 136-139.

Crump, W. A. L., Reavy, B., and Moore, N. F. (1983). Intra-cellular RNA expressed in black beetle virus infected *Drosophila* cells. J. Gen. Virol. 64, 717-721.

D'Arcy, C. J., Burnett, P. A., Hewings, A. D., and Goodman, R. M. (1981). Purification and characterization of a virus from the aphid *Rhopalosiphum padi*. *Virology 112*, 346-349.

Dearing, S. C., Scotti, P. D., Wigley, P. J., and Dhana, S. D. (1980). A small RNA virus isolated from the grass grub, *Costelytra zealandica* (Coleoptera: Scarabaeidae). *New Zeal. J. Zool. 7*, 267-269.

Devauchelle, G., Bergoin, M., and Vago, C. (1971). Etude ultrastructurale du cycle de réplication d'un Entomopoxvirus dans les hémocytes de son hôte. *J. Ultrastr. Res. 37*, 301-321.

Dobos, P. (1976). Size and structure of the genome of infectious pancreatic necrosis virus. *Nucleic Acids Res. 3*, 1903-1924.

Dobos, P. (1977). Virus-specific protein synthesis in cells infected by infectious pancreatic necrosis virus. *J. Virol.* *21*, 242-258.

Dobos, P. and Rowe, D. (1977). Peptide map comparison of infectious pancreatic necrosis virus-specific polypeptides. *J. Virol.* *24*, 805-820.

Dobos, P., Hallett, R., Kells, D. T. C., Sorenson, O., and Rowe, D. (1977). Biophysical studies of infectious pancreatic necrosis virus. *J. Virol.* *22*, 150-159.

Dobos, P., Hill, B. J., Hallett, R., Kells, D. T. C., Becht, H., and Teninges, D. (1979). Biophysical and biochemical characterisation of five animal viruses with bisegmented double stranded RNA genomes. *J. Virol.* *32*, 593-605.

Eaton, B. T., and Steacie, A. D. (1980). Cricket paralysis virus RNA has a 3' terminal poly (A). *J. Gen. Virol.* *50*, 167-171.

Ehresmann, D. W. and Schaffer, F. L. (1979). Calicivirus intracellular RNA: fractionation of 18 to 22S RNA, and lack of typical 5'-methylated caps on 36S and 22S San Miguel sealion virus RNA. *Virology* *95*, 251-255.

Finch, J. T., Crowther, R. A., Hendry, D. A., and Struthers, J. K. (1974). The structure of *Nudaurelia capensis* β virus: the first example of a capsid with icosahedral surface symmetry. *J. Gen. Virol.* *24*, 191-200.

Fretz, M. and Schaffer, F. L. (1978). Calicivirus proteins in infected cells: evidence for a capsid polypeptide precursor. *Virology* *89*, 318-321.

Friesen, P., Scotti, P., Longworth, J., and Rueckert, R. (1980). Black beetle virus: propagation in *Drosophila* line 1 cells and an infection-resistant subline carrying endogenous black beetle virus-related particles. *J. Virol.* *35*, 741-747.

Friesen, P. D., and Rueckert, R. R. (1981). Synthesis of black beetle virus proteins in cultured *Drosophila* cells: differential expression of RNAs 1 and 2. *J. Virol.* *37*, 876-886.

Friesen, P. D. and Rueckert, R. R. (1982). Black beetle virus: messenger for protein B is a subgenomic viral RNA. *J. Virol.* *42*, 986-995.

Furuichi, Y. (1974). "Methylation-coupled" transcription by virus associated transcriptase of cytoplasmic polyhedrosis virus containing double stranded RNA. *Nucleic Acids Res.* *1*, 809-822.

Furuichi, Y. (1978). Pretranscriptional capping in the biosynthesis of cytoplasmic polyhedrosis virus messenger RNA. *Proc. Nat. Acad. Sci. U.S.* *75*, 1086-1090.

Furuichi, Y. and Miura, K. (1975). A blocked structure at the 5' terminus of mRNA from cytoplasmic polyhedrosis virus. *Nature (London)* *253*, 374-375.

Furusawa, T. and Kawase, S. (1971). Synthesis of ribonucleic acid resistant to actinomycin D in silkworm midguts infected with the cytoplasmic polyhedrosis virus. *J. Invertebr. Pathol. 18*, 156-158.

Furusawa, T. and Kawase, S. (1973). Virus-specific RNA synthesis in the midgut of silkworm, *Bombyx mori* infected with cytoplasmic polyhedrosis virus. *J. Invertebr. Pathol. 22*, 335-344.

Garzon, S. and Kurstak, E. (1976). Ultrastructural studies on the morphogenesis of the densonucleosis virus (Parvovirus). *Virology 70*, 517-531.

Gateff, E., Gissmann, L., Shrestha, R., Plus, N., Pfister, H., Schröder, J., and Zurhausen, H. (1980). Characterization of two tumorous blood cell lines of *Drosophila melanogaster* and the viruses they contain. *In* "Invertebrate Systems *In Vitro*" (E. Kurstak, K. Maramorosch, and A. Dübendorfer, eds.), pp. 517-533. Elsevier/North-Holland Biomedical Press, Amsterdam.

Grace, T. D. C. and Mercer, E. H. (1965). A new virus of the saturniid *Antherea eucalypti* (Scott). *J. Invertebr. Pathol. 7*, 241-244.

Granados, R. R. (1973a). Insect poxviruses: Pathology, morphology and development. *Misc. Publ. Entomol. Soc. Amer. 9*, 73-94.

Granados, R. R. (1973b). Entry of an insect poxvirus by fusion of the virus envelope with the host cell membrane. *Virology 52*, 305-309.

Granados, R. R. (1978). The biology of cytoplasmic polyhedrosis viruses and entomopoxviruses. *In* "Viral Pesticides: Present Knowledge and Potential Effects on Public and Environmental Health" (M. D. Summers and C. Y. Kawanishi, eds.), pp. 89-102. U.S. Environmental Protection Agency, North Carolina.

Granados, R. R. (1981). Entomopoxviruses infections in insects. *In* "Pathogenesis of Invertebrate Microbial Diseases" (E. W. Davidson, ed.), pp. 101-126. Allanheld, Osmun, Totowa.

Granados, R. R. and Roberts, D. W. (1970). Electron microscopy of a pox-like virus infecting an invertebrate host. *Virology 40*, 230-243.

Granados, R. R., McCarthy, W. J., and Naughton, M. (1974). Replication of a cytoplasmic polyhedrosis virus in an established cell line of *Trichoplusia ni* cells. *Virology 59*, 584-586.

Greenwood, L. K. and Moore, N. F. (1981a). A single protein *Nudaurelia* β-like virus of the pale tussock moth, *Dasychira pudibunda*. *J. Invertebr. Pathol. 38*, 305-306.

Greenwood, L. K. and Moore, N. F. (1981b). The *Nudaurelia*-β group of small RNA viruses of insects. Serological comparison of four members. *Microbiologica 4*, 271-280.

Greenwood, L. K. and Moore, N. F. (1982a). The purification and partial characterization of a small RNA-virus from *Lymantria*, the identification of a Nodamura-like virus. *Microbiologica 5*, 49-52.

Greenwood, L. K. and Moore, N. F. (1982b). The *Nudaurelia*-β group of small RNA-containing viruses of insects: serological identification of several new isolates. *J. Invertebr. Pathol. 39*, 407-409.

Grylls, N. E. (1975). Leafhopper transmission of a virus causing maize wallaby ear disease. *Ann. Appl. Biol. 79*, 283-296.

Guarino, L. A. and Kaesberg, P. (1981). Isolation and characterization of an RNA-dependent RNA polymerase from black beetle virus-infected *Drosophila melanogaster* cells. *J. Virol. 40*, 379-386.

Guarino, L. A., Hruby, D. E., Ball, L. A., and Kaesberg, P. (1981). Translation of black beetle virus RNA and heterologous viral RNAs in cell-free lysates derived from *Drosophila melanogaster*. *J. Virol. 37*, 500-505.

Haars, R., Zentgraf, H., Gateff, E., and Bautz, F. A. (1980). Evidence for endogenous reovirus-like particles in a tissue culture cell line from *Drosophila melanogaster*. *Virology 101*, 124-130.

Harrap, K. A. and Payne, C. C. (1979). The structural properties and identification of insect viruses. *Advan. Virus Res. 25*, 273-355.

Hashimoto, Y. and Kawase, S. (1983). Characteristics of structural proteins of infectious Flacherie virus from the silkworm, *Bombyx mori*. *J. Invertebr. Pathol. 41*, 68-76.

Hatta, T. and Franchi, R. I. B. (1977). Morphology of Fiji disease virus. *Virology 76*, 797-807.

Hatta, T. and Franchi, R. I. B. (1982). Similarity in the structure of cytoplasmic polyhedrosis virus, leafhopper A virus and Fiji disease virus particles. *Intervirology 18*, 203-208.

Hayashi, Y. (1970). RNA in midgut of tussock moth, *Orgyia leucostigma*, infected with cytoplasmic polyhedrosis virus. *Can. J. Microbiol. 16*, 1101-1107.

Hayashi, Y. and Donaghue, T. P. (1971). Cytoplasmic polyhedrosis virus: RNA synthesised *in vivo* and *in vitro* in infected midgut. *Biochem. Biophys. Res. Commun. 42*, 214-221.

Hayashi, Y. and Kawarabata, T. (1970). Effect of actinomycin on synthesis of cell RNA and replication of insect cytoplasmic polyhedrosis viruses *in vivo*. *J. Invertebr. Pathol. 15*, 461-462.

Hayashi, Y. and Retnakaran, A. (1970). The site of RNA synthesis of a cytoplasmic polyhedrosis virus (CPV) in *Malacosoma disstria*. *J. Invertebr. Pathol. 16*, 150-151.

L'Héritier, Ph. and Teissier, G. (1937). Une anomalie physiologique héréditaire chez la *Drosophile*. *C. R. Scad. Sci.* *205*, 1099.

Heine, C. W., Kelly, D. C., and Avery, R. J. (1980). The detection of intracellular retrovirus-like entities in *Drosophila melanogaster* cell cultures. *J. Gen. Virol.* *49*, 385-395.

Hess, R. T., Summers, M. D., Falcon, L. A., and Stoltz, D. B. (1977). A new isosahedral insect virus: apparent mixed nuclear infection with the baculovirus of *Autographa californica*. *IRCS Med. Sci.* *5*, 562.

Hess, R. T., Summers, M. D., and Falcon, L. A. (1978). A mixed virus infection in midgut cells of *Autographa californica* and *Trichoplusia ni* larvae. *J. Ultrastr. Res.* *65*, 253-265.

Hill, B. J. (1976). Properties of a virus isolated from the bivalve mollusc *Tellina tenius* (da Costa). *In* "Wild Life Diseases" (L. A. Page, ed.), pp. 445-452. Plenum Press, New York.

Hillman, B., Morris, T. J., Kellen, W. R., Hoffman, D., and Schlegel, D. E. (1982). An invertebrate calici-like virus: evidence for partial virion disintegration in host excreta. *J. Gen. Virol.* *60*, 115-123.

Jousset, F.-X. (1976). Etude experiméntale du spectra d'hôtes du virus C de *Drosophila melanogaster* chez quelques diptères et lépidoptères. *Ann. Microbiol. Inst. Pasteur* *127A*, 529-544.

Jousset, F.-X., Plus, N., Croizier, G., and Thomas, M. (1972). Existence chez *Drosophila* de deux groupes de Picornavirus de propriétés sérologiques et biologiques différentes. *C.R. Acad. Sci. Paris* *D275*, 3043-3046.

Jousset, F.-X., Beroin, M., and Revet, B. (1977). Characterization of the *Drosophila* C virus. *J. Gen. Virol.* *34*, 269-285.

Juckes, I. R. M. (1970). Viruses of the pine emperor moth. *Bull. S. Afr. Soc. Plant Pathol. Microbiol.* *4*, 18.

Juckes, I. R. M. (1979). Comparison of some biophysical properties of the *Nudaurelia* β and ε viruses. *J. Gen. Virol.* *42*, 89-94.

Jutila, J. W., Henry, J. E., Anacker, R. L., and Brown, W. R. (1970). Some properties of a crystalline-array virus (CAV) isolated from the grasshopper *Melanoplus bivattatus* (Say) (Orthoptera: Acrididae). *J. Invertebr. Pathol.* *15*, 225-231.

Kawase, S. and Miyajima, S. (1969). Immunofluorescence studies on the multiplication of cytoplasmic polyhedrosis virus of the silkworm, *Bombyx mori*. *J. Invertebr. Pathol.* *13*, 330-336.

Kawase, S. and Furusawa, T. (1971). Effect of actinomycin D on RNA synthesis in the midguts of healthy and CPV-infected silkworms. *J. Invertebr. Pathol.* *18*, 33-39.

Kellen, W. R. and Hoffman, D. F. (1981). A pathogenic non-occluded virus in hemocytes of the Navel Orangeworm, *Amyelois transitella* (Pyralidae: Lepidoptera). *J. Invertebr. Pathol. 38*, 52-66.

Kelly, D. C. and Elliott, R. M. (1977). Polyamines contained by two densonucleosis viruses. *J. Virol. 21*, 408-410.

Kelly, D. C. and Bud, H. M. (1978). Densonucleosis virus DNA fine structure analysed by electron microscopy and agarose gel electrophoresis. *J. Gen. Virol. 40*, 33-43.

Kelly, D. C., Barwise, A. H., and Walker, I. O. (1977). DNA contained by two densonucleosis viruses. *J. Virol. 21*, 396-407.

Kelly, D. C., Ayres, M. D., Spencer, L. K., and Rivers, C. F. (1980a). Densonucleosis virus 3. A recent insect parvovirus isolate from *Agraulis vanillae* (Lepidoptera: Nymphalidae). *Microbiologica 3*, 455-460.

Kelly, D. C., Moore, N. F., Spilling, C. R., Barwise, A. H., and Walker, I. O. (1980b). Densonucleosis virus structural proteins. *J. Virol. 36*, 224-236.

Kelly, D. C., Ayres, M. D., Howard, S. C., Lescott, T., Arnold, M. K., Seeley, N. D., and Primrose, S. B. (1982). Isolation of a bisegmented double-stranded RNA virus from Thirlmere Reservoir. *J. Gen. Virol. 62*, 313-322.

King, A. M. Q., McCahon, D., Slade, W. R., and Newman, J. W. I. (1982). Recombination in RNA. *Cell 29*, 921-928.

Kobayashi, M. (1971). Replication cycle of cytoplasmic Polyhedrosis Virus of the Silkworm" (H. Aruga and Y. Tanada, eds.), pp. 103-128. University of Tokyo Press, Tokyo.

Kurstak, E. (1972). Small densonucleosis virus (DNV). *Advan. Virus Res. 17*, 207-241.

Kurstak, E. and Garzon, S. (1977). Entomopoxviruses (Poxviruses of invertebrates). *In* "The Atlas of Insect and Plant Viruses" (K. Maramorosch, ed.), pp. 29-66. Academic Press, New York

Kurstak, E., Vernoux, J. P., Niveleau, A., and Onji, P. A. (1971). Visualization de la densonucleose (VDN) a chaines monocatenaires complementaries de polarities inverses plus et moins. *C.R. Acad. Sci. 272*, 762-765.

Kurstak, E., Vernoux, J. P., and Brakier-Gingras, L. (1973). Etude biophysique de l'acide désoxyribonucléique du virus de la densonucléose (VDN). II. Extraction du DNA viral et mise en êvidence de la présence de chaines polynucléotidiques complémentaires, encapsidées séparément dans les virions VDN. *Arch. Ges. Virusforsch. 40*, 274-284.

Lipa, J. J. (1977). Electron microscope observations on the development of cytoplasmic polyhedrosis virus in *Scotogamma trifolii* Lepidoptera, Noctuidae. *Bull. Acad. Pol. Sci. Ser. Sci. Biol. 25*, 155-158.

Longworth, J. F. (1978). Small isometric viruses of invertebrates. *Advan. Virus Res. 23*, 103-157.

Longworth, J. F. (1981). The replication of a cytoplasmic polyhedrosis virus from *Chrysodeixis-eriosoma* Lepidoptera Noctuidae in *Spodoptera frugiperda* cells. *J. Invertebr. Pathol. 37*, 54-61.

Longworth, J. F., and Archibald, R. D. (1975). A virus of black beetle *Heteronychus arator* (F.) (Coleoptera: Scarabaeidae). *New Zeal. J. Zool. 2*, 233-236.

Longworth, J. F. and Carey, G. P. (1976). A small RNA virus with a divided genome from *Heteronychus arator* (F.) (Coleoptera: Scarabaeidae). *J. Gen. Virol. 33*, 31-40.

Longworth, J. F., Payne, C. C., and Macleod, R. (1973a). Studies on a virus isolated from *Gonometa podocarpi* (Lepidoptera: Lasiocampidae). *J. Gen. Virol. 18*, 119-125.

Longworth, J. F., Robertson, J. S., Tinsley, T. W., Rowlands, D. J., and Brown, F. (1973b). Reactions between an insect picornavirus and naturally occurring IgM antibodies in several mammalian species. *Nature (London) 242*, 314-316.

MacCallum, F., Brown, G., and Tinsley, T. W. (1979). Antibodies in human sera reacting with an insect pathogenic virus. *Intervirology 11*, 234-237.

McCrae, M. A. (1982). Coding assignments for the genes of a cytoplasmic polyhedrosis virus. *Proc. 3rd Intern. Colloq. Invertebr. Pathol.*, pp. 20-24.

McCrae, M., and Joklik, W. (1978). The nature of the polypeptide coded by each of the 10 double-stranded RNA segments of reovirus type 3. *Virology 89*, 578-593.

MacDonald, R. D. and Yamamoto, T. (1977). The structure of infectious pancreatic necrosis virus RNA. *J. Gen. Virol. 34*, 235-247.

MacDonald, R. D. and Dobos, P. (1981). Identification of the proteins encoded by each genome segment of infectious pancreatic necrosis virus. *Virology 114*, 414-422.

Matthews, R. E. F. (1982). Classification and nomenclature of viruses. Fourth Report of the International Committee on Taxonomy of Viruses. *Intervirology 17(1-3)*, 1-199.

Mertens, P. P. C. (1979). A study of the transcription and translation (*in vitro*) of the genomes of cytoplasmic polyhedrosis viruses types 1 and 2. D. Phil. Thesis, University of Oxford.

Mertens, P. P. C. and Dobos, P. (1982). Messenger RNA of infectious pancreatic necrosis virus is polycistronic. *Nature (London) 297*, 243-246.

Mertens, P. P. C. and Payne, C. C. (1978). *S*-adenosyl-*L*-homocysteine as a stimulator of viral RNA synthesis by 2 distinct cytoplasmic polyhedrosis viruses. *J. Virol. 26*, 832-835.

Mertens, P. P. C., Jamieson, P. J., and Dobos, P. (1982). *In vitro* RNA synthesis by infectious pancreatic necrosis virus-associated RNA polymerase. *J. Gen. Virol. 59*, 47-56.
Meynadier, G., Vago, C., Plantevin, G., and Atger, P. (1964). Virose d'un habituel chez le Lepidoptera *Galleria mellonella* L. *Rev. Zool. Agr. Appl. 63*, 207-209.
Milne, R. G. and Lovisolo, O. (1977). Maize rough dwarf and related viruses. *Advan. Virus Res. 21*, 267-341.
Moore, N. F. and Kelly, D. C. (1980). Interrelationships of the proteins of two insect parvoviruses (densonucleosis virus types 1 and 2). *Intervirology 14*, 160-166.
Moore, N. F. and Pullin, J. S. K. (1982). Plaque purification of Cricket paralysis virus using an agar overlay on *Drosophila* cells. *J. Invertebr. Pathol. 39*, 10-14.
Moore, N. F., Kearns, A., and Pullin, J. S. K. (1980). Characterization of Cricket paralysis virus-induced polypeptides in *Drosophila* cells. *J. Virol. 33*, 1-9.
Moore, N. F., Greenwood, L. K., and Rixon, K. R. (1981a). Studies on the capsid protein of several members of the *Nudaurelia* β group of small RNA viruses of insects. *Microbiologica 4*, 59-71.
Moore, N. F., McKnight, L., and Tinsley, T. W. (1981b). Occurrence of antibodies against insect virus proteins in mammals: simple model to differentiate between passive exposure and active virus growth. *Infect. Immun. 31*, 825-827.
Moore, N. F., Pullin, J. S. K., Crump, W. A. L., Reavy, B., and Greenwood, L. K. (1981c). Comparison of two insect picornaviruses, *Drosophila* C and Cricket paralysis viruses. *Microbiologica 4*, 359-370.
Moore, N. F., Pullin, J. S. K., and Reavy, B. (1981d). Inhibition of the induction of heat-shock proteins in *Drosophila melanogaster* cells infected with insect picornaviruses. *FEBS Lett. 128*, 93-96.
Moore, N. F., Pullin, J. S. K., and Reavy, B. (1981e). The intracellular proteins induced by Cricket paralysis virus in *Drosophila* cells: the effect of protease inhibitors and amino acid analogues. *Arch. Virol. 70*, 1-9.
Moore, N. F., Reavy, B., and Pullin, J. S. K. (1981f). Processing of Cricket paralysis virus-induced polypeptides in *Drosophila* cells: production of high molecular weight polypeptides by treatment with iodoacetamide. *Arch. Virol. 68*, 1-8.
Moore, N. F., Reavy, B., Pullin, J. S. K., and Plus, N. (1981g). The polypeptides induced in *Drosophila* cells by *Drosophila* C virus (strain Ouarzazate). *Virology 112*, 411-416.
Moore, N. F., Hunt, R. C., Pullin, J. S. K., and Marshall, L. M. (1982a). Cricket paralysis virus specific polypeptides in infected *Drosophila* cells analysed by isoelectric focusing and 2-D analysis. *Microbiologica 5*, 215-223.

Moore, N. F., Pullin, J. S. K., Crump, W. A. L., and Plus, N. (1982b). The proteins expressed by different isolates of *Drosophila* C virus. *Arch. Virol. 74*, 21-30.

Moore, N. F. and Pullin, J. S. K. (1983). Heat shock used in combination with amino acid analogues and protease inhibitors to demonstrate the processing of the proteins of an insect picornavirus (*Drosophila* C virus) in *Drosophila melanogaster* cells. *Ann. Virol. Inst. Pasteur 134E*, 285-292.

Moore, N. F. and Tinsley, T. W. (1982). The small RNA-viruses of insects. *Arch. Virol. 72*, 229-245.

Morris, T. J., Hess, R. T., and Pinnock, D. E. (1979). Physiochemical characterization of a small RNA virus associated with baculovirus infection in *Trichoplusia ni*. *Intervirology 11*, 238-247.

Morrison, T. G. (1980). Rhabdoviral assembly and intracellular processing of viral components. *In* "Rhabdoviruses (D. H. L. Bishop, ed.), Vol. II, pp. 95-114. CRC Press, Inc., Boca Raton, Florida.

Moussa, A. Y. (1978). A new virus disease in the housefly, *Musca domestica* (Diptera). *J. Invertebr. Pathol. 31*, 204-216.

Müller, H. and Becht, H. (1982). Biosynthesis of virus-specific proteins in cells infected with infectious bursal disease virus and their significance as structural elements for infectious virus and incomplete particles. *J. Virol. 44*, 384-392.

Murphy, F. A., Scherer, W. F., Harrison, A. K., Dunne, M. W., and Gary, G. W. (1970). Characterization of Nodamura virus, an arthropod transmissible picornavirus. *Virology 40*, 1008-1021.

Newman, J. F. E. and Brown, F. (1973). Evidence for a divided genome in Nodamura virus, an arthropod-borne picornavirus. *J. Gen. Virol. 21*, 371-384.

Newman, J. F. E. and Brown, F. (1976). Absence of poly (A) from the infective RNA of Nodamura virus. *J. Gen. Virol. 30*, 137-140.

Newman, J. F. E. and Brown, F. (1977). Further physicochemical characterization of Nodamura virus. Evidence that the divided genome occurs in a single component. *J. Gen. Virol. 38*, 83-95.

Newman, J. F. E., Matthews, T., Omilianowski, D. R., Salerno, T., Kaesberg, P., and Rueckert, R. (1978). *In vitro* translation of the two RNAs of Nodamura virus, a novel mammalian virus with a divided genome. *J. Virol. 25*, 78-85.

Nick, H., Cursiefen, D., and Brecht, H. (1976). Structural and growth characteristics of infectious bursal disease virus. *J. Virol. 18*, 227-234.

Ohanessian, A. (1971). Sigma virus multiplication in *Drosophila* cell lines of different genotypes. *Current Topics Microbiol. Immunol. 55*, 230-233.

Ohanessian, A. and Echalier, G. (1967). Multiplication of *Drosophila* hereditary virus (σ virus) in *Drosophila* embryonic cells cultivated *in vitro*. *Nature (London) 212*, 1049-1050.

Overton, H. A., Buck, K. W., Bailey, L., and Ball, B. V. (1982). Relationships between the RNA components of chronic bee-paralysis virus and those of chronic bee-paralysis virus associate. *J. Gen. Virol. 63*, 171-179.

Parker, L., Barker, E., and Stanley, N. F. (1965). The isolation of reovirus type 3 from mosquitoes and a sentinel infant mouse. *Aust. J. Exp. Biol. Med. Sci. 43*, 167-170.

Payne, C. C. and Harrap, K. A. (1977). Cytoplasmic polyhedrosis virus. *In* The Atlas of Insect and Plant Viruses" (K. Maramorosch, ed.), pp. 105-125. Academic Press, New York.

Payne, C. C. and Kalmakoff, J. (1973). The synthesis of virus-specific single stranded RNA in larvae of *Bombyx mori* infected with cytoplasmic polyhedrosis virus. *Intervirology 1*, 34-40.

Pfefferkorn, E. R. and Shapiro, D. (1974). Reproduction of togaviruses. *In* "Comprehensive virology" (H. Fraenkel-Conrat and R. R. Wagner, eds.), Vol. 2, pp. 171-230. Plenum Press, New York.

Plus, N. (1978). Endogenous viruses of *Drosophila melanogaster* cell lines: their frequency, identification and origin. *In Vitro 12*, 1015-1021.

Plus, N. (1980). Further studies on the origin of the endogenous viruses of *Drosophila melanogaster* cell lines. *In* "Invertebrate Systems *In Vitro*" (E. Kurstak, K. Maramorosch, and A. Dübendorfer, eds.), pp. 535-539. Elsevier/North-Holland Biomedical Press, Amsterdam.

Plus, N. and Duthoit, J. L. (1969). Un nouveau virus de *Drosophila melanogaster*, le virus P. *C.R. Acad. Sci. D268*, 2313-2315.

Plus, N., Croizier, G., Duthoit, J. L., David, J., Anxolabehere, D., and Periquet, G. (1975). Découverte, chez la *Drosophile*, de virus appartenant à trois nouveaux groupes. *C.R. Acad. Sci. 280*, 1501-1504.

Plus, N., Croizier, G., Veyrunes, J. C., and David, J. (1976). A comparison of buoyant density and polypeptides of *Drosophila* P, C and A viruses. *Intervirology 7*, 346-350.

Plus, N., Croizier, G., Reinganum, C., and Scotti, P. D. (1978). Cricket paralysis virus and *Drosophila* C virus: serological analysis and comparison of capsid polypeptides and host range. *J. Invertebr. Pathol. 31*, 296-302.

Plus, N., Gissmann, L., Veyrunes, J. C., Pfister, H., and Gateff, E. (1981). Reoviruses of *Drosophila* and *Ceratitis* populations and *Drosophila* cell lines: a possible new genus of the *Reoviridae* family. *Ann. Virol. Inst. Pasteur 132E*, 261-270.

Polson, A., Stannard, L., and Tripconey, D. (1970). The use of haemocyanin to determine the molecular weight of *Nudaurelia cytherea capensis* β virus by direct particle counting. *Virology 41*, 680-687.

Printz, P. (1973). Relationship of sigma virus to vesicular stomatitis virus. *Advan. Virus Res. 18*, 143-157.

Pudney, M., Newman, J. F. E., and Brown, F. (1978). Characterization of Kawino virus: an entero-like virus isolated from the mosquito *Mansonia uniformis* (Diptera: Culicidae). *J. Gen. Virol. 40*, 433-441.

Pullin, J. S. K., Moore, N. F., Clewley, J. P., and Avery, R. J. (1982). Comparison of the genomes of two insect picornaviruses, cricket paralysis virus and *Drosophila* C virus, by ribonuclease T_1 oligonucleotide fingerprinting. *FEMS Micro. Lett. 15*, 215-218.

Quiot, J. M. and Belloncik, S. (1977). Charactérisation d'une polyédrose cytoplasmique chez le lépidoptère *Euxoa scandens*, Riley (Noctuidae, Agrotinae. Etudes *in vivo* et *in vitro*. *Arch. Virol. 55*, 145-154.

Reavy, B. (1982). Replication of Small RNA-Viruses of Insects. D. Phil. Thesis, University of Oxford.

Reavy, B. and Moore, N. F. (1981a). Cell-free translation of Cricket paralysis virus RNA: analysis of the synthesis and processing of virus-specified proteins. *J. Gen. Virol. 55*, 429-438.

Reavy, B. and Moore, N. F. (1981b). *In vitro* translation of Cricket paralysis virus RNA. *Arch. Virol. 67*, 175-180.

Reavy, B. and Moore, N. F. (1982). The replication of small RNA-containing viruses of insects. *Microbiologica 5*, 63-84.

Reavy, B., Crump, W. A. L., Greenwood, L. K., and Moore, N. F. (1982). Characterization of a bipartite genome ribovirus isolated from *Lymantria ninayi*. *Microbiologica 5*, 305-319.

Reavy, B., Crump, W. A. L., and Moore, N. F. (1983). Characterization of Cricket paralysis virus and *Drosophila* C virus induced RNA species synthesised in infected *Drosophila melanogaster* cells. *J. Invertebr. Pathol. 41*, 397-400.

Reinganum, C. (1973). Studies on a non-occluded virus of the field crickets *Telleogryllus oceanicus and T. commodus*. M.Sc. Thesis, Monash University, Melbourne, Australia.

Reinganum, C., O'Loughlin, G. T., and Hogan, T. W. (1970). A non-occluded virus of the field crickets *Telleogryllus oceanicus* and *T. commodus* (Orthoptera: Gryllidae). *J. Invertebr. Pathol. 16*, 214-220.

Reinganum, C., Robertson, J. S., and Tinsley, T. W. (1978). A new group of RNA viruses from insects. *J. Gen. Virol. 40*, 195-202.

Repik, P. (1980). Adsorption, penetration, uncoating, and the *in vivo* mRNA transcription process. *In* "Rhabdoviruses".

(D. H. L. Bishop, ed.), Vol. II, pp. 1-33. CRC Press Inc., Boca Raton, Florida.
Revet, B. and Delain, E. (1982). The *Drosophila* X virus contains a 1-μ*M* double-stranded RNA circularized by a 67-kd terminal protein: high-resolution denaturation mapping of its genome. *Virology 123*, 29-44.
Rivers, C. F. and Longworth, J. F. (1972). A non-occluded virus of *Junonia coenia* (Nymphalidae: Lepidoptera). *J. Invertebr. Pathol. 20*, 369-370.
Schaffer, F. L. (1979). Caliciviruses. *In* "Comprehensive Virology" (H. Fraenkel-Conrat and R. R. Wagner, eds.), Vol. 14, pp. 249-284. Plenum Press, New York.
Schaffer, F. L., Ehresmann, D. W., Fretz, M. K., and Soergel, M. E. (1980). A protein, VpG, covalently linked to 36S calicivirus RNA. *J. Gen. Virol. 47*, 215-220.
Schaffer, F. L., Bachrach, H. L., Brown, F., Gillespie, J. H., Burroughs, J. N., Madin, S. H., Madeley, C. R., Povey, R. C., Scott, F., Smith, A. W., and Studdert, M. J. (1980). Caliciviridae. *Intervirology 14*, 1-6.
Scherer, W. F. and Hurlbut, H. S. (1967). Nodamura virus from Japan: a new and unusual arbovirus resistant to ether and chloroform. *Amer. J. Epidemiol. 86*, 271-285.
Scherer, W. F., Verna, J. E., and Richter, G. W. (1968). Nodamura virus, an ether- and chloroform-resistant arbovirus from Japan. *Amer. J. Trop. Med. Hyg. 17*, 120-128.
Schlesinger, R. W. (ed.), (1980). "The Togaviruses: Biology, Structure, Replication." Academic Press, New York.
Scott, M. P., Fostel, J. M., and Pardue, M. L. (1980). A new type of virus from cultured *Drosophila* cells: characterization and use in studies of the heat-shock response. *Cell 22*, 929-941.
Scotti, P. D. (1976). Cricket paralysis virus replicates in cultured *Drosophila* cells. *Intervirology 6*, 333-342.
Scotti, P. D. (1977). End-point dilution and plaque-assay methods for titration of Cricket paralysis virus in cultured *Drosophila* cells. *J. Gen. Virol. 35*, 393-396.
Scotti, P. D. and Longworth, J. F. (1980). Naturally occurring IgM antibodies to a small RNA insect virus in some mammalian sera in New Zealand. *Intervirology 13*, 186-191.
Scotti, P. D., Longworth, J. F., Plus, N., Croizier, G., and Reinganum, C. (1981). The biology and ecology of strains of an insect small RNA virus complex. *Advan. Virus Res. 26*, 117-143.
Scotti, P. D., Dearing, S., and Mossop, D. W. (1983). Flock House Virus: Nodavirus isolated from *Costelytra zealandica* (White) (Coleoptera: Scarabaeidae). *Arch. Virol. 75*, 181-189.

Shiba, T. and Saigo, K. (1983). Retrovirus-like particles containing RNA homologous to the transposable element *copia* in *Drosophila melanogaster*. *Nature (London) 302*, 119-124.

Shimotohno, K. and Miura, K. (1973). Transcription of double-stranded RNA in cytoplasmic polyhedrosis virus *in vitro*. *Virology 53*, 283-286.

Shimotohno, K. and Miura, K. (1976). The process of formation of the 5'-terminal modified structure in messenger RNA of cytoplasmic polyhedrosis virus. *FEBS Lett. 64*, 204-208.

Smith, R. E. and Furuichi, Y. (1980). Gene mapping of cytoplasmic polyhedrosis virus of silkworm by the full length mRNA prepared under optimized conditions of transcription *in vitro*. *Virology 103*, 279-290.

Smith, R. E. and Furuichi, Y. (1982). The double-stranded RNA genome segments of cytoplasmic polyhedrosis virus are independently transcribed. *J. Virol. 41*, 326-329.

Somogyi, P. and Dobos, P. (1980). Virus specific RNA synthesis in cells infected by infectious pancreatic necrosis virus. *J. Virol. 33*, 129-139.

Stoltz, D. B. and Hilsenhoff, W. L. (1969). Electron-microscopic observations on the maturation of a cytoplasmic-polyhedrosis virus. *J. Invertebr. Pathol. 14*, 39-48.

Stoltz, D. B. and Summers, M. D. (1972). Observations on the morphogenesis and structure of a hemocytic poxvirus in the midge *Chironomus attenuatus*. *J. Ultrastr. Res. 40*, 581-598.

Storer, G., Shepherd, M., and Kalmakoff, J. (1973/74). Enzyme activities associated with cytoplasmic polyhedrosis virus from *Bombyx mori*. I. Nucleotide phosphohydrolase and nuclease activities. *Intervirology 2*, 87-94.

Struthers, J. K. (1973). Physico-chemical and substructural studies on *Nudaurelia capensis* beta virus. M.Sc. Thesis. Rhodes University, South Africa.

Struthers, J. K. and Hendry, D. A. (1974). Studies on the protein and nucleic acid components of *Nudaurelia capensis* β virus. *J. Gen. Virol. 22*, 355-362.

Studdert, M. J. (1978). Caliciviruses. *Arch. Virol. 58*, 157-191.

Teninges, D. (1979). Protein and RNA composition of the structural components of *Drosophila* X virus. *J. Gen. Virol. 45*, 641-649.

Teninges, D. and Plus, N. (1972). P virus of *Drosophila melanogaster*, as a new picornavirus. *J. Gen. Virol. 16*, 103-109.

Teninges, D., Ohanessian, A., Richard-Molard, Ch., and Contamine, D. (1979a). Isolation and biological properties of the *Drosophila* X virus. *J. Gen. Virol. 42*, 241-254.

Teninges, D., Ohanessian, A., Richard-Molard, Ch., and Contamine, D. (1979b). Contamination and persistent infection of *Drosophila* cell lines by reovirus type particles. *In Vitro 15*, 425-428.

Teninges, D., Contamine, D., and Brun, G. (1980). *Drosophila* sigma virus. *In* "Rhabdoviruses" (D. H. L. Bishop, ed.), Vol. III, pp. 113-134. C.R.C. Press Inc., Boca Raton, Florida.

Tesh, R. B. (1980). Infectivity and pathogenicity of Nodamura virus for mosquitoes. *J. Gen. Virol. 48*, 177-182.

Tijssen, P., Van de Hurk, J., and Kurstak, E. (1976). Biochemical, biophysical and biological properties of densonucleosis virus. I. Structural proteins. *J. Virol. 17*, 686-691.

Tijssen, P., Tijssen-van Der Slikke, T., and Kurstak, E. (1977). Biochemical, biophysical and biological properties of densonucleosis virus (Parvovirus). II. Two types of infectious virions. *J. Virol. 21*, 225-231.

Tijssen, P. and Kurstak, E. (1979). A simple and sensitive method for the purification and peptide mapping of proteins solubilized from densonucleosis virus with sodium dodecyl sulfate. *Anal. Biochem. 99*, 97-194.

Tijssen, P., Kurstak, E., Su, T.-M., and Garzon, S. (1982). Densonucleosis viruses: unique pathogens of insects. Invertebrate pathology and microbial control. *Proc. 3rd Intern. Colloq. Invertebr. Pathol.*, pp. 148-153.

Tripconey, D. (1970). Studies on a non-occluded virus of the pine tree emperor moth. *J. Invertebr. Pathol. 15*, 268-275.

Underwood, B. O., Smale, C. J., Brown, F., and Hill, B. J. (1977). Relationship of a virus from *Tellina tenius* to infectious pancreatic necrosis virus. *J. Gen. Virol. 36*, 93-110.

Vernoux, J. P. and Kurstak, E. (1972). Etude biophysique de l'acide desoxyribonucleique du virus de la densonucleose (VDN). *Arch. Ges. Virusforsch. 39*, 190-195.

Watanabe, H. (1967). Site of viral RNA synthesis within the midgut cells of the silkworm, *Bombyx mori*, infected with cytoplasmic-polyhedrosis virus. *J. Invertebr. Pathol. 9*, 480-487.

Xeros, N. (1956). The virogenic stroma in nuclear and cytoplasmic polyhedroses. *Nature (London) 178*, 412-413.

Xeros, N. (1966). Light microscopy of the virogenic stromata of cytoplyhedroses. *J. Invertebr. Pathol. 8*, 79-87.

QUANTITATION OF INSECT VIRUSES

W. J. KAUPP AND S. S. SOHI

Forest Pest Management Institute
Canadian Forestry Service
Ontario, Canada

I. INTRODUCTION

The need for quantitation of active material in viral preparations cannot be overemphasized. The easiest insect viruses to quantify are those which can be seen with the light microscope, for example, the viruses in the families Baculoviridae, Spheruloviridae, and Reoviridae. It is mainly to these viruses that most of the techniques discussed in this chapter are applicable. Since most of the viruses that are being developed as insecticides fall in this category, they can be quantified by these methods.

The major problem in quantifying viral insecticides or any other viral preparation is the discrepancy between quantitative assessments and related biological activity. More often than not, the virus preparations used for laboratory and field experiments are enumerated as occlusion bodies (OBs) per milliliter, OBs per hectare, or OBs per gram of insect material with no mention of the biological activity of the preparation, making it difficult to compare the efficacy of different viral preparations and different isolates of a virus. It must be emphasized here that, in any assessment of viral preparations, it is very important to include information on their biological activity. It is the purpose of this chapter

ISBN 0-12-470295-3

to discuss the current methods of virus quantitation and to illustrate that techniques are now available to accurately determine both the number and biological activity of OBs in viral preparations.

II. DIRECT PHYSICAL QUANTITATION

Direct physical quantitation of viral insecticides is the most widely used method of assessment in which one determines the number of OBs in a preparation. Its universal application is based on the ease of observation of unstained OBs under the light microscope.

Most of the published work dealing with occluded insect viruses provides information on the number of OBs in a preparation, but quite often this is also mistakenly considered as an indication of the biological activity of the preparation. Direct physical quantitation is only useful in standardizing the concentration of OBs. Dose estimates based on OB counts have been viewed suspiciously due to their inaccuracy; they are misleading and meaningless when used for comparative purposes (Ignoffo, 1964, 1966; Martignoni and Iwai, 1968, 1978; Pinnock, 1975).

A. COUNTING CHAMBERS

Counting chambers are by far the most commonly used method for counting OBs. Either a hemacytometer or Petroff-Hauser bacterial counting chamber is used in which a known volume of virus suspension is trapped between a machined coverglass and a precision-made, ruled sunken platform on a glass slide. The loaded chamber is placed on the stage of a microscope and examined using phase contrast or dark field optics. According to Martignoni and Iwai (1978), while using a counting chamber with the new Neubauer ruling, the OBs should be allowed to settle for 8 minutes. To ensure accuracy, these authors suggested counting the OBs in the four corner 1-mm squares and in the central 1-mm square of one counting area of the chamber, and calculating the average number of OBs per square. This procedure is repeated for a second counting area, and if the difference between these two "counts" exceeds 10% of their mean, the chamber is washed and reloaded. Usually, the number of OBs in the stock suspension is calculated as follows:

$$\text{OBs per } \mu\text{l} = \frac{\text{Total number of OB counted} \times \text{dilution} \times 10}{\text{Number of 1-mm squares counted}} \quad (1)$$

The factor 10 in the numerator of Eq. (1) is derived from the fact that the 1-mm squares are only 0.1 mm deep. The practical procedure for determining the concentration of OBs in a preparation can vary depending on the type of counter used. Some methods require as many as 400 OBs to be counted, in four different grid areas of the counting chamber, to reduce error caused by an imperfect fit of the coverglass (Burges and Thomson, 1971).

Although a very simple technique, the use of counting chambers to quantify OBs has some serious drawbacks. First, inaccurate counts are often caused by improper filling of the chamber; excess liquid is trapped which results in high counts. Second, viral preparations in which the OBs have clumped are impossible to count by this procedure. Third, and most significantly, it is almost impossible to count suspensions which contain a high proportion of insect debris. Because of the relatively low magnification one has to use for examining the counting chamber, and the refractile nature of OBs and other extraneous material, it is very difficult to distinguish between OBs and debris. The usefulness of this technique is generally restricted to the quantitation of relatively pure suspensions of OBs. Further information on the precision of counting chambers is found in Burges and Thomson (1971).

B. ELECTRONIC PARTICLE COUNTER

The use of an electronic Coulter counter to count OBs has been described by Martignoni and Iwai (1968). Counts on highly purified and relatively impure (the number of like-size particles other than OBs varied between 1.5 and 7.9% of gross count) suspensions of *Orgyia pseudotsugata* nuclear polyhedrosis virus (NPV) were compared to counts done with the light microscope using a hemacytometer. Occlusion bodies were counted by first determining the total number of particles of a size range encompassing the OB size distribution in an aliquot of a suspension. A similar aliquot was treated with alkali and ultrasound to dissolve the OBs, and was counted to determine the number of undissolved particles which represent nonviral debris. The number of OBs in the suspension was determined by subtracting the second count from the first. Ten 50-μl samples of each of the untreated and treated suspensions were counted and each was corrected for background particles and coincidence loss. They were then averaged to obtain OB concentration.

The electronic Coulter counter produced results comparable to those obtained with the light microscope using the counting chamber. However, the electronic counter allowed for faster enumeration of the suspensions independent from the decision-making process of an observer in differentiating OBs and nonviral particles. Further, the large sample size enumerated by the Coulter counter reduced the standard error of the counts. Unfortunately, this procedure could not be used for counting suspensions of tissue homogenates due to the large amount of impurities present, and those preparations in which the OBs had clumped.

C. DRY-FILM COUNTING TECHNIQUES

Dry-film counting techniques are procedures in which OBs are deposited as films on a slide or membranes and are stained to differentiate between OBs and nonviral particles. These include the methods of Smirnoff (1962) and Morris (1973) in which OBs were deposited on filter membranes, stained, and were observed with the aid of a light microscope.

A dry-film method to distinguish between OBs and foreign material has been described by Burges and Thomson (1971). A more elaborate and statistically accurate procedure has been developed by Wigley (1976, 1980a). Using the latter method OBs can be quantified not only in purified virus preparations and triturates of infected insects, but also in soil extracts. Preparation for this technique requires that a 5-μl aliquot of a particular suspension, mixed with an equal volume of 10% Meyer's albumen, be carefully smeared over a 15-mm diameter circular area on a glass slide. Four such smears are prepared for each preparation to be counted. Once air-dried, the smears are fixed by dipping the slides in absolute alcohol for 1 minute. The slides are then rinsed in tap water, drained, and stained with naphthalene black 12B at 40°-45°C for 5 minutes. Occlusion bodies stain an intense blue-black and are counted in an area defined by an eyepiece grid. Counts are made at present intervals across four diametrically opposed radii and, because the distribution of OBs is not random, the counts are mathematically corrected for their relative position on the smear. The number of OBs on the smear is calculated and then used for determining their number in the preparation.

This procedure has been used extensively to accurately quantify preparations impossible to differentiate using counting chambers. These include soil extracts (Evans et al., 1980; Kaupp, 1981) and insect triturates (Kaupp, 1983; Kelly et al., 1978a; Wigley, 1976). Results from enumerating purified suspensions are comparable with those of the counting chamber (Wigley, 1976; Allaway, 1982), and tend to be twice as accurate when debris is present in the suspension (Cunningham, personal communication). Allaway (1982) preferred the counting chamber to the dry-film counting procedure in the studies to enumerate preparations of granulosis virus (GV), because of the lack of a suitable staining technique to differentiate the small OBs on the dry-film. Recent developments in staining techniques, however, now make it possible to use this dry counting method to quantify all types of occluded insect viruses (Kaupp and Burke, 1984).

The main advantage of the dry-film counting procedure is that the OBs are stained and fixed to a glass slide like a permanent mount. This allows the examiner to critically examine and differentiate between debris and OBs. The usefulness of this procedure is also increased by the development of differential staining methods which distinguish between OBs of entomopox virus (EPV), cytoplasmic polyhedrosis virus (CPV), NPV, and GV (Wigley, 1980b). Cunningham et al. (1981) were able to assess the ratio of NPV:CPV OBs as 161:1 in preparations of *Operophtera brumata* virus sprayed in control operations. This explained the appearance of CPV infection in larval populations in the experimental plots sprayed with NPV. Similar work resulted in the determination of the degree of CPV contamination in preparations of *Choristoneura fumiferana* NPV, originally deemed to be pure (Cunningham, 1982). Wider application of this procedure should reduce many of the problems associated with quantifying viral preparations.

D. QUANTITATION BY ELECTRON MICROSCOPY

The use of the electron microscope (EM) to count OBs has usually been restricted to those viruses, the OBs of which are difficult to detect with the light microscope. Occlusion bodies of a GV have been counted on grids under the electron microscope standardized with polystyrene-latex granules (Williams and Backus, 1949).

Recently, EM has been used to quantify the numbers of virus particles (VP) per OB in order to assess possible differences in virus isolates. In a study to determine if there were any morphological differences between isolates of *Lymantria dispar* NPV obtained from the U.S.S.R. (VIRIN-ENSh) and United States (Gypchek), Adams (1983) quantified differences in both the number of VP per OB and nucleocapsids per VP observed in these isolates. After examination of 116 OBs, Gypchek was found to have a mean of 23.07 ± 15.69 VP per OB while after observing 201 OBs of VIRIN-ENSh, an average of 13.50 ± 9.27 VP per OB was observed. However, the mean number of VP per OB for the VIRIN-ENSh isolate rose to 25.52 ± 15.71 when 189 OBs from its first passage in *L. dispar* larvae were examined. The VIRIN-ENSh isolate was found to have approximately twice as many nucleocapsids per VP (2.18 ± 1.32) as the Gypchek isolate (1.29 ± 0.09).

Similar techniques were used by Allaway (1982, 1983) in studies on the infectivity of two NPVs. On examination of thin sections of OBs of two isolates of *Agrotis segetum* NPV (AsNPV) and one isolate of *Mamestra brassicae* NPV (MbNPV), in which the VP had been sectioned transversely, it was found that the mean number of nucleocapsids per VP for the two AsNPV isolates were 4.04 ± 0.56 and 3.89 ± 0.42, respectively. The MbNPV isolate averaged 2.40 ± 0.26 nucleocapsids per VP. The mean number of VP per OB was calculated by a novel method which used the mean OB volume and the volume of the OB sampled by the thin section. The number of VP per OB thin section was determined for each isolate and converted to the mean number of VP per μm^2. The mean number of VP per μm^3 was found by dividing the number per μm^2 by the sample thickness. Subsequently, the mean number of VP and nucleocapsids per OB were calculated by multiplying the mean number of VP per μm^3 by the OB volume and by the mean number of nucleocapsids per VP. Results show that the two AsNPV isolates had an average of 168 ± 27.4 and 55.3 ± 9.6 VP per OB and 673.7 ± 144.6 and 215.0 ± 44.2 nucleocapsids per OB, respectively. The MbNPV isolate was observed to have 47.1 ± 8.7 VP per OB and 100.3 ± 24.1 nucleocapsids per OB. These studies indicate how useful a tool the EM can be in characterizing the internal morphology of OBs.

E. FILM IMPRESSION TECHNIQUES

The film impression technique was first employed by Entwistle et al. (1979) and Elleman et al. (1980) to quantify the persistence of a number of NPVs on various plant surfaces. The plant surface to be sampled, previously contaminated with virus, was pressed to one side of a strip of double sticky sellotape mounted on a glass slide. Sufficient pressure was applied being careful to avoid crushing the leaf tissue, and the specimen was slowly removed from the adhesive surface. After the impression had been made, the tape was stained with naphthalene black 12B for 5 minutes at 40°-45°C, rinsed in tap water, and air dried. The sellotape did not stain, while OBs stained an intense blue-black. The OBs were enumerated under oil immersion using the light microscope.

The lower limit of detection with this procedure for cotton, *Gossypium* sp., and cabbage, *Brassicae oleracea*, was found to be 3×10^4 OBs per cm^2 of leaf surface. For coniferous foliage, *Picea abies*, *P. sitchensis*, *Pinus contorta*, and *P. sylvestris*, the lowest concentration in which needles could be dipped yielding detectable levels of OBs on the tape was

5×10^5 OBs per ml. The accuracy of this method, at the upper limit of detection, was found to be influenced by the concentration of OBs and their behavior on the plant surface. Aggregation on the plant surface made the method impractical and inaccurate at levels higher than 100 OBs per area enumerated. Entwistle et al. (1979) removed NPVs of *Gilpinia hercyniae* and *Neodiprion sertifer* from *P. abies* and *P. sylvestris* needles, and found the film impression technique usable for needles dipped in suspensions of 5×10^5 to 10^8 OBs per ml. The efficiency of removal of purified OB preparations from cotton was 40% but increased to nearly 100% when OBs were pretreated with filtered triturates of healthy larvae. Differences in the brands of the sellotape have also been seen to influence the accuracy and application of this method (W. J. Kaupp, unpublished data). The use of this technique will provide invaluable information on the persistence of OBs in the host habitat.

F. *NOVEL METHODS OF QUANTITATION*

Although not dealing directly with the physical quantitation of viruses, several techniques have been developed which deserve mention here. The separation of OBs of CPV and NPV of *O. pseudotsugata* by Martignoni (1967) using isopycnic centrifugation demonstrated the potential of this technique for the identification, salvage, and quality control of insert virus preparations. The OBs of these two viruses were completely separated in the density gradient column of sucrose, with NPV OBs being contained in the lighter fraction (700-680 mg sucrose per ml) and CPV OBs contained in the lower fraction of the gradient (760 to 740 mg sucrose per ml) (Martignoni, 1967).

Buoyant density centrifugation of viral DNA has been used to quantify the approximate percentage of two baculoviruses of *O. pseudotsugata* in mixed preparations of these viruses (Rohrmann et al., 1978). After preparing DNA from a suspension containing both SNPV and MNPV, samples were centrifuged on CsCl gradient saturated at 25°C (r_f = 1.4185) in a Sorval TV865 rotor for 44 to 48 hours at 30,000 rpm. The contents of the tube were removed from the top and monitored for absorbance at 254 nm. The relative concentration of the DNA from each virus was quantified by cutting out and weighing the peaks from the absorbance scan. Results indicated that in the two preparations tested, 7.2 ± 1.2% and 10.8 ± 1.9% of the OBs were of the SNPV type. These values were in close agreement with results from examination using the light microscope. By this method the authors were able to detect the presence of SNPV when its DNA comprised as little as 5% of the total DNA present in the sample.

III. SEROLOGICAL QUANTITATION

The use of serological methods in insect virology has basically been restricted to the identification, classification, and assay of isolates. Latex agglutination tests and tube precipitation tests (Carter, 1973), complement fixation and immunodiffusion (Kelly et al., 1978b), and neutralization tests (Aizawa, 1954, 1962; Martignoni et al., 1980) all have been used to characterize occluded insect viruses. All these techniques show promise in quality control programs and in field monitoring of viral strains.

Application of the enzyme-linked immunosorbent assay (ELISA) system has shown promise in terms of enumerating OBs. The ELISA technique outlined by Engvall and Perlmann (1971, 1972) and based on the microplate method described by Voller et al. (1974) is well established as an assay for plant, animal, and insect viruses (Voller et al., 1976; Kelly et al., 1978a,b; Kahane et al., 1979; Crook and Payne, 1980; Kaupp, 1980, 1981; Longworth and Carey, 1980; Volkmann and Falcon, 1982). It involves the detection of virus antigen by specific antibodies to which enzymes have been chemically linked. Exposed to the proper antigen, the enzyme-linked antibody forms an antibody-antigen complex which is then detected by the degradation of a suitable substrate by the enzyme carried on the antibody. If no suitable antigen is present, the complex cannot be formed and enzymatic degradation of the substrate does not occur. The use of reactants in the ELISA system is more economical, is more versatile, and is easier to apply than most labeled antibody techniques.

The potential of the ELISA system for the quantitation of viruses has not gone unnoticed. Longworth and Carey (1980) indicated that the indirect sandwich ELISA method could detect as little as 20 ng of purified *Orytes* baculovirus but quantitative determination was difficult because of high background interference from the host protein. Caution is advised in the use of ELISA procedures because similar antigenic properties among baculoviruses may cause false positives (Bilmoria et al., 1974; Volkmann and Falcon, 1982; Naser and Miltenburger, 1983).

Development of the ELISA system to enumerate baculoviruses has been demonstrated by Kelly et al. (1978a) and Kaupp (1980). By transforming standard curves illustrating the relationship between OD_{405} for ELISA tests and quantities of VP protein and polyhedral protein (polyhedrin) into equations describing the number of *N. sertifer* OBs represented by these protein quantities, Kaupp (1980) found the indirect ELISA system more sensitive to polyhedrin if polyhedrin instead of VP protein was assayed. Similar results have been demonstrated by Volkmann and Falcon (1982) in detecting *Trichoplusia ni* SNPV. Using this technique, Kaupp (1980) was able to detect and quantify as low as 4×10^5 OBs in infected *N. sertifer* larvae.

Additional improvements in the ELISA system which will employ different procedures and highly specific monoclonal antibodies will undoubtedly improve the capability of this technique to quantify viruses. Presently, it is accepted as an assay for purified preparations of a given virus, but is limited as a quantitative technique because of nonspecific reactions involving host proteins.

IV. QUANTITATION OF BIOLOGICAL ACTIVITY

It is apparent that direct physical enumeration of preparations of OBs provides no real indication of the biological activity of the virus. Reliance on physical counts has led to difficulties in standardization of virus preparations, questionable reliability of these preparations, difficulty in comparing experimental results, and anomalies in the efficacy data generated from similar batches of virus. The inception of larval equivalents (L.E.) and viral units (V.U.) did not resolve these discrepancies since they also relied on the physical quantitation of OBs (Allen, 1967; Ignoffo, 1964; Pinnock, 1975; Woodall and Ditman, 1967). Physical quantitation only approximates biological activity. Only by biologically assaying in insect larvae or tissue cultures can the actual potency of viral preparations be measured as activity or infectivity.

The quantitation of biological activity of a virus preparation requires a system in which a dosage of the virus produces a detectable response. Both *in vivo* and *in vitro* systems are used for the bioassay of insect pathogenic viruses. In the *in vivo* system larvae are used for the end-point dilution assay where the calculated titer represents a statistical, biological unit. In the *in vitro* systems cell cultures are employed to quantitate viral infectivity by either (1) end-point titration, i.e., quantal assay, or (2) enumerative titration, i.e., plaque assay. Both *in vivo* and *in vitro* methods have statistical validity within the constraints of Poisson distribution.

The choice of method for virus titration depends on the type of virus, viral preparation, laboratory facilities available, and the nature of response of the test organism. For example, a preparation of OBs is assayed in larvae because the approximate 6.5 pH of the insect cell culture systems is too low to liberate the virions from the OBs. Consequently, no infection occurs in cell cultures. The high pH of the larval gut dissolves OBs and liberates virions which initiate infection. Commercial preparations of viral insecticides that have various adjuvants, such as spreaders, stickers, and uv protectants are also assayed in larvae because the adjuvants are toxic to cell cultures. Furthermore, OBs are the active ingredient in almost all the currently available viral insecticides which precludes the use of *in vitro* methods of titration. However, OBs can be purified from such preparations and dissolved in alkali to liberate virions, which, in turn, can be titrated *in vitro*. However, that method is time consuming, and alkali treatment can inactivate virions which consequently affects the accuracy of the assay. Thus, larval bioassay is the method of choice, if not the only one, for the quantitation of biological activity of viral insecticides.

A. *IN VIVO BIOASSAY*

The development of various techniques of bioassay to ascertain the response of the insect to virus preparations has progressed considerably in recent years. Difficulties in the precision and analysis of bioassays has been reviewed by Burges and Thomson (1971) and Finney (1971a,b). In an attempt to accurately dose the insect larvae, procedures have been developed to administer the inoculum by diet surface contamination (Martignoni and Iwai, 1981), by incorporation into synthetic diet (Ignoffo, 1964), by foliar contamination (Evans, 1981), and by direct ingestion of droplets (Hughes and Wood, 1981). All of these procedures rely on physical quantitation to standardize the concentration of OBs administered and express the LD_{50} dose in terms of numbers of OBs.

A method of measuring the biological activity of virus preparations, not based on any physical enumeration of OBs, has been developed by Martignoni and Iwai (1978). The concentration of viral preparations was measured as the activity titer or potency of the virus in terms of activity units (AU), and defined as the amount of activity per unit weight or volume of the preparation. For the activity standardization of preparations of *O. pseudotsugata* NPV, the AU is the LC_{50} dose expressed as nanograms of viral preparation per cup, and the potency or activity titer is the number of AU per gram of the preparation. In other words, potency is the number of LC_{50} doses, referred to as activity units (AU) per gram of the virus preparation. These authors reported an AU of 6.98 ng per cup and the potency was 1.43×10^8 AU per gm (10^9 ng ÷ 6.98 ng). Elaborate instructions on how to perform the bioassays to determine the activity titer of preparations of *O. pseudotsugata* NPV are given in Martignoni and Iwai (1977).

The advantage of measuring the biological activity as AU is that comparisons can be made between the activity of various viral preparations. By measuring the activity titer by this method, it was found that the response of the GL-1 strain of *O. pseudotsugata* larvae to technical preparations of NPV had not changed over three generations (Martignoni and Iwai, 1978). Similar procedures were used to measure the biological activity of three baculoviruses in *Choristoneura occidentalis* larvae (Martignoni and Iwai, 1981). Future developments in bioassay techniques should enable more accurate estimates of biological activity to be made, independent of and yet related to the number of OBs.

B. *IN VITRO BIOASSAY*

Granados (1976) and Knudson and Buckley (1977) have reviewed the earlier literature on the *in vitro* bioassay of insect viruses. As alluded to above, the *in vitro* titration methods are useful mainly for preparations of nonoccluded or free virions.

1. *Plaque Assay*

For the plaque assay, monolayers of susceptible cells in log phase of growth are inoculated with virus. The virus is removed after an adsorption period of about 1 hour and a suitable overlay is added to the monolayer to restrict the lateral movement of progeny virus. The monolayers are then stained with a dye, generally neutral red for insect pathogenic viruses, 3-10 days p.i. Neutral red stains the living cells red, but the plaques of infected cells appear as pale pink or unstained. Depending upon the sensitivity of the assay, two types of plaques have been observed in almost all of the baculoviruses that have been tested: one type that has cells with many polyhedra (MP), and the other type that has cells with few polyhedra (FP).

Hink and Vail (1973) were the first to develop a plaque assay for a pathogenic insect virus. They used a liquid overlay containing 0.6% methylcellulose (MC) for the *Trichoplusia ni* cells infected with *Autographa californica* MNPV (AcMNPV). This technique was further improved by Hink and Strauss (1977, 1979) by adding 2-(*N*-morpholino)ethanesulfonic acid (MES) buffer to the MC overlay to stabilize pH in the range 6.2-6.3. This produced larger and more plaques than the original technique.

Knudson (1976, 1979) reported plaques of AcMNPV, *Spodoptera exempta* MNPV, *S. exigua* MNPV, and *T. ni* MNPV (TnMNPV) in *S. frugiperda* cells using a solid overlay of 0.75% Seakem agarose. Brown and Faulkner (1977, 1978) obtained plaques of

AcMNPV and TnMNPV in *S. frugiperda* cells using a solid overlay of 1.5% Seaplaque or Seakem agarose. They found that this assay was more sensitive than that using an MC overlay but less sensitive than end-point dilution titration. *Trichoplusia ni* and *Bombyx mori* cells were not satisfactory for this assay. *Manduca sexta* cells could be used for AcMNPV but they were less sensitive than *S. frugiperda* cells. Wood (1977) obtained plaques of AcMNPV in *T. ni* cells using a solid overlay of 1.5% Seaplaque agarose. A mixture of AcMNPV inoculum and cells was centrifuged at 1000 *g* for 60 minutes to improve infection.

Plaque assay for a singly enveloped NPV of *Heliothis zea* (HzSNPV) was reported by Yamada and Maramorosch (1981) in *H. zea* cells using a solid overlay of a 1% mixture of Seakem and ultrapure agarose in 50:50 ratio.

Fraser and Hink (1982) compared several plaque assay techniques using *T. ni* and *S. frugiperda* cells for detecting and distinguishing MP and FP plaque variants of *Galleria mellonella* MNPV (GmMNPV). Both types of plaques were obtained in both the cell lines. The largest plaques in either of the cell lines were produced with the 1.0 or 0.75% agarose overlay, but MP and FP plaques were most easily distinguished with these overlays in *S. frugiperda* cells. The 0.9% MC overlay was the only one that did not detect FP plaques. The authors concluded that the FP variant of GmMNPV is not a host-dependent phenomenon, and that detection can be influenced by the overlay formulation. Fraser (1982) reported that an overlay of 0.75% agarose produced plaques of AcMNPV, GmMNPV, and *Choristoneura fumiferana* MNPV (CfMNPV) in *T. ni* and *S. frugiperda* cells, and also plaques of HzSNPV in a *H. zea* cell line. In addition, he observed that plaque variants that did not produce polyhedra could also be easily detected.

Brown et al. (1977) obtained plaques of iridescent type 22 (from *Simulium* sp.) in *S. frugiperda* cells using 0.75% Seaplaque agarose overlay and were able to titrate this virus using the plaque assay. Earlier, Bellet (1965) titrated iridescent virus type 2 (*Sericesthis* iridescent virus) in *Antheraea ecalypti* cells using fluorescent antibody. He reported that this technique was versatile, but more tedious and less precise than a plaque assay titration.

A plaque assay method for cricket paralysis virus (CrPV) titration in *Drosophila melanogaster* cells was reported by Scotti (1977).

2. *$TCID_{50}$*

The $TCID_{50}$ (tissue culture infectious dose $_{50}$) or quantal assay is based on the end-point dilution method of Reed and Muench (1938). Serial dilutions of virus preparation are inoculated into multiple cultures of susceptible cells until the dilution end-point is reached. The cells may be exposed to virus in tubes or microtiter plates. The latter has been the method of choice in recent years since it conserves labor and materials. The cultures are then scored for virus replication based on cytopathic effect (CPE) and polyhedra formation. The unit of activity determined by this method is called $TCID_{50}$ (a dose of virus that infects 50% of the cultures) and the titer of virus is expressed in terms of $TCID_{50}$ units per milliliter of the virus preparation.

The titer of baculoviruses has been determined using continuous cell lines by many researchers including Knudson and Tinsley (1974), Brown and Faulkner (1975), Wood (1977), Knudson (1979), Granados et al. (1981), Yamada and Maramorosch (1981), Yamada et al. (1982); of a cytoplasmic virus by Belloncik and Chagnon (1980); of iridescent viruses by Ohba and Aizawa (1978) and Webb et al. (1975); and of a picorna-like virus by Scotti (1977).

Both the $TCID_{50}$ and plaque assay methods are based on the Poisson distribution and the single-particle concept of active virus. Assuming that there are no errors in dilution, and the systems for assay by both the $TCID_{50}$ and plaque assay methods are equally sensitive, 0.69 plaque units equal one $TCID_{50}$ (Cunningham, 1973). Thus, either of the two *in vitro* bioassay methods can be used for the titration of insect viruses. In the case of baculoviruses, both these methods are based on the production of polyhedra in infected cultures. However, polyhedra production is not always a reliable indication of baculovirus infection and replication (Volkman, 1978; Duncan et al., 1983). Polyhedra production by a virus differs with the line and subline of the host cells. Cloned cell lines from the *T. ni* line TN-368 differed from one another in their capability to form plaques of AcMNPV (Volkman and Summers, 1975). Thus, in order to improve the *in vitro* titration of baculoviruses, we need to develop an assay method that is based on virus replication rather than polyhedra production.

V. CONCLUSION

It is evident that most viral preparations can now be accurately quantified as to OB numbers. Also, the biological activity of preparations of OBs and NOVs can be determined per unit weight or volume of the preparation. This permits comparisons between isolates of a virus and between different preparations of the same isolate. It is strongly recommended that procedures for the standardization of viral preparations should include a physical quantitation of OBs as well as some indication of their activity so that the results obtained from the application of viral insecticides can be interpreted more easily and fully, thus providing more meaningful information.

REFERENCES

Adams, J. R. (1983). Electron microscope examinations of Gypchek and Virin-EnSh. *In* "A Comparison of the US (Gypcheck) and USSR (Virin-ENsH) Preparations of the Nuclear Polyhedrosis Virus of the Gypsy Moth, *Lymantria dispar*" (C. M. Ignoffo, M. E. Martignoni, and J. L. Vaughn, eds.), pp. 11-20. Amer. Soc. Microbiol. Publ., Washington, D.C.

Aizawa, K. (1954). Immunological studies of the silkworm jaundice virus. I. Neutralization and absorption test of the silkworm jaundice virus. *J. Virol. 4*, 238-240.

Aizawa, K. (1962). Infection of the Greater Wax Moth, *Galleria mellonella* (Linnaeus), with the nuclear polyhedrosis virus of the silkworm, *Bombyx mori* (Linnaeus). *J. Insect Pathol. 4*, 122-127.

Allaway, G. P. (1982). Infectivity of some occluded insect viruses. Ph.D. Thesis, University of London, London, England. 274 pp.

Allaway, G. P. (1983). Virus particle packaging in baculovirus and cytoplasmic polyhedrosis virus inclusion bodies. *J. Invertebr. Pathol. 42*, 357-368.

Allen, G. E. (1967). Report of second work conference on the utilization of nuclear polyhedrosis virus for the control of *Heliothis* species. *J. Invertebr. Pathol. 9*, 447-448.

Bellett, A. J. D. (1965). The multiplication of *Sericesthis* iridescent virus in cell cultures from *Antheraea eucalypti* Scott. II. An *in vitro* assay for the virus. *Virology 26*, 127-131.

Belloncik, S., and Chagnon, A. (1980). Titration of a cytoplasmic polyhedrosis virus by a tissue microculture assay: Some applications. *Intervirology 13*, 28-32.

Bilmoria, S. L., Parkinson, A. J., and Kalmakoff, J. (1974). Comparative study of ^{125}I and [3H] acetate-labelled antibodies in detecting irridescent virus. *Appl. Microbiol. 28*, 133-137.

Brown, M., and Faulkner, P. (1975). Factors affecting the yield of virus in a cloned cell line of *Trichoplusia ni* infected with a nuclear polyhedrosis virus. *J. Invertebr. Pathol. 26*, 251-257.

Brown, M., and Faulkner, P. (1977). A plaque assay for nuclear polyhedrosis viruses using a solid overlay. *J. Gen. Virol. 36*, 361-364.

Brown, M., and Faulkner, P. (1978). Plaque assay of nuclear polyhedrosis viruses in cell culture. *Appl. Environ. Microbiol. 36*, 31-35.

Brown, D. A., Lescott, T., Harrap, K. A., and Kelly, D. C. (1977). The replication and titration of iridescent virus Type 22 in *Spodoptera frugiperda* cells. *J. Gen. Virol. 38*, 175-178.

Burges, H. D., and Thomson, E. M. (1971). Standardization and Assay of Microbial Insecticides. *In* "Microbial Control of Insects and Mites" (H. D. Burges and N. W. Hussey, eds.), 861 pp. Academic Press, London.

Carter, J. B. (1973). Detection and assay of *Tipula* irridescent virus by the latex agglutination test. *J. Gen. Virol. 21*, 181-185.

Crook, N. E., and Payne, C. P. (1980). Comparison of three methods of ELISA for baculoviruses. *J. Gen. Virol. 46*, 29-37.

Cunningham, C. H. (1973). Quantal and enumerative titration of virus in cell cultures. *In* "Tissue Culture, Methods and Applications" (P. F. Kruse, Jr., and M. K. Patterson, Jr., eds.), pp. 527-532. Academic Press, New York.

Cunningham, J. C. (1982). Field trials with baculoviruses: Control of forest insect pests. *In* "Microbial and Viral Pesticides" (E. Kurstak, ed.), pp. 335-386. Dekker, New York.

Cunningham, J. C., Tonks, N. V., and Kaupp, W. J. (1981). Viruses to control winter moth, *Operophtera brumata* (Lepidoptera: Geometridae). *J. Entomol. Soc. Br. Columbia 78*, 17-24.

Duncan, R., Chung, K. L., and Faulkner, P. (1983). Analysis of a mutant of *Autographa californica* nuclear polyhedrosis virus with a defect in the morphogenesis of the occlusion body macromolecular lattice. *J. Gen. Virol. 64*, 1531-1542.

Elleman, C. J., Entwistle, P. F., and Hoyle, S. R. (1980). Application of the impression film technique to counting inclusion bodies of nuclear polyhedrosis viruses on plant surfaces. *J. Invertebr. Pathol. 36*, 129-132.

Engvall, E., and Perlmann, P. (1971). Enzyme-linked immunosorbent assay (ELISA). Quantitative assays of immunoglobulin G. *Immunochemistry 8*, 871-874.

Engvall, E., and Perlmann, P. (1972). Enzyme-linked immunosorbent assay, ELISA. III. Quantitation of specific antibodies by enzyme-labelled anti-immunoglobulin in antigen-coated tubes. *J. Immunochem. 109*, 129-135.

Entwistle, P. F., Elleman, C. J., Hoyle, S. R., Evans, H. F., and Adams, P. H. W. (1979). Aspects of the physical behaviour of baculoviruses on plant surfaces. *Prog. Invertebr. Pathol. 1958-78, Intern. Colloq. Invertebr. Pathol., Prague, Czechoslovakia.*

Evans, H. F. (1981). Quantitative assessment of the relationships between dosage and response of the nuclear polyhedrosis virus of *Mamestra brassicae*. *J. Invertebr. Pathol. 37*, 101-109.

Evans, H. F., Bishop, J. M., and Page, E. A. (1980). Methods for the assessment of nuclear-polyhedrosis virus in soil. *J. Invertebr. Pathol. 35*, 1-8.

Finney, D. J. (1971a). "Probit Analysis," 3rd Ed., 33 pp. Cambridge University Press, Cambridge.

Finney, D. J. (1971b). "Statistical Method in Biological Assay," 2nd Ed., 668 pp. Griffin, London, UK.

Fraser, M. J. (1982). Simplified agarose overlay plaque assay for insect cell lines and insect nuclear polyhedrosis viruses. *J. Tissue Culture Methods 7*, 43-46.

Fraser, M. J., and Hink, W. F. (1982). Comparative sensitivity of several plaque assay techniques employing TN-368 and IPLB-SF 21AE insect cell lines for plaque variants of *Galleria mellonella* nuclear polyhedrosis virus. *J. Invertebr. Pathol. 40*, 89-97.

Granados, R. R. (1976). Infection and replication of insect pathogenic viruses in tissue culture. *Advan. Virus Res. 20*, 189-236.

Granados, R. R., Lawler, K. A., and Burand, J. P. (1981). Replication of *Heliothis zea* baculovirus in an insect cell line. *Intervirology 16*, 71-79.

Hink, W. F., and Strauss, E. M. (1977). An improved technique for plaque assay of *Autographa californica* nuclear polyhedrosis virus on TN-368 cells. *J. Invertebr. Pathol. 29*, 390-391.

Hink, W. F., and Strauss, E. M. (1979). Plaque assay of alfalfa looper nuclear polyhedrosis virus on the TN-368 cell line. *T.C.A. (Tissue Culture Assoc.) Manual 5*, 1033-1035.

Hink, W. F., and Vail, P. V. (1973). A plaque assay for titration of alfalfa looper nuclear polyhedrosis virus in a cabbage looper (TN-368) cell line. *J. Invertebr. Pathol. 22*, 168-174.

Hughes, P. R., and Wood, H. A. (1981). A synchronous peroral technique for the bioassay of insect viruses. *J. Invertebr. Pathol. 37*, 154-159.

Ignoffo, C. M. (1964). Bioassay technique and pathogenicity of a nuclear-polyhedrosis virus of the cabbage looper, *Trichoplusia ni* (Hübner). *J. Insect Pathol. 6*, 237-245.

Ignoffo, C. M. (1966). Standardization of products containing insect viruses. *J. Invertebr. Pathol. 8*, 547-548.

Kahane, S., Goldstein, V., and Sarov, I. (1979). Detection of IgG antibodies specific for measles virus by enzyme-linked immunosorbent assay (ELISA). *Intervirology 12*, 39-46.

Kaupp, W. J. (1980). Development of the enzyme-linked immunosorbent assay to detect a nuclear polyhedrosis virus in European pine sawfly larvae. *Can. Forest. Serv. Bi-mon. Res. Notes 36*, 10-11.

Kaupp, W. J. (1981). Studies on the ecology of the nuclear polyhedrosis virus of the European pine sawfly, *Neodiprion sertifer* (Geoff). Ph.D. Thesis, University of Oxford, Oxford, 363 pp.

Kaupp, W. J. (1983). Estimation of nuclear polyhedrosis virus produced in field populations of the European pine sawfly, *Neodiprion sertifer* (Geoff) (Hymenoptera: Diprionidae). *Can. J. Zool. 61*, 1857-1861.

Kaupp, W. J., and Burke, R. F. (1984). A staining technique for the inclusion bodies of a granulosis virus. *Can. Forest. Serv. F.P.M.I. Tech. Note 1*, 2 pp.

Kelly, D. C., Edwards, M. L., Evans, H. F., and Robertson, J. S. (1978a). The use of the enzyme-linked immunosorbent assay to detect a nuclear polyhedrosis virus in *Heliothis arigera* larvae. *J. Gen. Virol. 40*, 465-469.

Kelly, D. C., Edwards, M. L., and Robertson, J. S. (1978b). The use of enzyme-linked immunosorbent assay to detect and discriminate between small irridescent viruses. *Ann. Appl. Biol. 90*, 369-374.

Knudson, D. L. (1976). Plaque assay of baculoviruses: Visible plaques under solid agarose overlays. *Proc. First Intern. Colloq. Invertebr. Pathol.*, pp. 104-107. Kingston, Ontario, Canada.

Knudson, D. L. (1979). Plaque assay of baculoviruses employing an agarose-nutrient overlay. *Intervirology 11*, 40-46.

Knudson, D. L., and Buckley, S. M. (1977). Invertebrate cell culture methods for the study of invertebrate-associated animal viruses. *In* "Methods in Virology" (K. Maramorosch and H. Koprowski, eds.), Vol. 6, pp. 323-391. Academic Press, New York.

Knudson, D. L., and Tinsley, T. W. (1974). Replication of a nuclear polyhedrosis virus in a continuous cell culture of *Spodoptera frugiperda*: Purification, assay of infectivity, and growth characteristics of the virus. *J. Virol. 14*, 934-944.

Longworth, J. F., and Carey, G. P. (1980). The use of an indirect enzyme-linked immunosorbent assay to detect baculovirus in larvae and adults of *Oryctes rhinoceros* from Tonga. *J. Gen. Virol. 47*, 431-438.

Martignoni, M. E. (1967). Separation of two types of viral inclusion bodies by isopycnic centrifugation. *J. Virol. 1*, 646-647.

Martignoni, M. E., and Iwai, P. J. (1968). Determination of nucleopolyhedron counts and size frequency distributions by means of a coulter transducer. *U.S. Forest. Serv. Res. Note PNW-85*, 14 pp.

Martignoni, M. E., and Iwai, P. J. (1977). Peroral bioassay of technical-grade preparations of Douglas-fir Tussock Moth nucleopolyhedrosis virus *(Baculovirus)*. *USDA Forest. Res. Pap. PNW-222*, 12 pp. Pacific Northwest Forestry and Range Experimental Station, Portland, OR.

Martignoni, M. E., and Iwai, P. J. (1978). Activity standardization of technical preparations of Douglas-fir Tussock Moth *Baculovirus*. *J. Econ. Entomol. 71*, 473-476.

Martignoni, M. E., and Iwai, P. J. (1981). Peroral bioassay of nucleopolyhedrosis viruses in larvae of the western spruce budworm. *USDA Forest. Serv. Res. Pap. PNW-285*, 20 pp. Pacific Northwest Forestry and Range Experimental Station, Portland, OR.

Martignoni, M. E., Iwai, P. J., and Rohrmann, G. F. (1980). Serum neutralization of nuclear polyhedrosis viruses *(Baculovirus* Subgroup A) pathogenic for *Orgyia pseudotsugata*. *J. Invertebr. Pathol. 36*, 12-20.

Morris, O. N. (1973). A method of visualizing and assessing deposits of aerially sprayed insect microbes. *J. Invertebr. Pathol. 22*, 115-121.

Naser, W. L., and Miltenburger, H. G. (1983). Rapid baculovirus detection, identification, and serological classification by western blotting-ELISA using a monoclonal antibody. *J. Gen. Virol. 64*, 639-647.

Ohba, M., and Aizawa, K. (1978). Comparative titration of *Chilo* iridescent virus *in vivo* and *in vitro*. *J. Invertebr. Pathol. 32*, 394-395.

Pinnock, D. E. (1975). Pest populations and virus dosage in relation to crop productivity. *In* "Baculoviruses for Insect Pest Control: Safety Consideration" (M. Summers, R. Engler, L. A. Falcon, and P. V. Vail, eds.), pp. 145-154. Amer. Soc. Microbiol. Publ., Washington, D.C.

Reed, L. J., and Muench, H. (1938). A simple method of estimating fifty percent endpoints. *Amer. J. Hyg. 27*, 493-497.

Rohrmann, G. F., Martignoni, M. E., and Beaudreau, G. S. (1978). Quantification of two viruses in technical preparations of *Orgyia pseudotsugata* baculovirus by means of buoyant density centrifugation of viral deoxyribonucleic acid. *Appl. Environ. Microbiol. 35*, 690-693.

Scotti, P. D. (1977). End-point dilution and plaque assay methods for titration of cricket paralysis virus in cultured *Drosophila* cells. *J. Gen. Virol. 35*, 393-396.

Shapiro, M. (1983). Comparative infectivity of Gypchek L-79 and VIRIN-ENSh to *Lymantria dispar*. *In* "A Comparison of the U.S. (Gypchek) and USSR (VIRIN-ENSh) Preparations of the

Nuclear Polyhedrosis Virus of the Gypsy Moth, *Lymantria dispar*" (C. M. Ignoffo, M. E. Martignoni, and J. L. Vaughn, eds.), pp. 38-42. Amer. Soc. Microbiol. Publ., Washington, D.C.

Smirnoff, V. A. (1962). Detection of polyhedra from insect virus diseases on filter membranes. *Stain Technol.* *37*, 207-210.

Volkman, L. E. (1978). Cell culture studies: Standardization of biological activity. *In* "Viral Pesticides: Present Knowledge and Potential Effects on Public and Environmental Health" (M. D. Summers and C. Y. Kawanishi, eds.), pp. 135-150. U.S. Environmental Protection Agency, Research Triangle Park, North Carolina, No. EPA-600/9-78-026.

Volkman, L. E., and Falcon, L. A. (1982). Use of monoclonal antibody in an enzyme-linked immunosorbent assay to detect the presence of *Trichoplusia ni* (Lepidoptera: Noctuidae) S nuclear polyhedrosis virus in *T. ni* larvae. *J. Econ. Entomol.* *75*, 868-871.

Volkman, L. E., and Summers, M. D. (1975). Nuclear polyhedrosis virus detection: Relative capabilities of clones developed from *Trichoplusia ni* ovarian cell line TN-368 to serve as indicator cells in a plaque assay. *J. Virol.* *16*, 1630-1637.

Voller, A., Bidwell, D., Huldt, G., and Engvall, E. (1974). A microplate method of enzyme-linked immunosorbent assay and its application to malaria. *Bull. WHO* *51*, 209-211.

Voller, A., Bartlett, A., Bidwell, D. E., Clark, M. F., and Adams, A. N. (1976). The detection of viruses by enzyme-linked immunosorbent assay (ELISA). *J. Gen. Virol.* *33*, 165-167.

Webb, S. R., Paschke, J. D., Wagner, G. W., and Campbell, W. R. (1975). Bioassay of mosquito iridescent virus of *Aedes taeniorhynchus* in cell cultures of *Aedes aegypti*. *J. Invertebr. Pathol.* *26*, 205-212.

Wigley, P. J. (1976). The epizootiology of a nuclear polyhedrosis virus disease of the winter moth, *Operophtera brumata* L., at Wistmans Woods, Dartmoor. Ph.D. Thesis, University of Oxford, Oxford, 185 pp.

Wigley, P. J. (1980a). Counting micro-organisms. *In* "Microbial Control of Insect Pests" (J. Kalmakoff and J. F. Longworth, eds.), 102 pp. New Zealand Department of Science and Industrial Research Bulletin 228, Wellington, New Zealand.

Wigley, P. J. (1980b). Diagnosis of virus infections--staining of insect inclusion body viruses. *In* "Microbial Control of Insect Pests" (J. Kalmakoff and J. F. Longworth, eds.), 102 pp. New Zealand Department of Science and Industrial Research Bulletin 228, Wellington, New Zealand.

Williams, R. C., and Backus, R. C. (1949). Macromolecular weights determined by direct particle counting. I. The weight of the bushy stunt virus particle. *J. Amer. Chem. Soc. 71*, 4052-4057.
Wood, A. H. (1977). An agar overlay plaque assay method for *Autographa californica* nuclear-polyhedrosis virus. *J. Invertebr. Pathol. 29*, 304-307.
Woodall, K. L., and Ditman, L. P. (1967). Control of the cabbage looper and corn earworm with nuclear polyhedrosis virus. *J. Econ. Entomol. 70*, 243-246.
Yamada, K., and Maramorosch, K. (1981). Plaque assay of *Heliothis zea* baculovirus employing a mixed agarose overlay. *Arch. Virol. 67*, 187-189.
Yamada, K., Sherman, K. E., and Maramorosch, K. (1982). Serial passage of *Heliothis zea* singly embedded nuclear polyhedrosis virus in a homologous cell line. *J. Invertebr. Pathol. 39*, 185-191.

RECEPTORS IN THE INFECTION PROCESS

H. M. MAZZONE

U.S. Department of Agriculture--Forest Service
Hamden, Connecticut

I. INTRODUCTION

Historically, two divergent views have evolved concerning the specificity of virus infection of susceptible host cells (Boulanger and Philipson, 1981). One view, supported by Dales (1973), stresses the absence of precise structural requirements for cell-virus interactions. The second view proposes specific cellular receptor sites in the plasma membrane of host cells which recognize one or several attachment proteins for a virus (cf. Lonberg-Holm and Philipson, 1981). In the latter view cellular receptor sites, therefore, serve to specifically bind viruses as the first event in infection.

In support for the absence of precise structural requirements for cell-virus interactions, Dales (1973) cautions on ascribing cell-virus requirements for infection based on interpretations from electron microscope images. The problem of artifacts in electron microscopy, rightly noted by Dales, has frequently provided the researcher with incorrect conclusions. In cell-virus interactions the most common artifact develops as a result of inadequate preservation of membranes. As pointed out by Dales, such artifacts may, for example, show a virus particle existing free in the cytoplasmic matrix, when in reality, it is enclosed by an inconspicuous membrane around a phagocytic vacuole. Again, in thin sections in the range of

ISBN 0-12-470295-3

500 to 1000 nm in thickness, this measure is equal to or greater than the width of many viruses. In random slices through cell-virus complexes, virus particles that are attached at the surface where invagination had occurred may appear as if they had lost morphological integrity at the site of contact and had merged with the plasma membrane.

While Dales' arguments are correct and appropriate, the overwhelming evidence to date on prokaryotic and eukaryotic cells emphasizes the existence of virus-receptor recognition. In this regard our understanding of bacterial receptors for bacterial viruses is more profound than that of cell receptors for animal viruses (Philipson, 1981). The reasons for this situation are believed to be the difference in structural complexity of the two host types and the availability of mutants for bacteria but not for animal cells. The lack of virus receptor mutants among eukaryotic host cells is considered to have hampered identification of specific receptor molecules such as proteins. It is held that rapid development of methods to identify mutants in membrane components of mammalian cells, for example, would lead to progress in this area of research, and also aid in our understanding of the role of other surface proteins in the differentiation processes of mammalian and other eukaryotic cells (Philipson, 1981).

The purpose of this chapter is to present the evidence existing for receptors of insect viruses in the infection process, particularly, in the case of the baculoviruses, which offer great potential as biological insecticides. This chapter will first consider receptor-virus relationships and cell-virus interactions in bacterial and vertebrate virus systems. Such considerations will then be applied to invertebrate virus systems and, in particular, to insect-baculovirus relationships.

Of the insect viruses, the baculoviruses have some unique properties. Many baculoviruses are occluded within a protein matrix, a polyhedron or capsule. Moreover, as cell receptors are believed to exist for both enveloped (Holmes, 1981) and nonenveloped (Boulanger and Philipson, 1981) viruses infecting vertebrate hosts, the baculoviruses have both enveloped and nonenveloped forms participating in the total infectious reaction.

In a discussion of virus receptors we need to describe the three principal events in an infection of a cell by a virus: attachment (adsorption), entry (penetration), and release. Interactions between cell and virus are noted in these events by which structural features, e.g., peplomers, are imparted to the virus. Such structural features appear to be necessary for continuation of the virus infection cycle, particularly for baculoviruses, by serving as the region on the virus which attaches to the host cell (Adams et al., 1975, 1977; Kawamoto et al., 1977; Hess and Falcon, 1977).

II. HOST CELL-VIRUS RECOGNITION

For a viral infection to occur, the host cell and virus express some degree of complementarity or recognition. The recognition may involve both a structural (physical) and chemical expression in the cell-virus interaction. In host cell-virus systems, this recognition is expressed in one commonly held concept of cell receptors and viral attachment proteins. In this section we consider systems that have been delineated in terms of proteins involved in the attachment of viruses to sites on host cells.

The essential terminology for such a concept has been defined by Lonberg-Holm (1981): The virus attachment protein(s) is a virion structure(s) which can recognize a cellular receptor. A cellular receptor unit refers to cellular molecules recognizing one virus attachment protein. The cellular receptor site is a cellular structure containing one or more cellular receptor units which can effectively bind one virion.

A. *NATURE OF CELL RECEPTORS*

Korn (1975) has reviewed the essential features of the plasma membrane of cells. The plasma membrane is approximately 100 Å in width consisting of protein and lipid in a ratio of about 1.5 to 1. In agreement with all cell membranes, the lipid of the plasma membrane contains little glyceride but a high concentration of phospholipid. The plasma membrane is unusual among cell membranes in its high content of glycolipid and sterol.

It is generally believed that most of the phospholipids and sterols of the plasma membrane are in the form of a molecular bilayer oriented with the polar head groups forming hydrophilic regions at the inner and outer surfaces and the fatty acyl chains forming a hydrophobic interior. There is some suggestion of specific arrangements of particular phospholipids and sterol molecules within the bilayer. Experiments on model systems infer that lipid molecules are free to move rapidly within the plane of the bilayer (Kornberg and McConnell, 1971a) but these same lipid molecules are unable to flip from one side of the bilayer to the other (Kornberg and McConnell, 1971b).

Proteins are present at the outer and inner surfaces of the plasma membrane, and proteins also lie within the hydrophobic interior of the lipid bilayer. The "Fluid mosaic model" of Singer and Nicholson (1972a,b) suggests that many of the membrane proteins and glycoproteins exist as mobile islands within a hydrocarbon sea and are able to move laterally through this liquidlike area. This conclusion rests on observations by freeze-cleavage electron microscopy.

In host cell-virus systems, the molecules used as receptors by viruses are on the plasma membrane. The virus receptors recognize or are recognized by virus particules and provide specific points of attachment on the cell membrane. Such receptors are believed to be controlled by genetic information: in eukaryotic cells these controlling factors are located on one or more chromosomes. The poliovirus receptor gene, for example, is located on chromosome 19 of human-mouse hybrid cells (Miller et al., 1974). As noted, cell receptor sites effectively bind virions and contain one or more cellular receptor units. Receptor sites may be composed of a number of molecules, each of which recognizes attachment proteins possessed by the virions (Lonberg-Holm and Philipson, 1980).

A variety of macromolecular structures may serve as receptors for viral attachment proteins. Many of these structures are substrates for enzymes and many have antigenic properties. Still others are biological effector molecules such as toxins, neurotransmitters, and regulatory molecules (Holmes, 1981; Incardona, 1981). The density and distribution of the various receptors affect the rate of the binding reaction. In mammalian cells the plasma membranes contain specialized organelles which perform many functions essential for interaction with the microenvironment. The structure specializations of cell membranes include microvilli, pinocytic ruffles, vesicles, coated pits, and desmosomes (Holmes, 1981). In metabolically active cells, new molecules are transferred into the plasma membrane while other membrane molecules may be destroyed or transferred into the microenvironment (Morrè et al., 1979).

Although the cell surface molecules which serve as receptors for some bacterial viruses have been identified (Bassford et al., 1977), the chemical composition of receptors for animal viruses is largely undetermined. For both enveloped and non-enveloped animal viruses it seems likely that many cell receptor units are proteins or glycoproteins (Holmes, 1981; Boulanger and Philipson, 1981). A few glycoproteins present in large amounts in cell membranes have been isolated and characterized. Such glycoproteins are believed to act as receptors for viruses (Hughes, 1973, 1976; Hughes and Nain, 1978; Hennache and Boulanger, 1977; Marchesi and Andrews, 1971; Marchesi et al., 1976; Nakajo et al., 1979; Yamada and Olden, 1978). Methods have been devised for the analysis of glycoproteins present in limited amounts in cell membranes. Such procedures will aid in determining whether these glycoproteins also serve a receptor role for viruses (Kulczycki et al., 1979; Lotan and Nicolson, 1979; Vitetta et al., 1977).

Crowell, Landau, and Siak (1981) analyzed the pathogenesis of picornavirus receptors. They note that a number of studies with cultured cells have revealed that receptors are present on cells which are targets for virus replication, but absent on cells which are not susceptible. The authors advise that extrapolation of such data to intact organs *in vivo* is justified only when the karyotypic, histologic, and physiologic characteristics of the cultured cells are similar to their *in vivo* counterparts. Cultures of cells which more accurately reflect *in vivo* conditions are needed to analyze the factors which control expression of functionally active viral receptors. Such cultures might reveal a relationship between expression and defined stages in differentiation (Goldberg and Crowell, 1971; Chairez et al., 1978).

B. NATURE OF VIRUS ATTACHMENT PROTEINS

Bramhall and Wisnieski (1981) have reviewed the nature of the envelope possessed by many viruses. Enveloped viruses are coated with a lipid matrix of varying complexity. This is present in a bilayer and is an essential constituent since in some virus systems, treatment with lipid solvents, detergents, or lipase inactivates infectivity or other essential processes such as hemagglutination (Kuwert et al., 1968). With several groups of enveloped viruses it has been shown that the lipid composition of the virus reflects that of the plasma membrane from which the virus membrane is derived during virus assembly or other intimate processes such as budding (Klenk and Choppin, 1969; McSharry and Wagner, 1971; Renkonen et al., 1971). Differences noted in specific virion lipids from those in the host plasma membrane may be attributed to a number of factors: (1) the influence of viral proteins; (2) the absence of intimate processes between host cell and virus--some viruses do not bud from the plasma membrane; (3) environmental factors, e.g., the conditions for replication.

Magnetic resonance studies suggest that the viral lipid bilayer is generally more rigid than the host plasma membrane (Landsberger et al., 1973) and this feature is thought to be partially dependent on the relative cholesterol content of the two membranes (Lee et al., 1972). In addition to cholesterol, viral membrane proteins play some role in determining the fluidity of the membrane bilayer. In general, the distribution of proteins in viral envelopes appears to be more dense than in cellular membranes.

For nonenveloped viruses the conformation of the capsid polypeptides may play a role in the ability of the virion to attach to host cells. The ligand which binds to the cellular receptor unit is presumed to be some element of the capsid protein (Bramhill and Wisnieski, 1981). The corresponding component of an enveloped virus is normally a glycoprotein spike which projects from and is intimately associated with the envelope lipid matrix (Blough et al., 1977; Bramhall et al., 1979; Chen et al., 1971; Collins and Knight, 1978; Inuma. et al., 1971; Mountcastle et al., 1971; Mussgay et al., 1975). The glycoprotein spikes are virus specific. Even though the structure of their carbohydrate moieties is largely determined by the host cell (Klenk and Choppin, 1970), the amino acid sequence of the glycoproteins is specified by the genome (Compans and Choppin, 1975). The number of glycoproteins differs among viruses, with one glycoprotein in rhabdovirus, two in myxo- and paramyxoviruses, and as many as seven in pox virus (Scheid, 1981). Figure 1 illustrates some of these points.

The spike glycoproteins of most enveloped viruses, e.g., the paramyxo glycoproteins, are anchored in the lipid bilayer of the viral membrane by a hydrophobic portion of the protein. This mode of spike attachment has been inferred not only from the arrangement of the glycoproteins on the surface of the virion, but also by their solubility properties. They can be solubilized by non-ionic detergents such as Triton X-100 or NP-40, and on removal of the detergent, the proteins aggregate by hydrophobic regions into rosettelike clusters (Scheid et al., 1972; Shimizu et al., 1974). Studies with Sendai virus suggest that the glycoproteins extend through the entire depth of the lipid bilayer (Lyles, 1979). Peplomeric glycoproteins serving as virus attachment proteins have been isolated from a number of viruses including orthomyxoviruses (Collins and Knight, 1978), paramyxoviruses (Nagai et al., 1976; Scheid and Choppin, 1974), rhabdoviruses (Kelley et al., 1972), coronaviruses (Sturman et al., 1980), alphaviruses (Helenius and Soderlund, 1973; Simons et al., 1973), and retroviruses (Strand and August, 1976). (Refer to Fig. 1c.)

Meager and Hughes (1977) in their review on virus receptors summarize our present knowledge of the chemical nature of the components of the capsid (for nonenveloped viruses) and the envelope surface of viruses. These surface components are mainly proteins and possibly glycoproteins in nonenveloped viruses and glycoproteins and glycolipids in enveloped viruses. The virus surface, as the first component in virus attachment, is envisaged as an electrostatically charged surface, often covered with projections, composed of proteins, glycoproteins, and glycolipids, or combinations of these, arranged in a regular manner and having varying degrees of freedom of movement relative to one another. The virus surface should be regarded as being in a dynamic state rather than the static picture presented in electron micrographs.

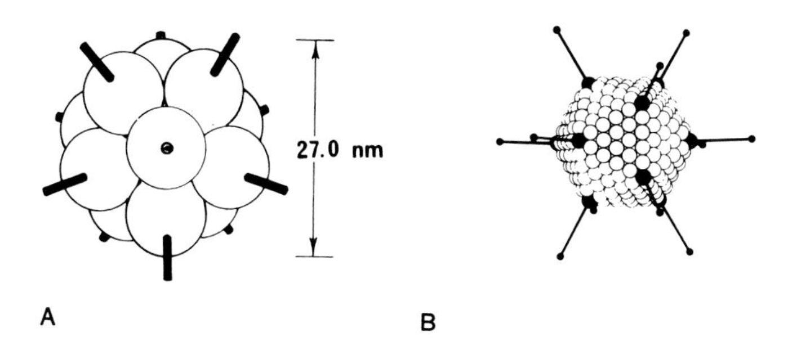

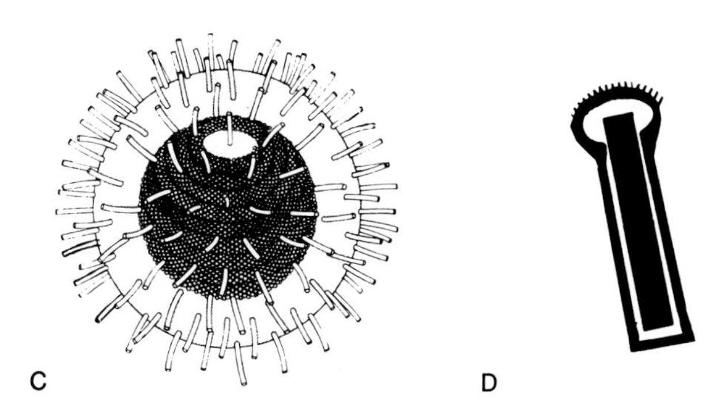

FIGURE 1. Models of viruses demonstrating surface projections. The projections are proteins and glycoproteins and are believed to be involved in the recognition between viruses and susceptible cells. As a result, the projections become attached to receptor sites on the cells. Models A, B, and C are from Horne (1974). Model D is based on reports by Adams et al. (1977) Kawamoto et al. (1977), and Hess and Falcon (1977). (A) Bacteriophage ϕX 174. The capsomeres have a spike projecting through the center. (B) Adenovirus capsid showing pentons, each with a base unit and a projecting shaft terminated by a knob. The projection is referred to as a fiber (refer to text). (C) The arrangement of components forming the spherical myxovirus structure. The surface envelope is covered with spikelike projections placed at regular spacings. The internal RNA helical nucleocapsid is coiled inside the outer envelope. (D) An enveloped nucleocapsid of a baculovirus (nuclear polyhedrosis virus). Spiked projections exist at one end only.

III. HOST CELL-VIRUS INTERACTIONS

Interactions between the host cell and virus involve three principal stages: attachment (adsorption), entry (penetration)--fusion or phagocytosis, and release--budding or lysis. Each stage, however, may include processes which overlap such divisions and most likely trigger the onset of the next stage. Such processes, for example, the assembly of virions and the acquisition of envelopes are intimately associated with events occurring between the penetration and release stages. Bramhall and Wisnieski (1981) have reviewed the principal stages of interaction between host cell and virus.

For enveloped viruses the location of the spike glycoproteins on the external surface of the virion predisposes them for a role in the early interaction between viruses and cells. As noted, the spike glycoproteins for most enveloped viruses are anchored in the lipid bilayer of the viral membrane by a hydrophobic portion of the protein. With several enveloped viruses, specific glycoproteins have been identified as being directly involved in the adsorption of the virus to the cell surface or in the penetration of the virus genome into the host cell (Scheid, 1981).

The nonenveloped animal viruses, in most cases, have icosahedral symmetry, with at least 12 identical sites for interaction with receptors on host cells. The multiple subunit structure of the virion capsid leads not only to multivalent receptor bonding, but may also confer allosteric properties to the virion (Boulanger and Lonberg-Holm, 1981).

A. ATTACHMENT (ADSORPTION)

In order to initiate infection a specific interaction takes place between viral attachment proteins and receptors on the cell surface. While cellular receptors serve to bind viruses specifically as the first event in infection, the exact location of receptor sites on the cell surface may determine the fate of the attached virions. Depending upon whether the virions attach to receptors on microvilli or the body of the cell, the virions may be processed differently (Roesing et al., 1975).

The cellular receptor unit for virus adsorption is not known in most virus systems. However, with myxoviruses and paramyxoviruses, which are examples of enveloped viruses, neuraminic acid has been identified as the attachment determinant. It has been shown that adsorption involved specific determinants on viral proteins that interact with neuraminic acid residues on the cell surface. Adsorption, as a prerequisite for infection, has been shown for several viruses to involve a specific host receptor, which is important for the cell, tissue, or organ tropism of the virus (Scheid, 1981).

Additional studies on the paramyxoviruses indicate that they possess HN glycoprotein spikes with both hemagglutination and neuraminidase activities (Seto et al., 1974). Electron microscopic observations have shown that prior to penetration, the viral envelope comes into close contact with the host plasma membrane. This attachment is mediated by the HN spike and the cellular receptor unit is believed to be a sialoglycoprotein or a ganglioside (Bramhall and Wisnieski, 1981).

As an example of a nonenveloped virus, adenoviruses are unique among animal viruses in their possession of apical projections which attach to cellular receptors (Boulanger and Lonberg-Holm, 1981). The adenovirus is an icosahedron, 70-90 nm in diameter, and is composed of 252 capsomeres. Of these, 240 have six neighbors and are called hexons. Each of the 12 apical capsomeres is surrounded by five neighboring hexons and is called a penton. The penton, 2 nm in diameter, is formed from a penton base unit and a projecting shaft terminated by a knob, 4 nm in diameter, called the fiber (Ginsberg et al., 1966; Horne et al., 1959; Valentine and Pereira, 1965; Wilcox et al., 1963). It is believed that the distal portion of the adenovirus fiber is the site of recognition for the cell receptors, and that this part of the fiber is attached to the cell plasma membrane at early stages of infection (Norrby and Skaaret, 1967). (Refer to Fig. 1b.)

For enveloped viruses it is unlikely that viral lipid composition plays any significant role in the attachment process. However, the host lipid composition may affect the ability of the cell to act as a target for viral attack, both from the point of view of the presence of receptors for attachment sites and also for the orientation and display of cellular receptors on the outer face of the plasma membrane (Bramhall and Wisnieski, 1981).

B. ENTRY (PENETRATION)

There are several ways for virions, particularly those which are enveloped in lipid membranes, to enter cells, including: fusion, phagocytosis, simple engulfment and partial fusion. Controversy surrounds the relative importance of phagocytosis and fusion events in the entry of membrane-enveloped viruses. For many such viruses evidence supports the concept of virus-host fusion (Heine and Schnaitman, 1971; Morgan and Rose, 1968; Morgan et al., 1968). However, host penetration by paramyxoviruses appears to be phagocytosis in some cell types (Hosaka and Koshi, 1968; Li et al., 1975), by fusion of virus and cell membranes (Dourmashkin and Tyrrell, 1970) or by a combination of fusion and phagocytosis (Morgan and Rose, 1968) in others. These differences probably result from variations in the lipid composition of the host membrane at the site of virus attachment (Haywood, 1975). In this regard, although viral lipids may be needed for fusion, they do not appear to have a modulating role at any stage in the infective process (Bramhall and Wisnieski, 1981).

Fusion of the Sendai virus envelope and the host plasma membrane allows the naked viral nucleocapsid to enter the cell cytoplasm (Morgan and Rose, 1968). It is apparent that prior to fusion of Sendai virus with host cell membrane, the viral envelope undergoes a dramatic change in structure (Knutton, 1976). Freeze-fracture studies of the viral membranes reveal a change in the orientation of transmembrane proteins together with a general rearrangement of viral membrane components to produce smooth elevated regions which appear to be initial sites of fusion with the target cell membrane. Membrane fusion occurred only after areas of two opposing bilayers were brought close enough to interact (Ahkong et al., 1975); the particle-denuded areas probably represent such areas (Deamer and Branton, 1967).

The capacity of viruses to fuse with cells has been attributed to various hypothetical fusion factors. Since lysolecithin induces fusion between cells when added exogenously (Cullis and Hope, 1978; Lucy, 1970), it has been postulated that viruses may possess a phospholipase activity which acts at the site of virus-cell contact, or that they may utilize virus-associated lysolecithin to effect fusion (Barbanti-Brodana et al., 1971).

Another hypothesis implicates cholesterol in the fusion process. It has been shown that an increase in host membrane cholesterol enhances fusion by Sendai virus, whereas a decrease in cholesterol depresses virus-mediated cell fusion (Hope et al., 1977). Akhong et al. (1975) suggest that cholesterol may enhance cell fusion by inducing protein-free areas in the lipid bilayer, which subsequently provides the sites for cell fusion. In addition, the physical state of the membrane hydrophobic phase may govern the extent of protein insertion, both processes being necessary for fusion.

Still another hypothesis is centered on the role of viral proteins in the fusion reaction. In some instances, notably the paramyxoviruses, it appears that membrane fusion is mediated by a viral glycoprotein. As with virus adsorption, the interaction between this protein and the host cell may determine cell, tissue, and organ tropism, possibly by a mechanism other than recognition of virus protein and cellular receptor (Scheid, 1981).

For nonenveloped viruses, Dales (1973) maintained that entry to the host cytoplasm is accomplished by a process of phagocytosis after attachment of the capsid to the cellular receptor sites. It is also possible that subsequent to attachment some nonenveloped viruses penetrate the plasma membrane directly, with or without being engulfed (Lonberg-Holm and Philipson, 1974).

Some enveloped animal viruses may be introduced into the cell by a mechanism related to endocytosis, and similar to the internalization of cellular lipoproteins (Philipson, 1981). The Semliki forest virus (SFV) is the model system for this type of penetration and internalization (Helenius et al., 1979). A detailed study of the penetration mechanism for SFV into BHK21 cells shows that the virus preferentially binds on microvilli and then probably migrates on the cell surface to the special structures referred to as coated pits, where epidermal growth factors and low-density lipoproteins are also bound. Subsequently, the virus is probably endocytosed, mainly in coated vesicles. The time the virus spends in the coated vesicles is short, and within minutes virus can be seen in larger uncoated vacuoles within the cytoplasm of the infected cells. These vacuoles may have arisen through fusion with coated vesicles. Some virus-containing vacuoles are ultimately transformed into secondary lysosomes. The final step of the penetration of SFV nucleoprotein containing the viral genome and the capsid protein probably occurs from the lysosome.

C. RELEASE

Many classes of lipid-enveloped viruses leave their host cells through a budding mechanism. A preliminary to budding is the appearance of viral-coded proteins in the plasma membrane of the host cell. These proteins, typically as with Newcastle disease virus and other paramyxoviruses, aggregate into localized clusters and are exposed at both the external and cytoplasmic faces of the plasma membrane. Viral capsids formed in the cytosol migrate toward these modified regions and the final interaction results in further membrane distortion and the ultimate budding of the capsid with the lipoprotein complex through the host membrane (Choppin et al., 1976; Compans et al., 1966). Complex variations on this general pattern exist (Bramhall and Wisnieski, 1981).

Electron microscopic studies have revealed that during the early stages of budding, the viral envelope appears as a continuation of the cellular membrane (Tooze, 1973). Later, both the nucleocapsid and the coat proteins appear to be present in extended membrane complexes which are mobile in the plane of the membrane (Dubois-Dalcq and Reese, 1975).

For other viruses such as herpes, the envelope is acquired at the inner nuclear membrane. During their development, nucleocapsids are seen close to protruding regions of the inner nuclear membrane that contain an underlying protein layer. These regions form the viral envelope. The enveloped virus travels within a vacuole from the perinuclear space probably to the cell surface, where it is released into the extracellular space by reverse phagocytosis (Rodriquez and Dubois-Dalcq, 1978). Other viruses, such as rabies, appear to form complete particles without cellular membrane involvement, i.e., prior to the arrival of any host cell membranes (Hummeler and Koprowski, 1969; Hummeler et al., 1967).

Assembly of virus particles is an integral part of cell-virus activity prior to release of virions. How the virion components arrive at the site of assembly is not well understood. Some viruses, such as paramyxo- and myxoviruses, togaviruses, and some rhabdoviruses, are known to use preexisting membranes as templates for assembly. The envelope proteins of such viruses may display specificity for particular fatty acyl chains during the transport and insertion process (Blough and Tiffany, 1973, 1975; Tooze, 1973; Dubois-Dalcq and Reese, 1975). Proteins that bind to the cytoplasmic side of the plasma membrane, such as those of the vesicular stomatitis virus, may assemble spontaneously after synthesis. These proteins may serve to attach the nucleocapsids to the virus-modified plasma membrane during the last stage of assembly (McSharry et al., 1975). The pox viruses are also assembled in the cytoplasm and appear to leave their host cell via the microvilli or similar specialized regions of the cell membrane (Stokes, 1976).

Most nonenveloped viruses commonly gain release from their host cells by promoting lysis of the host plasma membrane resulting in cell death. In contrast to release by budding, some enveloped viruses also gain their release from host cells following lysis of the host plasma membranes. The mechanism responsible for membrane damage is obscure. Cytocidal viruses appear to cause alterations in cellular membranes and, in many cases, have profound effects on host membrane lipid composition following infection (Blair and Brennan, 1972; Collins and Roberts, 1972; Pfefferkorn and Hunter, 1973; Poste, 1970). Many cytocidal viruses stimulate lipid metabolism in infected cells. Examples include adenoviruses (McIntosh et al., 1971), picornaviruses (Penman, 1965; Plagemann et al., 1970), paramyxoviruses (Gilbert, 1963), and pox viruses (Gaush and Younger, 1963). Exceptions have been reported for enveloped and nonenveloped viruses. The Sindbis virus causes a decrease in phospholipid synthesis (Pfefferkorn and Hunter, 1973). The temperature-sensitive mutants of SV40 virus inhibit phospholipid synthesis and this fact correlates with their ability to induce the release of cytoplasmic proteins from infected cells (Norkin, 1977).

IV. HOST CELL-BACULOVIRUS RELATIONSHIPS

A. THE INSECT CELL MEMBRANE

The plasma membrane of the insect cell, like that of other eukaryotic cells, is a complex and delicate structure. In their study on the fine structure of membranes and intercellular communication in insects, Satir and Gilula (1973) noted the fundamental similarity of the insect cell membrane to that of other animal cells. The limited information available suggests that with respect to composition, electron microscopic appearance (Smith, 1968), and permeability properties, including those for excitation of muscle and nerve (Usherwood, 1969), cellular membranes of insects fit the criteria for unit membranes (Robertson, 1969).

That the insect cell is capable of responding to foreign bodies present in its immediate environment has been demonstrated by numerous investigators. Insect blood cells, hemocytes, are extremely efficient at removing foreign particles, such as bacteria, fungi, and nematodes from the hemocoel, by either phagocytosis, nodule formation, or encapsulation (Salt, 1970).

For the insect cell, as for other eukaryotic cells, cell receptors have not been unequivocally identified in terms of detailed chemistry. In prokaryotic systems, such as the bacteriophage-bacterium system, the bacterial receptors for phage attachment are considered to be fixed (Meager and Hughes, 1977). In eukaryotic systems, cells are not only growing and dividing, but are also differentiating. Considerable changes occur in eukaryotic cells and accordingly, the cell membrane changes in composition, structure, and properties. Moreover, eukaryotic cells grown *in vitro* are not identical to those growing *in vivo*, and our understanding of receptors and cell-virus interactions in each situation may not coincide.

Receptor sites may have been identified for insect cell-baculovirus attachment. Tanada, Hess, and Omi (1975) in their study on infection of the armyworm by its NPV, reported that the surface of the microvilli of cells in the insect gut was covered with filaments, presumed to be composed of polysaccharide. Virus particles were frequently observed in contact with this coating; often, more than one location of the virus particles were in contact. In their study on NPV infection in the oriental tussock moth, *Euproctis subflava*, Kawamoto et al. (1977) noted that after cytoplasmic budding, enveloped nucleocapsids located on the basement membrane or freed in the hemocoel appeared to enter neighboring cells by phagocytosis (viropexis) with the spike end at the head. Dense materials were observed along the inner lining of the plasma membrane which had contact with the spikes of the entering virus particles.

B. PROPERTIES OF BACULOVIRUS NUCLEOCAPSIDS

In this section we consider properties of the baculoviruses which appear to be of importance in cell-virus recognition. What structural indicators are available which may be utilized for attachment of the virus to the cell? Does the biochemistry of baculovirus membranes play a part in initiating the process of infection?

The baculoviruses include the nucleopolyhedrosis viruses (NPVs) and the granulosis viruses (GVs). The nucleocapsids of both groups are rod shaped (baculo) and are subject to occlusion within paracrystalline inclusion bodies: polyhedra for the NPVs and capsules (granules) for the GVs. A third group of baculoviruses include rod-shaped nucleocapsids which are not occluded within inclusion bodies.

The nucleocapsids of the NPVs and the GVs are enveloped when contained in inclusion bodies. Ingested along with food, the inclusion bodies are broken down in the alkaline environment of the host's gut. During their cycle of infecting host cells, the nucleocapsids may exist with envelopes at various times and exist without envelopes at others. Much of our knowledge of the structural aspects of baculovirus nucleocapsids comes from electron microscopic observations on sectioned and alkali-dissolved inclusion bodies and infected cells.

On the basis of thin sectioning, Bergold (1963) described the structural features of the nucleocapsids of NPVs and GVs. The nucleocapsids of the NPVs exist as single rod forms or as groups of single virus particles, bundles. In each case an envelope, or in Bergold's terminology, a developmental membrane, surrounds the virus particles--the single forms and the bundles. The envelope or developmental membrane was reported by Bergold to be about 75 Å thick. Proceeding from the envelope inward, a space of about 60 Å and of lesser density than the envelope exists between it and an inner or intimate membrane. The inner membrane was reported to be about 40 Å thick. A layer of lesser density about 60 Å thick follows, and, finally, the central, dense corelike column of the virus proper with a diameter of about 300 Å. The average diameter of an enveloped nucleocapsid of the NPV of the silkworm, *Bombyx mori* L., was about 85 nm, and the average length was 330 nm. Cross and longitudinal sections through GV capsules revealed that there was almost exclusively one, and rarely two, virus particles in each capsule. Each GV nucleocapsid was also surrounded by a developmental and intimate membrane. The average diameter and length of an enveloped nucleocapsid within a capsule of the GV of *Cacoecia murinana* Hubner, was 47 Å × 260 Å. For the NPVs and GVs studied by Bergold (1963), there was no evidence of spherical or disk-shaped subunits, a central channel, or protrusions at one or either end of the nucleocapsid.

Subjecting inclusion bodies to weak alkali (Bergold, 1953) liberates the nucleocapsids. Alterations in morphology have been noted using this procedure. In many cases the developmental membrane becomes removed from the nucleocapsid and the inner membrane can be studied. Kozlov and Alexeenko (1967) in such studies concluded that the inner membrane was composed of subunits or capsomeres. They considered the inner membrane to be double layered and more dense than the developmental membrane. Teakle (1969) noted that virus particles from alkali-treated polyhedra of the NPV of *Anthela varia* possessed structures resembling a claw at each end and a nipplelike structure at one end. Other investigators have also occasionally observed protrusions from one or both ends of similarly prepared virus particles (Bird, 1957; Smith, 1962; Ponsen et al., 1965; Summers and Paschke, 1970; Mazzone and McCarthy, 1981). Among alkali-liberated nucleocapsids from polyhedra of NPVs, Harrap (1972a) observed enveloped, naked, and empty particles. He regarded the developmental membrane surrounding the nucleocapsid to be a three-layered virus envelope: an outer layer in which no detailed substructure could be resolved, a layer of hexagonally packed subunits referred to by Harrap as peplomers, 20 nm in diameter, and a flexible membrane or virus membrane composed of 4 nm subunits packed hexagonally. Moreover, Bergold's inner or intimate membrane was considered to be a capsid composed of subunits, 3 nm in diameter, arranged in a loose type of lattice. In this regard the notion of the inner membrane being a capsid has also been considered by other investigators (Hughes, 1958, 1972; Kozlov and Alexeenko, 1967; Arnott and Smith, 1968). In Harrap's observations, the capsid surrounds a core of internal component containing a central hole or channel, 10-15 nm in diameter. The less electron-dense region between the densely staining rod-shaped particle and the virus envelope seen in sectioned virus particles was believed to possibly represent condensed nucleoplasm of the virus-infected cell (Harrap, 1972b). From his studies, Harrap concluded that the capsid construction of the NPVs must be similar.

In a related study, Beaton and Filshie (1976) analyzed the capsid structure of two NPVs, that of *B. mori* and the cluster caterpillar, *Spodoptera litura* (F.), and two GVs, that of the potato tuberworm, *Phthorimaea operculla* (Zell) and the cabbage white butterfly, *Pieris rapae* (L.). Positive transparencies of electron micrographs of virus particles were prepared and analyzed by optical diffraction. Beaton and Filshie concluded, in agreement with Harrap (1972a), that the periodic lattice structure of the capsids of the two NPVs and the two GVs were indistinguishable. Each capsid was composed of stacked rings of subunits spaced 4.5 nm. They also were in agreement with Harrap (1972a) that the viral envelope was a triple-layered structure. To provide a more current terminology for baculovirus, Beaton and Filshie encouraged the use of the terms virus envelope and capsid for the older terms, developmental membrane and inner (intimate) membrane, respectively. They also pointed out that their results showing the close structural relationship of the capsids of NPVs and GVs supported Bellett's (1969) taxonomic observations on the close serological and genetic relationships between these two groups of baculoviruses.

The suggestion of peplomers in the envelope of the nucleocapsids serving as attachment proteins in infections (Harrap, 1972b; Harrap and Robertson, 1968; Summers, 1971) was substantiated by the work of Adams et al. (1975, 1977), Kawamoto et al. (1977), and Hess and Falcon (1977). They undertook an extensive electron microscope investigation on invasion and replication of insect NPVs *in vivo* and *in vitro*. The baculoviruses were observed possessing an envelope nucleocapsid with a peplomer structure restricted to one end of the envelope and specific to certain virus forms only (Fig. 1D). For the NPVs there appears to be two forms of baculovirus enveloped nucleocapsids (refer to the chapter on "Pathology Associated with Baculovirus Infection" by Mazzone in this volume). The invasive form is that of enveloped nucleocapsids present in nuclei and occluded in polyhedra. These nucleocapsids are involved in a primary infection and in the invasion events of entry of gut columnar cell microvilli. Adams and co-workers noted no special modification of the envelope of these nucleocapsids. However, the hemocoelic or spreading form of nucleocapsids are involved in: (1) secondary infection pathogenesis in insect cells other than gut columnar cells *in vivo*, and (2) in attachment, fusion, and penetration of insect cells *in vitro*. Whereas the invasive nucleocapsids acquire an envelope within the host cell nuclei, the hemocoelic type of nucleocapsids acquire an envelope upon budding through plasma membranes. Adams and co-workers believe that the envelope becomes modified with peplomers on one end of the nucleocapsids, presumably, by a virus-coded event.

The nature of the infective virion has been the subject of controversy, in terms of it requiring an envelope (Bergold, 1958; Summers and Volkman, 1976) or not (Bird, 1959; Stairs and Ellis, 1971; Kawarabata, 1974). The routes of infection may determine the necessity of a viral envelope. From their study on the NPV of the silkworm, Kawarabata and Aratake (1978) concluded that peroral infection is largely the result of an enveloped virion, the peroral infectious unit. Infection of cells in the hemocoel was largely the result of the virion without an envelope, the hemocoelic infectious unit. In the case of *in vitro* observations, non-enveloped virions are reported to be highly infectious to cell cultures (Raghow and Grace, 1974; Henderson et al., 1974; Knudson and Tinsley, 1974; Dougherty et al., 1975; Knudson and Harrap, 1976).

Summers and Volkman (1976) attempted to clarify the controversy of the infectious form of the nucleocapsid. They made biophysical and morphological comparisons of the infectious virions from insect hemolymph and from cell culture medium with virions derived from inclusion bodies after alkaline treatment. They concluded that the nucleocapsids from hemolymph and cell culture media were predominantly loosely fitting enveloped single nucleocapsids. These virus forms, from two different sources, were similar with regard to morphological and biophysical characteristics, but were quite different from the enveloped virus particles derived from alkali-treated polyhedra. Peplomers, observed on the surface of enveloped nucleocapsids from hemolymph and cell culture media, were not associated with polyhedra-derived virus. This study supports earlier reports of nucleocapsids with loose-fitting envelopes derived from hemolymph (Summers, 1971) and from cell culture media (Henderson et al., 1974).

Biochemically, the virus envelope of baculoviruses has been shown to have a phospholipid character with a role in infectivity. Yamamoto and Tanada (1977) working with the GV and NPV which infect the armyworm extracted fatty acids and phospholipids from polyhedra and capsules and from the isolated envelope virions from each type of inclusion body. Only the enveloped viruses from the GV and the NPV contained detectable phospholipids. Extracting fatty acids from the inclusion bodies with acetone and then inoculating them per os into armyworm larvae did not affect the infectivity of the viruses. Extracting phospholipids from inclusion bodies with chloroform-methanol and then inoculating them per os into armyworm larvae almost completely inactivated the infectivity of the viruses. The phospholipids extracted were believed to have originated from the enveloped viruses contained within the inclusion bodies (Yamamoto and Tanada, 1977).

The phospholipids extracted from the envelopes of the isolated viruses were identified as phosphatidylcholine, phosphatidylethanolamine, and an unidentified phospholipid (Yamamoto and Tanada, 1978b). A higher quantity of phosphatidylcholine was present in the enveloped virons of the NPV than in the GV. Isoelectric focusing of enveloped virions demonstrated that the total electric charge distributed on the surface of the envelopes of nucleocapsids was negative in neutral and alkaline solutions. Although there was little difference in charges between enveloped virions from the NPV and GV, the charge was less negative in the former than in the latter. When the charges were neutralized by cationic detergents, infectivity of the NPV was enhanced. Yamamoto and Tanada (1978b) hypothesized that phosphatidylcholine enhances the NPV infection by overcoming the negative potential of the viral envelope.

C. HOST CELL-BACULOVIRUS INTERACTIONS

The common route of infection for the baculoviruses is by the oral route into the host insect. The occluded baculoviruses used as viral insecticides are customarily sprayed as virus-containing inclusion bodies in an infested area at sometime just preceding or during the early stages of larval development. Insect larvae feeding on foliage contaminated with inclusion bodies, polyhedra, or capsules, ingest the virus material. In the gut the inclusion bodies are broken down by alkaline juices and most likely by enzymes, liberating free, enveloped virions. The liberated virions then interact with susceptible cells to cause infection.

For the baculoviruses, there are two phases of infection, a primary one occurring in the gut of the insect, and a secondary infection occurring in the tissues and organs in the hemocoel. Therefore, the three principal stages of infection, attachment (adsorption), entry (penetration), and release, for the baculoviruses, may be considered under two presumably different sets of circumstances within the host. This section will also note other processes which occur during baculovirus infection, such as virus assembly and envelope acquisition.

1. *Attachment (Adsorption)*

For the primary infection in the gut of the insect, enveloped virions released from inclusion bodies become attached to the cell membranes of the midgut microvilli (Harrap and Robertson, 1968; Harrap, 1969, 1970; Summers, 1969). Kawanishi et al. (1972) followed the infection of the NPV of *Rachiplusia ou* in the midgut columnar cells of the cabbage looper, *Trichoplusia ni*. They observed that enveloped viruses were adjacent to and closely appressed to the microvilli. The virions were not oriented in any specific manner to the microvilli, and both tip-to-tip and side-to-side interactions were observed. However, Adams and co-workers (1975, 1977) noted that in the invasion of tissue culture cells and target larval cells, the enveloped virions attached to the cell membranes in the area of the virus envelope which contained peplomers. The peplomers were believed to be only on one end of the virion.

In studies on NPV infection in the armyworm, Tanada et al. (1975) noted that the attachment of free enveloped virions to midgut cells was enhanced by a factor found in the capsule protein of the Hawaiian strain of a GV which was also infectious to the armyworm (Tanada and Hukhuhara, 1971; Tanada et al., 1973; Hara et al., 1976). The synergistic factor from the GV had a molecular weight of approximately 126,000, and contained polypeptides and phospholipids (Yamamoto and Tanada, 1978a). The phospholipid fraction appears to be essential for the enhancing activity. Yamamoto and Tanada (1978a) treated the synergistic factor with phospholipase C and with phospholipase A_2. Phospholipase C did not decompose the synergistic factor, but did destroy its capacity to enhance the NPV. In contrast, phospholipidase A_2 had no effect on the synergistic factor. The investigators believe that the different reactions of the two phospholipases on the synergistic factor indicates that the hydrophilic group of the phospholipid fraction was exposed to the action of phospholipase C and was associated with the synergistic activity.

For attachment events occurring within the host cell, Summers (1969) observed that in cabbage looper cells infected with a GV, 2-6 hours after infection, some virions were associated with the nuclear envelope in an apparent nonspecific manner. However, other virions appeared directly associated end-on with the nuclear pore. The attachment of capsids at the nuclear pore site in insect cells has also been reported in other studies, both *in vivo* (Summers, 1971; Kawanishi et al., 1972; Tanada and Hess, 1976) and *in vitro* (Raghow and Grace, 1974). Within the infected cells of the armyworm, Tanada and Hess (1976) noted that nucleocapsids associated with the virogenic stroma were usually aligned with one end either attached to or extremely close to the edge of the dense stroma. They point out that this observation had been previously described by other workers (refer to Bergold, 1958; Smith, 1967).

In the primary infection of gut columnar cells, the NPV nucleocapsids rarely take on polyhedra. At this point in the infection, the nucleocapsids do not have envelopes, a requisite for containment within inclusion bodies. However, on leaving the gut columnar cells, the nucleocapsids acquire envelopes. Robertson et al. (1974) point out that the nucleocapsids which become enveloped by the cell membrane play a specific role in the recognition of, and attachment to, susceptible sites on cell surfaces in the secondary infection. These nucleocapsids, therefore, cannot serve as sites for inclusion body deposition.

2. *Entry (Penetration)*

Fusion of the virus envelope of baculoviruses with plasma membranes of host cells has been reported in gut columnar cells of the microvilli of insects (Summers, 1969, 1971; Harrap, 1970). In such occurrences virus particles lacking envelopes were seen within the cells. Particles attached to the microvilli exhibited varying degrees of contiguity with the microvilli membranes. In some observations the juncture of the envelope and plasma membrane appeared indistinct while in others the two membrane elements were more distinctly confluent (Kawanishi et al., 1972). Single as well as bundles of nucleocapsids were encountered within microvilli at varying distances from the columnar cell body (Kawanishi and Pashke, 1970; Kawanishi et al., 1972). The virus envelope appeared to be lost at the cell surface since the virions were observed in the microvilli without the envelope.

Tanada et al. (1975) noted that virus particles also appeared to enter the cells at points other than the microvilli. Moreover, groups of single viruses from bundles appear to enter together into the cytoplasm of a microvillus.

Adams and co-workers (1975, 1977) in their extensive study on invasion and replication of baculoviruses *in vivo* and *in vitro* noted that fusion of virus envelope and plasma membrane occurred as one step of the infection process. However, entry of virus particles into *in vitro* cells probably occurred by phagocytosis. In this latter case, nucleocapsids were found in envelopes in vesicles and naked and in envelopes in the cytoplasm. Phagocytosis of many nucleocapsids was believed to involve lysosomes and/or microbodies. Those virus particles that reach the nuclear membrane may have gained entry by causing an inpouching and phagocytosis by the nuclear membrane in cell cultures, or through nuclear pores of cells *in vivo*. Virions may acquire an envelope within the nucleus but also from the nuclear membrane when exiting or from the plasma membrane at the time of release from the cell.

In infected hemocytes, Kislev et al. (1969) maintain that phagocytosis is the mechanism of penetration regardless of whether the virus is introduced into the body per os or by intrahemocoelic injection. Kislev et al. conducted electron microscopic studies on hemocytes of the Egyptian cottonworm, *Spodoptera littoralis* (Boisduval) infected with an NPV. Of the four major types of hemocytes differentiated in the blood of the insect, virus formation was found to occur mainly in the plasmatocytoids, and only to a much lesser extent in the granular hemocytes and oenocytoids. Plasmatocytoids were observed phagocytosing free virus particles as well as several whole polyhedra. Virus particles originating, presumably, from the nucleus of the cell were found in cytoplasmic extensions of infected plasmatocytoids. Kislev et al. noted many protuberances of the nuclear membrane of the hemocytes but did not report the budding of viruses from the nuclei.

From a study of baculovirus infection in the cabbage looper, Tanada and Leutenneger (1970) postulated an alternative major route of invasion of virus particles into the hemocoel. Virions released from polyhedra in the gut may move through the intercellular spaces of the gut columnar cells to the basal lamina and, ultimately, to the hemocoel. This mode of entry of virus particles into the hemocoel was also observed as occurring in the webbing clothes moth, *Tineola bisselliella*, by Hunter et al. (1973).

The uncoating or release of viral nucleic acid has been noted as occurring *in vivo* at the site of the nuclear pore of cells. Summers (1969) suggested that the virus genome was passed into the nucleus without the virions entering the nuclear region, thus resembling a mechanism similar to bacteriophage infection of bacteria. Uncoating was also stated as occurring by this procedure for infected cells *in vitro* (Raghow and Grace, 1974; Hirumi and Hirumi, 1975).

Adams et al. (1975, 1977) observed no peplomers on those virions which replicate in the nucleus and are occluded in polyhedra. It was uncertain whether these nucleocapsids in acquiring an envelope from the nuclear membranes had peplomers upon exit from the cells. The peplomer morphology was most evident on those nucleocapsids passing through the plasma membrane. These investigators believe that occluded virions may differ from nonoccluded virions as to the existence or timing of peplomer formation. Moreover, the peplomers may differ biochemically, perhaps being triggered upon release from the polyhedron prior to invasion of the microvilli of the gut.

The assembly of baculoviruses within the cytoplasm of insect cells has been observed in a number of studies (Huger,

1963; Arnott and Smith, 1968; Summers, 1971; Falcon and Hess, 1977). In cells infected with the NPV of the alfalfa looper, *Autographa californica* (Speyer), the cytoplasm-located viruses were frequently found in more than one cell in the same area. Interspersed with cell organelles such as mitochondria, glycogen, and rough endoplasmic reticulum were numerous vesicular membrane profiles which occurred much more frequently than any observed in the nucleus associated with virus development. These membranes were believed to be used in enveloping nucleocapsids (Falcon and Hess, 1977).

Nucleocapsids may acquire envelopes by a number of processes including: (1) *de novo* morphogenesis in the nucleus (Summers, 1971; Harrap, 1972 ; Adams et al., 1977); (2) from the inner layer of the nuclear membrane (Summers, 1971; Hughes, 1972; Stoltz et al., 1973; MacKinnon et al., 1974; Tanada and Hess, 1976); (3) budding from the nuclear envelope (Injac et al., 1971; MacKinnon et al., 1974; Nappi and Hammill, 1975; Adams et al., 1977); (4) *de novo* synthesis in the cytoplasm (Robertson et al., 1974); (5) from the endoplasmic reticulum membrane (Injac et al., 1971; Smith, 1971; MacKinnon et al., 1974); and (6) budding through the cell membrane (Summers, 1971; Robertson et al., 1974; Adams et al., 1977).

Kawamoto et al. (1977) followed the acquisition of envelopes by nucleocapsids via three processes in the oriental tussock moth: (1) *de novo* morphogenesis in the nucleus; (2) nuclear budding; and (3) cytoplasmic budding. The direction of nucleocapsids in the envelopes was the same in the three modes of envelopment. The envelopment seemed to occur from a nipple end which was at one end of the nucleocapsid.

In the latter two modes of envelope acquisition, nucleocapsids wrapped by these ways are not occluded and seemed to be released into extracellular space, such as, the hemocoel of insects or the culture medium of *in vitro* systems.

After the envelopment by the three processes, electron-dense materials were observed between the envelope and the nucleocapsid, although the contents and morphological features differed among the three types of envelopes. The authors believe that these materials may function similarly as mediator between the envelope and nucleocapsid, as have been observed in many vertebrate viruses which acquire envelopes. The function of the electron-dense material in the NPVs seems to correspond to that of the tegument of herpesviruses (Roizman and Furlong, 1974) or the membrane protein of myxo- and paramyxoviruses (Schulze, 1972; Yoshida et al., 1976).

A marked difference among the three types of envelopes was the characteristic cap-shaped structures with spikes which were seen only on the surface of the envelope derived from the plasma membrane (Fig. 1D). After cytoplasmic budding, nucleocapsids enveloped in this manner were located on the basement membrane or liberated in the hemocoel. They then appeared to enter neighboring healthy cells via phagocytosis (viropexis) with the spike end at the head. At the sites where these spikes came into contact with healthy cells, coated, vesicle-like structures were observed inside the plasma membrane. Occasionally, incomplete particles which lacked nucleocapsids were also budding through the plasma membrane and released into the extracellular space.

Release

Nucleocapsids may exit from infected nuclei by outpouching of the nuclear membrane and pinching off. Enveloped virions and nucleocapsids may exit through nuclear pores or ruptured areas in the nuclear membrane. Those nucleocapsids in vesicles that escape the cells' protective defenses may be ejected intact through the plasma membrane. Those nucleocapsids which reach the plasma membrane naked acquire an envelope as they pass through (Adams et al., 1975). At this stage some investigators have reported the modification of the anterior cap of the envelope with peplomers. This modification was also noted as occurring in the columnar cells of the insect gut as replicated nucleocapsids passed through the basement membrane (Adams et al., 1975).

As noted above, envelopes may not be needed for invasion of the virion into susceptible cells in the hemocoel (Stairs and Ellis, 1971; Kawarabata, 1974; Tanada and Hess, 1976). How nucleocapsids enter the hemocoel without envelopes is not clear. Harrap and Robertson (1968) suggest that virus particles released from the nucleus move along a gradient between the nucleus and its basement membrane. Summers (1969, 1971) reported virus particles engulfed in vesicles of unknown origin in the cytoplasm and suggested that they were transplanted by this means through the gut columnar cells to the basement membrane. Tanada and Leuteneger (1970) reported enveloped and unenveloped nucleocapsids in the endoplasmic reticulum and intercellular space and in the basement membrane, and they suggested that such nucleocapsids entered the hemocoel through this pathway. Hunter et al. (1973) similarly suggested that the virus particles moved through the intercellular space into the hemocoel.

The process of budding is of interest not only as a mode of viral release but also because it represents a nucleocapsid-plasma membrane interaction by which nucleocapsids acquire their envelopes. Nappi and Hammill (1975) studied the envelope acquisition of viral particles of the NPV of the gypsy moth (*Lymantria dispar*, Linnaeus) in host hemocytes. The most apparent pathological change was the development of numerous protrusions or buds of the nuclear membrane, many of which contained naked virus particles. The process of budding which was expressed by such protrusions involved an extension of both inner and outer lamellae of the nuclear membrane. The exact process by which viruses emerged from infected gypsy moth hemocytes was suggested as involving vesicles in cytoplasmic extensions of the hemocytes. The vesicles were formed as extensions of the nuclear envelope that had been pinched off into the cytoplasm. Enveloped nucleocapsids were observed as present in many of the vesicles. The nucleocapsids were believed to be transported within the vesicles to the periphery of the cell and released by exocytosis. During this process, the vesicular membrane, i.e., the outer lamellae of the nuclear envelope, could fuse with the plasma membrane to liberate the enveloped nucleocapsids into the hemolymph. This was believed to be one mode of transmission of the virus from cell to cell during the early stages of infection before the rupture of the nuclear membrane.

V. CONCLUSIONS

The baculovirus particles are considered to be nucleocapsids, which for the greater part of the infection cycle are surrounded by envelopes. In this regard, the older terms of intimate (inner) and developmental membranes to describe the structures surrounding the naked particle have been replaced by capsid and envelope, respectively (Harrap, 1972a; Beaton and Filshie, 1976). These changes in terminology give a somewhat better perception of how the baculovirus particle may fit into a concept of virus attachment proteins recognizing or being recognized by receptor sites on susceptible cells.

In its activities, expecially in lepidoptera, a baculovirus is generally involved in two types of infections (Harrap and Robertson, 1968; Harrap, 1969, 1970; Summers, 1969). After ingestion of inclusion bodies by an insect larva and liberation of enveloped nucleocapsids in the host gut, the virions commence a nonlethal infection of gut columnar cells. This infection involves one series of reactions between virus and cell, including attachment, entry, and release of enveloped virus particles. After this initial series of events, nucleocapsids reach the hemocoel by pathways which are not adequately delineated to engage in secondary, but lethal, infections of cells. In the hemocoel, a second series of reactions ensues involving, again, attachment, entry, and release of enveloped virus particles.

In terms of virus attachment proteins, the attachment reaction in the host gut is less clearly defined than is the attachment of virions to cells in the hemocoel. In the latter situation glycoprotein spikes in the form of peplomers are contained on one end of the enveloped nucleocapsid. Such structures are believed to be similar to the glycoprotein spikes of vertebrate viruses, serving the function of attaching the enveloped nucleocapsid to the cell (Adams et al., 1975, 1977). However, the virus receptor concept also requires sites on the cell which serve to bind the virus particles. In the primary infection occurring in the larval gut, receptor elements may have been identified on host cells. On the surface of the microvillus, filaments have been observed to which virus particles are attached (Tanada et al., 1975). Moreover, in the secondary infection occurring in cells of organs and tissues in the hemocoel, receptor sites may also have been observed. Dense materials have been noted along the inner lining of the plasma membrane which are in contact with spikes of the entering particles (Kawamoto et al., 1977).

Attachment of an NPV nucleocapsid to susceptible cells of the armyworm appears to be enhanced by a synergistic factor from a GV which also infects the insect (Tanada et al., 1975). The synergistic factor with an estimated molecular weight of 126,000 contains polypeptides and phospholipid (Hara et al., 1976; Yamamoto and Tanada, 1978a). The phospholipid is intimately related to the enhancement which is abolished by the action of phospholipase C but not phospholipase A_2. The different reactions of the two phospholipases on the synergistic factor suggested that the hydrophylic group of the phosphophipid was exposed to the action of phospholipase C and was associated with synergistic activity (Yamamoto and Tanada, 1978a). Attachment of the NPV virus particles also appears to be enhanced by a phospholipid, phosphatidylcholine, in the envelope of the nucleocapsid. Phosphatidylcholine is believed to enhance NPV infection by overcoming the negative potential of the viral envelope, thus increasing the attachment of virus to cell (Yamamoto and Tanada, 1978b).

In the penetration of the virus particle into the cell both fusion and phagocytosis have been observed. Phagocytosis appears to be especially frequent in hemocytes (Kislev et al., 1969) and in cultured cells (Raghow and Grace, 1974; Knudson and Tinsley, 1974; Hirumi and Hirumi, 1975; Knudson and Harrap, 1976). Injection of viral nucleic acid into the cell in a bacteriophage-bacterium type of reaction has been reported as occurring at the site of the nuclear pore of cells (Summers, 1969). Penetration does not appear to require any specific orientation of the virus particle to the cell. Thus, virus particles have been observed in their apposition to cells as side-to-side and tip-to-tip (Kawanishi et al., 1972); in other cases, only the end of the virus bearing peplomers, presumably glycoprotein spikes, appeared to attach to cells (Adams et al., 1975, 1977).

The mechanism of release of viral particles from the nucleus or from the plasma membrane in secondary infections results in the acquisition of envelopes for virus particles (Nappi and Hammill, 1975; Adams et al., 1977). The envelope acquired by the nucleocapsid in budding through the plasma membrane becomes modified with peplomers on one end. However, the nucleocapsids which acquire an envelope by *de novo* synthesis in the nucleus and are subsequently occluded within inclusion bodies do not have a peplomeric morphology on their envelopes (Adams et al., 1977). The peplomeric morphology was further delineated by Kawamoto et al. (1977). In the acquisition of envelopes by baculovirus nucleocapsids by *de novo* synthesis in the nucleus, by nuclear budding, and by cytoplasmic budding through the plasma membrane, the characteristic cap-shaped structure with spikes, at one end of the nucleocapsid, was seen only in the surface of the envelope derived from the plasma membrane.

While the present report brings out some strong support for a virus receptor concept in the baculovirus-host cell system, a number of observations require further elucidation. Since glycoprotein attachment units have not been reported for nucleocapsids which attach to gut columnar cells, what binding mechanisms do occur? Entry of the virus particles in the primary infection is reported to occur by fusion. Are the attachment proteins of the nucleocapsids the short projections observed on some virus particles in the gut (Harrap and Robertson, 1968; Adams et al., 1977) or the plaque structures referred to by Summers (1971)?

If the envelope is not required for virus particles to infect cells in the hemocoel (Bird, 1959; Stairs and Ellis, 1971; Kawarabata, 1974; Kawarabata and Aratake, 1978) or cells in culture (Dougherty et al., 1975), what is the mechanism of attachment of the naked capsid to the cell? Are there two functionally different forms of infectious units, i.e., enveloped nucleocapsids versus nonenveloped nucleocapsids? In this connection, one should note at this time that in some hymenoptera only midgut cells are infected by baculoviruses which results in the death of the hosts.

From the discussion presented it is clear that baculoviruses present many interesting and challenging topics for research. Elucidation of the virus receptor concept for the baculoviruses is one area of investigation which is equally as complex in terms of structure and function as any that is being pursued in other cell-virus systems.

REFERENCES

Adams, J. R., Goodwin, R. H., and Wilcox, T. A. (1975). Electron microscopic investigations on invasion and replication of insect viruses *in vivo* and *in vitro* (abstract). *Proc. 7th Annu. Meeting Soc. Invertebr. Pathol.* Oregon State University, Corvallis, Oregon.

Adams, J. R., Goodwin, R. H., and Wilcox, T. A. (1977). Electron microscopic investigations on invasion and replication of insect baculoviruses *in vivo* and *in vitro*. *Biol. Cell. 28*, 261-268.

Ahkong, Q. F., Fisher, D., Tampion, W., and Lucy, J. A. (1975). Mechanisms of cell fusion. *Nature (London) 253*, 194-195.

Arnott, H. J., and Smith, K. M. (1968). An ultrastructural study of the development of a granulosis virus in the cells of the moth *Plodia interpunctella* (Hbn.). *J. Ultrastr. Res. 21*, 251-268.

Barbanti-Brodana, G., Possati, L., and LaPlaca, M. (1971). Inactivation of polykaryocytogenic and hemolytic activities of Sendai virus by phospholipase B (lysolecithinase). *J. Virol. 8*, 796-800.

Bassford, P. J., Jr., Diedrich, D. L., Schnaitman, C. L., and Reeves, P. (1977). Outer membrane proteins of *Escherichia coli*. VI. Protein alteration in bacteriophage-resistant mutants. *J. Bacteriol. 131*, 608-622.

Beaton, C. D., and Filshie, B. K. (1976). Comparative ultrastructural studies of insect granulosis and nuclear polyhedrosis. *J. Gen. Virol. 31*, 151-161.

Bellett, A. J. D. (1969). Relationships among the polyhedrosis and granulosis viruses of insects. *Virology 37*, 117-123.

Bergold, G. H. (1953). Insect viruses. *Advan. Virus Res. 1*, 91-139.

Bergold, G. H. (1958). Viruses of insects. *In* Handbuch der Virusforschung" (C. Hallauer and K. F. Meyer, eds.), Vol. 4, pp. 60-142. Springer, Vienna.

Bergold, G. H. (1963). Fine structure of some insect viruses. *J. Insect Pathol. 5*, 111-128.

Bird, F. T. (1957). On the development of insect viruses. *Virology 3*, 237-242.

Bird, F. T. (1959). Polyhedrosis and granulosis viruses causing single and double infections in the spruce budworm, *Choristoneura fumiferana* (Clemens). *J. Insect Pathol. 1*, 406-430.

Blair, C. D., and Brennan, P. J. (1972). Effect of Sendai virus infection on lipid metabolism in chick fibroblasts. *J. Virol. 9*, 813-822.

Blough, H. A., and Tiffany, J. M. (1973). Lipids in viruses. *Advan. Lipid Res. 11*, 267-339.

Blough, H. A., and Tiffany, J. M. (1975). Theoretical aspects of structure and assembly of viral envelopes. *Curr. Topics Microbiol. Immunol. 70*, 1-30.
Blough, H. A., Tiffany, J. M., and Aaslestad, H. G. (1977). Lipids of rabies virus and BHK-21 cell membranes. *J. Virol. 21*, 950-955.
Boulanger, P., and Lonberg-Holm, K. (1981). Components of non-enveloped viruses which recognize receptors. *In* "Virus Receptors, Part 2: Animal Viruses" (K. Lonberg-Holm and L. Philipson, eds.), pp. 21-46. Series B, Vol. 8 of "Receptors and Recognition" (P. Cuatrecasas and M. F. Greaves, series eds.). Chapman and Hall, London.
Boulanger, P., and Philipson, L. (1981). Membrane components interacting with non-enveloped viruses. *In* "Virus Receptors, Part 2: Animal Viruses" (K. Lonberg-Holm and L. Philipson, eds.), pp. 117-139. Series B, Vol. 8 of "Receptors and Recognition" (P. Cuatrecasas and M. F. Greaves, series eds.). Chapman and Hall, London.
Bramhall, J., and Wisnieski, B. (1981). The role of lipids in virus-cell interactions. *In* "Virus Receptors, Part 2: Animal Viruses" (K. Lonberg-Holm and L. Philipson, eds.), pp. 141-153. Series B, Vol. 8 of "Receptors and Recognition" (P. Cuatrecasas and M. F. Greaves, series eds.). Chapman and Hall, London.
Bramhall, J. S., Shiflett, M. A., and Wisnieski, B. J. (1979). Mapping the membrane proteins of Newcastle-Disease virus with a photoreactive glycolipid probe. *Biochem. J. 177*, 765-768.
Chairez, R., Yoon, J-W., and Notkins, A. L. (1928). Virus-induced diabetes mellitus. X. Attachment of encephalomyocarditis virus and permissiveness of cultured pancreatic B cells to infection. *Virology 85*, 606-611.
Chen, C., Compans, R. W., and Choppin, P. W. (1971). Parainfluenza virus surface projections: Glycoproteins with hemagglutinin and neuraminidase activities. *J. Gen. Virol. 11*, 53-58.
Choppin, P. W., Klenk, H-D., Compans, R. W., and Caliguiri, L. A. (1976). *In* Perspectives in Virology (M. Pollard, ed.), pp. 127-156. Academic Press, New York.
Collins, J. K., and Knight, C. A. (1978). Purification of the Influenza hemagglutinin glycoprotein and characterization of its carbohydrate components. *J. Virol. 26*, 457-467.
Collins, F. D., and Roberts, W. K. (1972). Mechanism of mengo virus-induced cell injury in L cells: Use of inhibitors of protein synthesis to dissociate virus-specific events. *J. Virol. 10*, 969-978.
Compans, R. W., and Choppin, P. W. (1975). Reproduction of myxoviruses. *In* "Comprehensive Virology" (H. Fraenkel-Conrat and R. R. Wagner, eds.), pp. 179-252. Plenum Press, New York.

Compans, R. W., Holmes, K. V., Dales, S., and Choppin, P. W. (1966). An electron microscopic study of moderate and virulent virus-cell interactions of the Parainfluenza virus SV 5. *Virology 30*, 411-426.

Crowell, R. L., Landau, B. J., and Siak, J.-S. (1981). Picornavirus receptors in pathogenesis. *In* "Virus Receptors, Part 2: Animal Viruses" (K. Lonberg-Holm and L. Philipson, eds.), pp. 169-184. Series B, Vol. 8 of Receptors and Recognition (P. Cuatrecasas and M. F. Greaves, Series eds.). Chapman and Hall, London.

Cullis, P. R., and Hope, M. J. (1978). Effects of fusogenic agent on membrane structure of erythrocyte ghosts and the mechanism of membrane fusion. *Nature (London) 271*, 672-674.

Dales, S. (1973). Early events in cell-animal virus interaction. *Bacteriol. Rev. 37*, 103-135.

Deamer, D. W., and Branton, D. (1967). Fracture planes in an ice-bilayer model membrane system. *Science 158*, 655-657.

Dougherty, E. M., Vaughn, J. L., and Reichelderfer, C. F. (1975). Characteristics of the non-occluded form of a nuclear polyhedrosis virus. *Intervirology 5*, 109-121.

Dourmashkin, R. R., and Tyrrell, D. A. J. (1970). Attachment of two myxoviruses to ciliated epithelial cells. *J. Gen. Virol. 9*, 77-88.

Dubois-Daloq, M., and Reese, T. S. (1975). Structural changes in the membrane of vero cells infected with a paramyxovirus. *J. Cell Biol. 67*, 551-565.

Falcon, L. A., and Hess, R. T. (1977). Electron microscope study on the replication of Autographa nuclear polyhedrosis virus and Spodoptera nuclear polyhedrosis virus in *Spodoptera exigua*. *J. Invertebr. Pathol. 29*, 36-43.

Gaush, C. R., and Younger, J. S. (1963). Lipids of virus infected cells. II. Lipid analysis of HeLa cells infected with vaccinia virus. *Proc. Soc. Exp. Biol. Med. 112*, 1082-1085.

Gilbert, V. E. (1963). Enzyme release from tissue culture as an indicator of cellular injury. *Virology 21*, 609-616.

Ginsberg, H. S., Pereira, H. G., Valentine, R. C., and Wilcox, W. C. (1966). A proposed terminology for the adenovirus antigens and virion morphological subunits. *Virology 28*, 782-783.

Goldberg, R. J., and Crowell, R. L. (1971). Susceptibility of differentiating mouse cells of the fetal mouse in culture to coxsackievirus A13. *J. Virol. 7*, 759-769.

Granados, R. R., and Lawler, K. A. (1981). *In vitro* pathway of *Autographa californica* baculovirus invasion and infection. *Virology 108*, 297-308.

Hara, S., Tanada, Y., and Omi, E. M. (1976). Isolation and characterization of a synergistic enzyme from the capsule of a granulosis virus of the armyworm, *Pseudaletia unipuncta*. *J. Invertebr. Pathol. 27*, 115-124.

Harrap, K. A. (1969). Viruses of invertebrates. *Proc. 1st Intern. Congr. Virol., Helsinki, 1968*, p. 281.
Harrap, K. A. (1970). Cell infection by a nuclear polyhedrosis virus. *Virology 42*, 311-318.
Harrap, K. A. (1972a). The structure of nuclear polyhedrosis viruses. II. The virus particle. *Virology 50*, 124-132.
Harrap, K. A. (1972b). The structure of nuclear polyhedrosis viruses. III. Virus assembly. *Virology 50*, 133-139.
Harrap, K. A., and Robertson, J. S. (1968). A possible infection pathway in the development of a nuclear polyhedrosis. *J. Gen. Virol. 3*, 221-225.
Haywood, A. M. (1975). "Phagocytosis" of Sendai virus by model membranes. *J. Gen. Virol. 29*, 63-68.
Heine, J. W., and Schnaitman, C. A. (1971). Entry of vesicular stomatitis virus into L cells. *J. Virol. 8*, 786-795.
Helenius, A., and Soderlund, H. (1973). Stepwise dissociation of the Semliki Forest virus membrane with Triton X-100. *Biochim. Biophys. Acta 307*, 287-300.
Helenius, A., Morein, B., Fries, E., Simons, K., Robinson, P., Schirrmacher, V., Terhorst, C., and Strominger, J. L. (1978). Human(HLA-A and HLA-B) and murine (H-2K and H-2D) histocompatibility antigens are cell surface receptors for Semliki Forest virus. *Proc. Nat. Acad. Sci. U.S. 75*, 3846-3850.
Henderson, J. F., Faulkner, P., and MacKinnon, E. A. (1974). Some biophysical properties of virus present in tissue cultures infected with the nuclear polyhedrosis virus of *Trichoplusia ni*. *J. Gen. Virol. 22*, 143-146.
Hennache, B., and Boulanger, P. (1977). Biochemical study of KB-cell receptor for adenovirus. *Biochem. J. Cell. Aspects 166*, 237-247.
Hess, R. T., and Falcon, L. A. (1977). Observations on the interaction of baculovirus with the plasma membrane. *J. Gen. Virol. 36*, 525-530.
Hirumi, H., and Hirumi, K. (1975). Morphogenesis of a nuclear polyhedrosis virus of the cabbage looper in a continuous cabbage looper cell line. *Ann. N.Y. Acad. Sci. 266*, 302-326.
Holmes, K. V. (1981). The biology and biochemistry of cellular receptors for enveloped viruses. *In* "Virus Receptors, Part 2: Animal Viruses" (K. Lonberg-Holm and L. Philipson, eds.), pp. 85-115. Series B, Vol. 8 of "Receptors and Recognition" (P. Cuatrecasas and M. F. Greaves, series eds.). Chapman and Hall, London.
Hope, M. J., Bruchdorfer, K. R., Hart, C. A., and Lucy, J. A. (1977). Membrane cholesterol and cell fusion of hen and quinea-pig erythrocytes. *Biochem. J. Cell. Aspects 166*, 255-263.
Horne, R. W. (1974). "Virus Structure." Academic Press, New York.
Horne, R. W., Brenner, S., Waterson, A. P., and Wildy, P.

(1959). The icosahedral form of an adenovirus. *J. Mol. Biol. 1*, 84-86.

Hosaka, Y., and Koshi, Y. (1968). Electron microscopic study of cell fusion by HVJ virions. *Virology 34*, 419-434.

Huger, A. (1963). Granuloses of insects. *In* "Insect Pathology" (E. A. Steinhaus, ed.), Vol. I, pp. 531-575. Academic Press, New York.

Hughes, K. M. (1958). The question of plurality of virus particles in insect virus capsules and an attempt at clarification of insect virus terminology. *Trans. Amer. Microscop. Soc. 77*, 22-30.

Hughes, K. M. (1972). Fine structure and development of two polyhedrosis viruses. *J. Invertebr. Pathol. 19*, 198-207.

Hughes, R. C. (1973). Glycoproteins as components of cellular membranes. *Progr. Biophys. Mol. Biol. 26*, 189-268.

Hughes, R. C. (1976). "Membrane Glycoproteins: A Review of Structure and Function." Butterworths, London.

Hughes, R. C., and Nain, C. (1978). Labeling, solubilization, and peptide mapping of fibroblast surface glycoprotein. *Ann. N.Y. Acad. Sci. 312*, 192-206.

Hunter, D. K., Hoffman, D. K., and Collier, S. J. (1973). The histology and ultrastructure of a nuclear polyhedrosis virus of the webbing clothes moth, *Tineola bisselliolla. J. Invertebr. Pathol. 21*, 91-100.

Hummeler, K., Koprowski, H., and Witkor, T. J. (1967). Structure and development of rabies virus in tissue culture. *J. Virol. 1*, 152-170.

Hummeler, K., and Koprowski, K. (1969). Investigating the rabies virus. *Nature (London) 221*, 418-421.

Incardona, N. L. (1981). The chemical nature of virus-receptor interactions. *In* "Virus Receptors, Part 2: Animal Viruses" (K. Lonberg-Holm and L. Philipson, eds.), pp. 155-168. Series B, Vol. 8 of "Receptors and Recognition" (P. Cuatrecasas and M. F. Greaves, series eds.). Chapman and Hall, London.

Linuma, M., Yoshida, T., Nagai, Y., Maeno, K., and Matsumoto, T. (1971). Subunits of NDV. Hemagglutinin and Neuraminidase subunits of Newcastle Disease Virus. *Virology 46*, 663-677.

Injac, M., Vago, C., Duthoit, J. L., and Veyrunes, J. C. (1971). Liberation (release) des virions dans les polyhedrosis nucleaires. *C.R. Acad. Sci. D273*, 439-441.

Kawamoto, F., Suto, C., Kumada, N., and Kobayashi, M. (1977). Cytoplasmic budding of a nuclear polyhedrosis virus and comparative ultrastructural studies of envelopes. *Microbiol. Immunol. 5*, 255-265.

Kawanishi, C. Y., and Paschke, (1970). Density gradient centrifugation of the virions liberated from *Rachiplusia ou* nuclear polyhedra. *J. Invertebr. Pathol. 16*, 89-92.

Kawanishi, C. Y., Summers, M. D., Stoltz, D. B., and Arnott, H. J. (1972). Entry of an insect virus *in vivo* by fusion of viral envelope and microvillus membrane. *J. Invertebr. Pathol. 20*, 104-108.

Kawarabata, T. (1974). Highly infectious free virions in the hemolymph of the silkworm (*Bombyx mori*) infected with a nuclear polyhedrosis virus. *J. Invertebr. Pathol. 24*, 196-200.

Kawarabata, T., and Aratake, Y. (1978). Functional differences between occluded and nonoccluded viruses of a nuclear polyhedrosis of the silkworm, *Bombyx mori*. *J. Invertebr. Pathol. 31*, 329-336.

Kelley, M. J., Emerson, S. U., and Wagner, R. R. (1972). The glycoprotein of vesicular stomatitis virus is the antigen that gives rise to and reacts with neutralizing antibody. *J. Virol. 10*, 1231-1235.

Kislev, N., Harpaz, I., and Zelcer, A. (1969). Electron microscopic studies on hemocytes of the Egyptian cottonworm, *Spodoptera littoralis* (Boisduval) infected with a nuclear polyhedrosis virus, as compared to noninfected hemocytes. II. Virus-infected hemocytes. *J. Invertebr. Pathol. 14*, 245-257.

Klenk, H.-D., and Choppin, P. W. (1969). Lipids of plasma membranes of monkey and hamster kidney cells and of parainfluenza virions grown in these cells. *Virology 38*, 255-268.

Klenk, H.-D., and Choppin, P. W. (1970). Plasma membrane lipids and parainfluenza virus assembly. *Virology 40*, 939-947.

Knudson, D. L., and Harrap, K. A. (1976). Replication of a nuclear polyhedrosis virus in a continuous cell culture of *Spodoptera frugiperda*. Microscopy study of the sequence of events of the virus infection. *J. Virol. 17*, 254-268.

Knudson, D. L., and Tinsley, T. W. (1974). Replication of a nuclear polyhedrosis virus in a continuous cell culture of *Spodoptera frugiperda*. Purification, assay of infectivity, and growth characteristics of the virus. *J. Virol. 14*, 934-944.

Knutton, S. (1976). Changes in viral envelope structure preceding infection. *Nature (London) 264*, 672-673.

Korn, E. D. (1975). Biochemistry of endocytosis. *In* "Biochemistry of Cell Walls and Membranes" (C. F. Fox, ed.), pp. 1-26. Biochemistry Series One, Vol. 2. MTP Intern. Rev. Sci. (H. L. Kornberg and D. C. Phillips, Consultant eds.). Butterworths, London.

Kornberg, R. D., and McConnell, H. M. (1971a). Lateral diffusion of phospholipids in a vesicle membrane. *Proc. Nat. Acad. Sci. U.S. 68*, 2564-2568.

Kornberg, R. D., and McConnell, H. M. (1971b). Inside-outside transitions of phospholipids in vesicle membranes. *Biochemistry 10*, 1111-1120.

Kozlov, E. A., and Alexeenko, I. P. (1967). Electron-microscope investigation of the structure of the nuclear polyhedrosis virus of the silkworm, *Bombyx mori*. *J. Invertebr. Pathol. 9*, 413-419.

Kulczycki, A., Jr., Hempstead, B. L., Hoffman, S. L., Wood, E. W., and Parker, C. W. (1979). The cell surface receptor for immunoglobulin E. *J. Biol. Chem. 254*, 3194-3200.

Kuwert, E., Witkor, T. J., Sokol, F., and Koprowski, H. (1968). Hemagglutination by rabies virus. *J. Virol. 2*, 1381-1392.

Landsberger, F. R., Compans, R. W., Choppin, P. W., and Lenard, J. (1973). Organization of the lipid phase in viral membranes. Effects of independent variation of the lipid and the protein composition. *Biochemistry 12*, 4498-4502.

Lee, A. G., Birdsall, N. J. M., Levene, Y. K., and Metcalf, J. C. (1972). High resolution proton relaxation studies of lecithins. *Biochim. Biophys. Acta 255*, 43-56.

Li, K.-K., J. Williams, R. E., and Fox, C. F. (1975). Effects of temperature and host lipid composition on the infection of cells by Newcastle Disease virus. *Biochem. Biophys. Res. Commun. 62*, 470-477.

Lonberg-Holm, K. (1981). Attachment of animal virus to cells: An introduction. *In* "Virus Receptors, Part 2: Animal Viruses" (K. Lonberg-Holm and L. Philipson, eds.), pp. 1-20. Series B, Vol. 8 of "Receptors and Recognition" (P. Cuatrecasas and M. F. Greaves, series eds.). Chapman and Hall, London.

Lonberg-Holm, K., and Philipson, L. (1974). Early interactions between animal viruses and cells. *Monogr. Virol. 9*, 1-148.

Lonberg-Holm, K., and Philipson, L. (1980). *In* "Cell Membranes and Viral Envelopes" (H. A. Blough and J. Tiffany, eds.), Vol. II, pp. 789-848. Academic Press, New York.

Lonberg-Holm, K., and Philipson, L. (eds.) (1981). "Virus Receptors, Part 2: Animal Viruses." Series B, Vol. 8 of "Receptors and Recognition" (P. Cuatrecasas and M. F. Greaves, series eds.). Chapman and Hall, London.

Lotan, R., and Nicolson, G. (1979). Purification of cell membrane glycoproteins by lecithin affinity chromatography. *Biochim. Biophys. Acta 559*, 329-376.

Lucy, J. A. (1970). The fusion of biological membranes. *Nature (London) 227*, 815-817.

Lyles, D. S. (1979). Glycoproteins of Sendai virus are transmembrane proteins. *Proc. Nat. Acad. Sci. U.S. 76*, 5621-5625.

McIntosh, K., Payne, S., and Russell, W. C. (1971). Studies on lipid metabolism in cells infected with adenovirus. *J. Gen. Virol. 10*, 251-265.

McSharry, J. J., and Wagner, R. R. (1971). Lipid composition of purified vesicular stromatitis viruses. *J. Virol. 7*, 59-70.

McSharry, J. J., Compans, R. W., Lackland, H., and Choppin, P. W. (1975). Isolation and characterization of the nonglycosylated membrane protein and a nucleocapsid complex from the paramyxovirus SV 5. *Virology 67*, 365-374.

Marchesi, V. T., and Andrews, E. P. (1971). Glycoproteins: Isolation from cell membranes with lithium diiodosalicylate. *Science 174*, 1247-1258.

Marchesi, V. T., Furthmayr, H., and Tomita, M. (1976). The red cell membrane. *Annu. Rev. Biochem. 45*, 667-698.

MacKinnon, E. A., Henderson, J. F., Stoltz, D. B., and Faulkner, P. (1974). Morphogenesis of nuclear polyhedrosis virus under conditions of prolonged passage *in vitro*. *J. Ultrastruct. Res. 49*, 419-435.

Mazzone, H. M., and McCarthy, W. J. (1981). The gypsy moth nucleopolyhedrosis virus. Biochemistry and biophysics. *In* "The Gypsy Moth: Research Toward Integrated Pest Management" (C. C. Doane and M. L. McManus, eds.), pp. 487-495. Tech. Bull. 1584. U.S. Dept. of Agriculture Forest Service, Washington, D.C.

Meager, A., and Hughes, R. C. (1977). Virus Receptors. *In* "Receptors and Recognition" (P. Cuatrecasas and M. F. Greaves, eds.). Series A, Vol. 4, pp. 141-195. Chapman and Hall, London.

Miller, D. A., Miller, O. J., Dev, V. G., Hashmi, S., Santravahi R., Medrano, L., and Green, H. (1974). Human chromosome 19 carries a poliovirus receptor gene. *Cell 1*, 167-173.

Morgan, C., and Rose, H. M. (1968). Structure and development of viruses as observed in the electron microscope. VIII. Entry of influenza virus. *J. Virol. 2*, 925-936.

Morgan, C., Rose, H. M., and Mednis, B. (1968). Electron microscopy of Herpes Simplex virus. I. Entry. *J. Virol. 2*, 507-516.

Morré, D. J., Kartenbeck, J., and Franke, W. W. (1979). Membrane flow and interconversions among endomembranes. *Biochim. Biophys. Acta 559*, 71-152.

Mountcastle, W. E., Compans, R. W., and Choppin, P. W. (1971). Proteins and glycoproteins of paramyxoviruses: a comparison of Simian virus 5, Newcastle Disease virus, and Sendai virus. *J. Virol. 7*, 47-52.

Mussgay, M., Enzmann, P. J., Horzinek, M. C., and Weiland, E. (1975). Growth cycle of arboviruses in vertebrate and arthropod cells. *Progr. Med. Virol. 19*, 257-323.

Nagai, Y., Klenk, H.-D., and Rott, R. (1976). Proteolytic cleavage of the viral glycoproteins and its significance for the virulence of Newcastle Disease virus. *Virology 72*, 494-508.

Nakajo, S., Nakaya, K., and Nakamura, Y. (1979). Isolation

and partial characterization of the major glycoprotein from the plasma membranes of AH-66 hepatoma cells. *Biochim. Biophys. Acta 579*, 88-94.

Nappi, A. J., and Hammill, T. M. (1975). Viral release and membrane acquisition by budding through the nuclear envelope of hemocytes of the gypsy moth, *Porthetria dispar. J. Invertebr. Pathol. 26*, 387-392.

Norkin, L. C. (1977). Cell killing by Simian virus 40: Impairment of membrane formation and function. *J. Virol. 21*, 872-879.

Norrby, E., and Skaaret, P. (1967). The relationship between soluble antigens and the virion of adenovirus type 3. III. Immunological identification of fiber antigen and isolated vertex capsomer antigen. *Virology 32*, 489-502.

Ohba, M., and Tanada, Y. (1983). A synergistic factor enhances the *in vitro* infection of an insect baculovirus. *Naturwissenschaften 70*, 613-615.

Penman, S. (1965). Stimulation of the incorporation of choline in polio virus-infected cells. *Virology 25*, 148-152.

Pfefferkorn, E. R., and Hunter, H. S. (1973). The source of the ribonucleic acid and phospholipid of Sindbis virus. *Virology 20*, 446-456.

Philipson, L. (1981). Evaluation and Conclusions. *In* "Virus Receptors, Part 2: Animal Viruses" (K. Lonberg-Holm and L. Philipson, eds.), pp. 203-211. Series B, Vol. 8 of "Receptors and Recognition" (P. Cuatrecasas and M. F. Greaves, series eds.). Chapman and Hall, London.

Plagemann, P. G. W., Cleveland, P. H., and Shea, M. A. (1970). Effect of Mengovirus replication on choline metabolism and membrane formation in Novikoff hepatoma cells. *J. Virol. 6*, 800-812.

Ponsen, M. B., Henestra, S., and Van der Scheer, C. (1965). Electron microscopy of DNA-cores in nuclear polyhedrosis virus. *Netherlands J. Plant Pathol. 71*, 54-56.

Poste, G. (1970). Virus-induced polykaryocytosis and the mechanism of cell fusion. *Advan. Virus Res. 16*, 303-356.

Raghow, R., and Grace, T. D. C. (1974). Studies on a nuclear polyhedrosis virus in *Bombyx mori* cells *in vitro. J. Ultrastr. Res. 47*, 384-399.

Renkonen, O., Kääriäinen, L., Simons, K., and Gahmberg, C. G. (1971). The lipid class composition of Semliki Forest virus and of plasma membranes of the host cells. *Virology 46*, 318-326.

Robertson, J. D. (1969). Molecular structure of biological membranes. *In* "Handbook of Molecular Cytology" (A. Lima-de Faria, ed.), pp. 1403-1443. North-Holland, Amsterdam.

Robertson, J. S., Harrap, K. A., and Longworth, J. F. (1974). Baculovirus morphogenesis: the acquisition of the virus envelope. *J. Invertebr. Pathol. 23*, 248-251.

Rodriquez, M., and Dubois-Dalcq, M. (1978). Intramembrane changes occurring during maturation of Herpes Simplex Virus type 1: Freeze-fracture study. *J. Virol. 26*, 435-447.

Roesing, T. G., Toselli, P. A., and Crowell, R. L. (1975). Elution and uncoating of Coxsackievirus B3 by isolated HeLa cell plasma membranes. *J. Virol. 15*, 654-667.

Roizman, B., and Furlong, D. (1974). The replication of herpesviruses. *In* "Comprehensive Virology" (H. Fraenkel-Conrat and R. R. Wagner, eds.), Vol. 3, pp. 229-404. Plenum Press, New York.

Salt, G. (1970). "The Cellular Defense Reactions of Insects." Cambridge Monographs in Experimental Biology, No. 16, 118 pp. Cambridge Univ. Press, London.

Satir, P., and Gilula, N. B. (1973). The fine structure of membranes and intercellular communication in insects. *Annu. Rev. Entomol. 18*, 143-166.

Scheid, A. (1981). Subviral components of myxo- and paramyxo-viruses which recognize receptors. *In* "Virus Receptors, Part 2: Animal Viruses" (K. Lonberg-Holm and L. Philipson, eds.), pp. 47-62. Series B, Vol. 8 of "Receptors and Recognition" (P. Cuatrecasas and M. F. Greaves, series eds.). Chapman and Hall, London.

Scheid, A., and Choppin, P. W. (1974). The hemagglutinating and neuraminidase protein of a paramyxovirus: Interaction with neuraminic acid in affinity chromatography. *Virology 62*, 125-133.

Scheid, A., Caliguiri, L. A., Compans, R. W., and Choppin, P. W. (1972). Isolation of paramyxovirus glycoproteins. Association of both hemagglutinating and neuraminidase activities with the larger SV 5 glycoprotein. *Virology 50*, 640-652.

Schulze, I. T. (1972). The structure of influenza virus. II. A model based on the morphology and composition of subviral particles. *Virology 47*, 181-196.

Seto, J. T., Becht, H., and Rott, R. (1974). Effect of specific antibodies on biological functions of the envelope components of Newcastle Disease virus. *Virology 61*, 354-360.

Shimizu, K., Shimizu, Y. K., Kohama, T., and Ishida, N. (1974). Isolation and characterization of two distinct types of HVJ (Sendai virus) spikes. *Virology 62*, 90-101.

Simons, K., Helenius, A., and Garoff, H. (1973). Solubilization of the membrane proteins from Semliki Forest virus with Triton X-100. *J. Mol. Biol. 80*, 119-133.

Singer, S. J., and Nicholson, G. L. (1972a). The fluid mosaic model of the structure of cell membranes. *Science 175*, 720-731.

Singer, S. J., and Nicholson, G. L. (1972b). *In* "Membranes and Viruses in Immunopathology" (L. Day and R. A. Good, eds.), pp. 7-47. Academic Press, New York.

Smith, D. S. (1968). "Insect Cells. Their Structure and Function." Oliver and Boyd, Edinburgh.

Smith, K. M. (1962). The arthropod viruses. *Advan. Virus Res.* *9*, 195-238.

Smith, K. M. (1967. "Insect Virology." Academic Press, New York.

Smith, K. M. (1971). The viruses causing the polyhedroses and granuloses of insects. *In* "Comparative Virology" (K. Maramorosch and E. Kurstak, eds.), pp. 479-507. Academic Press, New York.

Stairs, G. R., and Ellis, B. J. (1971). Electron microscope and microfilter studies on infectious nuclearpolyhedrosis virus in *Galleria mellonella* larvae. *J. Invertebr. Pathol.* *17*, 350-353.

Stokes, G. V. (1976). High-voltage electron microscopy study of the release of vaccinia virus from whole cells. *J. Virol.* *18*, 636-643.

Stoltz, D. B., Pavan, C., and da Cunha, A. B. (1973). Nuclear polyhedrosis virus: A possible example of de novo intranuclear membrane morphogenesis. *J. Gen. Virol.* *19*, 145-150.

Strand, M., and August, J. T. (1976). Structural proteins of ribonucleic acid tumor viruses. Purification of envelope, core, and internal components. *J. Biol. Chem.* *251*, 559-564.

Sturman, L. S., Holmes, K. V., and Behnke, J. N. (1980). Isolation of coronavirus envelope glycoproteins and interaction with the viral nucleocapsid. *J. Virol.* *33*, 449-462.

Summers, M. D. (1969). Apparent *in vivo* pathway of granulosis virus invasion and infection. *J. Virol.* *4*, 188-190.

Summers, M. D. (1971). Electron microscopic observations on granulosis virus entry, uncoating and replication processes during infection of the midgut cells of *Trichoplusia ni*. *J. Ultrastr. Res.* *35*, 606-625.

Summers, M. D., and Paschke, J. D. (1970). Alkali liberated granulosis virus of *Trichoplusia ni*. I. Density gradient purification of virus components and some of their *in vitro* chemical and physical properties. *J. Invertebr. Pathol.* *16*, 227-240.

Summers, M. D., and Volkman, L. E. (1976). Comparison of biophysical and morphological properties of occluded and extracellular nonoccluded baculovirus from *in vivo* and *in vitro* host systems. *J. Virol.* *17*, 962-972.

Tanada, Y., and Hess, R. T. (1976). Development of a nuclear polyhedrosis virus in midgut cells and penetration of the virus into the hemocoel of the armyworm, *Pseudaletia unipuncta*. *J. Invertebr. Pathol.* *28*, 67-76.

Tanada, Y., and Hukuhara, T. (1971). Enhanced infection of a nuclear-polyhedrosis virus in larvae of the armyworm,

Pseudaletia unipuncta, by a factor in the capsule of a granulosis virus. *J. Invertebr. Pathol. 17*, 116-126.

Tanada, Y., and Leutenegger, R. (1970). Multiplication of a granulosis virus in larval midgut cells of *Trichoplusia ni* and possible pathways of invasion into the hemocoel. *J. Ultrastr. Res. 30*, 589-600.

Tanada, Y., Himeno, M., and Omi, E. M. (1973). Isolation of a factor, from the capsule of a granulosis virus, synergistic for a nuclear-polyhedrosis virus of the armyworm. *J. Invertebr. Pathol. 21*, 31-40.

Tanada, Y., Hess, R. T., and Omi, E. M. (1975). Invasion of a nuclear polyhedrosis virus in the midgut of the armyworm, *Pseudaletia unipuncta*, and the enhancement of a synergistic enzyme. *J. Invertebr. Pathol. 26*, 99-104.

Teakle, R. E. (1969). A nuclear-polyhedrosis virus of *Anthela varia* (Lepidoptera: Anthelidae). *J. Invertebr. Pathol. 14*, 18-27.

Tooze, J. (1973). *In* "The Molecular Biology of Tumor Viruses" (J. Tooze, ed.), pp. 1-743. Cold Spring Harbor Press, New York.

Usherwood, P. N. R. (1969). Electrochemistry of insect muscle. *Advan. Insect. Physiol. 6*, 205-278.

Valentine, R. C., and Pereira, H. G. (1965). Antigens and structure of the adenoviruses. *J. Mol. Biol. 13*, 13-20.

Vitetta, E. S., Uhr, J. W., Klein, J., Pazderka, F., Moticka, E. J., Ruth, R. F., and Capra, J. D. (1977). Homology of (murine) H-2 and (human) HLA with a chicken histocompatibility antigen. *Nature (London) 270*, 535-536.

Wilcox, W. C., Ginsberg, H. S., and Anderson, T. F. (1963). Structure of Type 5 adenovirus. *J. Exp. Med. 118*, 307-314.

Yamada, K. M., and Olden, K. (1978). Fibronectins--adhesive glycoproteins of cell surface and blood. *Nature (London) 275*, 179-184.

Yamamoto, T., and Tanada, Y. (1977). Possible involvement of phospholipids in the infectivity of baculoviruses. *J. Invertebr. Pathol. 30*, 279-281.

Yamamoto, T., and Tanada, Y. (1978a). Phospholipids, an enhancing component in the synergistic factor of a granulosis virus of the armyworm, *Pseudaletia unipuncta*. *J. Invertebr. Pathol. 31*, 48-56.

Yamamoto, T., and Tanada, Y. (1978b). Biochemical properties of viral envelopes of insect baculoviruses and their role in infectivity. *J. Invertebr. Pathol. 32*, 202-211.

Yoshida, T., Nagai, Y., Yoshii, S., Marno, K., Matsumoto, T., and Hoshino, M. (1976). Membrane (M) protein of HVJ (Sendai virus): Its role in virus assembly. *Virology 71*, 143-161.

ACKNOWLEDGMENT

For valuable discussion and suggestions, I am grateful to Professor Y. Tanada, Division of Entomology and Parasitology, College of Natural Resources, University of California, Berkeley.

NOTE ADDED IN PROOF

Recently, Ohba and Tanada (1983) reported on the enhancement of *in vitro* infection of an insect baculovirus by a synergistic factor (SF). The SF, derived from the Hawaiian strain of the GV of *P. unipuncta*, markedly enhanced infection of cells of *Leucania separata* by the typical NPV of *P. unipuncta*. At a concentration of 75 μg/ml of SF, the NPV infection was enhanced approximately 100 times over that observed when no SF was used. Moreover, the enhancement obtained under *in vitro* conditions was higher by 56 times that obtained under *in vivo* conditions. This is the first report of enhancement of insect viruses under *in vitro* conditions. The SF is believed to act *in vivo* and *in vitro* as an enhancer in the fusion of enveloped virions to the cell membrane.

In terms of alternate pathways of infection, Granados and Lawler (1981) observed that some nucleocapsids pass directly through the cytoplasm of midgut cells and into the hemocoel as early as 1/2 hr post infection. These nucleocapsids do not replicate in the midgut. In other studies, it has been noted that while uncoating of nucleocapsids may occur at the nuclear pores, the process may also take place in the nucleoplasm (Hirumi et al., 1975; Granados, 1978; Walker et al., 1982).

Granados, R. R. (1978). Early events in the infection of *Heliothis zea* midgut cells by a baculovirus. *Virology 90*, 170-174.

Hirumi, H., Hirumi, K., and McIntosh, A. H. (1975). Morphogenesis of a nuclear polyhedrosis virus of the alfalfa looper in a continuous cabbage looper cell line. *Ann. N.Y. Acad. Sci. 266*, 302-326.

Walker, S., Kawanishi, C. Y., and Hamm, J. J. (1982). Cellular pathology of a granulosis virus infection. *J. Ultrastruct. Res. 80*, 163-177.

MULTIPLE VIRUS INTERACTIONS

KENNETH E. SHERMAN

George Washington University School of Medicine
Washington, D.C.

I. INTRODUCTION

A. BACKGROUND AND SIGNIFICANCE

The interaction of a virus and its host involves a complex relationship which can be examined at the ecological, organismal, cellular, or molecular levels. The sum of such investigations provides us with some level of understanding regarding the ennumerable factors that influence a virus-host relationship. This chapter attempts to review one subset of such factors, namely, how viruses interact with each other in specific hosts, and how this interaction affects both the host and the course of viral infection. This chapter will focus on what is known regarding virus-virus relationships in invertebrate systems. It will also deal with explanations of the general types of viral interactions. Pertinent examples that illustrate the variety of interactions that have been recognized will be discussed. Later sections will concentrate on invertebrate systems.

The study of multiple virus interactions is particularly important when arthropod systems are being discussed. The reason for this is based on some of the basic biological differences between invertebrate and vertebrate species. When an

ISBN 0-12-470295-3

immunocompetent vertebrate is infected with a virus, the host generally responds by active production of specific antibodies directed against the invader. This highly effective immune system not only eliminates the infectious agent (with certain exceptions), but it often "remembers" the attacker and can prevent reinfection. The mosquito can be used as an example of an invertebrate host. If a female mosquito ingests a viremic blood meal, this blood bypasses the gut diverticula and goes directly to the midgut. Because proximal parts of the digestive tract are lined by a chitinous, cuticular lining, the midgut is the first place that viruses come in direct contact with an epithelial surface. The midgut epithelial cells secrete a viscous, lamellar secretion called the peritrophic membrane. This electrically charged barrier, as well as the action of some digestive juices, are the first line of defense against viral infection. In this respect, the invertebrates do not differ significantly from vertebrates, since skin and mucosal linings have similar protection. The odds that a virus particle will breach the gut membrane barrier is primarily a function of virion number. The higher the virus titer, the more likely that epithelial infection will occur (Orihel, 1975; Paschke and Summers, 1975; Tinsley, 1975).

The virus will multiply in the midgut epithelial cells and then spill out into the hemolymph. Invertebrate have not been shown to have an antibody-specific immune system. The only remaining defense are primitive phagocytic cells called hemocytes. Generally, the proliferation of virus in the midgut epithelial cells is so great that the phagocytic capacity of the invertebrate immune system is overwhelmed, and all somatic cells are infected (Jones, 1975; Murphy et al., 1975).

Based on current knowledge, the invertebrate immune system plays a relatively minor role in the mediation of viral infection. For this reason, the modulation of viral infection by viral interactions may assume a more significant role in invertebrate virus infection.

In 1975, Murphy published an article in which he described several categories of various types of viral interactions that modified the infection process and did not involve specific cellular or humoral immunity. The categories included interferon induction, homologous viral autointerference which covers defective interfering particles, mutant virus, and wild-type autointerference, and intrinsic interference in heterologous viral types (Murphy, 1975).

B. INTERFERON

Interferon has been demonstrated to be a powerful mediator of viral infection in many vertebrate systems. This has not necessarily been true in invertebrate cell studies, however. Most investigators have not noted any interferon production in insects or insect cell lines. There are, however, a few reports in the literature that are worth noting. Bergold and Ramirez (1972) described suppression of both yellow fever virus 17D and vesicular stomatitis virus (VSV) when 10 μg/ml poly(I:C) was added to cultured *A. aegypti* cells. A similar result was obtained using 50 μg/μl poly(I:C) in *A. albopictus* cells. They also tested live mosquitoes by injecting a mixture of 1000 plaque-forming units (PFU) VSV and 0.25 μg poly(I:C) per mosquito. At various time intervals, samples were titrated on BHK-21 cells to determine the number of PFU. The problem with this study is that the assay system is extremely sensitive to induction by both VSV and poly(I:C) (Fenner, 1974). Therefore, the suppression may be an artifact of the method used to assay suppression. A more recent report, using a very similar system found no evidence of interferon induction in mosquito cells (Novokhatskii and Berezina, 1978).

Mosquito cells replicate togaviruses and show no cytopathogenic effect. Enzman (1973) speculated that chronic togavirus infection may be mediated by an interferon-like gene product. Despite this theory, no investigator has characterized such a protein, or has any form of interferon been found.

Based on the best evidence to date, we can rule out interferon as a mediator of viral infections in invertebrate hosts.

C. HOMOLOGOUS AUTOINTERFERENCE

This category includes all interactions in which the infection is mediated by various forms of the same virus. The most study in this category has been done in the area of defective interfering (DI) particles. In 1954, von Magnus first reported that high multiplicity passage of influenza virus would cause a decrease in the number of virions produced in subsequent generations. Later, Huang and Baltimore coined the term "DI" particle and developed criteria for this definition. In summary, the virions have normal structural proteins, but the viral genome is incomplete. Reproduction must be aided by the gene products of intact helper viruses, and there is specific interference of homologous virus replication (Huang and Baltimore, 1970). DI particle interference has been reported in a wide variety of viral groups (Joklik et al., 1980). DI particles and viruses that infect invertebrates will be discussed later in this chapter.

The total observable effect of a virus on its host is the sum of the effects due to the virion population. If some percentage of a virus population mutates from the wild-type form, the potential exists for a modification of the infection process. These mutant virus particles can be distinguished from DI virus particles since they are capable of normal infection and multiplication under a given set of circumstances.

Viruses that infect both invertebrates and vertebrates merit particular attention with respect to the production of mutant viruses, in general, and of temperature-sensitive (*ts*) mutants, in particular. Shenk et al. (1974) described *ts* mutants of Sindbis virus, a prototype alphavirus. The nonpermissive temperature was 39.5°C. Since many togaviruses spend part of their life cycle in avian hosts, whose body temperature is greater than 39°C, the presence of nonreplicating *ts* mutants could radically alter the infection process. Likewise, invertebrates are poikilothermic creatures and environmental temperature variation may play a major role in the course of viral infection. The type of mutation does not have to be of the *ts* variety. Continuous passage of dengue type 2 in *A. albopictus* cells leads to an alteration of the virion surface and results in progeny with reduced virulence. This is not the same as DI particle interference (Sinarachatanant and Olsen, 1973). If insect pathogenic virus produce *ts* or otherwise restricted mutants, then environmental factors could emerge as a significant mediator of viral biological control agents. In point of fact, *ts* mutants among baculoviruses have been described.

The third type of homologous autointerference is described by Murphy as wild-type autointerference (Murphy, 1975). It is speculated that this type of interference may be responsible for the development of chronic infection where no cytopathogenic effect is observed. Fifteen to twenty percent of cultured mosquito cells, which have been infected with Venezuelan equine encephalitis virus, produced virus up to 24 hours postinfection. After 24 hours, only 0.01-0.05% of the total population produced virus. If nonproducing cells were subcultured, all were able to produce virus. Therefore, cell resistance was not involved. Interferon or other antiviral agents could not be found. For this reason, a viral modification of cellular genetic mechanisms which limit virion production was postulated (Esparaza and Sanchez, 1975). Raghow et al. (1973) provided evidence that there is an accumulation of viral protein in infected mosquito cells. This protein may be part of a regulatory mechanism. It is quite possible that the mechanism of noncytopathic viral chronicity is based solely on the characteristics of host cell viral restriction. If this is the case, the wild-type autointerference is not really a form of viral interference. More recent evidence, argueing against this and for a virally produced mediator is found in the work of Riedel and Brown in Sindbis virus persistently infected mosquito cell cultures. They describe a low-molecular weight substance which was found in the culture medium. This substance differed from interferon in that it was virus as well as host specific (Riedal and Brown, 1979).

D. HETEROLOGOUS VIRAL INTERFERENCE

During the World War II, considerable research was done in the area of insect-borne viruses, particularly in the Pacific theater of operations. It was observed that yellow fever was not a problem in areas where dengue virus was endemic. *Aedes aegypti* mosquitoes are the vectors for both viruses, and there was no apparent reason why India, Indonesia, and Australia should be spared significant yellow fever outbreaks. Experiments done with Rhesus monkeys revealed that inoculation with both viruses, within a 7-day period, protected the monkeys from fatal yellow fever infection. If yellow fever virus was injected 1 month after dengue infection, there was no cross-immunity and most monkeys died of yellow fever. Similar results were obtained when mosquitoes which were proved to be infected with dengue were found to not be susceptible to further infection with the virulent Asibi strain of yellow fever (Sabin, 1952). A similar effect was observed in *Culex tritaeniorhynchus* mosquitoes infected with two different viruses. If the mosquitoes were infected with Japanese encephalitis virus, Murray Valley encephalitis transmission was reduced (Altman, 1963).

Heterologous virus interference was given the name "intrinsic" interference by Marcus and Carver in 1965. They coined this term after discovering that rubella virus infection made primary African green monkey cells refractory to superinfection with Newcastle Disease virus. Intrinsic interference was first characterized in the experiments of Marcus and Carver. They found that only cells infected with active virus became refractory to NDV infection. This ruled out interferon as the mediator of the observed effect. If protein synthesis is blocked, intrinsic interference does not take place. Cells with normal protein synthesis that are refractory to superinfection with NDV are still susceptible to a broad range of other cytocidal viruses. The intrinsic interference to NDV can also be induced by Sindbis virus, West Nile virus, lactate dehydrogenase virus, poliovirus, and an adventitious murine leukemia virus. This can occur in many other cell lines besides African green monkey kidney (Marcus and Carver, 1967). Therefore, a broad definition of intrinsic interference would be as follows: A virus infecting a cell makes that cell refractory to superinfection by a heterologous virus type via the production or presence of virus or viral products of the initially infecting agent. There is speculation that in some specific types of intrinsic interference, the effect is due to interference with mRNA translation (Joklik et al., 1980).

Originally, intrinsic interference was used to develop a hemabsorption-negative plaque assay for enumeration of the noncytopathic rubella virus (Marcus and Carver, 1965). Other investigators followed this approach and intrinsic interference was noted between lymphocytic choriomeningitis virus and Sendai virus (Wainwright and Mims, 1967). Attempts were made to identify the hepatitis B virus using the hemoabsorption-negative assay system with NDV, but results were mixed (Carver and Seto, 1971; Berthold et al., 1973; Vordham et al., 1973).

Another example of intrinsic interference is seen if adenovirus type 2 is permitted to infect HeLa cells for greater than 18 hours. If, at this time, vaccinia virus is added to the culture, vaccinia replication is inhibited. Protein synthesis seems to be inhibited by late adenovirus protein. RNA synthesis is not inhibited (Giorno and Kates, 1971).

Heterologous virus infection is not limited to interference effects. If Shope fibroma virus (pox virus) is permitted to infect rabbit cells, a persistent, noncytopathic infection results. If these cells are now exposed to Sindbis virus, NDV, vesicular stomatitis virus (VSV), encephalomyocarditis virus, or Japanese B encephalitis virus, these RNA viruses replicate. Without the Shope fibroma virus, these rabbit cells are refractory to replication by any of the abovementioned viruses (Padgett and Walker, 1970; Tsuchiya and Tagaya, 1970). This process is called "facilitation." The mechanism of this facilitation is not known, but it does not seem to involve merely the suppression of interferon production (Abrams, 1976). The facilitation differs from "helper" functions discussed below in that the viruses that are facilitated are fully capable of independent replication in permissive cells.

Some viruses are incapable of replication, unless they coinfect with a helper virus. This is analogous to the previously described situation where DI particles need normal homologous visions to replicate. In this case, we are referring to heterologous helper virus functions. The prototype example of the viral interaction is adeno-associated virus (subgroup B) of the parvovirus group. These viruses need adenovirus to fulfill specific helper functions for replication (Fenner et al., 1974).

II. MULTIPLE VIRUS INTERACTIONS INVOLVING INVERTEBRATE VIRUSES

A. *HOMOLOGOUS-TYPE INTERACTIONS*

The fact that baculovirus virus production decreases with continued subcultivation in susceptible cells has been recognized for some time. This effect can be observed if cells are first infected with a high-passage strain of a baculovirus and then challenged with a low-passage, high-production strain. The measurable level of viral replication is consistently limited by the production capacity of the high-passage strain. This effect seems to be comparable to the classic von Magnus phenomenon (Mackinnon et al., 1974; Hink and Strauss, 1976; McIntosh et al., 1979). It is worthwhile to note that caution must be observed when one looks at experiments comparing the effects of homologous high- and low-passage virus. In one *in vitro* study, polyhedra from high-passage NPV did not cause significant mortality in *H. zea* larvae. This may have been due to a lack of occluded virions in the late-passage polyhedral inclusion bodies (PIBs) since there was no significant difference between the mortality between the high- and low-passage nonoccluded virions (Yamada et al., 1981). This implies that only the "packaged" form of the virus changed and not the actual virus or its inherent infectivity.

Viral persistence with little or no cytocidal activity is a common occurrence when we are speaking of togaviruses in invertebrate cells. This is not necessarily true for the various groups of important insect pathogenic viruses, since the inherent definition of biological control agents implies the ability to induce abnormal pathology in target cells. Mitsuhashi reported on the establishment of a persistent iridescent virus infection in hemocytes of *Chilo suppressalis* (Mitsuhashi, 1967). Granados et al. (1978) described baculovirus-like particles visualized by EM in a *H. zea* cell line. McIntosh and Ignoffo (1981) established a persistent infection of *Spodoptera frugiperda* NPV is *S. frugiperda* cells. Viral antigens were observed in 70% of the persistently infected cells. The yield of virus was one hundred times lower than the viral production level in wild-type virus and cells. The high percentage of antigen-containing cells may lead one to speculate that the mechanism of persistence is by the formation of incomplete, but antigenically reactive interfering particles. This does not seem to be true, however, because heterologous as well as homologous interference to superinfection was observed (McIntosh and Ignoffo, 1981). There is now evidence that a persistent nonoccluded baculovirus is present in at least one cell line of *H. zea* cells, and that this virus can be induced by superinfection with two different nuclear polyhedrosis viruses (Kelly et al., 1981; Langridge, 1981). Aruga et al. (1961) described cold induction and interference associated with two cytoplasmic polyhedrosis viruses in *Bombyx mori* larvae.

As in the togaviruses, temperature may play an important role in the modulation of agents used for the biological control of poikilothermic invertebrate species. The effects of temperature variation are often very difficult to evaluate, because the host cells, as well as the viral agent in question, have optimal temperature ranges. Although there are many reports of alterations of viral production in insects both *in vivo* and *in vitro* (Kobayashi et al., 1981), no information is available that discusses the role of production of various temperature-restricted mutants, and what effect these may have on modulating viral infection. At least in some cases, high temperature inhibition seems related to restriction of DNA production (Kobayashi et al., 1981). Investigators studying baculoviruses at the molecular level have produced and isolated temperature-sensitive mutants and placed them in complementation groups (Brown et al., 1979; Lee and Miller, 1979). Further studies in this area are desirable to better define the natural mutation rate and the environmental significance of such mutants.

B. HETEROLOGOUS VIRUS INTERACTION

Electron microscopy has provided the basis of several reports in the literature that describe multiple virus infection of invertebrate cells. In one study, *Galleria mellonella* cells were observed to contain a parvovirus, a baculovirus, and an iridovirus. Cytologic pathology particular to each specific agent was described (Odier, 1977). In 1977, Hess et al. described a 40-mm nonenveloped particle which occurred in association with the baculovirus, *Autographa californica* NPV (ACmNPV). Both viruses were observed in the nuclei of *T. ni* and *A. californica* cells. EM performed on cells not inoculated with ACmNPV had no evidence of either virus. In a later report, two viruses of similar appearance, as well as three other cytoplasmic viruses, were described (Hess et al., 1978). These studies provided visual evidence of multiple virus infection, but did not attempt to examine viral interaction or the biological effects of multiple infectivity.

Other studies have been performed at the organismal level. If third instar *Bombyx mori* larvae are fed both nuclear polyhedrosis virus and cytoplasmic polyhedrosis virus, the viruses interfered with one another. When mixed infection occurred in a single larva, no cell was observed to contain both kinds of polyhedra. If second or third instar larvae were fed NPV followed by CPV at a 1-4-day interval, CPV infection was not observed (Tanaka and Aruga, 1967). Double infection of NPV and CPV in silkworm larvae greatly enhances larval mortality rates (Piasecka-Serafin, 1977). Ritter and Tanada (1978) described the interaction between two different nuclear polyhedrosis viruses in the larvae of the armyworm, *Pseudaletia unipuncta*. If the larvae were fed a mixed inoculum simultaneously, only one type of NPV was symptomatically apparent. The only way to obtain mixed infection was to inoculate one virus 1-2 days after the other. The rate of infection may be a significant factor in this phenomenon (Ritter and Tanada, 1978). The concept that the time of multiple infection is important has been brought out in several other studies as well (Lowe and Paschke, 1968a; Hughes, 1979).

In a classical series of experiments reported in 1959, Bird described the relationship of polyhedrosis and granulosis virus double infection in the spruce budworm. Heavy infection by granulosis virus resulted in a reduction in the number of polyhedrosis virions, as well as morphological changes in developmental membranes and the number of polyhedra formed. Conversely, the faster acting polyhedrosis virus can prevent granulosis virus infection, In spite of this inhibition, which may be due to competition for individual host cells, mixed infection significantly increased larval mortality (Bird, 1959). Double viral infections of an NPV and a granulosis virus in *T. ni* larvae, provided evidence that simultaneous infection with both viruses resulted in neither interference nor enhancement. The authors concluded that the increased mortality seen in this dual infection was the additive effect of both viruses. However, if the NPV was administered 5-7 days after the granulosis virus, the polyhedrosis infection was interfered with. Cells appeared to be refractory to the NPV superinfection (Lowe and Paschke, 1968a,b). A larval mortality study carried out by Whitlock demonstrated that dual infection of an NPV and a granulosis virus in *Heliothis armigera* resulted in interference. This interference was manifested as a lower mortality rate for dual infection than for each virus separately. Further, this effect seems to require live virus, and does not occur in the presence of inactivated structural viral components (Whitlock, 1977).

If *S. frugiperda* cells are multiply infected with both an iridescent virus, and with an NPV, iridescent virus infection significantly reduced NPV replication. This phenomenon was affected by both time of infection and temperature. It appears that live virus is most effective at creating this interference phenomenon. Viruses were not observed to be in the same cells (Kelly, 1980).

A reovirus has been reported to inhibit the envelopment and occlusion of the *Spodotera litteralis* NPV. Both viruses were observed within the same cell (Quiot et al., 1980). This study, and many of the previously described reports, indicate that intrinsic interference may be involved in a variety of heterologous multiple infection phenomena.

In a series of experiments, the interaction between the baculovirus, ACmNPV, and Sindbis virus, which is a member of the alphavirus group of the Togaviridae, was examined (Sherman, 1980). The initial study began with the hypothesis that mosquitoes, like all other cellular organisms, are susceptible to specific viral agents which cause a deviation of normal cellular function. It was further postulated that such agents may be capable of modifying the normal biology of arboviruses which mosquitoes carry between vertebrate hosts.

Initially, the primary problem was one of defining a model system in which an insect pathogenic viral agent would replicate. Although there have been many reports of baculovirus infection in mosquitoes since Clark's original paper in 1969, these remain limited to EM studies and crude infectivity experiments (Clark and Chapman, 1969; Clark and Fukada, 1971; Chapman, 1974). No mosquito baculovirus has yet been cultured and characterized in *in vitro* systems. In 1979, we reported on the replication of the lepidopteran baculovirus ACmNPV in an *Aedes aegypti* mosquito cell line. A low level of viral replication was observed by utilizing radioactive labeling of progeny DNA. Immunoprecipitation of the radioactive product confirmed baculovirus production (Sherman and McIntosh, 1979). This artificial system provided the model for the further investigation of viral interactions (Sherman, 1980).

Table I shows the titer of Singbis virus grown singly and with ACmNPV coinfection at the times indicated. In this experiment, Sindbis virus was added at a multiplicity of infection (MOI) of 10 and ACmNPV at an MOI of 3. The ACmNPV inoculum was gradient purified, and EM showed a homogenous collection of nonoccluded ACmNPV virions. By 6 hours, coinfection significantly altered Sindbis virus production. If ACmNPV titer was reduced to an MOI of 1.0, Sindbis virus growth was not affected. Similarly, Sindbis viral production was unchanged if Sindbis was added to a cell at an MOI of 3 and ACmNPV at an MOI of 2. By the Poisson distribution, most cells are doubly infected in the first experiment presented. This was shown to result in strong inhibition of viral replication. When the ACmNPV MOI is lowered, no significant effect can be detected. This suggests that a competition for cell surface receptors may be responsible (Sherman, 1980). Such interference has been previously reported (Rubin, 1960, 1961).

The problem with the experiment described above, and with all such experiments in viral coinfections is that most studies rely on assaying one or both of the involved viruses in a biological assay system. Most investigators make the assumption

TABLE I. Sindbis Virus Titer: (Log_{10}) PFU/ml[a]

	Time (hrs)				
Virus	*0*	*6*	*10*	*26*	*48*
Sindbis	*3.7*	*5.0*	*5.6*	*6.0*	*6.1*
Sindbis + ACNPV	*3.7*	*4.3*	*4.5*	*4.9*	*3.8*

[a]*Aedes aegypti cells infected with Sindbis virus at an MOI of 10 or same cells infected with Sindbis virus at an MOI of 10 and Autographa californica NPV at an MOI at 3. Cells incubated for times indicated at 28°C. Sindbis virus was titrated by plaque assay in BHK-21 cells.*

that if one of the viruses is defined as nonpermissive in a given cell line by standard virological techniques, it is assumed that only the effects of one virus will be seen in the assay system. It was shown that this assumption is not necessarily true.

BHK-21 cells were inoculated with Sindbis virus alone or with Sindbis plus ACmNPV at an MOI of 5. After 1 hour, cells were washed and methionine-free MEM supplemented with 5% FBS and 20 μCi/ml [^{35}S]methionine were added. The cells were incubated 2 days at 28°C, and supernatant fluid was removed. This fluid was subjected to low-speed centrifugation to remove cellular debris, and then virus was pelleted by spinning at 55,000 *g* for 1.5 hours. The pellet was allowed to gently resuspend in Tris buffer (pH 7.4) for 12 hours. This suspension was layered on a 20-60% sucrose gradient and spun 1 hour at 80,000 *g*. Fractions were collected and spotted onto glass filter paper. Macromolecules were precipitated using 10% cold trichloroacetic acid (TCA). Free nucleic acids were solubilized by further treatment with hot TCA. The filter paper was then washed, dried, and placed in scintillation vials with 3 ml of scintillation counting fluid. The results (not shown) reveal a single major peak of radioactivity that occurs in both single and mixed cultures at a gradient density of approximately 1.23 gm/cm^3. The height of this peak is almost 30% greater in the single Sindbis infection versus the mixed Sindbis and ACmNPV peak. This is a statistically significant difference of about 60,000 cpm. These data strongly suggest that mixed infection decreases the level of new viral protein produced in the BHK-21 cell line (Sherman, 1980).

In an effort to examine this phenomenon further, peak fraction samples were run on a discontinuous SDS-polyacrylamide gel according to the method of Laemmli (1970). Only Sindbis virus proteins were observed when the autoradiographic results were examined. This supports the supposition that ACmNPV does not replicate in mammalian cells (Smith, 1976; McClelland and Collins, 1978; Tinsley, 1979), although even this premise has been challenged (McIntosh and Shamy, 1980). Virions produced in the same host under identical conditions should have similar composition. When the relative area of the densitometer scan peaks for Sindbis virus proteins E_1 and E_2 were compared to capsid (C) protein, a relative ratio of 1.52:1 was observed for single Sindbis virus infection. Surprisingly, the mixed infection peak had a E_1E_2:C ratio of 3.90:1. This difference in virion composition can be explained in two ways. One is that ACmNPV causes a decreased synthesis of capsid proteins, or that it selectively increases glycoprotein production. An alternate possibility is that capsid protein is excluded or extra glycoprotein is added at the level of virion assembly (Sherman, 1980). Measles virus occurs in a range of relative proportional differences in its capsid composition and can serve as a precedent for the structural variation noted here. The reason for structural variation in measles is unknown (Miller and Raine, 1979). The observed differences between mixed and single Sindbis infection in BHK-21 cells suggest that more is involved than simple competition for cell surface receptors. It clearly demonstrates that great care must be taken in interpreting coinfection data that is examined using biological assay systems. If ACmNPV is exerting influence over Sindbis production in BHK-21 cells via either preformed or posttranslational viral products, then this may well be classified as a form of intrinsic interference (Sherman, 1980).

Enhancement is another form of viral interaction. Wagner et al. described isolation of a picornavirus from *Aedes taeniorhyrchus* mosquito larvae. If larvae are coinfected with mosquito iridescent virus, there is a significant increase in the production of both viruses. The authors suggest that the picornavirus may indeed be a "helper" virus for mosquito iridescent virus, since they were not able to prove singly infected cultures did not contain the picornavirus (Wagner et al., 1974). Another form of enhancement is seen when the armyworm, *Pseudaletia unipuncta*, is infected with its homologous NPV and a granulosis virus. A direct increase in the number of larvae infected with NPV was noted as granulosis virus levels were increased. Live granulosis virus was not required for this effect and evidence was provided that the active component was a minor protein of the virus capsule. The mechanism seems related to enhancement of virus entry into the hemocoel (Tanada and Hukuhara, 1971). The specific site of action has been shown to be the cell membrane of the midgut microvilli (Tanada et al., 1980).

III. IMPLICATIONS OF MULTIPLE VIRAL INTERACTIONS

In terms of commercial application of viruses as invertebrate biological control agents, the implications surrounding multiple virus interactions must be considered. The pertinent areas of interest and concern fall into two major categories. First, we must consider safety in the production and use of all viral pesticides. Heterologous phenotypic viral mixing due to genetic recombination can radically affect the action of an agent in its host. In a morphological study of mixtures of different alphaviruses, cocultivation resulted in mixtures of morphologically different progeny (Gushchin et al., 1981). Mazurenko et al. (1980) described several cases of viral-viral cancerogenesis where a nononcogenic virus can activate a latent oncogenic virus. Laboratory studies involving restriction endonuclease mapping of baculovirus wild types suggest that genetic rearrangement may frequently occur among baculoviruses (Carstens, 1982). One *in vivo* study has demonstrated that *Galleria mellonella* larvae can be infected with a mixture of *G. mellonella* NPV and *A. californica* NPV, and that within four generations, parental DNA's have been replaced by new genetic rearrangements (Croizier et al., 1980). The long and tedious process of viral pesticide licensure with various government agencies is one method of assuring product safety. It seems clear that prepared commercial biological pesticides must not be permitted to become contaminated with more than one agent. The implications of a second contaminating agent are that the contaminant may itself be harmful, or that either genetic recombination or phenotypic mixing will adversely affect the safety of the agent in question. In this regard, Morris et al. describes their efforts to detect a 35 mm RNA virus that was found associated with preparations of *A. californica* NPV (Morris et al., 1981).

The practical considerations of multiple virus interaction are highly dependent on the combination of viruses and the specific host in question. Previous sections of this paper provided examples of viral interference, viral enhancement, as well as variations in *in vivo* mortality rates that were not necessarily related to the phenomenon observed on the cellular level. A preparation that inadvertently contains two viruses may cause a significantly lower mortality in the target population than each agent separately. Conversely, mixtures of viruses may purposely be used to enhance the effect of a viral pesticide formulation. At least one group is using mathematical modeling to determine the optimal mixtures of NPV and granulosis viruses under a given set of field conditions (Tarasevich et al., 1980).

Finally, there are the environmental implications of multiple virus interactions. Different viruses and mutants of homologous viruses clearly interact at one or more levels of complexity in a given organism. There are a wide variety of possibilities for these interactions, and the range of possibility is further extended when relative times of infection and environmental conditions are taken into account. It will undoubtably be many years before we begin to unravel the intricacies of how viruses interact with each other in the natural environment, and how this interaction modulates all the interdependent organisms in the ecological community.

REFERENCES

Abrams, H. D. (1976). Facilitation of Sindbis virus replication in a rabbit cell line by co-infection with poxviruses. *Abstr. Annu. Meeting Amer. Soc. Micro.*, p. 244.

Altman, R. M. (1963). The behavior of Murray Valley encephalitis virus in *Culex tritaeniorhynchus and Culex pipiens quinquefasciatus*. *Amer. J. Trop. Med. Hyg. 12*, 425-434.

Aruga, H., Hukuhara, T., Yoshitake, N., and Ayudhya, I. N. (1961). Interference and latent infection in the cytoplasmic polyhedrosis of the silkworm, *Bombyx mori*. *J. Insect. Pathol. 3*, 81-92.

Bergold, G. H. and Ramirez, N. (1972). "Moving Frontiers in Invertebrate Virology. Monographs in Virology, Vol. 6" (T. W. Tinsley and K. A. Harrap, eds.), pp. 56-59. Karger, Basel.

Berthold, H., Mielke, G., and Merk, W. (1973). Intrinsic interference caused by hepatitis sera. *Proc. Soc. Exp. Biol. Med. 143*, 698-700.

Bird, F. T. (1959). Polyhedrosis and granulosis viruses causing single and double infections in the spruce budworm, *Choristoneura fumiferana* Clemens. *J. Insect Pathol. 1*, 406-430.

Brown, M., Crawford, A. M., and Faulkner, P. (1979). Genetic analysis of a baculovirus, *Autographa californica* nuclear polyhedrosis virus I. Isolation of temperature-sensitive mutants and assortment into complementation groups. *J. Virol. 31*, 190-198.

Carstens, E. B. (1982). Mapping the mutation site of *Autographa californica* nuclear polyhedrosis virus polyhedron morphology mutant. *J. Virol. 43*, 809-818.

Carver, D. H. and Seto, D. S. Y. (1971). Production of hemabsorption-negative areas by serums containing Australia antigens. *Science 172*, 1265-1267.

Chapman, H. C. (1974). Biological control of insects. *Annu. Rev. Entomol. 19*, 33-59.

Clark, T. B. and Chapman, H. C. (1969). A polyhedrosis in *Culex salinarius* of Louisiana. *J. Invertebr. Pathol. 13*, 312.

Clark, T. B. and Fukada, T. (1971). Field and laboratory observations of two viral diseases in *Aedes sollicitans* in southwestern Louisiana. *Mosquito News 31*, 193-199.

Croizier, G., Godse, D., and Vlok, J. (1980). Selection of new types of virus in larvae of *Galleria mellonella* L. infected with two baculoviruses. *C.R. Acad. Sci. 290*, 579-582.

Enzman, P. J. (1973). Induction of an interferon-like substance in persistently infected *Aedes albopictus* cells. *Arch. Virusforsch. 40*, 382-389.

Esparaza, J. and Sanchez, A. (1975). Multiplication of Venezuelan equine encephalitis in cultured mosquito cells. *Arch. Virol. 49*, 273-280.

Fenner, F., McAuslan, B. R., Mims, C. A., Sambrook, J., and White, D. O. (1974). "The Biology of Animal Viruses." Academic Press, New York.

Giorno, R. and Kates, J. R. (1971). Mechanism of inhibition of vaccinia virus replication in adenovirus-infected HeLa cells. *J. Virol. 7*, 208-213.

Granados, R. R., Nguyen, T., and Cato, B. (1978). An insect cell line persistently infected with a baculovirus-like particle. *Intervirology 10*, 309-317.

Gushchin, V., Tsilinsky, Y. Y., Karpova, E. F., Gushchina, E. A., and Klimenko, S. M. (1981). Morphological study of mixed infection with alphaviruses. *Vopr. Virusol. 0(6)*, 728-731.

Hess, R. T., Summers, M. D., Falcon, L. A., and Stoltz, D. B. (1977). A new icosahedrol insect virus: apparent mixed nuclear infection with the baculovirus of *Autographa californica*. *ICRS Med. Sci. 5*, 562.

Hess, R. T., Summers, M. D., and Falcon, L. A. (1978). A mixed virus infection in midgut cells of *Autographa california* and *Trichoplusia ni* larvae. *J. Ultrastr. Res. 65*, 253-265.

Hink, W. F. and Strauss, E. (1976). Replication and passage of alfalfa looper nuclear polyhedrosis virus plaque variants in cloned cell cultures and larval stages of four host species. *J. Invertebr. Pathol. 27*, 49-55.

Huang, A. S. and Baltimore, D. (1970). Defective viral particles and viral disease processes. *Nature (London) 226*, 325-327.

Hughes, K. M. (1979). Some interactions of two baculoviruses of the Douglas-fir tussock moth. *Can. Entomol. 111*, 521-523.

Joklik, W. K., Willett, H. P., and Amos, D. B. (1980). "Microbiology." Appleton-Century-Crofts, New York.

Jones, J. C. (1975). Forms and functions of insect hemocytes. *In* "Invertebrate Immunity" (K. Maramorosch and R. Shope, eds.), pp. 119-128. Academic Press, New York.

Kelly, D. C. (1980). Suppression of baculovirus and iridescent virus replication in dually infected cells. *Microbiologica 3*, 177-185.

Kelly, D. C., Lescott, T., Ayres, M. D., Casey, D., Coutts, A., and Harrap, K. A. (1981). Induction of a nonoccluded baculovirus persistently infecting *Heliothis zea* cells by *Heliothis armigera* and *Trichoplusia ni* nuclear polyhedrosis viruses. *Virology 112*, 174-189.

Kobayashi, M., Inagaki, S., and Kawase, S. (1981). Effect of high temperature on the development of nuclear polyhedrosis virus in the silkworm, *Bombyx mori*. *J. Invertebr. Pathol. 38*, 386-394.

Laemmli, U. K. (1970). Cleavage of structural proteins during the assembly of the head of bacteriophage T4. *Nature (London) 227*, 680-685.

Langridge, W. H. R. (1981). Biochemical properties of a persistent nonoccluded baculovirus isolated from *Heliothis zea* cells. *Virology 112*, 770-774.

Lee, H. H. and Miller, L. K. (1979). Isolation, complementation and initial characterization of temperature-sensitive mutants of the baculovirus *Autographa californica* nuclear polyhedrosis virus. *J. Virol. 31*, 240-252.

Lowe, R. E. and Paschke, J. D. (1968a). Simultaneous infection with the nucleopolyhedrosis and granulosis viruses of *Trichoplusia ni*. *J. Invertebr. Pathol. 12*, 86-92.

Lowe, R. E. and Paschke, J. D. (1968b). Pathology of a double viral infection of *Trichoplusia ni*. *J. Invertebr. Pathol. 12*, 438-443.

McClelland, A. J. and Collins, P. (1978). U.K. investigates virus insecticides. *Nature (London) 276*, 548-549.

McIntosh, A. H. and Ignoffo, C. M. (1981). Establishment of a persistent baculovirus infection in a Lepidopteran cell line. *J. Invertebr. Pathol. 38*, 395-403.

McIntosh, A. H. and Shamy, R. (1980). Biological studies of a baculovirus in a mammalian cell line. *Intervirology 13*, 331-341.

McIntosh, A. H., Shamy, R., and Ilsley, C. (1979). Interference with polyhedral inclusion body (PIB) production in *Trichoplusia ni* cells infected with a high passage strain of *Autographa californica* nuclear polyhedrosis virus (NPV). *Arch. Virol. 60*, 353-358.

MacKinnon, E. A., Henderson, J. F., Stoltz, D. B., and Faulkner, P. (1974). Morphogenesis of nuclear polyhedrosis virus under

conditions of prolonged passage *in vitro*. *J. Ultrastr. Res.* *49*, 419-435.
Marcus, P. I. and Carver, D. H. (1965). Hemabsorption-negative plaque test: new assay for rubella virus revealing a unique interference. *Science 149*, 983-986.
Marcus, P. I. and Carver, D. H. (1967). Intrinsic interference: a new type of viral interference. *J. Virol. 1*, 334-343.
Mazurenko, N. P., Merekalova, Z. I., Jakouleva, L. S., Scherbak, N. P., Kurzman, M. J., Zueva, J. N., and Pavlish, O. A. (1980). Virus-viral co-cancerogenesis and the other viral interactions. *Arch. Geschwulstforsch. 50*, 399-407.
Miller, C. A. and Raine, C. S. (1979). Heterogeneity of virus particles in measles virus. *J. Gen. Virol. 45*, 441-453.
Mitsuhashi, J. (1967). Establishment of an insect cell strain persistently infected with an insect virus. *Nature (London) 215*, 863-864.
Morris, T. J., Vail, P. V., and Collier, S. S. (1981). An RNA virus in *Autographa californica* nuclear polyhedrosis preparation: detection and identification. *J. Invertebr. Pathol.* *38*, 201-208.
Murphy, F. A. (1975). Cellular resistance to arbovirus infection. *In* "Pathobiology of Invertebrate Vectors of Disease" (L. Bulla and T. C. Cheng, eds.), pp. 197-203. New York Academy of Science, New York.
Murphy, F. A., Whitfield, S. G., Sudia, W. D., and Chamberlain, R. W. (1975). Interactions of vector with vertebrate viruses. *In* "Invertebrate Immunity" (K. Maramorosch and R. Shope, eds.), pp. 25-48. Academic Press, New York.
Novokhatskii, A. S. and Berezina, L. H. (1978). Arbovirus multiplication in mosquito cells treated with an interferon inducer. *Vopr. Virusol. 3*, 357-359.
Odier, F. (1977). Mise en évidence et étude d'un complexe d'un maladies a parvovirus, baculovirus et iridovirus. *Entomophaga 22*, 397-404.
Orihel, T. C. (1975). The peritrophic membrane: its role as a barrier to infection of the arthropod host. *In* "Invertebrate Immunity" (K. Maramorosch and R. Shope, eds.), pp. 65-74. Academic Press, New York.
Padgett, B. L. and Walker, D. L. (1970). Effect of persistent fibroma virus infection on susceptibility of cells to other viruses. *J. Virol. 5*, 199-204.
Paschke, J. D. and Summers, M. D. (1975). Early events in the infection of the arthropod gut by pathogenic insect viruses. *In* "Invertebrate Immunity" (K. Maramorosch and R. Shope, eds.), pp. 75-112. Academic Press, New York.
Piasecka-Serafin, M. (1977). A double infection with nuclear and cytoplasmic polyhedrosis viruses in *Bombyx mori*. *Bull. Acad. Polon. Sci. 25*, 287-301.

Quiot, J. M., Vago, C., and Tchoukchry, M. (1980). Experimental study of the interaction of two invertebrate viruses in lepidopteran cell culture. *C.R. Acad. Sci. 290*, 199-201.

Raghow, R. S., Grace, T. D. C., Filshie, B. K., Barthey, W., and Dalgarno, L. (1973). Ross River virus replication in cultured mosquito and mammalian cells: virus growth and correlated ultrastructural changes. *J. Gen. Virol. 21*, 109-122.

Riedel, B. and Brown, D. T. (1979). Novel antiviral activity found in the media of Sindbis virus-persistently infected mosquito (*Aedes albopictus*) cell cultures. *J. Virol. 29*, 51-60.

Ritter, K. S. and Tanada, Y. (1978). Interference between two nuclear polyhedrosis viruses of the armyworm, *Pseudoletia unipuncta*. *Entomophaga 23*, 349-359.

Rubin, H. (1960). A virus in chick embryos which induces resistance *in vitro* to infection with Rous sarcoma virus. *Proc. Natl. Acad. Sci. U.S. 46*, 1105.

Rubin, H. (1961). The nature of a virus-induced cellular resistance to Rous sarcoma virus. *Virology 13*, 200.

Sabin, A. B. (1952). Research on dengue during WWII. *Amer. J. Trop. Med. 1*, 30-50.

Shenk, T. E., Kolshelnyn, K. A., and Stollar, V. (1974). Temperature-sensitive virus from *Aedes albopictus* cells chronically infected with Sindbis virus. *J. Virol. 13*, 439-447.

Sherman, K. E. (1980). Replication of a Lepidopteran baculovirus. Ph.D. dissertation (Rutgers Univ.). University Microfilms International, Ann Arbor.

Sherman, K. E. and McIntosh, A. H. (1979). Baculovirus replication in a mosquito (Dipteran) cell line. *Infect. Immun. 26*, 232-234.

Sinarachatanant, P. and Olsen, L. C. (1973). Replication of dengue virus type 2 in *Aedes albopictus* cell culture. *J. Virol. 12*, 275-283.

Smith, K. M. (1976). "Virus-Insect Relationships." Longman Green, London.

Tanada, Y. and Hukuhara, T. (1971). Enhanced infection of a nuclear polyhedrosis virus in larvae of the armyworm, *Pseudaletia unipuncta*, by a factor in the capsule of a granulosis virus. *J. Invertebr. Pathol. 17*, 116-126.

Tanada, Y., Inoue, H., Hess, R. T., and Omi, E. M. (1980). Site of action of a synergistic factor of a granulosis virus of the armyworm, *Pseudoletia unipuncta*. *J. Invertebr. Pathol. 34*, 249-255.

Tanaka, S. and Aruga, H. (1967). Interference between the midgut nuclear polyhedrosis virus and the cytoplasmic polyhedrosis virus in the silkworm, *Bombyx mori*. *J. Sericult. Sci. Japan 36*, 169-176.

Tarasevich, L. M., Kitik, U. S., and Mencher, E. M. (1980). Mathematical modeling of optimal conditions for efficiency of mixed viral infection of turnip moths. *Izv. Akad. Nauk SSSR, Ser. Biol. 3*, 387-394.

Tinsley, T. W. (1975). Factors affecting virus infection of insect gut tissue. *In* "Invertebrate Immunity" (K. Maramorosch and R. Shope, eds.), pp. 55-65. Academic Press, New York.

Tinsley, T. W. (1979). The potential of insect pathogenic viruses as pesticidal agents. *Annu. Rev. Entomol. 24*, 63-87.

Tsuchiya, Y. and Tagaya, I. (1970). Enhanced or inhibited plaque formation of superinfecting viruses in Yaba virus-infected cells. *J. Gen. Virol. 7*, 71-73.

Von Magnus, P. (1954). Incomplete forms of influenza virus. *Advan. Virus Res. 2*, 59-78.

Vordham, A. V., Murphy, B. L., Hollinger, F. B., and Maynard, J. E. (1973). Attempts to detect hepatitis B virus by negative hemabsorption assay. *Proc. Soc. Exp. Biol. Med. 143*, 395-399.

Wagner, G. W., Webb, S. R., Paschke, J. D., and Campbell, W. R. (1974). A picornavirus isolated from *Aedes taeniorhynchus* and its interaction with mosquito iridescent virus. *J. Invertebr. Pathol. 24*, 380-382.

Wainwright, S. and Mims, C. A. (1967). Plaque assay for lymphocytic choriomeningitis virus based on hemabsorption interference. *J. Virol. 1*, 1091-1092.

Whitlock, V. H. (1977). Simultaneous treatments of *Heliothis armigera* with a nuclear polyhedrosis and a granulosis virus. *J. Invertebr. Pathol. 29*, 297-303.

Yamada, K., Sherman, K. E., and Maramorsch, K. (1981). *In vivo* infectivity of early and late passaged *Heliothis zea* polyhedra produced in tissue culture. *Appl. Entomol. Zool. 16*, 504-505.

VI.

Production and Field Application

CONSIDERATIONS IN THE LARGE-SCALE AND COMMERCIAL PRODUCTION OF VIRAL INSECTICIDES

KENNETH E. SHERMAN

George Washington University School of Medicine
Washington, D.C.

I. INTRODUCTION

The true measure of a concept's worth becomes apparent when it can be taken from the realm of theories and experiments, and applied as a solution or alternative in a concrete manner. When we consider the viability of viruses as biological control agents, this axiom certainly applies. Much of this book is devoted to improving the understanding of the biology and chemistry of viral agents that have been found to be pathogenic to host invertebrate species. Where do we stand today, in terms of applying basic science to the population control of pest species? Has the use of pathogenic viruses been demonstrated to be a truly viable alternative to chemical agents?

This chapter will serve as a review of factors and considerations that must be taken into account when a viral insecticide is sought as a large-scale commercial alternative to other more traditional methods of pest control. There are many excellent papers and technical bulletins which detail the specifics of each industrial production step, and except as examples, those details are beyond both the scope and intent of this chapter. This chapter examines and evaluates methods of viral insecticide production, and highlights some of the salient points pertaining to safety and quality control.

ISBN 0-12-470295-3

There are currently several viral insecticide formulations that have been approved for use by the United States Environmental Protection Agency. A larger number of such formulations are in various stages of review and evaluation. Literature pertaining to these agents provides ample material for evaluation and review of the current state-of-the-art.

II. PRODUCTION

A. *In Vivo* PRODUCTION METHODS

Virus production in living host insects is the predominant method used today to obtain suitable material for viral insecticides. Infected hosts can be obtained in one of two ways. The first method involves identifying geographic locations where large numbers of the target populations occur. A seed inoculum of the specific desired virus agent can be introduced into this population, and infected individuals can be harvested This process can be made even simpler, if the specific virus sought is naturally occurring in a population. For example, during the last decade an epidemic wave of the gypsy moth (*L. dispar*), has been rolling through forests and devastating millions of acres of foliage in the eastern United States. Each time this forest pest establishes itself in a new uninfested area, its population explodes and excessive forest damage occurs. In general, however, these waves are followed within several years by increases in natural or introduced biological control agents and the gypsy moth population typically "crashes." Following this, a cycle defined by a sine wave rhythmicity is established and the ecosystem returns to a more normal state. If one was interested in obtaining a naturally occurring virus that was pathogenic to the gypsy moth and that could be harvested in relatively large numbers, insect larvae could be collected from the fringe zone where natural biological controls were becoming prevalent. Obviously, these methods are fraught with problems and uncertainties. Collection would be highly dependent upon the survey and identification of the population dynamics in a wide variety of locations. Collection would be seasonable at best. There would be great variation in yield as well as viral activity. Contamination with adventitious agents would pose a major problem in terms of safety and licensing. For all of these reasons, field collection of viral product is not a highly desirable method of production for biological control agents. Its usefulness seems to rest primarily on the value of field collection to provide new sources of inoculum which may have properties not available in laboratory strains.

The second primary *in vivo* production methodology employs the laboratory colonization of host insect species, and the controlled introduction of viral agents. It is in this area that the majority of research efforts have concentrated. In this setting, a variety of variables can be addressed and attempts can be made to optimize production in a cost-effective manner. Factors that have been considered include variations in host insect stock, numerous factors related to the host surroundings and environment, strain differences among viral pathogens, and critical time-related events in inoculation and infection.

1. *The Host Insect*

Rapid, vigorous growth of colonized hosts is an essential element in the development of a large-scale virus production program. In one study by Bell et al. (1980) the rate of population increase per generation was calculated by multiplying the potential rate of population increase from the parental to the first generation by the percentage hatch. They reared gypsy moths collected from different geographic locations. There was greater than a 2.5-fold difference in the actual rate of increase between insects collected from two of the selected sites. The authors felt that the presence of intrinsic pathogenic virus was the single most important factor in this observation (Bell et al., 1980). Several articles in the literature cite sex ratio variations that occur in wild populations of insects. This is important to consider because of differences in hatch times, maturation rates, food consumption, and rearing-container effects on growth between males and females (Campbell, 1963, 1967; Leonard, 1968; Bell et al., 1980). Therefore, potential laboratory stock insect hosts must be carefully evaluated before they are selected as production subjects.

In virus production facilities, there should be as much separation and isolation of different activities and processes as is feasible. Thus, there should be a work area set aside for the rearing of healthy, uninfected stock host insects. A separate area, to be used for mass rearing of insects is important, because healthy, uninfected stock strains must be carefully maintained on a small scale with constant selection for desirable growth characteristics. Virus inoculation and production and processing must be physically separated from all host-rearing activities.

2. *Environmental Factors*

The growth and vigor of the host insect, and its subsequent ability to produce harvestable virus depends on a large number of environmental factors. These include nutrition, container size, temperature, humidity, population density, the male:female sex ratio, and lighting, as well as other factors which have been less well defined. Insect species is an important variable. For example, Ignoffo notes that the cabbage looper feeds gregariously and is much easier to raise than the cannabalistic bollworm (Ignoffo, 1966). Therefore, optimization of growth would require different rearing environments for these two species. There has been considerable research in the evaluation of factors affecting insect rearing over the years. There are many excellent articles and pamphlets which depict in great detail the particular requirements of many insect species. What follows here is a brief discussion of several of these factors, with emphasis on insects that are of interest in large-scale virus production.

Feeding colonized insects their natural diet involves cultivating, collecting, and storing plant material. This process is generally tedious and costly and does not lend itself efficiently to large-scale production schemes. Artificial or seminatural diets have been used for many years in the domestic colonization of insects. Many of the issues and considerations concerning artificial diet development were discussed in a 1966 review article by Vanderzant, and are still relevant (Vanderzant, 1966). The first diet used in a laboratory setting for the production of insect viruses consisted of a combination of cassein, wheat germ, and cotton leaves. This base was supplemented with a vitamin stock as well as with antibiotics to reduce bacterial contaminants. An agar base was used to bring the mixture to a semisolid consistency. A hot liquid diet was dispensed into rearing containers where it cooled and solidified (Ignoffo, 1964, 1966). This diet, with minor modifications, as well as artificial diets made with varying protein or plant base ingredients have been used extensively over the last decade to feed insects being raised for viral production. Shapiro presents a comparison of gypsy moth diets which have been used in production of a nuclear polyhedrosis virus (NPV). He notes that a high wheat germ diet surpassed other diets tested, in terms of virus yield and relative cost per comparative unit of virus production (Shapiro, 1982). This underscores the fact that extensive testing is necessary for each species of insect being reared to assure that cost effectiveness is optimized.

Container size and population density are complementary concerns. As noted above, some insects are gregarious and others are cannibalistic. Because the container is a microenvironment, it has a great effect on almost all of the other environmental characteristics that seem to make a difference in terms of growth rate and eventual virus production. This is illustrated in one set of experiments by Bell et al. in which they investigated gypsy moth growth in 180-ml polyethylene cups. They reared between four and ten larvae per cup. The cups contained enough food so that competition for food would not be a factor. The results showed no difference in the rate of development rate, but fewer individuals reached the pupal stage in the higher density containers and the male pupal weight was decreased. The authors felt that this difference was due to competition for molting and pupation space (Bell et al., 1980). They did not investigate whether this difference was an inherent biological characteristic that is population density-dependent, or if it was related to differences in humidity, air flow, and/or temperature. Containment units have been made of plastic, glass, waxed cartons, paper, paraffin-coated bags, screen units, and Petri dishes (Ignoffo, 1964; Chauthani and Claussen, 1968; Vail et al., 1973). The containers should be large enough to permit optimal growth without the need for insect transfer and they should be inexpensive. Stacking the containers in a configuration which would permit optimal use of space without compromising air flow would also be desirable. Finally, harvest or inoculation should be simple, with easy access to the growing insects. The 180-ml plastic containers (Dixie type) with cardboard lids have been used extensively, but they lack some of the critical characteristics noted above.

Temperature and humidity are two important environmental factors that can alter both insect growth and the level of virus production. As with many of these factors, it must be borne in mind that we are dealing with two different, but interrelated, biological systems. There are temperature and humidity optimums for insect growth which become particularly important at the stage of up-scaling production of colony insects destined to serve as virus propagation stock. This temperature level is not necessarily the optimal temperature for virus production. Each insect species has its own optimal range, and this temperature should be carefully regulated for maximum cost-effectiveness. Temperature is critical in maximizing virus production and will be discussed later in this chapter. Humidity must be kept high enough so that larvae do not dehydrate. A relative humidity of about 50% is adequate for most species.

The photoperiod or relative light:dark cycle is important in the development and maturation of some insects. Gypsy moth larvae have been routinely reared in 24-hour cycles--16 hours light and 8 hours dark (Leonard and Doane, 1966; Odell and Rollinson; 1966). However, experiments by Bell et al. failed to show any difference in larval growth, virus-induced mortality, or virus yield when differing photoperiods were tried (Bell et al., 1980).

3. *Virus Inoculation, Growth, and Harvest*

The majority of research effort in the development of viral insecticides has concentrated on crop and forest pests that devour relatively large volumes of plant material. The best way to cause a viral infection in these pests is to associate the insecticidal agent with a food product. Various research teams have devised methods of spraying virus suspensions on the provided food source; virus has also been directly incorporated into food preparations. Shapiro properly notes that the ideal method of viral inoculation is variable and depends on the biology of the host and efficiency versus efficacy (Shapiro, 1982). As early as 1964, Ignoffo described a procedure by which he produced *Trichoplusia ni* nuclear polyhedrosis virus under *in vivo* laboratory conditions. Larvae and a semisynthetic diet were sprayed using a hand atomizer. The atomizer contained a suspension containing 5×10^6 polyhedra/ml (Ignoffo, 1964). Variations of this method utilizing surface spraying have been widely employed since then (Vail et al., 1973; Ignoffo, 1965; Bell et al., 1980). While other workers have incorporated virus into the diet (Lewis, 1971; Hedlund and Yendol, 1974), problems with temperature inactivation and ease of preparation (Shapiro, 1982) have made the spray method more attractive as the primary means of virus inoculation. The inoculation dose is expressed in units of PIB/ml, and the optimal dose varies with the virus and host being considered. In general, a dose above the optimally determined level causes early death of the production insects, thus reducing total virus output. Suboptimal inoculation means that all hosts do not become infected initially and final yield may be lower than input virus dose. Theoretically, the optimal dose can be calculated by application of the Poisson distribution, and by estimating the average number of viral particles needed to cause clinical infection in each individual. In practice, variations in virulence require each production facility to determine the optimal level for their system. Guidelines for optimal virus concentration of important viral agents are reviewed and presented in tabular form in Shapiro's review of *in vivo* production methods (Shapiro, 1982).

Virus growth and yield is highly dependent on environmental factors. These factors are often, but not always, coincident with conditions necessary for optimal growth of the host insect. Thus, in early studies, Ignoffo maintained the noninfected host insect at a somewhat higher temperature than

NPV-infected individuals. This temperature difference was necessary so that living, intact infected larvae could be harvested (Ignoffo, 1964, 1965). Studies on the gypsy moth NPV revealed that yield and virus activity were similar regardless of the temperature in the 23°-29°C range (Bell et al., 1980). However, growth at 29 C produced the highest yields in the shortest time (Bell et al., 1980). At higher temperatures, problems develop in terms of restriction of viral replication. *Bombyx mori* larvae inoculated with an NPV survive longer at 35°C than at lower temperatures. Thermal inhibition of these viruses may be at least partially due to a block in virus replication at the molecular level (Kobayashi et al., 1981). Such inhibition has been noted for a variety of insect pathogenic viruses (Thompson, 1959; Tanada and Chang, 1968; Watanabe and Tanada, 1972).

The yield of virus from a particular host is directly related to the body mass of potentially infected tissues in the individual. There are several very important factors which figure in the final viral output in a colonized, virus-infected population because of variation in available body mass. The sex ratio is one such factor to consider. In general, female larvae of lepidopterous insects tend to be larger than their male counterparts at each stage of development. Data presented by Bell et al. (1980) clearly demonstrate that female gypsy moth larvae produce more NPV than males, solely on the basis of increased body mass. The fact that population density and container size can affect the growth rate of colonized insect hosts was discussed in a previous chapter. Under suboptimal conditions for that particular species, growth rates and final body mass will be reduced and virus yield would suffer. Finally, the larval age at time of inoculation will affect the final virus yield. If larvae are infected too early in their life cycle, they will die well before they reach maximal size. If infection occurs at too late a stage, then not all tissues will be carrying a maximal virus yield before pupation. For this reason, most studies have utilized larvae from the third or fourth instar. Ignoffo was able to achieve yields of greater than 10,000 times inoculum in cabbage looper larvae (Ignoffo, 1966). Vail obtained a similar increase by growing *Autographa californica* NPV in third to fourth instar cabbage looper larvae (Vail et al., 1973). However, fifth instar larvae produced the highest yield of gypsy moth NPV (Bell et al., 1980). It should be apparent that, in general, larger size strains or varieties of the same species will produce more virus than smaller individuals, as long as other possible interfering factors such as endogenous virus are not present.

The final virus activity seems to be influenced by whether the infected larvae are alive or dead at the time of harvest. There is an increase in the activity of the *Heliothis zea* NPV if it is collected from virus-killed host insects (Ignoffo and Shapiro, 1978). Preliminary data published regarding the gypsy moth NPV demonstrates a similar relationship between percentage mortality and increased viral activity (Bell et al., 1980). It is unknown why this increase in activity occurs. It is known that dead larvae contain a higher intrinsic bacterial load than living infected larvae (Ignoffo and Heimpel, 1965; Ignoffo and Shapiro, 1978). It is possible that degradative enzymes or metabolic by-products released by saprophytic bacteria may have some role in the modification of viral activity that is observed. Under laboratory conditions, it has been observed that simple modification of the electrostatic interaction between the viral envelope and the cell membrane significantly alters viral infectivity (Yamamoto and Tanada, 1978).

Virus harvest involves the collection and storage of infected larvae which have been raised in containers containing just a few individuals. Obviously, this is a tedious and time-consuming step in the production of viral insecticides from host insects. Most workers have included some variation of a freezing step in which containers with larvae are placed in freezers. This accomplishes several objectives. Bacterial growth and degradation of carcasses is retarded, frozen larvae are easier to transfer than mushy, decaying bodies, and larvae could be held in storage until processing of large batches could begin (Ignoffo, 1966; Chauthani and Claussen, 1968; Bell et al., 1980).

Extraction of virus from infected larvae generally involves a blending step in which larval tissues containing the viral product are liquefied. In early research studies, three parts of water were mixed with one part infected larvae. Larvae were crushed and allowed to decompose. Varying grades of cheesecloth were then used to filter the suspension of large solids (Ignoffo, 1966). Work done by Bell et al. (1980) on the gypsy moth NPV showed that a 1:10 dilution was most effective in maximizing recovery of PIB's, and that the blending time was also another crucial factor in increasing product yield. Purification of PIB's from polyhedrosis viruses is generally accomplished by ultracentrifugation. Ignoffo notes that centrifugation should ensure maximal recovery without the packing of inclusion bodies (Ignoffo, 1966). Most workers have centrifuged PIB's at 5000 to 15,000 *g* for 15 to 20 minutes (Ignoffo, 1966; Smith et al., 1976; Bell et al., 1980). The supernatant is then discarded, and the pellet resuspended for further formulation. Current research efforts revolve around the elimination of centrifugation and the use of freeze-drying to isolate the infectious agent (Bell et al., 1980).

B. *In Vitro* PRODUCTION METHODS

As discussed above, there are many variables associated with the production of a viral insecticide in a living host insect. This problem is similar to that which faced early researchers who were searching for viral-containing material for the purpose of vaccine production. The classic work of Enders and his colleagues clearly demonstrated the feasibility of producing virus or virus products in tissue culture systems, and has become the basis of much of the commercial vaccine industry of today. Growth of virus in cell culture seems to have many advantages over *in vivo* production methodologies. Mass production techniques, many of which can be automated, lend themselves to cell culture techniques. Nutritional and environmental requirements can be very strictly monitored and regulated. Quality control would be theoretically easier because the entire system has been simplified to more basic components. Despite these advantages, to date, there has not been wide application of *in vitro* methods. However, many workers have examined factors involved in optimizing this production method, and in keeping the cost competitive with more widely practiced *in vivo* production schemes.

1. *Cell Culture Methods*

Permanent cell lines, derived from the cells of an insect species were first cultured in 1962, utilizing tissues from *Antheraea eucalypti* (Grace, 1962). Since that time a large number of arthropod cell lines have been established. Many of these cell lines support the growth of one or more viral agents. Several of the lepidopteran cell lines have been used extensively because of their ability to grow quantities of nuclear polyhedrosis, cytoplasmic polyhedrosis, and granulosis viruses which are being considered as possible insecticidal agents. The cell lines most frequently encountered in the literature represent several important agricultural pest species including the cabbage looper, *Trichoplusia ni*, the alfalfa looper, *Autographa californica*, the fall army worm, *Spodoptera frugiperda*, and the cabbage moth, *Mamestra brassicae*. The methodology involved in the primary culture of these and other insect cells is beyond the scope of this chapter, but certain generalizations can be made. The starting material is usually eggs or early stage larvae which have been surface sterilized with alcohol, hypochlorite or other disinfecting solutions. Eggs or larvae are physically macerated and then tissues are dissociated with an enzyme preparation such as trypsin or collagenase. The resultant individual cells and small tissue fragments are placed in tissue culture flasks or Leighton tubes with a small amount of growth medium containing fetal bovine serum and an antibiotic (e.g., streptomycin). Growth surfaces are microscopically examined for cellular outgrowth. Attempts are then made to subculture and eventually a cell line is established. Dilutional procedures can be used to select specific clones which may prove to be more efficient in the production of viruses.

Over the years, a variety of growth media have been developed or adapted for the purpose of growing insect cells in culture (Wyatt, 1956; Mitsuhashi and Maramorosch, 1964; McIntosh et al., 1973; Hink, 1976). Variations of these media, containing lower levels of ionized calcium are utilized to grow cells in suspension. Calcium is an important factor in the attachment and binding of cells to container surfaces.

Efficient scale-up of cell production requires that many cells grow in a single container. The use of tissue culture flasks (glass or plastic) is limited by the volume and surface area which can be effectively managed on a large scale. For this reason, efforts have focused on the use of spinner cultures and fermentation systems. In his review of work done by himself and co-workers, Hink describes how *Trichoplusia ni* cells were grown in 100-ml spin flasks. They found that important factors in optimization of cell growth included the oxygen tension, clumping, and pH. Aeration, addition of methylcellulose, and constant pH adjustment tended to resolve these respective problems and to increase cell yield (Hink, 1982). Further scale-up to 2-liter fermenters required adjustment of all the above factors (Hink and Strauss, 1980). The importance of these factors and others will be increased in the future, primarily because the twin goals of increased yield and decreased cost will eventually determine the practical viability of *in vitro* production alternatives.

2. *Virus Growth and Related Factors*

Since a viable insect pathogenic virus was first produced in a cultured insect cell line in 1970 (Goodwin et al., 1970), a number of other such systems have been demonstrated to be alternatives to virus production in living host insects. There are some key differences between the *in vivo* and *in vitro* infection processes. Virus trapped in polyhedral inclusion bodies has no effect on cells growing in culture in terms of replicative potential. Therefore, the starting inoculum must be either nonoccluded virions harvested from a tissue culture system, or free virus must be produced by dissolution of the polyhedron protein matrix. This methodology is discussed elsewhere in this book. Free virus particles seem to enter the cultured cells by phagocytosis and viropexis (Ragshow and Grace, 1974; Adams et al., 1977).

The phase of cell culture growth is one factor that influences virus yield. Studies by Volkman and Summers (1975) and Vaughn (1976) demonstrated that cells in the log phase of growth were more susceptible to infection and produced more polyhedra than cells on other growth phases. However, data presented by Lynn and Hink (1978) demonstrated that virus production was not a function of infectivity versus cell cycle. Yield of polyhedra is significantly influenced by the cell density. Both number of polyhedra per milliliter and the number of polyhedra per infected cell showed marked variation when cell density was considered in the production of the *Autographa californica* NPV (Hink, 1982).

The phenomenon of viral attenuation with continuous passage of virus in cell culture is important to consider in light of attempts to maximize *in vitro* production. MacKinnon et al. (1974) reported that the number of polyhedra per cell, as well as the titer of nonoccluded virus, declined significantly when the *T. ni* NPV was passaged more than 15 times. PIB production and the number of normal virions per PIB was reduced after prolonged passage of the *Autographa californica* NPV in *T. ni* cells (Hirumi et al., 1975). Long-term serial passage of the *Heliothis zea* NPV resulted in a decline in both the total number of polyhedral inclusion bodies produced and the level of nonoccluded infectious virus available (Yamada et al., 1982). Even more significant is evidence that the polyhedra from late passaged *H. zea* NPV are virtually unable to cause mortality in cotton bollworm larvae, although similar doses of early passaged PIB's cause significant mortality (Yamada et al., 1981). McIntosh and Ignoffo (1981) reported on the establishment of a cell line persistently infected with a baculovirus. Although virus was consistently produced, the number of inclusion bodies was reduced about 98% below primary infection in the parent line and infectious viral titer was about 100 times lower. Obviously, these data indicate a serious problem which must be considered when commercial-scale attempts are made in cell culture systems. Hink (1982) describes a system for the semicontinuous production of virus in fermentation-type cultures. While this system provides an excellent model for an efficient cost-effective production system, it too suffered from decreased yields after the first four harvest cycles.

III. CONSIDERATIONS IN PRODUCT FORMULATION

Thus far, this chapter has dealt primarily with the actual production of insect pathogenic viruses that might be used as a viral insecticide. For a commercial venture to be viable, a final usable product must be developed. This product should be formulated in such a way so that it can be easily distributed under field conditions, preferably utilizing existing technologies and methods. Further, the product should have a standardized level of activity so that accurate determinations about application levels could be made. Finally, the product should be formulated so that it maintains its maximal effectiveness under prevailing environmental conditions.

Traditional chemical insecticides are formulated in several different ways, including solutions, emulsifiable concentrates, wettable powders, dusts, granules, and baits (James and Harwood, 1969). Of these, wettable powder, dust, or bait formulations are most applicable to the use of insect pathogenic virions which are encapsulated in a natural protein matrix. Wettable powders consist of inert carriers which are impregnated with the viral agent. Fluid is added and the formulation is kept in suspension by the use of agitation in the applicator system. Dusts are generally applied in dry form and consist of the agent diluted out in an inert dry carrier. A bait is generally a preferred food product which contains the agent in question. The formulation should be noncorrosive, and should be evenly distributed so that it does not cause clogging of applicator equipment. A major problem in the use of wettable powders and dusts is equipment wear caused by the abrasive qualities of many formulations.

Sunlight is a significant factor in the loss of infectivity of baculoviruses which are placed under field conditions (Bullock, 1967; David et al., 1968; Jaques, 1977). This is probably a function of the intensity and wavelength of the ultraviolet radiation component of sunlight (Smirnoff, 1972). For this reason, the formulation of baculovirus-containing insecticidal agents should contain agents that retard the loss of infectivity due to sun exposure. Jaques (1971, 1972) investigated a number of additives and combinations of additives including protein products, stains and dyes, charcoal, and commercially prepared protective agents to determine their relative ability to protect *T. ni* NPV from uv inactivation. Several of these products were highly successful at providing protection for the periods of uv exposure tested. Similarly, studies on the *H. zea* NPV have demonstrated the value of charcoal in prolonging viral activity (Ignoffo and Batzer, 1971). Infectivity can also be affected by the final pH of the spray tank mixture, and by the presence of chlorine in the water used in final preparation.

As previously discussed, there are many variables that affect the number of polyhedra produced per batch, the number of viable virions per PIB, and the relative activity of the virions to produce disease. For these reasons, it is imperative that some effort be made to standardize preparations of viral insecticides that are to be used in field applications. A technical bulletin produced by the USDA Forest Service Science and Education Agency summarizes the methods used to evaluate activity of the Douglas-fir tussock moth NPV. Briefly, several dilutions of the virus are fed to second instar larvae and the median lethal concentration (LC_{50}) is calculated by minimum logit chi-square analysis. One activity unit is the weight of preparation that kills 50% of the test larvae (Martignoni, 1978). Similar methods are used to determine the activity of other agents being developed as viral insecticides because bioassay is the legally required method for lot evaluation in the United States.

IV. SAFETY AND QUALITY CONTROL

Obviously, safety is a primary consideration in the use of viral agents as insecticides that will be disseminated into the environment. Empirically speaking, viral agents isolated from environmental sources should be safe. Natural epizootics among insect populations occur frequently (Thomas, 1975). Studies in one state revealed that 68% of 116 untreated fields had active *T. ni* NPV (Jaques, 1975). Heimpel et al. (1973) report that up to 1.1×10^6 PIB/cm^2 of cabbage leaf surface was found in supermarket produce. This produce came from previously untreated fields and was a naturally occurring virus load. Despite this evidence, there has been considerable work done to more formally evaluate the safety of possible viral insecticides. Martignoni (1978) summarizes the tests required for each production lot of the Douglas-fir tussock moth NPV. Test of safety to vertebrates includes enumeration of coliform bacteria, detection of fecal coliform, or pathogenic Enterobacteriaciae, safety to mice by intraperitoneal injection, and safety to mice following peroral administration. In addition, dark-field and electron microscopy is performed to detect other viral or bacterial contaminants. This type of testing meets the standards of the United States Environmental Protection Agency. Further testing for dermal and eye toxicity is also required.

To prevent loss of the final product, steps must be taken at all phases in the production process to assure the highest purity of the final product. Therefore, it behooves the manufacturer to start with a pure viral inoculum and to infect viral- and bacterial-free hosts. This is often very difficult because of the prevalence of both recognized and unrecognized adventitious agents which can occur in living insects as well as tissue culture systems. Granados et al. (1978) described an established lepidopteran cell line which was persistently infected with a baculovirus-like particle. This is of concern because regulations require full characterization of biologically active agents in a commercial formulation. In addition, the presence of adventitious agents may adversely affect the level of viral yield by the process of interference. Time of harvest from dead larvae as well as temperature and storage environment of recently harvested product make a difference in terms of the growth of saprophytic bacteria that contaminate the final product.

V. THE FUTURE OF COMMERCIAL PRODUCTION OF VIRAL INSECTICIDES

At the outset of this chapter, it was noted that the final decision on the value of viral insecticides would be determined by their efficacy, ease of use, and cost relative to more traditional methods of pest control. Studies which have investigated several of the baculovirus agents have confirmed their value as biological control agents. Current formulations can be applied without the need to develop or purchase new technologies for their dispersal. The final question is one of cost. Bell et al. (1980) report that the use of fifth instar larvae reduced previous reported costs to about 2 cents per larvae and that modifications in processing would add about 1 cent per larvae, yielding a total cost of about 3 cents per infected larvae. Hink (1982) estimated that the production of one larval equivalent in tissue culture currently costs about 23 cents. Those involved with both *in vivo* and *in vitro* technologies believe that there is substantial room for cost improvement using either methodology. The future seems to hold great promise for the wide commercial application of viral insecticides and they will surely play a larger role in the future of pest management.

REFERENCES

Adams, J. R., Goodwin, R. H., and Wilcox, T. A. (1977). Electron microscope investigations on invasion and replication of insect baculoviruses *in vivo* and *in vitro*. *Biol. Cellulaire 28*, 261-268.

Bell, R. A., Owens, C. D., Shapiro, M., and Tardif, J. R. (1980). *In* "The Gypsy Moth: Research Towards Integrated

Pest Management" (C. C. Doane, ed.), pp. 599-655. USDA Tech. Bull. 1584.

Bullock, H. R. (1967). Persistence of *Heliothis* nuclear polyhedrosis virus on cotton foliage. *J. Invertebr. Pathol. 9*, 434-436.

Campbell, R. W. (1963). Some factors that distort the sex ratio of the gypsy moth, *Porthetria dispar*. *Can. Entomol. 95*, 465-474.

Campbell, R. W. (1967). Studies on the sex ratio of the gypsy moth. *Forest Sci. 13*, 19-22.

Chauthani, A. R., and Claussen, D. (1968). Rearing Douglas-fir tussock moth larvae on synthetic media for the production of a nuclear polyhedrosis virus. *J. Econ. Entomol. 61*, 101-103.

David, W. A. L., Gardiner, B. O. C., and Wollmer, M. (1968). The effects of sunlight on a purified granulosis virus of *Pieris brassicae* applied to cabbage leaves. *J. Invertebr. Pathol. 11*, 496-501.

Goodwin, R. H., Vaughn, J. L., Adams, J. R., and Louloudes, S. J. (1970). Replication of a nuclear polyhedrosis virus in an established insect cell line. *J. Invertebr. Pathol. 16*, 284-288.

Grace, T. D. C. (1962). Establishment of four strains of cells from insect tissues grown *in vitro*. *Nature (London) 195*, 788-789.

Granados, R. R., Nguyen, T., and Cato, B. (1978). An insect cell line persistently infected with a baculovirus-like particle. *Intervirology 10*, 309-317.

Hedlund, R. C., and Yendol, W. G. (1974). Gypsy moth nuclear polyhedrosis virus production as related to inoculating time, dosage, and larval weight. *J. Econ. Entomol. 67*, 61-63.

Heimpel, A. M., Thomas, E. D., Adams, J. R., and Smith, L. J. (1973). The presence of nuclear polyhedrosis viruses of *Trichoplusia ni* on cabbage from the market shelf. *J. Environ.Entomol. 2*, 72-75.

Hink, W. F. (1976). A compilation of invertebrate cell lines and culture media. *In* "Invertebrate Tissue Culture" (K. Maramorosch, ed.), pp. 358-369. Academic Press, New York.

Hink, W. F. (1982). Production of *Autographa californica* nuclear polyhedrosis virus in cells from large-scale suspension cultures. *In* "Microbial and Viral Pesticides" (E. Kurstak, ed.), pp. 493-506. Dekker, New York.

Hink, W. F., and Strauss, E. M. (1980). Semi-continuous culture of the TN-368 cell line in fermentors with virus production in harvested cells. "Invertebrate Systems *In Vitro*," pp. 27-33. Elsevier, North Holland, New York.

Hirumi, H., Hirumi, K., and McIntosh, A. H. (1975). Morphogenesis of a nuclear polyhedrosis virus of the alfalfa

looper in a continuous cabbage looper cell line. *Ann. N.Y. Acad. Sci. 266*, 302-326.
Ignoffo, C. M. (1964). Production and virulence of a nuclear-polyhedrosis virus from larvae of *Trichoplusia ni* reared on a semisynthetic diet. *J. Invertebr. Pathol. 6*, 318-326.
Ignoffo, C. M. (1965). The nuclear polyhedrosis virus of *Heliothis zea* and *Heliothis virescens. Part I. Virus propagation and its virulence. J. Insect Pathol. 7*, 209-216.
Ignoffo, C. M. (1966). Insect viruses. *In* "Insect Colonization and Mass Production" (C. N. Smith, ed.), pp. 501-530. Academic Press, New York.
Ignoffo, C. M., and Batzer, O. F. (1971). Microencapsulation and ultraviolet protectants to increase sunlight stability of an insect virus. *J. Econ. Entomol. 64*, 850-853.
Ignoffo, C. M., and Heimpel, A. M. (1965). The nuclear-polyhedrosis virus of *Heliothis zea* and *Heliothis virescens*. 5. Toxicity-pathogenicity of virus to white mice and guinea pigs. *J. Invertebr. Pathol. 7*, 329-340.
Ignoffo, C. M., and Shapiro, M. (1978). Activity of viral preparations processed from living and dead larvae. *J. Econ. Entomol. 71*, 186-188.
James, M. T., and Harwood, R. F. (1969). "Medical Entomology." The Macmillan Company, London.
Jaques, R. P. (1971). Tests on protectants for foliar deposits of a polyhedrosis virus. *J. Invertebr. Pathol. 17*, 9-16.
Jaques, R. P. (1972). The inactivation of foliar deposits of viruses of *Trichoplusia ni* and *Pieris rapae* and tests on protectant activities. *Can. Entomol. 104*, 1985-1994.
Jaques, R. P. (1975). Persistence, accumulation, and denaturation of nuclear polyhedrosis and granulosis viruses. *In* "Baculoviruses for Insect Pest Control: Safety Considerations" (M. D. Summers, R. Engle, L. A. Falcon, and P. G. Vail, eds.), pp. 90-99. Amer. Soc. for Micro. Publ., Washington, D.C.
Jaques, R. P. (1977). Stability of entomopathogenic viruses. *Misc. Publ. Entomol. Soc. Amer. 10*, 99-116.
Leonard, D. E. (1968). Sexual differentiation in time of hatch of eggs of the gypsy moth. *J. Econ. Entomol. 61*, 698-700.
Leonard, D. E., and Doane, C. C. (1966). An artificial diet for the gypsy moth *Porthetria dispar. J. Econ. Entomol. 59*, 462-464.
Lewis, F. B. (1971). Mass propagation of insect viruses with special reference to forest insects. *Proc. 4th Intern. Colloq. Insect Pathol. Microbiol. Control, College Park, Maryland.*
Lynn, D. E., and Hink, W. F. (1978). Infection of synchronized TN-368 cell cultures with alfalfa looper nuclear polyhedrosis virus. *J. Invertebr. Pathol. 32*, 1-5.

McIntosh, A. H., and Ignoffo, C. M. (1981). Establishment of a persistent baculovirus infection in a lepidopteran cell line. *J. Invertebr. Pathol. 38*, 395-403.

McIntosh, A. H., Maramorosch, K., and Rechtoris, C. (1973). Adaptation of an insect cell line in a mammalian cell culture medium. *In Vitro 8*, 375-378.

MacKinnon, E. A., Henderson, J. F., Stoltz, D. B., and Faulkner, P. (1974). Morphogenesis of nuclear polyhedrosis virus under condition of prolonged passage *in vitro*. *J. Ultrastr. Res. 49*, 419-435.

Martignoni, M. E. (1978). Production, activity, and safety. "The Douglas-fir Tussock Moth: A Synthesis." pp. 140-147. USDA Forest Service Sci. and Educ. Agency Tech. Bull.

Mitsuhashi, J., and Maramorosch, K. (1964). Leafhopper tissue culture: embryonic, nymphal, imaginal tissues from aseptic insects. *Contrib. Boyce Thompson Inst. 22*, 435-460.

Odell, T. M., and Rollinson, W. D. (1966). A technique for rearing gypsy moth, *Porthetria dispar* on an artificial diet. *J. Econ. Entomol. 59*, 741-742.

Ragshow, R., and Grace, T. D. C. (1974). Studies on a nuclear polyhedrosis virus in *Bombyx mori* cells *in vitro*. *J. Ultrastruct. Res. 47*, 384-399.

Shapiro, M. (1982). *In vivo* mass production of insect viruses for use as pesticides. *In* "Microbial and Viral Pesticides" (E. Kurstak, ed.), pp. 463-493. Dekker, New York.

Smirnoff, W. A. (1972). The effect of sunlight on the nuclear polyhedrosis virus of *Neodiprion swainei* with measurement of the solar energy received. *J. Invertebr. Pathol. 19*, 179-188.

Smith, R. P., Wraight, S. P., Tardiff, M. F., Hasenstab, M. J., and Simeone, J. B. (1976). Mass rearing of *Porthetria dispar* for in-host production of nuclear polyhedrosis virus. *J. N.Y. Entomol. Soc. 84*, 212.

Tanada, Y., and Chang, G. Y. (1968). Resistance of the alfalfa caterpillar, *Colias eurytheme*, at high temperatures to a cytoplasmic-polyhedrosis virus and thermal inactivation point of the virus. *J. Invertebr. Pathol. 10*, 79-83.

Thomas, E. D. (1975). Normal virus levels and virus levels added for control. *In* "Baculoviruses for Insect Pest Control: Safety Considerations" (M. D. Summers, R. Engle, L. A. Falcon, and P. G. Vail, eds.). pp. 87-89. Amer. Soc. for Micro. Publ., Washington, D.C.

Thompson, C. G. (1959). Thermal inhibition of certain polyhedrosis virus disease. *J. Insect. Pathol. 1*, 189-190.

Vail, P. V., Anderson, S. J., and Jay, D. L. (1973). New procedures for rearing cabbage loopers and other lepidopterous larvae for propagation of nuclear polyhedrosis viruses. *Environ. Entomol. 2*, 339-344.

Vanderzant, E. S. (1966). Defined diets for phytophagous insects. *In* "Insect Colonization and Mass Production" (C. N. Smith, ed.), pp. 273-303. Academic Press, New York.

Vaughn, J. L. (1976). The production of nuclear polyhedrosis viruses in large-volume cell cultures. *J. Invertebr. Pathol. 17*, 233-237.

Volkman, L. E., and Summers, M. D. (1975). Nuclear polyhedrosis virus detection: relative capabilities of clones developed from *Trichoplusia ni* ovarian cell line TN-368 to serve as indicator cells in a plaque assay. *J. Virol. 16*, 1630-1637.

Watanabe, H., and Tanada, Y. (1972). Infection of a nuclear-polyhedrosis virus in armyworm, *Pseudaletia unipuncta* HAWORTH reared at high temperature. *Appl. Ent. Zool. 7*, 43-51.

Wyatt, S. S. (1956). Culture *in vitro* of tissue from the silkworm, *Bombyx mori*. *J. Gen. Physiol. 39*, 841-852.

Yamada, K., Sherman, K. E., and Maramorosch, K. (1981). *In vivo* infectivity of early and late passaged *Heliothis zea* polyhedra produced in tissue culture. *Appl. Entomol. Zool. 16*, 504-505.

Yamada, K., Sherman, K. E., and Maramorosch, K. (1982). Serial passage of *Heliothis zea* singly embedded nuclear polyhedrosis virus in a homologous cell line. *J. Invertebr. Pathol. 39*, 185-191.

Yamamoto, T., and Tanada, Y. (1978). Biochemical properties of viral envelopes of insect baculoviruses and their role in infectivity. *J. Invertebr. Pathol. 32*, 202-211.

STRATEGIES FOR FIELD USE OF BACULOVIRUSES

J. D. PODGWAITE

U.S. Department of Agriculture, Forest Service
Center for Biological Control of Northeastern
Forest Insects and Diseases
Hamden, Connecticut

I. INTRODUCTION

In recent years, there has been increased awareness in maintaining the quality of the environment. This has led to the development and use of microbial agents as alternatives to chemicals for controlling noxious insect populations. The insect pathogens in the family Baculoviridae, by virtue of their specificity, virulence, and safety for nontarget species, have become logical candidates in this regard, and several have been registered with the United States Environmental Protection Agency (EPA) for use in the United States (Martignoni, 1978; Lewis et al., 1979; Ignoffo and Couch, 1981). In fact, these viruses, as well as members of the Reoviridae, have been in general use for some time, in one form or another, throughout the world (Bedford, 1981; Cunningham and Entwistle, 1981; Katagiri, 1981; Falcon, 1982; Huber, 1982).

Of the baculovirus products registered for use in the United States, only one, Elcar, *Heliothis* nucleopolyhedrosis virus (NPV), is in commercial production and in general use. The others, Gypchek, the gypsy moth, *Lymantria dispar*, NPV product, and TM-Biocontrol-1, the Douglas-fir tussock moth, *Orgyia pseudotsugata*, NPV product, are registered for use under the auspices of the USDA Forest Service (FS), although

VIRAL INSECTICIDES
FOR BIOLOGICAL CONTROL

ISBN 0-12-470295-3

at present neither is in commercial production nor in general use. Another FS product, Neochek-S, the European pine sawfly, *Neodiprion sertifer* (NPV), has only recently been registered for use in the United States (Podgwaite et al., 1983). This virus is commercially produced and has been used in Europe for some time (Cunningham and Entwistle, 1981).

The development of virus products and control strategies for their use, has been patterned after conventional pesticide use and technology, i.e., broadcast application, usually from aircraft, timed to coincide with the most susceptible stage of the insect, and in response to threatened damage from an insect population that has already reached the economic threshold. Although this may be the most desired approach with many pests, with few exceptions (most notably the sawfly viruses), viral insecticides have provided less than expected results when used as substitutes for chemical pesticides.

The following discussion is focused on past, present, and future strategies for baculovirus use, with examples of conventional and novel application tactics, and the conditions under which these strategies and tactics have been successful, fallen short, or are likely to emerge as important elements in future pest management systems.

II. STRATEGIES FOR USE OF BACULOVIRUSES

Strategies for use of baculoviruses are indicated by the pest-virus interaction and by the nature of the resource being protected. Overall objectives may differ, but efficacy is usually defined in terms of population reduction or foliage protection, or both. For example, a Christmas tree grower can tolerate little defoliation before his crop becomes unmarketable and severe economic loss occurs. Thus, his approach in the use of a viral insecticide and how he measures efficacy will differ from that of the manager of a hardwood timber stand, who, though suffering some loss due to reduced tree growth during periods of defoliation, generally is more tolerant of pest populations, since it takes 2-3 years of heavy defoliation to cause tree mortality and the associated severe economic loss. For the manager of a spruce-fir forest, the situation is quite different, since his resource does not refoliate and extensive mortality follows heavy defoliation. His objectives, strategies, and tactics will reflect the high risk of tree mortality. Scenarios could be similar for agricultural pests, but the point remains the same: strategies for virus use are largely dictated by the economics of the situation, and perhaps secondarily by the nature of the particular virus-insect-host plant system being manipulated.

A. DIRECT CONTROL OF OUTBREAK POPULATIONS

Most control attempts using baculoviruses are made when the target pest has approached the economic threshold or is in the outbreak phase and has already caused substantial losses in the previous generation. The purpose of introducing virus in such situations is to induce artificial epizootics to reduce the pest population below some economic threshold. To introduce effective amounts of virus to as many individuals of the target population as possible in the shortest time, broadcast application usually is chosen. This tactic takes advantage of the density of the host population in the rapid transmission of the disease and the initiation of a widespread epizootic.

There are problems associated with introducing viruses, the most important of which is their instability. Numerous studies have shown that baculoviruses lose most of their insecticidal activity within a few days, and sometimes hours, after application. This has been demonstrated for cabbage looper, *Trichoplusia ni*, NPV, imported cabbageworm, *Pieris rapae*, NPV, and large white butterfly, *Pieris brassicae*, granulosis virus on cabbage (David et al., 1968; Jaques, 1972), *Heliothis* NPV on several host plants (Bullock, 1967; Ignoffo et al., 1973, 1974; Ignoffo and Batzer, 1971), and for *Lymantria dispar* NPV on oak (Lewis and Yendol, 1981). This rapid loss of activity is most often ascribed to the virucidal action of uv irradiation (David, 1969; Bullock et al., 1970; Morris, 1971).

There are other factors that affect the stability of viruses, most notably moisture and leaf surface pH. The former may, through some undetermined process, enhance uv inactivation of baculoviruses (David, 1969; Jaques, 1967a). The latter may be particularly critical to the survival of *Heliothis* NPV on cotton, where the pH (9.0-9.9) has been shown to be high enough to affect the integrity of the polyhedral inclusion body (PIB) (Andrews and Sikorowski, 1973). The problem of virus instability has been addressed to some extent by adding uv screens to baculovirus formulations before application. Several of these adjuvants/extenders have been tested in the laboratory and in the field (Yendol and Hamlen, 1973; Bull, 1978; Couch and Ignoffo, 1981; Luttrell et al., 1982, 1983; Hostetter et al., 1982). These adjuvants, some of which have been commercially developed and marketed, afford varying degrees of protection, but rarely what is needed for the least efficacious baculoviruses to be cost effective, that is, the extension of maximum viral activity over the entire life of the target stage.

The problem of viral instability is compounded by the mechanisms of viral pathogenicity. Since baculoviruses must be ingested to kill, sufficient infectious material must be deposited before a susceptible insect and remain viable long enough for the insect to ingest a lethal dose. This is complicated by nonuniform hatch rates that lead to nonuniform susceptibility of the target pest. Also, the rate of foliage development and weather conditions at the time of application can be major factors which prevent sufficient foliar coverage. Further, as much as 70% of the biological activity of a virus preparation can be lost through drift or droplet evaporation before impact on the target foliage (Lewis and Etter, 1978). Thus, the application window for these agents is very small. Many physical and biological factors act in concert to make broadcast application a very complex, high-risk tactic, the results of which can be compromised by only minor errors in timing, dose rates, and estimates of insect and foliage development.

Incorporating antievaporants and gustatory stimulants into spray formulations may help prevent physical loss of virus and offset rapidly decreasing virus activity after the product has been deposited on target foliage. Luttrell et al. (1983) found that mortality of cotton bollworm, *H. zea*, in the field was increased significantly when Elcar was used with a high rate of the feeding stimulant Coax. Nonuniformity of insect and foliage development can be compensated for by staggered applications of virus during periods of early larval development (Wollam et al., 1978). This may increase efficacy, but it also increases cost.

Depending on the particular baculovirus used, it may be as many as 7-14 days after application before mortality is significant. During this time the insect may continue to defoliate, causing damage and loss that must be an accepted part of any direct control strategy. Again, timing of application is critically important so that the target is reached at as early a life stage as possible to minimize defoliation before peak insect kill.

Environmental acceptability is a major reason why microbials have reached at least an introductory level of prominence in pest management strategies. They have been introduced to the pest management arena after several decades of indiscriminant use of chemical pesticides, the results of which have rendered well over 200 insect pest species resistant to these chemicals (FAO, 1970). However, there is some evidence that insect populations develop resistance to baculoviruses (see Briese and Podgwaite, this volume). Since the use of baculoviruses in control programs is a relatively recent development, neither enough time has passed nor enough data been made available to determine whether insect resistance is evolving in areas that have been treated repeatedly with these agents. Nevertheless, users should be aware of the possibility of resistance and be advised that any strategy involving treatments on successive insect generations is likely to result in a certain degree of resistance.

A common argument against the use of chemical pesticides is the resurgence of target pests that often is seen in the year following application (Falcon, 1973). This necessitates repeated applications of pesticides in succeeding generations. Similarly, the use of baculoviruses may not be prudent in situations where natural population collapse is imminent, either from starvation or disease. Since the least efficaceous viruses, used at dosages that are economically competitive with chemicals, generally reduce populations by 50-90% (Lewis, 1981; Ignoffo and Couch, 1981), it should be recognized that population rebound is likely when baculoviruses are used as direct control agents, particularly at high population densities. As with chemicals, yearly applications of baculovirus products may be necessary to afford desired control and crop protection.

Finally, baculovirus products are expensive to produce and to use. However, added cost often can be rationalized on the basis of safety and specificity relative to chemicals. Unfortunately, specificity is one reason why more baculoviruses are not available for use. Commercial producers are understandably interested in developing products that satisfy a broad market, as opposed to investing substantive amounts on products that will be used against one insect and perhaps only sporadically. The following is a brief account of the status of various baculoviruses as direct control agents.

1. *Heliothis zea* NPV (Elcar)

This virus, researched intensively for over a decade before its registration with EPA in 1975, has been in operational use on a fairly wide scale since 1979. Ignoffo and Couch (1981) reported that 150-200 field tests were conducted with *H. zea* NPV between 1960 and 1980; about 90% of these were on cotton and corn. Although results have varied, generally this NPV has been as effective as insecticide on cotton when used at rates above ca. 5×10^{10} PIB/0.4 ha. At high infestation levels of *H. zea* on corn, control has been less than that obtained with standard insecticides. However, at light to moderate infestations, results have been somewhat better. Similarly, tests on soybean and sorghum have provided reasonable control--90-96% and 88% population reduction, respectively.

Desired levels of control of tobacco budworm, *H. virescens*, on tobacco and *H. zea* on tomato have not been demonstrated.

Bohmfalk (1982) reported that since 1976 when Elcar was commercialized by Sandoz, Inc., there has been a mixed response to the product, which has been reflected in sales. As a partial explanation he noted that most cotton producers in the United States are resistant to change and have grown accustomed to seeing dead bollworms immediately after treating their fields with chemical insecticides. Since no such response is seen immediately following treatment with Elcar, many growers have decided that this product is not effective. He also cited competition with chemicals (particularly pyrethrins), formulation and application problems, and the lack of sophistication of the user in recognizing differences between pests and beneficial insects as other roadblocks to the success of Elcar. He concluded that the industry is experiencing difficulties in introducing this product into general use despite concentrated research and development efforts.

2. *Lymantria dispar NPV (Gypchek)*

A widespread outbreak of gypsy moth throughout the Northeastern United States in the 1970's led to the intensification of ongoing research directed toward the registration of a gypsy moth NPV product. Before this, direct-broadcast spray tests had provided limited control or unacceptable results primarily due to problems with formulation and differences in susceptibility among gypsy moth populations (Rollinson et al., 1965; Magnoler, 1968). From 1972 to 1978, the year that Gypchek was registered, there were several field tests on variations in virus product, application hardware, dose rates, timing, and formulation. In 1972 a ground mistblower test at 1×10^{13} PIB/0.4 ha provided a 90% reduction in egg masses compared with a 65% increase in untreated plots (Yendol et al., 1977). In 1974, an aerial test at the rate of 1×10^{12} PIB/0.4 ha resulted in no significant differences between numbers of pretreatment and posttreatment egg masses in treated plots (Yendol et al., 1977). Results of single and double applications of NPV at 2.5×10^{12} PIB/ha in 1975 were compromised by natural NPV epizootics in both treated and control plots. However, second year evaluation of the test indicated the NPV application did not improve the population quality of the next generation as measured by no increase in egg mass size, reduced larval emergence, and prolonged incidence of virus in the treated population (Wollam et al., 1978). Field tests with lower dosages in 1976 and tests of different nozzling systems in 1977 resulted in acceptable foliage protection (<55% defoliation) and population reduction (57-87%). However, a 1978 test of new formulations was again compromised by a natural population collapse (Lewis and Yendol, 1981). In 1981, results of further tests with Gypchek in 4L and Protec formulation, although promising, were inconclusive (Lewis and Podgwaite, 1982, unpublished).

Lewis (1981) concluded that the cost benefit in using Gypchek is positive if environmental considerations are paramount, if damage to other naturally occurring biotypes must be minimized, if gypsy moth is the primary insect in the treatment area, and if careful timing and application are not serious constraints. He added that the cost benefit is negative if immediate pest knockdown and long residual pesticide activity are required. Unfortunately, the conditions for achieving a positive cost benefit often are not realized. Environmental considerations are rarely as important to users as protecting their investment, and "environmental conciousness" suddenly becomes secondary under the threat of economic loss. Further, it is rare when careful timing and application are not constraints when using microbials. Disconcertingly, 5 years after registration, Gypchek is not in commercial production.

3. *Orgyia pseudotsugata NPV (TM-Biocontrol-1)*

This virus was registered with EPA by the FS in 1976 (Martignoni, 1978). Extensive field tests have shown it is an effective direct-control agent (Stelzer and Neisess, 1978). In a small-scale aerial test in 1973, dosages of 2.5×10^{11} and 2.5×10^{12} PIB/ha in a formulation of Shade and molasses resulted in reductions of about 97 and 98%, respectively, in treated tussock moth populations 35 days after treatment. Control populations were reduced by 47% by a natural virus epizootic. Average defoliation was 25% in treated plots compared with 64% in the controls.

In a large-scale pilot test in 1974 (plots ca. 80-500 ha), a dose rate of 2.5×10^{11} PIB/ha resulted in a 90% reduction on treated plots. No survival to the pupal stage was noted in any of the virus-treated areas. Defoliation was 20% in treated areas and averaged 60% in control plots.

Like Gypchek, TM-Biocontrol-1 is not commercially available and there appears to be little interest in its development. One can only speculate that the primary reasons are cost, specificity, extended time between tussock moth outbreaks and, as with cotton growers and Elcar, the resistance of timber managers to use other than relatively inexpensive chemical controls.

4. *Neodiprion sertifer NPV (Neochek-S) and Other Sawfly Viruses*

Sawfly baculoviruses are the most efficacious of the insect viruses, and some can be substituted for chemicals in direct control strategies. They also are inexpensive to produce with costs ranging from $0.75 to $2.00/ha equivalent compared with $10 to $125/ha equivalent for lepidopteran baculoviruses (Cunningham and Entwistle, 1981). Many have the distinct advantage of being amenable to spot innoculation since virus is spread rapidly within and between sawfly colonies (Bird, 1961).

Of the sawfly baculoviruses, the NPV of *N. sertifer* has been the most widely used; at least 10,000 ha in 12 countries have been treated since the early 1950's (Cunningham and Entwistle, 1981). Podgwaite et al. (1983) found that Neochek-S at a dose rate of 2.5×10^9 PIB/ha provided acceptable control (>90% reduction) and foliage protection (ca. 94%) in moderate to dense populations of *N. sertifer*. This product is in production by the FS and, currently, ca. 12,000 ha equivalents--a 5-year supply at the current demand--are available for use.

Lecontvirus, the Canadian redheaded pine sawfly, *N. lecontei*, NPV product, is in the final stages of development and should soon be registered for use in the United States and Canada. For acceptable direct control of redheaded pine sawfly, Cunningham and Entwistle (1981) recommended a dose rate of 5.0×10^9 PIB/ha. They also reported that the NPVs of Swaine's jack pine sawfly, *N. swainei*, the Virginia pine sawfly, *N. pratti pratti*, the loblolly pine sawfly, *N. taedea linearis*, and the European spruce sawfly, *Gilpinia hercyniae*, are efficient alternatives to chemical insecticides, although more data must be accumulated before their registration as commercial insecticides. This may well be moot as economic losses caused by these pests are insigificant relative to losses caused by the major agricultural and forest pests (NAS, 1975). Limited markets do not encourage commercialization, so if the production of these NPVs is not subsidized by interested governments, it is likely that they will not be made available.

5. *Other Baculoviruses*

In addition to the baculoviruses already discussed, several others have been tested and at least developed partially as direct control agents (see review by Stairs, 1971). A few of these have shown considerable promise and are potential candidates for commercialization.

A granulosis virus (GV) of the codling moth, *Cydia pomonella*, has been tested extensively in apple orchards in the United States, Canada, Australia, and Europe (Huber, 1982). In European trials, dosages between 0.5 and 1 × 10^{14} PIB/ha resulted in reduction of 75 to 95% in unmarketable apples. In Australia, weekly applications at 1 × 10^{14} PIB/ha provided 98% protection against deep entries and unmarketable fruit (Morris, 1972). Huber (1982) reported that an added advantage of using this virus is that the natural enemies of other orchard pests, especially mites, aphids, and leaf hoppers, are unaffected and remain to exert control over their hosts.

Sandoz, Inc. has produced a NPV product (SAN 406) that has been used in certain of the tests mentioned. Falcon (1982) reported that this product was as effective as commonly used chemical insecticides in controlling codling moth in two commercial fruit orchards. Since codling moth is the key pest on crops on which it occurs, the results bode well for the eventual registration and commercialization of its baculovirus.

The NPVs of the alphalpha looper, *Autographa californica*, and the cabbage looper, *T. ni*, have been produced by Sandoz, Inc. as experimental preparations SAN 404 and SAN 405, respectively (Falcon, 1982). The development of the former has been of particular interest because of its wide host range, e.g., *T. ni*; beet armyworm, *Spodoptera exigua*; saltmarsh caterpillar, *Estigmene acrea*; *H. zea*; cotton leafperforator, *Bacculatrix thurberiella*; and diamond back moth, *Plutella xylostella* (Vail and Jay, 1973). In field studies, Falcon (1982) found that SAN 404 was effective in controling *T. ni*, but that it did not infect populations of *S. exigua* or *E. acrea*. Both products were more effective than *Bacillius thuringiensis* (*B.t.*) for controlling *H. zea* on cotton; one application of either provided acceptable control.

B. *COLONIZATION FOR LONG-TERM CONTROL*

Colonization of a biological control agent is most often thought of in the classical sense, i.e., the entomogenous parasite is introduced into a susceptible population, actively searches for primary and alternate hosts, gradually becomes established in close association with the primary host population, and causes some numerical responses in that population.

Since baculoviruses do not search out their hosts, the success of colonization depends heavily on both the biology and behavior of the host and on environmental factors that enhance viral persistence and transmission within and between insect generations. Because predisposing conditions to the successful use of this strategy are rare, it is not surprising that there have been relatively few cases where a single introduction of a baculovirus has resulted in acceptable control in any year other than the one in which it was introduced.

One such case involved *G. hercyniae* NPV, which was accidentally introduced, probably through contaminated parasites that had been imported from Europe to mitigate sawfly populations in North America (Balch and Bird, 1944). The effects of a single introduction (a few treated trees) of the virus in Ontario were followed over a 10-year period. Epizootics occurred each year and sawfly populations were held below the economic threshold (Bird and Burk, 1961). As a result of this and earlier introductions and parasites, *G. hercyniae* has remained enzootic in Eastern North America since the 1950's. Originally attributed to transovum transmission (Bird, 1953; Neilson and Elgee, 1968), the successful colonization of this virus is now thought to be due largely to overwintering persistence on foliage (Entwistle and Adams, 1977) and perhaps dispersal by birds (Entwistle et al., 1977a,b, 1978) which ultimately leads to infection of newly emerged larvae in the succeeding generation.

Introductions of *N. sertifer*, *N. lecontei*, and *N. swainei* NPVs can be expected to exert at least some degree of control in succeeding generations as these viruses persist and spread rapidly in the ecosystems of their respective hosts (Bird, 1961; Smirnoff et al., 1962). Thus, colonization strategies with NPVs of these species certainly merit consideration. Introductions can be single or multiple point, and should be precalculated to account for degree and rate of NPV dispersal (Cunningham and Entwistle 1981).

There have been successful colonizations of the GV of the rhinoceros beetle, *Orcytes rhinocerus*, a serious pest of palm in the South Pacific Islands. In Western Samoa, virus introduced into log heaps established in *O. rhinoceros* populations and spread throughout the islands (Marschall, 1970). On Wallis Island, the virus spread over the entire island 2 months after its introduction into artificial log heaps (Hammes, 1971). As a result, palm damage dropped by 70-90%. Similar releases of the virus in Fiji, Tongatapu, American Samoa, the Takelau Islands, and Mauritius, either in log or manure heaps or as infected beetles, have resulted in the gradual spread of the virus and, generally, a significant decline in palm damage throughout the area (Bedford, 1981).

It should be noted that in what are considered major colonization successes, i.e., with sawfly and *Orcytes* baculoviruses, mechanisms of pathogenicity and host behavior have played major roles. These viruses are replicated in the midgut epithelium and are shed in the feces, providing for rapid horizontal transmission. Further, they are transmitted at mating, in *O. rhinoceros* GV by consumption of feces in contaminated breeding sites (Zelazny, 1976), and in sawfly NPVs by contamination of foliage during oviposition. Finally, since most sawfly species oviposit on coniferous hosts, persistent

NPV is readily available on needles consumed by emerging larvae in the succeeding generation. Viruses of pests of deciduous trees and most agricultural pests are lost to the soil with leaffall and through tilling, and are not readily available to insects of the following generation. Most of these virus do not replicate in the gut and are not dispersed in feces, so their effective colonization is much more difficult to achieve.

C. AUGMENTING NATURALLY OCCURRING BACULOVIRUS MORTALITY

Augmentation strategies may be particularly applicable in row crops where naturally occurring or introduced viruses can be returned to plant surfaces from the soil through cultivation or by other means. There is evidence that viruses in soil naturally contaminate leaves of low-growing crops through wind or rain splash (Jaques, 1967b; Tanada and Omi, 1974). Ignoffo (1978) suggested that epizootics could be created in field pest populations by blowing residual NPV from soil onto plants. For this to be reasonably effective, growers would need to know with some certainty the types, amounts, and distributions of viruses in their field. Thus, this tactic perhaps would best be implemented as a supplement to other control methods.

Ruzicka (1924) demonstrated efficacy of the nun moth, *Lymantria monacha*, NPV by blowing naturally contaminated litter into tree crowns with trench mortars. This apparently resulted in a virus epizootic. There is considerable evidence that forest insect viruses persist in soil and litter for extended periods (Podgwaite et al., 1979; Thompson and Scott, 1979; Mohamed et al., 1982; Kaupp, 1983). It is possible that epizootics could be initiated, or at least naturally occurring mortality could be increased, if technology were available to easily distribute contaminated litter from the forest floor to the forest canopy.

Parasitic and predacious insects are effective mechanical vectors of baculoviruses (Thompson and Steinhaus, 1950; Smirnoff, 1959; Raimo et al., 1977), and have been implicated in the natural transmission of these agents in the field (Smith et al., 1956; Bird, 1961; Reardon and Podgwaite, 1976; Levin et al., 1983). Similarly, birds feed on insects that have been infected with or killed by viruses (Hostetter and Biever, 1970; Entwistle et al., 1977a), and passively transport baculoviruses within their home range. The best documented case is that of *G. hercynia*, whose baculovirus has been carried up to 6 km from NPV-contaminated areas by birds, and whose dynamics appears to be partly regulated by this activity (Entwistle et al., 1977b). Small mammals also may feed on infected insects and foodstuffs and distribute infectious NPV in the environment (Lautenschlager and Podgwaite, 1977, 1979; Lautenschlager et al., 1980).

Certain mechanical vectors, particularly parasites, have been released or managed in direct control or colonization strategies, but rarely increase baculovirus mortality (Mohamed et al., 1981; Raimo and Reardon, 1981). This concept merits consideration but has not been exploited.

D. *INDUCING BACULOVIRAL EPIZOOTICS*

Both naturally occurring and artificially induced epizootics ensue most often after the pest insect has caused considerable economic loss. In most cases, this is a result of insufficient viral inoculum in the ecosystem during the initial susceptible stages of the pest or a distribution of virus such that it is unavailable for consumption and subsequent transmission of disease. Thus, it may be advisable to introduce virus earlier in the insects' life cycle, or perhaps in the preceding generation, to induce epizootics earlier than they normally would occur. Stairs (1965) introduced NPV into virus-free populations of the forest tent caterpillar *Malacasoma disstria* and initiated carryover and slowly developing epizootics in the next generation. Jaques (1977) found that an early single application of *T. ni* NPV and *P. rapae* GV was as effective in reducing cabbage loss as a series of viral applications. Podgwaite et al. (1981) introduced gypsy moth NPV into areas supporting sparse, enlarging gypsy moth populations (<175 egg masses/ha) by treating egg masses. This resulted in a 20% incidence of polyhedrosis in 4th-6th stage larvae in the year of treatment, followed by an NPV epizootic in the succeeding generation.

Early introduction of baculoviruses has not been widely tested, but lack of implementation probably lies in this strategy being recognized as a preventive measure. Often it is difficult to convince resource managers, farmers, and foresters to treat what at first glance are insignificant insect populations. This speaks to our collective inability, in the vast majority of cases, to forecast pest numbers with the degree of precision necessary to evoke such decisions.

E. INTEGRATED PEST MANAGEMENT

As emphasized earlier, most baculoviruses cannot be substituted for chemical pesticides with the expectation that they will provide the same quickness of kill and degree of population reduction. Thus, most are best suited for integration into a pest management scheme whose overall strategy is to take advantage of a variety of complementary control tactics, the combined effects of which bring pest populations below the economic threshold.

Although antagonisms do exist, baculoviruses generally are compatible with chemical pesticides, either as mixtures or when used sequentially (see reviews by Benz, 1971; Jaques and Morris, 1981). Falcon (1973) reported that cotton yields were not significantly different between plots that had received one application of Azodrin and then five applications of *H. zea* NPV and those that had received four applications of the chemical alone. Mohamed et al. (1983) tested Elcar in combination with eight chemical pesticides against *H. virescens*. Synergistic responses were found with several combinations, but differences were seldom in excess of 15% of calculated expected mortality. Since dosages used in the test were greater than normal field dosages, the authors concluded that at least with these chemical-NPV combinations, *H. virescens* control would not be altered significantly in the field. Jaques (1977) demonstrated increased efficacy when *T. ni* NPV and *A. californica* NPV were mixed with chemicals for control of cabbage looper and imported cabbage worm on late cabbage. Morris (1977) concluded that NPV plus a low dose of Sumithion was effective in suppressing spruce budworm populations for 3 years after treatment; there were no concomitant deleterious effects on the incidence of budworm parasites.

Combinations of viruses, or combinations of viruses with other entomopathogens, especially (*B.t.*), have shown promise in reducing pest populations. Jaques (1977) found that either *T. ni* NPV or *A. californica* NPV in combination with formulations of *B.t.* generally increased crop protection over that resulting from the use of the viruses in combination without *B.t.* Sears et al. (1983) found that various combinations of *T. ni* NPV, *A. californica* NPV, and *B.t.* provided 90% marketable cabbage compared with 95% using either permethrin or methomyl. Bell and Romine (1980) reported adequate control of *H. virescens* on cotton with a combination of *B.t.* and *A. californica* NPV. However, Luttrell et al. (1982) found that combinations of *B.t.* with *H. zea* NPV and *A. californica* NPV against *H. zea* and *H. virescens* resulted in no greater mortality than that elicited by either virus alone. They concluded that the use of these combinations was not advantageous. Similarly, Johnson

(1982) found that combinations of *H. zea* NPV and *B.t.* on cotton, although suppressing pest populations by an average of 28% compared with control plots, were no more effective than either microbial used alone. Bell and Romine (1982) reported that a combination of *B.t.* and *A. californica* NPV resulted in a 95% reduction of *B. thurberiella* on cotton compared with untreated controls, but they speculated that the combination might not be cost effective when the cotton leafperforator is the only pest in the system.

Other integrated strategies have been tested. Campbell (1983) found that various combinations of gypsy moth NPV egg mass treatments, Disparlure, and mechanical methods were ineffective in reducing gypsy moth numbers in succeeding generations. However, parasite-NPV combinations have shown some promise for use in gypsy moth management systems. Raimo and Reardon (1981) found that the release of NPV-contaminated *Apanteles melanoscelus* females resulted in a 39% incidence of NPV in treated blocks compared with 22% in controls, while percentage parasitism was virtually the same in both treated and control blocks.

These examples, some positive, some not, reveal an increasing willingness toward integrating baculoviruses into pest management strategies through a variety of tactics.

III. APPLICATION TACTICS

Strategies discussed here can be implemented by a variety of tactics, some of which have been tested extensively. There is little question that direct control of outbreak populations is most often successful with broadcast application, either from aircraft or with ground spray equipment. Aerial application as it relates to viral control of forest pests is discussed in detail by Akesson and Yates (1978). Current aerial and ground technology (foggers, mistblowers, etc.) are assessed by Smith and Bouse (1981).

Other tactics have not been widely tested and are not in general use. Some of these have been suggested by Ignoffo (1978) and include (a) the release of both virus-infected and contaminated hosts (Evans and Allaway, 1983); (b) the release of contaminated parasitic and predacious insects; (c) the contamination, release, and management of birds and mammals, (d) the use of insect attractants to lure hosts to contaminated baits (Gard and Falcon, 1978); (e) the mechanical manipulation of the environment to make naturally occurring baculovirus more available for host consumption; (f) the induction of "latent" virus through the use of chemical stressors; and (g) the development of timed-release technology for distribution of virus.

The choice of tactical approaches for implementing control strategies that include the use of baculoviruses depends largely on the dynamics of the host-pest system to be managed, and the relative threat of economic damage. It may be possible to intercede with spot inoculation tactics early in the insects' developmental cycle, or in the preceding generation. However, waiting until pest numbers have reached the economic threshold almost certainly will require the use of broadcast application. What is critical to choice of tactics, and what is almost always lacking, is a clear understanding of host dynamics. Less clear is how introductions of baculoviruses affect the flux of succeeding generations, both quantitatively and qualitatively. If baculoviruses are to be fully exploited for more than knee-jerk responses to pest crises, these areas must be more fully understood.

IV. SUMMARY AND CONCLUSIONS

Pest managers appear to be at a crossroad in the collective development of baculoviruses as microbial insecticides. Although for several decades, baculoviruses have been recognized as potentially useful in controlling noxious pests, only few are available for use in pest management systems. The reasons for this are complex and interrelated but the primary reason involves efficacy. It is clear that if the marginally effective viruses, which are, ironically, pathogens of the most economically important insects, are to be seriously considered as alternatives to chemicals in any strategy that includes broadcast application, they must be formulated in such a manner as to extend or amplify their activity on foliage, or they must be manipulated genetically to enhance their virulence. Commercialization of baculovirus products has been hindered because these products are less marketable than chemicals, because of their specificity, and because of their unpredictability in the field. Specificity, of course, is a double-edged sword, cutting for the environmentalist but against the producer who would prefer a product that can be used against a broad spectrum of pests. That this has been recognized is evidenced by both the commercialization of *H. zea* NPV and increased interest in the development and commercialization of *A. californica* NPV, two "broad-spectrum" viruses which are effective against a variety of agricultural and forest pests. There is no logical argument against research on increasing the host range of the more specific baculoviruses through genetic manipulation. In the shorter term, methods research should focus intently on extending the activity of baculoviruses once applied to the host plant. This will be realized only through an intensive, systematic research and development program in the area of formulation and application.

The public at large, the users, generally equate the performance of microbials with that of chemicals--they expect equivalent results. The future of baculoviruses as pesticides may well lie in the collective abilities of scientists, educators, and producers to resolve this misconception by promoting the use of some baculoviruses as adjuncts to, rather than substitutes for, other pesticidal agents.

ACKNOWLEDGMENTS

The author thanks H. C. Coppel, D. L. Hostetter, and F. B. Lewis for their thoughtful reviews of the manuscript.

REFERENCES

Akesson, N. B., and Yates, W. E. (1978). Aerial application technology: Aircraft and spray systems. *In* "The Douglas-Fir Tussock Moth: A Synthesis" (M. H. Brookes, R. W. Stark, and R. W. Campbell, eds.), pp. 160-164. U.S. Dept. Agric. Tech. Bull. 1585.

Andrews, G. L., and Sikorowski, P. P. (1973). Effects of cotton leaf surface on the nuclear-polyhedrosis virus of *Heliothis zea* and *Heliothis virescens* (Lepidoptera: Noctuidae). *J. Invertebr. Pathol.* 22, 290-291.

Balch, R. E., and Bird, F. T. (1944). A disease of the European spruce sawfly, *Gilpinia hercyniae* (Htg), and its place in natural control. *Sci. Agr.* 25, 65-80.

Bedford, G. O. (1981). Control of the rhinoceros beetle by baculovirus. *In* "Microbial Control of Pests and Plant Diseases 1970-1980" (H. D. Burges, ed.), pp. 409-426. Academic Press, London.

Bell, M. R., and Romine, C. L. (1980). Tobacco budworm field evaluation of microbial control in cotton using *Bacillus thuringiensis* and a nuclear polyhedrosis virus with a feeding adjuvant. *J. Econ. Entomol.* 73, 427-430.

Bell, M. R., and Romine, C. L. (1982). Cotton leafperforator (Lepidoptera: Lyonetidae): Effect of two microbial insecticides on field populations. *J. Econ. Entomol. 75*, 1140-1142.

Benz, G. (1971). Synergism of micro-organism and chemical insecticides. *In* "Microbial Control of Insects and Mites" (H. D. Burges and N. W. Hussey, eds.), pp. 327-355. Academic Press, London.

Bird, F. T. (1953). The effect of metamorphosis on the multiplication of an insect virus. *Can. J. Zool. 31*, 300-303.

Bird, F. T. (1961). Transmission of some insect viruses with particular reference to ovarial transmission and its importance in the development of epizootics. *J. Invertebr. Pathol. 3*, 352-380.

Bird, F. T., and Burk, J. M. (1961). Artificially disseminated virus as a factor controlling the European spruce sawfly, *Diprion hercyniae* (Htg.) in the absence of introduced parasites. *Can. Entomol. 93*, 228-238.

Bohmfalk, G. T. (1982). Progress with the nuclear polyhedrosis virus of *Heliothis zea* by commercialization of Elcar. *Proc. 3rd Intern. Colloq. Invertebr. Pathol., Brighton, UK*, pp. 113-117.

Bull, D. L. (1978). Formulations of microbial insecticides: Microencapsulation and adjuvants. *Misc. Publ. Entomol. Soc. Amer. 10*, 11-20.

Bullock, H. R. (1967). Persistence of *Heliothis* nuclear-polyhedrosis virus on cotton foliage. *J. Invertebr. Pathol. 9*, 434-436.

Bullock, H. R., Hollingsworth, J. P., and Hartsack, A. W., Jr. (1970). Virulence of *Heliothis* nuclear polyhedrosis virus exposed to monochromatic ultraviolet irradiation. *J. Invertebr. Pathol. 16*, 419-422.

Campbell, R. W. (1983). Gypsy moth (Lepidoptera: Lymantriidae) control trials combining nucleopolyhedrosis virus, disparlure, and mechanical methods. *J. Econ. Entomol. 76*, 610-614.

Couch, T. L., and Ignoffo, C. M. (1981). Formulation of insect pathogens. *In* "Microbial Control of Pests and Plant Diseases 1970-1980" (H. D. Burges, ed.), pp. 622-634. Academic Press, London.

Cunningham, J. C., and Entwistle, P. F. (1981). Control of sawflies by baculovirus. *In* "Microbial Control of Pests and Plant Diseases 1970-1980" (H. D. Burges, ed.), pp. 379-407. Academic Press, London.

David, W. A. L. (1969). The effect of ultraviolet radiation of known wave-lengths on a granulosis virus of *Pieris brassicae*. *J. Invertebr. Pathol. 14*, 336-342.

David, W. A. L., Gardiner, B. O. C., and Woolner, M. (1968). The effects of sunlight on a purified granulosis virus of

Pieris brassicae applied to cabbage leaves. *J. Invertebr. Pathol. 11*, 496-501.

Entwistle, P. F., and Adams, P. H. W. (1977). Prolonged retention of infectivity in the nuclear polyhedrosis virus of *Gilpinia hercyniae* (Hymenoptera, Diprionidae) on foliage of spruce species. *J. Invertebr. Pathol. 29*, 392-394.

Entwistle, P. F., Adams, P. H. W., and Evans, H. F. (1977a). Epizootiology of a nuclear polyhedrosis virus in European spruce sawfly (*Gilpinia hercyniae*). The status of birds as dispersal agents during the larval season. *J. Invertebr. Pathol. 29*, 354-360.

Entwistle, P. F., Adams, P. H. W., and Evans, H. F. (1977b). Epizootiology of a nuclear polyhedrosis virus in European Spruce sawfly (*Gilpinia hercyniae*). Birds as dispersal agents of the virus during winter. *J. Invertebr. Pathol. 30*, 15-19.

Entwistle, P. F., Adams, P. H. W., and Evans, H. F. (1978). Epizootiology of a nuclear polyhedrosis virus in European Spruce sawfly (*Gilpinia hercyniae*): The rate of passage of infective virus through the gut of birds during cage tests. *J. Invertebr. Pathol. 31*, 307-312.

Evans, H. F., and Allaway, G. P. (1983). Dynamics of baculovirus growth and dispersal in *Mamestra brassicae* L. (Lepidoptera: Noctuidae) larval populations introduced into small cabbage plots. *Appl. Environ. Microbiol. 45*, 493-501.

Falcon, L. A. (1973). Biological factors that affect the success of microbial insecticides: development of integrated control. *Ann. New York Acad. Sci. 217*, 173-186.

Falcon, L. A. (1982). The baculoviruses of *Autographa, Trichoplusia, Spodoptera* and *Cydia*. *Proc. 3rd Intern. Colloq. Invertebr. Pathol., Brighton, UK*, pp. 125-128.

FAO. (1970). Pest resistance to pesticides in agriculture. Importance, recognition and countermeasure. Food and Agric. Organ., UN. Rome, Italy.

Gard, I. E., and Falcon, L. A. (1978). Autodissemination of entomopathogens: virus. *In* "Microbial Control of Insect Pests: Future Strategies in Pest Management Systems" (G. E. Allen, C. M. Ignoffo, and R. P. Jaques, eds.), pp. 46-51. NSF-USDA-U.FLA., Gainesville.

Hammes, C. (1971). Multiplication et introduction d'um virus d' *Oryctes rhinoceros* a l'ile Wallis. *C.R. Acad. Sci. (Paris) 273*, 1048-1050.

Hostetter, D. L., and Biever, K. D. (1970). The recovery of virulent nuclear polyhedrosis virus of the cabbage looper, *Trichoplusia ni*, from feces of birds. *J. Invertebr. Pathol. 15*, 173-176.

Hostetter, D. L., Smith, D. B., Pinnell, R. E., Ignoffo, C. M., and McKibben, G. H. (1982). Laboratory evaluation of adjuvants for use with *Baculovirus heliothis* virus. *J. Econ. Entomol. 75*, 1114-1119.

Huber, J. (1982). The baculoviruses of *Cydia pomonella* and other tortricids. *Proc. 3rd Intern. Colloq. Invertebr. Pathol., Brighton, UK*, pp. 119-124.

Ignoffo, C. M. (1978). Strategies to increase the use of entomopathogens. *J. Invertebr. Pathol. 31*, 1-3.

Ignoffo, C. M., and Batzer, O. F. (1971). Microencapsulation and ultraviolet protectants to increase sunlight stability of an insect virus. *J. Econ. Entomol. 64*, 850-853.

Ignoffo, C. M., and Couch, T. L. (1981). The nucleopolyhedrosis virus of *Heliothis* species as a microbial insecticide. *In* "Microbial Control of Pests and Plant Diseases 1970-1980" (H. D. Burges, ed.), pp. 329-362. Academic Press, London.

Ignoffo, C. M., Parker, F. D., Boening, O. P., Pinnell, R. E., and Hostetter, D. L. (1973). Field stability of the *Heliothis* nucleopolyhedrosis virus on corn silks. *Environ. Entomol. 2*, 302-303.

Ignoffo, C. M., Hostetter, D. L., and Pinnell, R. E. (1974). Stability of *Bacillus thuringiensis* and *Baculovirus heliothis* on soybean foliage. *J. Environ. Entomol. 3*, 117-119.

Jaques, R. P. (1967a). The persistence of the nuclear-polyhedrosis virus in the habitat of the host insect, *Trichoplusia ni*. I. Polyhedra deposited on foliage. *Can. Entomol. 99*, 785-794.

Jaques, R. P. (1967b). The persistence of a nuclear polyhedrosis virus in the habitat of the host insect, *Trichoplusia ni*. II. Polyhedra in soil. *Can. Entomol. 99*, 820-829.

Jaques, R. P. (1972). The inactivation of foliar deposits of viruses of *Trichoplusia ni* and *Pieris rapae* and tests on protectant activities. *Can. Entomol. 104*, 1985-1994.

Jaques, R. P. (1977). Field efficacy of viruses infectious to the cabbage looper and imported cabbageworm on late cabbage. *J. Econ. Entomol. 70*, 111-118.

Jaques, R. P., and Morris, O. N. (1981). Compatibility of pathogens with other methods of pest control and with different crops. *In* "Microbial Control of Pests and Plant Diseases 1970-1980" (H. D. Burges, ed.), pp. 695-715. Academic Press, London.

Johnson, D. R. (1982). Suppression of *Heliothis* spp. on cotton by using *Bacillus thuringiensis*, *Baculovirus heliothis*, and two feeding adjuvants. *J. Econ. Entomol. 75*, 207-210.

Katagiri, K. (1981). Pest control by cytoplasmic polyhedrosis viruses. *In* "Microbial Control of Pests and Plant Diseases 1970-1980" (H. D. Burges, ed.), pp. 433-440. Academic Press, London.

Kaupp, W. J. (1983). Persistence of *Neodiprion sertifer* (Hymenoptera: Diprionidae) nuclear polyhedrosis virus on *Pinus contorta* foliage. *Can. Entomol. 115*, 869-873.

Lautenschlager, R. A., and Podgwaite, J. D. (1977). Passage of infectious nuclear polyhedrosis virus through the alimentary tracts of two small mammal predators of gypsy moth, *Lymantria dispar*. *Environ. Entomol. 6*, 737-738.

Lautenschlager, R. A., and Podgwaite, J. D. (1979). Passage of nucleopolyhedrosis virus by avian and mammalian predators of the gypsy moth, *Lymantria dispar*. *Environ. Entomol. 8*, 210-214.

Lautenschlager, R. A., Podgwaite, J. D., and Watson, D. E. (1980). Natural occurrence of the nucleopolyhedrosis virus of the gypsy moth, *Lymantria dispar* (Lepidoptera: Lymantriidae) in wild birds and mammals. *Entomophaga 25*, 261-267.

Levin, D. B., Laing, J. E., Jaques, R. P., and Corrigan, J. E. (1983). Transmission of the granulosis virus of *Pieris rapae* (Lepidoptera: Pieridae) by the parasitoid *Apanteles glomeratus* (Hymenoptera: Braconidae). *Environ. Entomol. 12*, 166-170.

Lewis, F. B. (1981). Control of gypsy moth by a baculovirus. *In* "Microbial Control of Pests and Plant Diseases 1970-1980" (H. D. Burges, ed.), pp. 363-377. Academic Press, London.

Lewis, F. B., and Etter, D. O., Jr. (1978). Use of pathogens in forest pest management systems: Gypsy moth, *lymantria dispar* L. *In* "Microbial Control of Insect Pests: Future Strategies in Pest Management Systems" (G. E. Allen, C. M. Ignoffo, and R. P. Jaques, eds.), pp. 261-272. NSF-USDA. Univ. Florida, Gainesville.

Lewis, F. B., and Podgwaite, J. D. (1982). Unpublished data on Gypchek field tests. Center for Biological Control of Forest Insects and Diseases. Hamden, CT.

Lewis, F. B., and Yendol, W. G. (1981). Gypsy moth nucleopolyhedrosis virus: Efficacy. *In* "The Gypsy Moth: Research Toward Integrated Pest Management" (C. C. Doane and M. L. McManus, eds.), pp. 503-512. U.S. Dept. Agr. Tech. Bull. 1584.

Lewis, F. B., McManus, M. L., and Schneeberger, N. F. (1979). Guidelines for the use of Gypchek to control the gypsy moth. *USDA Forest Serv. Res. Paper NE-441*, pp. 9.

Luttrell, R. G., Young, S. Y., Yearian, W. C., and Horton, D. L. (1982). Evaluation of *Bacillus thuringiensis*-spray adjuvant-viral insecticide combinations against *Heliothis* spp. (Lepidoptera: Noctuidae). *Environ. Entomol. 11*, 783-787.

Luttrell, R. G., Yearian, W. C., and Young, S. Y. (1983). Effect of spray adjuvants on *Heliothis zea* (Lepidoptera:

Noctuidae) nuclear polyhedrosis virus efficacy. *J. Econ. Entomol. 76*, 162-167.
Magnoler, A. (1968). Laboratory and field experiments on the effectiveness of purified and nonpurified nuclear polyhedral virus of *Lymantria dispar* L. *Entomophaga 13*, 335-344.
Marschall, K. J. (1970). Introduction of a new virus disease of the coconut rhinoceros beetle in Western Samoa. *Nature (London) 225*, 288-289.
Martignoni, M. E. (1978). Production, activity and safety. *In* "The Douglas-Fir Tussock Moth: A Synthesis" (M. H. Brookes, R. W. Stark, and R. W. Campbell, eds.), pp. 140-147. U.S. Dept. Agr. Tech. Bull. 1585.
Mohamed, A. I., Young, S. Y., and Yearian, W. C. (1983). Effects of microbial agent-chemical pesticide mixtures on *Heliothis virescens* (F.) (Lepidoptera: Noctuidae). *Environ. Entomol. 12*, 478-481.
Mohamed, M. A., Coppel, H. C., Hall, D. J., and Podgwaite, J. D. (1981). Field release of virus-sprayed adult parasitoids of the European pine sawfly (Hymenoptera: Diprionidae) in Wisconsin. *Great Lakes Entomol. 14*, 171-175.
Mohamed, M. A., Coppel, H. C., and Podgwaite, J. D. (1982). Persistence in soil and on foliage of nucleopolyhedrosis virus of the European pine sawfly, *Neodiprion sertifer* (Hymenoptera: Diprionidae). *Environ. Entomol. 11*, 1116-1118.
Morris, D. S. (1972). A cooperative programme of research into the management of pome-fruit pests in southeastern Australia. III. Evaluation of nuclear granulosis virus for control of codling moth. *Abstr. 14th Intern. Congr. Entomol., Canberra*, p. 238.
Morris, O. N. (1971). The effect of sunlight, ultraviolet and gamma radiations and temperature on the infectivity of nuclear polyhedrosis virus. *J. Invertebr. Pathol. 18*, 292-294.
Morris, O. N. (1977). Long-term effects of aerial applications of virus-fenitrothion combinations against the spruce budworm, *Choristoneura fumiferana* (Lepidoptera: Tortricidae). *Can. Entomol. 109*, 9-14.
NAS. (1975). "Pest Control: An Assessment of Present and Alternative Technologies," Vol. 4, p. 170. "Forest Pest Control" Nat. Acad. Sci., Washington, D.C.
Neilson, M. M., and Elgee, D. E. (1968). The method and role of vertical transmission of a nucleopolyhedrosis virus in the European spruce sawfly, *Diprion hercyniae*. *J. Invertebr. Pathol. 12*, 132-139.
Podgwaite, J. D., Shields, K. S., Zerillo, R. T., and Bruen, R. B. (1979). Environmental persistence of the nucleopolyhedrosis virus of the gypsy moth, *Lymantria dispar*. *Environ. Entomol. 8*, 528-536.

Podgwaite, J. D., Smith, H. R., and Zerillo, R. T. (1981). Feasibility of integrating nucleopolyhedrosis virus treatment of egg masses with small mammal management for control of the gypsy moth, *Lymantria dispar* . *Abstr. 14th Annu. Meet. Soc. Invertebr. Pathol., Bozeman, Montana,* pp. 41-42.

Podgwaite, J. D., Rush, P., Hall, D., and Walton, G. S. (1984). Efficacy of the *Neodiprion sertifer* (Hymenoptera: Diprionidae) nucleopolyhedrosis virus (baculovirus) product, Neochek-S. *J. Econ. Entomol.*, in press.

Raimo, B. J., and Reardon, R. C. (1981). Preliminary attempts to use parasites in combination with pathogens in an integrated control approach. *In* "The Gypsy Moth: Research Toward Integrated Pest Management" (C. C. Doane and M. L. McManus, eds.), pp. 408-413. U.S. Dept. Agr. Tech. Bull. 1584.

Raimo, B., Reardon, R. C., and Podgwaite, J. D. (1977). Vectoring gypsy moth nuclear polyhedrosis virus by *Apanteles melanocelus* (Hym.: Braconidae). *Entomophaga 22*, 207-215.

Reardon, R. A., and Podgwaite, J. D. (1976). Disease-parasitoid relationships in natural populations of *Lymantria dispar* (Lep.: Lymantriidae) in the northeastern United States. *Entomophaga 21*, 333-341.

Rollinson, W. D., Lewis, F. B., and Waters, W. E. (1965). The successful use of a nuclear polyhedrosis virus against the gypsy moth. *J. Invertebr. Pathol. 7*, 515-517.

Ruzicka, J. (1924). Die neusten Erfahrungen uber die Nonne in Bohmen. *Cent. Ges. Forsta. 50*, 33-68.

Sears, M. K., Jaques, R. P., and Laing, J. E. (1983). Utilization of action thresholds for microbial and chemical control of lepidopterous pests (Lepidoptera: Noctuidae, Pieridae) on cabbage. *J. Econ. Entomol. 76*, 368-374.

Smirnoff, W. A. (1959). Predators of *Neodiprion swainei* Midd. (Hymenoptera: Tenthredinidae) larval vectors of virus diseases. *Can. Entomol. 91*, 246-248.

Smirnoff, W. A., Fettes, J. J., and Haliburton, W. (1962). A virus disease of Swaine's jack pine sawfly, *Neodiprion swainei* Midd. sprayed from an aircraft. *Can. Entomol. 94*, 477-486.

Smith, D. B., and Bouse, L. F. (1981). Machinery and factors that affect the application of pathogens. *In* "Microbial Control of Pests and Plant Diseases 1970-1980" (H. D. Burges, ed.), pp. 635-653. Academic Press, London.

Smith, O. J., Hughes, K. M., Dunn, P. H., and Hall, I. M. (1956). A granulosis virus disease of the western grape leaf skeletonizer and its transmission. *Can. Entomol. 88*, 507-515.

Stairs, G. R. (1965). Artificial initiation of virus epizootics in forest tent caterpillar populations. *Can. Entomol. 97*, 1059-1062.

Stairs, G. R. (1971). Use of viruses for microbial control of insects. *In* "Microbial Control of Insects and Mites" (H. D. Burges and N. W. Hussey, eds.). Academic Press, London.

Stelzer, M. J., and Neisess, J. (1978). Field efficacy tests. *In* "The Douglas-Fir Tussock Moth: A Synthesis" (M. H. Brookes, R. W. Stark, and R. W. Campbell, eds.), pp. 149-152. U.S. Dept. Agr. Tech. Bull. 1585.

Tanada, Y., and Omi, E. M. (1974). Persistence of insect viruses in field populations of alfalfa insects. *J. Invertebr. Pathol. 23*, 360-365.

Thompson, C. G., and Scott, D. W. (1979). Production and persistence of the nuclear polyhedrosis virus of the Douglas-fir tussock moth, *Orgyia pseudotsugata* (Lepidoptera: Lymantriidae), in the forest ecosystem. *J. Invertebr. Pathol. 33*, 57-65.

Thompson, C. G., and Steinhaus, E. A. (1950). Further tests using a polyhedrosis virus to control the alfalfa caterpillar. *Hilgardia 19*, 411-414.

Vail, P. V., and Jay, D. L. (1973). Pathology of a nuclear polyhedrosis virus of the alfalfa looper in alternate hosts. *J. Invertebr. Pathol. 21*, 198-204.

Wollam, J. D., Yendol, W. G., and Lewis, F. B. (1978). Evaluation of aerially-applied nuclear polyhedrosis virus for suppression of the gypsy moth, *Lymantria dispar* L. U.S. Dept. Agr. Forest. Serv. Res. Paper NE-396. p. 8.

Yendol, W. G., and Hamlen, R. A. (1973). Ecology of entomogenous viruses and fungi. *Ann. N.Y. Acad. Sci. 217*, 18-30.

Yendol, W. G., Hedlund, R. C., and Lewis, F. B. (1977). Field investigations of a baculovirus of the gypsy moth, *Lymantria dispar* L. *J. Econ. Entomol. 70*, 598-602.

Zelazny, B. (1976). Transmission of a baculovirus in populations of *Orcytes rhinoceros*. *J. Invertebr. Pathol. 27*, 221-227.

INDEX

A

Acid, inactivation of virus by, 325–327
Actin, 563–564
Activity titer, 684
Activity unit, 684–685, 769
Acylamine, 330
Adenovirus, 701, 703, 706
Adjuvants, 328–330, 777–778
Agarose gel electrophoresis, 56
Air current, man-made, dispersal of virus by, 256–257
Alkali, inactivation of virus by, 325–327
Amino acid analysis, 56, 58
Amyelois virus, 6, 12
Animals
  aquatic, 265–266
  baculovirus in, 260–262, 265–266
  domestic, 260–261
  granulosis virus in, 424–426
  nuclear polyhedrosis virus in, 408–414
  and persistence of virus in habitat, 337–338
  wild, 261–262
Antibody
  enzyme-labeled, 14
  monoclonal, *see* Monoclonal antibody
  as reagent for virus identification, 30–32
Antievaporant, 778
Antiserum, preparation of, 13–14, 30–31
Application study, 361
Application techniques, 788–789
Arbovirus, 29, 31, 104, 635–638
Arkansas bee virus, 237
Arthropod
  parasitic, 336–337
  predaceous, 336–337
Assembly of virus, 558–561, 606–617, 706, 715–716
Attachment of virus, 577, 584–589, 695–720
Attachment protein, viral, 697–702, 710, 719
Augmentation strategies, 785–786
Autointerference
  homologous, 737–739
  wild-type, 738–739
Aves, *see* Birds
Avidin–biotin immunohistochemical technique, 42–43

B

Bacteria, persistence in field habitat, 286–287
Bacteriophage ϕX 174, 701
Baculoviridae, 6–8, 19, 29, 249–250, 441, 675
Baculovirus, 71, 489–493
  alkaline protease of, 58–59, 100–101
  assembly of, 715–716
  attachment to host cell, 584–589, 713–714
  augmenting naturally occurring mortality, 785–786
  cell culture systems for, 571–580, 619, 767
  characteristics of, 6–8, 469–483
  classification of, 82–83, 494
  deoxyribonucleoproteins of, 473–474
  dispersal of, 249–276
    by agronomic practices, 251–256
    by aquatic animals, 260–266
    by insects, 266–274
    by meterological factors, 256–260
    by parasitoids, 273–274
    by predatory insects, 272–273
    by terrestrial animals, 260–266
  DNA of, 17, 85–86, 97–98, 470–472, 495–497, 593–594, 620–621
  environmental considerations in use of, 104–107
  enzyme-linked immunosorbent assay, 39–41
  epidemiology of, 104

Baculovirus *(continued)*
evolution of, 526–528
field use of, 775–790
FP variant, 619–621
genetic engineering of, 103
hemagglutinating activity of, 34
homologous-type interactions of, 741–742
host range of, 87–88, 266–272, 491, 570
infection by
changes in hemolymph, 91–93
events in hemocoel, 90–91
events in insect gut, 89–90
gross appearance of host, 91
in vitro, 93–95
in vivo, 88–93
infectious unit of, 580–584, 711
infectivity of, 95–98
as insecticide, 103, 105–107, 444–446, 775–790
interaction with host cell, 707–718
latent, 101–102
lipid of, 478, 711–712, 719
in mixed infection, 99–100, 742–744
mode of entry of, 88, 95–96, 104–105
morphological forms of, 95–98
MP variant, 619–621
mutants of, 616–617
nonoccluded, 86–87, 95, 102, 446, 469, 494, 498–499, 606–613
nucleocapsid of, 470–479, 494–495, 613–617, 708–718
formation of, 606–609
uncoating of, 587–589
occluded, 85–86, 613–617
occurrence of, 491, 494
pathology associated with, 81–107, 498–499
penetration of host cell, 582–589, 714–717
persistence of, 783–785
polyhedron of, 479–483
proteins of, 16–17, 58–67, 70, 497, 597–605
receptors for, 696, 707–708
recombinant, 530
relatedness of strains, 497
release from host cell, 717–718
replication of, 498–499, 569–623
assembly and release of progeny, 606–617
defective, 618–621
early events in, 589–593
nucleic acid synthesis during, 593–598
protein synthesis during, 597–605
resistance to, 102, 778–779
RNA of, 595–598
stability of, 777–778
structural relationships, 84–97
structure of, 469–483, 495
subviral entities, 98
transmission of, 104–105, 250–251, 580
virion of, 85, 95–96
wasp, 102
western blot, 41
Bait, virus formulated as, 315–317, 768
Bioassay
in vitro, 685–687
in vivo, 684–685
Biological activity of virus, quantitation of, 683–687
Biological control
safety of, 399–429
and viral resistance, 383–390
Birds
baculovirus in, 262–263
nuclear polyhedrosis virus in, 413
and persistence of virus in habitat, 337–338
Birnavirus, 636, 643–644
Black beetle virus, 650–651
Bombyx mori, see Silkworm
Boric acid, 328, 330
Borrelina, 4
Broadcast application, 776–778, 788–789
Broad-spectrum virus, 789
Budding, 705–706, 716–718, 720
Bundle, 84–85, 708
Bunyaviridae, 29, 638

C

Calciviridae, 6, 12, 29, 637
replication of, 652–653
RNA of, 653
Capsid, 169–170, 204, 474–477, 559, 588, 606–609, 699–710, 746
Capsule, 84, 469
Captan, 328
Carbaryl, 328
Carbon, as protectant, 313–315
Carbon bisulfide, 329
Carbon dioxide, sensitivity to, 12, 239, 654
Carry-over effect, 293–294, 339
Cell culture
baculovirus in, 93–95, 569–623
cell lines used, 571–574
effect of medium, 575–577
efficiency of infection, 577–580

Cell culture *(continued)*
for commercial preparation of virus, 765–766
continuous passage of virus in, 619
cytoplasmic polyhedrosis virus in, 145–147, 639
densovirus in, 221–223
granulosis virus in, 425–426
growth phase of, 577–579
iridescent virus in, 547
medium for, 575–577
multiplicity of infection, 579–580
nuclear polyhedrosis virus in, 408–414
Centrifugation
buoyant density, 681
isopycnic, 681
Charcoal, *see* Carbon
Chemicals
inactivation of virus by, 325–331
natural, 330–331
Chlordimeform, 329
Chloriridovirus, 168–169
Cholesterol, 704
Chromosome aberration, virus-induced, 414–415, 426–427
Chronic bee paralysis virus, 658
Chronic stunt virus, 653
Classification
of baculovirus, 82–83
of densovirus, 198–199
of insect viruses, 27–46
of iridescent viruses, 164–169
Coated pit, 705
Coax (feeding stimulant), 778
Colonization of virus, 783–785
Commercial production, 757–770, 779
cost of, 770
formulation of product, 767–769
in vitro methods, 765–767
in vivo methods, 758–764
quality control of, 769–770
safety of, 769–770
Complement fixation, 15–16, 34–35, 682
Complement receptor, 420, 428
Copper sulfate, 328
Cotton seed products, as protectants, 312
Counting chamber, 676–677
Cricket paralysis virus, 13, 69–70, 234–235, 419, 645–646, 686
Cross-over electrophoresis, 14–15
Crystalline-array virus, 240
Cultivation, dispersal of baculovirus by, 254–255
Cytocidal virus, 706
Cytoplasmic polyhedrosis virus, 71, 84, 501
A strain, 123–125, 127–129, 134–136
B strain, 123–124, 128–130, 134–136
B_1 strain, 123–124, 130–131, 134–136
B_2 strain, 123–124, 132–136
characteristics of, 6, 9, 636, 638
commercial production of, 765
C_1 strain, 123–125, 133–136
C_2 strain, 123–125
DNA of, 17
enzyme-linked immunosorbent assay, 40
epizootiological studies on, 455–456
H strain, 123–124, 126, 134–136
infection by
in cell culture, 145–147
effects on host, 140–144
in embryo culture, 145
histopathology of, 136–143
in larval stage, 122–142
in postlarval stage, 142–144
symptomatology of, 136–140
as insecticide, 443
I strain, 123–126, 134–136
in mixed infection, 335, 742–743
mutants of, 148
occlusion body of, 490–491
pathology associated with, 121–148
persistence in field habitat, 289
proteins of, 68, 641–642
P strain, 123–124, 126–127, 134–136
quantitation of, 679
receptor for, 137–138
replication of, 639–642
resistance to, 362–370
RNA of, 640
stability of, 310, 325
strains of, 122–136
transmission of, 333
Cytosine arabinoside, 601–602
Cytoskeleton, and viroplasmic center, 561–564

D

DDT, 328
Debris, stability of virus in, 302–308
Defective interfering particle, 404–405, 737–738
Densonucleosis virus, *see* Densovirus
Densovirus
characteristics of, 6, 10–11, 637
classification of, 198–199
DNA of, 205–208, 211, 655–656

Densovirus *(continued)*
  epizootiology of, 222–224
  hemagglutination, 214–215
  host range of, 215–216
  immunological characterization of, 214–215
  immunoperoxidase staining of, 43
  infection by
    in cell culture, 221–223
    cytopathology of, 218–221
    fluorescent antibody studies of, 221–222
    histopathological symptoms of, 199–201
    immunoperoxidase studies of, 221
    metabolism in infected tissue, 217
    in organ culture, 222–223
  in mixed infection, 134, 216–217
  morphogenesis of, 215–223
  morphology of, 201–205
  pathology associated with, 197–224
  polyamines of, 212–214
  proteins of, 69, 208–213, 655–656
  replication of, 215–223, 655–656
  resistance to, 215–216, 362–365, 369–370
  serological relationships between isolates, 214
  size of, 201–202
  stability of, 321
  structure of, 203–205
  temperature sensitivity of, 217, 321
Detection of virus, 27–46
Dew, 301, 323–324, 326
Diet, insect, 760
Digestive fluid, 331, 370–371
Dinocap, 328
Disinfectant, inactivation of virus by, 330
Dispersal of virus, *see also* specific viruses
  by agronomic practices, 251–256
  by aquatic animals, 260–266
  by insects, 266–274
  by terrestrial animals, 260–266
  via meterological factors, 256–260
DNA
  of baculovirus, 85–86, 97–98, 470–472, 495–497, 593–594, 620–621
  buoyant density centrifugation of, 681
  circular, 470–472
  of densovirus, 205–208, 211, 655–656
  electrophoresis of, 56
  of granulosis virus, 422–423
  groups of viruses containing, 6, 636–637
  host, 589–590
  infectious, 472, 618
  of iridescent virus, 172, 174–175, 547–559
  of nuclear polyhedrosis virus, 402–407
  recombinant, 530
  synthesis in virus-infected cells, 547–559
  viral
    packaging of, 594
    subgenomic fragments of, 554–555
DNA binding protein, 473
DNA polymerase, virus-induced, 592–593
Drosophila, reovirus in, 642–643
Drosophila A virus, 238–239
Drosophila C virus, 69–70, 234–235, 239, 645–646
Drosophila P virus, 238–239
Dry-film counting technique, 678–679
Dust, virus formulated as, 768

E

Ecosystem, role of viruses in, 441–459
Effectiveness of virus, 292–294, 339
Efficacy of virus, 789
Efficiency of infection, 577–580
Elcar, 775, 778–780, 787
Electronic particle counter, 677–678
Electron microscopy, quantitation of virus by, 679–680
Electrophoresis, 56–58
ELISA, *see* Enzyme-linked immunosorbent assay
Embryo culture, cytoplasmic polyhedrosis virus in, 145
Endocytosis, 705
Endosulfan, 329
Endrin, 328
Enhancement, 746
Entomopoxvirus, 71, 637
  characteristics of, 6, 8
  as insecticide, 443–444
  occlusion body of, 490–491
  persistence in field habitat, 286
  proteins of, 67
  quantitation of, 679
  replication of, 656–657
  stability of, 303, 310
  transmission of, 333
Envelope
  of nucleocapsid, 477–479, 497, 581–583, 609–613, 708–709, 716
  of polyhedron, 482–483, 499, 613–617
  transport, 607, 609
Environmental Protection Agency (EPA), United States
  criteria for microbial insecticides, 106
  guidelines for safety testing, 400, 769
Enzootic, 455

Enzyme-linked immunosorbent assay (ELISA), 15, 37–40, 682–683
  indirect, 37–39
  sandwich, 38–40
Enzymes, viral, 56, 58–59, 175, 640–641
EPA, *see* Environmental Protection Agency
Epidemiology, of baculovirus, 104
Epizootic, 446–458, 784–785
  artificial, 777
  densovirus, 222–224
  development of, 454–458
  effect of inoculum release on, 451–454
  geometry of
    spatial, 450–451
    temporal, 447–450
  induction of, 786
Epizootic wave, 447–449, 758
Establishment of virus in host population, 292–294
Ethylene dichloride, 329
Ethylene oxide, 329
Evolution
  of insects, 526–528
  of polyhedrin, 526–528

F

Facilitation, 740
Feeding stimulants, 329
Fenitrothion, 329
Fetal bovine serum, 575
Field collection of virus, 758–759
Filamentous virus, honey bee, 70
Film impression technique, 680–681
Fish
  baculovirus in, 265–266
  granulosis virus in, 426
  nuclear polyhedrosis virus in, 413–414
Flacherie virus, 69, 138–140, 197–198, 239, 647
  resistance to, 362–369, 371
Flock house virus, 237
Fluorescent antibody technique, 41
Foliage
  quantitation of virus on, 680–681
  stability of virus on, 297–301
Forest, as reservoir of baculovirus, 253
Formalin, 330
Frog virus 3, 546, 550, 553–555, 560–563
Fungicides, 328–329
Fungus, persistence in field habitat, 288
Fusion, virus–host, 703–704, 714, 719

G

Gamma radiation, 317
Gel immunodiffusion, 33
Genetic engineering, of baculovirus, 103
Genome, *see also* DNA
  viral, 16
    variation in, 658
Genomic mixing, 658, 747
Glycolipid, 700
Glycoprotein, 61, 587, 602–603, 655, 698, 704–705, 746
  spike, 700, 702–703
Gonometa virus, 234, 645, 647
Gradient electrophoresis, 57
Granule, 84–85, 469–470, 489, 495
Granulin, 58, 64, 71, 422–423, 426, *see also* Polyhedrin
Granulosis virus, 83, 249–250, 469, 493–494, 572, *see also* Baculovirus
  commercial production of, 765
  DNA of, 422–423
  effect of pesticides on, 328
  enzyme-linked immunosorbent assay, 37, 39
  host range of, 334, 421–423
  as insecticide, 445–446, 782–783
  interaction with mammalian serum, 427–428
  in mixed infection, 331, 743, 746
  neutralization test, 36
  persistence of, 286, 289–292, 332–338, 423, 784–786
  protectant additives for, 311–317
  proteins of, 64–67
  quantitation of, 679
  resistance to, 362–363, 371–374, 378–383, 421–422
  safety testing of, 421–428
  stability of
    effect of pH, 326–327
    effect of temperature, 319–323
    effect of water, 323–324
    on foliage, 298–301
    in soil, 302–304, 307
  storability of, 294–297
  ultraviolet-resistant strain of, 317
  western blot, 41
Gustatory adjuvant, 329, 778
Gut, inset
  baculovirus in, 89–90
  cytoplasmic polyhedrosis virus in, 136–140
Gypchek, 383, 679, 775, 780–781
Gypsy moth, 758–764
  viral resistance in, 383–387

H

Harvesting of virus, commercial, 762–764
Helper virus, 658, 737, 740, 746
Hemagglutination, by densovirus, 214–215
Hemagglutination inhibition, 34
Hemocoel, baculovirus in, 90–91
Hemocyte, 736
Hemolymph, in baculovirus infection, 91–93
Honey bee, small RNA virus, 239
Host cell–virus interaction, 702–718
Host cell–virus recognition, 697–701
Host insect, *see also* specific viruses
  alternate, 334–335
  in commercial production of virus, 759–762
  competition for, 335–336
  direct control of outbreak population, 777–783
  and persistence of virus in habitat, 291–292, 332–336
  rearing of, 759–762
Host range, 789–790, *see also* specific viruses
  of baculovirus, 87–88
  of densovirus, 215–216
  of iridescent virus, 176–177
Humidity, inactivation of virus by, 323–324
2-Hydroxy-4-methoxybenzophenone, 312, 315
Hydroxyurea, 602

I

ICNV, *see* International Committee on the Nomenclature of Viruses
Identification of virus, 27–46, 55–71
Immunity, insect, to baculovirus, 102
Immunoaffinity chromatography, 44–45
Immunodiffusion, 682
Immunoelectrophoresis, 14, 33
Immunofluorescent staining, 41–44
Immunoglobulins, interaction with baculovirus proteins, 417–420, 427–428
Immunoperoxidase staining, 41–44
Immunoprecipitation, 44–45
Inclusion body, 4–9, 469, *see also* Polyhedron
  of baculovirus, 84
  breakdown of, 89
  effect of pH, 325–327
  formation of, 135–136
  proteins of, 59, 64, *see also* Granulin; Polyhedral protein; Polyhedrin; Spheroidin
  stability of, 307
  storability of, 295–297, 320
India ink, as protectant, 313, 315
Induction of virus, 367
Infection
  frank, 5
  latent, 5
Infectivity
  of baculovirus, 95–98
  of iridescent virus, 181–182
Infrared light, 317
Inoculum
  for commercial production of virus, 762–764, 766–767
  primary, 447
  release of, 451–454
  secondary, 447
Insect
  evolution of, 526–528
  predatory, *see* Predator
  quantitation of virus in, 678–679
Insect viruses
  categories of, 29
  as pathogen, 442–446
  relationships between, determination of, 13–18
Integrated pest management, 787–788
Interference
  heterologous viral, 739–740, 742–746
  intrinsic, *see* Interference, heterologous viral
Interferon, 737
Intermediate filament, 561–562
International Committee on the Nomenclature of Viruses (ICNV), 4–5, 18–19
Invertebrate icosahedral cytoplasmic deoxyribovirus, *see* Iridescent virus
Iridescence, mechanism of, 170–171
Iridescent virus, 4, 35
  as biological control agent, 186–187
  characteristics of, 6, 9–10, 171–175
  classification of, 164–169
  DNA of, 172, 174–175, 547–559
  entry into host cells, 546–548
  enzyme-linked immunosorbent assay, 39–40
  homologous-type interactions of, 741
  host range of, 176–177
  immunoperoxidase staining, 43
  infection by
    cytopathology of, 178–179
    gross pathology of, 180–181
    pathway of, 182–184
    symptoms of, 177
  infectivity of, 181–182
  lipid of, 172
  macromolecular synthesis in infected cells, 547–553

Iridescent virus *(continued)*
 mechanism of iridescence, 170–171
 in mixed infection, 100, 216, 742–743, 746
 natural incidence of, 181–182
 pathobiology associated with, 163–187
 proteins of, 16, 68–69, 173, 175, 547–553
 quantitation of, 686–687
 relationships between, 175–176
 replication of, 178–179, 545–565
 RNA of, 547–553
 transmission cycle of, 183–186
 ultrastructure of, 169–170
 viroplasmic center, 559–565
Iridoviridae, 6, 9–10, 29, 163–187, 441–442
Irrigation water, dispersal of baculovirus by, 255
Isolation of virus, 5–7

L

Large white butterfly, viral resistance in, 371–374
Larva
 cytoplasmic polyhedrosis virus in, 122–142
 midgut epithelium, regeneration of, 138–140
Latency, 5, 101–102, 367, 404–405, 455–456, 788–789
Lecontvirus, 782
Lepidoptera, cell lines from, 571–574
Light-brown apple moth, viral resistance in, 374–378, 402–403
Lipid
 of baculovirus, 478, 711–712, 719
 of iridescent virus, 172
 of viral envelope, 699–700

M

Mammals
 baculovirus in, 260–262
 granulosis virus in, 424–426
 nuclear polyhedrosis virus in, 408–413
 and persistence of virus in habitat, 337–338
Marketing, dispersal of baculovirus by, 255–256
Messenger RNA
 of baculovirus, 595–598
 of cytoplasmic polyhedrosis virus, 640–641
 polyhedrin, 513–518
Methomyl, 329
3-(4-Methylbenzyliden)camphor, 315
Methyl bromide, 329
Methyl parathion, 328
Mevinphos, 329
Microfilament, 561–564
Microsporidia, persistence in field habitat, 288
Microtubule, 560–564
Microvillus, 585–586, 713–714, 719
Mite, in dispersal of baculovirus, 274
Molasses, as protectant, 312
Monoclonal antibody, 14, 31–34, 44–45
 against polyhedrin, 512
 neutralizing, 36
Morator, 4
Morphogenesis, of densovirus, 215–223
Multiple virus interactions, 735–748
Multiply enveloped virion, 84
Mutant virus, 738
Myxovirus, 701, 706

N

Naled, 329
Neochek-S, 776, 782
Neuraminic acid, 702–703
Neutralization test, 35–37, 682
Nodamura virus, *see* Nodavirus
Nodaviridae, 6, 11, 29, 235–237, 636, 657–658
Nodavirus, 29, 234, 638
 characteristics of, 6
 pathology associated with, 235–237
 proteins of, 650–652
 replication of, 649–652
 RNA of, 650–652
Nomenclature of viruses, 3–19
Nuclear membrane, 609–613, 717–718
Nuclear polyhedrosis virus, 3, 71, 83, 103, 249–250, 469, 492–494, 701, *see also* Baculovirus
 carry-over effect, 293
 commercial production of, 760, 762–767
 commercial use of, 779–782, 787
 DNA of, 402–407, 496
 effect of pesticides, 328–330
 enzyme-linked immunosorbent assay, 38–40
 epizootiological studies on, 455–458
 establishment in host population, 292–293
 evolution of, 526–528
 homologous-type interactions of, 741
 host range of, 334–335, 401–410
 immunofluorescence staining of, 43
 immunoprecipitation of, 45
 as insecticide, 444–445
 interaction with mammalian serum, 416–420
 in mixed infection, 134, 216, 238, 331, 335, 648, 742–746

Nuclear polyhedrosis virus *(continued)*
multiple nucleocapsid, 470
mutants of, 524–525
neutralization test, 35–37
persistence of, 286, 289–292, 332–338, 452–454, 784–786
production of, 452–454
protectant additives for, 311–317
proteins of, 33, 406–407
quantitation of, 677, 680–681, 684–687
resistance to, 362–364, 367–371, 374–378, 381–387, 402–403
retrovirus inducing activity of, 415–416
safety tests of, 401–420
single nucleocapsid, 470
stability of
effect of pH, 325–327
effect of temperature, 318–323
effect of water, 323–324
on foliage, 298–301
in soil, 302–308
and sunlight, 308–317
storability of, 294–297
ultraviolet-resistant strain, 317
western blot, 41
Nuclear pore, 588, 611, 713–715, 717
Nucleic acids, *see also* DNA; RNA
analysis of, 17–18
Nucleocapsid, 581–585, 701, 706
of baculovirus, 470–479, 494–495, 613–617, 708–718
formation of, 606–609
uncoating of, 587–589
cap of, 477, 588
claws of, 476–477, 709
envelope of, 477–479, 581–583, 609–613, 708–709, 716
movement across nuclear membrane, 609–612
movement across plasma membrane, 609–612
nipples of, 476–477, 709
structure of, 708–709
Nucleoprotein, 594, 473–474, 607–609
Nucleus
in iridescent virus infection, 553–559
skeletal proteins of, 564–565
in viral DNA synthesis, 553–559
Nudaurelia β virus, 6, 11, 234, 636
pathology associated with, 237–238
proteins of, 70
replication of, 647–649

O

Obligatory mutualism, 102
Occlusion body, *see also* Polyhedron
formation of, 616–617
identification of, 681
Occult virus, 101
Oncogenic virus, 415–416, 747
Orbivirus, 638
Orcytes virus, 249–250, 274–276, 319, 469, 572, 609, 612–613, 784–785
Organ culture
densovirus in, 222–223
granulosis virus in, 425

P

Paramyxovirus, 702–706
Parasite, and persistence of virus in habitat, 336–337, 785–786
Parasitoid, in dispersal of baculovirus, 273–274
Parvoviridae, 6, 10–11, 29, 198, 441–442, 637, 742
Penetration of virus into cell, 582–589, 696, 703–705, 714–717
Peplomer, 478, 583, 587, 610–611, 696, 700, 709–720
Peptide mapping, 16–17, 56, 58
Peritrophic membrane, 736
Peroxidase–antiperoxidase technique, 42–43
Persistence of virus, 285–340, 451–452, 459, 783–785, *see also* specific viruses
biotic factors affecting, 331–338
and development of epizootic, 455–458
and host feeding habits, 291–292
significance in integrated pest management, 338–340
Pesticides, 328–330, 399–400, 545, 778–779, 787–788
pH, and viral stability, 325–327, 777
Phagocytosis, 586, 703–705, 708, 714–717, 719
reverse, 706
Phosphine, 329
Phospholipase, 704
Phospholipid, 711–712
Phosphoprotein, 602–604
Phylogeny, baculovirus, 526–528
Picornaviridae, 6, 11, 29, 234–235, 441–442, 636
Picornavirus, 182, 233, 602, 706
interaction with human serum, 417
in mixed infection, 746

Picornavirus *(continued)*
pathology associated with, 234–235
proteins of, 69–70
quantitation of, 687
receptor for, 679
replication of, 645–647
Pink shrimp baculovirus, 104
Pisces, *see* Fish
Plaque assay, 685–687
Plasma membrane, 609–612
lysis of, 706
receptors on, 695–720
structure of, 697–698
Polyacrylamide gel electrophoresis, 16, 57–58
Polyamines, 212–214, 473, 594, 655
Polyhedral protein, 68, 71, 481–482
Polyhedrin, 33, 38–39, 58–60, 71, 403, 489–530, 598–601, 616–617
aggregation of, 502–509, 587
amino acid composition of, 500–502
amino acid sequence of, 504–510
antigenicity of, 510–512
crystallization of, 502–509
evolution of, 526–528
gene for, 512–525
codon usage in, 519, 522–523
localization of, 514–515
nucleotide sequence of, 519–524
glycosylation of, 502–503
isoelectric point of, 500–502
messenger RNA for, 595–596
structure of, 515–518
transcription of, 513–514, 530
translation of, 513–514
mutants of, 524–525
phosphorylation of, 502–503
size of, 500–502
solubility of, 503–509
structure of, 529
tryptic peptide analysis of, 509
viral origin of, 512–513
Polyhedron, 81, 84–85, 93–95, 99, 122–125, 469–470, 478, 495, 640, 656, *see also* Inclusion body; Occlusion body; Polyhedrin
alkaline protease of, 499–500
commercial production of, 764, 767
envelope of, 482–483, 613–617
mutations affecting, 524–525, 616–617
physical characteristics of, 482
small RNA virus in, 648
structure of, 479–483
Population infective unit, 453
Postapplication study, 361
Potato moth, viral resistance in, 378–383, 421–422
Powder, wettable, 768
Poxviridae, 6, 8, 29, 490, 558, 637, 706
Preapplication study, 361
Predator, 785–786
and persistence of virus in habitat, 336–337
transmission of baculovirus by, 272–273
Production of virus, commercial, *see* Commercial production of virus
Protease, alkaline, 58–60, 64–68, 71, 100–101, 498–500, 605
Protectant additive, 311–317, 329, 768, 777–778
Protein kinase, 175, 179, 603–604
Proteins
attachment, *see* Attachment protein
of baculovirus, 58–67, 70, 497, 597–605
of cytoplasmic polyhedrosis virus, 68, 641–642
of densovirus, 69, 208–213, 655–656
early, 591–593
of entomopoxvirus, 67
of granulosis virus, 64–67
host, 590
nuclear skeletal, 564–565
receptor, 698
of iridescent virus, 68–69, 173, 175, 547–553
of nodavirus, 650–652
of nuclear polyhedrosis virus, 406–407
of *Nudaurelia* β virus, 70, 647–649
of picornavirus, 69–70
separation of, 57–58
of sigmavirus, 655
synthesis in virus-infected cells, 547–553
viral
analysis of, 16–17
antiserum to individual, 31
DNA binding, 473
envelope, 478–479
in identification of viruses, 55–71
posttranslational modification of, 602–605
Pupae, cytoplasmic polyhedrosis virus in, 142–143

Q

Quality control, of commercial preparations, 769–770
Quantal assay, 686–687

Quantitation of virus, 675–688
  direct physical, 676–681
  serological techniques, 682–683

R

Radioimmunoassay, 15, 37–40
Rain
  dispersal of virus by, 257–259
  inactivation of virus by, 323–324
Receptor, *see also* specific viruses
  cellular, 695–720
    macromolecular structure of, 698
Release of virus, 606–617, 696, 705–707, 717–720
Reoviridae, 6, 9, 29, 441, 490, 636, 638, 675, 775
Reovirus, 642–643, 743–744
Replication of virus, 635–659, *see also* specific viruses
  early events in, 589–593
Resistance to virus, *see* Viral resistance
Restriction endonucleases, 17
Retrovirus, 637, 653–654, 700
  induction of, 415–416
Rhabdoviridae, 6, 11–12, 29, 441–442, 637
Rhabdovirus, 654–655, 700, 706
RNA
  of baculovirus, 595–598
  of calcivirus, 653
  of cytoplasmic polyhedrosis virus, 640
  electrophoresis of, 56
  groups of viruses containing, 6, 636–637
  of iridescent virus, 547–553
  messenger, *see* Messenger RNA
  of nodavirus, 650–652
  of *Nudaurelia* β virus, 647–649
  of picornavirus, 646
  of sigmavirus, 655
  synthesis in virus-infected cells, 513–514, 547–553
RNA polymerase, 513–514, 522–523, 530, 595, 640–641, 644, 649, 652

S

Safety tests, 747
  of commercial preparations, 769–770
  with granulosis virus, 421–428
  of insect viruses, 399–429
  with nuclear polyhedrosis virus, 401–420
San 285, 312
SAN 404, 783
SAN 405, 783
SAN 406, 783
Satellite virus, 12–13, 658
Scavengers, and persistence of virus in habitat, 337–338
Semliki forest virus, 705
Sendai virus, 704
Serological techniques, 13–16, 27–46
  for quantitation of virus, 682–683
Serum, mammalian
  interaction with granulosis virus, 427–428
  interaction with nuclear polyhedrosis virus, 416–420
Shade (commercial protectant), 312–315
Sigmavirus, 6, 11–12, 654–655
Silkworm, viral resistance, 362–371
Sindbis virus, 738–739, 744–746
Sister chromatid exchange, virus-induced, 414–415, 426–427
Skim milk, as protectant, 312–314
Small RNA virus, 12, 29, 69–70, 638
  characteristics of, 6, 11
  of *Drosophila,* 238–239
  enzyme-linked immunosorbent assay, 40
  of honey bee, 239
  pathology associated with, 233–240
  in polyhedron, 648
Sodium hypochlorite, 330
Soil
  pH of, 327
  quantitation of virus in, 678–679
  as reservoir of baculovirus, 251–254
  stability of virus in, 302–308
Spheroidin, 67, 71
Spheruloviridae, 441, 675
Spike, 700, 702–703, 708, 717–720
Stability of virus, 285–340
Storability of virus, 294–297
Stress fiber, 562–563
Subviral entity, in baculovirus infection, 98, 101
Sulfur, 328
Sunlight
  inactivation of virus by, 308–317
  protection of virus against, 311–317
Synergistic factor, 713, 719, 733

T

Taxonomy of viruses, 3–19
$TCID_{50}$, *see* Tissue culture infectious dose 50
Temperature
  effect on densovirus, 217
  inactivation of virus by, 317–323

Thin filament, 562–563
Thymidine kinase, 592–593
Tissue culture infectious dose 50 ($TCID_{50}$), 686–687
TM-Biocontrol-1, 775, 781
Togaviridae, 29, 638, 737–738, 741, 744
Toxaphene, 328
Transfection, 582
Transmission of virus, *see also* specific viruses
  adult, 267–272, 333
  ovarial, 266–267
  by parasitic arthropods, 336–337
  transovarial, 250, 266–267, 271, 451
  transovum, 250, 267–272, 333, 451
  transtadial, 250
Tube precipitation, 33, 682
Tubulin, 561, 563–564
Tumor, epidermal, 181
Two-dimensional electrophoresis, 57–58

U

Ultraviolet irradiation, inactivation of virus by, 309–311
Ultraviolet-resistant virus, 317

V

Viral resistance, 361–391, 402–403, *see also* specific viruses
  and biological control, 383–390
  development of, 361–391
  and development of epizootic, 455
  genetic mechanism of, 365–368, 373–383, 387–390, 421–422
  physiological mechanism of, 368–371, 378, 389
  research requirements in, 389–390
VIRIN-ENSh, 679–680
Virion
  assembly of, 559–561
  of baculovirus, 85, 95–96
Virogenic stroma, 90–95, 124–125, 218–219, 606–609, 640
Viroplasmic center, 90–91, 554, 559–561
  cytoskeleton involvement in, 561–564
  nuclearlike proteins in, 564–565
Virus-like particle, 618–619

W

Wasp, baculovirus of, 102
Water
  natural, dispersal of virus by, 259–260
  stability of virus in, 308, 323–324
Western blot, 40–41
Wind, dispersal of virus by, 256–257

X

X virus, 12, 643

Z

Zinc sulfate, 328
Zineb, 328